AF770652

Photoinduced Processes in Nucleic Acids and Proteins

Kerala, India

11–13 January 2018

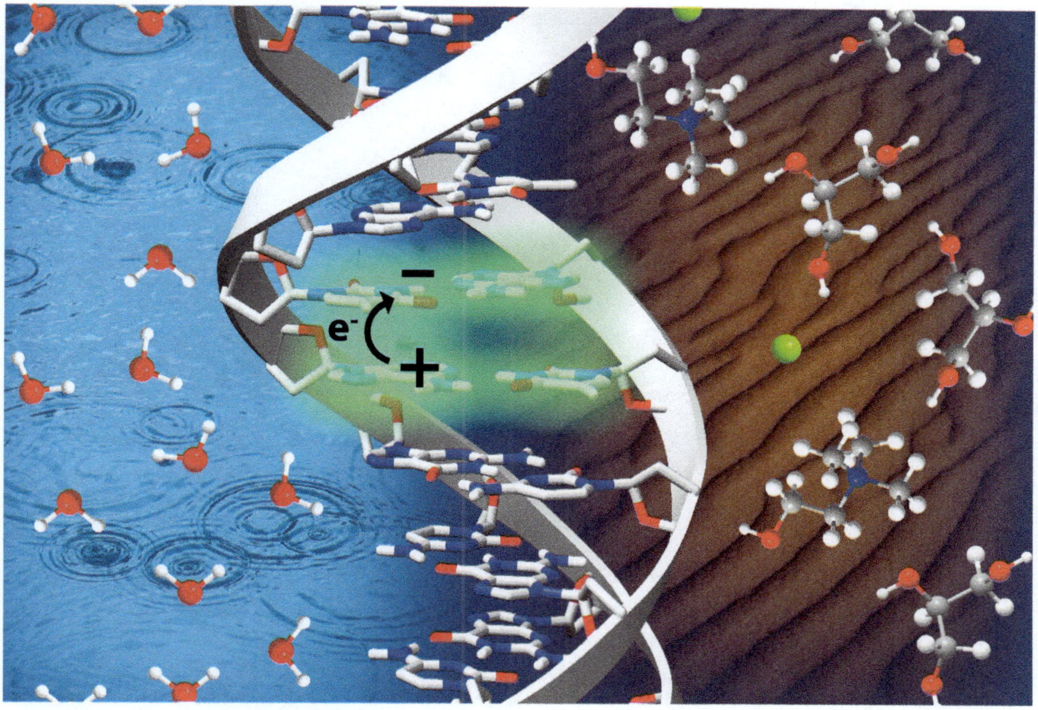

FARADAY DISCUSSIONS

Volume 207, 2018

ROYAL SOCIETY OF CHEMISTRY

The Faraday Division of the Royal Society of Chemistry, previously the Faraday Society, was founded in 1903 to promote the study of sciences lying between chemistry, physics and biology.

Editorial Staff

Executive Editor
Philippa Hughes

Deputy Editor
Heather Montgomery

Editorial Production Manager
Claire Darby

Publishing Editors
Jack Busby, Joseph Gibson

Editorial Assistant
Rebecca Withers

Publishing Assistants
Allison Holloway, Julie-Ann Roszkowski

Publisher
Jamie Humphrey

Faraday Discussions (Print ISSN 1359-6640, Electronic ISSN 1364-5498) is published 7 times a year by the Royal Society of Chemistry, Thomas Graham House, Science Park, Milton Road, Cambridge, UK CB4 0WF.

Volume 207 ISBN 13: 978-1-78801-158-7

2018 annual subscription price: print+electronic £1103 US $1940; electronic only £1050, US $1848.
Customers in Canada will be subject to a surcharge to cover GST. Customers in the EU subscribing to the electronic version only will be charged VAT.

All orders, with cheques made payable to the Royal Society of Chemistry, should be sent to the Royal Society of Chemistry Order Department, Royal Society of Chemistry, Thomas Graham House, Science Park, Milton Road, Cambridge, CB4 0WF, UK
Tel +44 (0)1223 432398; E-mail **orders@rsc.org**

If you take an institutional subscription to any Royal Society of Chemistry journal you are entitled to free, site-wide web access to that journal. You can arrange access via Internet Protocol (IP) address at **www.rsc.org/ip**

Customers should make payments by cheque in sterling payable on a UK clearing bank or in US dollars payable on a US clearing bank.

Whilst this material has been produced with all due care, the Royal Society of Chemistry cannot be held responsible or liable for its accuracy and completeness, nor for any consequences arising from any errors or the use of the information contained in this publication. The publication of advertisements does not constitute any endorsement by the Royal Society of Chemistry or Authors of any products advertised. The views and opinions advanced by contributors do not necessarily reflect those of the Royal Society of Chemistry which shall not be liable for any resulting loss or damage arising as a result of reliance upon this material. The Royal Society of Chemistry is a charity, registered in England and Wales, Number 207890, and a company incorporated in England by Royal Charter (Registered No. RC000524), registered office:
Burlington House, Piccadilly, London W1J 0BA, UK, Telephone: +44 (0) 207 4378 6556.

Printed in the UK

MIX Paper from responsible sources FSC® C013604 FSC www.fsc.org

Faraday Discussions are unique international discussion meetings that focus on rapidly developing areas of chemistry and its interfaces with other scientific disciplines.

Scientific Committee volume 207

Co-Chairs
Mahesh Hariharan, IISER-Thiruvananthapuram, India
Anthony Watts, University of Oxford, UK

Krishna N. Ganesh, IISER Pune, India
John Helliwell, The University of Manchester, UK
Judith A. K. Howard, Durham University, UK
Reji Varghese, IISER-Thiruvananthapuram, India

Faraday Standing Committee on Conferences

Chair
John M. Seddon, Imperial College London, UK

Secretary
Marie Cote, Royal Society of Chemistry, UK

Sharon Ashbrook, University of St Andrews, UK
Eleanor Campbell, University of Edinburgh, UK
Graeme Day, University of Southampton, UK
Susan Perkin, University of Oxford
Ivana Radosavljević Evans, Durham University, UK
Paul Raithby, University of Bath, UK
Pat Unwin, University of Warwick, UK

Advisory Board

Vic Arcus, The University of Waikato, New Zealand
Dirk Guldi, University of Erlangen-Nuremberg, Germany
Luiz Liz-Marzán, CIC biomaGUNE, Spain
Guy Lloyd-Jones, University of Edinburgh, UK
Marina Kuimova, Imperial College London, UK
Andrew Mount, University of Edinburgh, UK
Frank Neese, Max Planck Institute for Chemical Energy Conversion, Germany
Michel Orrit, Leiden University, The Netherlands
Zhong-Qun Tian, Xiamen University, China
Siva Umapathy, Indian Institute of Science, Bangalore, India
Mike Ward, University of Sheffield, UK
Bert Weckhuysen, Utrecht University, The Netherlands
Julia Weinstein, University of Sheffield, UK
Sihai Yang, University of Manchester, UK

This journal is © the Royal Society of Chemistry 2018.
Apart from fair dealing for the purposes of research or private study for non-commercial purposes, or criticism or review, as permitted under the Copyright, Designs and Patents Act 1988 and the Copyright and Related Rights Regulation 2003, this publication may only be reproduced, stored or transmitted, in any form or by any means, with the prior permission in writing of the Publishers or in the case of reprographic reproduction in accordance with the terms of licences issued by the Copyright Licensing Agency in the UK. US copyright law is applicable to users in the USA.

⊗ The paper used in this publication meets the requirements of ANSI/NISO Z39.48–1992 (Permanence of Paper).

Registered charity number: 207890

Photoinduced Processes in Nucleic Acids and Proteins

Faraday Discussions

www.rsc.org/faraday_d

A General Discussion on Photoinduced Processes in Nucleic Acids and Proteins was held in Kerala, India on the 11th, 12th and 13th of January 2018.

RSC Publishing is a not-for-profit publisher and a division of the Royal Society of Chemistry. Any surplus made is used to support charitable activities aimed at advancing the chemical sciences. Full details are available from www.rsc.org

CONTENTS

ISSN 1359-6640; ISBN 978-1-78801-158-7

Cover
See Kohler *et al.*, *Faraday Discuss.*, 2018, **207**, 267–282.

Exploring the effects of a non-aqueous solvent on excited electronic states in DNA.

Front cover image courtesy of Dr Kimberly de La Harpe, Department of Physics, United States Air Force Academy, USA.

INTRODUCTORY LECTURE

PAPERS AND DISCUSSIONS

IISER
THIRUVANANTHAPURAM

ANATEK
SERVICES PVT. LTD.

IUPAB
CONNECTING THE
WORLD OF BIOPHYSICS

CONCLUDING REMARKS

ADDITIONAL INFORMATION

This journal is © The Royal Society of Chemistry 2018

Faraday Discussions

PAPER

Spiers Memorial Lecture

Introductory lecture: the impact of structure on photoinduced processes in nucleic acids and proteins

Tatiana Domratcheva and Ilme Schlichting *

Received 6th March 2018, Accepted 7th March 2018
DOI: 10.1039/c8fd00058a

Light is an important environmental variable and most organisms have evolved means to sense, exploit or avoid it and to repair detrimental effects on their genome. In general, light absorption is the task of specific chromophores, however other biomolecules such as oligonucleotides also do so which can result in undesired outcomes such as mutations and cancer. Given the biological importance of light-induced processes and applications for imaging, optogenetics, photodynamic therapy or photovoltaics, there is a great interest in understanding the detailed molecular mechanisms of photoinduced processes in proteins and nucleic acids. The processes are typically characterized by time-resolved spectroscopic approaches or computation, inferring structural information on transient species from stable ground state structures. Recently, however, structure determination of excited states or other short-lived species has become possible with the advent of X-ray free-electron lasers. This review gives an overview of the impact of structure on the understanding of photoinduced processes in macromolecules, focusing on systems presented at this Faraday Discussion meeting.

1. Introduction

Light is an important environmental variable. High energy UV light can damage for example DNA while visible light provides energy for photosynthesis, producing biomass, enabling and sustaining oxygenic life on earth. Organisms have therefore evolved photosensory proteins that sense most of the spectral region, enabling phototaxis or photoavoidance, allowing us to see, and providing clues to *e.g.* synchronize daily rhythms. Understanding these processes not only has an impact on human health (melanoma, jet lag, photodynamic cancer therapy...) but also on sustainable energy sources by inspiring photovoltaic systems that use solar energy to produce electrical current. Moreover, the use of light-sensitive proteins has transformed cell- and neurobiology. High resolution imaging of live cells is made possible by fluorescent proteins, providing

Max Planck Institute for Medical Research, Jahnstr. 29, 69120 Heidelberg, Germany. E-mail: ilme.schlichting@mpimf-heidelberg.mpg.de

This journal is © The Royal Society of Chemistry 2018

unprecedented insight into various processes. Optogenetics allows manipulation of diverse processes *in vivo* ranging from vectorial ion transport and signalling by second messengers (Ca^{2+}, cyclic nucleotides) to changing oligomerization states of proteins with high temporal and spatial resolution.

For all these reasons, there is great interest in understanding photoinduced processes in nucleic acids and proteins. Recent developments in ultrafast spectroscopy, computational chemistry and in particular time-resolved crystallography have provided unprecedented and novel insights into the very early processes induced by photon absorption, which so far had been difficult to investigate. Taken together, this information provides a detailed understanding of how the early excited-state relaxation couples to the protein matrix and how protein dynamics shapes early photodynamics and the subsequent photochemical events. This Faraday Discussion meeting brings together experts who use different methods to provide detailed insights in the early events in the various photoinduced processes that can take place in nucleic acids and proteins. It will be interesting to see common themes and variations emerge.

2. Photoinduced DNA damage and repair by photoreactivation

Nucleobases strongly absorb UV photons. In single bases the absorbed energy is transformed into heat on a sub-picosecond time scale by nonradiative decay.[1] In contrast, the decay of excess electronic energy is significantly slower in DNA oligonucleotides[2,3] which are multichromic molecules with numerous overlapping transitions in the UV. These transitions populate different types of excited states including excitons, excimers, exciplexes, and charge transfer (CT) states delocalized on several nucleobases, in addition to the locally excited states of single nucleobases. Stacking and pairing of nucleobases in DNA duplexes favours population of CT states that correspond to intra-strand or inter-strand electron transfer between nucleobases thus resulting in the transient formation of a radical pair. The CT states have been identified as gateways to oxidative damage and photodamage (Banyasz *et al.*, DOI: 10.1039/c7fd00179g; Giussani *et al.*, DOI: 10.1039/c7fd00202e; Jian *et al.*[4]), but at the same time they mediate energy dissipation increasing DNA photostability (Martinez-Fernandez and Improta, DOI: 10.1039/c7fd00195a). The processes are typically characterized by time-resolved spectroscopic approaches[2,3,5–9] or computation (reviewed in ref. 10), inferring structural information on transient species from stable ground state structures. Known structures of DNA duplexes provide a basis for the development of highly accurate computational-chemistry models in order to study the reactivity of single nucleobases, dimers and higher-order assemblies in a complex DNA environment (Chakraborty *et al.*, DOI: 10.1039/c7fd00188f; Giussani *et al.*, DOI: 10.1039/c7fd00202e; Martinez-Fernandez and Improta, DOI: 10.1039/c7fd00195a). Ultimately, the excitation and relaxation processes of nucleic acids depend not only on oligomeric state and sequence, but also on conformational dynamics (Segarra-Martí *et al.*, DOI: 10.1039/c7fd00201g; Brown *et al.*, DOI: 10.1039/c7fd00186j) and solvent effects (Zhang *et al.*, DOI: 10.1039/c7fd00205j).

Although base stacking and pairing in single and double stranded DNA result in significantly longer lived excited states, they rarely cause photochemical

 This journal is © The Royal Society of Chemistry 2018

damage. In fact, the energy of 99.9% of the absorbed photons is transformed into heat *via* fast nonradiative decay. Nevertheless, in strong sunlight the far UV region of the solar spectrum can induce 100 000 lesions in DNA per exposed cell per hour. So-called direct DNA damage can occur for example when two adjacent pyrimidine bases react inside DNA upon UV-irradiation to form covalent pyrimidine dimers: either a cyclobutane–pyrimidine-dimer (CPD) (80%) or less frequently (20%) a (6-4) pyrimidine–pyrimidone photoproduct (6-4PP) (Scheme 1a). Both lesions are mutagenic and can eventually lead to cancer. Organisms have thus evolved means to revert this damage. In placental mammals lesion repair occurs *via* different processes including nucleotide excision. Most other organisms use photolyases, a distinct and ubiquitous class of flavoenzymes which are believed to be ancient DNA repair proteins as evidenced for example by their occurrence in giant DNA viruses.[11] Photolyases can reverse the UV-induced DNA damage using the energy of light, a process known as photoreactivation.[12] Notably, photoreactivation is also mediated by CT states formed between the enzyme flavin cofactor and the DNA photolesion, and thus involves transient formation of a radical pair that undergoes recombination after cleavage of the interbase covalent bonds. Such a complex repair mechanism critically depends on orchestration of the electron-transfer and bond-cleavage steps by coupling the protein dynamics with the photo-induced dynamics of the charge-transfer radical pair. Structure determination of photolyases complexed with lesion-containing DNA duplexes[13–15] greatly contributed to the understanding of the photoreactivation mechanism.

2.1 DNA repair by photolyase

CPD and 6-4PP lesions are repaired by specific photolyases which are therefore classified as CPD photolyases and 6-4 photolyases, respectively. They share similar

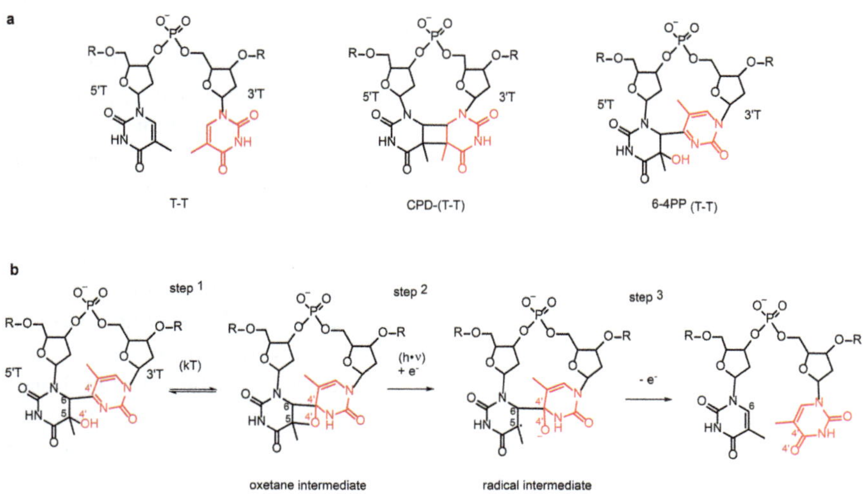

Scheme 1 (a) Chemical structure of the T–T dinucleotide and its CPD and 6-4 PP photolesions; (b) sequence of steps involved in the hypothesised repair of 6-4 PP photolesions by 6-4 photolyase *via* formation and photoreactivation of the oxetane intermediate. The atoms of the 3′ base (3′T) are shown in red.

primary sequences and structures but can only repair one kind of photoproduct, which they recognize with high specificity. Photolyases absorb blue light by the catalytically active flavin adenine dinucleotide (FAD) cofactor or through energy transfer from an excited antenna chromophore, either 5,10-methenyltetrahydrofolate (MTHF) or 8-hydroxy-5-deazariboflavin (F_0), which have absorption maxima of ~380 and ~440 nm, respectively. In CPD and 6-4 photolyases the catalytically active cofactor is a fully reduced $FADH^-$. It binds to the enzyme's active site in an unusual U-shaped configuration, with its isoalloxazine and adenine rings stacked in close distance to the lesion, which is flipped out from the double stranded DNA, pointing into the active site.[13–15] Upon photoactivation, the excited $FADH^-$ donates an electron to the lesion, initiating the repair reaction. Ultimately back transfer of the electron to the flavin completes the reaction cycle. Repair of the CPD lesion involves splitting of the cyclobutane ring and formation of the two canonical bases. Ultrafast spectroscopy was used to characterize the light-induced electron-transfer steps in CPD photolyase in great detail.[16,17] Some controversies remain[18,19] concerning the experimental details and the rate of the electron-back transfer.

Repair of the 6-4PP lesions to the undamaged T–T or T–C sequence is more complicated than the CPD repair which only involves bond cleavage. To recover the initial two-base sequence, 6-4 photolyase must control the timing of the repair steps which involve not only bond cleavage but also bond formation. The 6-4 PP lesion undergoes transfer of O and H atoms (shown in red in Scheme 1b) from the 5′ base (5′T) to the 3′ base (3′T) base in addition to the cleavage of the C6–C4′ covalent bond.

Before the crystal structure of the 6-4 photolyase complexed to a DNA duplex containing 6-4PP lesion was available[20,21] it was believed that 6-4PP rearranges to an oxetane intermediate upon binding to the enzyme (step 1 in Scheme 1b). The excited $FADH^-$ cofactor donates an electron to the oxetane intermediate (step 2 in Scheme 1b). Since the oxetane ring is not stable in the radical-anion form it undergoes rearrangement *via* cleavage of the C5–O4′ bond. Backward electron transfer (step 3) finalizes the enzymatic cycle by recovering reduced $FADH^-$ before the repaired DNA is released from the enzyme.

This reaction mechanism was shown to be incorrect, as the crystal structure of 6-4 photolyase complexed with a lesion containing DNA showed a 6-4 pyrimidine-pyrimidone photoproduct lesion instead of the proposed oxetane intermediate. The lack of observation of the oxetane intermediate was not due to a crystallization artefact[15] since the crystalline 6-4 photolyase was enzymatically active, repairing the lesion *in situ* upon illumination to yield the undamaged pyrimidine–pyrimidine bases. The absence of the thermally populated intermediate indicates that all repair steps are coordinated by the protein in the transiently populated radical-pair state with a lifetime of few nanoseconds,[22] making 6-4 photolyase an important example of functional coupling between protein dynamics and substrate photodynamics. The availability of a structure of the 6-4 photolyase–lesion containing DNA complex not only falsified the accepted mechanism of the 6-4PP rearrangement, but also spurred a number of new suggestions for the repair mechanism. Most of the mechanisms proposed on the basis of spectroscopic characterization and computational studies include a proton transfer step coupled to the initial photoinduced electron transfer.[15,22] According to a two photon mechanism, an oxetane intermediate formed after an initial photoinduced electron-transfer transfer cycle was proposed to undergo rearrangement

 This journal is © The Royal Society of Chemistry 2018

into undamaged bases *via* a second photoinduced electron transfer cycle.[23,24] An alternative mechanism that avoids proton transfer steps[25,26] suggests a direct transfer of the hydroxyl group (O4′H⁻) from the 5′ base to the 3′ base followed by the cleavage of the C6–C4′ bond.

To determine which one of the currently proposed reaction mechanisms correctly describes the function of 6-4 photolyase, structural information is required of the early intermediates that precede backward electron transfer. Several kinetic intermediates featuring the radical of the photoproduct with the lifetimes ranging from tens of picoseconds to nanoseconds were identified by transient spectroscopy.[22] The structure determination will be challenging as spectroscopic studies[22] indicate that the population of the nanosecond lifetime intermediate is low, as the photoreaction competes against the futile backward electron transfer from the 6-4 photoproduct anion-radical to the FADH radical, resulting in 10% photoreactivation yield. Despite this complication, new powerful X-ray sources delivering femtosecond pulses enable high resolution insight into the early events following photon absorption by time-resolved crystallography. This opens completely uncharted territory for structural biology that was previously only accessible by ultra-fast spectroscopy.

3. Time-resolved crystallography

Crystals are highly ordered ensembles of molecules. In particular, crystals of macromolecules are often catalytically active due to their high solvent content (typically 30–80%) and relatively few weak interactions between the molecules, providing crystal contacts which allow for reaction-induced conformational changes. In order to determine structures of reaction intermediates by time-resolved crystallography, the reaction needs to be initiated efficiently, gently and faster than the reaction – or step thereof – investigated, in order to resolve the structural changes "in step" and to not wash out the structural changes by averaging different states. Photoinduced reactions can be synchronized easily over macroscopic ensembles of molecules, facilitating the study of various processes at high temporal resolution by an optical pump and X-ray probe technique. However, due to the high molecular concentration in crystals (typically 10–60 mM in protein crystals), optical densities are often very high, complicating if not preventing efficient reaction triggering by light. It is therefore advantageous to use crystals that are as small as possible but still yield high resolution diffraction. Collecting diffraction data of very small crystals and other weakly scattering objects is made possible by X-ray free-electron lasers (XFELs), linear accelerator based X-ray sources that deliver femtosecond pulses with a peak brilliance exceeding that of synchrotron sources by nine orders of magnitude. An additional advantage over synchrotron sources is that the short pulse length of XFELs not only extends the time-resolution from 100 ps (limited by the electron bunch length at synchrotrons) to the chemical timescale of femtoseconds, but also allows outrunning most radiation damage effects.[27] Each microcrystal yields a single diffraction pattern upon XFEL pulse exposure before it is destroyed by subsequent damage effects (diffraction before destruction), necessitating high throughput serial data collection, an approach called serial femtosecond crystallography (SFX). Depending on crystallization conditions and the amount of material available, the microcrystals are delivered into the XFEL beam by jetting in thin liquid microstreams, streaming of viscous materials with embedded crystals or rastering of chips (fixed targets) containing the microcrystals.[28] All these approaches are compatible with a pump probe data collection scheme and allow data acquisition at ambient temperature, enabling time-resolved experiments on light-sensitive systems.

4. Double bond isomerization in photoactive proteins

Light-induced double bond isomerizations occur in the chromophores of various photoreceptors, including tetrapyrrole in phytochromes, *p*-coumaric acid in photoactive yellow protein, retinal in rhodopsins and 4-(*p*-hydroxybenzylidene) imidazolidin-5-one in fluorescent proteins (Scheme 2). In photoreceptor proteins, photoinduced change of the chromophore configuration facilitates rearrangement of the hydrogen-bonding network surrounding the chromophore and often causes protonation or deprotonation events that together trigger larger structural changes of the protein resulting in formation of the biologically active signalling state. In fluorescent proteins, change of protein–chromophore interactions induced by isomerization drastically changes the emission properties. Since many of these proteins have important applications in imaging or optogenetics, it is of great practical interest to understand the molecular origins of their colour tuning, specificity, photoactivation and fluorescence efficiency or nature of side-reactions in order to modify these properties in a rational way. To this end, the photoreaction itself needs to be characterized in depth (photophysics, photochemistry) as well as the interactions of the chromophore with the protein matrix. These interactions play a central role in excited-state evolution by introducing steric constraints as well as dynamic effects, controlling the timing of elementary steps and defining coupling between the cofactor photoreaction and protein motions.

4.1 Rhodopsins

Rhodopsins are α-helical membrane proteins that contain a retinal cofactor linked to the protein *via* a Schiff base with a lysine side chain. The most

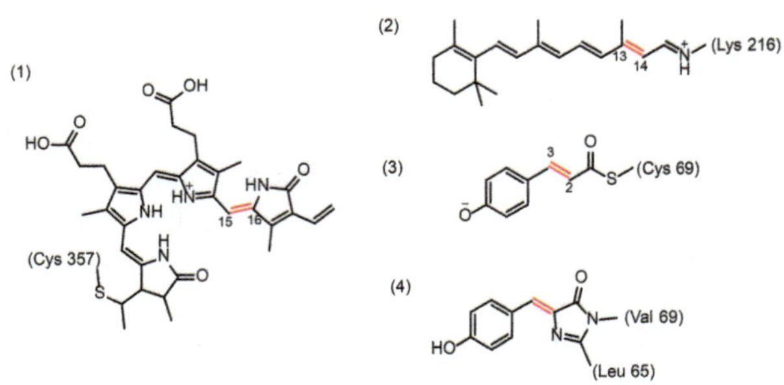

Scheme 2 Chemical structures of biological chromophores undergoing photoisomerization around a double bond (shown in red): (1) protonated tetrapyrrole in plant phytochrome, (2) protonated retinal Schiff-base in bacteriorhodopsins, (3) *p*-coumaric acid (anion) in photoactive yellow protein, and (4) 4-(*p*-hydroxybenzylidene)imidazolidin-5-one in fluorescent proteins.

prominent member of the family is rhodopsin, a G-protein coupled photoreceptor involved in visual phototransduction in animals. Microbial and algal rhodopsins are involved in directed ion transport and have prominent applications in optogenetics. While photoexcited retinal in solution isomerizes at several double bond positions, the protein matrix constrains this process. In visual rhodopsins, 11-*cis*-retinal is converted to all-*trans*-retinal whereas microbial rhodopsins convert all-*trans* retinal to 13-*cis*. The best characterized microbial rhodopsin is bacteriorhodopsin, a light-driven proton pump. Its photocycle is described by a sequence of distinct spectroscopic states: upon photon absorption a short-lived excited intermediate I decays within ~0.5 ps to the J intermediate. A hierarchical series of intermediates (J → K → L → M → N → O) with ps (J), μs (K, L) and ms (M, N, O) lifetimes follows before the resting state is reached. The crucial steps enabling proton transport take place at the active site, which comprises Wat402 forming hydrogen bonds with the protonated Schiff base and the anionic side chains of Asp85 and Asp212. Deprotonation of the Schiff base of the photoisomerized retinal and protonation of Asp85 during the L → M transition ultimately results in proton translocation to the extracellular side. Recently, time-resolved SFX was used to visualize the steps from the K-intermediates onward. The K-intermediate, captured 16 ns after photon absorption[29] shows a strained 13-*cis* isomer of retinal and disorder of Wat402 which is close to the Schiff base and part of a water-mediated hydrogen bonded network involved in proton translocation. Most likely, changes in Wat402 relate to the very early events in the photoisomerization reaction. Water-mediated hydrogen bonded networks are not only important for ion conductance in bR but also for stabilizing different conformations in the related retinal containing protein rhodopsin, a G protein coupled receptor (Lesca *et al.*, DOI: 10.1039/c7fd00207f).

Insight into the early events following photon absorption has been obtained by ultrafast spectroscopy (*e.g.* ref. 30–34 and Agathangelou *et al.*, DOI: 10.1039/c7fd00200a) and computation (reviewed in ref. 35 and see *e.g.* Borin *et al.*, DOI: 10.1039/c7fd00198c). Induced by photoexcitation, charge transfer is followed by bond length alteration (BLA) related to changes of bond character along the retinal polyene chain as well as activation of hydrogen-out-of-plane (HOOP) wagging motions which together facilitate twisting about the isomerizing bond of the chromophore. These molecular motions drive the excited chromophore toward the conical intersection between the S_1 and S_0 potential energy surfaces.[35] The site of retinal isomerization and quantum yield differ markedly between retinal protonated Schiff base bound to a protein and retinal in solution. Similar vibrational signatures of retinal in protein and solution, imply similar starting geometries,[36] suggesting that the differences in reactivity likely results from the activation of distinct nuclear motion by energy transfer from the initially populated coordinates.[37,38] It is to be expected that these high-frequency retinal modes couple with lower frequency modes of the protein already on an ultrafast timescale.

4.2 Fluorescent proteins

Fluorescent proteins (FP) have revolutionized cell biology by serving as genetically encoded markers that enable visualization of different constituents and processes *in vivo*. FPs derived from a green fluorescent protein (GFP) consist of a beta barrel that restricts access of water to the chromophore as well as its conformational flexibility. The chromophore is formed by oxidation of amino acids and consists of two π-conjugated phenol and imidazolinone rings linked by a methylene bridge. Variants of GFP can undergo a variety of light-induced transformations:[39] FPs can be activated by light from a dark non-fluorescent form to a bright fluorescent form or *vice versa*. Reversibly photoswitchable FPs can be toggled back and forth between dark and bright forms using light with different wavelength, whereas photoconvertible FPs can be switched between two fluorescent forms of different colours. Photoconvertible and photoswitchable FPs provide the basis for many super-resolution microscopy techniques.

Reversibly photoswitchable FPs include Dronpa,[40] Padron,[41] IrisFP,[42] asFP395,[43] and rsEGFP2.[44] The light-induced interconversion between dark and bright forms involves *cis* to *trans* isomerization and a change in protonation state of the chromophore. While the order of events is still under some debate, there is evidence[45–47] that isomerization occurs on a picosecond time scale, followed by protonation on a microsecond time scale. The first comprehensive study on the isomerization reaction was performed recently on rsEGFP2 (ref. 44) using a combination of transient absorption spectroscopy, computation and time-resolved crystallography at an XFEL.[48] rsEGFP2 can be switched from the fluorescent ON state (*cis* anionic chromophore) with 488 nm light to the non-fluorescent OFF state (*trans* protonated chromophore); OFF to ON switching can be done with 400 nm light.[44] Since the quantum yield for the ON to OFF reaction is an order of magnitude lower (0.04) than that for the OFF to ON reaction (0.4), only the latter has been investigated in detail[48] so far. Analysis of time-resolved absorption spectra showed that within 90 fs after absorption of a 400 nm photon two electronically excited intermediates are populated from the Franck Condon state. They decay with 0.9 ps and 3.7 ps time constants to the ground electronic state, either back to the protonated *trans* isomer or forward to the still-protonated *cis* isomer. Unlike GFP[49] and Venus (Dhamija *et al.*, DOI: 10.1039/c7fd00187h), rsEGFP2 does not undergo excited state proton transfer as indicated by similar results obtained when the measurements were performed in D_2O instead of H_2O. Instead, proton transfer occurs on a slower time scale in the ground state similar to that characterized by previous studies on Dronpa[45,46] and IrisFP.[47]

Time-resolved serial femtosecond crystallography measurements were performed to visualize the two short-lived intermediates by collecting diffraction data 1 ps and 3 ps after microcrystals of ON state rsEGFP2 had been excited by an optical pump pulse (see Fig. 1). The 1 ps pump probe delay data showed two low occupancy (~6–7% total occupancy) intermediates, one with a twisted conformation of the chromophore (T), the other with a planar conformation (P). In the 3 ps delay data the occupancy of the twisted intermediate T was lower, indicating decay of this intermediate. The planar intermediate P could not be identified in 3 ps data, either because it had been fully depleted or because of lower quality of the diffraction data. The 3 ps delay data also indicated population of the ground state *cis* isomer and repopulation of the OFF state, suggesting that a fraction of the chromophores has isomerized at 3 ps but not yet at 1 ps after photon absorption. In the T intermediate, the phenol and imidazolinone rings of the chromophore are oriented perpendicularly, halfway between the ON- and OFF state conformations. This distorted geometry is in line with the results from QM/MM and

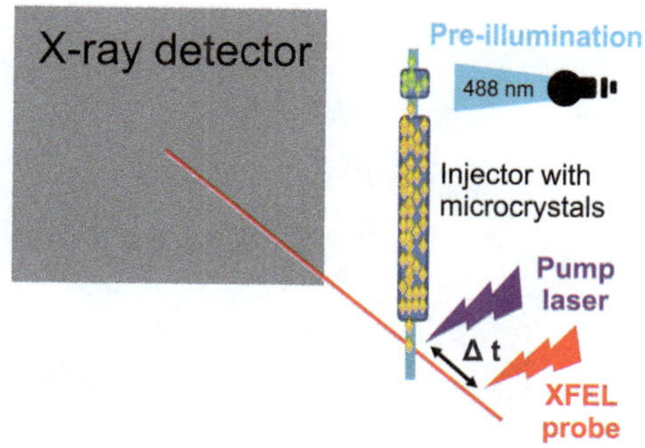

Fig. 1 Experimental setup of a time-resolved pump probe experiment to study photo-switching in a fluorescent protein.[48]

classical MD calculations for the S_1 excited-state species. The P intermediate was also characterized by calculations and likely corresponds to one of the short-lived intermediates observed by transient absorption spectroscopy.

The conformational changes of the chromophore during *trans* to *cis* isomerization are in line with a hula-twist motion.[50] Formation of the twisted intermediate is accommodated by changes in the chromophore binding pocket that involve residues Val151 and Thr204 lining chromophore phenol group. Notably, Thr204 has different rotamers in the OFF-, T intermediate-, and ON states, whereas a clash with the side chain of Val151 is avoided by a tilt of the chromophore imidazolinone ring which may cause a translation of the α-helix containing the chromophore. It thus appears as if chromophore twisting was constrained at the phenol side. Accommodation of a twisted chromophore by the protein is believed to be essential for photoswitching in reversibly switching FPs. Since Val151 seems to constrain the twisted conformation (see Fig. 2), it was hypothesized that a smaller amino acid might facilitate photoswitching. Indeed, the Val151Ala variant has a higher photoswitching quantum yield than the wildtype (Val151Ala (wt) OFF-to-ON: 0.77 (0.4); ON-to-OFF 0.064 (0.043)). Thus, detailed knowledge of the structure of the excited state intermediate was used for the first time to rationally engineer a fluorescent protein.[48]

4.3 Photoactive yellow protein

Photoactive yellow protein (PYP) is a blue light photoreceptor in purple photosynthetic bacteria. PYP has a *p*-coumaric acid (pCA) chromophore attached to a cysteine residue (Cys69) *via* a thioester bond. In the dark-adapted state, the deprotonated *trans* chromophore is stabilized by strong hydrogen bonds with Tyr42 and Glu46. Upon exposure to blue light, PYP undergoes a fully reversible branched photocycle from the dark state, through two early spectroscopic intermediates (I_0 and $I_0^{\ddagger}$)† that decay on a picosecond and nanosecond timescale to a red-shifted intermediate denoted pR (or I_1). On a microsecond timescale pR decays to a blue-shifted state denoted pB (or I_2), which then reverts to the dark state. The primary photochemical event activating PYP is the *trans* to *cis* isomerization of pCA around the central C3=C2 double bond.[51] The structural

† There is no commonly agreed-on nomenclature for the various intermediates.

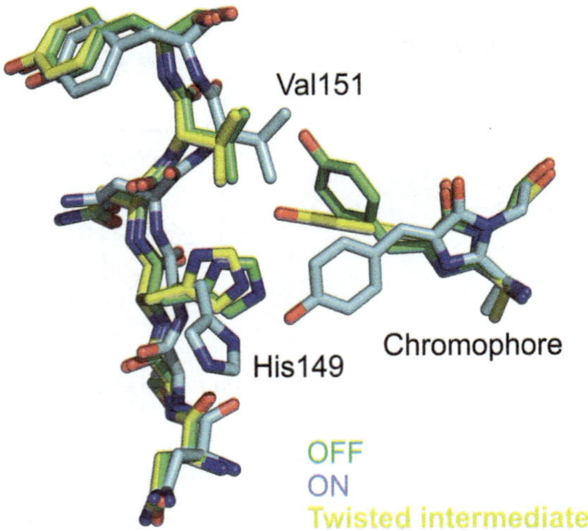

Fig. 2 Chromophore conformation in the non-fluorescent OFF (green) and the fluorescent ON (light-blue) state as well in the twisted intermediate of rsEGFP2 observed 1 ps after photoexcitation. The close proximity of the Val151 side chain to the twisted chromophore is noticeable. Indeed, the Val151Ala variant has a higher switching quantum yield.[48]

changes upon formation of the *cis*-like configuration (I_0) described as volume-saving hula-twist and bicycle bipedal isomerization pathways[52] induce rearrangement of hydrogen bonds. The chromophore adopts a planar *cis*-geometry at the cost of breaking the hydrogen bond between its carbonyl group and the Cys69 backbone. Subsequently, the side-chain configuration of Cys69 changes. Two short hydrogen bonds between the pCA phenolate and Tyr42 and Glu46 break, exposing the pCA phenolate to water and facilitating protonation, which is spectroscopically observed as a blue-shifted intermediate.[53] The rearrangement of the hydrogen-bonding network around isomerized pCA propagates the local photochemically-induced conformational changes to the N-terminal region of the protein, resulting in formation of the biologically active signalling state.

The structural changes occurring during PYP photoactivation have been followed by time-resolved crystallography from 100 picoseconds to several seconds[52–54] using synchrotron sources. Based on similar diffraction data different reaction schemes were proposed for the nanosecond and microsecond intermediates and the chromophore conformation in the early 100 ps structure was interpreted differently.[52,53] In the first intermediate (lifetime ~600 ps) the pCA chromophore is highly contorted, with its carbonyl group rotated nearly 90° with respect to the plane of the phenolate ring. A hydrogen bond between the pCA carbonyl and the Cys69 backbone constrains the chromophore in this unusual twisted conformation. There were, however, different interpretations[52,53,55,56] concerning the geometry of the twisted cofactor, in particular the dihedral angle involving the isomerizing C3=C2 double bond (30° (pR_0 state[53]) *versus* 90° (I_T state[52])). Recent ultrafast time-resolved crystallography resolved these issues, showing an angle of ~35° for a structure determined 3 ps after photoexcitation.[57] The initial steps along the isomerization were also resolved, indicating a change

 This journal is © The Royal Society of Chemistry 2018

of the double-bond configuration between 500 fs and 700 fs from a twisted *trans* conformation of the chromophore to a near *cis* form with the dihedral angle changing from ~140° to ~50°. In the twisted *trans* conformation the C2=C3 double bond is displaced by ~1 Å from the plane defined by the chromophore structure in the ground state. The hydrogen bonds between the phenolate ring and the Tyr42 and Glu46 side chains (or between the chromophore phenolate ring and the protein) are significantly elongated.

The time delays probed by SFX (0.1 to 3 ps) are within the vibrational dephasing time of the PYP S_1 state[58] and ground state modes in proteins. However, no oscillatory motions were observed for the protein residues or the cofactor. This could be due to sampling rate or to the inherently low quality of the extrapolated structure factors.

5. Ligand dissociation from myoglobin

Many gaseous ligand-binding heme proteins not only bind their natural ligand but also carbon monoxide (CO). This paves the way for time-resolved studies on ligand dissociation and re-binding exploiting the fact that the iron–CO bond is light-sensitive and dissociates upon absorption of a visible photon. The poster child for such studies is the oxygen storage protein myoglobin (Mb). It was the first protein whose three-dimensional structure was determined almost 60 years ago.[59] The structure of the α-helical protein showed that there is no direct path for ligands to bind from the solvent to the heme, implying transient structural changes. Mb has thus become a model system to study protein dynamics, including the vibrational energy flow in proteins.

Ligand binding and dissociation in myoglobin have been studied in great detail using numerous approaches. In synchrotron-based time-resolved X-ray crystallography,[53] the time-resolution was limited by the 100 ps duration of the X-ray pulse and the structural changes emanating from the breaking ligand bond had already spread throughout the protein, resulting in large scale helix movements.[53] It was therefore not possible to identify how and in which order these changes are induced by the breaking of the Fe–CO bond. These initial events were followed by sub-ps time-resolved SFX at an XFEL.[60] Within 100 fs, CO appears in the primary docking site and the iron moves 0.2 Å out of the heme plane, in line with prior spectroscopic evidence. Structural changes appear throughout the protein within 500 fs in a quake-like fashion[61] again in line with spectroscopic evidence, with the C-, F- and H-helices moving away from the heme and the E- and A-helices moving towards it. This is followed by further displacement of the E-helix and of the distal histidine. These collective movements are predicted by quantum mechanics/molecular mechanics (QM/MM) simulations and also observed in time-resolved solution studies.[62] Interestingly, also sub-picosecond dynamics of residues close to the heme were observed. The χ_1 torsion angles of Phe43, Val68, and Ile107 show an exponential change with time. Their side chains have a high probability of being hit by the dissociating CO[63], which would result in a large momentum transfer, causing a movement of the side chain, followed by an exponential relaxation. In contrast, the χ_1 and/or χ_2 torsion angles of a number of side chains in close proximity to the heme and of a number of hydrogen bond distances seem to oscillate with time at sub-ps time delays. The oscillations share a 500 ± 150 fs period, which is reminiscent of the doming motion of the heme photoproduct of 430 fs (75 cm^{-1}).[64] Based on fs coherence spectroscopy, it has been suggested that electronic rearrangements in the heme iron upon ligand photodissociation can simultaneously excite associated low-frequency vibrational modes.[65] The vibrational modes of the heme couple to collective global modes of the protein as suggested earlier,[66,67] resulting not only in an immediate collective response of the protein but also the later, large scale motions. This notion of a network of coupled modes is important for the understanding of the nature and time-scales of enzyme catalysis[68] and protein dynamics.[69,70] Excitation and coupling of collective motions results in anisotropic energy flow through the protein structure. This is likely a general allosteric mechanism for the transport of local binding energy to remote sites in proteins.[70]

6. Conclusions and challenges

The advent of femtosecond time-resolved crystallography at XFELs and multidimensional spectroscopies allows us to obtain detailed insights into ultrafast protein dynamics, including structures of early intermediates, concerted motions of the excited chromophore and the surrounding protein. This not only provides structural information on vibronic coupling in photosensitive biomolecules, but generally enables correlating information obtained from various time-resolved spectroscopy methods with direct structural information.

So far, all time-resolved crystallographic studies using femtosecond laser excitation used significantly higher pump energies than typically used for spectroscopic measurements. Comparing information derived from data that were obtained using very different excitation energies can, however, result in complications.[71,72] Goodno *et al.*[71] suggested that at high power excitation conditions, a larger fraction of the excess thermal energy will be transferred *via* the protein to the solvent *via* collective modes rather than through the slower diffusive channel. The reason would be a stronger anharmonic coupling between the collective modes of the protein and an increased spatial dispersion of the excess energy.[71]

Despite this challenge, it is enticing to correlate spectroscopic and structural data by comparing vibrational spectra obtained experimentally and derived computationally using structural models of short-lived species obtained by time-resolved crystallography. This will also facilitate development of computational methods towards an accurate description of interactions between the photoactive chromophore and its molecular environment (protein, DNA, solvent *etc.*) in particular in the excited state. An accurate description of intermolecular interactions at the quantum-mechanics level[73] is still challenging, but necessary since the vibrational frequencies, especially of low-frequency modes, are sensitive to these interactions. Another challenge will be to account for anharmonic coupling enabling energy transfer from the high-frequency modes directly activated by photoexcitation to the low-frequency modes driving the system towards a conical intersection. These modes are observed as vibrational coherences in photoactive proteins, *e.g.* in rhodopsin and bacteriorhodopsin,[37,74–76] but assignment of these coherences to specific vibrations of the chromophore coupled to the vibrations of the protein is still an open question. For a complex many-atom system, it is often unclear how to relate the computed frequencies to the experimental ones without additional experimental data, for instance the frequency shifts caused by site-directed isotope labeling.[77] Hence, visualizing the coupled chromophore and protein vibrations by means of femtosecond time-resolved crystallography at XFELs may indeed result in a breakthrough in the identification of collective motions of biomolecules, which would advance our knowledge on the photochemical and functional dynamics of DNA and proteins.

Conflicts of interest

There are no conflicts to declare.

Acknowledgements

Open Access funding provided by the Max Planck Society.

References

1 C. E. Crespo-Hernandez, B. Cohen, P. M. Hare and B. Kohler, Ultrafast excited-state dynamics in nucleic acids, *Chem. Rev.*, 2004, **104**(4), 1977–2019.

2 C. E. Crespo-Hernandez, B. Cohen and B. Kohler, Base stacking controls excited-state dynamics in A.T DNA, *Nature*, 2005, **436**(7054), 1141–1144.

3 C. E. Crespo-Hernandez, K. de la Harpe and B. Kohler, Ground-state recovery following UV excitation is much slower in G x C-DNA duplexes and hairpins than in mononucleotides, *J. Am. Chem. Soc.*, 2008, **130**(33), 10844–10845.

4 Y. Jian, E. Maximowitsch, D. Liu, S. Adhikari, L. Li and T. Domratcheva, Indications of 5′ to 3′ Interbase Electron Transfer as the First Step of Pyrimidine Dimer Formation Probed by a Dinucleotide Analog, *Chem.–Eur. J.*, 2017, **23**(31), 7526–7537.

5 D. Markovitsi, UV-induced DNA Damage: The Role of Electronic Excited States, *Photochem. Photobiol.*, 2016, **92**(1), 45–51.

6 G. W. Doorley, M. Wojdyla, G. W. Watson, M. Towrie, A. W. Parker, J. M. Kelly and S. J. Quinn, Tracking DNA Excited States by Picosecond-Time-Resolved Infrared Spectroscopy: Signature Band for a Charge-Transfer Excited State in Stacked Adenine–Thymine Systems, *J. Phys. Chem. Lett.*, 2013, **4**, 2739–2744.

7 D. B. Bucher, B. M. Pilles, T. Carell and W. Zinth, Charge separation and charge delocalization identified in long-living states of photoexcited DNA, *Proc. Natl. Acad. Sci. U. S. A.*, 2014, **111**(12), 4369–4374.

8 Y. Zhang, J. Dood, A. A. Beckstead, X. B. Li, K. V. Nguyen, C. J. Burrows, *et al.*, Efficient UV-induced charge separation and recombination in an 8-oxoguanine-containing dinucleotide, *Proc. Natl. Acad. Sci. U. S. A.*, 2014, **111**(32), 11612–11617.

9 Y. Y. Zhang, K. de La Harpe, A. A. Beckstead, R. Improta and B. Kohler, UV-Induced Proton Transfer between DNA Strands, *J. Am. Chem. Soc.*, 2015, **137**(22), 7059–7062.

10 R. Improta, F. Santoro and L. Blancafort, Quantum Mechanical Studies on the Photophysics and the Photochemistry of Nucleic Acids and Nucleobases, *Chem. Rev.*, 2016, **116**(6), 3540–3593.

11 M. G. Fischer, I. Kelly, L. J. Foster and C. A. Suttle, The virion of Cafeteria roenbergensis virus (CroV) contains a complex suite of proteins for transcription and DNA repair, *Virology*, 2014, **466–467**, 82–94.

12 R. Dulbecco, Reactivation of ultra-violet-inactivated bacteriophage by visible light, *Nature*, 1949, **163**(4155), 949.

13 A. Mees, T. Klar, P. Gnau, U. Hennecke, A. P. Eker, T. Carell, *et al.*, Crystal structure of a photolyase bound to a CPD-like DNA lesion after *in situ* repair, *Science*, 2004, **306**(5702), 1789–1793.

14 S. Kiontke, Y. Geisselbrecht, R. Pokorny, T. Carell, A. Batschauer and L. O. Essen, Crystal structures of an archaeal class II DNA photolyase and its complex with UV-damaged duplex DNA, *EMBO J.*, 2011, **30**(21), 4437–4449.

15 M. J. Maul, T. R. Barends, A. F. Glas, M. J. Cryle, T. Domratcheva, S. Schneider, *et al.*, Crystal structure and mechanism of a DNA (6-4) photolyase, *Angew. Chem., Int. Ed. Engl.*, 2008, **47**(52), 10076–10080.

16 V. Thiagarajan, M. Byrdin, A. P. Eker, P. Muller and K. Brettel, Kinetics of cyclobutane thymine dimer splitting by DNA photolyase directly monitored in the UV, *Proc. Natl. Acad. Sci. U. S. A.*, 2011, **108**(23), 9402–9407.

17 Z. Liu, C. Tan, X. Guo, Y. T. Kao, J. Li, L. Wang, *et al.*, Dynamics and mechanism of cyclobutane pyrimidine dimer repair by DNA photolyase, *Proc. Natl. Acad. Sci. U. S. A.*, 2011, **108**(36), 14831–14836.

18 K. Brettel and M. Byrdin, DNA photolyase: is the nonproductive back electron transfer really much slower than forward transfer?, *Proc. Natl. Acad. Sci. U. S. A.*, 2012, **109**(23), E1462.

19 D. Zhong, A. Sancar and A. Stuchebrukhov, Reply to Brettel and Byrdin: On the efficiency of DNA repair by photolyase, *Proc. Natl. Acad. Sci. U. S. A.*, 2012, **109**, E1463.

20 X. Zhao, J. Liu, D. S. Hsu, S. Zhao, J. S. Taylor and A. Sancar, Reaction mechanism of (6-4) photolyase, *J. Biol. Chem.*, 1997, **272**(51), 32580–32590.

21 K. Hitomi, H. Nakamura, S. T. Kim, T. Mizukoshi, T. Ishikawa, S. Iwai, *et al.*, Role of two histidines in the (6-4) photolyase reaction, *J. Biol. Chem.*, 2001, **276**(13), 10103–10109.

22 J. Li, Z. Liu, C. Tan, X. Guo, L. Wang, A. Sancar, *et al.*, Dynamics and mechanism of repair of ultraviolet-induced (6-4) photoproduct by photolyase, *Nature*, 2010, **466**(7308), 887–890.

23 J. Yamamoto, R. Martin, S. Iwai, P. Plaza and K. Brettel, Repair of the (6-4) photoproduct by DNA photolyase requires two photons, *Angew. Chem., Int. Ed. Engl.*, 2013, **52**(29), 7432–7436.

24 K. Sadeghian, M. Bocola, T. Merz and M. Schutz, Theoretical study on the repair mechanism of the (6-4) photolesion by the (6-4) photolyase, *J. Am. Chem. Soc.*, 2010, **132**(45), 16285–16295.

25 T. Domratcheva and I. Schlichting, Electronic structure of (6-4) DNA photoproduct repair involving a non-oxetane pathway, *J. Am. Chem. Soc.*, 2009, **131**(49), 17793–17799.

26 T. Domratcheva, Neutral histidine and photoinduced electron transfer in DNA photolyases, *J. Am. Chem. Soc.*, 2011, **133**(45), 18172–18182.

27 H. N. Chapman, C. Caleman and N. Timneanu, Diffraction before destruction, *Philos. Trans. R. Soc., B*, 2014, **369**, 20130313.

28 I. Schlichting, Serial femtosecond crystallography: the first five years, *IUCrJ*, 2015, **2**(Pt 2), 246–255.

29 E. Nango, A. Royant, M. Kubo, T. Nakane, C. Wickstrand, T. Kimura, *et al.*, A three-dimensional movie of structural changes in bacteriorhodopsin, *Science*, 2016, **354**(6319), 1552–1557.

 This journal is © The Royal Society of Chemistry 2018

30 R. A. Mathies, C. H. Brito Cruz, W. T. Pollard and C. V. Shank, Direct observation of the femtosecond excited-state *cis–trans* isomerization in bacteriorhodopsin, *Science*, 1988, **240**(4853), 777–779.

31 G. R. Loppnow and R. A. Mathies, Excited-state structure and isomerization dynamics of the retinal chromophore in rhodopsin from resonance Raman intensities, *Biophys. J.*, 1988, **54**(1), 35–43.

32 T. Kobayashi, T. Saito and H. Ohtani, Real-time spectroscopy of transition states in bacteriorhodopsin during retinal isomerization, *Nature*, 2001, **414**(6863), 531–534.

33 J. Herbst, K. Heyne and R. Diller, Femtosecond infrared spectroscopy of bacteriorhodopsin chromophore isomerization, *Science*, 2002, **297**(5582), 822–825.

34 V. I. Prokhorenko, A. M. Nagy, S. A. Waschuk, L. S. Brown, R. R. Birge and R. J. Miller, Coherent control of retinal isomerization in bacteriorhodopsin, *Science*, 2006, **313**(5791), 1257–1261.

35 S. Gozem, H. L. Luk, I. Schapiro and M. Olivucci, Theory and Simulation of the Ultrafast Double-Bond Isomerization of Biological Chromophores, *Chem. Rev.*, 2017, **117**(22), 13502–13565.

36 S. O. Smith, M. S. Braiman, A. B. Myers, J. A. Pardoen, J. M. L. Courtin, C. Winkel, J. Lugtenburg and R. A. Mathies, Vibrational analysis of the all-*trans*-Retinal Chromophore in light-adapted bacteriorhodopsin, *J. Am. Chem. Soc.*, 1987, **109**, 3108–3125.

37 M. Liebel, C. Schnedermann, G. Bassolino, G. Taylor, A. Watts and P. Kukura, Direct observation of the coherent nuclear response after the absorption of a photon, *Phys. Rev. Lett.*, 2014, **112**(23), 238301.

38 P. J. Johnson, A. Halpin, T. Morizumi, L. S. Brown, V. I. Prokhorenko, O. P. Ernst, *et al.*, The photocycle and ultrafast vibrational dynamics of bacteriorhodopsin in lipid nanodiscs, *Phys. Chem. Chem. Phys.*, 2014, **16**(39), 21310–21320.

39 A. Acharya, A. M. Bogdanov, B. L. Grigorenko, K. B. Bravaya, A. V. Nemukhin, K. A. Lukyanov, *et al.*, Photoinduced Chemistry in Fluorescent Proteins: Curse or Blessing?, *Chem. Rev.*, 2017, **117**(2), 758–795.

40 M. Andresen, A. C. Stiel, S. Trowitzsch, G. Weber, C. Eggeling, M. C. Wahl, *et al.*, Structural basis for reversible photoswitching in Dronpa, *Proc. Natl. Acad. Sci. U. S. A.*, 2007, **104**(32), 13005–13009.

41 M. Andresen, A. C. Stiel, J. Folling, D. Wenzel, A. Schonle, A. Egner, *et al.*, Photoswitchable fluorescent proteins enable monochromatic multilabel imaging and dual color fluorescence nanoscopy, *Nat. Biotechnol.*, 2008, **26**(9), 1035–1040.

42 V. Adam, M. Lelimousin, S. Boehme, G. Desfonds, K. Nienhaus, M. J. Field, *et al.*, Structural characterization of IrisFP, an optical highlighter undergoing multiple photo-induced transformations, *Proc. Natl. Acad. Sci. U. S. A.*, 2008, **105**(47), 18343–18348.

43 M. Andresen, M. C. Wahl, A. C. Stiel, F. Grater, L. V. Schafer, S. Trowitzsch, *et al.*, Structure and mechanism of the reversible photoswitch of a fluorescent protein, *Proc. Natl. Acad. Sci. U. S. A.*, 2005, **102**(37), 13070–13074.

44 T. Grotjohann, I. Testa, M. Reuss, T. Brakemann, C. Eggeling, S. W. Hell, *et al.*, rsEGFP2 enables fast RESOLFT nanoscopy of living cells, *eLife*, 2012, **1**, e00248.

45 M. M. Warren, M. Kaucikas, A. Fitzpatrick, P. Champion, J. T. Sage and J. J. van Thor, Ground-state proton transfer in the photoswitching reactions of the fluorescent protein Dronpa, *Nat. Commun.*, 2013, **4**, 1461.
46 D. Yadav, F. Lacombat, N. Dozova, F. Rappaport, P. Plaza and A. Espagne, Real-time monitoring of chromophore isomerization and deprotonation during the photoactivation of the fluorescent protein Dronpa, *J. Phys. Chem. B*, 2015, **119**(6), 2404–2414.
47 J. P. Colletier, M. Sliwa, F. X. Gallat, M. Sugahara, V. Guillon, G. Schiro, *et al.*, Serial Femtosecond Crystallography and Ultrafast Absorption Spectroscopy of the Photoswitchable Fluorescent Protein IrisFP, *J. Phys. Chem. Lett.*, 2016, **7**(5), 882–887.
48 N. Coquelle, M. Sliwa, J. Woodhouse, G. Schiro, V. Adam, A. Aquila, T. R. M. Barends, S. Boutet, M. Byrdin, S. Carbajo, E. De la Mora, R. B. Doak, M. Feliks, F. Fieschi, L. Foucar, V. Guillon, M. Hilpert, M. S. Hunter, S. Jakobs, J. E. Koglin, G. Kovacsova, T. J. Lane, B. Levy, M. Liang, K. Nass, J. Ridard, J. S. Robinson, C. M. Roome, C. Rueckbusch, M. Seaberg, M. Thepaut, M. Cammarata, I. Demachy, M. Field, R. L. Shoeman, D. Bourgeois, J.-P. Colletier, I. Schlichting and M. Weik, Chromophore twisting in the excited state of a photoswitchable fluorescent protein captured by time-resolved serial femtosecond crystallography, *Nat. Chem.*, 2018, **10**(1), 31–37.
49 D. Stoner-Ma, A. A. Jaye, P. Matousek, M. Towrie, S. R. Meech and P. J. Tonge, Observation of excited-state proton transfer in green fluorescent protein using ultrafast vibrational spectroscopy, *J. Am. Chem. Soc.*, 2005, **127**(9), 2864–2865.
50 R. S. H. Liu, Photoisomerization by hula-twist: A fundamental supramolecular photochemical reaction, *Acc. Chem. Res.*, 2001, **34**(7), 555–562.
51 D. S. Larsen and R. van Grondelle, Initial photoinduced dynamics of the photoactive yellow protein, *ChemPhysChem*, 2005, **6**(5), 828–837.
52 Y. O. Jung, J. H. Lee, J. Kim, M. Schmidt, K. Moffat, V. Srajer, *et al.*, Volume-conserving *trans–cis* isomerization pathways in photoactive yellow protein visualized by picosecond X-ray crystallography, *Nat. Chem.*, 2013, **5**(3), 212–220.
53 F. Schotte, H. S. Cho, V. R. Kaila, H. Kamikubo, N. Dashdorj, E. R. Henry, *et al.*, Watching a signaling protein function in real time *via* 100-ps time-resolved Laue crystallography, *Proc. Natl. Acad. Sci. U. S. A.*, 2012, **109**(47), 19256–19261.
54 H. Ihee, S. Rajagopal, V. Srajer, R. Pahl, S. Anderson, M. Schmidt, *et al.*, Visualizing reaction pathways in photoactive yellow protein from nanoseconds to seconds, *Proc. Natl. Acad. Sci. U. S. A.*, 2005, **102**(20), 7145–7150.
55 Y. O. Jung, J. H. Lee, J. Kim, M. Schmidt, K. Moffat, V. Srajer, *et al.*, Reply to 'contradictions in X-ray structures of intermediates in the photocycle of photoactive yellow protein', *Nat. Chem.*, 2014, **6**(4), 259–260.
56 V. R. Kaila, F. Schotte, H. S. Cho, G. Hummer and P. A. Anfinrud, Contradictions in X-ray structures of intermediates in the photocycle of photoactive yellow protein, *Nat. Chem.*, 2014, **6**(4), 258–259.
57 K. Pande, C. D. Hutchison, G. Groenhof, A. Aquila, J. S. Robinson, J. Tenboer, *et al.*, Femtosecond structural dynamics drives the *trans/cis* isomerization in photoactive yellow protein, *Science*, 2016, **352**(6286), 725–729.

 This journal is © The Royal Society of Chemistry 2018

58 R. Nakamura, N. Hamada, H. Ichida, F. Tokunaga and Y. Kanematsu, Coherent oscillations in ultrafast fluorescence of photoactive yellow protein, *J. Chem. Phys.*, 2007, **127**(21), 215102.

59 J. C. Kendrew, R. E. Dickerson, B. E. Strandberg, R. G. Hart, D. R. Davies, D. C. Phillips, *et al.*, Structure of myoglobin: A three-dimensional Fourier synthesis at 2 A. resolution, *Nature*, 1960, **185**(4711), 422–427.

60 T. R. Barends, L. Foucar, A. Ardevol, K. Nass, A. Aquila, S. Botha, *et al.* Direct observation of ultrafast collective motions in CO myoglobin upon ligand dissociation, *Science*, 2015, **350**(6259), 445–450.

61 A. Ansari, J. Berendzen, S. F. Bowne, H. Frauenfelder, I. E. Iben, T. B. Sauke, *et al.*, Protein states and proteinquakes, *Proc. Natl. Acad. Sci. U. S. A.*, 1985, **82**(15), 5000–5004.

62 M. Levantino, G. Schiro, H. T. Lemke, G. Cottone, J. M. Glownia, D. Zhu, *et al.*, Ultrafast myoglobin structural dynamics observed with an X-ray free-electron laser, *Nat. Commun.*, 2015, **6**, 6772.

63 R. Elber and M. Karplus, Enhanced Sampling in Molecular Dynamics: Use of the Time-Dependent Hartree Approximation for a Simulation of Carbon Monoxide Diffusion through Myoglobin, *J. Am. Chem. Soc.*, 1990, **112**, 9161–9175.

64 L. Zhu, J. T. Sage and P. M. Champion, Observation of coherent reaction dynamics in heme proteins, *Science*, 1994, **266**(5185), 629–632.

65 P. M. Champion, F. Rosca, D. Ionascu, W. Cao and X. Ye, Rapid timescale processes and the role of electronic surface coupling in the photolysis of diatomic ligands from heme proteins, *Faraday Discuss.*, 2004, **127**, 123–135.

66 R. J. D. Miller, Energetics and Dynamics of Deterministic Protein Motion, *Acc. Chem. Res.*, 1994, **27**, 145–150.

67 J. T. Sage, S. M. Durbin, W. Sturhahn, D. C. Wharton, P. M. Champion, P. Hession, *et al.*, Long-range reactive dynamics in myoglobin, *Phys. Rev. Lett.*, 2001, **86**(21), 4966–4969.

68 V. L. Schramm, Transition States and transition state analogue interactions with enzymes, *Acc. Chem. Res.*, 2015, **48**(4), 1032–1039.

69 H. Fujisaki and J. E. Straub, Vibrational energy relaxation in proteins, *Proc. Natl. Acad. Sci. U. S. A.*, 2005, **102**(19), 6726–6731.

70 G. Li, D. Magana and R. B. Dyer, Anisotropic energy flow and allosteric ligand binding in albumin, *Nat. Commun.*, 2014, **5**, 3100.

71 G. D. Goodno, V. Astinov and R. J. D. Miller, Femtosecond heterodyne-detected four-wave-mixing studies of deterministic protein motions. 2. Protein response, *J. Phys. Chem. A*, 1999, **103**, 10630–10643.

72 V. I. Prokhorenko, A. Halpin, P. J. Johnson, R. J. Miller and L. S. Brown, Coherent control of the isomerization of retinal in bacteriorhodopsin in the high intensity regime, *J. Chem. Phys.*, 2011, **134**(8), 085105.

73 J. Y. Hasegawa, K. Yanai and K. Ishimura, Quantum mechanical molecular interactions for calculating the excitation energy in molecular environments: a first-order interacting space approach, *ChemPhysChem*, 2015, **16**(2), 305–311.

74 P. J. Johnson, A. Halpin, T. Morizumi, V. I. Prokhorenko, O. P. Ernst and R. J. Miller, Local vibrational coherences drive the primary photochemistry of vision, *Nat. Chem.*, 2015, **7**(12), 980–986.

75 J. P. Kraack, T. Buckup, N. Hampp and M. Motzkus, Ground- and excited-state vibrational coherence dynamics in Bacteriorhodopsin probed with degenerate four-wave-mixing experiments, *ChemPhysChem*, 2011, **12**(10), 1851–1859.

This journal is © The Royal Society of Chemistry 2018

76 D. P. Hoffman and R. A. Mathies, Femtosecond Stimulated Raman Exposes the Role of Vibrational Coherence in Condensed-Phase Photoreactivity, *Acc. Chem. Res.*, 2016, **49**(4), 616–625.
77 T. Domratcheva, E. Hartmann, I. Schlichting and T. Kottke, Evidence for Tautomerisation of Glutamine in BLUF Blue Light Receptors by Vibrational Spectroscopy and Computational Chemistry, *Sci. Rep.*, 2016, **6**, 22669.

 This journal is © The Royal Society of Chemistry 2018

Faraday Discussions

PAPER

The role of water molecules in phototransduction of retinal proteins and G protein-coupled receptors

Elena Lesca,[*ab] Valérie Panneels[a] and Gebhard F. X. Schertler[*ab]

Received 9th November 2017, Accepted 20th November 2017
DOI: 10.1039/c7fd00207f

G protein coupled receptors (GPCRs) are a key family of membrane proteins in all eukaryotes and also very important drug targets for medical intervention. The extensively studied visual pigment rhodopsin is a prime example of a family A GPCR. Its chromophore ligand retinal is covalently linked to a lysine in helix seven forming a protonated Schiff base. Interestingly, this is the same situation in other-non-GPCR-retinal proteins, like the prototype light-driven microbial proton pump bacteriorhodopsin, albeit there is no (or only a very remote) phylogenetical link. Close to the retinal ligand, several water molecules help to organise a functionally important hydrogen bond network that undergoes significant changes during photo-activation. Such water-mediated networks are also critical for ligand binding of other GPCRs and they are becoming increasingly important in drug discovery. GPCRs also contain a partially conserved water mediated hydrogen bond network that stabilises the ground state of the receptor, and rearrangement of this network leads to the stabilization of the active state. Some water molecules have a specific role in this process to appropriately orient specific residues relative to the Schiff base, and to modulate the fine structure of the transmembrane bundle, particularly near the intracellular G protein binding site. While the atomic details of these mechanisms are still missing, the recent developments in free electron lasers (FELs) are enabling us to begin to observe the changes in waters and relevant side chains shortly after photo activation at an unprecedented level of spatial and temporal resolution.

Introduction

Retinal proteins catalyse proton or ion transport and they are the most important signal transducers in photoreception in multicellular organisms.[1] For example, the archaea proteins bacteriorhodopsin and halorhodopsin are a proton pump and chloride pump, and a sensory rhodopsin is important for phototaxis, respectively. Visual pigments of vertebrates and invertebrates have the key

[a]Division of Biology and Chemistry, Laboratory of Biomolecular Research, Paul Scherrer Institute, 5232 Villigen PSI, Switzerland. E-mail: elena.lesca@psi.ch; gebhard.schertler@psi.ch
[b]Department of Biology D-BIOL ETH Zurich, 8093 Zürich, Switzerland

conserved sequence signatures of family A GPCRs and they trigger the nucleotide exchange of heterotrimeric G proteins.[2,3] Visual pigments can be grouped in so-called monostable pigments, which lose the retinal upon photoactivation and need an external regeneration system, and bistable pigments, in which retinal remains covalently attached upon activation and a second light stimulus triggers the reverse reaction.[4]

Monostable GPCRs include the vertebrate pigment rhodopsin for dim light vision and the green, red, and blue cone pigments that mediate our colour vision. After their photoactivation (see ref. 5 for a review), which isomerises retinal from the inactive form with 11-*cis* to the active form with all-*trans* retinal, several intermediate states can be identified including an active state *meta*-rhodopsin II that is characterized by a deprotonated Schiff base (Fig. 1).[6] Like most GPCRs, activated rhodopsin interacts with and activates a heterotrimeric G protein (transducin) by catalysing nucleotide exchange (GDP for GTP). In vertebrate rhodopsins *meta*-rhodopsin II decays after some time, the Schiff base is hydrolysed, and the retinal is released. Retinal is then re-isomerised back to the 11-*cis* form in the pigment epithelium and then transported back to the sensory cells where it binds to opsins to form the regenerated pigment.

While bistable visual pigments[7] are commonly found in invertebrates, they also include melanopsins, which are found in the retinal ganglion cells of vertebrate eyes, where they serve as critical amongst others light sensors for the entrainment of the circadian rhythm.[8] For instance, melanopsin is responsible for bringing us back to our normal day and night rhythm after jetlag. In bistable

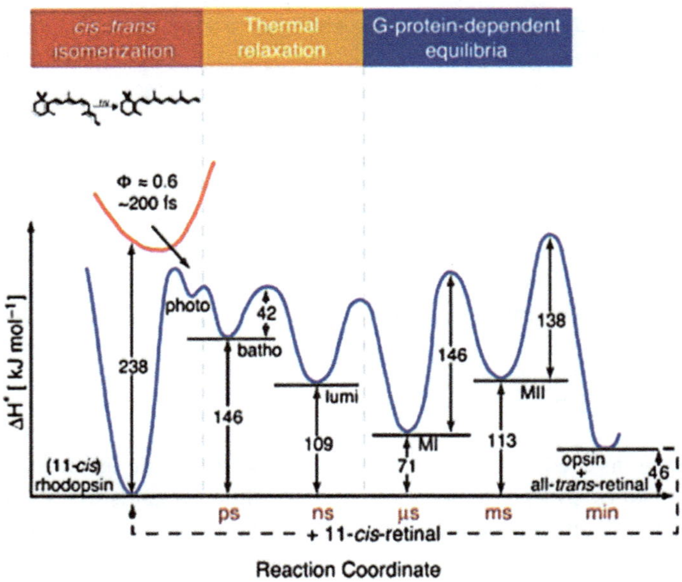

Fig. 1 Reaction coordinate for the activation of rhodopsin. 11-*cis* retinal in rhodopsin is isomerized after absorption of a photon. Then, the protein thermally relaxes over several intermediates until reaching an equilibrium between the Meta I (MI) and Meta II (MII) states. Meta-II represents the activated form of the receptor that binds to and activates the G protein. Taken from V. Lemaître and A. Watts, "Conformational Studies of Retinal in Rhodopsin" published on researchgate.

 This journal is © The Royal Society of Chemistry 2018

pigments the retinal remains attached to the receptor, the Schiff base link stays protonated[4] and retinal therefore does not leave the protein. In this case, a second photon is used to regenerate the 11-*cis* retinal from the all-*trans* retinal form.

During the last 20 years, substantial progress in the structure determination of membrane proteins has been made[9] and, by now, there are many structures of microbial retinal proteins available. In addition, over the last decade, more and more structures from GPCRs, including vertebrate and invertebrate rhodopsins, have been determined,[3] and some of them have high enough resolution that we can see water molecules embedded in specific regions of the transmembrane bundle.[10] As the resolution of these X-ray structures increased, several highly relevant water molecules inside the protein structure were discovered, and it became increasingly obvious that analyzing the role of these waters is critical for understanding these systems.[11–15] These water molecules have several roles including an essential function for the changes that occur after activation. Specifically, three major roles can be assigned to water molecules: (1) they can be important components of hydrogen bond networks that stabilise the overall structures of inactive and active states[11,16–19] (Fig. 2 and 3); (2) they can provide a local hydrophilic environment for polar ligands (Fig. 4), a property that is now being more and more exploited in structure-based drug discovery where water has an important modulating effect in ligand or cofactor binding;[20,21] and (3) they can play a similar role for polar amino acid side chains (Fig. 3), in which case the water molecule is really a well-ordered co-factor that can modulate the catalytic action of the protein (Fig. 5).

In this article, we will look at several X-ray structures of membrane proteins to exemplify these different roles of water molecules. The focus will be on retinal proteins and photoactivatable GPCRs.

Results and discussion

Stabilising the ground and active states of rhodopsin and family A GPCRs

The ground (inactive) and Meta II (active; *i.e.* able to activate transducin) states of vertebrate rhodopsin (Fig. 1) are energetically stable and have been solved by X-ray

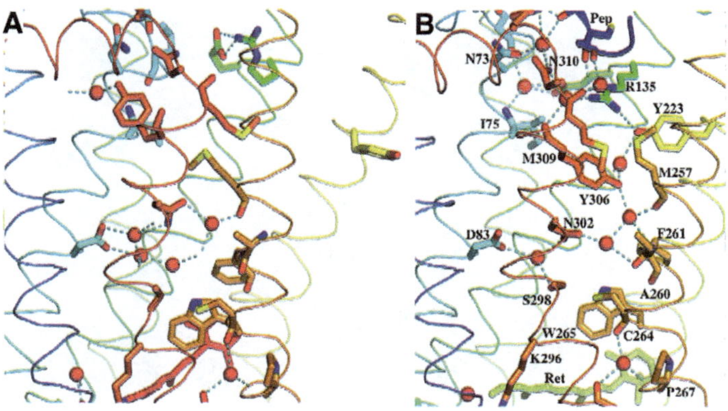

Fig. 2 The water-mediated hydrogen networks in rhodopsin are partially conserved in family A GPCRs. Panel (A) shows its ground inactive state. (B) shows the activated form of rhodopsin (Meta II-like state). Polar and highly conserved residues are highlighted. Red balls represent the oxygen atoms of water molecules.

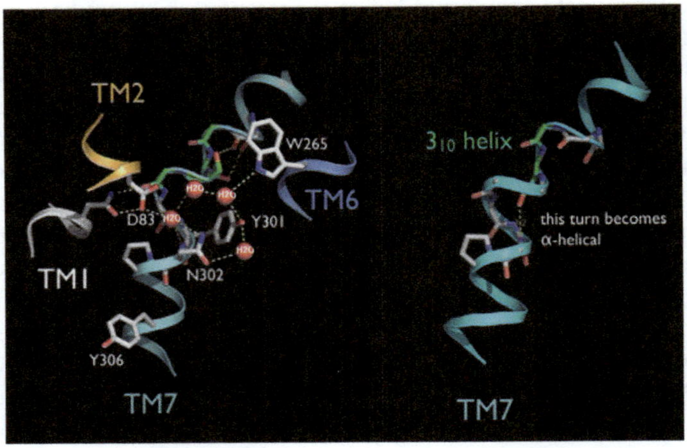

Fig. 3 A structural irregularity in transmembrane helix 7 is stabilized by water molecules. In the transition from the inactive (left) to the active (right) conformation of bovine rhodopsin, a 3–10 helix segment becomes alpha-helical, a transition that is mediated by the rearrangement of a water-mediated hydrogen bond network in this region.

crystallography. When we analyse these structures in detail (Fig. 2) we observe two distinct complex hydrogen bond networks that are strongly mediated by water molecules. Comparable states are found in many GPCRs and often include a sodium binding site.[10] In both networks, the water molecules help to maximize the completeness of the hydrogen bond networks and help to satisfy hydrogen bond donor and acceptor conditions.

Water molecules are often found at sites of helix irregularities,[12] for example at a segment of a 3_{10} helix in transmembrane helix 7, where the backbone has unsatisfied hydrogen bond potential that can be satisfied by coordinating water (see Fig. 3 for an example). Here, the water-mediated hydrogen network is linking

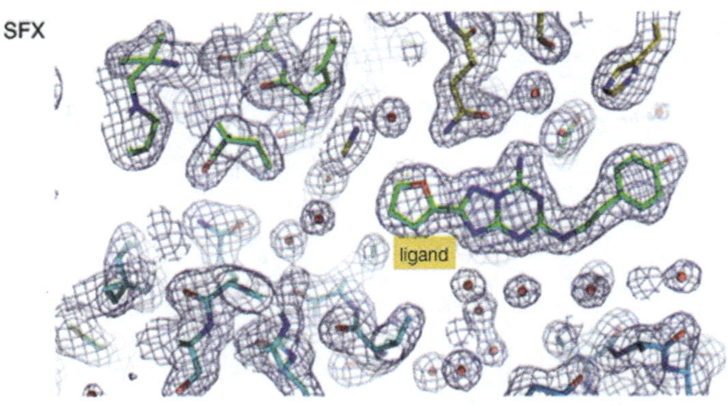

Fig. 4 Water molecules, labeled as red dots, in the pharmacologically relevant ligand binding site of the adenosine A2a receptor. The 1.7 Å electron density map was determined with an X-ray free electron laser. A model of the receptor and the ligand is shown in sticks (adapted from R. K. Y. Cheng, R. Abela and M. Hennig, X-ray free electron laser: Opportunities for drug discovery, *Essays Biochem.*, 2017, **61**, 529–542).

 This journal is © The Royal Society of Chemistry 2018

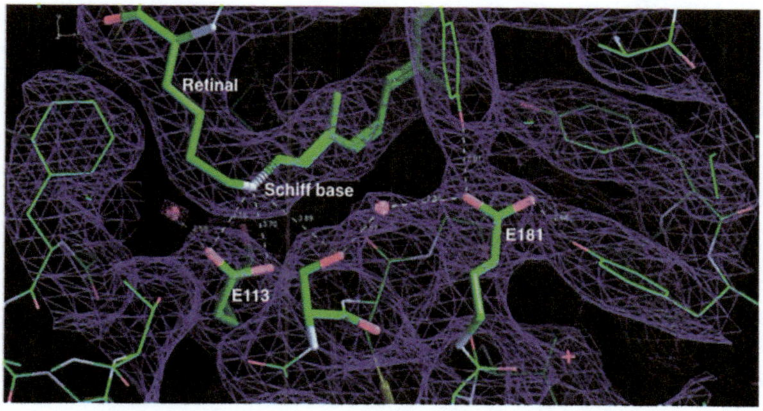

Fig. 5 Water molecules in the ground state of rhodopsin close to the Schiff base nitrogen. Amino acids in the local water-mediated hydrogen-bond network are highlighted. The carboxyl group of the counterion residue E113 is directly interacting with the Schiff base while the closest water is only interacting with the carboxyl group.

several helices together in the inactive state of rhodopsin, in a region that rearranges during activation (Fig. 2 and 3). This is an important factor for the stabilization of both inactive as well as active states. The existence of these two distinctive hydrogen bond networks in these structures is even more significant as they involve the most important conserved residues in family A GPCRs. The requirement of two distinct stabilising networks is likely the reason for the strong evolutionary conservation of many of the polar residues that are involved in these networks. It is interesting to realise that not only the amino acids in the polypeptide have been conserved, but also to some extent the positions where water molecules help in completing the network, a combination of (i) fulfilling the maximum number of hydrogen bonds by providing an adequate environment for polar side chains, and (ii) stereo-chemical constraints are important for reaching a stable receptor conformation in the inactive and active states. Therefore, water molecules play a significant role in the general allosteric activation mechanism, not only in vertebrate visual pigments but in all family A GPCRs.

Water molecules are important in ligand binding of photopigments and GPCRs

The analysis of high-resolution structures of ligand-binding sites in membrane proteins has led to the detection of water molecules for example in GPCRs such as the adenosine A2a receptor[22] (Fig. 4) or rhodopsin (Fig. 5 and 6), squid opsin,[23] and in bacteriorhodopsin[24] (Fig. 7). The differences in the electron density of water molecules could indicate different occupation and dynamics. Some of the waters are in positions where all polar interactions are very well fulfilled, and they are therefore low energy sites; others have less favourable positions and merely fulfil a 'filling' role for cavities in the ligand binding site by occupying space which is not occupied by the ligand.[20] Water molecules in the ligand binding pocket are also important for the 'stiffness' of the ligand binding site. Taken together, these waters contribute to ligand affinity, and potentially influence the on- and off-rate of ligands.[25] In drug design, it has been assumed that replacing water molecules

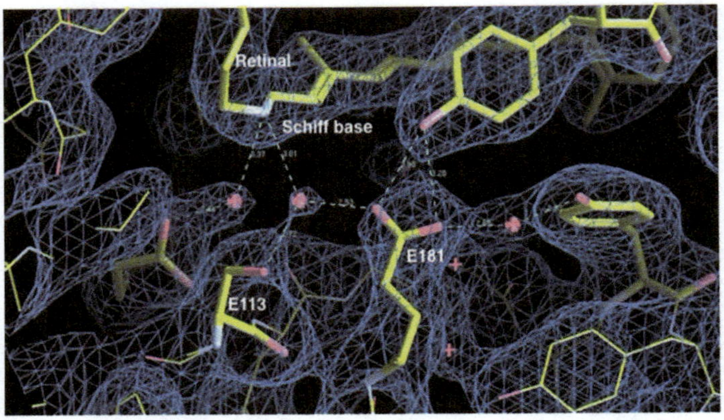

Fig. 6 Water molecules in the activated state of rhodopsin close to the Schiff base nitrogen. An extended water-mediated hydrogen-bond network forms in the Meta II state after photo-activation of rhodopsin. While the counterion residue 113 has moved further away from the Schiff base, the carboxyl group of E181 has moved closer to it, and, in addition, two water molecules are directly interacting with the Schiff base. Also, the tyrosine residues that 'hold' E181 further away from the Schiff base in the inactive state are now re-arranged, so additional water molecules can extend the hydrogen-bond network.

that are not in a low energy state can improve the properties of a ligand. Today, there are several methods that also can predict potential water binding positions even if the resolution of a receptor structure does not directly allow the determination of water positions experimentally.[21] Therefore, we speculate that understanding water in the context of ligand or co-factor binding will increasingly become important for understanding enzymatic mechanisms as well as substrate and ligand binding in proteins. Finally, in photo-pigments, water molecules situated close to a chromophore can also influence the polarizability of the chromophore binding site and thus influence the spectral tuning of chromophore absorption[26] and modulate dynamics of chromophore isomerisation.

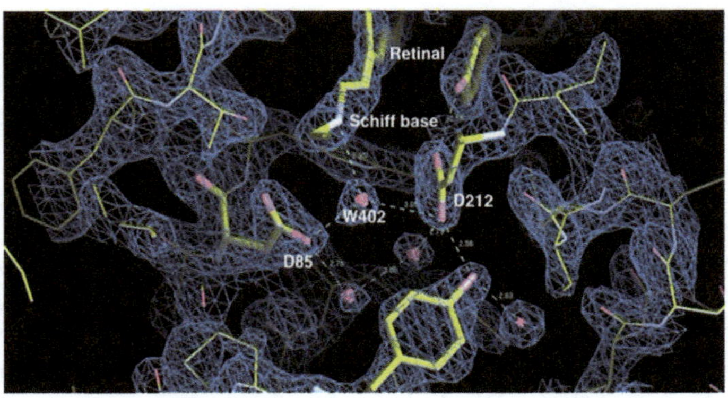

Fig. 7 The ground state structure of bacteriorhodopsin obtained with a room temperature serial crystallography experiment at the FEL is shown. Two carboxyl groups and three water molecules form a pentameric cluster close to the Schiff base, with one of the water molecules (W402) in direct contact with the Schiff base. Polar residues close to the Schiff base and the retinal are highlighted.

 This journal is © The Royal Society of Chemistry 2018

Water molecules are key cofactors in the catalytic action of retinal proteins

When we inspect the structures of rhodopsin (Fig. 5 and 6) and bacteriorhodopsin (Fig. 7) around the retinal binding site close to the Schiff base, we can identify several polar residues and water molecules in the ground state (Fig. 5 and 7). The water molecules and carboxyl groups neutralize the positively charged Schiff base and stabilize its proton by forming a hydrogen bond. After photo-activation, we can observe changes in the arrangement of the polar residues and of the position of ordered water molecules in the retinal binding pocket of rhodopsin (Fig. 6). These complex 'counterions' composed of negatively charged carboxyl groups and water molecules, play an important role in colour tuning of pigments and in influencing the quantum yield and speed of isomerisation upon photo-activation. Specifically, in the rhodopsin ground state the water molecules help to position the counterion carboxyl group E113 close to the protonated Schiff base.[17] On the other hand, in the activated state two water molecules are close to the nitrogen of the Schiff base which now is deprotonated. In this process, one water has inserted itself between the carboxyl group of E113 and the nitrogen of the Schiff base. The second water in the activated state is interacting with the carboxyl group of E181, which in the activated state of rhodopsin has moved closer to the Schiff base nitrogen after undergoing a rotamer change. This interplay between the E113 and E181 counterions is an interesting topic in the field. In invertebrate pigments, which are evolutionarily older, E181 has been postulated as the counterion.[4] However, in the ground state of vertebrate rhodopsin, the carboxyl group of E181 is engaged with several tyrosine residues in a water-mediated hydrogen bond network that we term the 'tyrosine cage'. Importantly, water molecules play a very direct role in orienting specific side chains (carboxyl groups or hydroxyl groups) relative to the Schiff base proton, a central functional element in all retinal proteins. Interestingly, in bistable pigments deprotonation of the Schiff base does not occur. In such pigments, no water molecules or negatively charged carboxyl groups that could act as proton acceptors are close to the protonated nitrogen of the Schiff base.[23] We speculate that the lack of deprotonation of the Schiff base in bistable pigments is due the lack of water molecules close to the Schiff base that could accept the proton and guide it away.

Time-resolved room temperature X-ray crystallography to study photo-activation of retinal proteins

A rapidly developing area in membrane protein crystallography is room temperature crystallography, using either X-ray free electron lasers (FELs) for serial femtosecond crystallography (SFX)[27] or high brilliance synchrotrons for serial microsecond crystallography (SMX).[28] With these powerful technologies, it has become possible to obtain full structural information by following fast photoreactions. For the first time, we have been able to observe early changes in the retinal protein bacteriorhodopsin in a series of structure snapshots that provide a movie-like view of the activation process.[24] Due to the high resolution, it was possible to monitor the rearrangement of side chains and, furthermore, the changes in the order and position of water molecules instants after photo activation; we were able to record snapshots as short as 16 ns after photo-activation. From a series of snapshots at 16 different time points, we assembled a movie showing electron density changes over time upon photo-activation. We found that, in bacteriorhodopsin, there is a cluster of three water molecules connected to two carboxyl groups. These water molecules are highly ordered in the ground state close to the Schiff base (Fig. 7). When the difference maps in the region of the Schiff base are inspected in detail, we observe that 16 ns after activation the water molecule closest to the Schiff base shows a negative difference density peak without a positive component. This most likely means that the water has already started to be less ordered 16 ns after light activation (Fig. 8). These measurements were carried out at room temperature using serial crystallography in a free electron laser set up with a lipidic cubic phase jet by measuring a large number of single crystal diffraction patterns for each time point. The negative density increases further over time and finally after 260 microseconds all three water molecules in the cluster close to the Schiff base have disordered and many side chains have rearranged.[29] These measurements are a first direct observation at room temperature of the behaviour of water molecules inside a retinal protein upon photoactivation. The data show how, in the dark state, the water molecule closest to the protonated Schiff base is stabilising its hydrogen, a factor that explains its high pK_a. Further along the activation reaction, in the active state there is a deprotonated Schiff base, which means that now its pK_a is lower. Therefore, disordering of a specific water molecule after photo-isomerisation is a key factor that influences the pK_a of the Schiff base and allows proton transfer.[24] Here, this water molecule is playing a key role in the catalytic photo-cycle of bacteriorhodopsin.

Similar experiments have not yet been carried out with monostable and bistable visual pigments, but it is likely that, also there, water molecules play an important role. However, in the ground state of these proteins, water molecules are only weakly connected to the Schiff base, and play the role of orientating the counterions relative to the Schiff base nitrogen. Furthermore, they also might be important to guide the proton away from the Schiff base.

Experimental

The program COOT[30] was used for the analysis and visualisation of maps and PDB coordinates. All coordinates and maps were imported as deposited in the Protein

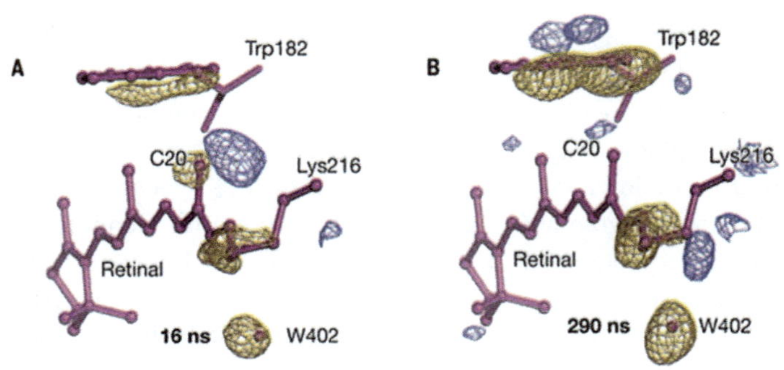

Fig. 8 Early changes in water occupancy in the photo-activation of rhodopsin. Difference electron density maps 16 (left) and 290 (right) ns after activation, obtained by time-resolved room temperature crystallography on highly ordered bacteriorhodopsin crystals. Positive and negative densities are indicated in blue and gold, respectively. It is apparent that, even after only 16 ns, water molecule W402 has been disordered, as shown by the negative difference density peak (adapted from ref. 24).

 This journal is © The Royal Society of Chemistry 2018

Data Bank (http://www.rcsb.org (ref. 31)). Coordinates, amplitudes and phases were imported in the EDS format into COOT to obtain electron density maps representations. Fig. 4–7 were prepared using PYMOL (The PyMOL Molecular Graphics System, Version 2.0 Schrödinger, LLC.) from the coordinates of inactive (PDB ID: 1GZM[17]) and active (PDB ID: 2X72[32]) bovine rhodopsin, of the adenosine A2A receptor solved by serial millisecond crystallography at room-temperature (PDB ID: 5NM4[22]), of the T94I rhodopsin mutant (5DYS[19]), and of bacteriorhodopsin in the ground state obtained with serial femtosecond crystallography (5J7A[29]).

Conclusions

Water plays an essential role in the function of membrane proteins and it is an integral part of their architecture. Specifically, water-mediated hydrogen bond networks close to the active site of retinal proteins are critical for their function, as they stabilise active and inactive conformations by complementing the existing intramolecular hydrogen bond networks in the receptor. Water molecules are also central in at least some of the key switch regions that are important for the allosteric mechanism of GPCR activation, and they can also modulate ligand binding in GPCRs. As a matter of fact, the study of the role of individual water molecules is becoming increasingly important for structure-based drug discovery. With the availability of X-ray free electron lasers (FELs), which are becoming powerful tools for time-resolved crystallography at room temperature, we will be able to analyse these key water interactions in much more detail. We are currently rapidly advancing our knowledge here in the field of retinal-binding membrane proteins, strengthening our ability to predict absorption maxima and quantum yields of these photoactivatable systems.

Conflicts of interest

Gebhard F. X. Schertler is cofounder of leadXpro and InterAx biotechnology companies.

Acknowledgements

G. F. X. Schertler acknowledges the support of all the members of his group over many years. Xavier Deupi prepared Fig. 3. The authors thank Gregor Cicchetti and Xavier Deupi for their input with the manuscript. Elena Lesca acknowledges support from the Peter and Traudl Engelhorn Foundation, and GFX Schertler acknowledges support from HFSP Research Grant RGP0034/2014, SNF grant No. 310030B_173335 and support through the ETH Zurich through the NCCR MUST and the ETH FAST programme.

Notes and references

1 O. P. Ernst, D. T. Lodowski, M. Elstner, P. Hegemann, L. S. Brown and H. Kandori, *Chem. Rev.*, 2014, **114**, 126–163.

2 A. J. Venkatakrishnan, X. Deupi, G. Lebon, F. M. Heydenreich, T. Flock, T. Miljus, S. Balaji, M. Bouvier, D. B. Veprintsev, C. G. Tate, G. F. Schertler and M. M. Babu, *Nature*, 2016, **536**, 484–487.

3 A. J. Venkatakrishnan, X. Deupi, G. Lebon, C. G. Tate, G. F. Schertler and M. M. Babu, *Nature*, 2013, **494**, 185–194.
4 M. Koyanagi and A. Terakita, *Biochim. Biophys. Acta*, 2014, **1837**, 710–716.
5 K. P. Hofmann, P. Scheerer, P. W. Hildebrand, H. W. Choe, J. H. Park, M. Heck and O. P. Ernst, *Trends Biochem. Sci.*, 2009, **34**, 540–552.
6 X. Deupi and B. K. Kobilka, *Physiology*, 2010, **25**, 293–303.
7 H. Tsukamoto and A. Terakita, *Photochem. Photobiol. Sci.*, 2010, **9**, 1435–1443.
8 S. Hattar, R. J. Lucas, N. Mrosovsky, S. Thompson, R. H. Douglas, M. W. Hankins, J. Lem, M. Biel, F. Hofmann, R. G. Foster and K. W. Yau, *Nature*, 2003, **424**, 76–81.
9 I. Moraes, G. Evans, J. Sanchez-Weatherby, S. Newstead and P. D. Stewart, *Biochim. Biophys. Acta*, 2014, **1838**, 78–87.
10 W. Liu, E. Chun, A. A. Thompson, P. Chubukov, F. Xu, V. Katritch, G. W. Han, C. B. Roth, L. H. Heitman, A. P. Ijzerman, V. Cherezov and R. C. Stevens, *Science*, 2012, **337**, 232–236.
11 T. Okada, Y. Fujiyoshi, M. Silow, J. Navarro, E. M. Landau and Y. Shichida, *Proc. Natl. Acad. Sci. U. S. A.*, 2002, **99**, 5982–5987.
12 L. Pardo, X. Deupi, N. Dolker, M. L. Lopez-Rodriguez and M. Campillo, *ChemBioChem*, 2007, **8**, 19–24.
13 T. E. Angel, M. R. Chance and K. Palczewski, *Proc. Natl. Acad. Sci. U. S. A.*, 2009, **106**, 8555–8560.
14 T. E. Angel, S. Gupta, B. Jastrzebska, K. Palczewski and M. R. Chance, *Proc. Natl. Acad. Sci. U. S. A.*, 2009, **106**, 14367–14372.
15 R. Nygaard, L. Valentin-Hansen, J. Mokrosinski, T. M. Frimurer and T. W. Schwartz, *J. Biol. Chem.*, 2010, **285**, 19625–19636.
16 H. Belrhali, P. Nollert, A. Royant, C. Menzel, J. P. Rosenbusch, E. M. Landau and E. Pebay-Peyroula, *Structure*, 1999, **7**, 909–917.
17 J. Li, P. C. Edwards, M. Burghammer, C. Villa and G. F. Schertler, *J. Mol. Biol.*, 2004, **343**, 1409–1438.
18 J. L. Miller-Gallacher, R. Nehme, T. Warne, P. C. Edwards, G. F. Schertler, A. G. Leslie and C. G. Tate, *PLoS One*, 2014, **9**, e92727.
19 A. Singhal, Y. Guo, M. Matkovic, G. Schertler, X. Deupi, E. C. Yan and J. Standfuss, *EMBO Rep.*, 2016, **17**, 1431–1440.
20 J. S. Mason, A. Bortolato, M. Congreve and F. H. Marshall, *Trends Pharmacol. Sci.*, 2012, **33**, 249–260.
21 A. Bortolato, B. G. Tehan, M. S. Bodnarchuk, J. W. Essex and J. S. Mason, *J. Chem. Inf. Model.*, 2013, **53**, 1700–1713.
22 T. Weinert, N. Olieric, R. Cheng, S. Brunle, D. James, D. Ozerov, D. Gashi, L. Vera, M. Marsh, K. Jaeger, F. Dworkowski, E. Panepucci, S. Basu, P. Skopintsev, A. S. Dore, T. Geng, R. M. Cooke, M. Liang, A. E. Prota, V. Panneels, P. Nogly, U. Ermler, G. Schertler, M. Hennig, M. O. Steinmetz, M. Wang and J. Standfuss, *Nat. Commun.*, 2017, **8**, 542.
23 M. Murakami and T. Kouyama, *Nature*, 2008, **453**, 363–367.
24 E. Nango, A. Royant, M. Kubo, T. Nakane, C. Wickstrand, T. Kimura, T. Tanaka, K. Tono, C. Song, R. Tanaka, T. Arima, A. Yamashita, J. Kobayashi, T. Hosaka, E. Mizohata, P. Nogly, M. Sugahara, D. Nam, T. Nomura, T. Shimamura, D. Im, T. Fujiwara, Y. Yamanaka, B. Jeon, T. Nishizawa, K. Oda, M. Fukuda, R. Andersson, P. Bath, R. Dods, J. Davidsson, S. Matsuoka, S. Kawatake, M. Murata, O. Nureki, S. Owada,

 This journal is © The Royal Society of Chemistry 2018

T. Kameshima, T. Hatsui, Y. Joti, G. Schertler, M. Yabashi, A. N. Bondar, J. Standfuss, R. Neutze and S. Iwata, *Science*, 2016, **354**, 1552–1557.

25 A. Bortolato, F. Deflorian, D. R. Weiss and J. S. Mason, *J. Chem. Inf. Model.*, 2015, **55**, 1857–1866.

26 K. Katayama, Y. Nonaka, K. Tsutsui, H. Imai and H. Kandori, *Sci. Rep.*, 2017, **7**, 4904.

27 R. Neutze, G. Branden and G. F. Schertler, *Curr. Opin. Struct. Biol.*, 2015, **33**, 115–125.

28 P. Nogly, D. James, D. Wang, T. A. White, N. Zatsepin, A. Shilova, G. Nelson, H. Liu, L. Johansson, M. Heymann, K. Jaeger, M. Metz, C. Wickstrand, W. Wu, P. Bath, P. Berntsen, D. Oberthuer, V. Panneels, V. Cherezov, H. Chapman, G. Schertler, R. Neutze, J. Spence, I. Moraes, M. Burghammer, J. Standfuss and U. Weierstall, *IUCrJ*, 2015, **2**, 168–176.

29 P. Nogly, V. Panneels, G. Nelson, C. Gati, T. Kimura, C. Milne, D. Milathianaki, M. Kubo, W. Wu, C. Conrad, J. Coe, R. Bean, Y. Zhao, P. Bath, R. Dods, R. Harimoorthy, K. R. Beyerlein, J. Rheinberger, D. James, D. DePonte, C. Li, L. Sala, G. J. Williams, M. S. Hunter, J. E. Koglin, P. Berntsen, E. Nango, S. Iwata, H. N. Chapman, P. Fromme, M. Frank, R. Abela, S. Boutet, A. Barty, T. A. White, U. Weierstall, J. Spence, R. Neutze, G. Schertler and J. Standfuss, *Nat. Commun.*, 2016, **7**, 12314.

30 P. Emsley, B. Lohkamp, W. G. Scott and K. Cowtan, *Acta Crystallogr., Sect. D: Biol. Crystallogr.*, 2010, **66**, 486–501.

31 H. M. Berman, J. Westbrook, Z. Feng, G. Gilliland, T. N. Bhat, H. Weissig, I. N. Shindyalov and P. E. Bourne, *Nucleic Acids Res.*, 2000, **28**, 235–242.

32 J. Standfuss, P. C. Edwards, A. D'Antona, M. Fransen, G. Xie, D. D. Oprian and G. F. Schertler, *Nature*, 2011, **471**, 656–660.

This journal is © The Royal Society of Chemistry 2018

Faraday Discussions

PAPER

Probing the excited state dynamics of Venus: origin of dual-emission in fluorescent proteins

Shaina Dhamija, [a] Bhisham Thakur, [b] Purnananda Guptasarma [b] and Arijit K. De *[a]

Received 4th October 2017, Accepted 6th November 2017
DOI: 10.1039/c7fd00187h

Fluorescent proteins exhibit interesting excited state photochemistry, leading to bright fluorescence emission that renders their versatile biological role and wide use as biomarkers. A molecular-level mechanism of the excited state dynamics is desirable to pinpoint the origin of the bright fluorescence of these proteins. Here we present studies on a yellow fluorescent protein variant, Venus, and investigate the photophysics behind the dual fluorescence emission upon UV excitation. Based on our studies, we propose that the unique nature of the potential energy surface is responsible for the observation of minor fluorescence in Venus which is not seen in wild type GFP.

1 Introduction

Photoinduced processes in proteins drive fundamental life processes such as light harvesting in photosynthesis and photo-reception in vision. Light absorption by protein complexes, for example by pigment–protein complexes (PPCs), initiates a set of downstream cascade processes. While light harvesting by photosynthetic PPCs involves photo-excitation, energy transfer between multiple chromophores and charge separation,[1] photoreception by visual PPCs involves *cis*/*trans* isomerization leading to the opening/closing of membrane proton pumps to generate a visual sensory signal. Photoinduced processes are also encountered in many natural systems other than PPCs, for example in fluorescent proteins (FPs). There are a number of light-induced phenomena in fluorescent proteins such as photoactivation, photoswitching, photoconversion, *etc.*[2,3] Some phototransformations can be reversible, giving rise to reversibly switchable fluorescent proteins.[4]

[a]*Department of Chemical Sciences, Indian Institute of Science Education and Research (IISER) Mohali, Knowledge City, Sector 81, SAS Nagar, Mohali, Punjab 140306, India. E-mail: akde@iisermohali.ac.in; arijitb4u@gmail.com*

[b]*Department of Biological Sciences, Indian Institute of Science Education and Research (IISER) Mohali, Knowledge City, Sector 81, SAS Nagar, Mohali, Punjab 140306, India*

The green fluorescent protein (GFP) from the jellyfish *Aequorea victoria* has an enormous role in biology; the bright fluorescence from GFP is utilized by jellyfish to perplex its prey or by anthozoa within coral to protect algae from harmful UV excitation.[5] GFP and its analogs are widely used as fluorescent biomarkers. Wild-type GFP (wt-GFP) consists of 238 amino acid residues forming a β-can structure with the chromophore located at the centre. By altering the side-chains in the vicinity of the chromophore, various GFP variants have been prepared that cover a range of wavelengths ranging from blue to red.

A yellow variant of GFP, enhanced yellow fluorescent protein (E-YFP), has four mutations (S65G, V68L, S72A and T203Y). A photoactive YFP is known to undergo structural rearrangement for optimum photoreception efficiency.[6] Venus, another YFP variant, contains five other mutations (F46L, F64L, M153T, V163A and S175G). Venus has an improved maturation rate at 37 °C because of its additional mutations in comparison with E-YFP.[7] While Venus does not undergo photo-switching, it undergoes photoconversion *via* excited state proton transfer (ESPT) like wt-GFP.[8,9] The photochemistry of Venus is insensitive to experimental conditions such as changes in pH and buffer content.[7,10]

Using picosecond fluorescence lifetime measurements and femtosecond pump-probe spectroscopy, in this paper we discuss the excited state dynamics of Venus upon UV photo-excitation leading to dual fluorescence emission. We compare and contrast the nature of its photophysics with those of wt-GFP.

2 Experiments

2.1 Cloning, expression and purification of the N-terminally 6xHis tagged Venus protein

A genetic construct (Venus/pCS2) was obtained from the laboratory of T. Nagai, with the gene encoding Venus present in the vector, pCS2, cloned between the BamHI and EcoRI restriction sites. Due to the unavailability of an expression host for this plasmid, the gene was sub-cloned in fusion with a 6xHis tag into the pET28a vector between the BamHI and XhoI restriction sites. The ligated pET28a plasmid containing the Venus-encoding gene was transformed into XL-I blue *E. coli* cells. Positive clones were screened for by colony PCR, using T7 universal primers. Positive clones were grown overnight at 37 °C in LB (Luria Bertani) media with shaking at 220 rpm. The pET28a plasmid was purified from these cells, using the Qiagen plasmid purification kit, and transformed into BL21(DE3)PlysS* *E. coli* cells. The protein was produced from these cells by inducing secondary cultures (at an OD of 0.6, with 1 mM IPTG) grown in the presence of kanamycin (50 μg mL^{-1}) and chloramphenicol (35 μg mL^{-1}) for 6–7 hours at 37 °C, with shaking at 220 rpm. The cells were sedimented by centrifugation at 6000 rpm for 10 minutes and then resuspended in a lysis buffer (20 mM PBS, 10 mM imidazole, 1 M NaCl, pH 7.3). The cells were lysed by sonication for 30 min and the cell lysate was separated from the debris by centrifugation at 11 000 rpm for 70 min. The released protein was purified by affinity chromatography on a Ni-NTA column, using the standard Qiagen protein affinity-purification protocol(s). Protein bound to the Ni-NTA column was eluted in 20 mM PBS, with 250 mM imidazole. The eluted protein was dialyzed against 20 mM PBS to remove the imidazole.

 This journal is © The Royal Society of Chemistry 2018

The amino acid sequence for Venus is:

MGSSHHHHHHSSGLVPRGSHMASMTGGQQMGRGSMVSKGEELFTGVVPILVELDGDV
NGHKFSVSGEGEGDATYGKLTLKLICTTGKLPVPWPTLVTTLGYGLQCFARYPDHMKQ
HDFFKSAMPEGYVQERTIFFKDDGNYKTRAEVKFEGDTLVNRIELKGIDFKEDGNILGH
KLEYNYNSHNVYITADKQKNGIKANFKIRHNIEDGGVQLADHYQQNTPIGDGPVLLPDN
HYLSYQSKLSKDPNEKRDHMVLLEFVTAAGITLGMDELYKstop

The following parameters apply to the protein Venus (as derived from sequence analysis using the software ProtParam tool[11]):

(1) Molar extinction coefficient: 23 505 M^{-1} cm^{-1}.
(2) Number of amino acids: 273.
(3) Molecular weight: 30 436.33 Daltons.
(4) Theoretical pI: 6.34.

2.2 Structural perturbation and protein stability studies

The effects of dithiothreitol (DTT), urea and temperature on the structure and local environment of the chromophore in Venus[12] were explored, for different combinations of urea or DTT concentrations and at either 25 °C (either before or after 10 minutes of heating at 95 °C) or 95 °C. Fluorescence emission spectra were recorded on a spectrofluorimeter (Cary Eclipse, Agilent Technologies), or on a Fluoromax-4C spectrofluorimeter (Horiba Scientific) using 400 nm excitation. Circular dichroism (CD) spectra were recorded with an MOS-500 CD spectrometer (BioLogic Science Instruments) or with a Chirascan CD spectrometer (Applied Photophysics). For all of the experiments, a known volume (200 μL) of 20 μM Venus was treated with different volumes of urea or DTT, and the spectra were corrected for dilution. The final concentrations of DTT that were used were either 3 mM or 6 mM, and the concentration of urea that was used was 8 M.

2.3 Steady-state and lifetime measurements

Absorption spectra were recorded on a UV-VIS spectrophotometer (Cary 5000 UV-VIS-NIR spectrophotometer, Agilent Technologies).

Fluorescence (emission) spectra and fluorescence excitation spectra were collected using a steady-state fluorimeter (FLUOROMAX-4C, Horiba Scientific).

Fluorescence lifetimes were measured using a 374 nm diode for excitation (pulse width < 200 ps), using a time-correlated single photon counting (TCSPC) fluorimeter (Deltaflex, Horiba Scientific). Lifetime analysis was carried out using the DAS6 software. The instrument response function (IRF) was ~260 ps (by measuring the FWHM of the scattering time trace using ~0.1% ludox solution, Sigma-Aldrich) which gives a sub-100 ps resolution after numerical deconvolution.

2.4 Pump-probe spectroscopy

The laser used was a Ti-sapphire amplified beam with an ~800 nm central wavelength with a ~50 fs pulse-width (FWHM) and a 1 kHz repetition-rate. A pair of beam samplers was used to split off the pump and probe beams from the main laser beam. The pump beam diameter was further reduced by 2.5 times using a convex/concave lens combination. We used a pair of half-wave plates in the pump and probe arms in order to have vertical polarization for the probe beam and to keep the pump polarization at the magic angle (54.7 deg) with respect to the probe polarization. Both of the pump and probe beams were routed to a custom-made femtosecond transient absorption spectrometer (TAS, Newport Corp) which was further re-designed according to the experiment. The pump and probe power values before TAS were both ~15 mW.

We used variable neutral density filters to control the intensity of the probe pulse at the sample. A type-I beta-barium borate crystal (0.2 mm thick) was used to generate the second-harmonic pump pulse (with a ~400 nm central wavelength) and a calcium fluoride crystal (placed on a motorized translation stage) was used to generate a broadband white-light probe. The 400 nm pump and white-light probe overlapped inside the sample; the transmitted probe was sent to a spectrometer (Oriel MS260i, Newport Corp) where it was dispersed with a grating (Richardson grating 74064, Newport Corp) and the spectrum was recorded with a linear array detector (CMOS S10453-1024Q, Hamamatsu Photonics).

The time delay between the pump and probe could be varied up to ~4 ns with ~0.13 fs step-size using a single mechanical delay line (ILS300LM, Newport Corp). A mechanical chopper (New Focus 3502, with a 7/5 slot wheel, Newport Corp) with a 500 Hz repetition-rate blocked every alternate pump pulse, and for every pump-probe delay a differential absorption spectrum (a spectrum taken with the pump beam blocked by the chopper subtracted from that taken with the pump beam unblocked) was recorded.

3 Results and discussion

Fig. 1 shows a ribbon diagram of the molecular structure of the polypeptide backbone of Venus and its chromophoric unit consisting of the amino acids, glycine, tyrosine and glycine.

3.1 SDS-PAGE analysis of the purity and size of affinity-purified Venus

As shown in Fig. 2, the Ni-NTA purified Venus was electrophoresed using sodium dodecylsulfate (SDS) polyacrylamide gel electrophoresis (PAGE) on a 15% polyacrylamide gel, in order to verify the protein's purity as well as its molecular weight. As can be seen in the figure, the protein was of greater than 95 percent purity after affinity purification.

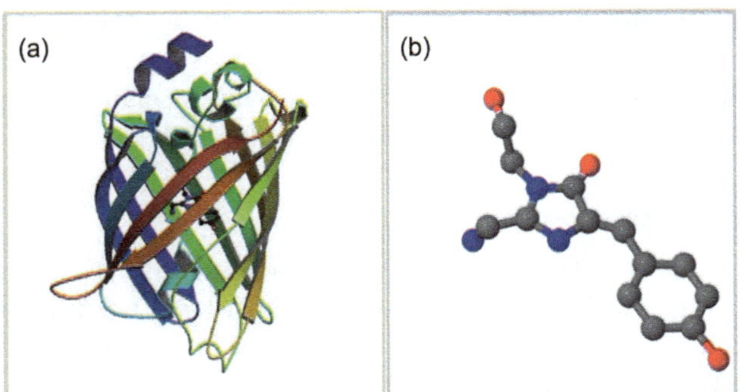

Fig. 1 (a) Molecular structure of Venus (PDB ID: 1MYW) and (b) the chromophoric unit within it (adapted from https://www.rcsb.org/pdb/explore/explore.do?structureId=1MYW).

 This journal is © The Royal Society of Chemistry 2018

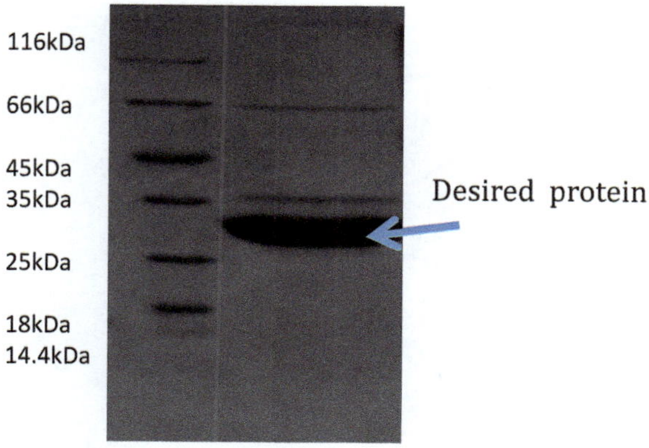

Fig. 2 The right lane shows the purified Venus protein appearing as a band at a ~30 kDa molecular weight against the left lane which shows a standard set of SDS-PAGE molecular weight markers.

3.2 Gel-filtration chromatographic analysis of the association state of Venus

Gel-filtration chromatography was performed on a Superdex-200 Increase 10/300 GL column equilibrated with PBS (20 mM phosphate buffered saline, pH 7.3) using an AKTA Purifier 10 (GE) instrument as shown in Fig. 3, in order to examine the association state (oligomeric status) of the purified Venus after Ni-NTA affinity chromatography. Fractions were collected and the protein was observed to elute from this column at 16 mL, corresponding to a molecular weight of 30 kDa, using the calibration curve shown in Fig. 4 (Table 1). The chromatographic analysis confirmed that the protein is monomeric.

3.3 Steady-state and time-resolved fluorescence measurements

3.3.1 Assignment of the major absorption/emission peaks. As shown in Fig. 5, the absorption spectrum of the purified Venus protein has a dominant maximum at ~516 nm.

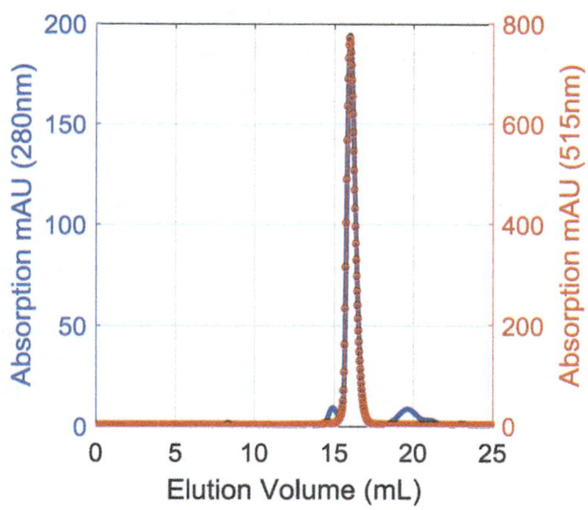

Fig. 3 Gel filtration profile for Venus.

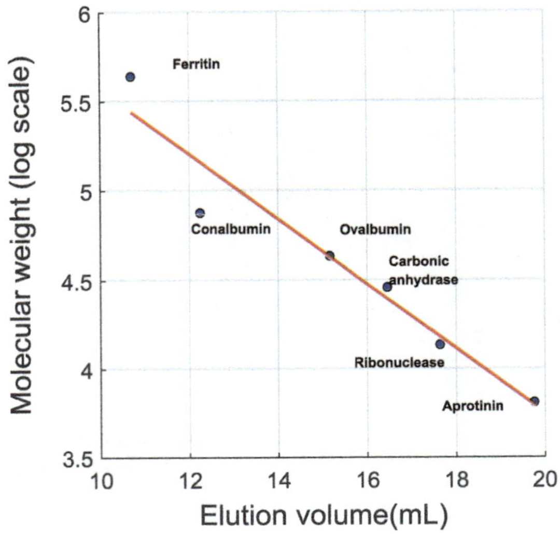

Fig. 4 Calibration curve for the Superdex 200 column.

Fig. 6 shows the fluorescence emission spectra of the purified Venus protein collected using three different excitation wavelengths, 374 nm, 400 nm and 516 nm (which correspond to the TCSPC diode excitation, pump excitation and UV-VIS excitation maximum, respectively). The spectra for both the 374 nm and 400 nm excitations show a small peak at ~462 nm and a major peak at ~529 nm with a shoulder at ~551 nm, while the spectrum for the 516 nm excitation shows the standard ~529 nm emission of Venus.

Fig. 7 shows the fluorescence excitation spectrum collected at the red-edge of the fluorescence spectrum (using an emission monochromator setting of 580 nm) along with the fluorescence lifetime data recorded at the major peak (~529 nm) and the shoulder (~551 nm).

The excitation spectrum is 'structured' and indicates strong vibrational manifolds and/or the presence of multiple absorbing species (arising most likely from static heterogeneity); notably, there is a small absorption at ~350 nm (zoomed within the inset in the left panel of Fig. 7).

Time constants for multiexponential fits to the fluorescence lifetime data collected at two different wavelengths, 529 nm and 551 nm (which correspond to

Table 1 Molecular weight markers with their actual molecular masses and elution volumes during gel filtration

Elution volume (mL)	Molecular weight (Da)	Molecular marker
10.697	440 000	Ferritin
12.246	75 000	Conalbumin
15.16	43 000	Ovaalbumin
16.46	29 000	Carbonic anhydrase
17.648	13 700	Ribonuclease A
19.759	6500	Aprotinin

 This journal is © The Royal Society of Chemistry 2018

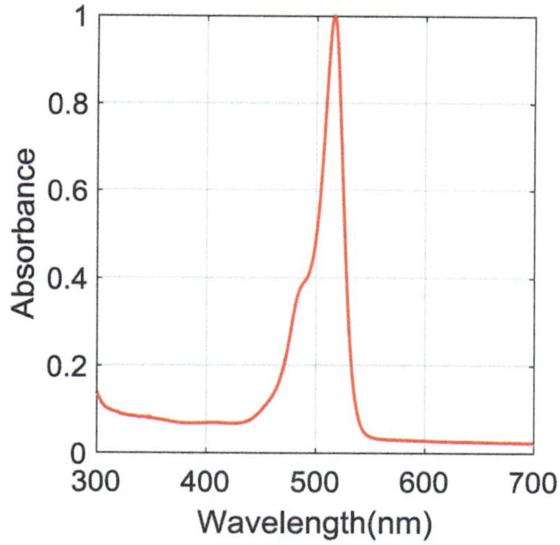

Fig. 5 Absorption spectrum.

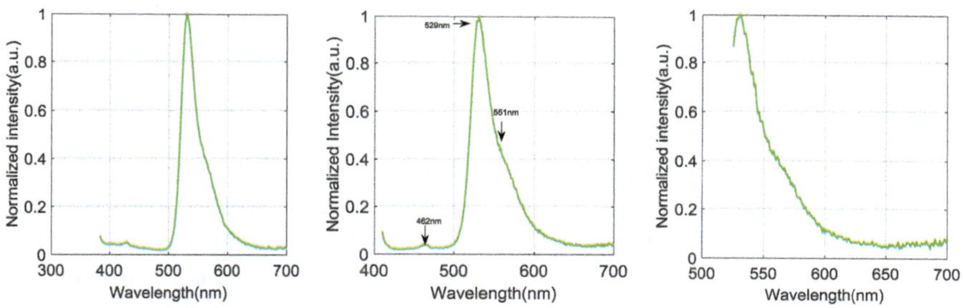

Fig. 6 Fluorescence emission spectra recorded using excitation wavelengths of 374 nm (left panel), 400 nm (middle panel) and 515 nm (right panel).

the fluorescence maximum and the shoulder, respectively) are shown in Table 2; the number of exponentials were chosen according to the goodness of the fitting.

It is interesting to compare the behavior of Venus with that of the wt-GFP from which Venus has been derived through multiple mutations. The absorption

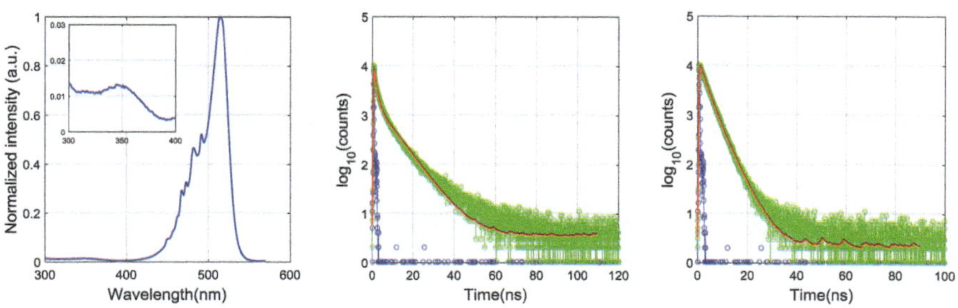

Fig. 7 Fluorescence excitation spectrum recorded at a 580 nm emission wavelength (left panel), and fluorescence lifetime data (green) along with multi-exponential fits (red) recorded at 529 nm (middle panel) and 551 nm (right panel), with the instrument response shown in blue.

Table 2 Lifetime data at 529 nm and 551 nm

Wavelength (nm)	α_1	α_2	τ_1 (ns)	τ_2 (ns)	τ_{avg} (ns)	χ^2
529	0.81	0.19	2.65	4.51	3.01	1.27
551	0.76	0.24	2.49	4.49	2.97	1.62

spectrum of wt-GFP has one minor peak at ~475 nm (apart from the major peak at ~395 nm), which arises because of the ionic (de-protonated) chromophore; the fluorescence emission spectrum of wt-GFP exhibits a peak at ~509 nm with a shoulder at ~540 nm.[13] It is known that the structure of the delocalized ionic chromophore is the same for different fluorescent proteins (FPs), although the absorption/emission tends to be red-shifted in E-YFP and Venus, relative to GFP, due to a change in the immediate local environment of the chromophore brought about by the mutations introduced to create these FP variants.[10] Therefore, by comparing the fluorescence behavior of Venus with that reported for wt-GFP, the electronic state in Venus displaying the ~516 nm/~529 nm absorption/emission pairs can be assigned to a state corresponding to the ionic chromophore which fluoresces in wt-GFP as the ~475 nm/~509 nm absorption/emission pairs; the shoulder at ~551 nm in Venus can be related to that at ~540 nm in wt-GFP. Therefore, these excitation/emission pairs of the ionic chromophore are red-shifted in Venus in comparison with wt-GFP.

3.3.2 Assignment of the minor absorption/emission peaks. Fig. 8 shows the fluorescence emission spectrum over the wavelength range of 440–490 nm, recorded with a small step-size of 0.05 nm. To explore the origin of the fluorescence emission peak at ~462 nm, the absorption spectrum was recorded with a small step-size (0.02 nm) over the wavelength range of 380–430 nm, revealing a weakly absorbing peak at ~400 nm; the optical density (OD) at 400 nm appears to be $\sim 10^{-3}$ after subtracting the baseline (OD ~0.0326).

To further confirm the origin of the ~462 nm emission peak, which appears to derive from the ~400 nm absorption, we recorded the fluorescence excitation spectrum and fluorescence lifetime data at a 462 nm emission wavelength, as shown in Fig. 9.

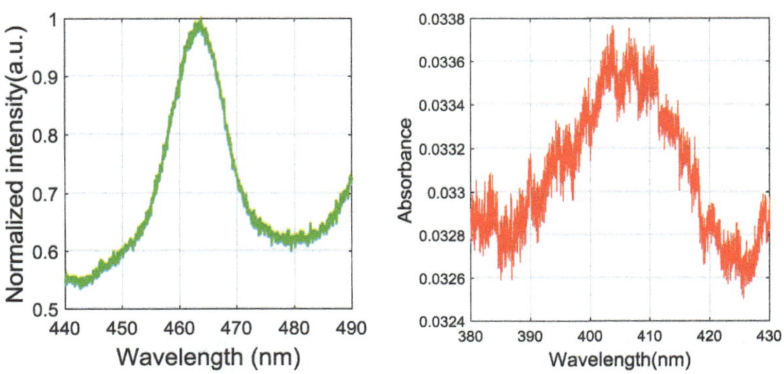

Fig. 8 Fluorescence emission spectrum recorded at a 400 nm excitation wavelength (left panel) and the absorption spectrum for the wavelength range of 380–430 nm.

 This journal is © The Royal Society of Chemistry 2018

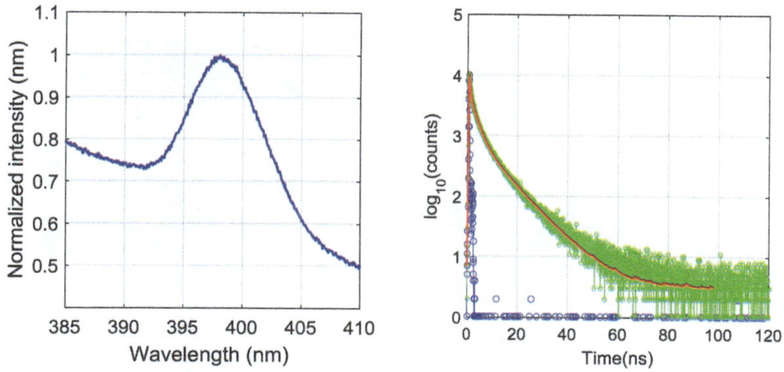

Fig. 9 Fluorescence excitation spectrum (left panel) recorded at a ~462 nm emission wavelength, and fluorescence lifetime data (green, right panel) along with a multi-exponential fit (red) recorded at a 462 nm emission wavelength, with the instrument response shown in blue.

The fluorescence excitation spectrum clearly reveals that the origin of the ~462 nm (21 645 cm^{-1}) emission peak is due to excitation at a wavelength just below ~400 nm (25 000 cm^{-1}), corresponding to the previously-mentioned absorption at ~400 nm. The 3355 cm^{-1} Stokes shift was initially suggestive of a possible origin due to Raman scattering (predominantly from the water solvent). However, this possibility is ruled out by the lifetime data (Table 3) on the nanosecond timescale (as scattering would have led to an extremely small lifetime).[14]

The major absorption peak at ~395 nm in wt-GFP is slightly red-shifted to a weakly absorbing peak at ~400 nm in Venus which means the energy levels of neutral (protonated) chromophore is not perturbed much by the mutations. In wt-GFP, this ~395 nm peak is thought to arise from the neutral (protonated) chromophore.[8] Quite interestingly, in wt-GFP excitation at 395 nm leads to the major emission peak at 509 nm (with the shoulder at 540 nm), whereas in Venus excitation at 400 nm leads to the major emission peak at 529 nm (with the shoulder at 551 nm) as well as the small emission peak at 462 nm (as shown in Fig. 6). In what follows, we focus on the excited state dynamics, following excitation at 400 nm, of Venus and compare it with that of wt-GFP.[8]

3.3.3 Pump-probe studies. Fig. 10 shows the spectrally resolved pump-probe data (a & b) along with a spectral slice taken at 530 nm (c & d). The poor signal-to-noise is due to the very weak absorption of Venus at 400 nm (OD is ~10^{-3}). A sigmoidal fit to the rise time (c) reveals an unusual fast growth with ~18 fs time constant (Table 4) which is presumably due to the overwhelming scattering signal (coherent artifacts), as also evident from the strong interference signal near time zero in (a). A bi-exponential fit to the decay (d) more or less correlates with the fluorescence lifetime (Table 5) which arises due to a stimulated emission.

Table 3 Lifetime data recorded at 462 nm

Wavelength (nm)	α_1	α_2	α_3	τ_1 (ns)	τ_2 (ns)	τ_3 (ns)	τ_{avg} (ns)	χ^2
462	0.26	0.68	0.06	2.49	9.74	0.34	1.48	1.59

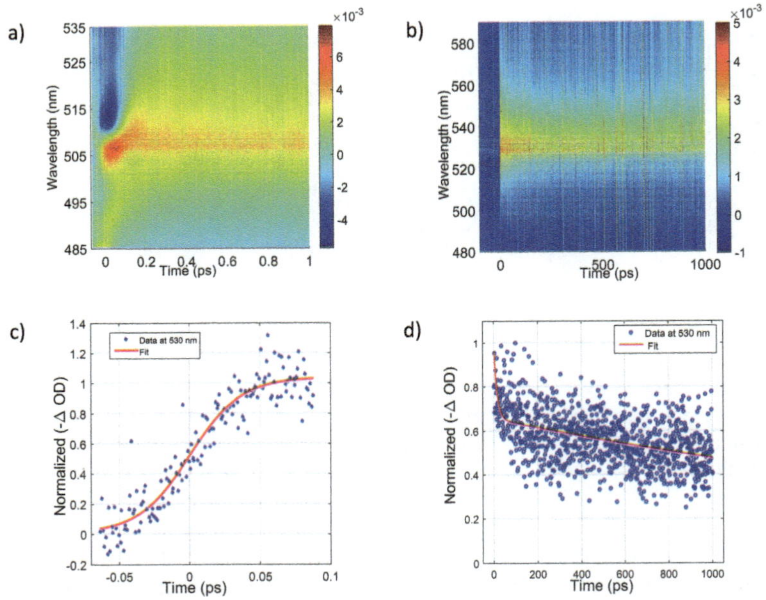

Fig. 10 Spectrally resolved pump-probe data for (a) −0.1 ps to 1 ps range and (b) −100 ps to 1000 ps range. (c) The rise of the signal (scatter, blue) along with the sigmoidal fit (line, red) and (d) the decay of the signal (scatter, blue) along with bi-exponential fit (line, red). The choice of time zero, baseline correction and normalization were done according to the sigmoidal fit.

Table 4 Fitting parameters for the rise in signal (sigmoidal fit) at 530 nm

$$y = A_2 + \frac{(A_1 - A_2)}{1 + e^{\frac{(t-t_0)}{\tau}}}$$

A_1	A_2	t_0 (ps)	τ (ps)	R^2
-8.51×10^{-5}	1.03	-1.82×10^{-6}	0.0183	0.91

Table 5 Fitting parameters for the decay in signal (bi-exponential fit) at 530 nm

$$y = A_1 e^{\left(-\frac{t}{\tau_1}\right)} + A_2 e^{\left(-\frac{t}{\tau_2}\right)}$$

A_1	A_2	τ_1 (ps)	τ_2 (ps)	τ_{avg} (ps)	R^2
0.35	0.65	14.27	3109.57	2026.21	0.23

3.3.4 Origin of the dual fluorescence in Venus. By comparing the behavior of Venus with that of wt-GFP, the electronic state in Venus that exhibits an absorption/emission pair at ~400 nm/~462 nm can potentially be attributed to a state corresponding to the neutral (protonated) chromophore. What is clear from the data presented in this paper is that the electronic state displaying the ~400 nm absorption is emissive for Venus, whereas the presumed electronic state that displays the corresponding ~395 nm absorption in wt-GFP is non-emissive.

 This journal is © The Royal Society of Chemistry 2018

Upon photo-excitation with ~400 nm light, ESPT generates the ionic (de-protonated) emissive species from the neutral (protonated) non-emissive/weakly-emissive species in both wt-GFP/Venus.[8]

In the case of wt-GFP, the major absorption peak corresponds to the neutral (protonated) form of the chromophore and excitation at both 395 nm and 475 nm leads to an emission at ~509 nm, thus suggesting the occurrence of ESPT.[8,10] In the case of Venus, unlike in GFP, the major absorption peak at 516 nm corresponds to the ionic (de-protonated) form of the chromophore. This suggests that in wt-GFP the ground state of the neutral (protonated) form is more stable compared with that of the ionic (de-protonated) form, while the opposite is observed for Venus. For Venus, excitation at 400 nm (or at 374 nm) leads to a branching of the excited state population from the Frank–Condon region into two potential wells: the locally excited (LE) state of the neutral (protonated) form *via* vibrational relaxation and the fully relaxed (FR) state of the ionic (de-protonated) form *via* ESPT. Radiative decay from the LE state gives rise to the ~462 nm emission peak, while that from the FR state gives rise to the ~529 nm emission peak (with a shoulder at ~551 nm).

Notably, dual-emission is observed in some GFP chromophore analogues and also in the GFP found in sea cactuses,[15,16] but not in wt-GFP. We conjecture that such a dual-emission pathway could also be present in wt-GFP but that it fails to manifest due to more efficient ESPT.

The contrasting nature of photoconversion between wt-GFP and Venus can be explained based on a schematic potential energy surface, as is depicted in Fig. 11.

In the following section, we conjecture that there may exist several alternate possibilities for explaining why the ~462 nm emission is seen in Venus and not in wt-GFP.

(1) It is possible that the ~395 nm excitation seen in GFP is immediately followed by efficient ESPT, as proposed earlier, which results in the formation of an ionic chromophore which then emits at ~509 nm. In Venus also, the ~400 nm

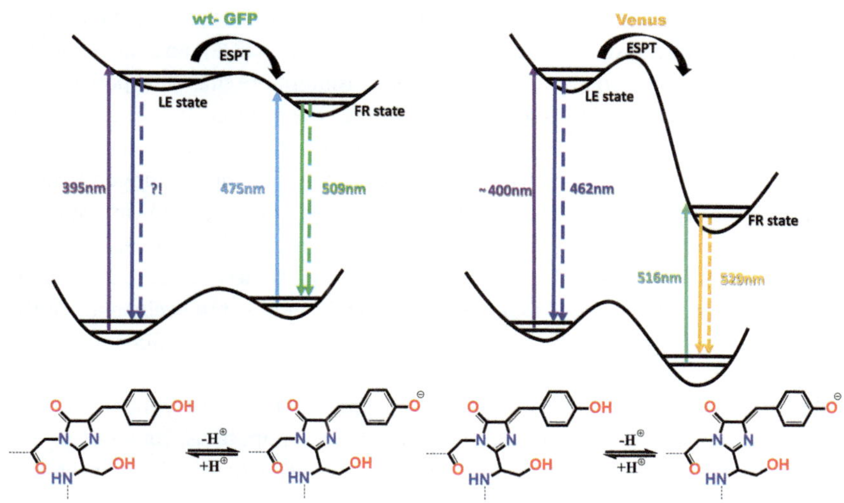

Fig. 11 Schematic of the excited state dynamics of wt-GFP (left panel) and Venus (right panel).

excitation is presumably followed by a similar occurrence of ESPT, leading to the formation of the ionic chromophore which then emits at ~529 nm. Indeed, excitation of Venus at ~400 nm also leads to emission at ~529 nm. As depicted schematically in Fig. 11, the barrier to ESPT is likely to be higher in Venus than in wt-GFP which may lead to less efficient ESPT in Venus. Also, there may exist a steeper decent immediately after ESPT in Venus which may render the ESPT too fast to be captured (see Fig. 10c).

(2) There is a second possibility to which we also allude. Since there is a significant overlap between the emission resulting from ~395 nm/~400 nm excitation (seen at ~462 nm, at least in Venus) and the excitation of the FR state of GFP (~475 nm), it is entirely possible that in wt-GFP radiative de-excitation (fluorescence emission) from the LE state (of the neutral chromophore) is re-absorbed by the FR state (of the anionic chromophore) of neighboring molecules; non-radiative de-excitation (internal conversion) from the LE state to the ground state may also be more efficient in wt-GFP. In contrast, for Venus the ~462 nm radiative de-excitation (fluorescence emission) from the LE state (of the neutral chromophore) is spectrally isolated from the ~516 nm absorption by the FR state (of the anionic chromophore).

(3) In recent years, it has been shown that hydrogen-bonded networks of peptide bonds in proteins exhibit fluorescence with an excitation close to ~400 nm and emission spanning the ~462 nm region, all the way up to above 550 nm; the peaks seen in the fluorescence envelope are dependent on the protein's secondary structural content.[17] The fluorescence is attributed to the excitation of a lone pair of electrons on the carbonyl oxygen of the peptide bond, which is sufficiently delocalized by the effects of hydrogen bonding to other peptide bonds (or to water) to display such a long wavelength excitation close to ~400 nm, with emission from a diffused fluorophore that is distributed over the secondary structural elements present in the structure of any protein.[18] For the purpose of this study, we presume that this fluorescence from hydrogen-bonded peptide networks, which has been referred to as 'protein intrinsic blue fluorescence', would also be seen in both wt-GFP and in Venus, since these fluorescent proteins (FPs) possess highly hydrogen-bonded, beta sheet-based, cylinders in the form of 'beta cans'. Therefore, we conjecture that the ~395 nm excitation in wt-GFP and the ~400 nm excitation in Venus could also derive from the chromophore that is responsible for the proteins' intrinsic blue fluorescence. In FPs, the peptide bonds are mostly networked into a beta sheet-based beta-can structure. It may be postulated that in wt-GFP there is very efficient FRET which occurs between this (hydrogen-bonded peptide network) chromophore, which is distributed over the beta-can, and the central chromophore residing inside the can that is responsible for the ~395 nm excitation. This may be why the hydrogen-bonded peptide network chromophore is not observed to fluoresce in wt-GFP. Presumably, in Venus, the occurrence of the FRET is far less efficient, thus resulting in the observation of fluorescence at ~462 nm.

There are four mutations in E-YFP compared to wt-GFP. In Venus, which is a mutated form of E-YFP, there are an additional five mutations, thus raising the total number of residue changes between Venus and wt-GFP to nine. It is conceivable that these nine changes significantly perturb the local environment of the protein and its structure in the vicinity of the FP chromophore, thus explaining why there is a red-shifting of both the absorption spectrum and the emission spectrum of the FP

 This journal is © The Royal Society of Chemistry 2018

between wt-GFP and Venus. Our speculation is that the same perturbations also result in a relative 'disconnect' between the beta-can and the FP chromophore, presumably through the placing of strain on the chromophore in addition to changes in its local chemical environment. We propose that this results in the less efficient occurrences of intramolecular FRET, leading to the observation of fluorescence from the LE state.

It thus becomes necessary to examine whether the proposed perturbations in the local structure and chemical environment of the FP chromophore, upon mutating wt-GFP into Venus, result in any significant alterations in the structural characteristics or structural stability (and resilience to denaturation by thermal or chemical means) of Venus. It is known that GFP is a very structurally-stable protein, requiring a combination of heat and chemical denaturants to exhibit any structural alterations whatsoever. Below, we show that Venus remains a highly structured and structurally-stable protein, which suggests that the perturbations eliciting less efficient ESPT and FRET in Venus must be due to very subtle changes, and not to any gross changes.

3.4 Studies of the structure and structural stability of Venus

Fig. 12 shows the mean residual ellipticity (MRE) and fluorescence emission intensity of Venus, as a function of wavelength. The CD spectrum shows that Venus is highly structured. Furthermore, it is evident that the structure of Venus is dominated by a beta sheet secondary structure from the consideration of two facts: (i) Venus has a CD band minimum between 210 and 220 nm, with a trough at ~216–218 nm, and (ii) the MRE value at 216 nm is only a few thousand degrees per cm^2 per dmol since (as is well known) even a small component of helical structure raises the negative MRE value to a few tens of thousands of degrees per cm^2 per dmol. The CD and fluorescence spectra in the presence of varying concentrations of the reducing agent dithiothreitol (DTT) (which have been corrected for dilution due to the addition of DTT) show that there is no significant effect of DTT on the structure and spectral characteristics of Venus.

Fig. 13 shows that there is no significant effect of heating on the CD spectrum of Venus upon the addition of DTT (right top panel), and upon the addition of 8 M

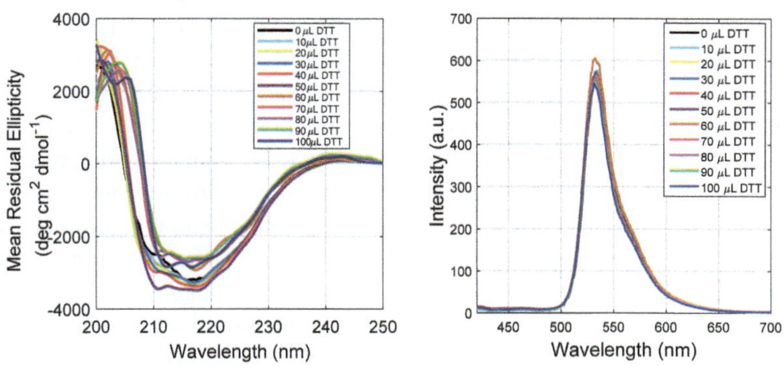

Fig. 12 Circular dichroism measurements showing a mean residual ellipticity *versus* wavelength plot (left panel) and fluorescence emission spectra of 200 μL of 20 μM Venus for different volumes of 20 mM dithiothreitol (DTT) (right panel).

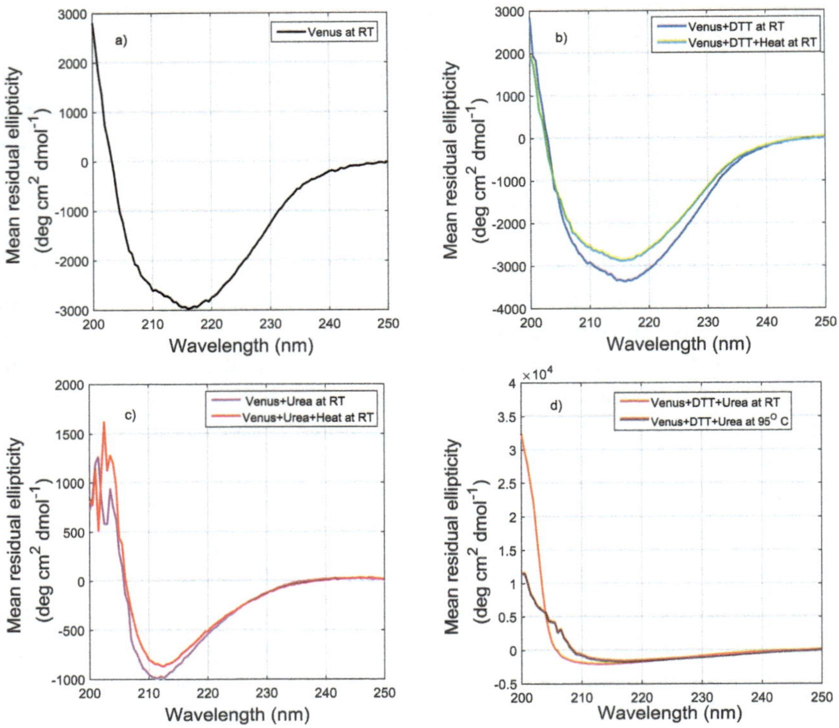

Fig. 13 Circular dichroism measurements showing mean residual ellipticity *versus* wavelength, recorded at room temperature (at 25 °C) before and after heating (at 95 °C) for (b) Venus + DTT and (c) Venus + urea. For comparison, the same measurement recorded at room temperature for (a) Venus is shown. Also shown are the results for (d) Venus + DTT + urea, recorded at room temperature and at 95 °C.

urea (left bottom panel). It should be noted that all of these spectra were collected at 25 °C, *i.e.*, after heating at 95 °C for 10 min and cooling back to 25 °C, if and when heating was carried out. The spectra show that Venus is highly structurally-stable and that there is no significant change in its CD spectrum upon the addition of reducing agents like DTT or hydrogen-bond destabilizing and denaturing agents like urea (8 M), either singly (DTT or urea) or in combination (both DTT and urea), or upon heating (up to 95 °C before cooling back down to 25 °C).

Furthermore, we also obtained CD measurements at 95 °C after 10 minutes of incubation at this temperature. The CD spectra obtained at 25 °C and 95 °C are shown in Fig. 13 (right bottom panel). The lack of change in the secondary structure indicates that even a combination of denaturant and heat causes no alteration in Venus over short timescales. It should be noted, however, that prolonged heating can cause some transient and partial denaturation of Venus, with refolding observed upon cooling.[12] This is understandable, given the known structural stability of wt-GFP. What we effectively show here is that Venus is highly-structurally stable like GFP (with no gross alterations in its structure or structural stability, with respect to wt-GFP), which can shed light on the poorer ESPT involving the FP chromophore seen in Venus *vis a vis* wt-GFP.

 This journal is © The Royal Society of Chemistry 2018

4 Conclusion

To summarize, we have investigated the excited state dynamics of the fluorescent protein Venus upon UV photo-excitation. By combining steady-state and lifetime measurements and pump-probe spectroscopy, we have proposed a general model for fluorescent proteins that satisfactorily explains the origin of dual fluorescence in Venus. The absence of dual fluorescence emission in wt-GFP can be due to several possibilities: (1) efficient ESPT from LE state leading to FR state, (2) re-absorption of LE state fluorescence by FR state and (3) intramolecular FRET occurring between the hydrogen-bonded peptide network and the FR state of the chromophore.

Conflicts of interest

The authors declare no conflicts of interest.

Acknowledgements

This work is supported by the start-up grant from IISER Mohali (to AKD) and by the Center of Excellence grant (for a Centre for Protein Science, Design and Engineering, CPSDE) from MHRD (MHRD-14-0064, to PG). SD thanks IISER Mohali and BT thanks CSIR for graduate fellowships.

References

1 R. E. Blankenship, *Molecular mechanisms of photosynthesis*, Wiley-Blackwell, 2nd edn, 25 April 2014.

2 X. X. Zhou and M. Z. Lin, Photoswitchable Fluorescent Proteins: Ten Years of Colorful Chemistry and Exciting Applications, *Curr. Opin. Chem. Biol.*, 2013, **17**, 682–690.

3 A. Acharya, A. M. Bogdanov, B. L. Grigorenko, K. B. Bravaya, A. V. Nemukhin, K. A. Lukyanov and A. I. Krylov, Photoinduced Chemistry in Fluorescent Proteins: Curse or Blessing?, *Chem. Rev.*, 2017, **117**, 758–795.

4 N. C. Shaner, P. A. Steinbach and R. Y. Tsien, A guide to choosing fluorescent proteins, *Nat. Methods*, 2005, **2**(12), 905–909.

5 A. Salih, A. Larkum, G. Cox, M. Kuèhl and O. Hoegh-Guldberg, Fluorescent pigments in corals are photoprotective, *Nature*, 2000, **408**, 850–853.

6 H. Kuramochi, S. Takeuchi, K. Yonezawa, H. Kamikubo, M. Kataoka and T. Tahara, Probing the early stages of photoreception in photoactive yellow protein with ultrafast time domain Raman spectroscopy, *Nat. Chem.*, 2017, **9**, 660–666.

7 S. T. D. Hsu, C. Behrens, L. D. Cabrita and C. M. Dobson, ^{1}H, ^{15}N and ^{13}C assignments of yellow fluorescent protein (YFP) Venus, *Biomol. NMR Assignments*, 2009, **3**, 67–72.

8 M. Chattoraj, B. A. King, G. U. Bublitz and S. G. Boxer, Ultra-fast excited state dynamics in green fluorescent protein: Multiple states and proton transfer, *Biophysics*, 1996, **93**, 8362–8367.

9 T. Fujisawa, H. Kuramochi, H. Hosoi, S. Takeuchi and T. Tahara, Role of Coherent Low-Frequency Motion in Excited-State Proton Transfer of Green

This journal is © The Royal Society of Chemistry 2018

Fluorescent Protein Studied by Time-Resolved Impulsive Stimulated Raman Spectroscopy, *J. Am. Chem. Soc.*, 2016, **138**, 3942–3945.

10 A. Rekas, J. R. Alattia, T. Nagai, A. Miyawaki and M. Ikura, Crystal Structure of Venus, a Yellow Fluorescent Protein with Improved Maturation and Reduced Environmental Sensitivity, *J. Biol. Chem.*, 2002, **277**(45), 50573–50578.

11 http://web.expasy.org/protparam/.

12 G. J. Kremers, J. Goedhart, E. B. van Munster and T. W. J. Gadella, Cyan and Yellow Super Fluorescent Proteins with Improved Brightness, Protein Folding, and FRET Forster Radius, *Biochemistry*, 2006, **45**, 6570–6580.

13 M. Chalfie, GreenFluorescent Protein, *Photochem. Photobiol.*, 1995, **62**, 651–656.

14 Y. Silori, S. Dey and A. K. De, How to study picosecond solvation dynamics using fluorescent probes with small Stokes shifts, *Chem. Phys. Lett.*, DOI: 10.1016/j.cplett.2017.12.031.

15 T. Chatterjee, M. Mandal, A. Das, K. Bhattacharyya, A. Datta and P. K. Mandal, Dual Fluorescence in GFP Chromophore Analogues: Chemical Modulation of Charge Transfer and Proton Transfer Bands, *J. Phys. Chem. B*, 2016, **120**, 3503–3510.

16 K. Ogoh, T. Kinebuchi, M. Murai, T. Takahashi, Y. Ohmiyab and H. Suzukia, Dual-color-emitting green fluorescent protein from the sea cactus Cavernularia obesa and its use as a pH indicator for fluorescence microscopy, *Luminescence*, 2013, **28**, 582–591.

17 A. Shukla, S. Mukherjee, S. Sharma, V. Agrawal, K. V. Radha Kishan and P. Guptasarma, A novel UV laser-induced visible blue radiation from protein crystals and aggregates: scattering artifacts or fluorescence transitions of peptide electrons delocalized through hydrogen bonding?, *Arch. Biochem. Biophys.*, 2004, **428**(2), 144–153.

18 P. Guptasarma, Solution-state characteristics of the ultraviolet A-induced visible fluorescence from proteins, *Arch. Biochem. Biophys.*, 2008, **478**(2), 127–129.

 This journal is © The Royal Society of Chemistry 2018

Faraday Discussions

PAPER

Effect of point mutations on the ultrafast photo-isomerization of Anabaena sensory rhodopsin†

D. Agathangelou,[a] Y. Orozco-Gonzalez,[ag] M. del Carmen Marín,[b] P. P. Roy,[c] J. Brazard,[a] H. Kandori,[d] K.-H. Jung,[e] J. Léonard,[a] T. Buckup,[c] N. Ferré,[f] M. Olivucci[bgh] and S. Haacke*[a]

Received 27th September 2017, Accepted 13th November 2017

DOI: 10.1039/c7fd00200a

Anabaena sensory rhodopsin (ASR) is a particular microbial retinal protein for which light-adaptation leads to the ability to bind both the all-*trans*, 15-*anti* (AT) and the 13-*cis*, 15-*syn* (13C) isomers of the protonated Schiff base of retinal (PSBR). In the context of obtaining insight into the mechanisms by which retinal proteins catalyse the PSBR photo-isomerization reaction, ASR is a model system allowing to study, within the same protein, the protein–PSBR interactions for two different PSBR conformers at the same time. A detailed analysis of the vibrational spectra of AT and 13C, and their photo-products in wild-type ASR obtained through femtosecond (pump-) four-wave-mixing is reported for the first time, and compared to bacterio- and channelrhodopsin. As part of an extensive study of ASR mutants with blue-shifted absorption spectra, we present here a detailed computational analysis of the origin of the mutation-induced blue-shift of the absorption spectra, and identify electrostatic interactions as dominating steric effects that would entail a red-shift. The excited state lifetimes and isomerization reaction times (IRT) for the three mutants V112N, W76F, and L83Q are studied experimentally by femtosecond broadband transient absorption spectroscopy. Interestingly, in all three mutants, isomerization is accelerated for AT with respect to wild-type ASR, and this the more, the shorter the wavelength of maximum absorption. On the contrary, the 13C photo-reaction is slightly slowed down, leading to an

[a]*University of Strasbourg, CNRS, Inst. de Physique et Chimie des Matériaux de Strasbourg, 67034 Strasbourg, France. E-mail: haacke@unistra.fr*

[b]*Department of Biotechnology, Chemistry and Pharmacy, University of Siena, Via A. Moro 2, I-53100 Siena, Italy*

[c]*Physikalisch-Chemisches Institut, Ruprecht-Karls Universität Heidelberg, 69120 Heidelberg, Germany*

[d]*Department of Frontier Materials, Nagoya Institute of Technology, Showa-ku, Nagoya 466-8555, Japan*

[e]*Department of Life Science, Institute of Biological Interfaces, Sogang University, Shinsu-Dong 1, Mapo-Gu, Seoul 121-742, South Korea*

[f]*Aix-Marseille Univ., CNRS, ICR, Marseille, France*

[g]*Department of Chemistry, Bowling Green State University, Bowling Green, Ohio 43403, USA*

[h]*USIAS Institut d'Études Avanceés, University of Strasbourg, 67083 Strasbourg, France*

† Electronic supplementary information (ESI) available. See DOI: 10.1039/c7fd00200a

inversion of the ESLs of AT and 13C, with respect to wt-ASR, in the blue-most absorbing mutant L83Q. Possible mechanisms for these mutation effects, and their steric and electrostatic origins are discussed.

Introduction

Light-induced processes in retinal proteins (RPs) have attracted vivid attention over the past two decades from experimentalists and theoreticians alike.[1–7] While initial work revealed the time scales for isomerization of the PSBR, and how they depend on the protein species or on point mutations, more attention was recently devoted to the mechanism, by which the protein steers this ultrafast reaction. New femtosecond spectroscopic[8–19] and quantum chemistry tools were devised[20–26] with the aim of drawing a very detailed picture of the excited state multi-dimensional potential energy surface (PES) including the conical intersection (CI) seam, of the time-resolved motion of PSBR at the atomic level, the reaction-driving vibrational modes and the dynamic changes in the charge distribution on PSBR and in the protein electrostatics. Additional, though partial, insight on these aspects comes from comparative studies of PSBR in solution[27–34] and of proteins reconstituted with non-isomerizing PSBR derivatives.[35–37]

While the relatively high complexity of the system impairs the ability to single out specific molecular or structural factors, it is generally accepted that the protein globally acts by limiting the isomerization to one specific double bond of the PSBR backbone and to increase the reaction quantum yield by steric and electrostatic constraints, preventing ultrafast non-productive routes[38] or possibly modifying the conical intersection (CI) topology.[22,39,40]

Nevertheless, the process by which the protein environment tunes the excited state PES or CI so as to shorten the S_1 lifetime and accelerate the reaction remains an open question. The recently discovered Anabaena Sensory Rhodopsin (ASR) offers a very interesting test ground in that context.[7,41–44] It bears a mixture of all-*trans*, 15-*anti* (AT) and 13-*cis*, 15-*syn* (13C) PSBR isomers, the relative content of which depends on the illumination, so-called 'light-adaptation' conditions. When left in the dark, the dark-adapted (DA) form adopts 97% AT, which drops to less than 40% under illumination with orange light, at neutral pH.[42,45] ASR therefore allows to study the PSBR protein interactions for two PSBR conformations, but with the same environment.

In this context, the observation of strikingly different excited state lifetimes (ESL) for AT (~0.7 ps (ref. 46 and 47)), and 13C (approx. 150 fs (ref. 44 and 47)) may be related to the different PSBR conformations,[48] but also to different electrostatic or dynamic interactions with the protein charge distribution. These findings were in qualitative agreement with quantum chemistry computations of the excited state PES along the minimum energy paths in ASR,[43] and with excited state trajectory calculations.[7] There, it was found that the S_1 PES of 13C is barrierless, while that of the AT form displays a plateau or a slight barrier that could account for a short resting period of the excited state population. Importantly, both reaction paths lead to a conical intersection seam with the ground state PES. The precise origin of these different shapes of the S_1 PES for both isomers remains unclear, however. As we reported recently,[48] a precise analysis of the nanosecond isomer difference spectra confirmed that AT has a 2.7 times higher reaction

 This journal is © The Royal Society of Chemistry 2018

quantum yield (QY) than 13C, and that ASR has the lowest QY ever reported for RPs (<0.2). The former result is intuitively surprising, a shorter ESL would imply a larger QY,[49] but is related to the fact that both quantities are fundamentally independent since determined by separate, independent regions of the PES. While the ESL reports about population decay close to the Franck–Condon region, the QY is a matter of transition probabilities and wave packet trajectories around the CI.[47]

Here, we report on ultra-broadband probed transient absorption spectroscopy (TAS) and transient Raman experiments performed with <30 fs, respectively 20 fs time resolution. The latter reports new and precise data for the C=C, C–C stretch and methyl rocking frequencies for AT, 13C ground states and their primary photoproducts K_{AT} and K_{13C}. As an extension of previous reports on ASR wild-type, we are focussing here on the effect of point mutations on the ESL of AT and 13C isomers. The mutants investigated – V112N, L83Q and W76F – were chosen due to their high level of expression in *E. coli* and the relatively large blue shifts their DA and LA ground state spectra afford, indicating significant changes of the electrostatic protein–PSBR interactions with respect to wild-type. Structural details of the mutated PSBR environment are displayed in Fig. S0.† Interestingly, for AT, the blue shifts are correlated with a reduction of the ESL, attaining 120 fs for the blue-most shifted mutant L83Q. On the contrary, the ESL of 13C is barely affected or at most increased by a factor of two *via* mutations, as compared to wt-ASR.

Materials and methods

ASR mutants were expressed in *E. coli*, prepared according to reported protocols[24,31] and solubilized in buffer (150 mM NaCl, 50 mM Tris–HCL, 0.02% DDM, pH 7.0). Steady state absorption spectra of dark-adapted (DA) and light-adapted (LA) protein samples were obtained using a Perkin Elmer "Lambda 950" UV/vis spectrometer equipped with integrating sphere detection. DA samples were prepared by incubation in the dark for at least 12 h at room temperature. LA samples were obtained by constant illumination for 30 min using Luxeon LEDs, LXHL-NB98 (460 ± 15 nm), or LXHL-PL01 (590 ± 10 nm). The former is referred to as blue-adaptation (BA) and the latter as orange-adaptation (OA). The procedure for determining the absorption maxima of all-*trans*, 15-*anti* (AT) and 13-*cis*, 15-*syn* (13C) isomers was already described elsewhere.[47] We measure the absorption spectra of the DA and LA proteins, determine the isomer content in both states after retinal oxime extraction by high performance liquid chromatography (HPLC),[50] and use the fact that the absorption spectra of DA and LA are linear combinations of AT and 13C absorbances according to the equations:

$$\begin{bmatrix} \mathrm{AT} \\ \mathrm{13C} \end{bmatrix} = R^{-1} \times \begin{bmatrix} A_{\mathrm{DA}} \\ A_{\mathrm{LA}} \end{bmatrix} \text{ with } R = \begin{bmatrix} r_{\mathrm{DA}}^{\mathrm{AT}} & r_{\mathrm{DA}}^{\mathrm{13C}} \\ r_{\mathrm{LA}}^{\mathrm{AT}} & r_{\mathrm{LA}}^{\mathrm{13C}} \end{bmatrix} \quad (1)$$

TAS was carried out with an improved time resolution (IRF 40 ± 5 fs), using a home-built non-collinear OPA with typically 30 fs pulse width as a pump source. The spectral widths of the pulses were deliberately limited in order to reduce the spectral region of pump light scattering. The pump wavelength was adjusted to be on the low-energy side of the absorption maxima, *i.e.* 530 nm for L83Q, 535 nm for V112N, and 540 nm for W76F. The excitation density was kept low enough in order to be in a linear excitation regime. Broadband probing was achieved with white-light pulses (330–1000 nm) generated by 40 fs pulses at 800 nm in a 2 mm CaF_2 window. Acquisition of single TAS spectra on a 1 kHz rate, ensured high sensitivity ($\Delta A_{min} \approx 5 \times 10^{-5}$), which in the present case is rather limited by the scattering properties of the protein samples. Data are averaged from 14 time scans (1400 laser pulses). All data are corrected for signal contributions from the solvent or flow cell and for the white-light group velocity dispersion (20 fs accuracy).[47] All decay traces are fitted by multi-exponential functions convolved with the IRF approximated as a Gaussian.[51]

For the transient vibrational spectroscopy, DFWM and pump-DFWM experiments were employed as described in detail in ref. 52–54. In short, two home-built non-collinear optical parametric amplifiers (1 kHz) were used to generate ultrashort excitation and probing pulses with typical durations below 15 fs. The spectrum of actinic-pump (AP) was made resonant to the molecular ground state absorption, while the DFWM spectrum was spectrally overlapped with the photo-induced, ground state photo-product absorption of ASR, respectively (Fig. 3a and b). Energy of the AP beam was 100 nJ. DFWM energies were set to 50 nJ for the pump and Stokes and 10 nJ for the probe. A rapid scan measurement was realized with a piezo, which continuously scans the delay between Stokes and probe pulses (τ) (Fig. 3c). The delay between pump and Stokes (τ_{12}) was kept zero in all measurements. Vibrational spectra were obtained by Fourier transformation, zero padding and windowing the oscillatory signal after the initial 50 fs. The optical density (OD) of the ASR sample was around 0.9 at 545 nm. The sample was circulated through a fused silica flow cell with a path length of about 1 mm by a motor-driven pump.

Results and discussion

Effect of light- and dark-adaptation on steady-state absorption spectra of the mutants: experiments and computational results

The ground state absorption spectra of the LA and DA forms are reported in Fig. 1 for the 3 mutants studied and for wt-ASR as a reference (see Fig. S0† for positions of mutations). All DA spectra are normalized to 1.0, since extinction coefficients were not available, and the LA spectra are plotted with the experimental relative reduction in absorbance. Table 1 summarizes the wavelengths of absorption maxima, the isomer composition, as well as the absorption maxima of the isomer specific spectra (not shown) obtained after linear decomposition.[47] Light-adaptation was carried out either with an orange (OA) or a blue (BA) LED (see Materials and methods), and the light-adaptation with the highest 13C isomer ratio was retained. V112N and W76F exhibit a blue-shift under LA as wt-ASR and many other mutants. The absorption spectrum of the 13C isomer of these mutants shifts by 2 nm (V112N) and 8 nm (W76F). L83Q, the blue-most shifted mutant, surprisingly shows the opposite effect, with the 13C isomer being 6 nm red-shifted with respect to AT. Finally, we note that mutations induce 18, 21 and 33 nm blue-shifts, *i.e.* between 615 (76 meV) and 1160 cm^{-1} (145 meV) spectral shifts.

 This journal is © The Royal Society of Chemistry 2018

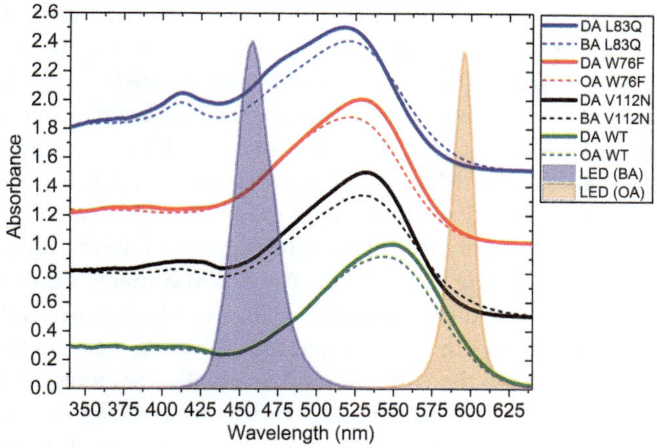

Fig. 1 Normalized ground state absorption spectra of light- and dark-adapted wt-ASR and the three mutants V112N, W76F and L83Q. Spectra are vertically shifted by multiples of 0.5 units. The spectral intensity distributions of the OA and BA LED's used for light-adaptation are displayed in orange and blue, respectively.

We report now the results obtained with QM/MM models of Anabaena Sensory Rhodopsin (ASR) and three of their mutants (V112N, W76F and L83Q) computed by the semi-automatic ARM protocol.[55] ARM employs the multi-configurational complete active space self-consistent field (CASSCF) method to obtain ground state geometries. Excitation energies are then computed using multi-configurational second-order perturbation theory (CASPT2) to recover the missing dynamical electron correlation associated with the CASSCF description.

Table 1 Isomer content of wt-ASR and the three mutants in the DA, OA and BA conditions, together with the measured maxima of absorption, highlighting the effect of light-adaptation. The maxima of isomer-specific ground state absorption spectra are calculated according to the linear decomposition method given in ref. 47

Sample	State	AT (%)	13C (%)	Abs max (nm)
WT	DA	97 ± 1	3 ± 1	550 ± 1
	OA	42 ± 1	58 ± 1	543 ± 1
	AT			550 ± 1
	13C			537 ± 1
V112N	DA	98 ± 1	2 ± 1	532 ± 1
	BA	42 ± 1	58 ± 1	530 ± 1
	AT			532 ± 1
	13C			527 ± 1
W76F	DA	96 ± 1	4 ± 1	529 ± 1
	OA	42 ± 1	58 ± 1	521 ± 1
	AT			529 ± 1
	13C			514 ± 1
L83Q	DA	98 ± 1	2 ± 1	517 ± 1
	BA	44 ± 1	56 ± 1	520 ± 1
	AT			517 ± 1
	13C			523 ± 1

The computed values obtained with ARM, are reported in Table S1 (*cf.* ESI†) and plotted in Fig. 2, where they are compared to the experimental values determined above. Apart from the previously documented[55] ≈2–3 kcal mol^{-1} overestimation of the computed values, the experimental trend is well reproduced for all mutants except for L83Q, where the ARM model fails to reproduce the observed λ_{max} values, and predicts the opposite trend of LA. Further investigation reveals that the discrepancy with respect to experiments is related to the orientation of the O–H dipole moment of the Glu83 residue towards the protonated Schiff base nitrogen (see Fig. S2 in ESI†). The QM/MM model generated by the ARM protocol orients this dipole with the positive end facing the SB (Fig. S2A†), leading to destabilization of the ground state (S_0). However, a second lower energy minimum (see Table S2 in the ESI†) is found, manually, for an orientation of the O–H dipole, with the negative end pointing toward the SB nitrogen (see Fig. S2B†), thus stabilizing S_0, and generating a blue-shift of λ_{max} similar to the experimental value. Nevertheless, while the L83Q$^*_{AT}$ appears to provide a λ_{max} substantially in line with the general trend, L83Q$^*_{13C}$ is clearly too blue shifted and, thus, the LA effect is not predicted in the right order (see also Table S1†). Within a new computational protocol analyzing the total free energy of the protein,[82] we are presently exploring a possible isomer-dependence of the protonation states of other amino acids in the retinal binding pocket, and orientation of Glu83 and Lys. We conclude that LA, in L83Q, causes a structural modification that cannot be correctly captured by the present modelling technique. Accordingly, the L83Q computational data will not be discussed/analyzed in the following.

We now focus on the differences between the WT model and the V112N and W76F mutant models. Mutants are usually challenging to model due to the smaller changes in the observed $\Delta E_{S_1-S_0}$ values and the absence of crystallographic data or other structural information. In spite of this difficulty, the computed WT, V112N and W76F $\Delta E_{S_1-S_0}$ trend parallels the observed values with only a relatively

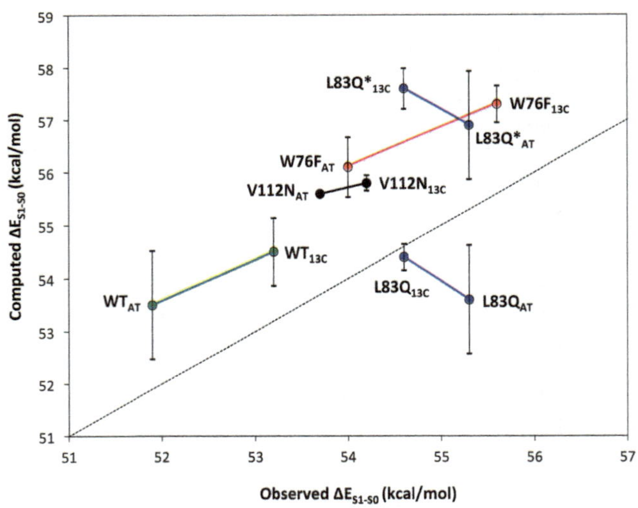

Fig. 2 Comparison between computed and observed $\Delta E_{S_1-S_0}$ (kcal mol^{-1}) values from ARM models of ASR WT (green), V112N (black), W76F (red), L83Q and modified L83Q* (blue) mutants. The error bars of the standard deviation are shown in black.

small blue-shifted deviation in the absolute $\Delta E_{S_1\text{-}S_0}$ values. The mutants studied show significant blue-shifts with respect to the WT reference, *e.g.* the computed $\Delta E_{S_1\text{-}S_0}$ change between WT and the most blue-shifted mutant (W76F) is 2.6 kcal mol^{-1} for the AT and 2.8 kcal mol^{-1} for the 13C isomer. We have computationally disentangled the electrostatic and steric effects causing the total blue shift by computing the $\Delta E_{S_1\text{-}S_0}$ values of the three chromophores in the gas-phase (keeping the chromophore geometries fixed at their optimized protein structure). In fact, the gas-phase $\Delta E_{S_1\text{-}S_0}$ value reflects the change in chromophore geometry induced by only the steric interaction with the protein cavity. By comparing the electrostatic and steric effects of the different chromophore cavities, we show that the electrostatic effect is the dominant one (Table 2, last column). These effects are due to the different stabilization of the chromophore ground state (S_0) and first excited state (S_1), which are characterized by different electronic distributions (Fig. S1†), due to the change in the protein charges and positions induced by the mutation. It is well-known that PSBR's photo-excited state S_1 is characterized by a partial charge transfer (*ca.* 30%)[56] electronic configuration. On the other hand, S_0 is dominated by a covalent electronic configuration, with the positive charge localized at the protonated Schiff base. The interaction of both electronic configurations with the mutated PSBR binding cavity explains the observed blue-shifts: for both the V112N and W76F mutants, a change from valine and tryptophan to asparagine and phenylalanine respectively, near the β-ionone ring of the chromophore produces a destabilization of the S_1 rather than a stabilization of S_0. The observed blue-shift is thus a combined effect of S_1 up-shift and S_0 down-shift, but the former is larger in amplitude than the latter, which may be important information for the mutation effects on the S_1 kinetics (see Table S3 in the ESI†).

Notice that, as usually found,[57] the chromophores separated from their protein environments present $\Delta E_{S_1\text{-}S_0}$ values about 12–20 kcal mol^{-1} ($\Delta E_{S_1\text{-}S_0}$ vacuum) lower than for the chromophore inside the protein ($\Delta E_{S_1\text{-}S_0}$ protein environment) (Table 2). Interestingly, the mutation-induced changes of the PSBR configurations cause, with respect to the WT, a red-shift and not a blue-shift in all $\Delta E_{S_1\text{-}S_0}$ values. However, as is shown in Table 2, the electrostatic contribution of the cavity dominates, causing a global blue-shift for the present mutants, with the strongest

Table 2 Comparison of the vertical excitation energies $\Delta E_{S_1\text{-}S_0}$ (kcal mol^{-1}) computed at the CASPT2/CASSCF(12,12)/6-31G*/AMBER level of theory for ASR WT and the mutants V112N and W76F when the retinal chromophore is inside the protein ($\Delta E_{S_1\text{-}S_0}$ protein environment) and in vacuum ($\Delta E_{S_1\text{-}S_0}$ vacuum), supposing the same conformation as in the protein. This allows to separate the dominant electrostatic effect of the mutations from conformational changes

Sample	State	$\Delta E_{S_1\text{-}S_0}$ protein environment	$\Delta E_{S_1\text{-}S_0}$ vacuum (steric contribution)	Effect of mutation: e.-static contribution
WT	AT	53.5	41.8	
	13C	54.5	43.1	
V112N	AT	55.6	40.2 (−1.6)	+3.7
	13C	55.8	40.9 (−2.2)	+3.5
W76F	AT	56.1	36.3 (−5.5)	+8.1
	13C	57.3	39.1 (−4.0)	+6.8

effect observed for the W76F mutant. In conclusion, the observed blue-shifts of the V112N and W76F are dominated by the change in the electrostatics of the chromophore hosting cavity. In Fig. S3† we provide a qualitative explanation of the electrostatic effects, which clearly show how the V112N mutation creates a new dipole moment destabilizing S_1 with respect to S_0. In contrast, the W76F mutation removes a dipole moment vector, which, in wild-type, stabilizes the S_1 with respect to S_0.

Photo-isomerization kinetics of wild-type ASR studied by femtosecond vibrational spectroscopy

Understanding the ultrafast vibrational evolution is a central aspect in the investigation of the reaction dynamics of the retinal chromophore.[58–63] Time resolved vibrational spectroscopies, like degenerate four wave mixing (DFWM),[53,64] pump-degenerate four wave mixing (pump-DFWM),[52,54,65–69] impulsive vibrational spectroscopy (IVS),[67,70–74] or femtosecond stimulated Raman (FSRS),[75,76] allow to resolve such structural changes in the ground and electronically excited states up to the formation of photoproducts. A detailed study of the evolution and origin of vibrational modes in the ground and excited state has been done, for example, in the case of bacteriorhodopsin (bR).[53,64,74] Excitation of vibrational coherences in bR has been shown to be highly dependent on the excitation wavelength in the spectral range from 520 up to 620 nm.[53] The "out-of-plane" modes of polyene-chain substituents and low frequency modes have been found to be very intense in the electronically excited state of bR. Contrasting to that, all remaining experimentally observed high frequency modes above 1010 cm^{-1} have to be assigned to the ground state manifold. A similar picture has been also observed for all-*trans*-retinal protonated Schiff-bases (RPSB).[64] While the origin of the experimental vibrational spectra and the vibrational dynamics of other retinal based systems, like bR and RPSB, has been already established, very little is known for ASR. In particular, it is still unknown how the retinal isomerization and environment alters the topology of PES and hence vibrational dynamics.[77] In this section, non-resonant DFWM and pump-DFWM (Fig. 3) is applied to DA and LA ASR. In particular, all the high frequency modes (>800 cm^{-1}) of ground states and photoproducts were compared both in DA and LA conditions.

Electronic ground state of AT and 13C. Non-resonant DFWM transients detected at 610 nm in DA and LA conditions display strong high frequency oscillations (Fig. 4a and b). The respective FFT spectra show characteristic localized high frequency vibrational modes: DA ASR shows four main peaks at 1003, 1165, 1230 and 1530 cm^{-1}; LA ASR shows only three major peaks at 1007, 1180 and 1539 cm^{-1} and a smaller one at 1300 cm^{-1}. The mode at about 1003/1007 cm^{-1} is assigned to the CH_3 rocking, while the mode at 1530/1539 cm^{-1} is assigned to the C=C stretching.[53,64–66] The blue shift of the frequency of the C=C stretching mode and the CH_3-rocking mode (smaller shift) due to light adaptation agrees with previous experimental observations.[74] The modes at 1165 and 1230 cm^{-1} observed for the DA ASR are characteristic for the C–C single bond stretch of the AT retinal isomer. For bR, these two frequencies are at 1165 cm^{-1} and 1210 cm^{-1}. The respective C–C mode for the LA ASR is at 1180 cm^{-1}. In bR, this mode is at 1190 cm^{-1} and is commonly attributed to the formation of the 13C isomer.[78]

 This journal is © The Royal Society of Chemistry 2018

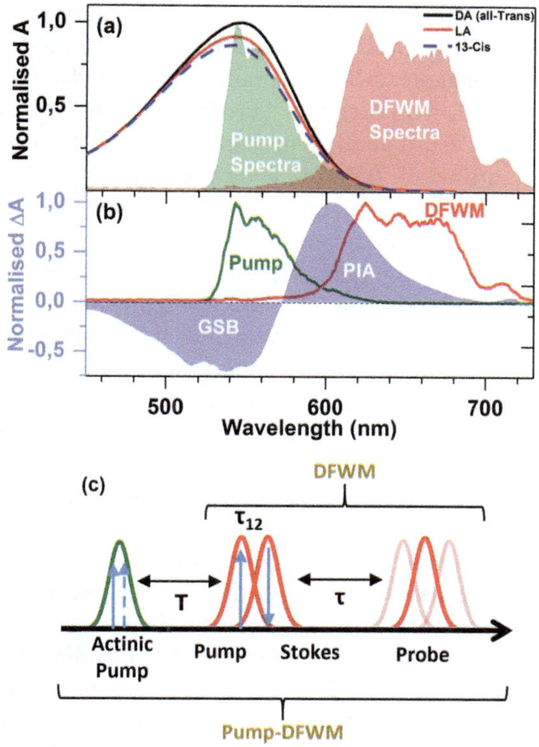

Fig. 3 (a) Ground state absorption spectra of ASR in DA and LA conditions. (b) Normalized difference absorption spectra at a pump-probe delay of 100 ps. Green and red (solid in (a) and line in (b)) curves represent the spectrum of AP and degenerate P. (c) Schematic of the pulse sequence and definition of the relevant delay times.

The contributions of each isomer to the DA and LA signals can be separated by fitting the individual FFT spectra with Gaussian line shapes and considering the isomer content in each case. Since DA ASR contains almost pure-AT isomer, a fitting of the respective FFT spectra delivers the pure vibrational frequencies

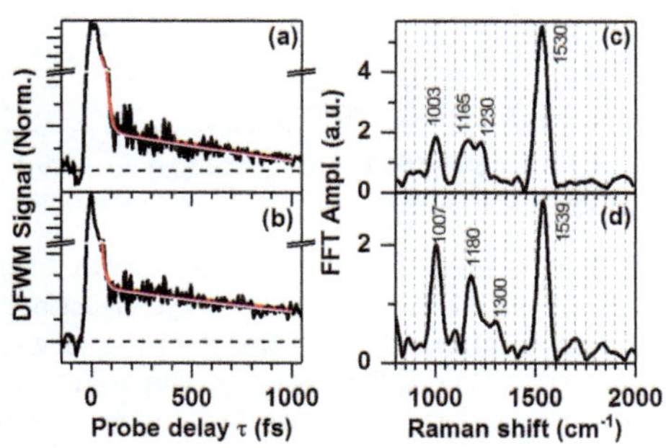

Fig. 4 Non-resonant DFWM transients probed at 610 nm: (a) DA ASR, (b) LA ASR. (c) FFT spectra of oscillatory signal of (a). (d) FFT spectra of oscillatory signal of (b). Non-oscillatory contribution (red line) was fitted by exponential decay and subtracted before FFT.

and widths of AT ground state modes. This information can be used to disentangle the pure-13C spectra by using a constrained double Gaussian fit for LA ASR, here orange-adapted, which contains both AT and 13C isomers. This is illustrated for the C=C stretching mode (Fig. 5a). In this case, the centre frequency and width of one Gaussian was fixed to the value obtained by single Gaussian fitting of DA ASR (=AT) (see Table S3 in the ESI† for fitting parameters). As the second Gaussian basically represents only the signature of 13C, the pure contribution of 13C can be retrieved by using HPLC isomeric ratio (58% 13C + 42% AT, Table 1). A comparison of AT and reconstructed-13C spectra shows that the C=C line shape is narrower and lower in amplitude. A similar approach can also be applied to the CH_3 rocking mode to separate the AT and 13C-isomeric contributions (see ESI†). The frequency difference of this mode for AT and 13C is about 5 cm^{-1}, while the widths are comparable within the fitting error. Separation of the isomeric contribution to the peaks in the fingerprint region 1100–1400 cm^{-1} is more challenging due to the strong spectral overlap between Raman lines and will be presented elsewhere.

The different Gaussian widths obtained for the 13C (52 cm^{-1}) and AT (68 cm^{-1}) above suggests that the C=C mode undergoes different dephasing dynamics for each isomer, *i.e.* the C=C mode for the 13C isomer dephases much slower than for AT. This can be further corroborated by directly comparing the dephasing rates *via* a sliding window FFT along several probe delays τ.[53] Fig. 5b compares the decay of the amplitude of the C=C mode obtained this way for the AT (=DA GS) and 13C (retrieved from LA data). While the C=C stretching dephases with a lifetime of about 750 fs in the AT GS, the same mode in the ground state of the 13C GS dephases slower with a time constant of 1030 fs. Moreover, since the LA form contains both AT and 13C, a constrained bi-exponential fitting reproduces the LA spectra well by fixing the time constants

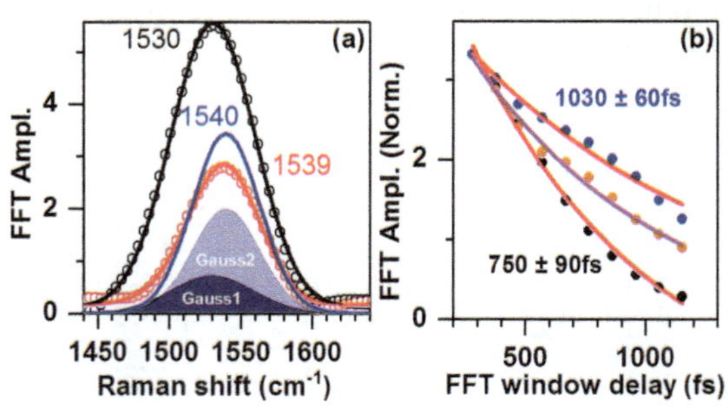

Fig. 5 (a) Gaussian fit of C=C stretching mode of GS ASR in DA (black curve) and LA (red curve) conditions. The DA GS peak was fitted with a single Gaussian. For LA GS, a double Gaussian constrained fit was performed fixing the centre and width of first Gaussian (Gauss1) as obtained by fitting the DA GS. The Gaussians required to fit the LA GS line shape are shown as filled curves. The reconstructed pure-13C spectrum is represented as a blue line. (b) Exponential fit of dephasing of GS C=C stretching mode obtained by sliding window FFT. Dephasing of DA (=AT), LA and extracted 13C isomer are shown by black, orange and blue dots. Red and pink solid lines represent the exponential and constrained bi-exponential fitting, respectively.

 This journal is © The Royal Society of Chemistry 2018

(750 fs and 1030 fs) obtained for AT and 13C, but keeping the amplitudes free (Fig. 5b).

K photoproducts. In order to compare the K photoproduct's vibrational signature with the electronic ground state of each isomer (see above), a pump-DFWM experiment was carried out. A resonant AP pulse was used to initiate the isomerization reaction. The DFWM pulse sequence, which was set spectrally resonant to the photoproduct absorption, was delayed by $T = 100$ ps. At this delay, the isomerization of $C_{13}{=}C_{14}$ is completed and a vibrationally relaxed K-photoproduct is already obtained.[60,61] The pump-DFWM signal measured shows similar vibrations as DFWM but with slightly shifted frequencies. FFT spectra of respective transients show for DA ASR four peaks at 1003, 1170, 1245 and 1535 cm^{-1}, while for LA ASR the peak frequencies are at 1007, 1174, 1305 and 1532 cm^{-1} (Fig. 6). Table 2 in the ESI† compares all values for the ground states of DA, LA and at $T = 100$ ps.

Discussion. Comparison of the FFT spectra obtained for the ground state of AT and 13C isomers and the respective photoproducts shows three major features regarding frequency shifts. The first one is the blue shift of the C=C stretching band (1530 to 1535 cm^{-1}) from GS to $T = 100$ ps in DA conditions, whereas it gets red shifted (1539 to 1532 cm^{-1}) in LA conditions (Fig. 4 and Table S4 in the ESI†). These differences can be rationalized by the structural differences at the C_{13} and C_{15} double bonds for the ground states and the respective photoproducts. For the two isomers in the ground state, both double bonds are *trans* (AT) or *cis* (13C) isomerized, which results in a 10 cm^{-1} blue shift for this mode (Fig. 3a). For the photoproducts, K_{AT} and K_{13C}, one of the two double bonds is in *trans*- while the other is in *cis*-conformation. In this situation one expects a frequency for this mode between 1530 cm^{-1} (AT) and 1540 m^{-1} (13C). Moreover, since at $T = 100$ ps the signal contains contributions of ground states as well as from the respective photoproducts, the frequency of C=C stretching appears at a frequency between the frequency of the AT and 13C, *i.e.*, at 1535 cm^{-1} in DA and 1532 cm^{-1} in LA conditions. The second aspect is the absence of any shift for the CH_3-rocking mode when the ground state is compared to the respective photoproduct (Table S4 in the ESI†). For example, for DA ground state as well as at $T = 100$ ps the frequency is 1003 cm^{-1}. This contrasts with the frequency shift observed for the ground states of AT (1003 cm^{-1}) and 13C (1007 cm^{-1}), and suggests that the CH_3

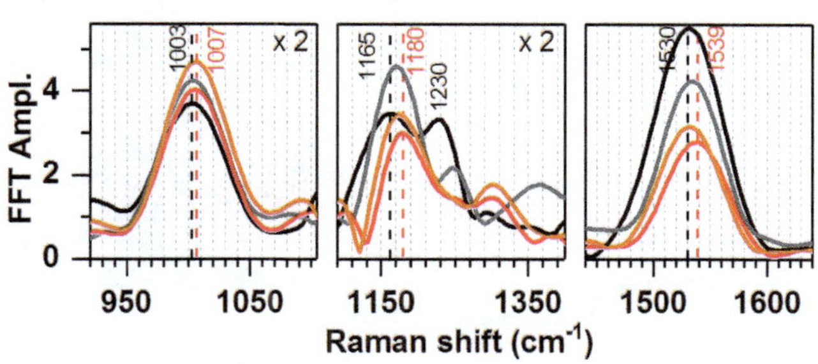

Fig. 6 Comparison of FFT spectra of DA and LA ASR for different vibrational modes. Black and red curves show FFT spectra of ground states of ASR in DA and LA conditions. Grey and orange curves represent the FFT spectra obtained from pump-DFWM transient at $T = 100$ ps for DA and LA conditions, respectively.

rocking mode is more sensitive when isomerization at both C_{13}- and C_{15}-positions takes place. The last and more complex aspect of the frequency shifts observed is in the fingerprint region between 1100 and 1400 cm^{-1}. Although many vibrational modes are overlapping in this region, large changes in frequency from GS to T = 100 ps are quite evident. Similar to the C=C stretching mode, the 1165 cm^{-1} mode blue shifts to 1170 cm^{-1} for DA ASR, whereas in LA conditions 1180 cm^{-1} red shifts to 1174 cm^{-1}. In addition, a band at 1230 cm^{-1} observed for DA GS is reduced at T = 100 ps. This band vanishes for LA ASR whereas a new peak around 1300 cm^{-1} appears. It is very similar to the 13C isomer of bR[79,80] and Channelrhodopsin,[73] where it has been assigned to CH_3 rocking and marked as a signature of *cis*-configuration.

Effects of point mutations on the photo-reaction kinetics of all-*trans*, 15-*anti* PSBR

The reaction kinetics of the all-*trans* species are retrieved from TAS data measured on DA mutants, as described in the Materials and methods section. Fig. 7 reports the time-resolved TAS spectra for W76F and L83Q, at time delays between 40 fs and 1.0 ps, and long-time difference spectra (90 ps) representing the AT-K_{AT} difference spectra.

Until 100–150 fs, the spectra show a prominent excited state absorption (ESA) in the range from 420 to 620 nm, and stimulated emission (SE) beyond 630 nm. During this early time period, significant dynamic spectral shifts are observed redistributing the ESA spectral intensity to shorter wavelengths, and allowing for the ground state bleach (GSB) to show up progressively as a trough in the ESA (L83Q-AT) or a negative signal (W76F-AT). This is most obvious for L83Q-AT, where the ESA maximum shifts from 550 to 540 nm within 60 fs. This initial phase of relaxation (<0.1 ps) is attributed to excited state relaxation, most likely due to C=C bond length alternation and IVR. GSB is observed at all delay times at wavelengths shorter than 400 nm.

After 100 fs, the ESA and SE globally decay in amplitude, indicating excited state decay due to isomerization. In contrast, the spectral range between 570 and 670 nm, now displays an increasing differential absorption, consistent with the

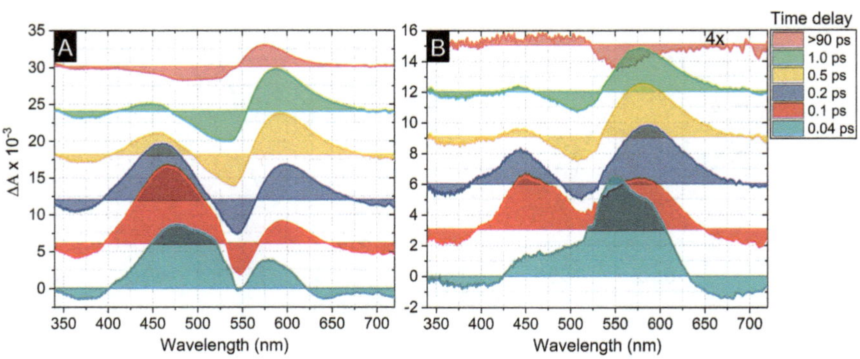

Fig. 7 Time-resolved TAS spectra for the AT forms of W76F (A) and L83Q (B) for highlighting delays up to 1.0 ps and a long-time, 90 ps, GS-K difference spectrum. Note the prominent spectral shifts of ESA, GSB and SE in the first 200 fs.

 This journal is © The Royal Society of Chemistry 2018

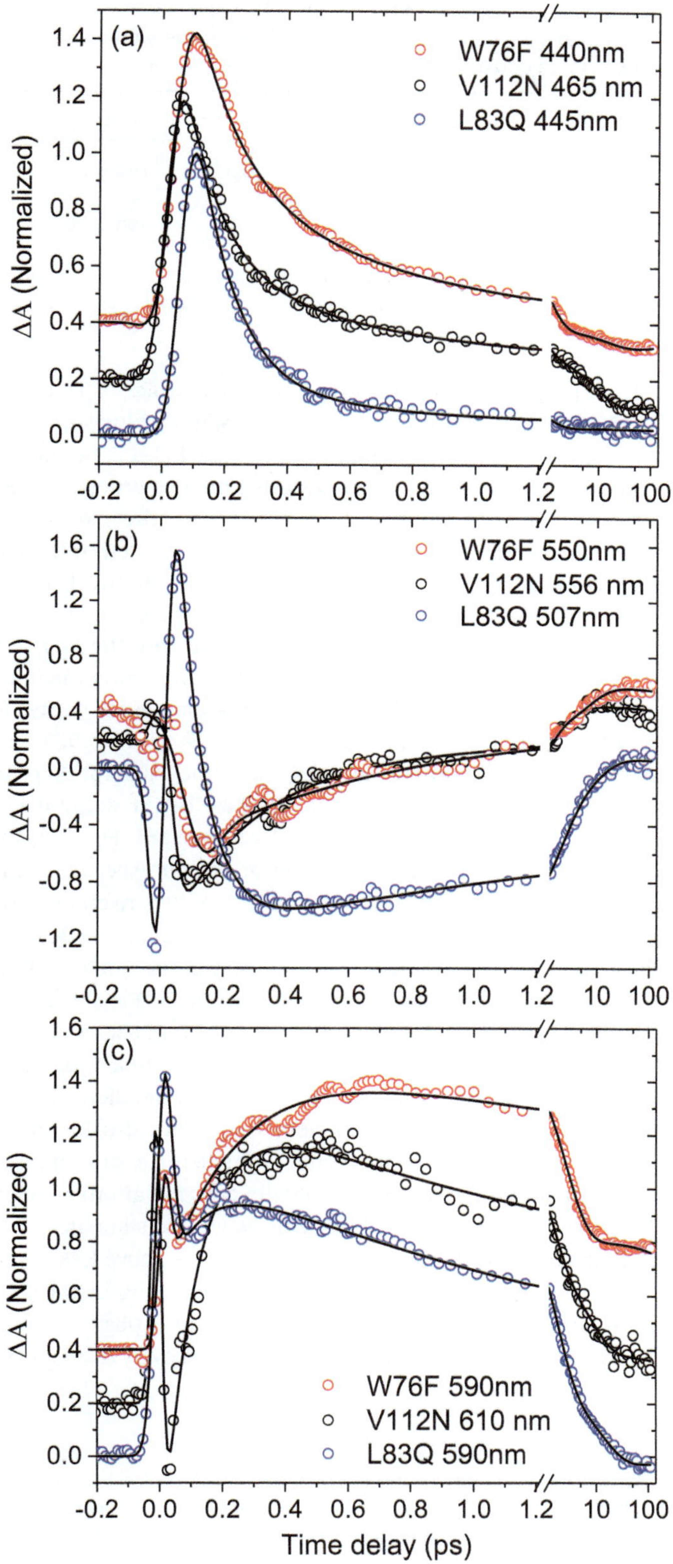
(a)
W76F 440nm
V112N 465 nm
L83Q 445nm
ΔA (Normalized)
(b)
W76F 550nm
V112N 556 nm
L83Q 507nm
ΔA (Normalized)
(c)
W76F 590nm
V112N 610 nm
L83Q 590nm
ΔA (Normalized)
Time delay (ps)

Table 3 Isomerization reaction times for AT and 13C isomers in the 3 mutants studied. Times are determined from ESA decay, PP rise and GSB recovery times. A2 ≠ 0 refers to bi-phasic excited state population decay. See text for details

Sample	Isomer	A_1	τ_1 (fs)	A_2	τ_2 (fs)	IRT (fs)
W76F	AT	0.67	140 ± 10	0.33	700 ± 30	340 ± 40
	13C	1.0	120 ± 30			120 ± 30
V112N	AT	0.72	110 ± 10	0.28	490 ± 80	230 ± 20
	13C	1.0	290 ± 50			290 ± 50
L83Q	AT	1.0	120 ± 30			120 ± 30
	13C	1.0	200 ± 40			200 ± 40

formation of the J_{AT} photo-product, as reported for wt-ASR.[44,47] The high-energy part of ESA, 400–500 nm, continues decaying and slightly blue-shifting until 1.0 ps. It is not clear, whether the remaining amplitude at 430–450 nm is ESA or part of the GS-J_{AT} difference spectrum, slowly decaying on a ps timescale (see below).

The spectral overlap of ESA and J_{AT}, throughout the entire 420–620 nm range, requires a careful analysis of the TAS transients for the determination of the excited state lifetime (ESL) and isomerisation reaction time (IRT). The SE decay reflects the ESL on a zero background and is therefore of particular relevance (see below). In addition, ESL and IRT have to be consistent with the PP rise time.

For L83Q, the ESA decay in the blue part of the spectrum after the initial 100 fs spectral shift can be fitted at 445 nm with a bi-exponential decay, featuring a dominant 120 ± 10 fs component and a 10-times smaller 0.76 ± 0.13 ps component (Fig. 8a). For shorter wavelengths, where the J absorption is negligible (424 nm, not shown) the slower component vanishes. The SE decay at >700 nm is purely mono-exponential with a 110 ± 10 fs decay (see ESI, Fig. S5†) indicating clearly that the S_1 population decays on this ultrafast timescale, *i.e.* 7 times faster than AT in wt-ASR.[47] This is corroborated by the 110 ± 10 fs risetime of the J_{AT} PP absorption followed by a 1.54 ± 0.06 ps decay due to $J_{AT} \rightarrow K_{AT}$ relaxation (Fig. 8c). The GSB in the 507 nm trace sets in only after the initial ESA decay and cannot inform about the isomerization-related GS recovery (Fig. 8b). Nevertheless, from the ESA, SE and PP kinetics it is clear that the IRT for L83Q-AT is 110 ± 10 fs.

For W76F, the ESA decay at 440 nm shows a bi-exponential decay with a dominant 140 ± 10 fs component and a 2.0 times smaller 0.70 ± 0.03 ps component (Fig. 8a). The slower component is present also in ESA traces at shorter wavelengths where the J contribution is smaller (*e.g.* 415 nm, not shown) even with a smaller amplitude. The SE at >700 nm does not afford a high SNR (see ESI, Fig. S5†), but a mono-exponential fit yields an excited state lifetime of 360 ± 20 fs, in good agreement with a weighted average of the above ESA components. We thus conclude that for W76F the S1 decay is biphasic. In the PA trace at 590 nm (Fig. 8c), only a fast 240 ± 20 fs rise component is observed, followed by a slower 2.55 ps $J_{AT} \rightarrow K_{AT}$ relaxation decay. The GSB recovery displays significant

Fig. 8 Normalized kinetic traces of the AT isomer in the 3 mutants, and probed at the wavelengths indicated. (a) ESA decay, (b) GSB recovery, (c) PP formation. Solid lines represent fits, according to the parameters discussed in the text. The W76F and V112N traces are displaced vertically.

 This journal is © The Royal Society of Chemistry 2018

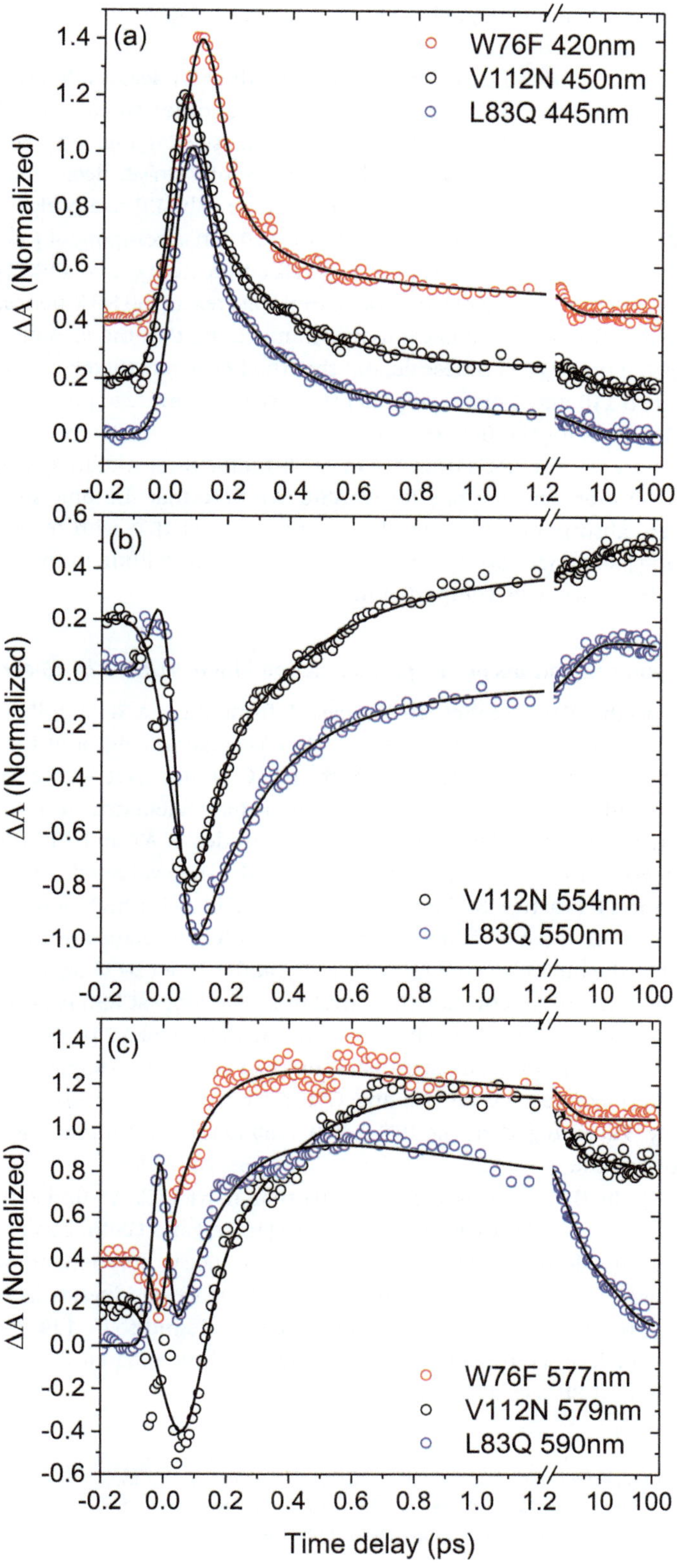
(a)
W76F 420nm
V112N 450nm
L83Q 445nm
ΔA (Normalized)
(b)
V112N 554nm
L83Q 550nm
ΔA (Normalized)
(c)
W76F 577nm
V112N 579nm
L83Q 590nm
ΔA (Normalized)
Time delay (ps)

non-exponential features (Fig. 8b), but the non-oscillating part is consistent with the above biphasic isomerization reaction (Table 3).

The situation is not as clear for V112N. Here, the ESA decay is bi-exponential with a dominant 110 $\pm$ 10 fs component and a 2.6-times smaller 490 $\pm$ 80 fs. SE is weaker at >700 nm than for the other mutants (see ESI, Fig. S5†), and shows a dominant 110 $\pm$ 10 fs component. The existence of a slower decay component cannot be excluded but is hidden in the noise floor. The PP rise is clearly faster than for V112N and shows only the fastest 110 $\pm$ 10 fs rise component followed by a slower 1.1 ps decay due to $J_{AT} \rightarrow K_{AT}$ relaxation decay (Fig. 8b). The GSB recovery, at 556 nm, also displaying oscillatory features as the PP kinetics, can be best fitted by a fast 250 $\pm$ 30 fs component, indicating that the isomerization is not completed on a 110 fs timescale, but that the IRT is most likely composed of the above two 110 and 490 fs components. A weighted average gives 210 fs. We retain 230 $\pm$ 20 fs for the IRT (see Table 3).

In summary, we find IRT's from 110 to 340 fs for the three mutants, which are 2 to 7 times shorter than the one for wt-ASR. We note that the blue-shift of the ground state absorption of AT correlates with a shorter IRT, with respect to wt-ASR. Notably, a similar correlation, though with less amplitude in the IRTs, was predicted theoretically by comparing different RPs.[81]

Effects of point mutations on the photo-reaction kinetics of 13-*cis*, 15-*syn* PSBR

The 13C-specific TAS kinetics are extracted from the LA data following the decomposition procedure used in ref. 47. Here we take into account the ground state isomer composition determined by HPLC under exactly the same LA conditions, and the excited state isomer composition calculated from the pump laser spectrum and the relative absorption amplitudes of AT and 13C (Fig. 1).

The effects of the mutations on the ESL and IRT are evaluated from selected ESA, GSB and PP kinetics. SE kinetics are not available with high enough signal-to-noise ratio for the 13C isomers. The time-resolved spectra are qualitatively equivalent to the ones observed for AT (Fig. 7), and will not be shown here. Fig. 9 displays the relevant kinetics for the 3 mutants. The longest ESL is observed for V112N (240 $\pm$ 30 fs) in the ESA, which displays an additional 5.7 ps decay time, 12 times smaller in amplitude, most probably due to a longer-lived signal assigned to the J_{13C} intermediate. The PP rise and GSB recovery times are 310 and 340 fs, respectively. Taken together, the IRT is 290 $\pm$ 50 fs for this mutant, *i.e.* twice as long as for wt-ASR.

For L83Q, the IRT of 13C is slightly longer than in wt-13C. We find 200 $\pm$ 40 fs, as a combination of ESA decay (220 fs), PP rise (160 fs) and GSB recovery (220 fs). ESA shows an additional 8-times smaller 4.3 ps decay component, which is attributed again to the J_{13C} intermediate. The 13C reaction kinetics in W76F do not seem to be affected by the mutations, since it is found to be 120 $\pm$ 30 fs, as a result of ESA decay (120 $\pm$ 30 fs, 8 times larger amplitude than a spurious 1.5 ps component) and PP rise (110 $\pm$ 10 fs).

Fig. 9 Normalized kinetic traces of the 13C isomer in the 3 mutants, and probed at the wavelengths indicated. (a) ESA decay, (b) GSB recovery, (c) PP formation. Solid lines represent fits, according to the parameters discussed in the text. The W76F and V112N traces are displaced vertically.

 This journal is © The Royal Society of Chemistry 2018

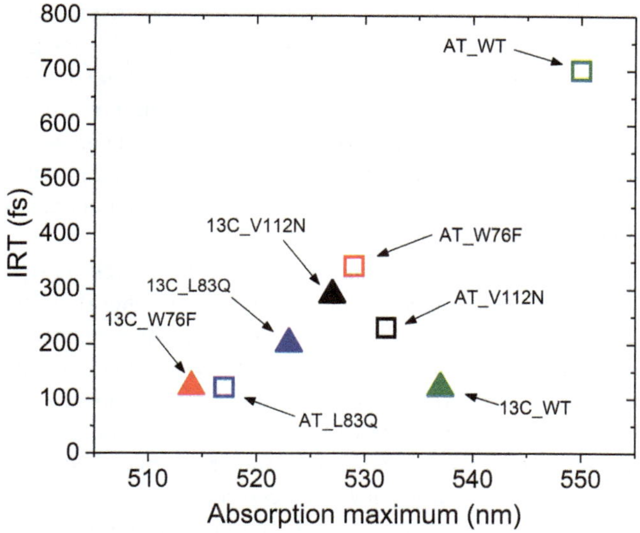

Fig. 10 Isomerization reaction times with respect to absorption maxima of AT (open squares) and 13C isomers (triangles) in wt-ASR and three mutants.

Interestingly, mutations with hydrophilic groups tend to slow down the IRT of the 13C isomer, while it significantly accelerates isomerization for the ATs (Table 3). For L83Q, the IRT of 13C is now slower than for AT, while the former is 7 times shorter than the latter in wt-ASR. Nevertheless, disregarding wt-ASR, the same trend as for AT is observed: a larger blue-shift of the absorption spectrum leads to a shorter IRT (Fig. 10).

Since the strongest effect of an enhanced IRT is observed for V112N, with a mutation in the vicinity of the ionone ring, *i.e.* far away from the C12=C13 double bond, the mechanism is most likely due to an electrostatic rather than a steric effect. Detailed excited state trajectory computations are needed, but the experimental results suggest that the 13C-PSBR excited state dipole moment increases during the reaction in a manner that reduces the slope of the S_1 PES, so as to slow down the excited state decay, with respect to the situation in wt-13C.

Conclusions and outlook

This paper reports the first detailed study on two aspects of the spectroscopy of Anabaena sensory rhodopsin, namely the vibrational spectra of the primary photoproducts K_{AT} and K_{13C} of wt-ASR and the effects of point mutations of ASR on the excited state decay and isomerization reaction times of the mutants V112N, W76F and L83Q. The ground state absorption spectra of these mutants are all blue-shifted with respect to wt-ASR and our quantum chemistry computations show that the electrostatic interactions in the PSBR binding pocket are responsible for the colour tuning in V112N and W76F. The transient absorption experiments show quite dramatic shortenings of the S_1 lifetimes and IRT's of the AT isomer with respect to wt-ASR. To the best of our knowledge, a 7-fold reduction of the excited state lifetime, as observed here for L83Q-AT is unique for a retinal protein, for which mutations generally induce slower and less efficient isomerization reactions. Such a slow-down is observed for the 13C. It is interesting to note that for AT, the shorter IRT's go along with the blue-shifted absorption spectra. Our strategy to proceed both experimentally and theoretically for providing a detailed comprehensive understanding of the mutation effects and of PSBR–protein interactions in general has made early progress, but it hinges presently on obtaining an adequate model for L83Q, and for realistic excited state trajectories for the three mutants. The time-resolved vibrational spectra will provide additional experimental data, to be tested by the QM/MM simulations.

The vibrational spectra of 13C and AT of wt-ASR, as well as the respective photoproducts, show all features well-known for bR and other rhodopsin systems in the region between 1000–1600 cm^{-1}. In particular, the separation of the 13C and AT contributions was demonstrated for the C=C and C–C stretching modes. The dephasing of the C=C for 13C is slower than for AT (1030 *vs.* 750 fs), hinting at a weaker coupling of this mode with the environment on the electronically ground state in respect to AT. Moreover, the frequency shifts detected and assigned for each isomer clearly show the potential of pump-DFWM to separate and resolve the small spectral changes due to mutation of ASR in the future.

Last, the specific biological function of ASR, as a photo-receptor for light-intensity level, requires differences in the absorption spectra of AT and CIS and differences in their isomerization quantum yield, so as to allow for sizeable changes of the isomer content upon light- or dark-adaptation. It will therefore be extremely important in the future to determine experimentally and understand theoretically the effect of the mutations on the reaction quantum yield.

Conflicts of interest

There are no conflicts to declare.

Acknowledgements

The Strasbourg team is grateful to Dr A. Cheminal for his contribution to the initial phase of the experiments, and to Dr O. Crégut and J.-P. Vola for technical assistance. The work was supported by the French-German ANR-DFG grant Femto-ASR (ANR-14-CE35-0015-01) and by the Labex NIE (ANR-11-LABX-0058_NIE). The France-Korea collaboration is supported by CNRS in the framework of the "LIA NanoFunc". M. O. is grateful for a USIAS 2015 Fellowship, and funding through the NSF Grant No. CHEM 1710191. D. A. thanks the "region Alsace" for partial funding of his PhD contract.

Notes and references

1 H. J. Polland, M. A. Franz, W. Zinth, W. Kaiser, E. Kölling and D. Oesterhelt, *Biophys. J.*, 1986, **49**, 651–662.

2 J. W. Petrich, J. Breton, J. L. Martin and A. Antonetti, *Chem. Phys. Lett.*, 1987, **137**, 369–375.

3 J. Herbst, K. Heyne and R. Diller, *Science*, 2002, **297**, 822–825.

4 R. Diller, in *Biological and Medical Physics, Biomedical Engineerin*, ed. M. Braun, P. Gilch and W. Zinth, Springer, Berlin Heidelberg, 2008, pp. 243–273.

5 J. Briand, J. Leonard and S. Haacke, *J. Opt.*, 2010, **12**, 084004.

 This journal is © The Royal Society of Chemistry 2018

6 A. Wand, I. Gdor, J. Y. Zhu, M. Sheves and S. Ruhman, *Annu. Rev. Phys. Chem.*, 2013, **64**, 437–458.
7 I. Schapiro and S. Ruhman, *Biochim. Biophys. Acta, Bioenerg.*, 2014, **1837**, 589–597.
8 S. Ruhman, B. X. Hou, N. Friedman, M. Ottolenghi and M. Sheves, *J. Am. Chem. Soc.*, 2002, **124**, 8854–8858.
9 G. I. Groma, A. Colonna, J. C. Lambry, J. W. Petrich, G. Váró, M. Joffre, M. H. Vos and J.-L. Martin, *Proc. Natl. Acad. Sci. U. S. A.*, 2004, **101**, 7971–7975.
10 S. Schenkl, F. van Mourik, G. van der Zwan, S. Haacke and M. Chergui, *Science*, 2005, **309**, 917–920.
11 P. Kukura, D. W. McCamant, S. Yoon, D. B. Wandschneider and R. A. Mathies, *Science*, 2005, **310**, 1006–1009.
12 D. W. McCamant, P. Kukura and R. A. Mathies, *J. Phys. Chem. B*, 2005, **109**, 10449–10457.
13 J. P. Kraack, T. Buckup, N. Hampp and M. Motzkus, *ChemPhysChem*, 2011, **12**, 1851–1859.
14 J. P. Kraack, T. Buckup and M. Motzkus, *Phys. Chem. Chem. Phys.*, 2012, **14**, 13979–13988.
15 V. I. Prokhorenko, A. M. Nagy, S. A. Waschuk, L. S. Brown, R. R. Birge and R. J. D. Miller, *Science*, 2006, **313**, 1257–1261.
16 G. Vogt, P. Nuernberger, T. Brixner and G. Gerber, *Chem. Phys. Lett.*, 2006, **433**, 211–215.
17 P. J. M. Johnson, A. Halpin, T. Morizumi, V. I. Prokhorenko, O. P. Ernst and R. J. Dwayne Miller, *EPJ Web Conf.*, 2013, **41**, 07020.
18 D. Polli, P. Altoe, O. Weingart, K. M. Spillane, C. Manzoni, D. Brida, G. Tomasello, G. Orlandi, P. Kukura, R. A. Mathies, M. Garavelli and G. Cerullo, *Nature*, 2010, **467**, 440–U488.
19 T. Sovdat, G. Bassolino, M. Liebel, C. Schnedermann, S. P. Fletcher and P. Kukura, *J. Am. Chem. Soc.*, 2012, **134**, 8318–8320.
20 L. M. Frutos, T. Andruniow, F. Santoro, N. Ferre and M. Olivucci, *Proc. Natl. Acad. Sci. U. S. A.*, 2007, **104**, 7764–7769.
21 T. Andruniów, N. Ferré and M. Olivucci, *Proc. Natl. Acad. Sci. U. S. A.*, 2004, **101**, 17908–17913.
22 M. Ben-Nun, F. Molnar, K. Schulten and T. J. Martínez, *Proc. Natl. Acad. Sci. U. S. A.*, 2002, **99**, 1769–1773.
23 F. Molnar, M. Ben-Nun, T. J. Martínez and K. Schulten, *J. Mol. Struct.*, 2000, **506**, 169–178.
24 A. Cembran, F. Bernardi, M. Olivucci and M. Garavelli, *Proc. Natl. Acad. Sci. U. S. A.*, 2005, **102**, 6255–6260.
25 I. Schapiro, M. N. Ryazantsev, L. M. Frutos, N. Ferre, R. Lindh and M. Olivucci, *J. Am. Chem. Soc.*, 2011, **133**, 3354–3364.
26 I. Rivalta, A. Nenov, O. Weingart, G. Cerullo, M. Garavelli and S. Mukamel, *J. Phys. Chem. B*, 2014, **118**, 8396–8405.
27 P. Hamm, M. Zurek, T. Röschinger, H. Patzelt, D. Oesterhelt and W. Zinth, *Chem. Phys. Lett.*, 1996, **263**, 613–621.
28 S. L. Logunov, M. A. ElSayed, L. Song and J. K. Lanyi, *J. Phys. Chem.*, 1996, **100**, 2391–2398.
29 O. Bismuth, N. Friedman, M. Sheves and S. Ruhman, *J. Phys. Chem. B*, 2007, **111**, 2327–2334.

30 O. Bismuth, N. Friedman, M. Sheves and S. Ruhman, *Chem. Phys.*, 2007, **341**, 267–275.
31 G. Zgrablić, S. Haacke and M. Chergui, *Chem. Phys.*, 2007, **338**, 168.
32 G. Zgrablic, S. Haacke and M. Chergui, *J. Phys. Chem. B*, 2009, **113**, 4384–4393.
33 G. Zgrablic, K. Voitchovsky, M. Kindermann, S. Haacke and M. Chergui, *Biophys. J.*, 2005, **88**, 2779–2788.
34 G. Bassolino, T. Sovdat, M. Liebel, C. Schnedermann, B. Odell, T. D. W. Claridge, P. Kukura and S. P. Fletcher, *J. Am. Chem. Soc.*, 2014, **136**, 2650–2658.
35 T. Ye, N. Friedman, Y. Gat, G. H. Atkinson, M. Sheves, M. Ottolenghi and S. Ruhman, *J. Phys. Chem.*, 1999, **103**, 5122.
36 S. Haacke, S. Schenkl, S. Vinzani and M. Chergui, *Biopolymers*, 2002, **67**, 306–309.
37 S. Haacke, S. Vinzani, S. Schenkl and M. Chergui, *ChemPhysChem*, 2001, **2**, 310–315.
38 G. Zgrablic, S. Haacke and M. Chergui, *J. Phys. Chem. B*, 2009, **113**, 4384–4393.
39 G. Zgrablić, A. M. Novello and F. Parmigiani, *J. Am. Chem. Soc.*, 2011, **134**, 955–961.
40 I. Burghardt and J. T. Hynes, *J. Phys. Chem. A*, 2006, **110**, 11411–11423.
41 L. Vogeley, O. A. Sineshchekov, V. D. Trivedi, J. Sasaki, J. L. Spudich and H. Luecke, *Science*, 2004, **306**, 1390–1393.
42 A. Kawanabe, Y. Furutani, K.-H. Jung and H. Kandori, *J. Am. Chem. Soc.*, 2007, **129**, 8644–8649.
43 A. Strambi, B. Durbeej, N. Ferré and M. Olivucci, *Proc. Natl. Acad. Sci. U. S. A.*, 2010, **107**, 21322–21326.
44 A. Wand, R. Rozin, T. Eliash, K.-H. Jung, M. Sheves and S. Ruhman, *J. Am. Chem. Soc.*, 2011, **133**, 20922–20932.
45 R. Rozin, A. Wand, K. H. Jung, S. Ruhman and M. Sheves, *J. Phys. Chem. B*, 2014, **118**, 8995–9006.
46 A. Wand, B. Loevsky, N. Friedman, M. Sheves and S. Ruhman, *J. Phys. Chem. B*, 2013, **117**, 4670–4679.
47 A. Cheminal, J. Leonard, S.-Y. Kim, K.-H. Jung, H. Kandori and S. Haacke, *Phys. Chem. Chem. Phys.*, 2015, **17**, 25429–25439.
48 A. Cheminal, J. Leonard, S. Y. Kim, K. H. Jung, H. Kandori and S. Haacke, *Chem. Phys. Lett.*, 2013, **587**, 75–80.
49 R. M. Weiss and A. W. Warshel, *J. Am. Chem. Soc.*, 1979, **101**, 6131–6133.
50 Y. Wada, A. Kawanabe, Y. Furutani, H. Kandori and H. Ohtani, *Chem. Phys. Lett.*, 2008, **453**, 105–108.
51 J. Briand, O. Braem, J. Rehault, J. Leonard, A. Cannizzo, M. Chergui, V. Zanirato, M. Olivucci, J. Helbing and S. Haacke, *Phys. Chem. Chem. Phys.*, 2010, **12**, 3178–3187.
52 J. Hauer, T. Buckup and M. Motzkus, *J. Phys. Chem. A*, 2007, **111**, 10517–10529.
53 J. P. Kraack, T. Buckup, N. Hampp and M. Motzkus, *Chem. Phys. Chem.*, 2011, **12**, 1851–1859.
54 T. Buckup and M. Motzkus, *Annu. Rev. Phys. Chem.*, 2015, **65**, 39–57.
55 F. Melaccio, M. Del Carmen Marin, A. Valentini, F. Montisci, S. Rinaldi, M. Cherubini, X. Yang, Y. Kato, M. Stenrup, Y. Orozco-Gonzalez, N. Ferre, H. L. Luk, H. Kandori and M. Olivucci, *J. Chem. Theory Comput.*, 2016, **12**, 6020–6034.

 This journal is © The Royal Society of Chemistry 2018

56 A. Zen, E. Coccia, S. Gozem, M. Olivucci and L. Guidoni, *J. Chem. Theory Comput.*, 2015, **11**, 992–1005.
57 L. Andersen, I. Nielsen, M. Kristensen, M. El Ghazaly, S. Haacke, M. Nielsen and M. Petersen, *J. Am. Chem. Soc.*, 2005, **127**, 12347–12350.
58 R. A. Mathies, C. H. B. Cruz, W. T. Polard and C. V. Shank, *Science*, 1988, **240**, 777–779.
59 Q. Wang, R. W. Schoenlein, L. A. Peteanu, R. A. Mathies and C. V. Shank, *Science*, 1994, **266**, 422–424.
60 A. Kawanabe, Y. Furutani, K.-H. Jung and H. Kandori, *J. Am. Chem. Soc.*, 2007, **129**, 8644–8649.
61 A. Kawanabe and H. Kandori, *Sensors*, 2009, **9**, 9741–9804.
62 O. A. Smitienko, M. N. Mozgovaya, I. V. Shelaev, F. E. Gostev, T. B. Feldman, V. A. Nadtochenko, O. M. Sarkisov and M. A. Ostrovsky, *Biochemistry*, 2009, **75**, 34–45.
63 A. V. Sharkov, A. V. Pakulev, S. V. Chekalin and a. Y. A. Matveetz, *Biochim. Biophys. Acta*, 1985, **808**, 94–102.
64 J. P. Kraack, T. Buckup and M. Motzkus, *Phys. Chem. Chem. Phys.*, 2011, **13**, 21402–21410.
65 J. P. Kraack, T. Buckup and M. Motzkus, *Phys. Chem. Chem. Phys.*, 2012, **14**, 13979–13988.
66 J. P. Kraack, T. Buckup and M. Motzkus, *J. Phys. Chem. Lett.*, 2013, **4**, 383–387.
67 J. P. Kraack, A. Wand, T. Buckup, M. Motzkus and S. Ruhman, *Phys. Chem. Chem. Phys.*, 2013, **15**, 14487–14501.
68 M. S. Marek, T. Buckup and M. Motzkus, *J. Phys. Chem. B*, 2011, **115**, 8328–8337.
69 T. Miki, T. Buckup, M. S. Krause, J. Southall, R. J. Cogdell and M. Motzkus, *Phys. Chem. Chem. Phys.*, 2016, **18**, 11443–11453.
70 S. Fujiyoshi, S. Takeuchi and T. Tahara, *J. Phys. Chem. A*, 2003, **107**, 494–500.
71 M. Liebel and P. Kukura, *J. Phys. Chem. Lett.*, 2013, **4**, 1358–1364.
72 M. Liebel, C. Schnedermann and P. Kukura, *Phys. Rev. Lett.*, 2014, **112**, 198302.
73 C. Schnedermann, V. Muders, D. Ehrenberg, R. Schlesinger, P. Kukura and J. Heberle, *J. Am. Chem. Soc.*, 2016, **138**, 4757–4762.
74 A. Kahan, O. Nahmias, N. Friedman, M. Sheves and S. Ruhman, *J. Am. Chem. Soc.*, 2007, **129**, 537–546.
75 P. Kukura, D. W. McCamant, S. Yoon, D. B. Wandschneider and R. A. Mathies, *Science*, 2005, **310**, 1006–1009.
76 P. Kukura, D. W. McCamant and R. A. Mathies, *Annu. Rev. Phys. Chem.*, 2007, **58**, 461–488.
77 I. Schapiro, *J. Phys. Chem. A*, 2016, **120**, 3353–3365.
78 S. Shim, J. Dasgupta and R. A. Mathies, *J. Am. Chem. Soc.*, 2009, **131**, 7592–7597.
79 M. Braiman and R. Mathies, *Proc. Natl. Acad. Sci. U. S. A.*, 1982, **79**, 403–407.
80 S. Smith, J. A. Pardoen, J. Lugtenburg and R. A. Mathies, *J. Phys. Chem.*, 1987, **91**, 804–819.
81 S. Rinaldi, F. Melaccio, S. Gozem, F. Fanelli and M. Olivucci, *Proc. Natl. Acad. Sci. U. S. A.*, 2014, **111**, 1714–1719.
82 Y. Orozco-Gonzalez, M. Manathunga, M. del Carmen Marín, D. Agathangelou, K.-H. Jung, F. Melaccio, N. Ferré, S. Haacke, K. Coutinho, S. Canuto and M. Olivucci, *J. Chem. Theory Comput.*, 2017, **13**, 6391–6404.

Faraday Discussions

PAPER

Computational studies on ground and excited state charge transfer properties of peptidomimetics†

Sherin Joy, Vommina V Sureshbabu and Ganga Periyasamy *

Received 20th August 2017, Accepted 13th September 2017
DOI: 10.1039/c7fd00183e

Chemical modifications at various peptide positions result in peptidomimetics with unique physical and chemical properties that can be used for a range of applications. Among many peptidomimetics, ureidopeptides are interesting due to their ability to act as donor–bridge–acceptor systems through which charge transfer occurs in one direction and can be triggered by an electrochemical pulse without perturbing the nuclear position. In this regard, some UP mimetics with different chromophoric units are studied in this work to understand their role using DFT based methods. Computational results and natural charge analysis provide evidence for the extensive contribution of the substituents to the excitation and hole migration dynamics. Further, the results show that the UP backbone preserves its uni-directional charge transfer phenomenon from the ureido to carboxylate terminal irrespective of the terminal groups and position. However, the substituent affects the excitation energies and the time scales of the hole migration. Among the substituents studied here, fluorine migrates to the hole within a shorter time scale while phenyl groups take longer.

1. Introduction

Recent advances in attosecond spectroscopy have provided access to explore the minute details of electron dynamics, which involve ultrafast charge migrations including Auger decays and intra Coulomb delays.[1,2] At the molecular level, the electronic dynamics are driven by either excitation to a higher energy state or ionization by a photon, which can be purely electron dynamics or can be coupled with electronuclear dynamics.[3,4] This complex phenomenon has been decoupled in order to achieve a suitable output for the process.[5] Lasers producing pulses lasting for attoseconds give access to pure electronic motion before nuclear

Department of Chemistry, Central College, Bangalore University, Bangalore, India. E-mail: ganga.periyasamy@gmail.com

† Electronic supplementary information (ESI) available: Comparison of structural parameters using different functionals, comparison of bond lengths and bond angles of UP, TUP and SUP at neutral and cationic states, excitation wavelengths and orbital pictures of F, Br, CN, NO_2 substituted UP models. See DOI: 10.1039/c7fd00183e

relaxation for many complex reactions.[5–7] This has revolutionised studies regarding DNA damage caused by light since ultraviolet radiation can mutate human DNA, causing cancer.[8] Biomolecules like proteins acting as units for signal transduction can also be associated with the same phenomenon.[9] Charge transfer through biomolecules has been studied in detail both experimentally and theoretically.[10,11] This is due to the fact that these studies have the potential to affect the areas of therapeutics and molecular electronics positively.

Polypeptide chains are widely used as models to study charge transfer phenomena in proteins.[12–14] Various studies have reportedly found that peptide sequences follow a range of mechanisms depending upon the nature of the substituents, the environment, components and so forth.[15,16] Citing an example, some molecular structures act as donor–bridge–acceptor systems to exhibit charge transfer through them.[17]

One electron oxidation of tryptophyl amino acid residues is found to produce radicals that undergo intramolecular electron transfer, which has been probed using hydrated electron spectroscopy.[18] Similarly, Schlag *et al.* experimentally studied electron or hole transfer through polypeptide chains with natural amino acids and natural chromophores as donors.[12] Remacle *et al.* described sub-femtosecond hole/charge migration in tetrapeptides using a computational method, where the charge migrates from the N-terminal through the entire backbone and returns back in a few femto seconds.[13] In natural peptides, the charge transfer occurs in both directions, where the carboxyl terminal to the amino end has an obvious advantage over the other.[14]

Following that, peptidomimetics have been introduced to tune the mechanism of charge transfer expediently for the required application. The peptidomimetics are biologically active peptides, structurally and functionally modified so that they have better bioavailability and less susceptibility to enzymatic degradation compared to natural peptides.[19] Use of the same motifs in therapeutics,[20] and mediating protein–protein interactions in signalling complexes are developing areas due to their receptor specificity.[21] Thus it is predicted that many of the properties of natural peptides can be tuned by modifying them in a beneficial manner, since protein signalling and other biologically important processes are based on electron transfer. Peptidomimetics can act as potential candidates for charge transfer by proper modification of the peptide backbone. Further, the reported computational studies on various peptidomimetics like ureidopeptides (UPs),[22] thioureidopeptides (TUPs) and selenoureidopeptide (SUPs),[23] and thioxo (TP) and selenoxo peptides (SP)[24] show charge transfer mechanisms with plausible applications, where the source can be either an ionizing or excitation pulse. Amongst several peptidomimetics, ureidopeptides are found to be important for the phenomenon of charge transfer with minimal structural relaxation after ionization along with characteristic one directional charge transfer from the ureido to carboxylate end within a 2–5 fs time scale. This encouraged us to explore the ureidopeptide family of peptidomimetics to a larger extent. This work focuses on various substituted ureidopeptides (Fig. 1), where the ligands at various positions are chemically modified. The electronic, optical and redox properties of these molecules are studied to understand the nature of charge transfer and time scales.

 This journal is © The Royal Society of Chemistry 2018

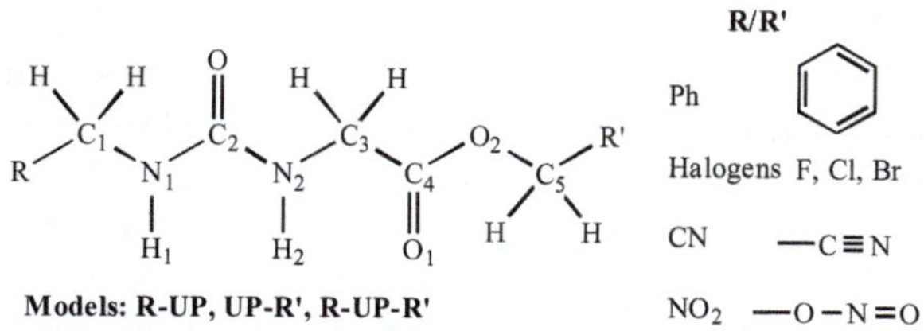

Fig. 1 Schematic representation of the different models of peptidomimetics studied in this work. R and R′ represent the substitution at the ureido end and carboxylate end respectively.

2. Computational methods

All the models reported in this work are optimized at the ground and first ionized states using the density functional theoretical (DFT) method with CAM-B3LYP/6-311++G** (ref. 25) as included in the Gaussian 09 package.[26] This functional is found to give appropriate long range electron–electron correlation interactions for UP, SUP, TUP, TP, and SP molecules.[22–24] Herein, the models have been selected in such a way that the substituent groups (R/R′ = Ph, F, Cl, Br, CN, and NO_2) are positioned at the ureido end (R–UP), carboxylate end (UP–R′) or simultaneously at both ends (R–UP–R′) (Fig. 2). TUP and SUP in place of UP refer to the substituted thioureidopeptides and selenoureidopeptides respectively. The hole migration dynamics have been carried out using the same functional since it represents the electron dynamics without any self-interaction error.[27] The

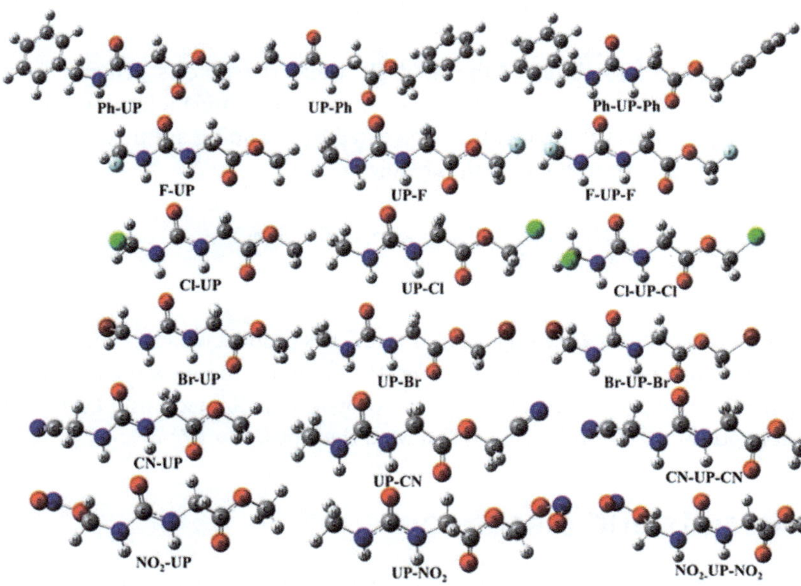

Fig. 2 Optimized geometries of the lowest energy conformers studied in this work using the CAM-B3LYP functional. Note that deviation from dihedral 180° is apparent at the ureido end substitution in all of the models.

structures are characterized as minima or maxima based on frequency analysis. The presence of all real frequencies confirms the structure as minima. Further, the methods are validated by reproducing the energies and structural parameters of different models with the wB97XD[28] and M06-HF[29] functionals and are found to give similar results (Table S1 in ESI†). Natural population analysis (NPA)[30] has been used to calculate the electron density distribution, mainly to describe the charge distribution on the donor and acceptor groups. Time dependent density functional theory (TDDFT) has been used to study the excited state geometries and properties like excitation energies.[7] For similar types of molecules like di-peptides, β-dipeptides and tri-peptides, CAM-B3LYP is proven to give a better description of the excitation energies without any correlation between the error and spatial orbital overlap (λ values).[31] Thus the same functional has been used here. A local vertical ionization at the specific donor unit introduces a hole in the electronic cloud which induces its migration throughout the peptide backbone.

Vertical ionization potential (VIE) and adiabatic ionization potential (AIE) are defined based on the equations given below.

$\text{VIE} = E'_{(N-1)} - E_{(N)}$, and $\text{AIE} = E_{(N-1)} - E_{(N)}$, where $E_{(N)}$ is the energy of the optimized neutral molecule with N electrons, $E'_{(N-1)}$ is the energy of the cation present in neutral (N electron) geometry and $E_{(N-1)}$ is the energy of the optimized cation with $(N-1)$ electrons.

The structural reorganisation energy (λ_{re}) is calculated as the difference between the vertical ionization potential and adiabatic ionization potential, which essentially gives the extent of nuclear relaxation after ionization.

Eqn (1) defines the initial non-stationary hole orbital $|\psi(0)|$, where N represents the number of atomic basis orbitals, which are stationary states of λth molecular orbitals.

$$|\psi(0)\rangle = \sum_{\lambda=1}^{N} \langle \phi_\lambda | \psi_{\text{MO}} | \phi_\lambda \rangle \tag{1}$$

ψ_{MO} is the molecular orbital of the neutral, where the hole is generated (in maximum cases it is the HOMO), whereas ϕ_λ is the λth cationic molecular orbital. After time, (t) the hole orbital can be represented by including a time dependent phase factor, as shown in the equation below,

$$|\psi(t)\rangle = \sum_{\lambda=1}^{N} \langle \phi_\lambda | \psi_{\text{MO}} \rangle \exp(-iE_\lambda t/\hbar) | \phi_\lambda \rangle \tag{2.}$$

The time scale for the hole migration dynamics is probed, which occurs before nuclear rearrangement, with a one electron DFT approach to follow the motion of an electronic wave packet. This method was proposed by Remacle and Levine on natural peptides and we have closely followed their procedure.[13]

3. Results and discussion

All models reported in this work are optimized and the energetics are analyzed. The relaxation shows that the lowest energy conformers are the ones with all the bonds oriented trans to each other as in natural peptides and unsubstituted UPs.[22] However the substitution of –R at the ureido end makes it oriented

 This journal is © The Royal Society of Chemistry 2018

perpendicular to the plane of the peptide backbone making a dihedral angle (C_2–N_1–C_1–R) of around 85° where the substitution at the carboxylate end (C_4–O_2–C_5–R′) follows the same trans orientation as other bonds (173°).

Investigation of the structural parameters indicates that the substitution of halogens at the ureido, the carboxylate or both ends shortens the C=X and C–N bond distances in UP by 0.005 Å and 0.012 Å (Table 1) respectively compared to the unsubstituted UP, which is attributed to its electronegative nature. The same phenomenon can be seen in the computed partial negative natural charge values. However, other groups like Ph and CN have minimal change in the structural parameters, because of their relatively less polarizing nature. All the substituents display their electron withdrawing ability (negative natural charge value) when substituted at the ureido end. The groups substituted at the ureido end act as electron rich groups irrespective of the models used here. The range of charge polarization between the ureido and carboxylate end varies with respect to the substituents, where the fluoride substitution has a maximum charge polarization with almost 0.4 e difference when it presents at the ureido end (Table 2). The electronegativity of the substituent is directly proportional to the amount of charge accepted by the same. Irrespective of the position of the substituents, the strength of electron acceptance follows the order F < NO_2 < Cl < Br < CN < Ph, as seen in the natural charge values (Table 2). The counteractive effect has been seen at the other end, where a more electropositive charge has been localized. Thus the

Table 1 Structural parameters (bond length (Å) and bond angles (°)) of different models studied at the neutral state using CAM-B3LYP/6-311++G**

Models	$r_{C–R/R'}$	$r_{C–X}$	$r_{C–N}$	$\angle_{N–C–R/R'}$
Free UP	—	1.226	1.373	—
Substitution at ureido end				
Ph–UP	1.517	1.228	1.371	114
F–UP	1.407	1.222	1.361	111
Cl–UP	1.833	1.222	1.360	113
Br–UP	1.994	1.222	1.360	113
CN–UP	1.475	1.225	1.361	113
NO_2–UP	1.462	1.221	1.360	108
Substitution at carboxylate end				
UP–Ph	1.512	1.227	1.372	114
UP–F	1.362	1.226	1.378	105
UP–Cl	1.768	1.226	1.378	107
UP–Br	1.915	1.226	1.375	109
UP–CN	1.463	1.225	1.379	107
UP–NO_2	1.402	1.222	1.377	107
Substitution at both ends				
Ph–UP–Ph	1.513	1.228	1.371	114
F–UP–F	1.406	1.221	1.363	111
Cl–UP–Cl	1.831	1.222	1.363	113
Br–UP–Br	1.992	1.222	1.361	113
CN–UP–CN	1.475	1.224	1.365	113
NO_2–UP–NO_2	1.461	1.221	1.362	108

Table 2 Computed NPA charges (e) at the substituted (R/R′), ureido and carboxylate ends using CAM-B3LYP/6-311++G**

Models	R (ureido)	Ureido	Carboxylate	R′ (carboxylate)
Free UP	—	−0.202	0.001	—
Ph–UP	−0.022	0.020	0.002	—
UP–Ph	—	0.002	0.421	−0.399
Ph–UP–Ph	−0.023	0.023	0.001	0.002
F–UP	−0.418	0.411	0.005	—
UP–F	—	0.002	0.362	−0.376
F–UP–F	−0.416	0.423	0.371	−0.374
Cl–UP	−0.124	0.117	0.008	—
UP–Cl	—	0.014	0.012	−0.026
Cl–UP–Cl	−0.120	0.122	0.015	−0.022
Br–UP	−0.082	0.073	0.008	—
UP–Br	—	0.014	−0.055	0.042
Br–UP–Br	−0.078	0.085	−0.051	0.046
CN–UP	−0.053	0.046	0.008	—
UP–CN	—	0.019	0.010	−0.029
CN–UP–CN	−0.043	0.061	0.015	−0.026
NO_2–UP	−0.353	0.346	0.008	—
UP–NO_2	—	0.007	0.275	−0.288
NO_2–UP–NO_2	−0.351	0.356	0.279	−0.286

donor end has been changed from a ureido group to carboxylate in the substituted UPs and the acceptor end has also been tuned accordingly. The changes in bond distances are reflected in their respective vibrational frequencies (Table 3). Further, the highest energy molecular orbitals are plotted to assess the

Table 3 Computed important vibrational frequencies (cm^{-1}) for the lowest energy conformers using the CAM-B3LYP functional

Models	$\upsilon_{C\text{–}R/R'}$	$\upsilon_{C\text{–}X}$	$\upsilon_{C\text{–}N}$	$\upsilon_{C\text{–}O}$
Free UP	—	1783	1593	1830
Ph–UP	1247	1773	1587	1831
UP–Ph	1256	1783	1593	1823
Ph–UP–Ph	1248	1772	1586	1830
F–UP	1156	1801	1589	1832
UP–F	1155	1796	1593	1830
F–UP–F	1155	1798	1585	1831
Cl–UP	977	1801	1587	1832
UP–Cl	837	1786	1593	1833
Cl–UP–Cl	970	1800	1586	1831
Br–UP	593	1801	1586	1832
UP–Br	651	1785	1594	1835
Br–UP–Br	590	1800	1585	1832
CN–UP	1140	1789	1591	1831
UP–CN	990	1787	1593	1840
CN–UP–CN	1137	1788	1590	1830
NO_2–UP	1430	1800	1589	1832
UP–NO_2	1390	1789	1593	1835
NO_2–UP–NO_2	1428	1799	1588	1830

 This journal is © The Royal Society of Chemistry 2018

involvement of the substituents and are given in Fig. 3 for the Ph substituted UPs. The plots clearly reflect that the substituent entries at both ends become localized within the orbital at the specific end, increasing the energy gap between the HOMO and LUMO compared to the unsubstituted one (from 9 eV in free UPs to 9.5–10 eV in substituted UPs) (Table 4). TDDFT calculations show that the excitation gives two major transitions occurring either from n to π^* or π to π^* (Table 5). The type of transition and its energies are decided by the nature of the substituent and its position in the models as seen in Fig. 3 and S1 to S4 in the ESI.† However, the transition occurs predominantly from the localized molecular orbitals.

Following that, the behavioral changes in the ionization processes have been studied. Similar to unsubstituted UP, SUP and TUP,[22,23] the ionization did not alter the structure much from the ground state geometry for the substituted models, which is reflected in the reorganization energy (Table 4). Comparison of the structural parameters (Table S2 in the ESI†) between the neutral and first ionized optimized geometries further validates minimal change in the nuclear position during ionization. Moreover, ionization removes electrons from the localized donor groups as predicted by the natural charge distribution and as reported for the unsubstituted UP models. This is shown also in the spin density plots of Ph–UP, UP–Ph and Ph–UP–Ph (Fig. 4). Hence, the hole migration is carried out using the fixed nuclei approach. The time scale for this process after vertical ionization in the ground state geometry is found to be lengthier or shorter in the case of substituted UPs. The contributions of individual constituents have been discussed in detail in the section below.

(i) Phenyl substituted UPs (Ph–UP, UP–Ph and Ph–UP–Ph)

Aromatic rings have been found to play crucial roles when associated with the peptide backbone. The interactions can contribute to the conformer stability of peptides and proteins. They have been added as a side chain or can be included as related amino acids like tyrosine.[32] There are no major structural changes observed resulting from the substitution of phenyl rings at the ureido or carboxylate ends of UPs. The phenyl substituted UPs (Ph–UP, UP–Ph and Ph–UP–Ph) are found to be chemically stable with larger HOMO–LUMO values as in the unsubstituted UP. TDDFT calculations show that the substitution of phenyl rings at various positions gives two intense peaks corresponding to π to π^* transitions as in Fig. 3. Note that the phenyl ring substitutions eliminate the n (lone pair) to π^* transitions, which was observed in the unsubstituted UP. This is due to the larger negative charge of the phenyl ring in the models studied here. The phenyl ring as a chromophore makes the excitation wavelengths red shift by $\sim$200 nm compared to the unsubstituted UP, since excitation transfers an electron from the π cloud of the phenyl ring. Substitution at the ureido end or both ends (Ph–UP and Ph–UP–Ph in Fig. 3a and c) gives a peak due to the excitation of an electron from the phenyl unit at the ureido end to the carboxylate group. This is apparent from the orbital picture, where the HOMOs of Ph–UP and Ph–UP–Ph are localised at the phenyl ring and the vacant π^* LUMO present at the carboxylate end.

However, Ph substitution at the carboxylate end (UP–Ph) shows that the orbitals involved in the transitions are localised at the same end (Fig. 3b), where both the π and π^* orbitals are localised at the phenyl ring. Hence, end-to-end

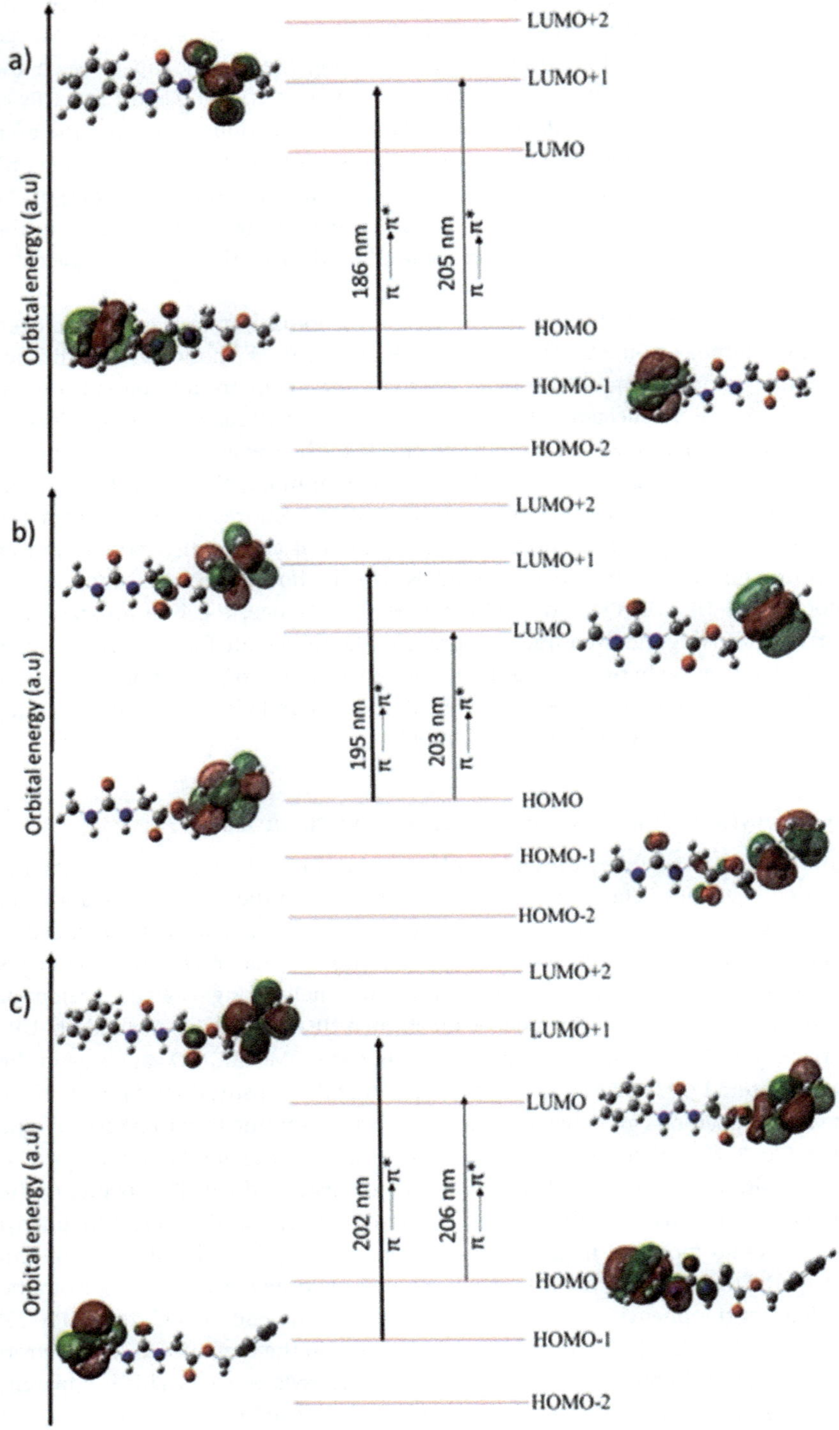

Fig. 3 Excitation wavelengths (nm) and corresponding orbitals involved in the transition for the (a) Ph–UP, (b) UP–Ph and (c) Ph–UP–Ph models plotted with the isocontour value 0.03 Å^{-3}.

 This journal is © The Royal Society of Chemistry 2018

Table 4 Computed VIE (eV), AIE (eV), ΔHOMO − LUMO (eV), λ_{re} and time scale for hole migration (fs) for the models studied using CAM-B3LYP/6-311++G**

Models	VIE	AIE	ΔHOMO − LUMO	λ_{re}	Time scale (fs)
Free UP	9.208	8.933	9.000	0.275	2.9
Ph–UP	8.716	8.963	8.802	0.247	4.8
UP–Ph	8.543	8.762	8.808	0.219	—
Ph–UP–Ph	8.683	8.896	8.480	0.213	4.0
F–UP	9.801	10.102	9.650	0.301	2.4
UP–F	9.393	9.692	9.219	0.299	2.3
F–UP–F	10.006	10.309	9.712	0.303	2.4
Cl–UP	9.681	9.805	9.556	0.124	2.3
UP–Cl	9.224	9.368	9.191	0.144	2.2
Cl–UP–Cl	9.712	9.844	9.577	0.132	2.2
Br–UP	9.537	9.768	9.273	0.231	2.7
UP–Br	9.431	9.650	8.988	0.219	2.5
Br–UP–Br	9.676	9.933	9.203	0.257	2.4
CN–UP	9.809	10.030	9.532	0.221	2.9
UP–CN	9.486	9.725	9.077	0.239	3.2
CN–UP–CN	9.819	10.060	9.551	0.241	3.0
NO_2–UP	9.668	9.969	9.665	0.301	3.6
UP–NO_2	9.379	9.678	9.568	0.299	3.4
NO_2–UP–NO_2	9.818	10.104	10.754	0.286	3.5

charge transfer in the UP–Ph model after excitation is not possible. Following that the wave packet propagation is studied using a frozen nuclei approach. The migration occurs from the ureido end to carboxylate end in both the Ph–UP and Ph–UP–Ph models. In Ph–UP, the hole created after ionization at the phenyl ring

Table 5 Excitation wavelengths (nm) and corresponding energies (eV) of the different models computed using TDDFT

Models	n to π^*	π to π^*
Free UP	173 (7.166)	184 (6.738)
Ph–UP	—	205 (6.048) 186 (6.665)
UP–Ph	—	203 (6.107) 195 (6.358)
Ph–UP–Ph	—	206 (6.018) 202 (6.137)
F–UP	176 (7.044)	179 (6.926)
UP–F	172 (7.208)	190 (6.525)
F–UP–F	165 (7.514)	185 (6.701)
Cl–UP	180 (6.888)	179 (6.926)
UP–Cl	170 (7.293)	172 (7.208)
Cl–UP–Cl	179 (6.926)	185 (6.701)
Br–UP	181 (6.850)	193 (6.424)
UP–Br	184 (6.738)	191 (6.491)
Br–UP–Br	210 (5.904)	193 (6.424)
CN–UP	168 (7.380)	179 (6.926)
UP–CN	172 (7.208)	192 (6.457)
CN–UP–CN	186 (6.665)	190 (6.525)
NO_2–UP	—	187 (6.630)
UP–NO_2	—	195 (6.358)
NO_2–UP–NO_2	—	196 (6.325)

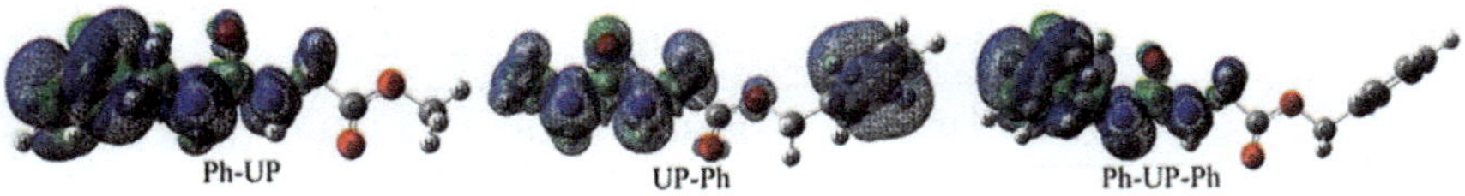

Fig. 4 Spin density plots of various Ph substituted UPs plotted with an isocontour value of 0.02 $Å^{-3}$. Note that the spin density in UP–Ph is localised equally on both ends eliminating the possibility for a charge transfer from the phenyl to ureido end though there is a charge polarisation.

migrates and reaches the carboxylate end in 3.2 fs, then bounces back from there in 4.8 fs which is found to be the largest time scale among all the models studied in this work (Fig. 5). The presence of the constituents at both ends decreases the time scale for hole migration from 4.8 fs to 4 fs. As suggested by the results of the TDDFT calculations, in the UP–Ph models, the hole created at the carboxylate end phenyl unit is not propagating towards the ureido end. Moreover, Ph–UP and Ph–UP–Ph do not move the charge from the carboxylate to ureido end. This suggests that Ph–UP and Ph–UP–Ph undergo hole migration in only one direction as reported for the unsubstituted UP.

(ii) Halogen substituted UPs

Fluorines are substituted into the peptides experimentally to stabilize the helical conformations.[33] Hence, the fluorines along with chlorine and bromine (iodine has been excluded for being radioactive) are incorporated in the UPs to tune their specific properties. The bond lengths between the halogens and UP backbone increase considerably (1.407–1.902 Å) down the group, which is due to a decrease in their electronegativity. The general tendency to donate the lone pair has made the carbonyl bonds of the ureido group short compared to the same group in the simple UP model. Thus halogens alter the electronic distribution on the complete peptide bonds. The same has been replicated in natural charge values, where the negative charge indicates that halogens can act as donors regardless of their

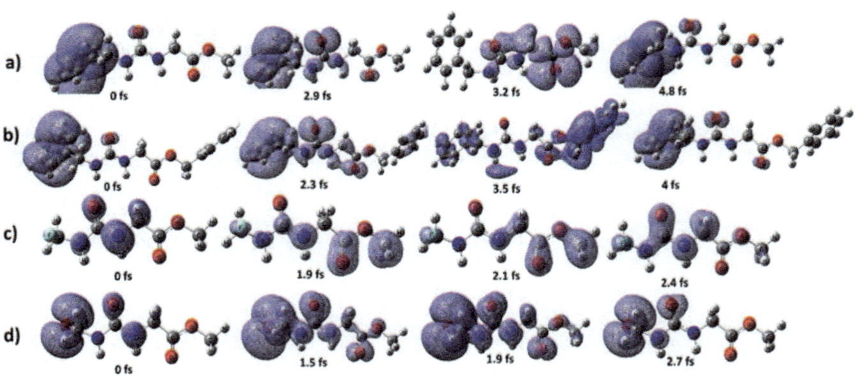

Fig. 5 Snapshots of the hole migration dynamics from the holes generated at the HOMO orbitals of (a) Ph–UP, (b) Ph–UP–Ph, (c) F–UP and (d) Br–UP plotted with an isocontour value of 0.002 $Å^{-3}$.

 This journal is © The Royal Society of Chemistry 2018

position. More than one halogen in the model combines with the ureido group and makes the ureido end more electronegative. In all of the chlorine-/fluorine-substituted models, the HOMOs and LUMOs are localized at the ureido and carboxylate group respectively (Fig. S1 in the ESI†). Thus the electronic transition wavelengths have not varied much, which is similar to that of the unsubstituted UP. Nevertheless, the trend with the bromine substituted UPs (Br–UP, UP–Br, and Br–UP–Br) is different, which is due to the presence of a loosely bound lone electron pair. The HOMOs of Br–UP and Br–UP–Br are largely localised at the bromine centre and the excitation transfers the non-bonded electrons from the same end towards the π^* orbital at the carboxylate end. In UP–Br, the transition is from the π orbital on the ureido group towards the π^* orbital located on the carboxylate end (Fig. S2 in the ESI†). The excitation wavelengths of the bromine substituted UPs are red-shifted compared to the unsubstituted UP (193–210 nm).

The ionization removes an electron from the HOMO, which is localised at the ureido group in the case of the F and Cl substituted UPs. As a consequence, hole migration is observed from the ureido to carboxylate group as observed in the unsubstituted UP. The time scale for such hole migration is shorter and falls in the range of 2.2–2.4 fs, whereas for the bromine substituted UPs, Br–UP, UP–Br and Br–UP–Br, the hole created in the HOMO localised on the bromine atom is found to take 2.4 to 2.7 fs for its migration (Fig. 5) from one end to the other. This computed time scale is shorter than the reported period for the unsubstituted UP. Ionization of UP–Br removes the electron density localised on the ureido end and the hole migrates from there to the bromine centre at the other end. Br–UP–Br exhibits hole migration from the bromine at the ureido end to the carboxylate end. Similar to phenyl substituted UP models, there is no charge transfer from the carboxylate to ureido end in any halogen-substituted models.

(iii) CN and NO_2 substituted UPs

Both CN and NO_2 groups are found to be electron withdrawing when associated with linear or cyclic conjugated systems. Cyanylation of cysteine residues is reported to promote selective bond cleavage. Similarly, nitrotyrosine containing peptides have also been formulated for peptide identification.[34] In this section both groups are coupled since they show similar behaviour when substituted on UPs. They show a C–R bond length of ~1.462–1.471 Å and are closer to the unsubstituted UP. It can also be attributed to the resonance structures of both of the groups in relation to the carboxylate partial double bond nature. For the NO_2 groups, being ambidendate ligands, both nitro (attachment through a nitrogen atom) and nitrito (attached through oxygen) structures have been analysed since there are instances of photo induced conversions of the linkage isomers. The UPs with ONO coordinated structures are found to be energetically stable since they are aligned in a bent formation out-of-plane from the peptide backbone (Fig. 2).

As exhibited by other groups, the donor and acceptor ends get tuned according to the substituent position. The HOMOs and LUMOs of the CN substituted UPs are localized more at the ureido and carboxylate terminals respectively (Fig. S4 in the ESI†). Thus charge transfer after excitation is observed from the ureido to carboxylate end as in the unsubstituted UP, whereas for the NO_2 substituted UPs, the amount of localisation differs depending upon its position. Localization is higher in the UP–NO_2 and NO_2–UP–NO_2 models compared to the NO_2–UP model,

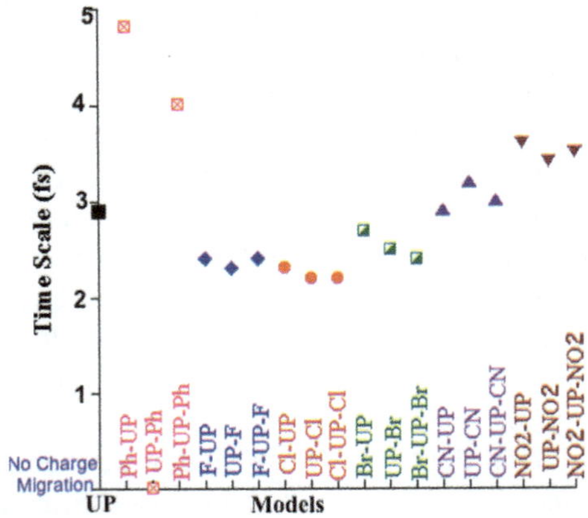

Fig. 6 Time scales for hole migrations for the models studied in this work.

where the π orbital localisation is seen at the NO_2 group and a vacant π* orbital at the carboxylate end (Fig. S3 in the ESI†). Hole migration after ionization follows from the ureido group in the CN substituted UPs, making the time scale similar to the bromine substituted UPs. End to end hole migration in NO_2 substituted UPs, specifically UP–NO_2, shows a longer time scale compared to CN as seen in Table 4.

Following that, the time scales for hole migration in the various models are compared and given in Fig. 6. The plot clearly shows that hole migration is favourable and happens faster in the F–UP, UP–F, F–UP–F, Cl–UP, UP–Cl, Cl–UP–Cl, Br–UP, UP–Br and Br–UP–Br models when compared to unsubstituted UP, which is due to the increased electronegative nature. Further CN and NO_2 group containing model (CN–UP, UP–CN, CN–UP–CN, NO_2–UP, UP–NO_2 and NO_2–UP–NO_2) migration dynamics are a little lengthier than for the unsubstituted UP. On the other hand, phenyl substitution in UPs is either forbidden or takes longer for migration of the hole from the ureido to carboxylate end.

Conclusions

The substitutions of UPs with different functional units (phenyl, F, Cl, Br, CN, and NO_2) are found to change the electronic structure of the peptide backbone considerably along with its polarization. As a result, the negative charge polarisation from the ureido group in the unsubstituted UP was shifted towards the end where the substituent was present. The excitation calculations show two intense peaks corresponding to the n to π* or π to π* transitions, where n and π are localised at different energy levels depending upon the substituent groups. The presence of NO_2 and phenyl substituents in UP dominates with π to π* transitions, while the F, Cl, Br and CN groups prefer both n to π* and π to π* transitions. The excitation studies show that for all the models, intense peaks appear from transition of the orbitals present at one end to the other, except for the UP–Ph model. In the UP–Ph model, end to end charge transfer is not possible because the HOMO and LUMO are located at the same terminal. Hole migration dynamics have been followed after immediate ionization for all of the substituted models. Among the many models, Ph–UP does not show any hole migration, while all of the other models show the dynamics with reasonable time scales. The F–UP, UP–F, and F–UP–F molecules show shorter time scales for hole migration compared to the other models studied in this work. Further, the migration dynamics provide evidence that, irrelevant of the substituent, the hole migration happens only in one direction, from the ureido to carboxylate end as in the unsubstituted UP.

Conflicts of interest

There is no conflict of interest.

Acknowledgements

We thank the UGC-Major Research Project (UGC-MRP), DST-SERB, Junior Research Fellowship (JRF), and Faculty Recharge Programme (FRP) for funding.

References

1 P. H. Bucksbaum, *Science*, 2007, **317**, 766–769.
2 M. Vacher, L. Steinberg, A. J. Jenkins, M. J. Bearpark and M. A. Robb, *Phys. Rev. A*, 2015, **92**, 040502.
3 O. M. Kirkby, M. Sala, G. Balerdi, R. de Nalda, L. Banares, S. Guerin and H. H. Fielding, *Phys. Chem. Chem. Phys.*, 2015, **17**, 16270–16276.
4 G. Sansone, F. Kelkensberg, J. F. Perez-Torres, F. Morales, M. F. Kling, W. Siu, O. Ghafur, P. Johnsson, M. Swoboda, E. Benedetti, F. Ferrari, F. Lepine, J. L. Sanz-Vicario, S. Zherebtsov, I. Znakovskaya, A. L'Huillier, M. Y. Ivanov, M. Nisoli, F. Martin and M. J. J. Vrakking, *Nature*, 2010, **465**, 763.
5 G. Periyasamy, R. D. Levine and F. Remacle, *Chem. Phys.*, 2009, **366**, 129–138.
6 F. Remacle, R. Kienberger, F. Krausz and R. D. Levine, *Chem. Phys.*, 2007, **338**, 342–347.
7 B. Mignolet, R. D. Levine and F. Remacle, *J. Phys. B: At., Mol. Opt. Phys.*, 2014, **47**, 124011.
8 J. P. Hall, F. E. Poynton, P. M. Keane, S. P. Gurung, J. A. Brazier, D. J. Cardin, G. Winter, T. Gunnlaugsson, I. V. Sazanovich, M. Towrie, C. J. Cardin, J. M. Kelly and S. J. Quinn, *Nat. Chem.*, 2015, **7**, 961–967.
9 R. Zhang, J. Liu, H. Yang, S. Wang, M. Zhang and Y. Bu, *ChemPhysChem*, 2015, **16**, 436–446.
10 H. Kim and E. Sim, *Curr. Appl. Phys.*, 2009, **9**, E276–E279.
11 H. S. Mandal and H. B. Kraatz, *Chem. Phys.*, 2006, **326**, 246–251.
12 E. W. Schlag, S. Y. Sheu, D. Y. Yang, H. L. Selzle and S. H. Lin, *Proc. Natl. Acad. Sci. U. S. A.*, 2000, **97**, 1068–1072.
13 F. Remacle and R. D. Levine, *Proc. Natl. Acad. Sci. U. S. A.*, 2006, **103**, 6793–6798.
14 Y. F. Wang, Z. Y. Yu, J. Wu and C. B. Liu, *J. Phys. Chem. A*, 2009, **113**, 10521–10526.
15 J. Gao, P. Mueller, M. Wang, S. Eckhardt, M. Lauz, K. M. Fromm and B. Giese, *Angew. Chem., Int. Ed.*, 2011, **50**, 1926–1930.
16 G. Jones, L. N. Lu, H. Fu, C. W. Farahat and C. Oh, *J. Phys. Chem. B*, 1999, **103**, 572–581.

This journal is © The Royal Society of Chemistry 2018

17 B. Giese, M. Wang, J. Gao, M. Stoltz, P. Mueller and M. Graber, *J. Org. Chem.*, 2009, **74**, 3621–3625.
18 W. A. Prutz, E. J. Land and R. W. Sloper, *J. Chem. Soc., Faraday Trans. 1*, 1981, **77**, 281–292.
19 J. Gante, *Angew. Chem., Int. Ed.*, 1994, **33**, 1699–1720.
20 M. Rubinstein and M. Y. Niv, *Biopolymers*, 2009, **91**, 505–513.
21 I. Baran, R. S. Varekova, L. Parthasarathi, S. Suchomel, F. Casey and D. C. Shields, *J. Chem. Inf. Model.*, 2007, **47**, 464–474.
22 S. Joy, V. V. Sureshbabu and G. Periyasamy, *J. Phys. Chem. B*, 2016, **120**, 6469–6478.
23 S. Joy, V. V. Sureshbabu and G. Periyasamy, *J. Phys. Org. Chem.*, 2017, e3693.
24 S. Joy, V. V. Sureshbabu and G. Periyasamy, *ChemistrySelect*, 2017, **2**, 4261–4266.
25 T. Yanai, D. P. Tew and N. C. Handy, *Chem. Phys. Lett.*, 2004, **393**, 51–57.
26 M. J. Frisch, G. W. Trucks, H. B. Schlegel, G. E. Scuseria, M. A. Robb, J. R. Cheeseman, G. Scalmani, V. Barone, B. Mennucci, G. A. Petersson, H. Nakatsuji, M. Caricato, X. Li, H. P. Hratchian, A. F. Izmaylov, J. Bloino, G. Zheng, J. L. Sonnenberg, M. Hada, M. Ehara, K. Toyota, R. Fukuda, J. Hasegawa, M. Ishida, T. Nakajima, Y. Honda, O. Kitao, H. Nakai, T. Vreven, J. A. Montgomery Jr, J. E. Peralta, F. Ogliaro, M. Bearpark, J. J. Heyd, E. Brothers, K. N. Kudin, V. N. Staroverov, R. Kobayashi, J. Normand, K. Raghavachari, A. Rendell, J. C. Burant, S. S. Iyengar, J. Tomasi, M. Cossi, N. Rega, J. M. Millam, M. Klene, J. E. Knox, J. B. Cross, V. Bakken, C. Adamo, J. Jaramillo, R. Gomperts, R. E. Stratmann, O. Yazyev, A. J. Austin, R. Cammi, C. Pomelli, J. W. Ochterski, R. L. Martin, K. Morokuma, V. G. Zakrzewski, G. A. Voth, P. Salvador, J. J. Dannenberg, S. Dapprich, A. D. Daniels, Ö. Farkas, J. B. Foresman, J. V. Ortiz, J. Cioslowski and D. J. A. Fox, *Gaussian 09, Revision D.01*, Gaussian, Inc., Wallingford, CT, 2009.
27 S. Chandra, G. Periyasamy and A. Bhattacharya, *J. Chem. Phys.*, 2015, **142**, 244309.
28 J. D. Chai and M. Head-Gordon, *Phys. Chem. Chem. Phys.*, 2008, **10**, 6615–6620.
29 Y. Zhao and D. G. Truhlar, *Theor. Chem. Acc.*, 2008, **120**, 215–241.
30 J. P. Foster and F. Weinhold, *J. Am. Chem. Soc.*, 1980, **102**, 7211–7218.
31 M. J. G. Peach, P. Benfield, T. Helgaker and D. J. Tozer, *J. Chem. Phys.*, 2008, **128**, 044118.
32 N. Y. Palermo, J. Csontos, M. C. Owen, R. F. Murphy and S. Lovas, *J. Comput. Chem.*, 2007, **28**, 1208–1214.
33 E. Danelius, H. Andersson, P. Jarvoll, K. Lood, J. Grafenstein and M. Erdelyi, *Biochemistry*, 2017, **56**, 3265–3272.
34 C. Batthyany, S. Bartesaghi, M. Mastrogiovanni, A. Lima, V. Demicheli and R. Radi, *Antioxid. Redox Signaling*, 2017, **26**, 313–383.

 This journal is © The Royal Society of Chemistry 2018

Faraday Discussions

PAPER

Protein charge transfer absorption spectra: an intrinsic probe to monitor structural and oligomeric transitions in proteins†

Mohd. Ziauddin Ansari, [a] Amrendra Kumar, [a] Dileep Ahari, [a] Anurag Priyadarshi, [a] Padmavathi Lolla,[b] Rashna Bhandari [b] and Rajaram Swaminathan *[a]

Received 14th September 2017, Accepted 13th November 2017
DOI: 10.1039/c7fd00194k

Protein Charge Transfer Spectra (ProCharTS) originate when charged amino/carboxylate groups in the side chains of Lys/Glu act as electronic charge acceptors/donors for photoinduced charge transfer either from/to the polypeptide backbone or to each other. The absorption band intensities in ProCharTS at wavelengths of 250–800 nm are dependent on the 3D spatial proximity of these charged functional groups across the protein. Intrinsically disordered proteins (IDPs) are an important class of proteins involved in signalling and regulatory functions in the eukaryotic cell. IDPs are rich in charged amino acids, but lack structure-promoting intrinsic spectral probes like Tyr or Trp in their sequences, making their structural characterisation difficult. Here, we exploit the richness of charged amino acid populations among IDPs (like the PEST fragment of human c-Myc, its mutant and dehydrin from maize) to sense structural transitions in IDPs using ProCharTS absorption spectra. Conformational changes induced in the protein by altering the pH and temperature of the aqueous medium were monitored by ProCharTS and confirmed by CD spectra. Further, the utility of ProCharTS to detect protein aggregation was examined using Hen Egg-White Lysozyme (HEWL) protein. The results revealed that in the presence of Trp/Tyr, ProCharTS absorbance was substantially reduced, specifically at wavelengths where the absorption by Trp or Tyr was near its maximum. Significant changes in the ProCharTS spectra were observed with changing pH in the range of 3–11, which correlated with changes in the secondary structure of the PEST fragment. Importantly, the absorbance at 280 nm, which is often employed as a measure of protein concentration, was profoundly altered by changes in ProCharTS intensity in response to changing the pH in dehydrin. The ProCharTS intensity was sensitive to temperature-induced changes in the secondary structures of the PEST fragments between 25–85 °C. The presence of 0.25 M NaCl or KCl in the

[a]*Department of Biosciences & Bioengineering, Indian Institute of Technology Guwahati, Guwahati 781039, Assam, India. E-mail: rsw@iitg.ernet.in*
[b]*Laboratory of Cell Signalling, Centre for DNA Fingerprinting and Diagnostics (CDFD), Hyderabad 500001, India*

† Electronic supplementary information (ESI) available. See DOI: 10.1039/c7fd00194k

medium also altered the ProCharTS spectrum. Finally, an increase in ProCharTS absorbance with time in HEWL at pH 2 directly correlated with the growth of HEWL aggregates and amyloid fibrils, as confirmed by the increasing thioflavin T fluorescence. Taken together, our work highlights the utility of ProCharTS as a label-free intrinsic probe to monitor changes in protein charge, structure and oligomeric state.

1 Introduction

Intrinsic spectral probes enable the structure and dynamics of biomolecules to be investigated in their natural state. Intrinsic UV-visible spectral chromophores present in the side chains of amino acids like tryptophan and tyrosine have in the past been extensively used for such purposes in numerous proteins.[1–8] However, Trp by itself is of limited utility in protein structure investigations as its fluorescence is mostly sensitive to the immediate local environment surrounding the indole chromophore and requires another external acceptor for yielding 3D distance information across the protein, using FRET.[9] Further, several important protein classes like intrinsically disordered proteins (IDPs)[10–12] lack these intrinsic chromophores in their sequences.[10,13,14] This imposes severe constraints on monitoring structural changes in IDPs[15] by a label-free approach using UV-vis spectroscopy. To overcome these constraints, there is a need for alternative intrinsic chromophores to investigate the structures and dynamics of IDPs.

IDPs are rich in charged amino acids including Glu and Lys.[10,13,14] Earlier work from our lab revealed that a concentrated aqueous solution of L-Lysine·HCl[16] and dilute aqueous solutions of lysine-rich proteins, like poly-L-lysine·HCl or human serum albumin, display noticeable UV-vis absorption spectra in the 250–350 nm wavelength range, although their origin has remained unexplained.[17] Recent work from our group using the synthetic protein α_3C, which is rich in charged amino acids like Lys and Glu but lacks aromatic amino acids, has suggested that such UV-vis absorption spectra in the 250–800 nm region arise from charge transfer transitions.[18] Specifically, photoinduced charge transfer from the polypeptide backbone to NH_3^+ in lysine, COO^- in glutamate to the polypeptide backbone, and COO^- in glutamate to NH_3^+ in lysine, was chiefly implicated to be the origin of the observed protein charge transfer spectra (ProCharTS).[18] From a protein structure perspective, it is intriguing to note that the ProCharTS intensity and wavelength were expected to be dependent on the 3D spatial proximity between charges in the Lys–Lys, Glu–Glu and Lys–Glu side-chain headgroup pairs. Thus, while the close proximity (~3 Å) of like charges in the Lys–Lys and Glu–Glu pairs enhanced the ProCharTS intensity at longer visible wavelengths, it diminished the intensity for the oppositely-charged Lys–Glu pair. However, moderate spatial separation (~6 Å) between the oppositely-charged Lys–Glu pair enhanced the ProCharTS intensity and extended its presence to longer wavelengths.[18] α_3C, which has several such pairs of Lys/Glu in favorable proximity, displays intense ProCharTS. In this context, it would be worthwhile to investigate whether ProCharTS can provide insight into the otherwise elusive structures of IDPs that are rich in charged amino acids.

Protein aggregation is a symptom of several diseases like Alzheimer's, type 2 diabetes and so on.[19–25] Monitoring the formation of early oligomers in protein aggregation is experimentally difficult, as dyes like thioflavin T and Congo Red are more sensitive to late-stage ordered aggregates like amyloids.[26–29] Intrinsic chromophores like tryptophan are of limited utility for monitoring aggregation.[8] Intermolecular events like protein aggregation that bring together protein chains and generate new close molecular contacts present another opportunity to validate ProCharTS, which is dependent on the proximity between charged amino acid residues. New spatially proximal pairs of like or oppositely-charged Lys/Glu arising in protein aggregates are expected to display ProCharTS. It will be worthwhile to determine whether ProCharTS intensity can detect the time-dependent growth of protein aggregates in a charged protein. In this report, we attempt to employ ProCharTS for monitoring protein aggregation and tracking changes in the structures of multiple IDPs.

The work presented in this paper is divided into two parts. The first part is focused on measuring ProCharTS absorption spectra among three IDPs in aqueous media, whose 3D structures are as yet unknown, namely wild-type human c-Myc PEST fragment (Wt), mutant human c-Myc PEST fragment (M1) with inserted Trp, and dehydrin protein (DHN1) from *Zea mays*. c-Myc oncoprotein acts as a transcription factor with involvement in cell growth, differentiation, proliferation and apoptosis.[30–32] c-Myc has a PEST region in its center, enriched in proline (P), glutamic acid (E), serine (S) and threonine (T). The PEST region serves as a proteolytic signal for the rapid destruction of c-Myc.[33–35] Dehydrin, an IDP from *Zea mays*, belongs to the Late Embryogenesis Abundant (LEA) protein family. Proteins of this family are known to accumulate under abiotic stress conditions like water desiccation, salt and cold stress.[36,37] The characteristic presence of ProCharTS was revealed among all the IDPs studied, even when aromatic residues like Trp or Tyr were present in the sequence. Subsequently, the effects of changing the pH, salt concentration and temperature of the aqueous medium on the ProCharTS spectra of the three IDPs are presented. The increased intensities of the ProCharTS spectra with rising temperature in the PEST fragments correlated with temperature-induced changes in protein structure. A modest increase in ProCharTS intensity at acidic pH compared to neutral and alkaline pH suggested an alternative structure in the PEST fragment at acidic pH.

In the second part, ProCharTS absorption was employed to investigate the aggregation of the moderately-charged protein Hen Egg-White Lysozyme (HEWL), which shows negligible ProCharTS in its monomeric form, as demonstrated previously.[18] HEWL aggregation *in vitro*, monitored by conventional techniques, at both acidic[38–42] and alkaline[43–47] pH, has been well documented previously. Further, different manoeuvres to arrest the aggregation of HEWL using DTT,[48,49] iodoacetamide[45,50] and competitive inhibitors like tri-*N*-acetyl chitotriose[51] have also been demonstrated previously. Our results revealed a marked and consistent increase in ProCharTS absorption among HEWL aggregates in comparison to the monomer in both acidic and alkaline conditions, which are known to promote aggregation. The increased ProCharTS absorption at alkaline pH was clearly diminished when HEWL aggregation was inhibited by blocking reactive thiols using iodoacetamide, establishing that the ProCharTS absorption was indeed tracking HEWL aggregation.

ProCharTS provides a new label-free, intrinsic spectral marker to detect early oligomer and late fibril formation during protein aggregation. It also facilitates the easy monitoring of structural transitions in IDPs in response to changes in temperature.

2 Materials

Biochemicals and reagents like ampicillin (A8351); nickel affinity gel (H0537); TCEP (C4706); iodoacetamide (I1149); and Hen Egg-White Lysozyme (L6876) protein were purchased from Sigma Aldrich Chemicals Pvt. Limited, Bengaluru, India. Luria broth (M1245) and imidazole (GRM1864) were procured from HiMedia Laboratories, India. His Trap column (17-5255-01) was purchased from GE Healthcare. Other chemicals, including buffer salts, were purchased from Merck India Limited. All the chemicals used were analytical grade with purity ≥98%.

Amino acid sequences of all the proteins employed in the current study are shown in Fig. 1. Uniprot IDs of the human c-Myc PEST region (residues 201–268) and HEWL are P01106 and P00698, respectively. PDB ID of α_3C is 2LXY.

3 Experimental methods

3.1 Cloning, expression and purification of recombinant proteins

3.1.1 PEST wild-type (Wt) and PEST mutant (M1). The disorder prediction profiles for human c-Myc PEST Wt and its mutant (M1) are displayed in Fig. S1A and B in the ESI,† respectively. The human c-Myc region from amino acid 201 to 268 (PEST Wt) and its mutant (PEST M1) were cloned in a pET21b expression vector between the *Nde*1 and *Xho*1 restriction sites, encoding the PEST Wt and PEST M1 residues with a hexahistidine tag fused with a C-terminus. 5′-TGACTCATATGGACAGCAGCTCGCCCAAGTC-3′ was used as the forward primer. 5′-CATGACTCGAGTTCCACAGAAACAACATCG-3′ and 5′-CATGACTCGAGCCATTCCACAGAAACAACATCG-3′ were used as reverse primers for PEST Wt and PEST M1, respectively. The recombinant PEST Wt had three extra residues at positions 1, 70

[A] **PEST Wt [77]** (Mass: 8342 Da)
MDSSSPKSCA SQDSSAFSPS SDSLLSSTES SPQGSPEPLV LHEETPPTTS
SDSEEEQEDE EEIDVVSVEL EHHHHHH

[B] **PEST M1 [78]** (Mass: 8528 Da)
MDSSSPKSCA SQDSSAFSPS SDSLLSSTES SPQGSPEPLV LHEETPPTTS
SDSEEEQEDE EEIDVVSVEW LEHHHHHH

[C] **DHN1 [167]** (Mass: 16955 Da)
MEYGQQGQHG HGATGHVDQY GNPVGGVEHG TGGMRHGTGT GGMGQLGEHG
GAGMGGGQFQ PAREEHKTGG ILHRSGSSSS SSSEDDGMGG RRKKGIKEKI
KEKLPGGHKD DQHATATTGG AYGQQGHTGS AYGQQGHTGG AYATGTEGTG
EKKGIMDKIK EKLPGQH

[D] **α_3C [67]** (Mass: 7460 Da)
GSRVKALEEK VKALEEKVKA LGGGGRIEEL KKKCEELKKK IEELGGGGEV
KKVEEEVKKL EEEIKKL

[E] **HEWL [129]** (Mass: 14313 Da)
KVFGRCELAA AMKRHGLDNY RGYSLGNWVC AAKFESNFNT QATNRNTDGS
TDYGILQINS RWWCNDGRTP GSRNLCNIPC SALLSSDITA SVNCAKKIVS
DGNGMNAWVA WRNRCKGTDV QAWIRGCRL

Fig. 1 Protein sequences: charged and aromatic amino acids are highlighted in bold using black and red coloured fonts, respectively.

 This journal is © The Royal Society of Chemistry 2018

and 71 other than the hexahistidine, while the mutant M1 had four at positions 1, 70, 71 and 72. The *E. coli* BL21 (DE3) strain was used for overexpression of PEST Wt and PEST M1. The cultures were grown in LB media, containing 100 μg mL^{-1} ampicillin (at 37 °C, 180 rpm) in a shaking incubator. At A_{600} of 0.6–0.7, the cultures were induced with isopropyl β-D-1-thiogalactopyranoside (IPTG, 1 mM) at 37 °C for 3 hours. The induced cells were centrifuged at 5000 rpm for 15 minutes at 4 °C to collect the pellets. The collected pellets were resuspended in ice cold lysis buffer (25 mM Tris·HCl (pH 8.0); 100 mM NaCl; 2 mM benzamidine; 10 μg mL^{-1} soybean trypsin inhibitor) and incubated at 4 °C for 10 minutes. Subsequently, the resuspended cells were disrupted by sonication at 33% power output with cycles of 4 s on and 10 s off for 30 minutes at 4 °C. To remove the cell debris, the cell lysates were centrifuged at 12 000 rpm for 30 minutes at 4 °C. The supernatants were filtered with a 0.45 μm filter. The filtered supernatants were mixed with 2 mL Ni-NTA agarose beads (for 1 L culture) pre-equilibrated with binding buffer (25 mM Tris·HCl (pH 8.0); 100 mM NaCl) and incubated for 4–5 hours with gentle shaking at 4 °C. The Ni-NTA beads bound with lysate were washed with wash buffer A (25 mM Tris·HCl (pH 8.0); 600 mM NaCl; 0.1% Triton-X-100), wash buffer B (25 mM Tris·HCl (pH 8.0); 600 mM NaCl; 0.1% Triton-X-100; 7.5 mM imidazole) and wash buffer C (25 mM Tris·HCl (pH 7.4); 100 mM NaCl). The recombinant PEST fragment proteins were eluted with an elution buffer (25 mM Tris·HCl (pH 7.4); 100 mM NaCl; 150 mM imidazole). The purity of the proteins was checked by SDS polyacrylamide gel electrophoresis (15% acrylamide) (Fig. S2†). The fractions containing pure proteins were pooled and dialyzed against dialysis buffer (25 mM Tris·HCl (pH 7.4); 25 mM NaCl) to remove imidazole. The masses of the purified PEST Wt and M1 were 8341.102 and 8525.045 Da, respectively (Fig. S3†).

3.1.2 Dehydrin (*Zea mays*, DHN1). The disorder prediction profiles for DHN1 are displayed in Fig. S1C.† The dehydrin protein was expressed and purified as described previously[52] with slight modifications. Overexpression of the DHN1 gene was done in the *E. coli* BL21 CodonPlus (DE3)-RIPL strain. The DHN1 culture was induced with 0.2 mM IPTG at A_{600} of ~0.6 and incubated for 3 hours in a shaking incubator (180 rpm) at 37 °C. The cells were harvested by centrifugation at 4000*g* for 10 min at 4 °C. The pellet was resuspended in lysis buffer (50 mM sodium phosphate (pH 7.4); 300 mM NaCl; 10 mM imidazole; 3 mM β-ME; 1 mM PMSF) and disrupted by sonication with a cycle of 4 s on and 10 s off for 20 minutes. The lysed cells were centrifuged at 12 000*g* for 30 minutes and the collected supernatant was heated at 95 °C for 30 minutes. Subsequently, the heated supernatant was ultracentrifuged (Beckman Coulter, Optima TLX) using a type 70 Ti rotor at 117 000*g* for 30 minutes at 4 °C. Finally the supernatant was allowed to bind His Trap FF SP column for 4 hours at room temperature (298 K). After binding, the column was washed with wash buffer (50 mM sodium phosphate (pH 7.4), 300 mM NaCl, 20 mM imidazole) followed by elution with an elution buffer (50 mM sodium phosphate (pH 7.4), 300 mM NaCl and 250 mM imidazole). The purity of DHN1 was ascertained by SDS polyacrylamide gel electrophoresis (15% acrylamide) (Fig. S4†). The mass of purified DHN1 was 16 951.451 Da (Fig. S5†).

3.1.3 α_3C. Overexpression of α_3C was done in the BL21 (DE3) strain of *E. coli*. The purification method reported earlier was followed.[18,53] The protein purity was monitored by SDS polyacrylamide gel electrophoresis and the mass determined using mass spectrometry was 7460.055 Da. The lyophilized protein was stored at −20 °C for further experiments.

3.2 Hen lysozyme aggregation

A stock solution (0.4–4.5 mM) of HEWL monomer was made in deionized water. The desired concentrations of HEWL in the chosen buffer were prepared by diluting this stock. All the HEWL concentrations mentioned refer to monomer concentration only.

3.2.1 Incubation in alkaline pH. HEWL (120 μM; 1 mL) was prepared in 50 mM sodium phosphate buffer (pH 12.2) and subsequently incubated at 23–25 °C without shaking for 10 days.[45] At the end of each day, an aliquot of sample was stored at 4 °C for further analysis or instantly diluted in 0.1 M bicarbonate buffer (pH 9.3), yielding 6–36 μM concentrations for absorbance measurements.

3.2.2 Incubation in acidic pH

3.2.2.1 pH 5. Aggregation of HEWL (100 μM; 1 mL) was carried out in 0.1 M citrate buffer (pH 5.0), incubated at 65 °C with intermittent mixing for 5 days.[54] At the end of each day aliquots of sample were stored at 4 °C for further analysis.

3.2.2.2 pH 2. Aggregation of HEWL (1.36 mM; 1 mL) was carried out in 10 mM glycine buffer (pH 2.0), incubated at 65 °C with intermittent mixing for 5 days.[39,41] At the end of each day 100 μL aliquots of sample were collected and stored at 4 °C for further analysis. For collecting the absorbance spectrum, the incubated sample was diluted to 100 μM in the same buffer.

3.2.3 Inhibition of HEWL aggregation. Inhibition of HEWL aggregation *at alkaline pH* was done in the presence of iodoacetamide as reported earlier.[50] Iodoacetamide was added at 2, 6, 12 and 24 hour time points. The final concentration of iodoacetamide after 24 hours in the incubated sample was 11.9 mM.

3.3 Spectroscopic techniques

3.3.1 Mass spectrometry. The masses of the proteins were determined using a MALDI-TOF mass spectrometer (Daltonics Bruker, Germany). The matrix was prepared by saturating a solution of Sinapic acid in TA-30 solvent. 0.1% trifluoroacetic acid and acetonitrile were mixed in a 7 : 3 ratio for preparing the TA-30 solvent. A further saturated solution of Sinapic acid was sonicated for 30 minutes, followed by centrifugation at 12 000 rpm at room temperature (25 °C) for 15 minutes. The collected supernatant was mixed with protein samples in 2 : 1 ratios and spotted on the target plate and subsequently allowed to dry at room temperature for 2 hours. Data acquisitions of all samples were done in linear mode with flexControl and analysis was done with flexAnalysis software from Bruker Daltonics.

3.3.2 Circular dichroism. CD spectra of proteins were recorded on a CD spectrometer (Make: Jasco, Model J-1500, Jasco Inc., Maryland, USA), equipped with a Peltier thermoregulator. For acquiring pH-dependent CD spectra, 20 μM of PEST Wt and M1, 12.5 μM DHN1 and 30 μM α_3C were dissolved in different 5 mM aqueous buffers. The PEST proteins contained 1 mM Tris (2-carboxyethyl)phosphine·HCl (TCEP) to prevent disulfide linked dimer formation. To maintain a constant pH, citric acid monohydrate (pH 3), sodium acetate (pH 5), sodium phosphate (pH 7), Tris·HCl (pH 9) and sodium phosphate dibasic (pH 11) buffers

 This journal is © The Royal Society of Chemistry 2018

were used. For measuring CD spectra at various temperatures (25–85 °C), 20 μM of PEST proteins were mixed in deionized water containing 1 mM TCEP. All the CD spectra were collected from 190 to 260 nm, at a scan speed of 100 nm minute^{-1}, bandwidth of 2 nm and data pitch of 0.1 nm. All the spectra were collected using a 2 mm quartz cuvette (Starna Scientific Ltd). The contributions of deionized water and the respective buffers were subtracted from the acquired spectra during data acquisition. For each protein sample at least three independent scans were carried out. The CD spectra from 190 to 260 nm were analyzed using the DichroWeb server. The CDSSTR program was used to calculate secondary structure content with reference set 7 and the experimental curves were well superimposed on the reconstructed curves.[55–58]

3.3.3 Thioflavin T fluorescence. For the determination of fibril formation under acidic conditions, a ThT assay was done as indicated previously, with some modifications.[50,59] Briefly, 10μM of aggregated protein was mixed with 20 μM of ThT in aqueous medium buffered at pH 8.5. The sample was excited at 450 nm (slit width 1 nm) and the emission was collected between 470 nm and 550 nm (slit width 5 nm). The area under the emission spectrum was calculated between 470 nm and 550 nm after subtraction of the blank spectrum without ThT. All fluorescence measurements were done in triplicate in a 2 mL quartz cuvette with a 10 mm path length [Hellma (Z600253)] using a Fluoromax-3 spectrofluorometer (Jobin-Yvon Horiba Inc., USA).

3.3.4 UV-visible absorption spectra. The absorption spectra of proteins were recorded using a double beam Lambda-25 UV-vis spectrophotometer (Perkin Elmer, USA) at room temperature (25 °C). A UV-transmissible quartz cuvette of 10 mm path length (Hellma; Z600210) was used for all absorption measurements. The absorption spectra were acquired with multiple scans (3–5) between 250 nm and 800 nm (1 nm bandwidth; 480 nm minute^{-1} scan speed) and subsequently averaged. All the spectra shown are the averages of at least two independent measurements. For measuring the molar extinction coefficients, PEST Wt, PEST M1 and DHN1 were dissolved in deionised water. The PEST samples always contained 5 mM TCEP in all measurements. Blanks were identical to the samples in all respects with the exception that they contained no protein. The protein concentrations were measured by the Lowry method.

For recording pH-dependent absorption spectra, the proteins were dissolved in different buffers (25 mM) as described for CD. To monitor the effects of salts on the absorption spectra, the proteins were dissolved in 250 mM KCl or NaCl.

To probe the temperature dependence of the absorption spectra, the proteins were dissolved in deionized water. The temperature of the protein samples was maintained using a circulating water bath (Julabo; FPW50-HL). The cuvette containing the sample was sealed with parafilm and incubated in a water bath for at least 20 minutes at the chosen temperature prior to collecting spectra. Initially the absorption spectra were measured at 25 °C. Subsequently, absorption spectra were collected by heating the sample to 85 °C. As the sample cooled, absorption spectra were recorded at 65 °C, 45 °C and 25 °C. All the spectral acquisition parameters were kept identical for each temperature.

Spectra for the HEWL aggregates formed at alkaline pH (12.2) were recorded after dilution in 0.1 M sodium bicarbonate buffer (pH 9.3) to keep the lysine side chain charged and protonated for ProCharTS. For some experiments, aggregated samples were centrifuged at 12 000*g* for 10 minutes prior to absorbance measurements to remove insoluble aggregates that can cause scattering. For the HEWL aggregates formed at acidic pH (2 and 5), the spectra were recorded in the same buffer as for incubation.

4 Results

4.1 Protein absorption spectra

Fig. 1 shows that selected proteins for the study are rich in charged amino acids with a few sparsely distributed aromatic amino acids. PEST Wt and PEST M1 are abundant in Glu, His and Asp; DHN1 has an abundance of His, Lys and Glu; whereas α_3C is rich in Lys and Glu. The mutant M1 was constructed to observe whether the insertion of tryptophan alters the ProCharTS absorption spectrum of Wt PEST.

Fig. 2 shows the absorption spectra observed for the PEST fragments and DHN1. The observed spectra are not due to scattering, as demonstrated by the poor overlap of the spectra with simulated Rayleigh scattering profiles (Fig. S6†). From the inset in Fig. 2A, it is noticeable that the influence of Trp absorption in PEST M1 is diminished after 315 nm, when both Wt and the mutant spectra of the PEST fragments nearly merge together. It is clear that the stronger Trp absorption in the mutant can significantly mask the weaker ProCharTS spectrum in the same protein between 250 and 315 nm.

Similarly, the inset in Fig. 2B displays the domination of tyrosine absorption over ProCharTS until 300 nm. Thus the absorption intensity arising from ProCharTS can be unambiguously identified only beyond these wavelengths, as shown in Table 1 and Fig. 2A and B, when all the other amino acid chromophores are silent. Therefore, the molar extinction coefficients observed at and beyond 315 nm arise solely from ProCharTS. Both PEST fragments and DHN1 spectra reveal a distinctive broad tail of ProCharTS absorption between 325–800 nm, as observed earlier with α_3C.[18]

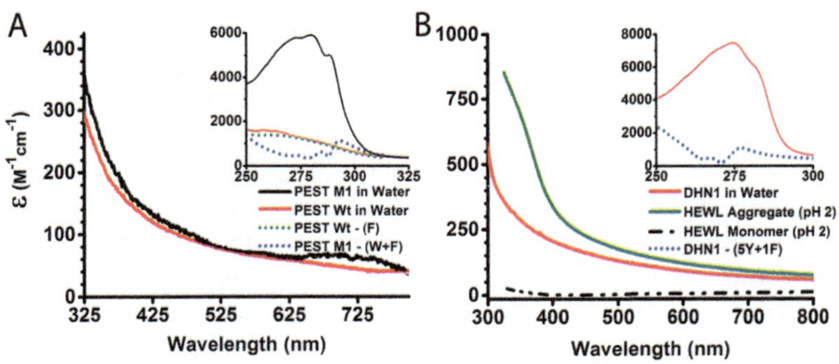

Fig. 2 Absorption spectra of proteins: [A] c-Myc PEST Wt and its mutant PEST M1 in deionized water containing 5 mM TCEP and [B] DHN1 in deionized water; the HEWL aggregate (100 μM) formed in glycine buffer (pH 2.0) and its monomer at the same concentration are displayed. Insets show the absorption spectra at shorter wavelengths in expanded form. The spectra left behind after subtracting the contribution of the aromatic chromophore are also shown in the insets. The reported extinction coefficients are averages of multiple concentrations between 25–75 μM for PEST Wt, 20–40 μM for PEST M1, 50–100 μM for DHN1 and 100 μM for HEWL.

 This journal is © The Royal Society of Chemistry 2018

Table 1 ProCharTS absorption: molar extinction coefficients of absorption arising from ProCharTS among the studied proteins. Other conditions are similar to Fig. 2. The numbers in brackets indicate standard deviations for $n = 2–3$

Protein	Charged residue fraction	ε (M^{-1} cm^{-1}) at different wavelengths in nm												
		315	320	325	350	400	450	500	550	600	650	700	750	800
PEST Wt	35%	343	316	293	214	138	103	82	67	59	52	44	39	37
		(57)	(57)	(49)	(45)	(20)	(9)	(5)	(2)	(1)	(1)	(1)	(1)	(1)
PEST M1	35%	430	392	357	255	157	112	90	73	65	64	64	60	37
		(11)	(9)	(10)	(9)	(8)	(8)	(19)	(22)	(26)	(27)	(27)	(19)	(4)
DHN1	29%	396	370	351	284	198	154	125	103	87	74	65	57	53
		(72)	(68)	(64)	(54)	(34)	(21)	(15)	(10)	(8)	(5)	(4)	(4)	(5)
HEWL	21%	1052	912	851	696	336	224	172	141	119	103	89	81	74
aggregate		(54)	(47)	(42)	(31)	(11)	(4)	(4)	(4)	(4)	(6)	(5)	(5)	(5)

Fig. 2A and B reveal that when the strong contribution from Trp/Tyr is deducted from the PEST M1 and DHN1 spectra near 250–290 nm using the molar extinction coefficients reported previously,[60] a weakened spectrum of ProCharTS is left behind. Interestingly, this loss in ProCharTS intensity is shifted with shifts in the peak absorption of stronger chromophores, as revealed for Tyr in DHN1 (Fig. S8†) at strongly alkaline pH. Further, the weak chromophore in Phe has negligible influence on the ProCharTS of PEST Wt (Fig. 2A), highlighting the importance of chromophore strength in this phenomenon. This needs further investigation. The additional absorption contributed by ProCharTS at 280 nm, if unaccounted for, can lead to significant overestimation of the protein concentration based on Trp/Tyr absorption. However, at wavelengths beyond 315 nm, the ProCharTS spectra of Wt and the mutant PEST appear nearly comparable (Fig. 2A). This indicates that the presence of extra tryptophan in the PEST mutant has minimal bearing on the ProCharTS spectrum in this region. Fig. S7† indicates that the absorbances of the PEST fragments (Wt and M1) and DHN1 at different wavelengths vary linearly with concentration, arguing against any contribution to the spectra arising from intermolecular interactions.

Table 1 lists the numerical values of the molar extinction coefficients observed for all the proteins shown in Fig. 2. DHN1 consistently shows marginally higher values compared to the PEST fragments at several of the wavelengths listed, although it has a smaller fractional population of charged residues. Perhaps DHN1 has a more favorable proximity between the charged Glu and Lys residues in the folded 3D structure compared to the PEST proteins, enhancing ProCharTS absorption. Additionally, the sequence of DHN1 appears richer in local clusters of three or more charged amino acids in tandem compared to PEST.

To evaluate the sensitivity of ProCharTS spectra to structural transitions in the proteins, several manoeuvres were attempted. In the first, the pH of the medium was varied from pH 3 to pH 11, so as to both alter the ionization state of the charged amino acid and gently perturb the structure of the protein, as discussed in the following section.

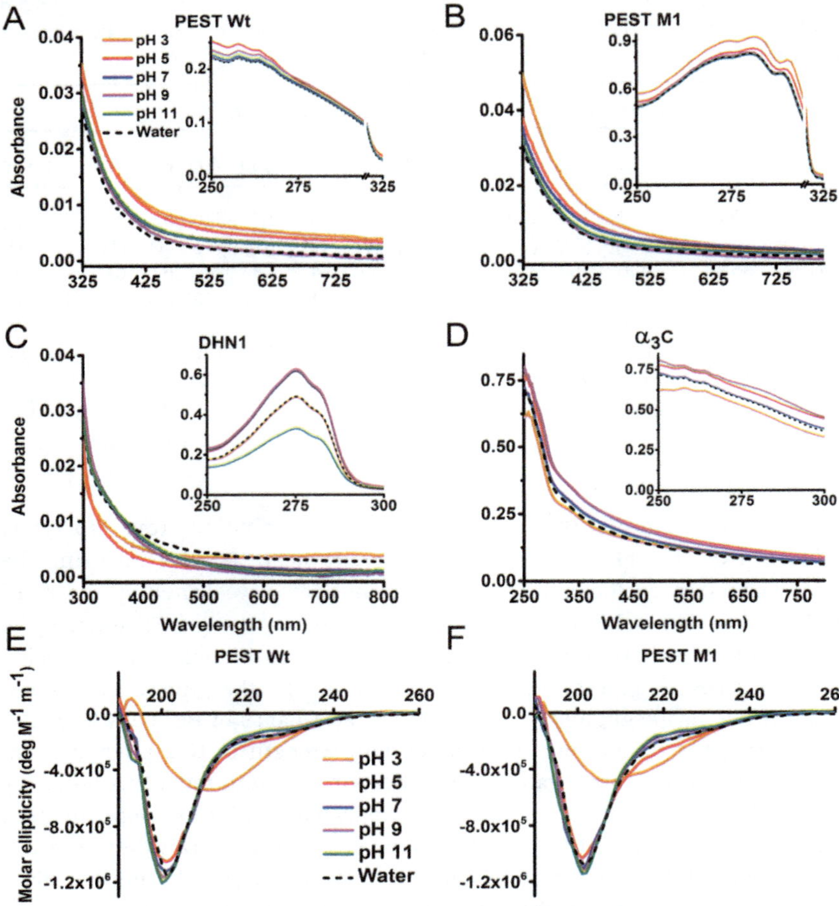

Fig. 3 Effect of pH on ProCharTS and CD: the effects of changing pH on the absorption spectra of [A] 150 μM PEST Wt, [B] 150 μM PEST M1, [C] 125 μM DHN1 and [D] 100 μM α_3C are shown. Insets show the absorption spectra at shorter wavelengths. The inset scales of [A] and [B] have a break between 294 and 295 nm. Solid lines and dashed lines show spectra in buffer and water, respectively. The effects of pH on the CD spectra for [E] 20 μM PEST Wt and [F] 20 μM PEST M1 in different buffers and water are also shown.

4.2 Effect of pH

Fig. 3A–D and S9† depict the effects of varying pH on the ProCharTS absorption spectra of different proteins. Far UV CD spectra for the PEST fragments are also shown for comparison in Fig. 3E and F. Comparing the insets in Fig. 3A and B and S9A and B,† it is noticeable that PEST M1 has uniformly higher absorption near 250–290 nm at pH 3, while PEST Wt also displays a consistent but weak increase between 250–270 nm at same pH. The modest increase over and above Trp (indole) absorption in PEST M1 must arise from an increase in ProCharTS absorption, because Trp absorption is barely sensitive to changes in pH in the acidic range. Further, Trp absorption is known to increase between 260–280 nm and decrease between 282–300 nm with increasing acidity.[60]

 This journal is © The Royal Society of Chemistry 2018

The uniform increase observed in the absorption of PEST M1 at pH 3 in the 250–290 nm wavelength range runs contrary to this reversal, thus ruling out the role of tryptophan. Fig. 3A and B and Fig. S9† show that both PEST Wt and PEST M1 exhibit increased absorbance at pH 3 starting from 325 nm, but PEST M1 has a slightly higher absorption. These results hint at an alternative structure of both PEST fragments at pH 3, and to a lesser extent at pH 5, in comparison to other pH conditions where no significant changes are apparent. The proposed structural change at pH 3 coincides with major changes in the far UV CD spectrum for both PEST Wt and M1 at pH 3 (Fig. 3E and F). Minor changes in the CD spectrum at pH 5 are also evident, in comparison to the remaining pH somewhat matching the observations from ProCharTS. Moving to higher pH, the trends in both PEST Wt and M1 appear nearly similar but are not identical. The absorption spectra for pH 7 and 11 are superimposed for Wt PEST, while for PEST M1, the pH 7 spectrum is marginally higher compared to that at pH 11. The spectra at pH 9 and in pure water appear quite similar for both PEST variants. However, the absorption of PEST Wt and PEST M1 was less in deionised water in comparison with buffers.

The inset in Fig. 3C and S9C† display changes in absorption predominantly arising from five tyrosine residues in DHN1 with a peak near 275 nm. However, a significant contribution from ProCharTS is evident here based on the spectral perturbations observed with changing pH. A prominent dip at pH 3 is seen in the spectrum between 250–290 nm, while a further decline is visible at pH 11 in comparison to the spectra at pH 5–9. No change is observed at 292 nm at pH 11, where the tyrosinate anion can increase the absorption significantly if present. In the main part of Fig. 3C, DHN1 shows moderately higher ProCharTS absorption in deionised water as compared with different pH values, unlike PEST fragments. Aside from the spectrum at pH 3, the spectra at the remaining pH values remain low in intensity and similar to each other at longer wavelengths, unlike those for the PEST proteins. Fig. S12A† reveals minor changes at pH 3 and 5 and no significant changes with pH in the CD spectra of DHN1, in line with similar ProCharTS spectra in the long wavelength region. The DHN1 results highlight the role of ProCharTS in altering the apparent tyrosine peak absorption intensity at 275 nm with changes in pH.

Fig. 3D and S9D† reveal the UV-visible absorption spectra of α_3C at different pH values. In the inset of Fig. 3D, a marginal rise in absorption is visible at pH 5 and 9, while the absorption at pH 7 closely matches that for water. A low intensity is evident at pH 3. In the main figure, the low intensity at pH 3 persists. The spectra in water and those at pH 3 and pH 7 lie nearly above each other, while the spectra at pH 5 and 9 are a shade higher than others. The collection of spectra at pH 11 for α_3C was not possible owing to precipitation. Overall, no major changes are observed from the ProCharTS spectrum in α_3C, which is consistent with the absence of any change in the CD spectra of α_3C in the same pH range (Fig. S12B†). Thus ProCharTS appears to be sensitive to pH-induced structural transitions in proteins, as shown for the PEST fragment, while it remains nearly unchanged in the long wavelength region in the absence of major structural transitions in α_3C and DHN1. In the next approach, we raised the temperature of the medium to disrupt the protein structure and proceeded to evaluate its effect on the ProCharTS spectrum.

4.3 Effect of temperature

Fig. 4A and B (with insets) show a prompt rise in ProCharTS absorption in PEST Wt and M1 when the temperature is raised from 25 to 85 °C. The increase in absorbance is quantitatively more for the Wt protein compared to the mutant M1 (Fig. S13A and B†).

Further, the proportional rise in absorbance appears non-uniform across the entire spectral range (Fig. S13A and B†) for both the Wt and the mutant, ruling out the possibility of an increase in absorbance as a result of evaporation. As the protein samples cool from 85 °C to lower temperatures (65, 45 and 25 °C), the absorbance values decrease. At 45 and 25 °C, the spectra appear nearly identical (including the insets). The spectrum after cooling to 25 °C appears almost identical to that at the beginning before heating (Fig. S13A and B†), suggesting

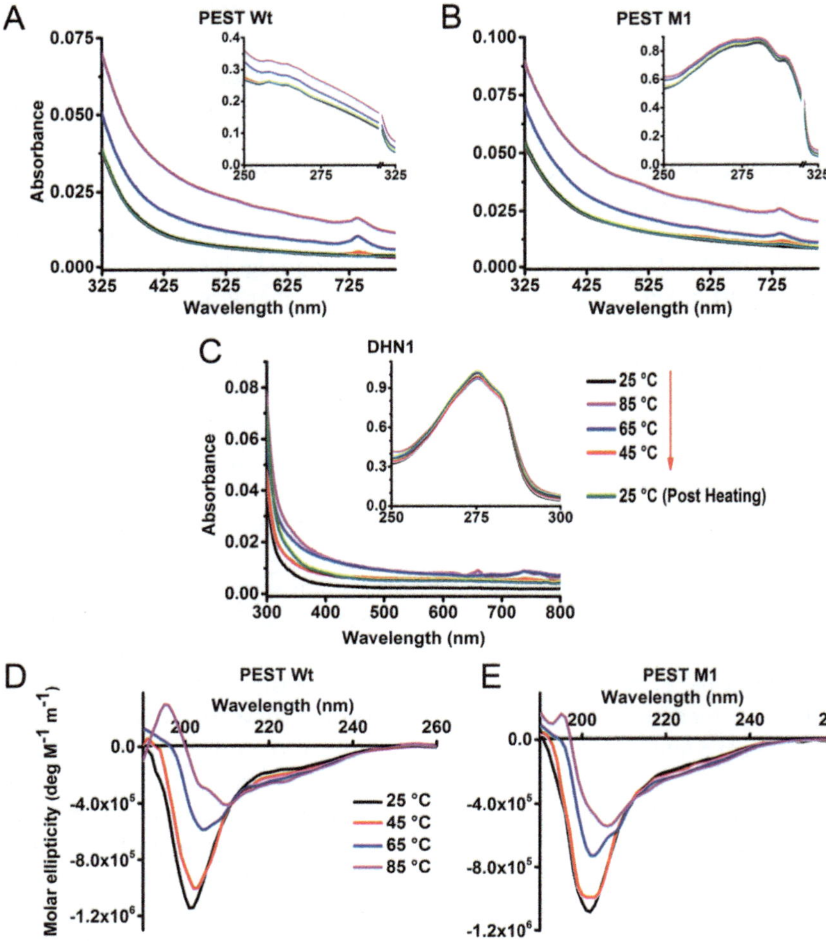

Fig. 4 Effect of temperature on ProCharTS and CD: the effect of temperature on the absorption spectra of [A] 150 µM PEST Wt, [B] 150 µM PEST M1 and [C] 125 µM DHN1 in water. The inset scales of [A] and [B] have a break between 294 and 295 nm. The effect of temperature on the CD spectra for [D] 20 µM PEST Wt and [E] 20 µM PEST M1 at different temperatures is also shown.

complete reversibility in the process of regaining the original conformation. Interestingly, at all higher temperatures (85, 65 and 45 °C), a significant peak in the tail of the ProCharTS spectrum near ~735 nm is visible for both PEST Wt and M1. The CD spectra for PEST Wt and M1 are shown in Fig. 4D and E, respectively. It is observed that for both PEST fragments, the individual CD spectra at 25 and 45 °C are nearly identical, matching the trend in the ProCharTS spectra. At 85 °C, a major change in the CD spectrum is observed for both PEST fragments, while at 65 °C a moderate change is noticeable compared to the spectrum at 25 °C for both PEST fragments. These structural changes observed in the CD spectra almost mirror the changes noticed in the ProCharTS spectra for the same proteins (Fig. S13†). Hence, ProCharTS spectra appear truly sensitive to heat-induced structural transitions in the PEST Wt and M1 proteins.

The DHN1 absorption spectra in Fig. 4C reveal a marginal increase in absorption at high temperatures (85 and 65 °C), hinting at only minor alterations in the structure of the protein. No significant changes are seen at shorter wavelengths (inset of Fig. 4C). Interestingly, the spectra at 65 and 85 °C appear nearly the same, hinting at similar conformations. With cooling to 45 and 25 °C, the absorption decreases, but interestingly the ProCharTS spectrum obtained at 25 °C after cooling is significantly different from the spectrum of the original sample (25 °C before heating). Clearly, the reversibility observed in the PEST fragments is absent in DHN1. This hints that upon cooling, DHN1 may have acquired an alternate conformation to the initial one.

Overall, heat-induced changes monitored using ProCharTS spectra could identify structural transitions in both PEST fragments at 85 °C. The absence of reversibility in the heat-induced structural change in DHN1 was also evident using ProCharTS.

In the next section we investigate the role of excess salt on the ProCharTS spectra of the chosen proteins. The objective of this experiment was to perturb the electrostatic interactions between like and opposite charges in the protein by introducing salt counterions. As ProCharTS spectra are dependent on the charges localized on amino acid side chains and their interactions with those in neighbouring side chains, stabilizing the charges by introducing excess salt counterions could influence the spectra. These changes in ProCharTS spectra could in turn provide insights into the multiple side chain interactions contributing to the spectra.

4.4 Effect of salt

In Fig. 5A and B and S16A and B,† the absorption of PEST Wt and M1 is significantly higher in the presence of NaCl and KCl at wavelengths longer than 325 nm. No significant changes are noticed at shorter wavelengths (insets of Fig. 5A and B). This trend is consistent with the higher ProCharTS observed in the presence of buffer salts (Fig. 3A and B) compared to pure water at shorter wavelengths. Interestingly, the larger potassium ion consistently appears to cause more change in ProCharTS compared to sodium in both of the PEST proteins.

Fig. 5C and S16C† reveal that a different trend is observed with DHN1 protein. The inset of Fig. 5C reveals a major increase in the ProCharTS absorption at shorter wavelengths (250–300 nm) in the presence of NaCl and KCl. At longer wavelengths, a perceptible drop in absorption is noticeable, more so with NaCl compared to KCl. Here, the rise and dip of ProCharTS at shorter and longer

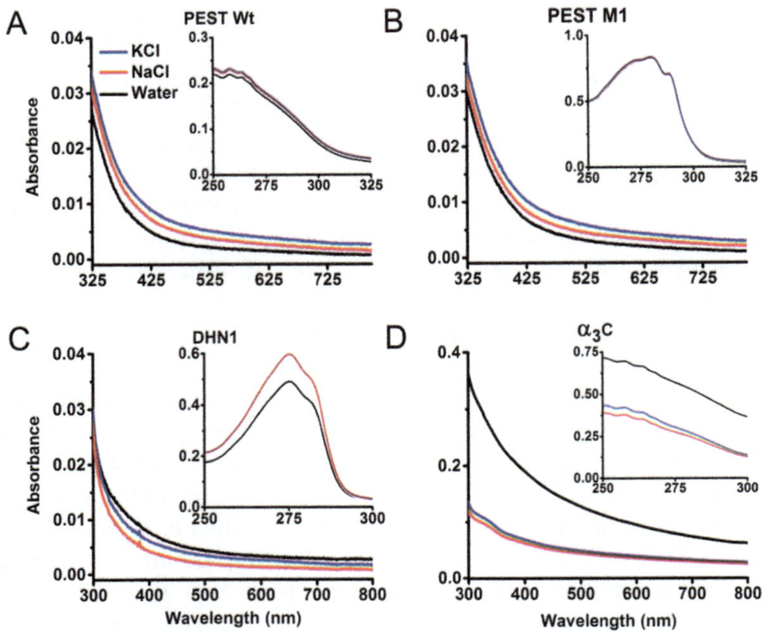

Fig. 5 Effect of excess salt on ProCharTS: the effects of 250 mM KCl and 250 mM NaCl on the absorption spectra of [A] 150 μM PEST Wt, [B] 150 μM PEST M1, [C] 125 μM DHN1 and [D] 100 μM α_3C dissolved in water.

wavelengths, respectively, mirrors a similar trend of ProCharTS in the presence of buffer salts compared to water (Fig. 3C).

Fig. 5D and Fig. S16D† display a unique trend observed with α_3C in the presence of the salts employed. Both NaCl and KCl substantially diminish the ProCharTS absorption observed between 250–800 nm in pure water. Interestingly, the effects of both sodium and potassium ions appear nearly indistinguishable. The effects observed in α_3C appear far more perturbative than the minor changes seen in the presence of buffer salts (Fig. 3D), indicating that major disruptions in electrostatic interactions occur in α_3C in the presence of excess salt. Taken together, the results from multiple proteins suggest that ProCharTS spectra are sensitive to the presence of excess salt in the medium.

In the following section, we examine how ProCharTS can arise when new intermolecular contacts between charged amino acid residues are established during protein aggregation. For this purpose, we chose HEWL, which has previously established conditions in acidic and alkaline pH that promote *in vitro* protein aggregation. HEWL, the sequence of which contains 21% charged amino acids[61] (Fig. 1E), displays an insignificant amount of ProCharTS absorption as a monomer (Fig. 2B).

4.5 HEWL aggregation at pH 12.2

Fig. 6 compares the absorption spectra of HEWL aggregates with those of its monomer under different concentrations after ten days of incubation at alkaline pH.

 This journal is © The Royal Society of Chemistry 2018

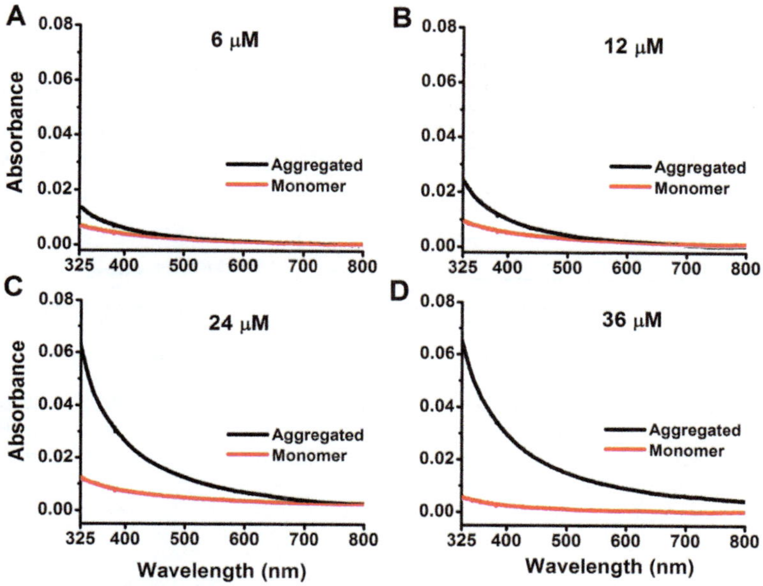

Fig. 6 ProCharTS of HEWL aggregates formed at alkaline pH: ProCharTS spectra for HEWL aggregates incubated at alkaline pH (12.2) for 10 days are shown alongside the freshly prepared monomer at identical concentrations. The spectra of [A] 6 μM, [B] 12 μM, [C] 24 μM and [D] 36 μM HEWL aggregates are shown. Each sample was centrifuged at 12 000*g* for 10 min and the supernatant was used for collecting absorption spectra.

A gradual increase in absorption with an increase in monomer concentration is evident among the HEWL aggregates. This indicates a major rise in the number of new contacts made between the side chains of charged residues in the aggregate compared to the monomer. Fig. S17† shows the difference spectra at each concentration. The ProCharTS intensity shows a clear rise as a function of concentration, suggesting a marked increase in the number and possibly the density of intermolecular contacts between charged residues in the aggregate population with a gradual increase in the monomer concentration. To ascertain whether the increase in ProCharTS intensity arises solely from the formation of HEWL aggregates, we proceeded to arrest HEWL aggregation at alkaline pH by adding iodoacetamide, which prevents the formation of intermolecular disulfide bonds, as shown previously.[50]

4.6 Inhibition of HEWL aggregation by iodoacetamide at pH 12.2

In the absence of iodoacetamide (Fig. 7A), an increase in ProCharTS absorption is evident from 6 to 7 day old and 8 to 9 day old aggregates. This indicates growth of the aggregates and some structural changes that further enrich the molecular contacts between the charged side chains, contributing to ProCharTS. A dip in absorption between 7 and 8 days suggests the loss of protein by proteolysis, due to prolonged exposure at alkaline pH. The spectra of the 9 and 10 day old aggregates reveal clear changes in shape, suggesting restructuring of the aggregates, which possibly affects the nature of the molecular contacts giving rise to ProCharTS absorption. In the presence of added iodoacetamide (Fig. 7B), a profound dip in

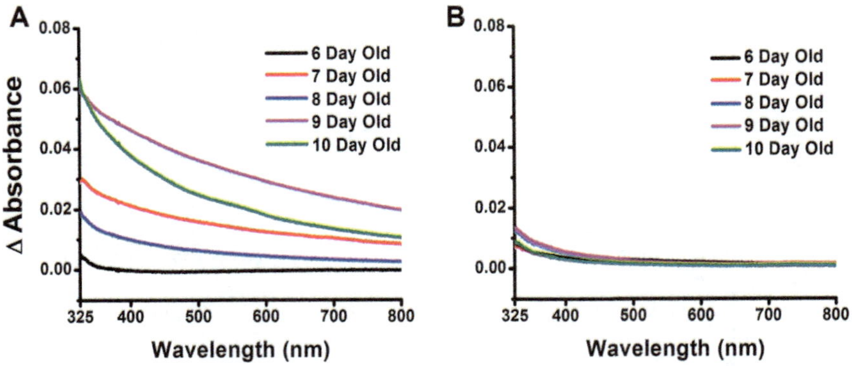

Fig. 7 Effect of inhibition of HEWL aggregation on ProCharTS: difference absorption spectra of different day old HEWL aggregates in the absence [A] and presence [B] of iodoacetamide. The aggregates (12 μM) were transferred to 0.1 M sodium bicarbonate buffer (pH 9.3) before recording the spectra.

the ProCharTS absorption is noticeable, compared to in its absence. This implies a dramatic decline in the population of intermolecular contacts between charged residues in HEWL contributing to ProCharTS. It is most likely that the population and density of intermolecular contacts in the HEWL aggregates that contributed to ProCharTS in the absence of iodoacetamide are now diminished in its presence. Earlier work using atomic force microscopy has shown that HEWL aggregates are markedly reduced in size and possess weak protein–protein interactions in the presence of iodoacetamide.[50] Thus the ProCharTS absorption intensity in the 325–800 nm wavelength range appears to directly track the formation of HEWL aggregates.

Next we investigated the time-dependent progress of HEWL aggregation at acidic pH values of 2 and 5. Here ProCharTS absorption can be directly measured in the aggregating samples without the need for transfer to another pH, avoiding dilution of the sample.

4.7 HEWL aggregation in acidic pH

Fig. 8A and C display the ProCharTS absorption from HEWL samples aggregating at pH 2. Increasing absorption, with notably large amplitudes at shorter wavelengths (305–325 nm) compared to longer wavelengths (400–700 nm) is evident with lengthening duration of sample incubation from 1 day to 3 days.

The increase most likely originates from the growth of HEWL aggregates with time that quantitatively contributes more new molecular contacts among the charged HEWL amino acid side chains, favouring ProCharTS. Subsequently, at day 4 and day 5, the intensity values show saturation, implying a halt in growth thereafter. Fig. 8E reveals that the growth kinetics of the HEWL aggregates, as measured from the thioflavin T fluorescence, almost mirror the trend observed with ProCharTS (Fig. 8C), indicating that ProCharTS is indeed monitoring the growth of HEWL aggregates.

Fig. 8B depicts the aggregation of HEWL at pH 5, which is a less harsh condition compared to pH 2 and is likely to yield aggregates of a different makeup than pH 2. Fig. 8D reveals that the rate of increase in ProCharTS absorption is

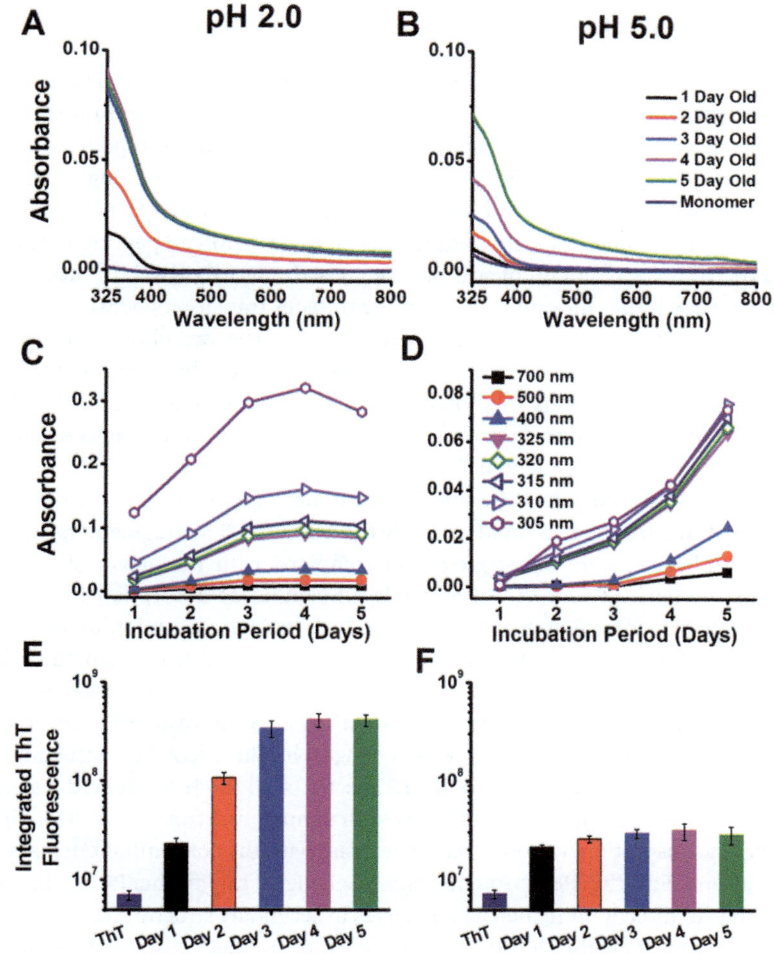

Fig. 8 ProCharTS of HEWL aggregates formed at acidic pH: absorption spectra for different day old HEWL aggregates (100 μM) formed at [A] pH 2.0 and [B] pH 5.0. [C] and [D] display the changes in absorbance as a function of the incubation period for different wavelengths among the HEWL aggregates formed at pH 2.0 and 5.0, respectively. [E] and [F] display similar variation with the integrated thioflavin T fluorescence intensity observed for the HEWL aggregates formed at pH 2.0 and pH 5.0, respectively.

significantly slower in comparison to pH 2. Interestingly, no saturation of absorption is evident here, unlike at pH 2. The low absorbance values compared to pH 2 hint at a reduced aggregate population and/or smaller aggregates, resulting in a poor density of molecular contacts contributing to ProCharTS in comparison to pH 2. Fig. 8F shows that the thioflavin T intensities are significantly lower at pH 5, which is consistent with the reduced ProCharTS. This marks a decreased population of ordered aggregates like amyloid fibrils at pH 5 compared to pH 2. These experiments validate that ProCharTS absorbance can successfully track the growth of HEWL aggregates at both pH 2 and pH 5 by a label-free approach.

5 Discussion

Intramolecular charge transfer with alternating donor/acceptor repeat units to reduce band gaps in conjugated polymers has been demonstrated previously.[62] This has been extended further to control the true colours of organic chromophores in π-conjugated polymers using the concept of 'spectral engineering'.[63] Charge transfer optical spectral bands in proteins have in the past been associated with cytochromes,[64] plastocyanin,[65] gas phase protein and peptide ions in an ion trap[66] and covalent conjugates of protein with methylglyoxal.[67,68] Recently it has been shown that intermolecular charge transfer transitions between the YFP chromophore and tyrosine aromatic ring enhance the two-photon absorption cross-section of YFP.[69] To the best of our knowledge there has been no report of charge transfer transitions in proteins devoid of aromatic amino acids, metal centres, prosthetic groups or external adducts, making the ProCharTS discussed here unique and unprecedented.[18]

In the present study, the UV-visible charge transfer spectra (ProCharTS) of three intrinsically unstructured proteins, namely the wild-type PEST fragment, its mutant M1 and dehydrin, were investigated and compared with the previously studied synthetic protein α_3C. All the IDPs displayed significant absorption in the ProCharTS spectra up to 800 nm. Interference from strong absorption arising from aromatic amino acids was clearly absent in the 325–800 nm wavelength region, making this a universal absorption band for ProCharTS. The ProCharTS spectra of all the IDPs displayed significant changes with pH. In the case of the PEST fragments, these changes correlated with changes in the secondary structure, as monitored by CD spectra. The ProCharTS spectra of all the IDPs showed a significant increase in absorption with an increase in temperature from 25 to 85 °C. Upon cooling, the absorption was observed to decrease with the concomitant decrease in temperature. For the PEST fragments the changes in ProCharTS at different temperatures directly correlated with changes in secondary structure, as revealed by CD spectra. Further, the ProCharTS spectra of all the IDPs displayed significant changes in absorption intensity in the presence of either 0.25 M KCl or 0.25 M NaCl.

In the next part of the study, ProCharTS was employed to monitor the aggregation of hen egg-white lysozyme under previously established pH conditions. At pH 12.2, the absorption arising from ProCharTS showed an increase in intensity among 10 day old aggregates, with an increase in the initial monomer concentration. When the aggregation at alkaline pH was inhibited by the addition of iodoacetamide, the increase previously observed in ProCharTS was clearly abolished. The growth of HEWL aggregates at pH 2 also resulted in increased ProCharTS intensity. This increase reached saturation after day 3 of incubation, which was consistent with similar saturation kinetics in thioflavin T fluorescence for the same aggregates. The growth of HEWL aggregates at pH 5 also mirrored the rise in ProCharTS intensity. However, unlike at pH 2, the magnitude of the increase in ProCharTS was significantly less with the growth of aggregates at pH 5, which was consistent with the reduced thioflavin T fluorescence of the same aggregates.

5.1 Effect of pH on ProCharTS

The pH effects observed with ProCharTS (Fig. 3 and S9†) can be analysed collectively to identify common features. At pH 3, the carboxylate groups in both

Glu and Asp are likely to be protonated and uncharged, so at this pH all the proteins are likely to have a net positive charge from His, Lys and Arg residues only. This also means that salt bridge interactions between oppositely-charged residues cannot exist at pH 3. α_3C, which has several such salt bridges, as seen from the NMR structure[53] and also from Radial Distribution Function plots,[18] shows a diminished ProCharTS at pH 3, unlike PEST. DHN1, which is rich in Glu/Asp, also shows a dip at pH 3 near 275 nm. The higher ProCharTS in the PEST fragments (Wt and M1) at pH 3 appear to correlate with increased helix and strand population, along with more ordered structures compared to those at other pH values (Fig. S10 and S11†), as revealed by CD. Interestingly, pH 5, 7 and 9 appear generally healthy and mostly uneventful for ProCharTS in nearly all the proteins (Fig. 3), perhaps due to the rich presence of negative and positive charges in all the charged residues in this pH range. As the protonated imidazole side chain of His has a pK_a close to 6.0, it is surprising that both PEST and DHN1, which are rich in His residues, reveal little change in ProCharTS between pH 5–9. DHN1 has 13 Lys residues, compared to one in PEST. Thus, the large drop in absorption around 275 nm in DHN1 at pH 11 from pH 7 could arise from the loss of a proton and positive charge from lysine, disrupting salt bridges with Glu or Asp, if they exist. Previous work has also shown a reduced absorption at both shorter and longer wavelengths in α_3C at pH 13.[18] Overall, pH-related changes in ProCharTS appear complex due to concomitant changes in: (a) the ionization state of charged residues; (b) protein conformation and (c) salt concentration. The interpretation of changes at extremes, like pH 3 and 11, appears to be easier.

5.2 Effect of temperature on ProCharTS

The effect of increasing temperature (25 to 85 °C) is manifested as increased ProCharTS in the PEST fragments and DHN1, starting from 325 nm and going to higher wavelengths (Fig. 4 and S13†). This increase is non-uniform against wavelength, similar to that observed with α_3C.[18] The CD spectra (Fig. S14 and S15†) reveal increased fractions of the helix and strand populations along with ordered structures at 85 °C in comparison to lower temperatures for the PEST fragments. Interestingly, while the 77/78 residue PEST fragment reverted back to the original ProCharTS spectrum upon cooling to 25 °C from 85 °C, the 167 residue DHN1 protein did not (Fig. S13†). This could imply the presence of an alternate structure (a trapped local thermodynamic minimum) upon cooling, which is more likely in a larger protein like DHN1 compared to a PEST fragment. Taken together, ProCharTS appears to be sensitive to thermally-induced structural transitions in charged proteins.

It may be added that ProCharTS may be sensitive to subtle changes in tertiary structure which are not discernible by CD. Such changes are however, difficult to verify in the absence of 3D structural data.

5.3 Effect of salt on ProCharTS

The addition of 0.25 M NaCl/KCl appears to have conflicting effects on the ProCharTS spectra (Fig. 5 and S16†). For the PEST fragments a steady increase is evident from 325 nm onwards, while both DHN1 and α_3C display a prominent decline in ProCharTS from 325 nm. Salt counter-ions are expected to stabilise protein intrinsic charges and weaken the CT bands. α_3C, which contains multiple salt bridge interactions, expectedly displays a profound decrease. A similar trend in DHN1 appears to hint at a significant perturbation of salt bridge interactions.

The dramatic dip in ProCharTS with excess salt in α_3C, unlike other proteins, could be linked to a large population of Lys–Glu salt bridge interactions through the 3D space in the protein. It is likely that stabilization of the charge by salt counter ions alters the electrostatics between Lys and Glu significantly, so as to diminish ProCharTS. This possibility is corroborated by a similar reduction in ProCharTS at extremes of pH when charges are removed.[18] Overall, the results suggest that salt ions can alter the ProCharTS spectra significantly.

5.4 Growth of protein aggregates monitored by ProCharTS

Fig. 2B shows the absorption spectrum of HEWL aggregates formed at pH 2 in comparison with HEWL monomers. A substantially increased molar absorptivity is noticed in HEWL with the formation of aggregates in comparison to the monomer, indicating that a dramatic rise in new molecular contacts between charged residues created due to the aggregation in HEWL contribute substantially to ProCharTS. An increase in absorption due to aggregation, arising presumably from ProCharTS, was previously observed for calf thymus histone protein, which is rich in charged residues.[17] Recently, ordered aggregates like protein fibrils have been shown to exhibit unique near-UV absorption and visible fluorescence arising from proton transfer across hydrogen bonds in the fibril.[70]

As ProCharTS originates from interactions between charged functional groups in proteins, protein–protein interactions that create new molecular contacts present an ideal opportunity for ProCharTS investigation without the use of extrinsic probes like thioflavin T or FRET donor/acceptor pairs. Unlike monomeric proteins, which require a large fractional population of charged amino acids (Table 1) to be visible in the ProCharTS spectrum, here the requirement of charged residues in the monomeric proteins is less stringent if the eventual protein cluster has a rich population of molecular contacts involving charged amino acids in comparison to the monomer, as shown in this work for hen lysozyme. Fig. 8C and D show that the sensitivity of ProCharTS in detecting protein aggregates is higher at wavelengths below 325 nm (305–320 nm), where the intensity from ProCharTS is higher but the likelihood of minor interference from Trp absorption is higher. An absence of Trp and Tyr could clearly amplify the sensitivity of ProCharTS, as wavelengths as low as 280 nm can yield rich spectra originating purely from ProCharTS. This can further ease the minimal requirement of charged residues and make the application of ProCharTS to monitor protein–protein interactions more robust. The pH and temperature dependence of the observed ProCharTS from a protein aggregate can further yield vital information on the molecular architecture and stability of the aggregate that is otherwise not available from thioflavin T fluorescence data. Thus the application of ProCharTS as a label-free approach to track the growth of protein aggregates appears to be promising indeed.

Our preliminary investigation revealed that the charge transfer states contributing to ProCharTS are emissive. This work is in progress.

6 Conclusions

Protein charge transfer spectra were distinctly observed in the 325–800 nm wavelength range among multiple IDPs in the presence of aromatic amino acids

 This journal is © The Royal Society of Chemistry 2018

like Phe, Trp and Tyr (Fig. 2). However, in the 250–315 nm region, strong absorption by Trp/Tyr suppressed ProCharTS absorption significantly (Fig. 2). Changes in ProCharTS absorbance with increasing temperature (Fig. 4) correlated with changes in protein secondary structure, as observed by CD. The ProCharTS intensity was sensitive to the presence of salts in the medium (Fig. 5). Further, the ProCharTS absorption in some proteins was also sensitive to changes in pH between 3 and 11 (Fig. 3). Aside from changes in protein structure, this sensitivity to pH, on a cautionary note, can also arise from alterations in protein charge and the addition of buffer salts. Finally, ProCharTS absorption serves as an ideal label-free intrinsic probe to monitor protein aggregation in moderately-charged proteins (Fig. 6 and 8).

Acknowledgements

RS acknowledges financial support from the Department of Biotechnology (NER Division), New Delhi under project BT/409/NE/U-Excel/2013. RB acknowledges funding from Human Frontier Science Program (RGP0025/2016). The authors acknowledge the service and support from Mass Spectrometry facilities at CIF and CD measurement facilities at IIT Guwahati. The authors thank Timothy J. Close at the University of California, Riverside for the DHN1 plasmid construct and Cecilia Tommos at the University of Pennsylvania, USA for the α_3C plasmid construct. MZA thanks Sayan Roy Chowdhury for help with CD measurements. RS thanks R. Venkatramani at TIFR, Mumbai for stimulating discussions. RS dedicates this paper to his parents on their golden anniversary.

References

1 J. M. Antosiewicz and D. Shugar, *Biophys. Rev.*, 2016, **8**, 163–177.

2 J. M. Antosiewicz and D. Shugar, *Biophys. Rev.*, 2016, **8**, 151–161.

3 M. R. Eftink, in *Methods of Biochemical Analysis*, John Wiley & Sons, Inc., 1991, vol. 35, ch. Fluorescence Techniques for Studying Protein Structure, pp. 127–205.

4 A. B. T. Ghisaidoobe and S. J. Chung, *Int. J. Mol. Sci.*, 2014, **15**, 22518–22538.

5 J. R. Lakowicz, in *Principles of Fluorescence Spectroscopy*, Springer, 3rd edn, 2006, ch. Protein Fluorescence, pp. 530–577.

6 B. R. Rami, G. Krishnamoorthy and J. B. Udgaonkar, *Biochemistry*, 2003, **42**, 7986–8000.

7 J. B. A. Ross, W. R. Laws, K. W. Rousslang and H. R. Wyssbrod, in *Topics in Fluorescence Spectroscopy: Biochemical Applications*, Plenum Press, New York, 1992, vol. 3, ch. Tyrosine Fluorescence and Phosphorescence from Proteins and Polypeptides, pp. 1–63.

8 R. Swaminathan, N. Periasamy, J. B. Udgaonkar and G. Krishnamoorthy, *J. Phys. Chem.*, 1994, **98**, 9270–9278.

9 A. M. Saxena, J. B. Udgaonkar and G. Krishnamoorthy, *J. Mol. Biol.*, 2006, **359**, 174–189.

10 A. K. Dunker, J. D. Lawson, C. J. Brown, R. M. Williams, P. Romero, J. S. Oh, C. J. Oldfield, A. M. Campen, C. M. Ratliff, K. W. Hipps, J. Ausio, M. S. Nissen, R. Reeves, C. Kang, C. R. Kissinger, R. W. Bailey, M. D. Griswold, W. Chiu, E. C. Garner and Z. Obradovic, *J. Mol. Graphics Modell.*, 2001, **19**, 26–59.

11 A. Tantos, K.-H. Han and P. Tompa, *Mol. Cell. Endocrinol.*, 2012, **348**, 457–465.
12 P. Tompa, *FEBS Lett.*, 2005, **579**, 3346–3354.
13 V. N. Uversky, *Int. J. Biochem. Cell Biol.*, 2011, **43**, 1090–1103.
14 P. Romero, Z. Obradovic, X. Li, C. E. Garner, C. Brown and A. K. Dunker, *Proteins: Struct., Funct., Genet.*, 2001, **42**, 38–48.
15 P. Mateos-Gil, A. Tsortos, M. Velez and E. Gizeli, *Chem. Commun.*, 2016, **52**, 6541–6544.
16 L. Homchaudhuri and R. Swaminathan, *Chem. Lett.*, 2001, **30**, 844–845.
17 L. Homchaudhuri and R. Swaminathan, *Bull. Chem. Soc. Jpn.*, 2004, **77**, 765–769.
18 S. Prasad, I. Mandal, S. Singh, A. Paul, B. Mandal, R. Venkatramani and R. Swaminathan, *Chem. Sci.*, 2017, **8**, 5416–5433.
19 G. M. Ashraf, N. H. Greig, T. A. Khan, I. Hassan, S. Tabrez, S. Shakil, I. A. Sheikh, S. K. Zaidi, M. A. Wali, N. R. Jabir, C. K. Firoz, A. Naeem, I. M. Alhazza, G. A. Damanhouri and M. A. Kamal, *CNS Neurol. Disord.: Drug Targets*, 2014, **13**, 1280–1293.
20 V. Kumar, N. Sami, T. Kashav, A. Islam, F. Ahmad and M. I. Hassan, *Eur. J. Med. Chem.*, 2016, **124**, 1105–1120.
21 A. Mukherjee, D. Morales-Scheihing, P. C. Butler and C. Soto, *Trends Mol. Med.*, 2015, **21**, 439–449.
22 H. Olzscha, S. M. Schermann, A. C. Woerner, S. Pinkert, M. H. Hecht, G. G. Tartaglia, M. Vendruscolo, M. Hayer-Hartl, F. U. Hartl and R. M. Vabulas, *Cell*, 2011, **144**, 67–78.
23 J. T. Pedersen and N. H. H. Heegaard, *Anal. Chem.*, 2013, **85**, 4215–4227.
24 C. A. Ross and M. A. Poirier, *Nat. Rev. Mol. Cell Biol.*, 2005, **6**, 891–898.
25 A. Mukherjee and C. Soto, *Cold Spring Harbor Perspect. Med.*, 2017, **7**, a024315.
26 J. Lee, E. K. Culyba, E. T. Powers and J. W. Kelly, *Nat. Chem. Biol.*, 2011, **7**, 602–609.
27 D. J. Lindberg, A. Wenger, E. Sundin, E. Wesén, F. Westerlund and E. K. Esbjörner, *Biochemistry*, 2017, **56**, 2170–2174.
28 B. Demeule, R. Gurny and T. Arvinte, *Int. J. Pharm.*, 2007, **329**, 37–45.
29 R. Khurana, V. N. Uversky, L. Nielsen and A. L. Fink, *J. Biol. Chem.*, 2001, **276**, 22715–22721.
30 K. M. Ryan and G. D. Birnie, *Biochem. J.*, 1996, **314**, 713–721.
31 M. Eilers, S. Schirm and J. M. Bishop, *EMBO J.*, 1991, **10**, 133–141.
32 D. S. Askew, R. A. Ashmun, B. C. Simmons and J. L. Cleveland, *Oncogene*, 1991, **6**, 1915–1922.
33 M. A. Gregory and S. R. Hann, *Mol. Cell. Biol.*, 2000, **20**, 2423–2435.
34 M. Rechsteiner and S. W. Rogers, *Trends Biochem. Sci.*, 1996, **21**, 267–271.
35 S. Rogers, R. Wells and M. Rechsteiner, *Science*, 1986, **234**, 364–368.
36 I. Amara, A. Odena, E. Oliveira, A. Moreno, K. Masmoudi, M. Pagès and A. Goday, *Plant Cell Physiol.*, 2012, **53**, 312–329.
37 A. Tunnacliffe and M. J. Wise, *Naturwissenschaften*, 2007, **94**, 791–812.
38 L. N. Arnaudov and R. de Vries, *Biophys. J.*, 2005, **88**, 515–526.
39 A. Chaari, C. Fahy, A. Chevillot-Biraud and M. Rholam, *PLoS One*, 2015, **10**, e0142095.
40 S. E. Hill, J. Robinson, G. Matthews and M. Muschol, *Biophys. J.*, 2009, **96**, 3781–3790.

 This journal is © The Royal Society of Chemistry 2018

41 M. R. H. Krebs, D. K. Wilkins, E. W. Chung, M. C. Pitkeathly, A. K. Chamberlain, J. Zurdo, C. V. Robinson and C. M. Dobson, *J. Mol. Biol.*, 2000, **300**, 541–549.
42 R. Mishra, K. Sörgjerd, S. Nyström, A. Nordigården, Y.-C. Yu and P. Hammarström, *J. Mol. Biol.*, 2007, **366**, 1029–1044.
43 M. Hameed, B. Ahmad, K. M. Fazili, K. Andrabi and R. H. Khan, *J. Biochem.*, 2007, **141**, 573–583.
44 L. Homchaudhuri, S. Kumar and R. Swaminathan, *FEBS Lett.*, 2006, **580**, 2097–2101.
45 V. K. Ravi, T. Swain, N. Chandra and R. Swaminathan, *PLoS One*, 2014, **9**, e87256.
46 A. J. Sophianopoulos and K. E. Van Holde, *J. Biol. Chem.*, 1961, **236**, PC82–PC83.
47 A. J. Sophianopoulos and K. E. Van Holde, *J. Biol. Chem.*, 1964, **239**, 2516–2524.
48 S. Kumar, V. K. Ravi and R. Swaminathan, *Biochem. J.*, 2008, **415**, 275.
49 S. S. S. Wang, K.-N. Liu and B.-W. Wang, *Eur. Biophys. J.*, 2010, **39**, 1229–1242.
50 V. K. Ravi, M. Goel, H. C. Kotamarthi, S. R. K. Ainavarapu and R. Swaminathan, *PLoS One*, 2014, **9**, e87012.
51 S. Kumar, V. K. Ravi and R. Swaminathan, *Biochim. Biophys. Acta, Proteins Proteomics*, 2009, **1794**, 913–920.
52 S. G. Jepson and T. J. Close, *Protein Expression Purif.*, 1995, **6**, 632–636.
53 C. Tommos, K. G. Valentine, M. C. Martínez-Rivera, L. Liang and V. R. Moorman, *Biochemistry*, 2013, **52**, 1409–1418.
54 M. Ahn, C. L. Hagan, A. Bernardo-Gancedo, E. De Genst, F. N. Newby, J. Christodoulou, A. Dhulesia, M. Dumoulin, C. V. Robinson, C. M. Dobson and J. R. Kumita, *Biophys. J.*, 2016, **111**, 2358–2367.
55 A. Lobley, L. Whitmore and B. A. Wallace, *Bioinformatics*, 2002, **18**, 211–212.
56 N. Sreerama and R. W. Woody, *Anal. Biochem.*, 2000, **287**, 252–260.
57 L. Whitmore and B. A. Wallace, *Nucleic Acids Res.*, 2004, **32**, W668–W673.
58 L. Whitmore and B. A. Wallace, *Biopolymers*, 2008, **89**, 392–400.
59 J. Wall, C. L. Murphy and A. Solomon, *Methods Enzymol.*, 1999, **309**, 204–217.
60 G. D. Fasman, *Practical handbook of biochemistry and molecular biology*, CRC Press, New York, 1992.
61 R. E. Canfield, *J. Biol. Chem.*, 1963, **238**, 2698–2707.
62 Q. T. Zhang and J. M. Tour, *J. Am. Chem. Soc.*, 1998, **120**, 5355–5362.
63 P. M. Beaujuge, C. M. Amb and J. R. Reynolds, *Acc. Chem. Res.*, 2010, **43**, 1396–1407.
64 D. Verbaro, A. Hagarman, J. Soffer and R. Schweitzer-Stenner, *Biochemistry*, 2009, **48**, 2990–2996.
65 A. A. Gewirth and E. I. Solomon, *J. Am. Chem. Soc.*, 1988, **110**, 3811–3819.
66 R. Antoine and P. Dugourd, *Phys. Chem. Chem. Phys.*, 2011, **13**, 16494–16509.
67 S. Bone, T. J. Lewis, R. Pethig and A. Szent-Györgyi, *Proc. Natl. Acad. Sci. U. S. A.*, 1978, **75**, 315–318.
68 J. A. McLaughlin, R. Pethig and A. Szent-Györgyi, *Proc. Natl. Acad. Sci. U. S. A.*, 1980, **77**, 949–951.
69 M. T. P. Beerepoot, D. H. Friese and K. Ruud, *Phys. Chem. Chem. Phys.*, 2014, **16**, 5958–5964.
70 D. Pinotsi, L. Grisanti, P. Mahou, R. Gebauer, C. F. Kaminski, A. Hassanali and G. S. Kaminski Schierle, *J. Am. Chem. Soc.*, 2016, **138**, 3046–3057.

Faraday Discussions

PAPER

Optical backbone-sidechain charge transfer transitions in proteins sensitive to secondary structure and modifications†

I. Mandal, S. Paul and R. Venkatramani *

Received 2nd October 2017, Accepted 30th October 2017
DOI: 10.1039/c7fd00203c

The absorption of light by proteins can induce charge transfer (CT) transitions in the UV-visible range of the electromagnetic spectrum. Metal–ligand complexes or active site prosthetic groups which absorb in the visible region exhibit prominent CT transitions. Furthermore, the protein backbone also exhibits CT transitions in the far UV range. In this manuscript, we present a detailed computational study of new near UV-visible CT transitions that involve amino acids with charged side chains. Specifically, using time dependent density functional theory calculations, we examine the absorption spectra of naturally charged amino acids (Lys, Glu, Arg, Asp and His), extracted from solution phase protein structures generated by classical molecular dynamics simulations, and phosphorylated amino acids (Tyr, Thr and Ser) from experimentally determined protein structures. We show that amino acids with charged sidechains present a directed electronic donor–bridge–acceptor paradigm, with the lowest energy optical excitations demonstrating peptide backbone-sidechain charge separations. The UV-visible spectral range of the backbone-sidechain CT transitions is determined by the chemical nature of the donor, bridge and acceptor groups within each amino acid, amino acid conformation and the protein secondary structure where the amino acids are located. Photoinduced CT occurs in opposite directions for the anionic and cationic amino acids along the ground state dipole moment vector for the chromophores. We find that photoinduced charge separation is more facile for the anionic amino acids (Asp, Glu, pSer, pThr and pTyr) relative to that for the cationic amino acids (Lys, Arg and Hsp). Our results provide a foundation for the development of spectroscopic markers based on the recently proposed Protein Charge Transfer Spectra (ProCharTS) which are relevant for the study of DNA-binding or intrinsically disordered proteins that are rich in charged amino acids.

Introduction

The absorption spectra of proteins in the ultraviolet and visible (UV-vis) region of the electromagnetic spectrum are primarily characterized in terms of aromatic

Department of Chemical Sciences, Tata Institute of Fundamental Research, Dr. Homi Bhabha Road, Colaba, Mumbai 400005, India. E-mail: ravi.venkatramani@tifr.res.in

† Electronic supplementary information (ESI) available. See DOI: 10.1039/c7fd00203c

amino acid sidechain excitations (Tyr, Trp and Phe) between 250–300 nm [1–3] and excitations of peptide bonds around 190–220 nm.[4–7] Such excitations are local in nature, consisting of n → π* transitions near 220 nm and π → π* transitions near 190 nm for the peptide bond and π → π* transitions near 270 nm for the aromatic amino acids. Additionally, light absorption in proteins leads to charge transfer (CT) transitions in the peptide backbone below 180 nm, which show signatures in both the electronic absorption and circular dichroism spectra of proteins.[8,9] Other prominent CT transitions in proteins involve metal–ligand complexes or active site prosthetic groups which absorb in the visible region.[2,10,11] CT transitions, wherein photoexcitation directly transfers electrons between covalently or non-covalently linked donor (D) and acceptor (A) groups, are sensitive to the chemical nature of the D and A groups and the bridging medium (B) separating them. For instance, CT transitions in metal–ligand complexes are sensitive markers of the metal-coordination geometry and can be used to track chemical reactions that are catalysed by such centres in enzymes.[10] Computational studies have proposed that the electronic spectra of proteins below 180 nm are sensitive to the conformation and secondary structure of the polypeptide chain due to the presence of peptide backbone CT transitions.[8,9,12,13]

Recently, we and our collaborators have reported new photoinduced CT transitions in proteins arising from Lys and Glu amino acids in the near UV-vis region (250–800 nm) of the electromagnetic spectrum.[14] Through computational and experimental studies of α_3C, a synthetic protein that is rich in charged amino acids, we demonstrated that interactions between charged Lys amino and Glu carboxylate side-chains can produce distinct spectral signatures well beyond 250 nm in charged amino acid rich proteins even if they lack aromatic amino acids. In the past decade and a half, there have been reports of non-aromatic near-UV/visible absorption (emission) extending up to 450 nm (600 nm) in high concentration protein solutions and peptide aggregates.[15–17] However, our studies indicate that even at low concentrations, monomeric proteins that are rich in charged amino acids should also exhibit significant absorption in the near UV-vis region. Indeed, monomeric α_3C (5–105 μM) solutions demonstrated distinctive absorption features spanning 250–800 nm ($\varepsilon = 7338 \pm 191$ M^{-1} cm^{-1}, 964 ± 129 M^{-1} cm^{-1}, and 501 ± 66 M^{-1} cm^{-1} at 250, 450 and 800 nm, respectively). The discovery of new photoinduced CT transitions for Lys and Glu amino acids adds a new spectral window (250–800 nm) which we call ProCharTS (Protein Charge Transfer Spectra), for label free tracking of biomolecular structure and dynamics.

Our previous study[14] also showed that ProCharTS signatures are also prominently present in natural proteins that are rich in charged amino acids containing aromatic chromophores. While ProCharTS features extend up to 800 nm, well beyond the absorption range of aromatic amino acids (up to 320 nm), the presence of ProCharTS has implications even for spectral ranges below 300 nm. Near-UV absorbance at 280 nm, which is typically only assumed to be from aromatic amino acids, and which can be used to quantify protein solutions (*e.g.* for estimating concentrations), should account for significant ProCharTS absorption when the protein is rich in charged amino acids. For instance, the α_3C protein which lacks aromatic amino acids still exhibits significant ProCharTS absorption ($\varepsilon = 4531 \pm 133$ M^{-1} cm^{-1}) at 280 nm. A better understanding of the ProCharTS spectral features below 300 nm, which is the range where monomer charged amino acids are expected to contribute, can thus significantly improve the

 This journal is © The Royal Society of Chemistry 2018

quantification of protein solutions. Finally, the broad ProCharTS profile will always have some overlap with the emission spectra of natural fluorescent chromophores such as Trp or synthetic dyes. This overlap may produce multi-exponential fluorescence lifetime decay kinetics which can be interpreted in terms of the dynamics and interactions of charged amino acids within protein folds. Charged amino acids such as Lys and Arg are integral constituents of DNA and RNA binding proteins such as histones, spliceosomal proteins and transcription factors.[18] Furthermore, intrinsically disordered proteins, which are implicated in both cell regulatory events as well as pathological disorders, are also rich in charged amino acids.[19] In order to develop the potential of ProCharTS to examine such proteins of biological and biomedical relevance, it is imperative to understand the distinct spectral ranges and features of photoinduced transitions in natural charged amino acids.

In our previous study,[14] we uncovered two distinct photoinduced CT transitions involving Lys and Glu amino acids within the helical protein α_3C: (1) Peptide Backbone-Sidechain (PBS) and (2) Sidechain–Sidechain (SS) CT transitions. PBS-CT transitions lead to charge separation between the charged carboxylate/amino sidechain groups of Glu/Lys and the peptide backbone. In contrast, SS-CT transitions occur between the Glu carboxylate and the Lys amino sidechain groups. We systematically showed that the broad 250–800 nm ProCharTS spectrum of α_3C could be explained by considering the sensitivity of PBS-CT and SS-CT transitions to the inter-residue contacts imposed by protein tertiary folds and their dynamics, solvation of the chromophores, and the protonation states of the amino acids. Based on these results, we hypothesize that charged amino acids, in isolation or in collective assemblies, form electronic D–B–A molecular complexes that exhibit photoinduced D to A CT transitions. Here, we validate our hypothesis by conducting a systematic investigation of the charged chromophores that are available in protein sidechains (Fig. 1).

In this manuscript, we present a detailed computational study and comparison of the absorption spectra of all amino acids that are charged under physiological conditions: Lys, Glu, Arg, Asp and His (Fig. 1). While the absorption spectra for the Lys and Glu amino acids located within α-helices were explored in our previous study, here we present new calculations for these amino acids within other secondary structure motifs (β-sheet and random coil). Furthermore, we also examine the absorption spectra of phosphorylated forms of the amino acids Ser, Thr and Tyr (pSer, pThr and pTyr), wherein negatively charged phosphate groups are post-translationally attached to these otherwise neutral amino acids (Fig. 1). We carry out Time Dependent Density Functional Theory (TDDFT) based spectra calculations on the charged amino acids extracted from a microsecond (μs) long classical molecular dynamics (MD) trajectory of the protein ubiquitin and from experimentally derived protein data bank structures of phosphorylated proteins. Our studies are aimed at demonstrating the presence of facile PBS-CT transitions in the absorption spectra of charged amino acid chromophores and comparing the properties of these transitions as a function of the following amino acid properties: (1) chemical structure, (2) conformation and (3) location in distinct secondary structural elements.

Our results have a direct bearing on the design of optical probes based on ProCharTS for monitoring the structure, dynamics and function of proteins that are rich in charged amino acids or that undergo post-translational modifications.

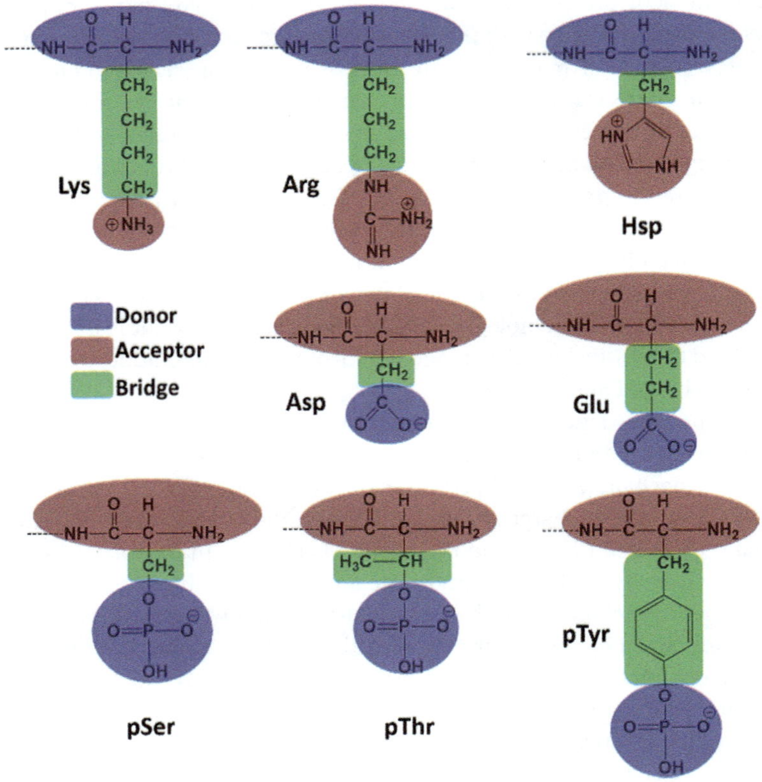

Fig. 1 Charged amino acids considered in this study. The colored segments indicate the electronic donor (D)–bridge (B)–acceptor (A) segments for the PBS-CT transitions expected in these systems.

Methods

The goal of the present manuscript is to examine the nature of PBS-CT transitions in the absorption spectra of charged and post-translationally modified amino acids located within protein folds. To this end, we carried out TDDFT calculations on amino acid conformations sampled from a classical atomistic MD trajectory of fully solvated ubiquitin, a globular protein with a β-grasp fold. Furthermore, we extracted multiple post-translationally modified amino acid conformers from the protein data bank. We then carried out a detailed analysis of the ground state electronic properties and UV-vis absorption spectra for each charged amino acid chromophore ensemble. Our methodology is described in the following subsections.

Molecular dynamics simulations of ubiquitin

Ubiquitin is a small 76 residue globular protein with a β-grasp topology consisting of 5 β-strands, a single α-helix, and a 3–10 helix (Fig. 2). In a biological context, the protein serves as a post-translational molecular tag for the regulation of cellular processes. The protein contains 7 Lys residues which play a critical role in the

 This journal is © The Royal Society of Chemistry 2018

function of the protein. The sequence of ubiquitin consists of 23 amino acids (7 Lys, 6 Glu, 5 Asp, 4 Arg and 1 His) whose sidechains exist in charged forms at physiologically relevant pH values. Fig. 2A shows the location of these charged amino acids in the ubiquitin sequence and their structural arrangement over the ubiquitin fold. Initial coordinates for ubiquitin were taken from a crystal structure in the protein data bank (PDB code: 1UBQ). We added hydrogen atoms and solvated the protein in a cubic box of explicit water molecules with 10 Å solvent padding. The solvated models were then subjected to an equilibration protocol and subsequent equilibrium MD production runs using GROMACS version 5.0 [20] and the CHARMM36 force field with the TIP3P water model for the solvent.[21] Our simulations employed periodic boundary conditions (PBC) with Particle Mesh Ewald (PME) calculations of long range electrostatic interactions and a cut-off of 1.0 nm for short range van der Waals and electrostatic interactions. The equilibration protocol consisted of a minimization step and thermal equilibration at

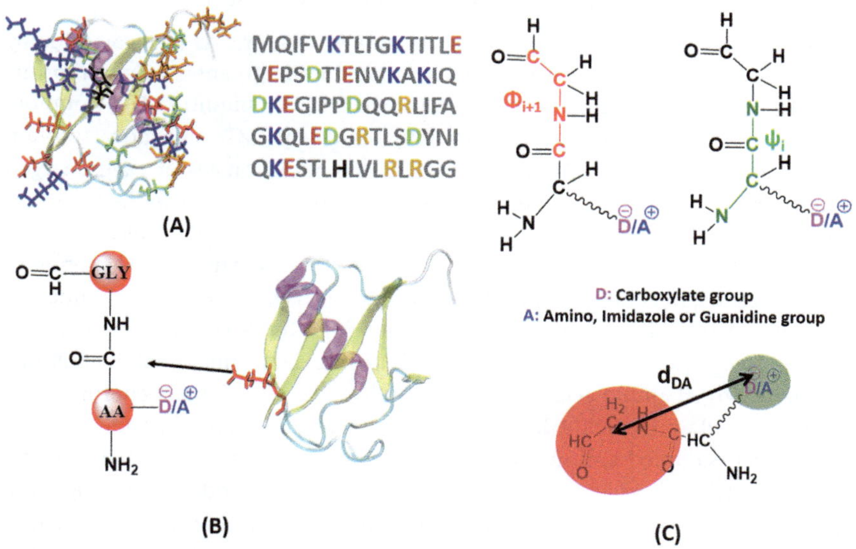

Fig. 2 (A) The structure (PDB code: 1UBQ) and amino acid sequence of ubiquitin. Ubiquitin has 23 charged amino acids whose position in the sequence and structure are indicated by colored single letter codes and colored licorice representations, respectively. (B) A schematic representation of the procedure that was used to create charged amino acid monomer models for our electronic structure calculations. We extracted a dimer (AA–XX) containing one of the 23 charged amino acids (AA = Lys/Glu/Asp/Arg/His), mutated the neighbouring amino acid (XX) to Gly and capped the amino and carbonyl ends with hydrogens. (C) We use two geometric parameters to describe the variation in conformation of the charged amino acid monomer models sampled from the MD trajectory of ubiquitin: (top) Ramachandran torsion angles (Φ_{i+1} and Ψ_i) and (bottom) the donor–acceptor distance (d_{DA}). Here, Ψ_i is the Ramachandran dihedral angle for the charged amino acid at the ith position in the sequence, calculated using the C–C$_\alpha$–N–C atoms (highlighted in green), and Φ_{i+1} is the Ramachandran dihedral angle for the neighbouring residue (represented by Gly) at the $i + 1^{th}$ position, calculated using the N–C$_\alpha$–C–N atoms (highlighted in red) of the dimer fragment. The distance d_{DA} is measured between the geometric centers of the extended backbone of the dimer fragment (red shaded region) and the sidechain headgroup of the charged amino acid (green shaded region).

300 K for 100 ps. Both of these steps were carried out initially with the protein's heavy atoms fixed, later constraining them harmonically around their structural positions with a spring constant of 25 kcal mol^{-1} $Å^{-2}$. The solvated protein system was then equilibrated further under constant temperature and pressure (NPT) conditions at 300 K and 1 atm. We carried out successive NPT equilibration runs (150 ps each), during which harmonic constraints on the heavy atoms were gradually reduced from 25–5 kcal mol^{-1} $Å^{-2}$. Finally, after an unconstrained 1 ns NPT equilibration run, an NVT production trajectory of 1 μs was generated. The μs long simulation of the ubiquitin structure in explicit solvent provided us with a solvated form for the protein with inter-residue sidechain interactions matching that of NMR derived solution structures (Fig. S1 of the ESI†). We uniformly sampled 100 charged amino acid conformations from the protein snapshots in the 1 μs trajectory for further analysis and spectra calculations. The sampling scheme is described in detail in the next subsection.

TDDFT calculations for amino acids with charged sidechains

In order to calculate the absorption spectra of Lys, Glu, Arg, Asp, His and doubly protonated His (Hsp), we extracted chromophore fragments from uniformly sampled protein snapshots from the 1 μs MD run of ubiquitin. The sampling covered multiple conformations (14–100) along the MD trajectory (protein snapshots captured every 70–10 ns) for all possible instances of each charged amino acid in the protein sequence. For instance, there are 7 Lys in the ubiquitin sequence (Fig. 2A). We therefore uniformly sampled 15 conformations for 2 of the 7 Lys and 14 conformations for the remaining 5 Lys to arrive at 100 monomer conformations. In contrast, since there is only one His in the ubiquitin sequence, we sampled 100 His monomer conformations from the MD trajectory. Through this sampling scheme we captured both the thermal conformational fluctuations for each charged amino acid monomer as well as the variations in conformation of the charged amino acids due to their placement in the protein fold.

For each sampled amino acid monomer conformation, we created a model with an extended backbone which mimics the polypeptide chain (Fig. 2B). Specifically, we extracted a dimer fragment containing the charged amino acid of interest (Lys/Arg/His/Glu/Asp) and sidechain atoms of the residue adjacent to the charged amino acid were deleted in order to mutate the residue to Gly. This procedure created an extended backbone monomer model. Then the dangling bonds on the backbone were capped with hydrogens to create a modified C terminus (CHO group) and N terminus (NH_2 group) using the psfgen module in VMD.[22] This capping strategy has been previously used by us before,[14] and we discuss its suitability along with control calculations in the last subsection of the "Methods" section. We generated 100 conformations for each charged amino acid (Lys/Glu/Asp/Arg/His) by sampling multiple MD conformations for all 23 charged amino acids in ubiquitin. We considered two protonation states for His, with the neutral form protonated at the epsilon nitrogen (His) and the doubly protonated form with protons at both the epsilon and delta nitrogens (Hsp) of the imidazole ring. For the 100 conformations of each charged amino acid, we calculated the absorption spectra using TDDFT with a CAM-B3LYP[23] functional and a 6–31++G(d) basis set on all atoms in the Gaussian 09 program.[24] The first 100 lowest energy electronic transitions were calculated in all TDDFT calculations.

Difference electron density plots were calculated using the Multiwfn 3.3.8 software[25] and visualised (isovalue set to 0.0004) in GaussView 5.0.[26]

Our previous studies on Lys/Glu indicated that the photoinduced PBS-CT transitions should be sensitive to the geometric (structural) parameters of the amino acids which influence the coupling between electronic donor–acceptor groups or the delocalization of the electronic donor and acceptor states. Two such geometric parameters (Fig. 2C) are the backbone-charged sidechain headgroup distance (d_{DA}) and the backbone torsion/Ramchandran angles for our monomer fragment models (Fig. 2C). Fig. S2–S5 of the ESI† show the variation in the geometric parameters (d_{DA} and torsion angles) defined in Fig. 2C over the 100 amino acid conformations sampled from the ubiquitin MD trajectory. In later sections we examine the sensitivity of the excitation energies and oscillator strengths for the PBS-CT transitions in charged amino acid monomer models to the geometric parameters (Fig. 2C) and the chemical nature of the electronic donor–bridge–acceptor complex.

TDDFT calculations for post-translationally modified amino acids

The post-translationally modified amino acid fragments of Ser, Thr and Tyr were extracted from 10 different experimental structures of proteins containing those residues from the protein data bank (https://www.rcsb.org/).[27] Post-translationally modified amino acid fragments include mono-anionic (singly protonated) phosphorylated Ser, Thr and Tyr residues. Although di-anionic (unprotonated) phosphate group forms exist under physiological conditions, we did not consider them here in our spectra calculations. The spectra of the mono-anionic forms can be more directly compared with the spectra of the singly charged amino acids (Lys, Glu, Arg, Asp and His). The pSer residues were extracted from 5 experimental structures (PDB codes: 4JFZ, 2FWN, 4KK4, 5TR5 and 2LO6), the pThr residues were extracted from 5 experimental structures (PDB codes: 2ERK, 3Q52, 3A62, 2JQL and 3OB1), and the pTyr residues were extracted from 6 experimental structures (PDB codes: 2ERK, 3OBI, 2DVJ, 2IVT, 1IR3, 1TRN and 1VX3). Specifically, we extracted a dimer fragment containing the phosphorylated version of the amino acid of interest (pThr/pSer/pTyr) and the sidechain atoms of the adjacent residue were deleted in order to mutate the residue to Gly. This procedure created an extended backbone monomer model with dangling bonds capped with hydrogens, as described in the previous subsection. TDDFT absorption spectra were computed (first 100 lowest energy electronic transitions), as described in the previous section, for 10 conformations of each phosphorylated amino acid.

Characterization of transitions

We computed two measures in order to characterize the transitions as charge transfer (CT) transitions or non-CT transitions. One measure is the average hole–electron separation distance:[25,28]

$$\Delta r = \frac{\sum_{i,j} \left(K_i^j\right)^2 \left| \langle \varphi_j | r | \varphi_j \rangle - \langle \varphi_i | r | \varphi_i \rangle \right|}{\sum_{i,j} \left(K_i^j\right)^2} \tag{1}$$

where φ is the molecular orbital (MO) and the indexes i and j go over all occupied and vacant MOs, respectively. Here $K_i^j = P_i^j + Q_i^j$, where P_i^j and Q_i^j are the excitation ($i \rightarrow j$) and de-excitation ($j \leftarrow i$) configuration coefficients. The second measure[25,29] is the distance between the centroid of the hole and electron distribution (D_{CT}), defined as

$$D_{\mathrm{CT}} = \sqrt{(D_{\mathrm{CT},X})^2 + (D_{\mathrm{CT},Y})^2 + (D_{\mathrm{CT},Z})^2} \quad (2)$$

with

$$D_{\mathrm{CT},\alpha} = \left| \int \alpha\rho^{\mathrm{electron}}(r)\mathrm{d}r - \int \alpha\rho^{\mathrm{hole}}(r)\mathrm{d}r \right| \quad (3)$$

where the index α represents the Cartesian components (X, Y, Z) and $\rho^{\mathrm{electron/hole}}$ is the electron/hole density distribution. The two measures were calculated using Multiwfn 3.3.8 [25] and then the following conditions were used to categorize the transitions: (i) CT transition when $\Delta r > 2$ Å and $D_{CT} > 1$ Å, (ii) non-CT transition when $\Delta r < 2$ Å or if $\Delta r > 2$ Å and $D_{CT} < 1$ Å. Furthermore, transitions for which Δr was within 5% of the threshold (2 Å) value were classified as borderline transitions. In the manuscript we only show the classification of transitions as either CT or non-CT transitions.

Critical discussion of modelling assumption and caveats

We have previously explored the solvation effects and inter-residue sidechain interactions for Lys and Glu chromophores using full QM treatments on systems of varying size and sampling conformations with different charge states.[14] Our previous study showed that the essential features of photoinduced CT transitions for Lys/Glu were present in the monomer spectra of these amino acids. Thus, in the present study, we examine only the monomer charged amino acid spectra with the goal of comparing the features of photoinduced CT transitions for anionic (Glu, Asp, pThr, pSer and pTyr) and cationic (Lys, Arg and His) amino acids. Since His can occur in both neutral (singly protonated) and charged (doubly protonated) states under physiological conditions, here we computed the electronic structures of both the singly protonated neutral (His) and doubly protonated cationic (Hsp) forms.

In the present study, we carry out spectra calculations on a large number of monomer amino acid fragments using the TDDFT method, which has been shown by us and others to be the method of choice for molecular systems with a good balance of cost and accuracy.[14,15,30–32] Furthermore, in our TDDFT calculations, we employ the CAM-B3LYP range corrected hybrid exchange-correlation (XC) functional which provides a reasonable description of CT states.[23,33,34]

Our goal in the present study is to examine the spectral features of amino acid chromophores embedded within protein folds. To this end, we constructed amino acid fragment models with extended backbones which mimic the polypeptide chain. This was accomplished by extracting a dimer fragment (AA–XX), where AA was one of the 23 charged amino acids in the ubiquitin sequence and XX was its neighboring residue. We mutated XX to Gly and capped the amino and carboxylate ends with hydrogens. We have critically examined the capping strategies for amino acids used in our electronic structure calculations by carrying out control calculations on extended peptide backbones and through utilizing

 This journal is © The Royal Society of Chemistry 2018

alternative capping models using methyl groups.[14] Based on these calculations, we concluded that extending the polypeptide backbone by adding an extra Gly unit to the C-terminal end of amino acid fragments provided robust converged spectra above 250 nm. Our control calculations (using Glu as a reference) show that further extending the backbone by adding Gly units or using methyl capping groups only alters the backbone transitions around 200 nm. However, backbone transitions and signature CT transitions involving charged amino acid sidechains/backbones near and above 250 nm are not altered by either extending the backbone or changing the capping groups. Here, we present new control calculations on Lys/Arg chromophores (Fig. S6 of ESI†) with more extended peptide backbones. These calculations show that using a more extended polypeptide backbone slightly increases the range of Lys and Arg excitation wavelengths towards lower energies by 30 and 10 nm, respectively. However, relative trends in the ranges for the cationic systems (Lys *vs.* Arg) or between cationic and anionic systems (Lys/Arg data from Fig S6† *vs.* Glu data from ref. 14) remain unaltered. Furthermore, the nature of the transitions is also preserved, with the lowest energy transitions demonstrating distinct PBS-CT character. Thus, the present capping scheme provides converged trends with respect to the observations reported and discussed in this study. We note that the capping scheme used here does produce a small number of fictitious low intensity non-CT transitions above 300 nm which are localized on our capping groups (see Fig. 6).[14] However, we were able to cleanly identify and separate these non-CT transitions above 300 nm from our characteristic CT transitions of interest. Thus our results and conclusions on PBS-CT transitions are free of any artefacts introduced by the capping scheme.

Results and discussion

Ground state electronic structures of charged amino acids present a directed D–B–A paradigm

The absorption of light by a molecular scaffold can create charge separation, provided the scaffold possesses distinct electron donor (D) and electron acceptor (A) moieties. Such groups may be coupled to each other within the molecular scaffold *via* a bridging medium (B) with less propensity (relative to D/A) to accept/donate electrons.[35,36] Fig. 3 shows the ground state energy gap distributions and representative localizations of the HOMO and LUMO for cationic Arg/Lys/Hsp, neutral His, and anionic Glu/Asp/pSer/pThr/pTyr. For the cationic species we find the HOMO to be localized on the peptide backbone, thus marking this group as the electronic D unit. In contrast, the LUMO is localized on the sidechain guanidine, amino and imidazole moieties of Arg, Lys and Hsp, respectively, thus marking these groups as electronic A units. The alkyl sidechain which possesses σ-character connects these D/A groups to form the B unit for the cationic amino acids. In contrast, for the anionic species the peptide backbone reverses its role to form the electronic A unit with the LUMO localized there. Here, the sidechain carboxylate moieties of Glu/Asp and phosphate moieties of the phosphorylated residues form the electronic D unit, with the HOMO localized on these groups. Except for the case of pTyr, where the sidechain B unit has a distinctive π-character due to the presence of the aromatic ring, the B unit for all other anionic amino acids is an alkyl chain with σ-character. Given that for both anionic and

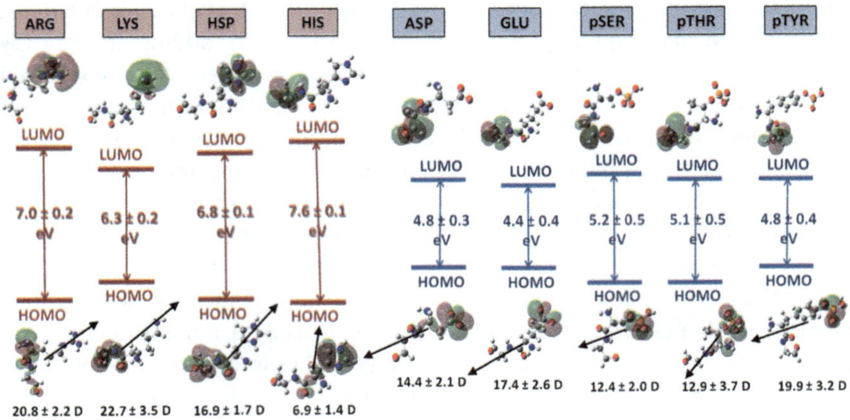

Fig. 3 Average ground state HOMO–LUMO gaps and dipole moment values along with their standard deviation from 100 conformations of all of the charged amino acids. The representative molecular structures at the top and bottom of the energy gap diagrams are identical, showing HOMO (bottom structures) and LUMO (top structures) contours plotted with an isovalue of 0.02. Ground state dipole vectors (arrows going from the negative end to the positive end on the bottom structures) for each system are also shown.

cationic systems, the D and A groups are populated with filled and empty frontier orbitals, respectively, we anticipate the lowest energy photoinduced electronic transitions in these ensembles to be PBS-CT transitions which transfer charge from D to A groups. The anionic systems exhibit considerably smaller (by 1–3 eV) HOMO–LUMO energy gaps relative to those of the cationic systems, thus indicating much more facile charge transfer and lower energy PBS-CT transitions for the former relative to the latter. Since the anionic (excluding pTyr) and cationic systems possess the same peptide backbone and a σ-bridge, the differences in ground state energy gaps can be attributed to the stronger electronic character of the anionic carboxylate and phosphate A groups relative to the guanidine, amino and imidazole D groups of the cationic systems.

For both anionic and cationic systems, the charged state of the sidechains polarizes the filled and empty orbitals in opposite directions. For the neutral His system, the frontier filled and empty orbitals are distributed indiscriminately over the peptide backbone and the imidazole ring. In the representative structure shown in Fig. 3, the HOMO is located mostly on the imidazole ring while the LUMO is localized on the peptide backbone. In contrast, for the doubly protonated HSP chromophore, the excess positive charge on the imidazole ring polarizes the molecular orbitals to shift the empty orbitals to the imidazole ring, thus transforming it to an electronic A group. The filled orbitals are in turn shifted to the peptide backbone, thus transforming it to an electronic D group. The dipole moment of HSP is also more well defined (about 10 Debye larger) relative to that of neutral His. Both cationic and anionic systems exhibit well defined dipole moments pointing from their D to A groups indicating the direction of photoinduced CT (*vide infra*). However, we find no systematic correlation between the dipole moment magnitude and the energy gap which could serve as a ground state proxy for predicting PBS-CT excitation energies. The dipole moment of cationic systems is larger than that for anionic systems, despite the latter

 This journal is © The Royal Society of Chemistry 2018

possessing significantly smaller energy gaps than the former. Within the cationic species, Lys has the largest dipole moment and the lowest energy gap. However, the dipole moment of Arg is slightly greater than that of Hsp, while both amino acids display similar energy gaps. Overall, the average energy gaps (E_{gap}) and the ground state dipole moment magnitudes (μ_{gs}) are both well resolved for the cationic species and follow the trends E_{gap}(Lys) < E_{gap}(Arg) ~ E_{gap}(Hsp) and μ_{gs}(Lys) > μ_{gs}(Arg) > μ_{gs}(Hsp). Within the anionic species, pTyr possesses the largest dipole moment but its energy gap is comparable to that of Asp whose dipole moment is significantly smaller. The variance in the energy gaps is large (0.3–0.5 eV) for the anionic species, and differences in their average energy gaps cannot be resolved. The dipole moment magnitudes for the anionic species follow the trend μ_{gs}(Glu) ~ μ_{gs}(pTyr) > μ_{gs}(pAsp) ~ μ_{gs}(pSer) ~ μ_{gs}(pThr).

To summarize this section, our ground state analysis of both cationic and anionic species reveals the significant polarization of orbitals due to excess charge in the side-chains, which enables us to clearly identify the electronic D–B–A groups for each chromophore. The ground state analysis reveals that the electronic D and A states are populated by frontier filled and empty orbitals, respectively, thus suggesting that the lowest energy transitions in these systems would be PBS-CT transitions. Furthermore, all of the charged amino acids examined here possess well defined dipole moments vectors pointing from D to A indicating the direction of photoinduced charge transfer.

The lowest energy transitions of anionic amino acids are PBS-CT transitions directed from the amino acid sidechains to the peptide backbone

We computed the optical absorption spectra (first 100 transitions) for all of the anionic amino acids using TDDFT electronic structure calculations. The top row of panels in Fig. 4A shows the computed spectra above 200 nm for 100 conformations of Asp and Glu sampled from the ubiquitin MD trajectory. Fig. 4B shows the spectra above 200 nm for 10 conformations of the three phosphorylated residues, pSer, pThr and pTyr, extracted from experimentally determined protein structures (see the Methods section). The insets in the top panels of Fig. 4A and B show the lowest energy transitions for each species in terms of difference density plots which present the photoexcited hole (pink) and electron (blue) densities on each amino acid fragment. In the lower panels of Fig. 4A and B, we present a classification of the transitions into CT or non-CT transitions based on two measurements of charge separation (see the Methods section). These panels show the percentage of CT and non-CT transitions within 5 nm windows over the whole absorption spectral range for each of the charged chromophores.

The computed spectrum for Glu in ubiquitin extends up to 483 nm, which is about ~34 nm higher relative to the computed spectral range for Glu residues from α_3C obtained in our previous study.[14] We attribute this decrease in the PBS-CT excitation energies to the difference in secondary structure folds of ubiquitin and α_3C (*vide infra*). The difference density plots for the lowest energy transition for Glu show well separated photoexcited electron and hole densities, which reside on the peptide backbone and Glu carboxylate groups, respectively. This transition is therefore a PBS-CT transition from D (carboxylate) to A (peptide backbone), in accordance with our ground state analysis (previous subsection). The difference density plots for the lowest energy transitions of Asp mirror that of

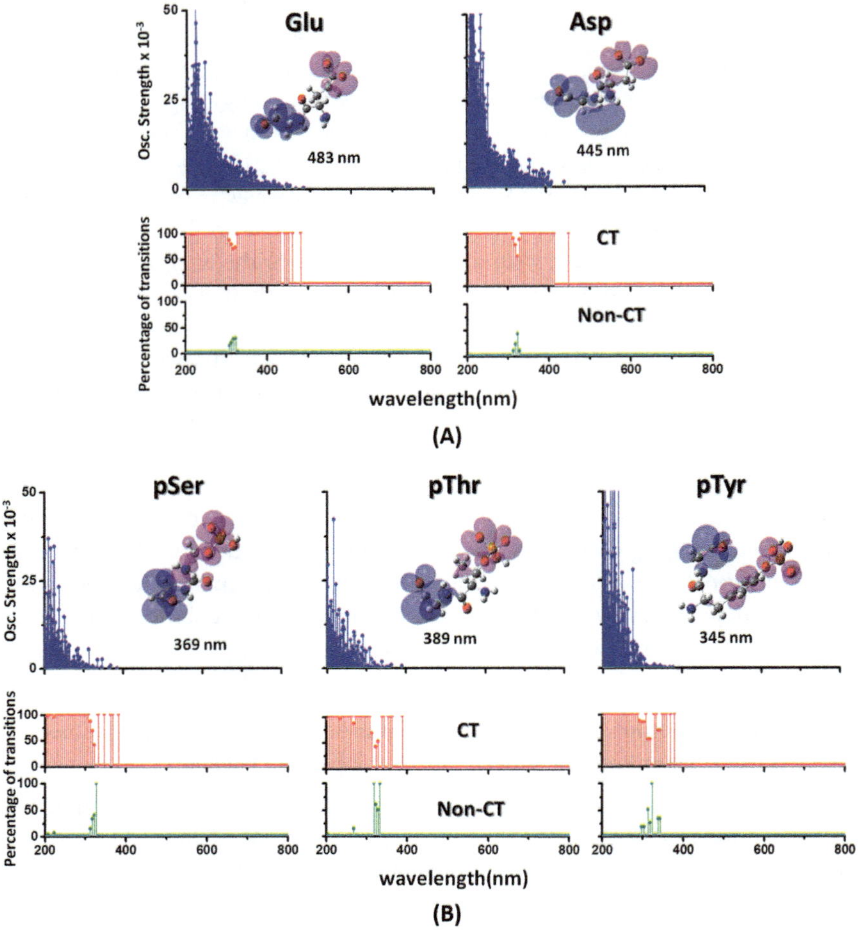

Fig. 4 Simulated absorption spectra (wavelength *vs.* oscillator strength) for the anionic amino acid monomers. The top panels show spectra for (A) 100 conformations of the Glu and Asp amino acids sampled from a 1 μs MD trajectory of ubiquitin and for (B) 10 different conformations of pSer, pThr and pTyr taken from crystal and NMR structures from the Protein Data Bank. Each amino acid fragment (chemical structures shown in Fig. 1) has an extended backbone capped with hydrogen atoms (see the Methods section). The lower panels in (A) and (B) show the assignment of transitions in the spectra to CT *vs.* non-CT transitions across the absorption profile. Difference density plots (the pink lobes represent hole density and the blue lobes represent electron density after photoexcitation) for the lowest energy transitions for each spectrum are shown in the insets of the top panels in (A) and (B).

Glu, showing a PBS-CT transition from D (carboxylate) to A (peptide backbone). Glu and Asp share the same electronic D and A states but differ in the length (1 methylene unit for Asp *vs.* 2 for Glu) of the sidechain B group. Intuitively, the shorter bridge length in Asp should lead to a stronger coupling of D and A states, thus leading to a more facile photoinduced CT relative to Glu. However, Fig. 4 shows that the spectral range of Asp (extending up to 445 nm) is in fact slightly lower than that observed for Glu. The shorter bridge length for Asp neither lowers the excitation energy of the PBS-CT transitions nor does it increase the oscillator strength for the low energy transitions relative to that of Glu. These results are

 This journal is © The Royal Society of Chemistry 2018

consistent with the ground state calculations (Fig. 3) which show that the average HOMO–LUMO gaps for Asp and Glu are comparable and that the lowest energy gap value for Asp is about 0.3 eV higher than that for Glu. The lower panels in Fig. 4 for Asp and Glu indicate that the spectra above 200 nm for these charged species are dominated by CT transitions. The few low intensity non-CT transitions above 300 nm are spurious transitions localized on the capping group,[14] and we do not ascribe any physical significance to them.

The computed spectra (Fig. 4) for the phosphorylated residues pSer, pThr and pTyr all lie below 400 nm, with the lowest energy transitions in these systems appearing at 369 nm, 389 nm and 345 nm, respectively. The difference density plots for the lowest energy transitions in all three systems show well separated photo-excited hole and electron densities on the phosphate and peptide backbone groups, thus indicating that these are PBS-CT transitions. The significantly higher excitation energies (0.4–0.8 eV) required for PBS-CT in phoshorylated amino acids relative to Asp/Glu loosely correlate with our ground state analysis (Fig. 3) which shows that the lowest energy gaps for the phosphorylated species are 0.1–0.9 eV higher than for Asp/Glu. The pSer and Asp chromophores share the same A (peptide backbone) and B (methylene) bridge, but differ in the nature of the D (carboxylate for Asp *vs.* phosphate for pSer) group. Thus, a comparison of the Asp and pSer spectra suggests that the phosphate group is a weaker electron donor than the carboxylate group. The phosphorylated amino acids share the same D and A group but differ in their B unit. Both pSer and pThr have similar B units which possess σ character. Accordingly, the ground state analysis reveals very similar energy gaps for pSer and pThr (Fig. 3), which correlates well with the similar spectral features observed for the two systems (Fig. 4). The pTyr chromophore, on the other hand, is distinct from the other anionic species in that the B unit possesses π character. In fact, the hole density in Fig. 4 shows significant delocalization on the phosphate group and the Tyr aromatic ring to form an extended D state. While the presence of a π bridge does not significantly change the average ground state energy gap (Fig. 3) relative to that of the other phosphorylated systems, the lowest energy gap for pTyr is still 0.4–0.5 eV lower than that for pSer and pThr (Fig. 4). However, in contrast with the ground state energy gap observations, the lowest energy PBS-CT transition for pTyr is 0.2–0.4 eV higher than that observed for pSer and pThr. This case illustrates the complexities in comparing PBS-CT excitation energies for systems with distinct B units. The pTyr spectra shows more intense transitions in the 200–300 nm window relative to those of the other phosphorylated species. The lower panels in Fig. 4B show that the entire spectral range above 200 nm for the phosphorylated species is dominated by CT transitions.

For a D–B–A complex, the lower bound of the TDDFT excitation energies for CT transitions (E_{CT}) is given by the relation:[37]

$$E_{CT} = (IP - EA) - \frac{1}{R} \tag{4}$$

where IP and EA represent the ionization potential of the donor and electron affinity of the acceptor, respectively, and the last term represents the electrostatic binding energy of the electron–hole pair created by the excitation and separated by a distance *R*. An orbital decomposition of the lowest energy PBS-CT transitions for Glu and Asp (Fig. S7 of ESI†) reveals that these transitions are predominantly (>96%) composed of the HOMO–LUMO D–A pairs. Thus, for these systems, we can approximate eqn (4) for the lower bound of the PBS-CT excitation energies in terms of the HOMO–LUMO energy gap (ΔE) and the charge separation index[34] D_{CT} (eqn (2)) as:

$$E_{CT} = \Delta E - \frac{1}{D_{CT}} \tag{5}$$

Fig. 5 shows that the excitation energies for the lowest energy PBS-CT transitions from 100 conformations of Glu and Asp correlate strongly with the HOMO–LUMO energy gaps and moderately with the hole–electron binding energy. For the phosphorylated systems, we observe a greater degree of mixing of the higher lying orbitals (*e.g.* LUMO+1) in the composition of the PBS-CT transitions. We thus expect the correlation of excitation energies with HOMO–LUMO gaps to be weaker for these systems.

The lowest energy transitions of cationic amino acids are PBS-CT transitions directed from the peptide backbone to the amino acid sidechains

We next examine the computed optical absorption spectra (100 lowest energy transitions) for the three cationic amino acids. The top panels in Fig. 6 show the TDDFT spectra beyond 200 nm for 100 conformations of the Lys, Arg and Hsp monomers sampled from the 1 μs MD trajectory of ubiquitin. As with the anionic species (previous subsection), we display difference density plots for the lowest energy transitions (insets) and a classification of the transitions as CT and non-CT transitions (lower panels).

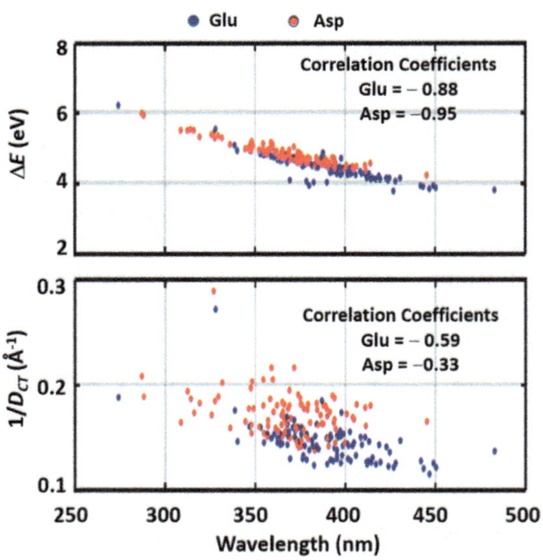

Fig. 5 (Top panel) Correlation between the HOMO–LUMO energy gap and the excitation wavelength for the lowest energy transitions of Asp and Glu. (Bottom panel) Correlation between the inverse charge separation distance (eqn (2)) and the excitation wavelength for the lowest energy transitions of Asp and Glu. The data was generated from 100 MD conformations of Asp and Glu.

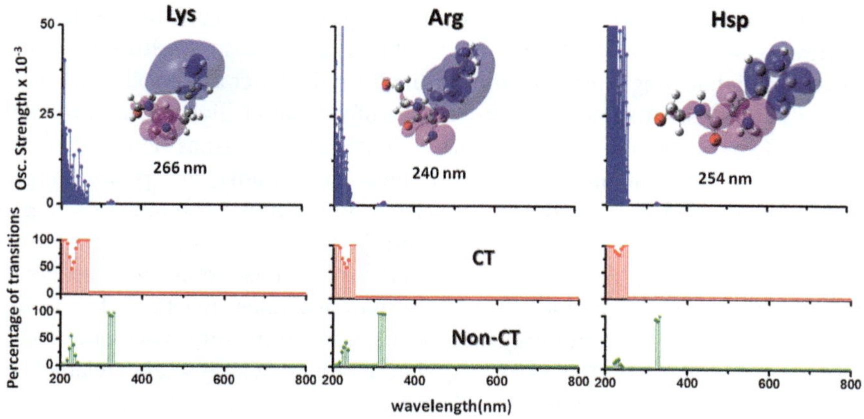

Fig. 6 Simulated absorption spectra (wavelength *vs.* oscillator strength) for the cationic amino acid monomers. The top panels show spectra for 100 conformations of the Lys, Arg and Hsp amino acids sampled from a 1 μs MD trajectory of ubiquitin. Each amino acid fragment (chemical structures shown in Fig. 1) has an extended backbone capped with hydrogen atoms (see the Methods section). The lower panel for each amino acid fragment shows the assignment of transitions in the spectra to CT *vs.* non-CT transitions across the absorption profile. Difference density plots (the pink lobes represent hole density and the blue lobes represent electron density after photoexcitation) for the lowest energy transitions for each spectrum are shown in the insets of the top panel.

The spectra for all three cationic species lie below 300 nm. The Lys spectra extends up to 266 nm which matches the range obtained by us earlier for amino acid conformers in α_3C.[14] The spectral range of the Arg and Hsp chromophores is similar to that of Lys, extending up to 240 nm and 254 nm, respectively. The difference density plots reveal clear backbone to sidechain PBS-CT transitions for the lowest energy excitations of the Lys, Arg and Hsp ensembles. For all three chromophores, the lowest photoexcited transition shows the hole density to be mostly localized on the peptide backbone, with some delocalization on the alkyl sidechain. On the other hand, the electron density for the lowest energy transition is mostly localized on the sidechain amino (Lys), guanidine (Arg) and imidazole (Hsp) groups, with some delocalization on the alkyl chain. Thus, for the lowest energy PBS-CT transitions in the cationic systems, we see more extended D/A units and less well-defined B units. In contrast, the anionic systems (Fig. 4) show more localized and well defined D, B and A units. Overall, the lowest PBS-CT excitation energy of the conformational ensemble of cationic amino acids is ~1.0–2.5 eV higher than that of the anionic amino acids. This correlates well with our ground state analysis (Fig. 3) which shows that the HOMO–LUMO energy gaps for the anionic systems are up to 2.5 eV lower than those for the cationic systems.

The spectra for Arg and Hsp show significantly higher oscillator strengths between 200-240 nm relative to Lys. This is likely due to the more extended conjugation in the Arg guanidine and Hsp imidazole moieties relative to the Lys amino group. On the other hand, the excess positive charge on the guanidine and imidazole moieties is more delocalized and negatively impacts the polarization and stabilization of the ground state orbitals. For instance, in the ground state picture (Fig. 3), the HOMO–LUMO gaps for Arg and Hsp are 0.5–0.7 eV higher than that for Lys. Furthermore, a decomposition of the orbitals for the lowest energy PBS-CT transitions (Fig. S8 of ESI†) of Lys, Arg and Hsp confirms a greater mixing of higher-lying filled and empty orbitals for the latter two systems. In fact, even the Lys system shows a greater degree of mixing of higher lying orbitals (LUMO+1) in the composition of the lowest energy PBS-CT transition relative to that of the anionic Glu and Asp chromophores. Furthermore, the spectral range for the PBS-CT transitions of the Arg and Hsp conformational ensembles overlaps with that of the localized backbone transitions (210–230 nm). In our calculations, we find that for 50% of Arg conformations and for 6% of His conformations, PBS-CT transitions are not the lowest energy excitations. Taken together, the above observations suggest that for cationic systems, the correlation between the excitation energy and the HOMO–LUMO gap should be significantly weaker relative to that for the anionic systems. Indeed, our analysis reveals significantly weak correlation between the ground state energy gap and $1/D_{CT}$ with the excitation wavelength for the lowest energy transitions extracted from 100 conformations of Lys, Arg and Hsp (Fig. S9 of ESI†).

Sensitivity of CT transitions to protein secondary structure

In this section, we present evidence for the sensitivity of the lowest energy PBS-CT transitions of charged amino acid residues to the protein secondary structure. The charged amino acid chromophores studied here were sampled from a 1 μs MD trajectory of ubiquitin, a globular protein with 5 β-strands wrapped around a single α-helix. The charged amino acid chromophores sampled in our study therefore come from different secondary structure elements including β-strands, α-helices, and loop regions of the protein. Fig. 7 shows the variation in the lowest energy PBS-CT excitation wavelength as a function of the Ramachandran torsion angles (Φ_{i+1} and Ψ_i) defined in Fig. 2C for conformations of the 7 Lys (left panel) and 6 Glu (right panel) amino acids in ubiquitin sampled from the MD trajectory. The correspondence of these torsion angles with the Ramachandran angles (Φ_I, Ψ_i) for the Lys and Glu residues is shown in Fig. S2 and S4 of ESI.†

The Lys conformational ensemble in Fig. 7 shows a single diffuse cluster in the β-sheet region of the plot (labelled as β_{Lys}), 2 localized clusters in the α-helical region (labelled $\alpha1_{Lys}$ and $\alpha2_{Lys}$), and one cluster (F_{Lys}) in the forbidden region (Φ_{i+1} = 60 to 90 degrees, Ψ_i = 120 to 150 degrees) for Lys. By examining the excitation wavelength of the lowest energy PBS-CT transitions (colour code shown in the left panel of Fig. 7) of each conformation as a function of the torsion angles (Φ_{i+1} and Ψ_i), we find distinct wavelength ranges for each conformational cluster. For instance, conformations in the $\alpha1_{Lys}$ cluster show relatively high energy PBS-CT transitions below ~245 nm. On the other hand, the lowest wavelength transitions (~245–265 nm) come from conformations in the β_{Lys} cluster. Conformations in the $\alpha2_{Lys}$ and F_{Lys} clusters show intermediate wavelength range PBS-CT transitions between 245–255 nm. For the Glu conformational ensemble, we identify 2 distinct clusters of conformations in the β-sheet region ($\beta1_{Glu}$ and $\beta2_{Glu}$), one cluster in the α-helical region (α_{Glu}), and 2 clusters in the forbidden region of the plot: $F1_{Glu}$ ($\Phi_{i+1} = -90$ to -60 degrees, $\Psi_i = 0$ to 30 degrees) and $F2_{Glu}$ (Φ_{i+1} = 60 to 120 degrees, $\Psi_i = -60$ to 0 degrees). Upon examining the variation in the lowest energy excitation wavelengths of the PBS-CT transitions over the conformational ensemble (colour code shown in the right panel of Fig. 7),

 This journal is © The Royal Society of Chemistry 2018

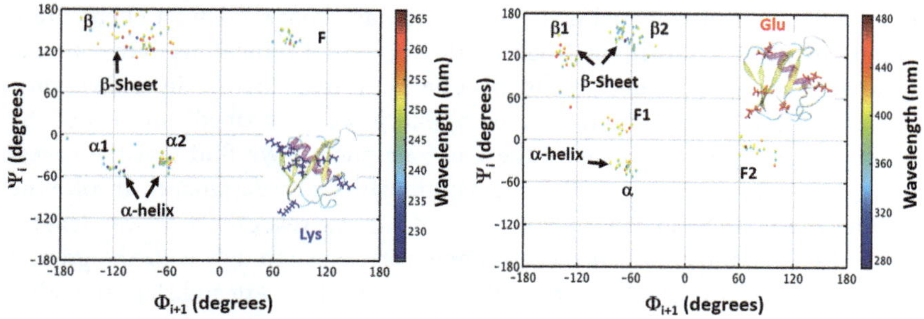

Fig. 7 Ramachandran dihedral angle (ϕ, ψ) maps for the Lys (left panel) and Glu (right panel) residues. Each point on the map represents a residue conformation sampled from 100 snapshots of a 1 μs ubiquitin trajectory (see the Methods section). Each point is colour coded according to the excitation wavelength of the lowest energy PBS-CT transition for that conformation. The wavelength range is indicated by a colour scale bar to the right of each panel. Ubiquitin has 7 Lys and 6 Glu residues and we sample all of them in multiple conformations in our analysis. The distribution of Lys and Glu residues over the ubiquitin β-grasp fold sampled in our calculations is shown in the inset of each panel.

we again find a sensitivity of the excitation energies to the clustering of conformations. Conformations in $\beta1_{\text{Glu}}$ and $\beta2_{\text{Glu}}$ show the lowest (~280–400 nm) and highest (~360–480 nm) ranges of excitation wavelengths for the lowest energy PBS-CT transitions, respectively. On the other hand, conformations within α_{Glu}, F1_{Glu} and F2_{Glu} mostly exhibit transitions in the intermediate wavelength (~360–440 nm) range. Interestingly, one of the Glu conformations with the lowest excitation energy PBS-CT transition (~480 nm) lies in the F2_{Glu} cluster.

To summarize, the data in this subsection show that the excitation wavelengths of the lowest energy PBS-CT transitions in charged amino acid monomers are sensitive to the secondary structure fold where the individual charged amino acids are located. These results motivate more detailed investigations of the spectra of interacting charged amino acid conformations sampled from different secondary structure elements in protein folds with the goal of connecting ProCharTS spectral features to the underlying protein structure and dynamics.

Conclusions

In this manuscript, we provide a detailed computational study and comparison of new optical (UV-vis) CT transitions in charged amino acid monomers. Previously, we had shown that such CT transitions from Lys and Glu amino acids could produce significant absorption in proteins that are rich in these charged amino acids well beyond 250 nm.[14] Here we computationally examine the UV-vis spectra of six other charged amino acid chromophores, Arg, Hsp, Asp and phosphorylated sidechains of Ser, Thr and Tyr (pSer, pThr, and pTyr), and compare them with those of Lys and Glu. For all eight amino acids we carried out electronic structure and TDDFT calculations on multiple conformations sampled from a classical MD trajectory and structures from the protein data bank. Our analysis reveals that the excess charge on the sidechains of charged amino acids polarizes the frontier orbitals to create well defined electronic D–B–A complexes in these systems. TDDFT calculations show that the UV-vis spectra of all charged amino acid chromophores is dominated by CT transitions and that the lowest energy transitions are PBS-CT transitions. The direction of PBS-CT transitions in the anionic systems is opposite (sidechain to peptide backbone) to that in the cationic systems (peptide backbone to sidechain) and is well described in terms of the ground state dipole moments of these chromophores. We find that the lowest energy PBS-CT in the anionic Glu and Asp transitions predominantly involve the HOMO and LUMO and consequently the red edge (lowest excitation wavelength) of the PBS-CT transitions in these systems correlates strongly with the ground state HOMO–LUMO gap. In contrast, for the cationic (Lys, Arg and Hsp) and other anionic (pSer, pThr and pTyr) species, we find that higher lying filled and empty orbitals contribute to the lowest energy PBS-CT transitions. Thus, for these systems the red-edge does not correlate significantly with the ground state HOMO–LUMO gap. Finally, we presented evidence for the sensitivity of the UV-vis spectral range associated with PBS-CT transitions to the protein secondary structure fold. These results advance our understanding of the range of CT transitions present in non-aromatic amino acids. The PBS-CT transitions studied here are sensitive to sidechain interactions and the protein fold[14] and produce novel spectra termed ProCharTS which can be used to probe the structure and dynamics of proteins that are rich in charged amino acids. The systematic investigation of charged amino acids presented here builds a foundation for understanding distinct features in the ProCharTS absorption spectra of proteins.

Conflicts of interest

There are no conflicts to declare.

Acknowledgements

The authors thank the Tata Institute of Fundamental Research (Department of Atomic Energy, India) for their financial support.

References

1 M. C. Otey and J. P. Greenstein, Studies on polycysteine peptides and proteins. II. Apparent dissociation constants, and ultraviolet and infrared absorption spectra of isomeric cystinylcystine peptides, *Arch. Biochem. Biophys.*, 1954, **53**, 501–513.

2 C. R. Cantor and P. R. Schimmel, *Biophysical Chemistry Part 2: Techniques for the study of biological structure and function*, W. H. Freeman and Company, New York, 1980.

3 G. D. Fasman, *Practical Handbook of Biochemistry and Molecular Biology*, CRC Press, New York, 1992.

4 H. D. Hunt and W. T. Simpson, Spectra of Simple Amides in the Vacuum Ultraviolet, *J. Am. Chem. Soc.*, 1953, **75**, 4540–4543.

5 K. Rosenheck and P. Doty, The far ultraviolet absorption spectra of polypeptide and protein solutions and their dependence on conformation, *Proc. Natl. Acad. Sci. U. S. A.*, 1961, **47**, 1775–1785.

6 D. B. Wetlaufer, Ultraviolet spectra Of Proteins and Amino Acids, in *Advances in Protein Chemistry*, 1963, vol. 17, pp 303–390.

 This journal is © The Royal Society of Chemistry 2018

7 J. S. Ham and J. R. Platt, Far U.V. spectra of peptides, *J. Chem. Phys.*, 1952, **20**, 335–336.
8 M. T. Oakley and J. D. Hirst, Charge-transfer transitions in protein circular dichroism calculations, *J. Am. Chem. Soc.*, 2006, **128**, 12414–12415.
9 L. Serrano-Andres and M. P. Fulscher, Theoretical study of the electronic spectroscopy of peptides. III. Charge-transfer transitions in polypeptides, *J. Am. Chem. Soc.*, 1998, **120**, 10912–10920.
10 E. I. Solomon, D. E. Heppner, E. M. Johnston, J. W. Ginsbach, J. Cirera, M. Qayyum, M. T. Kieber-Emmons, C. H. Kjaergaard, R. G. Hadt and L. Tian, Copper Active Sites in Biology, *Chem. Rev.*, 2014, **114**, 3659–3853.
11 D. L. Ma, M. D. Wang, C. F. Liu, X. M. Miao, T. S. Kang and C. H. Leung, Metal complexes for the detection of disease-related protein biomarkers, *Coord. Chem. Rev.*, 2016, **324**, 90–105.
12 L. Serrano-Andres and M. P. Fulscher, Charge transfer transitions in neutral and ionic polypeptides: A theoretical study, *J. Phys. Chem. B*, 2001, **105**, 9323–9330.
13 A. T. B. Gilbert and J. D. Hirst, Charge-transfer transitions in protein circular dichroism spectra, *J. Mol. Struct.: THEOCHEM*, 2004, **675**, 53–60.
14 S. Prasad, I. Mandal, S. Singh, A. Paul, B. Mandal, R. Venkatramani and R. Swaminathan, Near UV-Visible electronic absorption originating from charged amino acids in a monomeric protein, *Chem. Sci.*, 2017, **8**, 5416–5433.
15 D. Pinotsi, L. Grisanti, P. Mahou, R. Gebauer, C. F. Kaminski, A. Hassanali and G. S. Kaminski Schierle, Proton Transfer and Structure-Specific Fluorescence in Hydrogen Bond-Rich Protein Structures, *J. Am. Chem. Soc.*, 2016, **138**, 3046–3057.
16 P. Guptasarma, Solution-state characteristics of the ultraviolet A-induced visible fluorescence from proteins, *Arch. Biochem. Biophys.*, 2008, **478**, 127–129.
17 A. Shukla, S. Mukherjee, S. Sharma, V. Agrawal, K. V. Radha Kishan and P. A. Guptasarma, novel UV laser-induced visible blue radiation from protein crystals and aggregates: scattering artifacts or fluorescence transitions of peptide electrons delocalized through hydrogen bonding?, *Arch. Biochem. Biophys.*, 2004, **428**, 144–153.
18 S. Jones, P. van Heyningen, H. M. Berman and J. M. Thornton, Protein–DNA interactions: A structural analysis, *J. Mol. Biol.*, 1999, **287**, 877–896.
19 R. van der Lee, M. Buljan, B. Lang, R. J. Weatheritt, G. W. Daughdrill, A. K. Dunker, M. Fuxreiter, J. Gough, J. Gsponer, D. T. Jones, P. M. Kim, R. W. Kriwacki, C. J. Oldfield, R. V. Pappu, P. Tompa, V. N. Uversky, P. E. Wright and M. M. Babu, Classification of intrinsically disordered regions and proteins, *Chem. Rev.*, 2014, **114**, 6589–6631.
20 M. J. Abraham, T. Murtola, R. Schulz, S. Páll, J. C. Smith, B. Hess and E. Lindahl, GROMACS: High performance molecular simulations through multi-level parallelism from laptops to supercomputers, *SoftwareX*, 2015, **1–2**, 19–25.
21 R. B. Best, X. Zhu, J. Shim, P. E. M. Lopes, J. Mittal, M. Feig and A. D. MacKerell, Optimization of the Additive CHARMM All-Atom Protein Force Field Targeting Improved Sampling of the Backbone phi, psi and Side-Chain chi(1) and chi(2) Dihedral Angles, *J. Chem. Theory Comput.*, 2012, **8**, 3257–3273.

22 W. Humphrey, A. Dalke and K. Schulten, VMD: visual molecular dynamics, *J. Mol. Graphics*, 1996, **14**, 33–38.

23 T. Yanai, D. P. Tew and N. C. Handy, A new hybrid exchange-correlation functional using the Coulomb-attenuating method (CAM-B3LYP), *Chem. Phys. Lett.*, 2004, **393**, 51–57.

24 M. J. Frisch, G. W. Trucks, H. B. Schlegel, G. E. Scuseria, M. A. Robb, J. R. Cheeseman, G. Scalmani, V. Barone, B. Mennucci, G. A. Petersson, H. Nakatsuji, M. Caricato, X. Li, H. P. Hratchian, A. F. Izmaylov, J. Bloino, G. Zheng, J. L. Sonnenberg, M. Hada, M. Ehara, K. Toyota, R. Fukuda, J. Hasegawa, M. Ishida, T. Nakajima, Y. Honda, O. Kitao, H. Nakai, T. Vreven, J. A. Montgomery Jr, J. E. Peralta, F. Ogliaro, M. J. Bearpark, J. Heyd, E. N. Brothers, K. N. Kudin, V. N. Staroverov, R. Kobayashi, J. Normand, K. Raghavachari, A. P. Rendell, J. C. Burant, S. S. Iyengar, J. Tomasi, M. Cossi, N. Rega, N. J. Millam, M. Klene, J. E. Knox, J. B. Cross, V. Bakken, C. Adamo, J. Jaramillo, R. Gomperts, R. E. Stratmann, O. Yazyev, A. J. Austin, R. Cammi, C. Pomelli, J. W. Ochterski, R. L. Martin, K. Morokuma, V. G. Zakrzewski, G. A. Voth, P. Salvador, J. J. Dannenberg, S. Dapprich, A. D. Daniels, Ö. Farkas, J. B. Foresman, J. V. Ortiz, J. Cioslowski and D. J. Fox, *Gaussian 09*, Gaussian, Inc., Wallingford, CT, USA, 2009.

25 T. Lu and F. Chen, Multiwfn: a multifunctional wavefunction analyzer, *J. Comput. Chem.*, 2012, **33**, 580–592.

26 R. Dennington, T. Keith and N. J. Millam, *GaussView, Version 5*, Semichem Inc., Shawnee Mission, KS, 2009.

27 H. M. Berman, J. Westbrook, Z. Feng, G. Gilliland, T. N. Bhat, H. Weissig, I. N. Shindyalov and P. E. Bourne, The Protein Data Bank, *Nucleic Acids Res.*, 2000, **28**, 235–242.

28 C. A. Guido, P. Cortona, B. Mennucci and C. Adamo, On the Metric of Charge Transfer Molecular Excitations: A Simple Chemical Descriptor, *J. Chem. Theory Comput.*, 2013, **9**, 3118–3126.

29 T. Le Bahers, C. Adamo and I. Ciofini, A Qualitative Index of Spatial Extent in Charge-Transfer Excitations, *J. Chem. Theory Comput.*, 2011, **7**, 2498–2506.

30 D. M. Rogers, N. A. Besley, P. O'Shea and J. D. Hirst, Modeling the absorption spectrum of tryptophan in proteins, *J. Phys. Chem. B*, 2005, **109**, 23061–23069.

31 P. Stepanek and P. Bour, Multi-scale modeling of electronic spectra of three aromatic amino acids: importance of conformational averaging and explicit solute-solvent interactions, *Phys. Chem. Chem. Phys.*, 2014, **16**, 20639–20649.

32 D. Jacquemin, E. A. Perpete, G. E. Scuseria, I. Ciofini and C. Adamo, TD-DFT performance for the visible absorption spectra of organic dyes: Conventional *versus* long-range hybrids, *J. Chem. Theory Comput.*, 2008, **4**, 123–135.

33 M. J. G. Peach, P. Benfield, T. Helgaker and D. J. Tozer, Excitation energies in density functional theory: An evaluation and a diagnostic test, *J. Chem. Phys.*, 2008, **128**, 044118.

34 M. Campetella, F. Maschietto, M. J. Frisch, G. Scalmani, I. Ciofini and C. Adamo, Charge transfer excitations in TDDFT: A ghost-hunter index, *J. Comput. Chem.*, 2017, **38**, 2151–2156.

35 M. N. Paddon-Row, Investigating long-range electron-transfer processes with rigid, covalently linked donor–(norbornylogous bridge)–acceptor systems, *Acc. Chem. Res.*, 1994, **27**, 18–25.

 This journal is © The Royal Society of Chemistry 2018

36 O. S. Wenger, How Donor–Bridge–Acceptor Energetics Influence Electron Tunneling Dynamics and Their Distance Dependences, *Acc. Chem. Res.*, 2011, **44**, 25–35.

37 A. Dreuw, J. L. Weisman and M. Head-Gordon, Long-range charge-transfer excited states in time-dependent density functional theory require non-local exchange, *J. Chem. Phys.*, 2003, **119**, 2943–2946.

Faraday Discussions

PAPER

A QM/MM study of the initial excited state dynamics of green-absorbing proteorhodopsin†

Veniamin A. Borin, Christian Wiebeler and Igor Schapiro *

Received 23rd September 2017, Accepted 3rd November 2017
DOI: 10.1039/c7fd00198c

The primary photochemical reaction of the green-absorbing proteorhodopsin is studied by means of a hybrid quantum mechanics/molecular mechanics (QM/MM) approach. The simulations are based on a homology model derived from the blue-absorbing proteorhodopsin crystal structure. The geometry of retinal and the surrounding sidechains in the protein binding pocket were optimized using the QM/MM method. Starting from this geometry the isomerization was studied with a relaxed scan along the $C_{13}{=}C_{14}$ dihedral. It revealed an "aborted bicycle pedal" mechanism of isomerization that was originally proposed by Warshel for bovine rhodopsin and bacteriorhodopsin. However, the isomerization involved the concerted rotation about $C_{13}{=}C_{14}$ and $C_{15}{=}N$, with the latter being highly twisted but not isomerized. Further, the simulation showed an increased steric interaction between the hydrogen at the C_{14} of the isomerizing bond and the hydroxyl group at the neighbouring tyrosine 200. In addition, we have simulated a nonadiabatic trajectory which showed the timing of the isomerization. In the first 20 fs upon excitation the order of the conjugated double and single bonds is inverted, consecutively the $C_{13}{=}C_{14}$ rotation is activated for 200 fs until the S_1–S_0 transition is detected. However, the isomerization is reverted due to the specific interaction with the tyrosine as observed along the relaxed scan calculation. Our simulations indicate that the retinal – tyrosine 200 interaction plays an important role in the outcome of the photoisomerization.

Introduction

In 2000 Béjà *et al.*[1,2] discovered a protein in γ-proteobacteria that later turned out to be the most abundant member of the rhodopsin protein family.[3] Accordingly, the protein is named proteorhodopsin (PR). Rhodopsins are light-sensitive proteins that consist of the apoprotein opsin which is characterized by seven transmembrane helices and a prosthetic group retinal that serves as

Fritz Haber Center for Molecular Dynamics, Institute of Chemistry, The Hebrew University of Jerusalem, Jerusalem 91904, Israel. E-mail: igor.schapiro@mail.huji.ac.il

† Electronic supplementary information (ESI) available. See DOI: 10.1039/c7fd00198c

a chromophore. This chromophore is covalently bound to the apoprotein through a Schiff base bond to the lysine sidechain making it effectively a photoreceptor protein. This composition forms the structural basis for the members of the rhodopsin family that absorb light in the UV/Vis range. Upon light absorption, the retinal undergoes isomerization that represents the first step in a series of conformational changes summarized in a photocycle that ultimately leads to a molecular response. The exact function of this light-induced activation and the fine details vary within the rhodopsin family. The family is distinguished in two groups:[4] the vertebrate rhodopsins (type II) and microbial rhodopsins (type I). Type I rhodopsin starts from an all-*trans* retinal protonated Schiff base (rpSb) and has a closed photocycle, in which the retinal chromophore is not released from the protein and the reisomerization occurs thermally. Type II rhodopsin binds 11-*cis* rpSb and has an open photocycle, where the retinal can be released from the protein and reisomerised by enzymes or *via* a photochemical reaction. PR is categorized as type I rhodopsin and is closely related to its well-studied subfamily-representative bacteriorhodopsin (bR).[5]

The overall sequence identity between PR and bR is less than 30%,[1] nevertheless most of the functional amino acids are conserved.[6,7] Among them are R94 and D227 (corresponding to Arg82 and D212 in bR) that form the counterion complex. The primary proton acceptor is D97 in PR and D85 in bR while the primary proton donor is homologue Glu108 and D96, respectively. These amino acids are essential for the function of bR as a proton pump that translocates protons from the cytoplasmic to the extracellular side across the membrane. Hence, it is not surprising that PR was confirmed to be a proton pump as well.[1,2,6]

PR has some characteristics that make it distinct from bR despite the similar functional role and the preserved key amino acids for the pumping. One of them is the unusually high pK_a value of 7.5 of the primary proton acceptor D97 (D82 in bR has a pK_a of 2.5) which means that it is protonated to a large degree at physiological conditions (pH 7–8). Since the primary proton acceptor is protonated, the proton transfer from the retinal pSB is delayed or even aborted at very low pH values (<2.5).[8] That means PR is operating at alkaline pH values. The protonation state also has an influence on the photochemistry of PR. The deprotonation of D97 leads to a spectral shift from 544 nm at pH 5 to 516 nm at pH 9.[9] This effect of the electrostatic environment on the wavelength or colour of the light absorbed by the chromophore can be explained by the point charge model.[10] Briefly, the chromophore absorption is fine-tuned by the protein surroundings, in particular by the immediate environment in the so-called protein binding pocket. Due to the intramolecular charge-transfer of retinal after excitation, a differential effect can be achieved by introduction of amino acid mutations that will influence the ground state to a different extent than the excited state resulting in an overall spectral shift.[11] There are 23 amino acids which have at least one atom with 5 Å distance from the chromophore. Interestingly, one of these amino acids (position 105) is found to be responsible for colour tuning of 35 nm in PR. Therefore, the PR family is divided into two subgroups depending on whether this residue is a leucine or a glutamine. The PR subfamily with L105 absorbs green light (GPR), while the other group with Q105 absorbs blue light (BPR).[12] This is due to difference in the depth of the ocean where the bacteria were found. As blue light penetrates deeper into water, the bacteria that are found at lower levels of the ocean have adapted to absorb this

 This journal is © The Royal Society of Chemistry 2018

wavelength of light.[3,13] A single point mutant L105Q in GPR recovered 25 nm of the shift, although 4 of the 23 amino acids in close proximity are different (F137, A151, F152, and C156 in GPR and G137, W151, W152, and M156 in BPR). In this sense, proteorhodopsin serves as an ideal system to study spectral tuning.

However, the primary dynamics following the excitation have also been of interest in order to see, for example, if the excited state properties can be affected by mutations.[14–19] A relative shift between the ground and excited state potential energy surfaces could lead to changes in the excited state lifetimes or the quantum yield of the photoreaction. To study such ultrafast dynamics, time-resolved absorption and infrared spectroscopy measurements have been conducted mainly on GPR.[20–26] The ultrafast dynamics of native GPR at pH 9 was found to be best described by an excited state relaxation and a biexponential decay with time constants of 140 fs, 280 fs and 9.5 ps, respectively.[21] However, the underlying molecular mechanism of the isomerization has not been explored by computational methods, mainly due to the lack of a crystal structure of GPR.

In order to perform computational investigations of PRs, a structural model of the protein is required. So far, the crystal structure of GPR is not solved and only recently was this achieved for BPR.[27] Therefore, previous studies have used homology models of BPR and GPR based on the crystal structures of bR or SRII as a template.[28,29] Excited state properties have been determined based on a semi-empirical method by employing cluster models (rpSb and a few adjacent amino acids) from classical molecular dynamics (MD) simulations. Furthermore, the crystal structure of BPR served as a starting point to investigate ground state properties *via* classical MD.[30] Finally, Buda *et al.*[31] have also used cluster models to calculate the properties of excited states based on TDDFT and to perform *ab initio* molecular dynamics. For this purpose, they have relied on a previously reported homology model[32] and extracted the positions of the retinal atoms and its nearest neighbours. The focus of the present work is on QM/MM nonadiabatic molecular dynamics based on a multiconfigurational method. The purpose of this work is to make an initial step in closing the gap between the ultrafast spectroscopic studies and the computer simulations.

The paper is organized as follows: first we present the methodological details of the present work, then the computed excitation energies to validate our model. Starting from this optimized QM/MM model, a relaxed scan and an excited state trajectory are reported and discussed in context of other retinal proteins and the ultrafast measurements.

Computational methodology

BPR is known for its high sequence identity with GPR,[33] which makes it the best suitable rhodopsin as a template. To generate a homology model of GPR we have taken the sequence of amino acids from a solution NMR study (PDB code: 4L6X).[34] Several templates were evaluated by creating homology models using the Swiss Model[35,36] webserver and assessing the quality *via* MolProbity[37,38] as well as visual inspection. Based on this assessment, chain B of the BPR crystal structure (PDB code 4JQ6)[27] was rated as the best template: it has a sequence identity of *ca.* 63% and the resulting homology model has a MolProbity score of 1.34. The latter is based on clashscore, rotamer and Ramachandran evaluations and normalized to be on the same scale as X-Ray resolution. Further, the resulting model did not show any Ramachandran outliers. Therefore, this homology model was used as the starting point in the present work.

In a first step, we added the retinal chromophore to the homology-derived model and linked it to the residue K231. In order to assign the protonation states of the titratable residues at pH 8, we employed the PDB2PQR web server.[39,40] This pH value was used because it is close to that of the natural environment. The imidazole sidechain of H75 was reoriented to be inside the protein moiety according to the experimental study of Hempelmann *et al.*[16] Further, 10 water molecules were added to the protein using the DOWSER program.[41] The positions of all hydrogen atoms were optimized with the Amber16 program package.[42]

The QM/MM optimization of the homology model was carried out using Amber16 interfaced to the Orca 4.0 program.[43,44] The QM part, consisting of the retinal molecule and the lysine sidechain, was described with the BP86 functional[45,46] in conjunction with the cc-pVDZ basis set.[47] The hydrogen link atom was placed at the QM/MM boundary between C_α and C_β atoms of the K231 residue. The remaining protein was calculated with the Amber ff14SB force field,[48] while the water molecules were described using the TIP3P model.[49] All sidechains with at least one atom within 5 Å from the QM part were relaxed during the optimization.

The TD-BP86,[45,46] TD-B3LYP,[50,51] and TD-CAM-B3LYP[52] vertical excitation energies were obtained using Orca 4.0 [44] and the RI-CC2 [53,54] and RI-ADC(2)[55] calculations were realized with Turbomole 7.1.[56] All of the aforementioned calculations were carried out for the first five excited states with the cc-pVDZ basis set.[47] The excited state relaxed scan and nonadiabatic molecular dynamics simulations were performed for the first bright excited state using the Complete Active Space Self-Consistent Field (CASSCF)[57] method as implemented in MOLCAS 8.1 [58] developers version interfaced with Tinker 6.3.[59] All six bonding and six antibonding π-type orbitals were included into the active space (Fig. S2†). Vertical excitation energies were also computed using the complete active space second order perturbation theory in the single state (SS-CASPT2), multistate (MS-CASPT2), and extended multistate (XMS-CASPT2) formulations with an imaginary shift of 0.2 Hartree and without an IPEA shift. The single-point energy correction along the relaxed scan was done using the MS-CASPT2 method. In order to reduce the computational cost of the more demanding CASSCF method the QM region was reduced. Therefore, the QM/MM boundary was placed between the C_δ and C_ε atoms of the K231 residue. The geometry was reoptimized using CASSCF on the ground state and keeping all MM atoms frozen at their positions. The relaxed scan on the excited state was computed by twisting the C_{12}–C_{13}–C_{14}–C_{15} dihedral in steps of 5 degrees and relaxing the remaining degrees of freedom. The scan was computed for both senses of rotation of the C_{13}–C_{14} bond. One excited state trajectory starting in the S_1 state without initial velocities, termed "FC trajectory" or "0k trajectory" in the literature,[60–62] was propagated in time for 270 fs. The molecular dynamics was calculated using the Velocity-Verlet algorithm[63] and a time step of 1 fs. Surface hopping was realized *via* Tully's Fewest Switches scheme[64] using nonadiabatic coupling matrix elements based on wave function overlaps.[65] The integration of the electronic degrees of freedom was done employing 200 substeps and a decoherence correction of 0.1 Hartree was used.[66]

Results and discussion

Geometry optimization and excitation energies

The homology model was subject to a QM(BP86/cc-pDVZ)/MM geometry optimization that included relaxation of the sidechains in the protein binding pocket. These sidechains were selected if at least one of their atoms is within 5 Å distance of the QM region. This procedure was chosen because the retinal geometry is refined in the crystallographic structure using force field parameters which are less accurate then the electronic structure methods. A similar approach was applied to other retinal proteins in QM/MM studies of the initial photochemistry[67–69] and is justified by the fact that no conformational changes of the protein are expected on the time scale of the retinal photoisomerization. Indeed, a recent study on PR confirmed that first changes of the protein backbone appear after 500 fs.[22]

In the optimized geometry of GPR (Fig. 1) the Schiff base group is oriented towards the D227, which is part of the counterion complex. The distance to D227 of 2.8 Å (measured between iminium N and carboxyl C) is significantly shorter compared to the 3.9 Å of the primary proton acceptor D97. This arrangement of the salt-bridge is adapted from the BPR crystal structure that was a template for the homology model. It was noted that BPR is missing a water molecule typically found in the counterion complex of bR (W402). Instead D227 forms a direct hydrogen bond to the iminium hydrogen. This attraction leads to an *M* helicity of

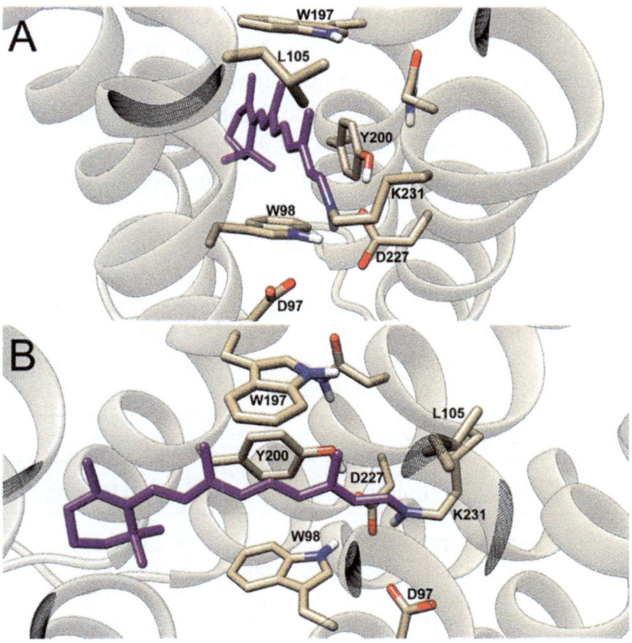

Fig. 1 Optimized QM/MM model of GPR. The retinal depicted in purple is shown together with important amino acids of the binding pocket in liquorice representation. (A) View along the Schiff base and the counter ion complex. (B) View perpendicular to the retinal polyene chain.

the retinal backbone, characterized by −161° torsion of the N=C bond, slowly relaxing along the polyene chain with twisted C_{13}=C_{14} (−169°) and C_{11}=C_{12} (−178°).

Based on this model, we have calculated the vertical excitation energies using various electronic structure methods (Table 1). The protonation state of our model corresponds to an alkaline environment, hence we take an experimental absorption maximum at 2.40 eV (516 nm) that was determined at pH 9.2.[9] An experimental absorption maximum cannot be compared directly to a vertical excitation energy, hence we are not expecting to have a one to one correspondence with the computed values. Nevertheless, we observe that all methods yield excitation energies within a range of ±0.25 eV. Even canonical TD-DFT methods like the GGA functional BP86 and the hybrid B3LYP yield results close to the measured value (2.26 eV and 2.42 eV, respectively), although both are notorious for their failure in describing charge transfer states. We have not found any dark states between the ground and the first excited state (Tables S1–S5 in ESI†). The long-range corrected functional CAM-B3LYP produces a slightly higher excitation energy of 2.56 eV. These results are in agreement with a recent study by Buda *et al.*[31] where B3LYP (2.43 eV) and CAM-B3LYP (2.72 eV) were employed as well. However, it should be noted that instead of a full QM/MM protein model a reduced cluster model was used in that study which contained rpSb with lysine, the counterion complex D97, D227, H75 and three water molecules which were treated at the DFT level and TZP basis set. Further, in that model the iminium moiety is hydrogen bonded to D97, instead of D227 as found in the BPR crystal structure. The comparison to the calculations in our study shows that this structural change has a small effect on the excitation energy.

In addition, we have tested RI-ACD(2) and RI-CC2 methods which have been previously applied for other retinal proteins.[70,71] These methods result in 2.30 eV and 2.46 eV excitation energies, for RI-ACD(2) and RI-CC2, respectively. Further we have computed multiconfigurational and multireference energies. Surprisingly, SA2-CASSCF(12,12) results in 2.46 eV which is close to the experimental value. This is a result of a fortuitous error compensation as the multireference

Table 1 Excitation energies in eV (nm) and oscillator strengths obtained from different quantum chemical methods for models optimized *via* DFT and CASSCF as the quantum chemical part in the QM/MM approach. Experimentally, the first absorption maximum for GPR is found at *ca.* 2.40 eV (516 nm) at pH 9.2 (ref. 9)

QM method	BP86		CASSCF	
Method	E_{Vert} [eV] ([nm])	f	E_{Vert} [eV] ([nm])	f
TD-BP86	2.26 (549)	1.41	2.91 (426)	0.94
TD-B3LYP	2.42 (512)	1.90	2.32 (534)	1.22
TD-CAM-B3LYP	2.56 (484)	2.15	2.70 (459)	1.72
CASSCF	2.46 (504)	1.90	3.36 (369)	1.01
SS-CASPT2	2.65 (468)	—	3.01 (412)	—
MS-CASPT2	2.27 (546)	2.23	2.64 (470)	1.22
XMS-CASPT2	2.66 (466)	2.31	3.09 (401)	1.37
RI-ADC(2)	2.30 (539)	1.81	2.37 (523)	1.26
RI-CC2	2.45 (506)	2.20	2.46 (504)	1.72

perturbation theory to second order produces 2.65 eV, 2.27 eV and 2.66 eV for single state (SS), multistate (MS) and extended multistate (XMS) implementations of CASPT2. These methods should give more reliable values because they include dynamic electron correlation. The MS-variant is below the experimental value, while SS and XMS-CASPT2 are above. The values for the latter should be more accurate because it is well known that MS leads to an artificially overestimated mixing of the electronic states at higher energy differences. MS-CASPT2 was originally designed to describe near-degeneracy situations and later replaced by XMS-CASPT2 that produces smoother surfaces close to degenerate energies.[72] Although an extensive benchmark for XMS-CASPT2 vertical excitation energies is lacking, the agreement with SS-CASPT2 indicates less pronounced mixing. However, we should note that the difference is still within 0.25 eV of the experimentally determined 2.40 eV absorption maximum. An assessment of excitation energies from Thiel and co-workers has shown that CASPT2 energies typically deviate by 0.2 eV from higher order methods.[73]

Since we are relying on CASSCF for the calculation of the photoisomerization reaction, we have re-optimized the geometry of retinal using this method in the adapted QM/MM scheme as described in the Computational methodology section. The resulting geometry has a significantly more pronounced bond length alternation pattern as documented in the literature (Fig. S1†).[69,74–76] It means the double bonds are shorter while the single bonds are more extended compared to BP86. This leads to altered excitation energies (Table 1). The CASSCF-optimized geometry resulted in a large shift of 0.65 eV for TD-BP86, while the B3LYP and CAM-B3LYP results have changed by 0.10 and 0.15 eV, respectively. Surprisingly, RI-ADC(2) and RI-CC2 methods are less sensitive to the geometry and show a change below 0.1 eV. For CASSCF the excitation energies are shifted by almost 1 eV. The subsequent CASPT2 correction decreases the change in energy due to the different geometry but the difference is between 0.35 and 0.40 eV with respect to the first model. Hence, the structural changes have a notable effect on the excitation energies. However, in order to describe the photoisomerization that involves evolution on the ground and excited state, as well as passage close to a CI, we have to use a multiconfigurational method such as CASSCF.

Relaxed scan along the $C_{13}{=}C_{14}$ dihedral

The relaxed scan was generated by twisting the dihedral C_{12}–$C_{13}{=}C_{14}$–C_{15} in steps of 5° in both senses of rotation starting from the FC point (Fig. 2). The energy for each optimized geometry has been subsequently corrected by MS-CASPT2 calculations. The counter clockwise path leads to the energetically accessible S_0/S_1 conical intersection that is encountered shortly after −120° through a low barrier of 8 kcal mol^{-1} (MS-CASPT2). It should be noted that the transition state characterizing this barrier is energetically below the FC point, which means that a sufficient amount of kinetic energy should be available to overcome the barrier on the excited state. The torsion approaching the S_0–S_1 degeneracy is smaller than the anticipated 90° from minimal conical intersection optimization. However, it is in line with previous computational studies that confirmed a 70° twist from planarity (starting from *cis* or *trans*). The nearly barrierless counter-clockwise path has been also discovered for other retinal proteins such as bR, channelrhodopsin and *Anabaena* sensory rhodopsin.[68,69,77–79] The

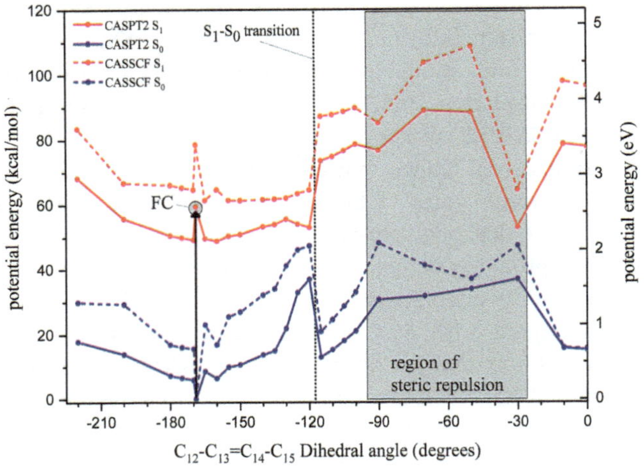

Fig. 2 QM/MM relaxed scan along the $C_{13}{=}C_{14}$ dihedral of retinal in GPR. CASSCF (dashed line) and MS-CASPT2//CASSCF (solid lines) energies are shown along a $C_{13}{=}C_{14}$ torsion. The transition from S_1 to S_0 is indicated by a dotted line.

structural changes associated with the excited state relaxation of the retinal are characterized by two dihedral angles with opposite senses of rotation. It can be seen in Fig. 3 that the $C_{15}{=}N$ bond is rotating in the opposite direction as

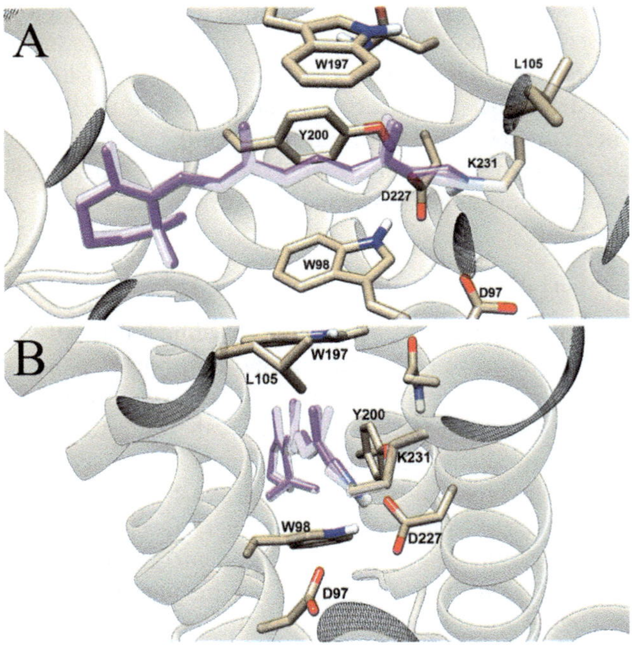

Fig. 3 Geometrical changes along the relaxed scan. Three characteristic geometries are shown. The structures are decreasing in occupancy in the following order: Franck Condon point, 145° twist and 120° (close to the conical intersection). (A) View along the Schiff base and the counter ion complex. (B) View perpendicular to the retinal polyene chain.

 This journal is © The Royal Society of Chemistry 2018

compared to the $C_{13}{=}C_{14}$ bond, therefore reducing the space required to undergo a *trans–cis* isomerization. Such a mechanism was first proposed by Warshel as the so-called bicycle pedal isomerization.[80] However, similar to its refined "aborted bicycle pedal" mechanism formulated for bR, we also see only a high twist but not a full isomerization of the $C_{15}{=}N$ bond.[81]

The constrained excited state geometry optimization for torsions above $-120°$ did not converge which is due the S_1 energy falling below the one of S_0. Hence, starting from $-115°$ we have continued the optimization on the ground state (dotted line in Fig. 2). The portion of the scan that follows the ground state after 90° shows the presence of a relatively high barrier due to steric repulsion of the hydrogen atom at C_{14} and the hydroxyl hydrogen of the Y200 residue (gray area in Fig. 2). However, towards the 13-*cis* isomer the retinal structure becomes planar again, the potential energy decreases and adapts a ground state configuration. This can also be seen from the evolution of the bond length in this region (ESI†). In comparison to the geometry at the FC point, a significant shortening of the single and elongation of the double bonds can be seen from the beginning of the scan due to the redistribution of the π-electron density in the excited state. However, after the S_1–S_0 transition and further optimization on the ground state the π-system recovers the alternation of the bond lengths, resembling the pattern of the ground state equilibrium geometry. The iminium hydrogen of the Schiff-base hydrogen bonded to the D227 residue throughout the scan and all retinal atoms, except for C_{14} and C_{15}, remain close to their initial positions.

The clockwise path shows a decrease in the excited state energy in the first 10°, but reveals a significant potential barrier after further 30° that is *ca.* 10 kcal mol^{-1} (MS-CASPT2) higher than the FC point. Thus, the clockwise isomerization path is not accessible which is why we decided to stop the scan in this direction.

In recent studies of retinal proteins, the two senses of isomerization directions were attributed to the biexponential nature of the excited state decay.[77] The faster time constant was assigned to the path with a smaller energy barrier while the slower time constant was interpreted as the path with a higher barrier. However, although a biexponential decay has been also confirmed for GPR, our data indicates a unidirectional isomerization in GPR, similar to what has been described in the case of *Anabaena* sensory rhodopsin.[78]

Excited state trajectory

In order to gain more insight into the excited state relaxation mechanism in GPR, we have carried out a nonadiabatic QM/MM molecular dynamics simulation. The trajectory was launched from the same FC point as in the case of the relaxed scan (Fig. 4). We have used zero initial velocities to run what has been referred to as a 0k-trajectory.[60–62] A single trajectory is not sufficient to make statistics about lifetimes and the quantum yield of the photoisomerization, but it can provide some preliminary insight into the molecular mechanism. The evolution of the excited state trajectory follows a well known pattern.[60,67,82–84] The first twenty femtoseconds of the MD can be characterized by an inversion of single and double bond length alternation at nearly planar geometry (Fig. 4B). Once the double bonds are becoming effectively like single bonds, an onset of the rotation about the double bonds is observed at around 30 fs of the dynamics (Fig. 4C). The largest twist is observed for the C_{12}–$C_{13}{=}C_{14}$–C_{15} and C_{14}–$C_{15}{=}N$–C_ε dihedral

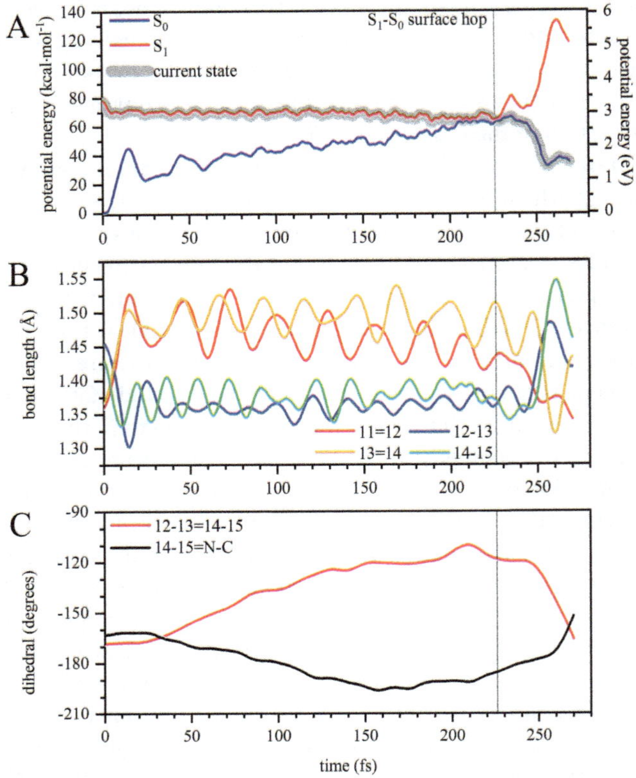

Fig. 4 QM/MM nonadiabatic trajectory of GPR. Shown are the evolution of the energy (A); the bond lengths (B); and the dihedral angles (C).

angles as already noticed in the relaxed scan. Although the evolution of the S_1 potential energy is flat, the potential energy is slightly decreasing until the hop from S_1 to S_0 is observed at 226 fs. At the time of the S_1–S_0 transition, the value of the C_{13}=C_{14} dihedral is −120°, in agreement with the distortion found in the relaxed scan. The C=N bond is twisting noticeably less, already rotating back to the *trans*-configuration at the time of the hop. In the following 30 fs it becomes apparent that both isomerizations are aborted, returning to an all-*trans* retinal geometry. Inspecting the structural changes we have identified a collision between the hydrogen at C_{14} and the OH group of Y200 (Fig. 5) that is responsible for an aborted isomerization. However, we cannot exclude the fact that our computational model with fixed amino acids in the binding pocket is responsible for this relaxation process on the ground state. However, we observe a space-saving rotation of the C_{13}=C_{14} and C_{15}=N double bonds of retinal in the excited state which leads to a conical intersection at 226 fs. This aborted bicycle pedal isomerization is consistent with the relaxed scan described above. The transition time of the 0k trajectory is on a similar time scale as the shortest of the two time constants for excited state decay (200 fs) from the pump–probe experiments by Wachtveitl and co-workers.[20]

Interestingly, Jung and coworkers[85] have recently characterized 18 variants of proteorhodopsin from the arctic ocean. They found 10 of them possessed an

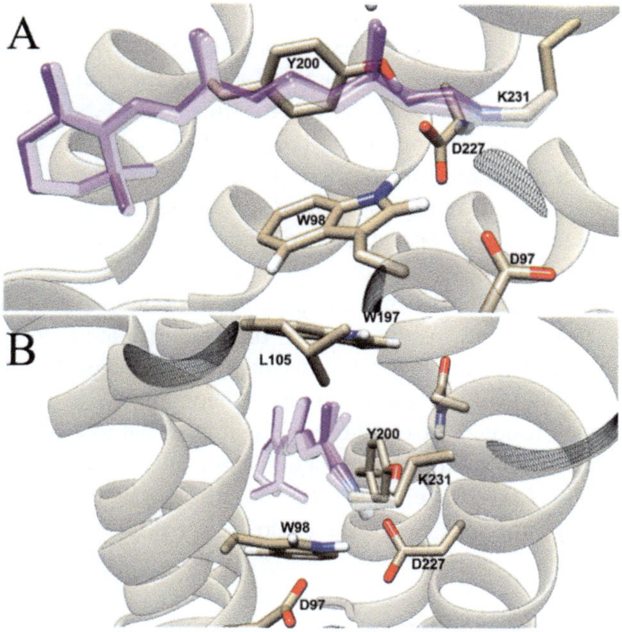

Fig. 5 Geometrical changes along the QM/MM trajectory. Four characteristic geometries at different times are shown. The structures are decreasing in occupancy in the following order: Franck Condon point, 70 fs, 140 fs and 227 fs (transition to the ground state). (A) View along the Schiff base and the counter ion complex. (B) View perpendicular to the retinal polyene chain.

N200Y mutant and slower photocycle. However, this study has focussed on the late intermediates of the photocycle and didn't address the ultrafast isomerization.

Conclusions

We have presented a computational study of the initial photochemistry in GPR. For this purpose a homology model was constructed based on the available crystallographic structure of the related BPR. This model was refined with a QM/MM geometry optimization that provided a starting point for an excited state relaxed scan and a trajectory simulation. The relaxed scan along the $C_{13}{=}C_{14}$ located a conical intersection for a 70° distortion from planarity. Further, it revealed a detailed insight into the molecular level rearrangements: the $C_{13}{=}C_{14}$ rotation is accompanied by an anticlockwise twist of the $C_{15}{=}N$ double bond, also known as a bicycle pedal mechanism. This simultaneous motion in opposite directions minimizes the space required for the *trans–cis* isomerization in a tight protein pocket. Nevertheless, we found a high barrier while continuing the rotation on the ground state due to steric interaction between the hydrogen at C_{14} and the Y200 sidechain. This repulsion is enhanced in our model where the binding pocket is held fixed. In a more flexible model such an interaction would probably lead to a delay of the isomerization without necessarily aborting it, which we plan to explore in a follow-up investigation.

The nonadiabatic QM/MM trajectory was computed starting from the same ground state geometry as the relaxed scan. The structural changes along the trajectory confirmed a bicycle pedal motion found in the relaxed scan computation. In addition, the trajectory revealed the timing of the events in the isomerization. The C_{13}=C_{14} and C_{15}=N double bond rotations start after the inversion of the bond length alternation in the first 20 fs. After 226 fs a transition to the ground state is observed, at this stage C_{13}=C_{14} is twisted at $-120°$. Eventually, the repulsion between the C_{14}–H and hydroxyl group of Y200 prevents an isomerization to 13-*cis*-rpSb that leads to the starting geometry.

The presented work is the first computational study of the ultrafast photochemistry in GPR. It confirms the course of the isomerization reported for other, related rhodopsins. However, it gives insight into the specific molecular level events in GPR, like the increased repulsion between C_{14} hydrogen with the adjacent tyrosine residue. This understanding might lead to a design of specific Y200 mutants that could further delay or accelerate the reaction as well as impact the success rate of the isomerization. The PR variants with a N200Y mutation characterized by Jung and coworkers could be a natural example of such an adaptation.

Conflicts of interest

There are no conflicts to declare.

Acknowledgements

This project has received funding from the European Research Council (ERC) under the European Union's Horizon 2020 research and innovation program (grant agreement 678169 "PhotoMutant"). Further, we would like to thank the Regional Computing Center of the University of Cologne (RRZK) for providing CPU time on the DFG-funded supercomputer CHEOPS and for their support. C. W. acknowledges support by the German Research Foundation (DFG) *via* a research scholarship (reference number: WI 4853/1-1).

References

1 O. Béjà, L. Aravind, E. V. Koonin, M. T. Suzuki, A. Hadd, L. P. Nguyen, S. B. Jovanovich, C. M. Gates, R. A. Feldman, J. L. Spudich, E. N. Spudich and E. F. DeLong, *Science*, 2000, **289**, 1902–1906.
2 O. Béjà, E. N. Spudich, J. L. Spudich, M. Leclerc and E. F. DeLong, *Nature*, 2001, **411**, 786–789.
3 C. Bamann, E. Bamberg, J. Wachtveitl and C. Glaubitz, *Biochim. Biophys. Acta, Bioenerg.*, 2014, **1837**, 614–625.
4 O. P. Ernst, D. T. Lodowski, M. Elstner, P. Hegemann, L. S. Brown and H. Kandori, *Chem. Rev.*, 2014, **114**, 126–163.
5 J. K. Lanyi, *Annu. Rev. Physiol.*, 2004, **66**, 665–688.
6 T. Friedrich, S. Geibel, R. Kalmbach, I. Chizhov, K. Ataka, J. Heberle, M. Engelhard and E. Bamberg, *J. Mol. Biol.*, 2002, **321**, 821–838.
7 R. A. Krebs, D. Dunmire, R. Partha and M. S. Braiman, *J. Phys. Chem. B*, 2003, **107**, 7877–7883.

8 M. Lakatos, J. K. Lanyi, J. Szakács and G. Váró, *Biophys. J.*, 2003, **84**, 3252–3256.
9 T. Köhler, I. Weber, C. Glaubitz and J. Wachtveitl, *Photochem. Photobiol.*, 2017, **93**, 762–771.
10 B. Honig, U. Dinur, K. Nakanishi, V. Balogh-Nair, M. A. Gawinowicz, M. Arnaboldi and M. G. Motto, *J. Am. Chem. Soc.*, 1979, **101**, 7084–7086.
11 F. Melaccio, N. Ferré and M. Olivucci, *Phys. Chem. Chem. Phys.*, 2012, **14**, 12485.
12 D. Man, W. Wang, G. Sabehi, L. Aravind, A. F. Post, R. Massana, E. N. Spudich, J. L. Spudich and O. Beja, *EMBO J.*, 2003, **22**, 1725–1731.
13 M. T. Suzuki, O. Beja, L. T. Taylor and E. F. Delong, *Environ. Microbiol.*, 2001, **3**, 323–331.
14 M. O. Lenz, A. C. Woerner, C. Glaubitz and J. Wachtveitl, *Photochem. Photobiol.*, 2007, **83**, 226–231.
15 M. K. Verhoefen, K. Neumann, I. Weber, C. Glaubitz and J. Wachtveitl, *Photochem. Photobiol.*, 2009, **85**, 540–546.
16 F. Hempelmann, S. Hölper, M. K. Verhoefen, A. C. Woerner, T. Köhler, S. A. Fiedler, N. Pfleger, J. Wachtveitl and C. Glaubitz, *J. Am. Chem. Soc.*, 2011, **133**, 4645–4654.
17 J. Herz, M. K. Verhoefen, I. Weber, C. Bamann, C. Glaubitz and J. Wachtveitl, *Biochemistry*, 2012, **51**, 5589–5600.
18 J. Mao, N. N. Do, F. Scholz, L. Reggie, M. Mehler, A. Lakatos, Y. S. Ong, S. J. Ullrich, L. J. Brown, R. C. D. Brown, J. Becker-Baldus, J. Wachtveitl and C. Glaubitz, *J. Am. Chem. Soc.*, 2014, **136**, 17578–17590.
19 C. E. Eckert, J. Kaur, C. Glaubitz and J. Wachtveitl, *J. Phys. Chem. Lett.*, 2017, **8**, 512–517.
20 R. Huber, T. Köhler, M. O. Lenz, E. Bamberg, R. Kalmbach, M. Engelhard and J. Wachtveitl, *Biochemistry*, 2005, **44**, 1800–1806.
21 M. O. Lenz, R. Huber, B. Schmidt, P. Gilch, R. Kalmbach, M. Engelhard and J. Wachtveitl, *Biophys. J.*, 2006, **91**, 255–262.
22 J. J. Amsden, J. M. Kralj, L. R. Chieffo, X. Wang, S. Erramilli, E. N. Spudich, J. L. Spudich, L. D. Ziegler and K. J. Rothschild, *J. Phys. Chem. B*, 2007, **111**, 11824–11831.
23 K. Neumann, M.-K. Verhoefen, I. Weber, C. Glaubitz and J. Wachtveitl, *Biophys. J.*, 2008, **94**, 4796–4807.
24 A. Rupenyan, I. H. M. van Stokkum, J. C. Arents, R. van Grondelle, K. Hellingwerf and M. L. Groot, *Biophys. J.*, 2008, **94**, 4020–4030.
25 A. Rupenyan, I. H. M. Van Stokkum, J. C. Arents, R. Van Grondelle, K. J. Hellingwerf and M. L. Groot, *J. Phys. Chem. B*, 2009, **113**, 16251–16256.
26 E. Bühl, M. Braun, A. Lakatos, C. Glaubitz and J. Wachtveitl, *Biol. Chem.*, 2015, **396**, 1109–1115.
27 T. Ran, G. Ozorowski, Y. Gao, O. A. Sineshchekov, W. Wang, J. L. Spudich and H. Luecke, *Acta Crystallogr., Sect. D: Biol. Crystallogr.*, 2013, **69**, 1965–1980.
28 J. R. Hillebrecht, J. Galan, R. Rangarajan, L. Ramos, K. McCleary, D. E. Ward, J. A. Stuart and R. R. Birge, *Biochemistry*, 2006, **45**, 1579–1590.
29 R. Rangarajan, J. F. Galan, G. Whited and R. R. Birge, *Biochemistry*, 2007, **46**, 12679–12686.
30 J. Feng and B. Mertz, *Biochemistry*, 2015, **54**, 7132–7141.
31 F. Buda, T. Keijer, S. Ganapathy and W. J. de Grip, *Photochem. Photobiol.*, 2017, **93**, 1399–1406.

32 S. Ganapathy, O. Bécheau, H. Venselaar, S. Frölich, J. B. van der Steen, Q. Chen, S. Radwan, J. Lugtenburg, K. J. Hellingwerf, H. J. M. de Groot and W. J. de Grip, *Biochem. J.*, 2015, **467**, 333–343.
33 S. Y. Kim, S. A. Waschuk, L. S. Brown and K. H. Jung, *Biochim. Biophys. Acta, Bioenerg.*, 2008, **1777**, 504–513.
34 S. Reckel, D. Gottstein, J. Stehle, F. Löhr, M. K. Verhoefen, M. Takeda, R. Silvers, M. Kainosho, C. Glaubitz, J. Wachtveitl, F. Bernhard, H. Schwalbe, P. Güntert and V. Dötsch, *Angew. Chem., Int. Ed.*, 2011, **50**, 11942–11946.
35 K. Arnold, L. Bordoli, J. Kopp and T. Schwede, *Bioinformatics*, 2006, **22**, 195–201.
36 M. Biasini, S. Bienert, A. Waterhouse, K. Arnold, G. Studer, T. Schmidt, F. Kiefer, T. G. Cassarino, M. Bertoni, L. Bordoli and T. Schwede, *Nucleic Acids Res.*, 2014, **42**, 252–258.
37 I. W. Davis, A. Leaver-Fay, V. B. Chen, J. N. Block, G. J. Kapral, X. Wang, L. W. Murray, W. B. Arendall, J. Snoeyink, J. S. Richardson and D. C. Richardson, *Nucleic Acids Res.*, 2007, **35**, 375–383.
38 V. B. Chen, W. B. Arendall, J. J. Headd, D. A. Keedy, R. M. Immormino, G. J. Kapral, L. W. Murray, J. S. Richardson and D. C. Richardson, *Acta Crystallogr., Sect. D: Biol. Crystallogr.*, 2010, **66**, 12–21.
39 T. J. Dolinsky, J. E. Nielsen, J. A. McCammon and N. A. Baker, *Nucleic Acids Res.*, 2004, **32**, W665–W667.
40 T. J. Dolinsky, P. Czodrowski, H. Li, J. E. Nielsen, J. H. Jensen, G. Klebe and N. A. Baker, *Nucleic Acids Res.*, 2007, **35**, W522–W525.
41 L. Zhang and J. Hermans, *Proteins: Struct., Funct., Genet.*, 1996, **24**, 433–438.
42 D. A. Case, R. M. Betz, D. S. Cerutti, T. E. Cheatham, T. A. Darden, R. E. Duke, T. J. Giese, H. Gohlke, A. W. Goetz, N. Homeyer, S. Izadi, P. Janowski, J. Kaus, A. Kovalenko, T. S. Lee, S. LeGrand, P. Li, C. Lin, T. Luchko, R. Luo, B. Madej, D. Mermelstein, K. M. Merz, G. Monard, H. Nguyen, H. T. Nguyen, I. Omelyan, A. Onufriev, D. R. Roe, A. Roitberg, C. Sagui, C. L. Simmerling, W. M. Botello-Smith, J. Swails, R. C. Walker, J. Wang, R. M. Wolf, X. Wu, L. Xiao and P. A. Kollman, *AMBER*, University of California, San Francisco, 2016.
43 A. W. Götz, M. A. Clark and R. C. Walker, *J. Comput. Chem.*, 2014, **35**, 95–108.
44 F. Neese, *Wiley Interdiscip. Rev.: Comput. Mol. Sci.*, 2012, **2**, 73–78.
45 A. D. Becke, *Phys. Rev. A: At., Mol., Opt. Phys.*, 1988, **38**, 3098–3100.
46 J. P. Perdew, *Phys. Rev. B: Condens. Matter Mater. Phys.*, 1986, **33**, 8822–8824.
47 T. H. Dunning Jr, *J. Chem. Phys.*, 1989, **90**, 1007.
48 J. A. Maier, C. Martinez, K. Kasavajhala, L. Wickstrom, K. E. Hauser and C. Simmerling, *J. Chem. Theory Comput.*, 2015, **11**, 3696–3713.
49 W. L. Jorgensen, J. Chandrasekhar, J. D. Madura, R. W. Impey and M. L. Klein, *J. Chem. Phys.*, 1983, **79**, 926–935.
50 A. D. Becke, *J. Chem. Phys.*, 1993, **98**, 5648–5652.
51 C. Lee, W. Yang and R. G. Parr, *Phys. Rev. B: Condens. Matter Mater. Phys.*, 1988, **37**, 785–789.
52 T. Yanai, D. P. Tew and N. C. Handy, *Chem. Phys. Lett.*, 2004, **393**, 51–57.
53 C. Hättig and F. Weigend, *J. Chem. Phys.*, 2000, **113**, 5154–5161.
54 C. Hättig and A. Köhn, *J. Chem. Phys.*, 2002, **117**, 6939–6951.
55 C. Hättig, *Adv. Quantum Chem.*, 2005, **50**, 37–60.

 This journal is © The Royal Society of Chemistry 2018

56 F. Furche, R. Ahlrichs, C. Hättig, W. Klopper, M. Sierka and F. Weigend, *Wiley Interdiscip. Rev.: Comput. Mol. Sci.*, 2014, **4**, 91–100.
57 B. O. Roos, in *Advances in Chemical Physics: Ab Initio Methods in Quantum Chemistry Part 2*, ed. K. P. Lawley, John Wiley & Sons, Inc., Hoboken, NJ, USA, 1987, vol. 69, pp. 399–445.
58 F. Aquilante, J. Autschbach, R. K. Carlson, L. F. Chibotaru, M. G. Delcey, L. De Vico, I. F. Galván, N. Ferré, L. M. Frutos, L. Gagliardi, M. Garavelli, A. Giussani, C. E. Hoyer, G. Li Manni, H. Lischka, D. Ma, P. Å. Malmqvist, T. Müller, A. Nenov, M. Olivucci, T. B. Pedersen, D. Peng, F. Plasser, B. Pritchard, M. Reiher, I. Rivalta, I. Schapiro, J. Segarra-Martí, M. Stenrup, D. G. Truhlar, L. Ungur, A. Valentini, S. Vancoillie, V. Veryazov, V. P. Vysotskiy, O. Weingart, F. Zapata and R. Lindh, *J. Comput. Chem.*, 2016, **37**, 506–541.
59 J. W. Ponder and F. M. Richards, *J. Comput. Chem.*, 1987, **8**, 1016–1024.
60 L. M. Frutos, T. Andruniów, F. Santoro, N. Ferré and M. Olivucci, *Proc. Natl. Acad. Sci. U. S. A.*, 2007, **104**, 7764–7769.
61 C. García-Iriepa, M. Marazzi, F. Zapata, A. Valentini, D. Sampedro and L. M. Frutos, *J. Phys. Chem. Lett.*, 2013, **4**, 1389–1396.
62 M. Manathunga, X. Yang, H. L. Luk, S. Gozem, L. M. Frutos, A. Valentini, N. Ferrè and M. Olivucci, *J. Chem. Theory Comput.*, 2016, **12**, 839–850.
63 W. C. Swope, H. C. Andersen, P. H. Berens and K. R. Wilson, *J. Chem. Phys.*, 1982, **76**, 637–649.
64 J. C. Tully, *J. Chem. Phys.*, 1990, **93**, 1061–1071.
65 S. Hammes-Schiffer and J. C. Tully, *J. Chem. Phys.*, 1994, **101**, 4657–4667.
66 G. Granucci and M. Persico, *J. Chem. Phys.*, 2007, **126**, 134114.
67 I. Schapiro, M. N. Ryazantsev, L. M. Frutos, N. Ferré, R. Lindh and M. Olivucci, *J. Am. Chem. Soc.*, 2011, **133**, 3354–3364.
68 I. Schapiro and S. Ruhman, *Biochim. Biophys. Acta, Bioenerg.*, 2014, **1837**, 589–597.
69 I. Dokukina and O. Weingart, *Phys. Chem. Chem. Phys.*, 2015, **17**, 25142–25150.
70 K. Sneskov, J. M. H. Olsen, T. Schwabe, C. Hättig, O. Christiansen and J. Kongsted, *Phys. Chem. Chem. Phys.*, 2013, **15**, 7567–7576.
71 C.-M. Suomivuori, A. P. Gamiz-Hernandez, D. Sundholm and V. R. I. Kaila, *Proc. Natl. Acad. Sci. U. S. A.*, 2017, **114**, 7043–7048.
72 T. Shiozaki, C. Woywod and H.-J. Werner, *Phys. Chem. Chem. Phys.*, 2013, **15**, 262–269.
73 M. Schreiber, M. R. Silva-Junior, S. P. A. Sauer and W. Thiel, *J. Chem. Phys.*, 2008, **128**, 134110.
74 F. Blomgren and S. Larsson, *J. Comput. Chem.*, 2005, **26**, 738–742.
75 M. Wanko, M. Hoffmann, P. Strodel, A. Koslowski, W. Thiel, F. Neese, T. Frauenheim and M. Elstner, *J. Phys. Chem. B*, 2005, **109**, 3606–3615.
76 O. Valsson, C. Angeli and C. Filippi, *Phys. Chem. Chem. Phys.*, 2012, **14**, 11015–11020.
77 Y. Hontani, M. Marazzi, K. Stehfest, T. Mathes, I. H. M. van Stokkum, M. Elstner, P. Hegemann and J. T. M. Kennis, *Sci. Rep.*, 2017, **7**, 7217.
78 A. Strambi, B. Durbeej, N. Ferré, M. Olivucci and J. Michl, *Proc. Natl. Acad. Sci. U. S. A.*, 2010, **107**, 21322–21326.
79 P. Altoè, A. Cembran, M. Olivucci and M. Garavelli, *Proc. Natl. Acad. Sci. U. S. A.*, 2010, **107**, 20172–20177.
80 A. Warshel, *Nature*, 1976, **260**, 619–621.

81 A. Warshel and N. Barboy, *J. Am. Chem. Soc.*, 1982, **104**, 1469–1476.
82 I. Schapiro, O. Weingart and V. Buss, *J. Am. Chem. Soc.*, 2009, **131**, 16–17.
83 D. Polli, P. Altoè, O. Weingart, K. M. Spillane, C. Manzoni, D. Brida, G. Tomasello, G. Orlandi, P. Kukura, R. A. Mathies, M. Garavelli and G. Cerullo, *Nature*, 2010, **467**, 440–443.
84 D. Polli, O. Weingart, D. Brida, E. Poli, M. Maiuri, K. M. Spillane, A. Bottoni, P. Kukura, R. A. Mathies, G. Cerullo and M. Garavelli, *Angew. Chem., Int. Ed.*, 2014, **53**, 2504–2507.
85 J. Y. Jung, A. R. Choi, Y. K. Lee, H. K. Lee and K. H. Jung, *FEBS Lett.*, 2008, **582**, 1679–1684.

 This journal is © The Royal Society of Chemistry 2018

Faraday Discussions

DISCUSSIONS

Light induced charge and energy transport in nucleic acids and proteins: general discussion

Amitabha Chattopadhyay, Richard Cogdell, Carlos E. Crespo-Hernández, Ankona Datta, Arijit De, Stefan Haacke, Mahesh Hariharan, John Helliwell, Ashley Hughes, Roberto Improta, Mike Jones, Joshy Joseph, Tolga Karsili, Bern Kohler, Retheesh Krishnan, Anvy Kuriakose, Mahil L, Dimitra Markovitsi, Himani Medhi, Ganga Periyasamy, P. I. Pradeepkumar, Priyadarshi Roy Chowdhury, Manas Sarangi, Igor Schapiro, Gebhard F. X. Schertler, Ilme Schlichting, Javier Segarra-Martí, Rajaram Swaminathan, Vishnu V, Rienk van Grondelle, Ravi Kumar Venkatraman, Ravindra Venkatramani and Anthony Watts

DOI: 10.1039/C8FD90004C

Amitabha Chattopadhyay opened the discussion of the paper by Gebhard F. X. Schertler: Do you have an idea of the timescale of the dynamics of these water molecules in GPCR structures?

Gebhard F. X. Schertler replied: This depends on the water molecules that you are referring to. The water molecules close to the retinal will respond very fast. From looking at the time resolved measurements with bacteriorhodopsin, we know that it is faster than 16 nanoseconds. I expect it to be on the picosecond timescale. Water molecules in the hydrogen bond network connecting to the G protein binding site will rearrange in the same time frame as the conformational changes. In rhodopsin this ranges from microseconds to milliseconds.

Richard Cogdell remarked: I agree that water molecules are important. Why do you think you would use water in this space rather than a larger amino acid? Water gives you a much greater structural flexibility and a faster timescale. Do you think that is a reasonable way to think about it?

Gebhard F. X. Schertler responded: As I pointed out, water molecules can have many roles inside a protein. I focused on the highly occupied water sites. This gives little information about the dynamics and exchange rates of the water molecules. It also does not mean that less ordered water molecules inside the protein are not important and so they could exist if the space allows it. I think you are implying that water molecules are important for the transition between conformational states and their dynamics, which I think is a reasonable assumption.

Ankona Datta asked: You mentioned that the water molecules in the ligand binding sites provide some stiffness; could you elaborate on that? Do you mean that water is providing structure to the active sites?

Gebhard F. X. Schertler answered: In the figures that I have included in the manuscript it is clear that some water molecules in the ligand binding site exhibit a structural role, but I am also convinced that some can play a role in the chemical reaction mechanism in the active site. There are programs that are used today in structure based drug discovery to evaluate and predict the role of water molecules in ligand binding sites.[1]

1 A. Bortolato, B. G. Tehan, R. T. Smith and J. S. Mason, *Methods Mol. Biol.*, 2018, **1705**, 207–232.

Rienk van Grondelle queried: I was wondering about the role of the water molecules in these systems. For instance, is there any evidence of charge transfer in the excited system?

Gebhard F. X. Schertler responded: At this moment all of our measurements are long after the charge separation, which happens upon the photo-excitation that puts the molecule in the exited state molecule landscape. In order to observe this, free electron laser measurements would be ideal, but we have not yet carried out these ultrafast measurements. Theory and spectroscopy puts these events between 100 and 500 fs.

Stefan Haacke commented: There has been a lot of work on dehydrated membranes but at the time it was not known which water molecules were removed by dehydration. Can you speculate what dehydration does to the molecules?

Gebhard F. X. Schertler replied: Dehydration clearly influences the ability of the protein to reach its full active conformation. This has been shown by neutron scattering and other methods for retinal proteins and a number of other membrane proteins. The exact mechanism of this observation is not completely understood. In my opinion even the removal of a sufficient number of exchangeable surface water molecules could be enough for the stabilisation of the active conformation to be inhibited. In my opinion, in my work it is less likely that the internal water molecules were affected as they would have required more severe dehydration which would have likely led to protein denaturation, which was not observed in these experiments.

Dimitra Markovitsi enquired: How do you expect the behaviour of the water molecules to change as you go from an ordered crystal to a macroscopically less ordered system, as may be the case for a biological system?

Gebhard F. X. Schertler answered: The outer hydration shell of the protein will increase a little bit, however one must consider that the protein contacts in a crystal only affect a small part of the surface. I expect the internal water molecule dynamics to be very similar.

Anthony Watts remarked: GPCRs are very mobile and going from R to R* traverses many energy minima. Are you suggesting that there are different water networks for each minimum? You have said on page 5 of your paper that this implies just two types of networks.

Gebhard F. X. Schertler replied: This is a very important question. I am implying that the clearly distinct water mediated hydrogen bond networks in the ground state and the fully activated *meta* II state make an important contribution to the minimal energy of these states in the energy landscape of the reaction. It is also likely that other intermediate states see important contributions to their minimal states from similar hydrogen-bond networks. It is important to mention here that not only our observed water molecules but also less ordered, exchangeable water molecules could contribute to the stabilization of an intermediate. In addition, we believe that the exchangeability of water molecules in the hydrogen-bond networks is important for the rates of the transitions between states and is therefore important for shaping the energy landscape of GPCRs.

Mike Jones asked: How tightly locked in are these water molecules? Can you exchange them with D_2O and do you see any effects from that?

Gebhard F. X. Schertler responded: Many water molecules inside a protein are exchangeable, with different rates observed. However, I have not done any experiments using D_2O.

John Helliwell communicated: With reference to the sentence beginning "After photo-activation, we can..." in the sub-section "Water molecules are key..."; a quantification of the atoms' positional errors is needed in order to statistically judge their dynamical movement. A web tool for such a calculation based on the Cruickshank diffraction precision index (dpi), *i.e.* when full matrix inversion isn't viable due to the atomic resolution diffraction data not being measurable, is available at http://cluster.physics.iisc.ernet.in/dpi/.[1]

1 K. S. D. Kumar, M. Gurusaran, S. N. Satheesh, P. Radha, S. Pavithra, K. P. S. Thulaa Tharshan, J. R. Helliwell and K. Sekar, *J. Appl. Crystallogr.*, 2015, **48**, 939–942.

Gebhard F. X. Schertler communicated in reply: This is a very good suggestion.

John Helliwell communicated: In the conclusions in your paper you have mentioned that you will be able to analyse key water interactions in "much more detail". Since the diffraction resolution of the crystals is intrinsically at a limit does "much more detail" refer to time-resolution? What time-resolution improvement is still needed for this important molecular system?

Gebhard F. X. Schertler communicated in response: What I mean is mainly time resolution, as in the bacteriorhodopsin case we have already reached 1.5 Å in spatial resolution. The next step is to carry out experiments with the FEL in the picosecond and subpicosecond time range. This will first reveal the role of the water molecules in the very first intermediate after isomerization. The sub-femtosecond measurements can even access the exited state of the protein and the transition state and formation of the first intermediate. I speculate that in all of these states we will see changes in the closest water molecules. I see them as part of the active site.

Priyadarshi Roy Chowdhury communicated: What are the effects of varying the timescale on the dynamics of water molecules in the photo transduction of retinal proteins?

Gebhard F. X. Schertler communicated in reply: As pointed out in the paper, there are different classes of water molecules. Some are in the orthosteric binding site of ligands, influencing the mobility and selectivity of the binding pocket. The second class of waters can be directly involved in enzymatic mechanisms, as pointed out in the paper. Both classes will respond very fast to photo-activation changes. The third class of water molecules between the allosteric binding site and the ligand binding site are important for the conformational change and for shaping the energy landscape that defines the change between inactive and active states. These water molecules will respond more slowly, often within microseconds to milliseconds, coinciding with the conformational change dynamics.

Priyadarshi Roy Chowdhury further communicated: In the case of a biological system, what is the most convenient method to study the fluctuation behaviour of water molecules?

Gebhard F. X. Schertler communicated in response: Time-resolved difference Fourier transform infrared methods have been widely used. The broadness of the infrared peaks is a serious limitation. New possibilities carrying out fast FEL measurements at room temperature are an amazing new tool that enable us to learn more about the dynamics, at least of the highly ordered water molecules in a structure.

Tolga Karsili opened the discussion of the paper by Arijit De: For Venus, the returned electronic absorption and emission spectra exhibit a more intense emission peak in comparison to the absorption peak, which is in obvious contrast to what is observed for wild type-GFP. The authors attribute this observation to a more stable anionic form of the chromophore (*cf.* the protonated form) which results in a greater initial Boltzmann population in the anionic form in the electronic ground state. Although plausible, the differing local protein environment in wild type-GFP *vs.* Venus may lead to modifications in the energetics of the bright and dark vertically excited states of the chromophore, thereby modulating the intensities observed in the absorption spectrum. Are the authors able to rule this out?

Arijit De replied: We have not ruled this possibility out. The schematic of the PES that we qualitatively proposed (Fig. 11) describes only one of the three possibilities (see Section 3.3.4), which is based on inverting the relative ground state Boltzmann population in the two forms (protonated form and neutral form) of the chromophore in the two proteins (Venus *vs.* wt-GFP) due to perturbation by local environment. It is certainly possible that the same local environment tunes the

Frank–Condon factor of the chromophores in these two proteins, leading to very different absorption intensities. In addition, there could also exist very different rates of non-radiative relaxation leading to differences in fluorescence intensities; we have suggested this as an alternate mechanism (see the paragraph beginning "(2) There is a second..." in Section 3.3.4).

Tolga Karsili asked: The authors propose a slower rate for ESIPT in the chromophore of Venus in comparison to that in wild type-GFP. Are the authors able to quantity this by a kinetic isotope effect? Such a study may also reveal if the rate is limited by a simple thermal motion over the barrier or H-atom tunneling through the barrier.

Arijit De answered: Yes, that is a good suggestion, thank you. In fact, the very first work on ESPT in wt-GFP carried out by Boxer *et al.*[1] discussed similar isotope effect experiments.

1 M. Chattoraj, B. A. King, G. U. Bublitz and S. G. Boxer, *Proc. Natl. Acad. Sci. U. S. A.*, 1996, **93**, 8362–8367.

Bern Kohler remarked: I would like to hear more about your reasons for ruling out Raman scattering from the solvent (water) as the source of the weak peak at 462 nm seen in the emission spectrum. A peak at 462 nm is labeled in the emission spectrum recorded at an excitation wavelength of 400 nm in the middle panel of Fig. 6, but although the text states that this feature is also seen when exciting at 374 nm, the only peak in the left panel of Fig. 6 between 400 and 500 nm is clearly shifted to a shorter wavelength, and thus appears to maintain the same constant frequency offset of approximately 3400 cm^{-1} relative to the excitation wavelength expected for Raman scattering by water. Raman peaks are, of course, difficult to detect in the presence of strong emission, but the weak absorbance of your sample at 374 and 400 nm coupled with the visible noise in the emission spectra in Fig. 6, and especially in Fig. 9, suggest that the emission is sufficiently weak for the water Raman band to be detectable. You write that you rule out Raman scattering based on the observation of a nanosecond lifetime, but such a response could arise from fluorescent impurities and could be present alongside the prompt scattering that originates from the Raman signal.

Arijit De replied: It is not the case that the Raman contribution is ruled out but, in addition to Raman scattering, there is also a (weak) fluorescence contribution leading to a longer lifetime component, as Raman scattering would yield a very short lifetime which is on the order of the instrument response,[1] as we discussed in our paper. We have further pursued this issue recently. Fig. 1 below shows the fluorescence/scattering spectra of Venus (left panel) and pure buffer (middle panel); in the third panel we show the fluorescence spectrum after subtraction of the scattering spectrum (followed by normalizing the intensity). It is clear that the Raman peak shifts with excitation wavelength, which is contributed to by water ($\sim$3400 cm^{-1}). However, in addition to the Raman peak, there is broad fluorescence in this wavelength region. The fluorescence (462 nm) excitation spectrum (Fig. 9, left panel), as well as the UV-Vis absorption spectrum (Fig. 8, right panel), confirms that there exists an excitation/emission pair (400 nm/462 nm), as we discussed in our

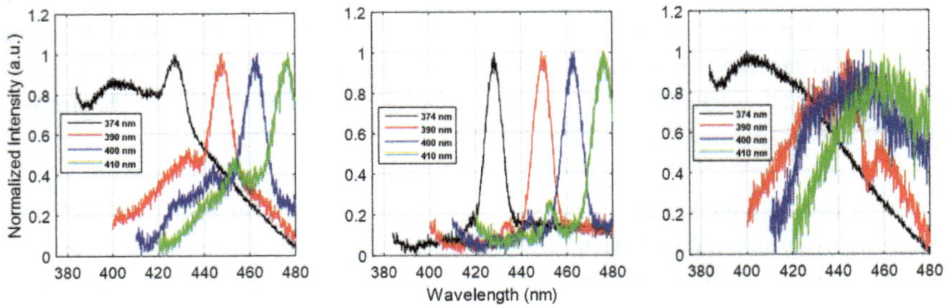

Fig. 1 Fluorescence/scattering spectra of Venus (left panel) and pure buffer (middle panel). Right panel: Fluorescence spectrum after subtraction of the scattering spectrum (followed by normalizing the intensity).

paper. Now, in order to estimate the fluorescence lifetime (at 462 nm), the correct approach would have been to excite the protein with a ~400 nm laser diode instead of a 374 nm laser diode. Quite interestingly, the 9.74 ns component is most likely due to emission from tryptophan[2] which absorbs over a broad region with a peak at ~280 nm (the 374 nm laser diode excites the red-edge of the tryptophan absorption). In addition, as shown in the right panel of the attached figure, there is also an emission at ~400 nm (apart from the absorption at ~400 nm), which most likely arises from a broad absorption with a peak at ~350 nm (inset in Fig. 7, left panel). These findings further complicate the assignment of the electronic states, which is what we are presently working on.

1 Y. Silori, S. Dey and A. K. De, *Chem. Phys. Lett.*, 2018, **693**, 222–226.
2 E. Gudgin, R. Lopez-Delgado and W. R. Ware, *Can. J. Chem.*, 1981, **59**, 1037–1044.

Roberto Improta asked: Can you please give some additional explanation about the origin of protein-backbone fluorescence? Is it necessarily associated with CT states?

Arijit De† answered: The origin of the protein-backbone fluorescence is believed to be due to the excitation of a lone pair of electrons on the carbonyl oxygen of the C=O group that is present in the peptide bond, according to Guptasarma and coworkers and numerous other groups (*e.g.* Kaminski and coworkers, Dobson and coworkers). It is thought that this lone pair can jump across to other hydrogen bonded peptide bonds (in protein secondary structures), leading to their substantial delocalization and a consequent long-wavelength excitation (peaking at ~387 nm but extending up to well past 400 nm) and long wavelength emission (peaking, sometimes with multiple peaks, in a protein secondary structure content-dependent fashion between 425 and 550 nm). Recently, some evidence was reported by the groups of Muschol and Uversky and colleagues in 2017, which suggested that some amino acids can also exhibit this fluorescence. It is considered likely that this lone pair can also exhibit fluorescence through hydrogen bonding to water (*i.e.* the solvent surrounding proteins), albeit with a much poorer intensity than what is seen with proteins. Thus, some (though not all, *e.g.* not proline) amino acids display this fluorescence, although they possess no peptide bonds, because of the behavior of the carbonyl oxygen and the presumed exchange of the electrons in the lone pair with the water molecules with which they are hydrogen-bonded. The extinction coefficients and quantum yields, however, are very low. There is a possibility that the mechanism also involves some form of excited state proton transfer, as demonstrated by Kaminksi and colleagues in 2015. Furthermore, Venkatramani has suggested in a theoretical treatment that such a low intensity fluorescence could also involve charge transfer (DOI: 10.1039/c7fd00203c). However, neither work has excluded the possibility of the charge transfer (whether with or without associated proton transfer) involving the lone pair of electrons on the carbonyl oxygen of the peptide bond, as originally suggested by Guptasarma and coworkers.[1,2]

1 A. Shukla, S. Mukherjee, S. Sharma, V. Agrawal, K. V. Radha Kishan and P. Guptasarma, *Arch. Biochem. Biophys.*, 2004, **428**, 144–153.
2 P. Guptasarma, *Arch. Biochem. Biophys.*, 2008, **478**(2), 127–129.

Anthony Watts queried: Are you sure that your protein is pure? The small impurities shown in Fig. 2 and 3 may be the origin of the low intensity fluorescence.

Arijit De† responded: The two minor impurities seen in Fig. 2 are known contaminants of Imidazole Metal Affinity Chromatographic (IMAC) protein purification; the contaminant of ~66 kDa size is the protein known as ARNA and the ~35 kDa contaminant is the protein known as SlyD. Both proteins are typically present at less than 2 percent of the total population, and often at less than 1 percent. They are both standard proteins with no known special spectroscopic characteristics other than those that are seen in all proteins. In Fig. 2, the minor contaminants consist of higher molecular weight (*e.g.* aggregates) as well as lower molecular weight (*e.g.* degraded) species of a hydrodynamic volume, which differs to that of the major species (Venus) undergoing purification. Typically, contaminants showing up in SDS-PAGE are removed upon gel filtration chromatography because the probability of two proteins of different polypeptide mass having the same hydrodynamic volume is negligible, except where one (or both) of them happens to be multimeric, with coincident multimeric molecular weights. In this case, the Venus protein is monomeric and the other two proteins are also monomers. Therefore, as gel filtration chromatography was performed after SDS-PAGE electrophoretic evaluation of Venus, there is no chance of either of the two contaminating proteins observed during electrophoresis also happening to contaminate Venus by co-eluting with Venus during chromatography. As the spectroscopy experiments were performed after chromatographic purification which would remove these contaminants (and also because the contaminants themselves have no scope for displaying a Venus-like fluorescence) there is no likelihood whatsoever of them being responsible for the low intensity fluorescence. However, one possible additional fact needs to be considered, and discarded (if necessary), before this question is fully answered, and this relates to the possibility of blue (peptide bond-derived) fluorescence. All proteins exhibit this fluorescence, with different quantum yields, and so if Venus displays this

† This response was prepared in collaboration with one of the authors of the paper, Purnananda Guptasarma, Department of Biological Sciences, Indian Institute of Science Education and Research (IISER) Mohali.

fluorescence, so would every other protein, irrespective of whether or not it contains any aromatic or other (*e.g.* GFP-like) chromophore/fluorophore. From this viewpoint, any/all contaminants would exhibit low intensity fluorescence in proportion to their presence and quantum yields. Once again, as the contaminants are present in low amounts, any contribution to blue fluorescence would be correspondingly low and, therefore, the low intensity fluorescence would still be largely derived from Venus, even if it was due to blue fluorescence from peptide bonds, because Venus is the major component of the population.

Retheesh Krishnan remarked: It seems that for Venus the 462 nm peak exhibits a lifetime of 9.74 ns. This lifetime seems to be too high for the argument you put forward, namely ESPT. However a further examination of Table 3 shows that the average reported lifetime is 1.48 ns. Can you please comment on this?

Arijit De replied: This 9.74 ns component is most likely due to emission from tryptophan[1] which is present at the 57th position in the amino acid sequence (see Section 2.1 in the paper).

1 E. Gudgin, R. Lopez-Delgado and W. R. Ware, *Can. J. Chem.*, 1981, **59**, 1037–1044.

Ravi Kumar Venkatraman noted: It is quite interesting that the protein's environment can selectively stabilize the protonated or deprotonated form of the chromophore embedded in it. As you propose that the relative intensities of dual fluorescence arise from the Boltzmann distribution of populations, would you expect that a temperature dependent experiment would reveal the mechanism?

Arijit De answered: I don't think that a temperature dependent study would reveal this as the required elevated temperature would be so high that the protein would unfold (or even degrade).

Arijit De opened the discussion of the paper by Stefan Haacke: (1) In 1991, Shank, Mathies and co-workers[1] showed (using pump–probe spectroscopy) that the initial isomerization (from rhodopsin to photorhodopsin) was complete after 200 fs. Later on, in 2005 Mathies and co-workers[2] showed (using FSRS) that initial 200 fs isomerization proceeds *via* a conical intersection and that the C–H HOOP (and not the C=C torsion) drives this rapid internal conversion. Subsequently, in 2010 Cerullo, Mathies and co-workers[3] demonstrated (using pump-probe spectroscopy) that the fast internal conversion takes place *via* the C=C torsion (and not the C–H HOOP), which is contradictory to the HOOP driven model. Is there any general consensus on how to identify the reaction coordinate (C–H HOOP *vs.* C=C torsion)?

(2) If you were to carry out a time–frequency analysis of this oscillatory data (along the waiting/population time), do you think you would see high frequency oscillations evolving into low frequency ones, thus revealing mode–mode coupling?

1 R. W. Schoenlein, L. A. Peteanu, R. A. Mathies and C. V. Shank, *Science*, 1991, **254**, 412–415.
2 P. Kukura, D. W. McCamant, S. Yoon, D. B. Wandschneider and R. A. Mathies, *Science*, 2005, **310**, 1006–1009.
3 D. Polli, P. Altoè, O. Weingart, K. M. Spillane, C. Manzoni, D. Brida, G. Tomasello, G. Orlandi, P. Kukura, R. A. Mathies, M. Garavelli and G. Cerullo, *Nature*, 2010, **467**, 440.

Stefan Haacke replied: (1) For rhodopsin, the group of Richard Mathies (UC Berkeley) in collaboration with M. Olivucci and I. Schapiro, have very recently shown[1] that the phase of the HOOP of the hydrogens attached to C11 and C12 is critical to the isomerisation quantum yield. This was demonstrated by a decrease in the quantum yield when one of the hydrogens was deuterated, leading to a phase shift and thus to an unfavourable arrival time at the conical intersection. I believe that this experimental evidence, clearly supported by QM/MM modelling, establishes the HOOP mode to be the driving mode that provides the swing to the torsional motion.

(2) Such a time-dependent frequency analysis, for example using a sliding window FFT, is difficult to perform on the present data since the low-frequency vibrational modes are critically damped and we do not observe any more than 2 or 3 periods of oscillations. I would therefore refrain from carrying out such an analysis on the present data of ASR.

1 C. Schnedermann, X. Yang, M. Liebel, K. M. Spillane, J. Lugtenburg, I. Fernández, A. Valentini, I. Schapiro, M. Olivucci, P. Kukura and R. A. Mathies, *Nat. Chem.*, 2018, DOI: 10.1038/s41557-018-0014-y.

Roberto Improta asked: How do you explain the effect of the point mutations you discussed on the excited state lifetime?

Stefan Haacke responded: As I indicated in the talk, two possible explanations can be invoked, but neither can explain the trends we observe in particular for the dramatic shortening of the excited state lifetime in L83Q. The response of the protein environment to the charge translocation on the PSBR builds up a large excited state difference dipole moment, with the positive end pointing towards the ionone ring. We know from the blue-shift of the S0–S1 transition in L83Q that the negative end of the Q dipole points towards the Schiff base, and thus rather opposes a repulsive interaction on the excited state, and even more so on the torsional motion, because the latter is predicted to increase the difference dipole moment.[1]

Another explanation relates to the mixing of S1 and S2, which, in the case of them being degenerate, was predicted to slow down the isomerisation.[2,3] According to those studies, when we extrapolate their findings to AT PSBR in solution, the stabilisation of the SB positive charge due to Q83 would lead to a slowing down of the reaction rather than to the shortened excited state lifetimes that we have observed. Therefore, our understanding is clearly incomplete at this point.

1 F. Melaccio, N. Calimet, I. Schapiro, A. Valentini, M. Cecchini and M. Olivucci, *J. Phys. Chem. Lett.*, 2016, **7**, 2563–2567.
2 B. Demoulin, S. F. Altavilla, I. Rivalta and M. Garavelli, *J. Phys. Chem. Lett.*, 2017, **8**, 4407–4412.
3 M. Manathunga, X. Yang, Y. Orozco-Gonzalez and M. Olivucci, *J. Phys. Chem. Lett.*, 2017, **8**, 5222–5227.

Tolga Karsili remarked: Upon point mutation, the modifications of the gradients (and thus the topographies) of the potential energy surfaces associated with *trans-cis* isomerization may also lead to modifications in the local topography around the conical intersections associated with such an isomerization. As a result, do the authors expect point mutation to have an effect on the branching into the *cis* and *trans* ground state minima following internal conversion at these conical intersections?

Stefan Haacke replied: Yes, we do expect changes, and it would be very interesting to look into this. The method that we used to determine the isomerization quantum yield in the wt, as explained by A. Cheminal *et al.*,[1] will be applied to the mutants as well. It should be noted that we have found remarkably low quantum yields of below 20% for the wt. More interestingly, it was found that the slower AT → 13C isomerization has a higher quantum yield than the reverse process. This is required in order to make light-adaptation happen, as is needed for the biological function of the protein, which is a light-intensity sensor that controls light-level dependent protein expression.

1 A. Cheminal, J. Léonard, S-Y. Kim, K-H. Jung, H. Kandori and S. Haacke, *Phys. Chem. Chem. Phys.*, 2015, **17**, 25429–25439.

Rienk van Grondelle queried: Why is the role of vibrations ignored in the explanations that you have shown? I could imagine some vibrations in the few hundred per cm range coupled to the isomerization.

Stefan Haacke answered: I agree that these low-frequency vibrations are related to the isomerization reaction, and we can even show in the case of the most rapid reaction that they are launched by the latter. Therefore, they are a consequence of the ultrafast speed of the reaction which, in turn, for reasons that we don't quite understand for now, is enabled by specific mutations, in particular for L83Q.

Igor Schapiro noted: The most blue shifting mutant that you have found for *Anabaena* sensory rhodopsin is L83Q. This corresponds to the L105Q mutant in a related retinal protein that we have studied: proteorhodopsin. In proteorhodopsin there are two variants that have different amino acids at position 105 because the two variants are present at different depths of the ocean. Therefore, it seems like nature has selected the same mutant to adapt the bacteria in the ocean.

It is interesting that you have found an effect of this mutant on the excited state lifetime. I'm wondering, have you looked at the effect of the long lived intermediates? Is there any effect in this mutant on proton pumping?

Stefan Haacke responded: This is indeed an interesting question, and I thank you for pointing out the analogy with proteorhodopsin.

We have not studied the long-lived intermediates thus far. I will ask my partners (Prof. K. H. Jung and H. Kandori) whether they have any information on these for L83Q.

Ravindra Venkatramani opened the discussion of the paper by Ganga Periyasamy: I have two questions regarding your calculations of charge transport in peptidomimetics.

(1) Have you considered the influence of peptide conformational change/dynamics on their charge transport properties?

(2) In your calculations of the hole transfer kinetics you neglect reorganization energy effects and justify this by arguing that the peptide geometry does not change significantly with the charge state. However, the dominant

contribution to reorganization energy should come from the outer-sphere component which includes the solvent and this cannot be neglected. Can you comment on how you could include the outer sphere reorganization energy effects and how this might influence the trends in the charge transport kinetics for your peptidomimetics?

Ganga Periyasamy replied: (1) Yes, we have indeed carried out a conformational search for the neutral and cation geometries and have considered the lowest energy conformer for further studies into hole migration dynamics.

(2) The reorganization energy has been calculated in both gas and PCM media and was found to vary negligibly. Furthermore, the computed structural parameters show an insignificant structural change between the neutral and optimized cation geometries.

We agree that the dominant contribution of reorganization comes from the outer-sphere component, which has not been considered in the dynamics study. The outer-sphere mechanism due to the solvation environment in the hole migration dynamics might alter the timescale, which is not included. Moreover, we have discussed only the relative timescale with respect to unsubstituted ureido peptides and have not commented on the absolute timescale.

P. I. Pradeepkumar asked: What would the effect of the secondary structures (helix/sheet) of proteins be on the charge transfer process?

Ganga Periyasamy answered: The structure decides the hole migration timescale and the presence of weak interactions in the helix and sheet could increase the time scale for hole migration, as we have observed for one of the bent peptide structures in our previous work.[1]

1 S. Joy, V. V. Sureshbabu and G. Periyasamy, *J. Phys. Chem. B*, 2016, **120**, 6469–6478.

Ankona Datta enquired: Could you provide any chemical insights into why hole migration is faster in the halogen substituted chromophores compared with those substituted with nitro groups? Do you observe any trends in hole migration rates for the halogen substituted chromophores?

Ganga Periyasamy responded: The change in electronegativity is inversely proportional to the hole migration time scale, where halogens are highly electronegative compared to NO_2 and benzene substituents. This is clearly reflected in our time scale.

Rienk van Grondelle asked: What is the physical basis that determines this extremely fast timescale?

Ganga Periyasamy replied: The appropriate orbital overlap between donor and acceptor units is the reason for the faster timescale. This is made apparent by the correlation between the timescale and functional groups.

Javier Segarra-Martí remarked: In the paper you report values for the first ionisation potentials of the different systems; did you check the higher-energy doublet states as well? If they are close in energy, they can modulate the coherence formed upon absorption.

Ganga Periyasamy answered: We have computed the higher energy doublet states and found a larger energy difference between the two states. Hence we have considered only one state in our hole migration dynamics.

John Helliwell communicated: With regards to the anticipated usage of the authors' chosen peptide in molecular electronics, I estimate that the speed of an electron in their peptide is $\sim 2 \times 10^5$ m s^{-1}. By contrast, the electron speed in graphene has been measured by Chuang *et al.*[1] who observed an electron velocity at the Dirac point of 1.093×10^6 m s^{-1}, which is the fastest velocity recorded for all known carbon materials. Therefore, what advantages does your peptide have for molecular electronics applications?

1 K-C. Chuang, R. S. Deacon, R. J. Nicholas, K. S. Novoselov and A. K. Geim, *Philos. Trans. R. Soc., A*, 2008, **366**, 237–243.

Ganga Periyasamy communicated in reply: Numerous candidates in different areas have been investigated in order to improve them for their use in molecular electronics. Graphene and peptides are used in two different areas. However they are interlinked, as peptides have been used as functionalizing units for graphene.[1] Our work highlights the preliminary studies into how charge transfer, through various functionalizing units containing peptido-mimetics, can be triggered by an electrochemical and optical pulse, which might be anchored to any metal or graphene surface. Charge transfer in peptidomimetics occurs through the backbone and the peptidomimetics used in this work have the advantage of being directional. Furthermore, the timescale of peptides can be appropriately controlled using functional units and conformers.

1 J. Katoch, S. N. Kim, Z. Kuang, B. L. Farmer, R. R. Naik, S. A. Tatulian and M. Ishigami, *Nano Lett.*, 2012, **12**, 2342–2346.

Priyadarshi Roy Chowdhury communicated: How is the substituent effect on the timescale of hole migration evaluated?

Ganga Periyasamy communicated in reply: The substituent changes the wave function (orbital) patterns, which have been used for following the evolution of the created hole. As a result the timescale of hole migration changes according to the nature of the functional groups at the donor and acceptor ends.

Rienk van Grondelle opened a general discussion of Stefan Haacke, Ganga Periyasamy, Arijit De and Gebhard F. X. Schertler's papers: Can the speakers first tell us what they believe to be the general theme of their presentations? What was the strongest example of a biologically relevant photophysical reaction?

Stefan Haacke responded: In my opinion, a common denominator was the quest to highlight and understand the influence of the protein environment on the photo-reaction.

 This journal is © The Royal Society of Chemistry 2018

Ganga Periyasamy replied: In this session, we have discussed photo physical processes on the ultrafast timescale.

Charge and energy transfer with fast timescales are important and can be affected by various parameters such as the structure, solvent and counter ion. To study these properties, a wide range of experimental and computational methodologies have been used and integrated appropriately. The studies have helped to address many of these faster timescale processes.

There are many examples of photophysical reactions in nature. In my opinion the strongest example is the 'photosystem', where the charge transfer and energy transfer dynamics occur in a controlled and selective manner. Photosystems contain a unique protein, with nature choosing the chlorophyll structure and dynamics according to its environment in order to obtain the most efficient charge transfer dynamics.

Arijit De answered: All four of us talked about the dynamics of chromophores inside proteins upon light absorption and how the local protein environment fine-tunes these dynamics. I think the strongest example of a biologically relevant photophysical reaction is photosynthetic light harvesting.

Rienk van Grondelle asked Stefan Haacke, Arijit De and Gebhard F. X. Schertler: How does the environment dynamically control the speed of a reaction? In your papers you observe very different speeds and coherences. There are big differences between the different systems; could this be due to different dynamic control by the environment?

Stefan Haacke answered: As I have discussed in my answer to a previous question, there are two explanations for how the protein environment may modify the excited state potential energy surface (PES) in such a way that the excited state lifetime is shortened. The basic picture we have in mind is that a larger negative slope of the PES will better accelerate the wavepacket, thus removing it faster from the FC zone and promoting it faster along the reaction coordinate towards the intersection with the photoproduct ground state. However, on the basis of the standard models of retinal-chromophore interactions (*e.g.* dipole-charge), an opposite trend would be expected, *i.e.* a longer excited state lifetime for L83Q and the other mutants.

One may now speculate about steric effects, like the mutations leading to a pre-twist of the AT retinal in the ground state, which would accelerate the reaction. Clearly, we need to obtain more structural information about the mutants.

Arijit De replied: It certainly depends on the (dynamic) coupling of the chromophores to their (immediate) environment, but I have a different view on the control of the speed of energy transfer. I think there is no simple answer to it as we simply do not know what biological systems have been trying to optimize over the course of evolution. It may be the case that photosynthetic organisms have optimized something else (*e.g.* fast adaptation to changing sunlight flux, which is more crucial for their survival) by tuning the local environment around the chromophores (pigments), and so the difference in the speed of energy transfer is just a by-product of this process. In other words, the environment evolved in such a way not to control the speed of energy transfer but to control something else.

Gebhard F. X. Schertler replied: For retinal proteins we are only just beginning to understand how retinal isomerization is coupled to the earliest changes in the protein. It is clear that the protein environment is the source of the increased quantum yield and stereoselectivity of the retinal isomerization. From analysing the Raman spectroscopy results, it seems very likely that particular vibrational modes of the chromophore are essential for both effects. We also have observed using dynamic X-ray scattering and time resolved crystallography that correlated motions are induced very early in the protein. It is, however, unclear whether these are either directly coupled to the first conformational changes or are a mechanism for dissipating the surplus energy from the absorbed photon to prevent damage.

Mahesh Hariharan returned to the discussion of the paper by Gebhard F. X. Schertler: You showed a couple of proteins where you had isolated and clustered water molecules within the protein pockets. Do you see any structural generalisation in the protein pocket that allows for single water molecules *vs.* clusters of water molecules?

Gebhard F. X. Schertler responded: There are a number of bioinformatics programs which are already used in drug discovery that predict water molecules and their energy states.[1] These predictions have already been utilized in drug discovery. At this point I have not yet included these bioinformatic approaches but I have instead only relied on experimental observations of the X-ray densities.

1 A. Bortolato, B. G. Tehan, R. T. Smith and J. S. Mason, *Methods Mol. Biol.*, 2018, **1705**, 207–232.

Richard Cogdell remarked: I would like to ask about water clusters as well. We have been looking at the high resolution structures of proteins; on the surfaces we sometimes find pentamers of water that interact/propagate. When you have looked at similar structures have you seen any correlation between what amino acids you observe and whether or not you observe pentamers?

Gebhard F. X. Schertler replied: We have not specifically looked at the arrangement of water molecules on the surfaces of proteins. Our study focuses on the highly occupied, well-ordered molecules inside the protein that have clear electron densities, as shown in the figures. There are always water molecules on the surfaces of the proteins, and some get added towards the end of the refinement of the X-ray structure. Their significance is hard to judge as they are very dependent on the refinement strategy, but I certainly believe that more specific water clusters on the surfaces of proteins exist.

Anthony Watts addressed Gebhard F. X. Schertler and Stefan Haacke: How does the dielectric constant affect each of your systems, and how can it be accommodated in each system?

Gebhard F. X. Schertler answered: This is an excellent question that is not addressable by structural biology at the moment, and which can only be answered using simulations and calculations, which are now more enabled by fast electron laser measurements. Dielectric constants are to some extent bulk properties

 This journal is © The Royal Society of Chemistry 2018

which are hard to define for a single protein molecule. It is even more difficult for membrane proteins, where we have to consider the local dielectric constants of polar groups embedded in the lipid bilayer and the hydrophobic membrane protein itself.

Stefan Haacke replied: The dielectric constant is a central quantity for the photoisomerization of the PSBR (Protonated Schiff Base of Retinal) since the first action of light absorption is a large charge translocation leading to a dipole moment change of approximately 20 D. However, the relevant dielectric constant to be considered is not a global one, but that of the amino acids forming the retinal binding pocket. We have to think here of a dielectric constant that is heterogeneous since it varies strongly on an atomic scale, reflecting the hydrophobic and hydrophilic character of the side chains, and of course the nearby H_2O molecules. It should be noted that the binding pocket is hydrophilic close to the Schiff base and hydrophobic at the opposite ionone ring end of the PSBR. This can be calculated or mapped, but what is really important is the high frequency limit, in the 10 THz range, since this is the relevant timescale for the retinal binding pocket being polarized by the charge translocation of the PSBR. I'm not sure that we have any precise information about this quantity. We showed, in collaboration with Majed Chergui (EPFL) in the mid-2000s for bacteriorhodopsin, that the PSBR charge translocation induces an intra-protein Stark effect that can be measured by monitoring the absorption shift of W86.[1,2] Interestingly, the effect persists over timescales of >10 ps, *i.e.* after isomerization, thus indicating a long-lived change of the dielectric constant, which is possibly associated with the rupturing of the H-bonding network or the dislocation of water molecules close to the moving Schiff base end.

1 S. Schenkl, F. van Mourik, G. van der Zwan, S. Haacke and M. Chergui, *Science*, 2005, **309**, 917–920.
2 J. Léonard, E. Portuondo-Campa, A. Cannizzo, F. van Mourik, G. van der Zwan, J. Tittor, S. Haacke and M. Chergui, *Proc. Natl. Acad. Sci. U. S. A.*, 2009, **106**, 7718–7723.

Amitabha Chattopadhyay addressed Arijit De, Gebhard F. X. Schertler and Anthony Watts: I would like to extend Anthony Watts's question a bit more. He is not only talking about a dielectric medium, he is talking about a dielectric gradient across the membrane leaflet experienced by a membrane protein, which is a hallmark of biomembrane organization. In the long run, we need to incorporate this parameter into the system.

Arijit De responded: That is a very important point, thank you. I would like to make a comment that the dielectric constant is a bulk property whereas the local electrostatic contribution from the nearby residues on the chromophore is a nano-scale phenomenon. Therefore, the spatial variation of a local electric field, rather than the dielectric constant, is perhaps a more relevant quantity to look at.

Gebhard F. X. Schertler answered: I agree.

Anthony Watts replied: That is correct and thank you for this comment. Some attempts have been made at incorporating the dielectric gradient across and within membranes, but it is incredibly complex; it is not a smooth linear gradient anywhere (protein and membrane landscapes are highly contorted), and it can vary over very short ranges from 2 to 80. Furthermore, such changes are not permanent and vary temporally, on timescales that are wide ranging (fs–ms, or slower). Some of these changes will indeed be coupled to photoinduced conformational changes in receptors, which is the topic of the meeting.

Rienk van Grondelle asked Anthony Watts: We know that there are other more efficient proton pumps out there. Why do you think the efficiency is so different between these fairly homologous proteins?

Anthony Watts answered: My guess is that there are differences in the energetics of charge separation in different systems. There is not much else that can be different. The differences are as a result of the positions and number of water molecules in the proton channel, and the residues inducing the charge separation that leads to H^+ and OH^- production following retinal isomerization. The data that Gebhard F. X. Schertler presented seems to support such a notion.

Rienk van Grondelle returned to the discussion of the paper by Ganga Periyasamy: If this energy landscape is so determined by the interaction with water, then water will play a role in the dynamics. If you were to add water to your model, what do you think would happen?

Ganga Periyasamy answered: The determination of the energy landscape with water can be considered in two ways, either implicitly or explicitly considering water molecules. Implicit solvation might only alter the timescale, while explicit solvation could alter the mechanism, depending upon its position at the donor/acceptor end. The explicit presence of water molecules in our structure requires a detailed study starting from a conformational search, which we are currently working on.

Anthony Watts addressed Stefan Haacke and Ganga Periyasamy: How does your work inform our understanding of photocurrent generation by photoreceptors? Very close analogues can generate very different photocurrents.

Stefan Haacke replied: Perhaps my understanding is wrong, but in my mind we can treat the photoisomerization sub-ps dynamics as being disconnected from the slower micro- and milli-second processes of proton translocation. The time scales are so different that they can be treated as independent. However, the quantum yield of the initial photoisomerization of course modulates the amount of photocurrent for a given incident power.

Ganga Periyasamy responded: The dynamics after excitation might be similar to the hole migration dynamics. However it purely depends on the physical and chemical nature of the photoreceptor. The generation of photocurrent in photoreceptors is influenced by many factors. A minimal change in the stoichiometry or substituent could trigger an inner and outer sphere structure reorganization mechanism. Similar to peptides, I expect that a conformational change will alter the timescale and the direction of photocurrent.

 This journal is © The Royal Society of Chemistry 2018

Ankona Datta opened the discussion of the paper by Rajaram Swaminathan: The changes of the Prochar spectra with pH would be due to the protonation/deprotonation of amino acids. In order to study aggregates you are changing the pH of the sample. The changes observed in the spectra could therefore be due to protonation/deprotonation, rather than aggregation. Can you clarify this point?

Do you see any shifts in your spectra with aggregation or pH changes?

Rajaram Swaminathan answered: HEWL aggregation was monitored with respect to time at a constant pH. Therefore, changes in the ProCharTS due to protonation/deprotonation were ruled out while observing the aggregation kinetics at constant pH. There was no shift in the ProCharTS spectra during HEWL aggregation in the wavelength range of 325–800 nm. There was a shift in the spectra under alkaline conditions, due to the presence of tyrosine, in the wavelength range of 250–325 nm. There were no marked shifts in the ProCharTS spectra with pH.

Dimitra Markovitsi remarked: We have shown that the irradiation of DNA duplexes in the UVA spectral domain, where they exhibit very weak absorption related to charge transfer transitions, induces photochemical reactions.[1,2] Do you expect that, analogous to DNA, peptides may also be damaged when irradiated at the red edge of their absorption spectra? If so, could this property be promising for therapeutical reasons, such as amyloid destruction?

1 S. Mouret, C. Philippe, J. Gracia-Chantegrel, A. Banyasz, S. Karpati, D. Markovitsi and T. Douki, *Org. Biomol. Chem.*, 2010, **8**, 1706–1711.
2 A. Banyasz, I. Vayá, P. Changenet-Barret, T. Gustavsson, T. Douki and D. Markovitsi, *J. Am. Chem.-Soc.*, 2011, **133**, 5163–5165.

Rajaram Swaminathan replied: This is an interesting thought. It is definitely worthwhile to explore photoinduced damage in proteins. We look forward to pursuing these experiments to obtain the answers.

Bern Kohler noted: It can be difficult to distinguish weak absorption by a solute from light scattering by aggregates. What evidence do you have that scattering can be neglected, and did you check that the weak absorbance measured at long wavelengths varies linearly with protein concentration?

Rajaram Swaminathan responded: The poor overlap of the observed spectra with the simulated Rayleigh scattering profiles (Fig. S6 in the ESI†) provides the evidence that the observed spectra are not due to scattering. In order to avoid scattering from aggregates, the samples were centrifuged at 12000*g* for 10 min and the supernatant was used to record the absorption spectra.

With regard to your second question, Fig. S7 in the ESI† shows that the weak absorption measured at long wavelengths (400–700 nm) increases linearly with protein concentration.

Rienk van Grondelle commented: With regard to the proteins that at high concentrations develop these 'charge transfer' bands, what happens to the fluorescence of these proteins under these conditions?

Rajaram Swaminathan answered: The charge transfer bands are emissive. Work relating to the fluorescence features of these protein charge transfer bands is in progress.

Richard Cogdell remarked: Unfolded regions are key for many proteins. Do you think that the small size of the signals you are measuring means that you will be able to use them as a probe for unfolded regions?

Rajaram Swaminathan replied: Yes indeed. This is something that we wish to do. In order to amplify the low absorbance, we intend to increase the pathlength of the observation volume.

Ashley Hughes commented: One could use a 10 cm pathlength, low volume cell here.

Ilme Schlichting asked: What happens in D_2O?

Rajaram Swaminathan replied: We have compared the absorption of lysine solution in both pure water and D_2O. There is no change observed in the ProCharTS absorption intensity in D_2O in comparison with that in water even after 24 h of incubation at room temperature.[1]

1 S. Prasad, PhD thesis, IIT Guwahati, 2017.

Himani Medhi communicated: What is nickel affinity gel (H0537) and what is the role of this material when used in the ProCharTS absorption spectra?

Rajaram Swaminathan communicated in reply: Nickel affinity gel is an immobilized metal-ion affinity chromatography product consisting of a proprietary quadridentate chelate on highly cross-linked beaded agarose charged with nickel (source: https://www.sigmaaldrich.com/). This gel was used to purify the hexa-histidine tagged proteins by affinity chromatography. There is no role of this gel in the ProCharTS absorption spectra.

Mahil L communicated: In aggregated protein is it possible to observe absorption due to charge transfer and due to aromatic amino acids separately? Can you please comment on the intensity of the absorption spectrum due to both charge transfer and due to aromatic amino acids? Why does the aggregated protein exhibit a higher intensity in the absorption spectrum compared to that of the normal one?

Rajaram Swaminathan communicated in response: Yes, we can separately observe the absorption due to charge transfer in aggregated proteins in the wavelength range of 325–800 nm. However, it is not possible to separately observe the absorption arising from aromatic amino acids in the wavelength range of 250–325 nm in the charge transfer spectra. This is due to a significant overlap of the ProCharTS spectrum with the aromatic amino acid absorption spectra in this wavelength region. The intensity of the absorption spectrum (325–800 nm) due to charge transfer is quite low in comparison with that of the

 This journal is © The Royal Society of Chemistry 2018

aromatic amino acids (250–325 nm). This is because aromatic chromophores possess much higher molar extinction coefficients in comparison with ProCharTS. The aggregation of protein creates new intermolecular contacts between the charged side chains of amino acids. This increase in the population of charged side chains that are in close proximity increases the ProCharTS absorption intensity in the aggregate.

Priyadarshi Roy Chowdhury communicated: What challenges do you face during an investigation of the structural aspects using protein charge transfer spectra?

Rajaram Swaminathan communicated in reply: Our work has shown that the changes in the protein charge transfer spectra correlate well with the changes observed in the circular dichroism data. However, at the present moment we are unable to derive specific structural signatures of the protein from the ProCharTS data. One challenge is to predict a low resolution 3D structure of a protein from its ProCharTS spectrum. Work in this direction is under progress in our lab.

Arijit De opened the discussion of the paper by Ravindra Venkatramani: As you have just mentioned, (intermolecular) solvation is important for estimating electron transfer rates, which you have not considered thus far (as this is a gas phase calculation). However, even in the absence of solvent, large (intramolecular) conformational change can also influence the rates of electron transfer, somewhat like I proposed as a model in my talk (DOI: 10.1039/c7fd00187h), by driving one locally excited (LE) state to one (twisted) intramolecular charge transfer ((T)ICT) state. Have you considered any such effect in your (gas phase) calculations?

Ravindra Venkatramani responded: While we have not considered solvation in the present manuscript, we have carried out a detailed analysis of the effect of the protein and solvent environment on the absorption spectra of Lys/Glu amino acids in our previous study.[1] Here, our calculations focus on the photoinduced direct CT transitions in the absorption profile of amino acids with charged sidechains. We have not investigated the subsequent excited state dynamics or relaxation pathways of the electrons in our systems.

1 S. Prasad, I. Mandal, S. Singh, A. Paul, B. Mandal, R. Venkatramani and R. Swaminathan, *Chem. Sci.*, 2017, **8**, 5416–5433.

Manas Sarangi remarked: This sort of charge transfer phenomenon has been very well reported elsewhere in the literature. What makes it special for these charge transport states to absorb over such a large range? How fast is this charge transfer?

Ravindra Venkatramani replied: While CT transitions in metal–ligand complexes and protein backbones have been examined previously, we and our collaborators are the first to report photoinduced CT processes involving charged amino acids.[1] In fact, our present study examines the absorption spectral range for CT transitions in charged amino acid monomers and attributes it to the backbone conformational flexibility of chromophores (Fig. 7). On the other hand, our previous study[1] showed that the interactions between charged amino acid sidechains in a protein can further significantly extend the absorption range of the CT transitions to visible wavelengths. The charge separations we examine are photoinduced vertical transitions, which are virtually instantaneous (on the order of femtoseconds) relative to any nuclear motion.

1 S. Prasad, I. Mandal, S. Singh, A. Paul, B. Mandal, R. Venkatramani and R. Swaminathan, *Chem. Sci.*, 2017, **8**, 5416–5433.

Carlos E. Crespo-Hernández asked: Have you investigated how the level of theory used (*i.e.* other density functionals that can be used to describe charge transfer states such as LC-ωPBEh or wave-based *versus* density-based theories) may affect the excitation energies calculated in this study?

Ravindra Venkatramani answered: We have compared the computed TDDFT/CAM-B3LYP/6-31++G(d) spectra of charged amino acid monomers (Lys cation/Glu anion) with those computed with two other range-corrected hybrid functionals (LC-ωPBE or wB97XD) for multiple amino acid conformations. For Lys, the CAM-B3LYP excitation energies are on average ~1.6 eV lower than those obtained from LC-ωPBE and ~0.5 eV lower than those obtained from wB97XD. For Glu, the CAM-B3LYP excitation energies are on average ~1.4 eV lower than those obtained from LC-wPBE and ~0.5 eV lower than those obtained from wB97XD. The CT excitation energies obtained from all three range-corrected hybrids are much higher (on average ~1.8–3.4 eV higher for Lys and ~2.2–3.4 eV higher for Glu) than those obtained from B3LYP. We have not yet compared the TDDFT results of our paper with high level wavefunction based approaches such as coupled cluster methods.

Mike Jones remarked: You briefly mentioned the fate of the electron and damage. Could the formation of these states actually have a beneficial physiological effect?

Ravindra Venkatramani responded: At present we do not have any computational or experimental data on the lifetime of the charge transfer excitons in our systems or on the probabilities of charge separation and recombination after photoexcitation. However, given that these transitions occur in charged amino acid rich proteins, we can speculate that an irreversible exchange of electrons between the peptide backbone and charged amino acid sidechains to create free charges could alter the surface electrostatics and sidechain interactions. The results of such photoexcitations could either be detrimental (*e.g.* disruption of DNA–protein interactions) or beneficial (*e.g.* hindering pathological aggregation of intrinsically disordered proteins), depending on the protein involved and the biological context. On the other hand, fast recombination of the electron–hole pair could efficiently quench the photoexcitations leading to no perceivable effect on either the protein structure or dynamics.

Joshy Joseph asked: Can you please comment on the fluorescence probability of these systems?

Ravindra Venkatramani replied: Thus far, we have only examined the absorption spectra for charged amino acids and not the fate of the excited electrons and their relaxation pathways. Thus, we are unable to comment on the fluorescence in these systems at present.

Dimitra Markovitsi addressed Rajaram Swaminathan and Ravindra Venkatramani: Does excitation at the red edge of the peptide absorption spectrum give rise to fluorescence? In such a case, a time-resolved study, such as those performed for DNA with UVA excitation,[1,2] could help better characterize the corresponding electronic transitions.

1 A. Banyasz, I. Vayá, P. Changenet-Barret, T. Gustavsson, T. Douki and D. Markovitsi, *J. Am. Chem. Soc.*, 2011, **133**, 5163–5165.
2 A. Banyasz, T. Gustavsson, D. Onidas, P. Changenet-Barret, D. Markovitsi and R. Improta, *Chem.–Eur. J.*, 2013, **19**, 3762–3774.

Rajaram Swaminathan responded: Our initial results show that ProCharTS emits fluorescence upon excitation at the red edge (600 nm). Time-resolved studies would indeed prove useful to better understand these transitions. This work is currently under active progress in our lab.

Ravindra Venkatramani responded: We have not yet computationally examined the relaxation pathways or dynamics of the excited electron in our charged amino acid spectra. However, experimentally exciting at the red-edge of the charged amino acid rich protein spectra to check for fluorescence would be very interesting. Thank you for your helpful suggestion for further characterizing the spectra and for your references to your previous studies on DNA UVA excitation which were relevant to this discussion.

Ravi Kumar Venkatraman commented: It is quite remarkable that one can sample the various conformational degrees of freedom of amino acids in native protein structures using classical-molecular dynamics simulations (c-MDS) and then measure their inhomogeneity in the charge-transfer (CT) absorption spectra using time-dependent density functional theory calculations. However, my question is regarding the computational methodology that was adopted in your paper. Generally one calculates the vertical excitation using the same level of theory for both the ground and excited electronic states. In this paper, the ground state structures were obtained from either c-MDS or X-ray crystal structures. Would you expect that this methodology will modify your results?

It is well known that many functionals fail to predict the CT energies owing to over-stabilization of the ground state structure.[1] Therefore, predicting the ground state structure is critical. To mitigate this problem we have adopted a methodology in our recent paper,[2] where instead of sampling intra-molecular conformational degrees of freedom we have sampled intermolecular (solute–solvent) conformational degrees of freedom. The low frequency conformational degrees of freedom sampled by c-MDS were frozen and a relaxed optimization along other high frequency vibrational modes was carried out using DFT calculations.

1 J. J. Eriksen, S. P. A. Sauer, K. V. Mikkelsen, O. Christiansen, H. J. A. Jensen and J. Kongsted, *Mol. Phys.*, 2013, **111**, 1235–1248.
2 V. R. Kumar, C. Verma and S. Umapathy, *J. Chem. Phys.*, 2016, **144**, 064302.

Ravindra Venkatramani replied: Vertical excitations by definition involve the photoinduced promotion of electrons to excited states at the ground state nuclear geometry of molecules. Thus, the initial and final electronic states of the transitions for our monomer amino acid fragments are computed at a single level of theory (TDDFT/CAM-B3LYP/6-31++G(d)) in the absorption spectra calculations. To sample the ground state nuclear geometries for our monomer amino acid fragments, we use classical molecular dynamics simulations on entire solvated proteins and extract individual amino acid conformations from the simulated structures. This method samples monomer amino acid conformations within an explicit solvated protein environment which is the system of interest in our present study. We do not carry out any additional DFT optimizations on the extracted amino acid fragments, as such calculations will lead to unnatural amino acid conformations in the absence of the protein environment. Constrained DFT optimization strategies such as those employed by Kumar *et al.*[1] also cannot account for the environment of the protein during geometry relaxation. We note that Kumar *et al.*[1] have applied their DFT optimizations to a neutral molecule (benzophenone) in a homogenous solvent environment which can be mimicked with a mixed explicit + implicit solvation model. In contrast, our aim is to examine charged amino acid chromophores which interact with other charged and polar sidechains in the protein as well as with water. In this case, classical molecular dynamics simulations appear to be the best way to sample the ground state amino acid conformations.

With regards to the comment on the TDDFT methodology, we have employed the CAM-B3LYP functional which provides a reasonable description of CT transitions in organic molecules. A discussion of this and accompanying references can be found in the subsection "Critical discussion of modelling assumption and caveats" in the "Methods" section of the present manuscript. We do not perform geometry optimizations using DFT/TDDFT as carried out by Eriksen *et al.*[2] which should reduce errors as the ground state geometries of our chromophores do not suffer from DFT limitations. Furthermore, in the present work we compare relative trends in the absorption profiles for a range of chromophores and not absolute energy gaps, which was the objective of Eriksen *et al.*[2]

1. V. R. Kumar, C. Verma and S. Umapathy, *J. Chem. Phys.*, 2016, **144**, 064302.
2 J. J. Eriksen, S. P.A. Sauer, K. V. Mikkelsen, O. Christiansen, H. J. A. Jensen and J. Kongsted, *Mol. Phys.*, 2013, **111**, 1235–1248.

Anvy Kuriakose remarked: The interactions between lysine and lysine and lysine and glutamate were discussed in the paper, but glutamate-glutamate interactions were never mentioned. I wondered why these were not mentioned.

Ravindra Venkatramani answered: I presume you are referring to the systems considered in our previous study.[1] In that study, Glu–Glu interactions have also been considered and discussed. Please see Fig. 5 (panel e), Fig. 6 (middle row b) and ESI Fig. S17† along with the associated discussion on pages 5426–5427 in that study.

1 S. Prasad, I. Mandal, S. Singh, A. Paul, B. Mandal, R. Venkatramani and R. Swaminathan, *Chem. Sci.*, 2017, **8**, 5416–5433.

Vishnu V communicated: Why do only lysine and glutamate show charge transfer?

Ravindra Venkatramani communicated in reply: Lys and Glu possess charged sidechain groups (NH_3^+/COO^-) which polarize the filled and empty orbitals across the chromophores (Fig. 3 of the present manuscript). Thus, during photoexcitation, electrons are spatially separated from the hole that they leave behind, giving the transitions a backbone–sidechain CT character. Our present study shows that such photoinduced CT transitions are not limited to only Lys/Glu, but are intrinsic properties of all amino acids with charged sidechains. We list such chromophores in Fig. 1 of the present manuscript.

John Helliwell communicated: At physiological pH the authors assume a particular (charged) protonation state for Asp and Glu. For His they rightly consider the two possible protonation states. Due to the local structure environment the pK_a values of Asp and Glu can vary from the values for the free amino acids.[1] Therefore, in addition to His, both protonation states of Asp and Glu must be accounted for. Do the authors plan to extend their calculations to do that?

1 S. J. Fisher, J. Wilkinson, R. H. Henchman and J. R. Helliwell, *Crystallogr. Rev.*, 2009, **15**, 231–259.

Ravindra Venkatramani communicated in reply: I fully agree with Professor Helliwell that the pK_a values of amino acids can deviate significantly within proteins from their free solution values and that Glu/Asp residues may be uncharged. We considered the effect of different charged states on the spectra of Lys and Glu in our previous study[1] and showed that in their uncharged states the lowest energy transitions for these amino acids do not possess backbone–sidechain CT character. We anticipate the same trends to hold for the charged/uncharged states of all of the amino acids considered in the present manuscript including Asp. Thus, choosing the right protonation states for the amino acids sampled from protein folds is very important to quantitatively interpret the spectra arising from proteins rich in charged amino acids. We thank Professor Helliwell for bringing our attention to his paper, which was relevant for this discussion.

1 S. Prasad, I. Mandal, S. Singh, A. Paul, B. Mandal, R. Venkatramani and R. Swaminathan, *Chem. Sci.*, 2017, **8**, 5416–5433.

Tolga Karsili opened the discussion of the paper by Igor Schapiro: Upon comparing the relaxed CASSCF and CASPT2 potential energy profiles in Fig. 2, I noted that the CASPT2 profiles show an initial rise in potential energy followed by a subsequent decline. The inevitable barrier returned from this rise and decline in potential energy may indicate a change in or mixing of the electronic configuration(s). This same profile is however unobserved in the CASSCF relaxed potential energy profiles. Since the QM/MM surface-hopping simulations are based on on-the-fly CASSCF energies, are the authors confident that the obtained timescale for isomerization is not underestimated?

Igor Schapiro answered: There is indeed a small barrier in the case of the relaxed scan that was re-computed using CASPT2. However, the height of this barrier is significantly smaller than the excess energy that becomes available to the retinal chromophore upon relaxation from the Franck–Condon point.

Therefore, I believe there will be no significant difference in the timescale of the isomerization.

Tolga Karsili asked: Is there a future plan to propagate more trajectories in order to obtain better statistics? If so, how will the initial conditions of such trajectories be defined?

Igor Schapiro answered: Yes, we plan to run more trajectories in the future to obtain statistically significant results. Up to now we have computed 10 trajectories (Fig. 2). The initial conditions were generated by computing a long classical trajectory which sampled the chromophore and the protein phase space. Then we took snapshots from this trajectory and used the geometries for the ground state QM/MM trajectories. Those were subsequently used for the excited state QM/MM trajectories after a sampling period of half a picosecond.

John Helliwell communicated: How do the you assess the statistical significance of the structural changes along your calculated trajectories, such as in Fig. 5 in your paper?

Igor Schapiro communicated in reply: We are in the process of running more trajectories to address the statistical relevance of the so called "0 K trajectory" that is presented in the manuscript. The sampling of the initial conditions and the

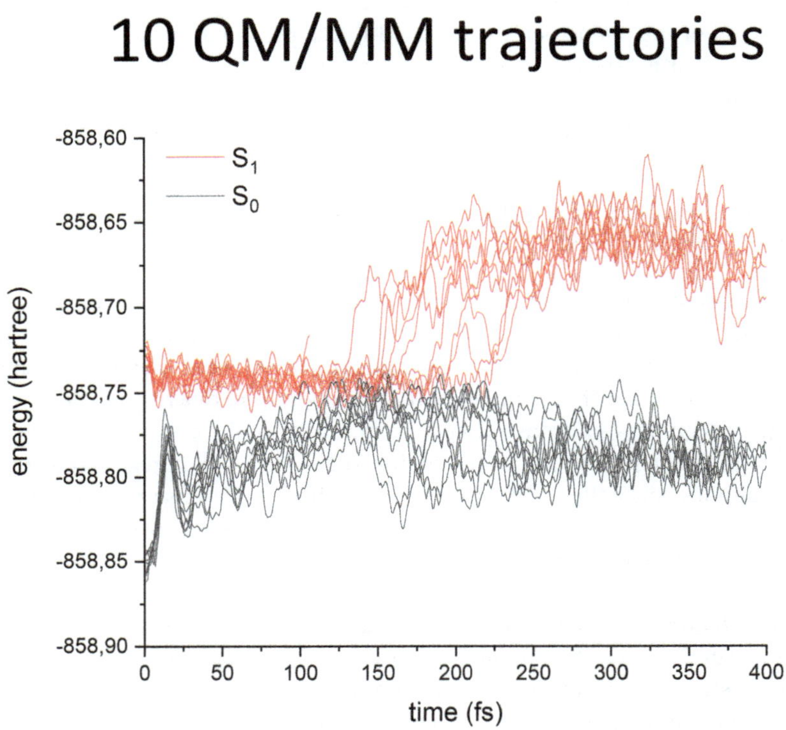

Fig. 2 10 QM/MM trajectories starting from different initial conditions.

 This journal is © The Royal Society of Chemistry 2018

excited state dynamics are computationally demanding. At the moment we have computed 10 trajectories, which can be seen in Fig. 2 in my response to the previous question. These 10 trajectories show the same mechanism of isomerization but have varying excited state lifetimes.

Carlos E. Crespo-Hernández commented: Does the size of the QM region in your QM/MM calculations affect the excited state dynamics predicted for the green-absorbing proteorhodopsin? In other words, what is the minimum number of moieties (*i.e.* retinal chromophore, side chains and water molecules) that needs to be included in the QM region in order to properly describe the dynamics of green-absorbing proteorhodopsin when carrying out QM/MM calculations?

Igor Schapiro responded: The QM region in our setup contains the full retinal chromophore and a part of the lysine sidechain, which together account for more than 50 atoms. This is close to the limit of what we can do in excited state dynamics simulations. Hence, it is not possible to extend the QM region much further. We did however carry out some tests where we extended the QM region for the calculations of the excitation energies. In these single point calculations we did not observe any significant change if neighbouring side chains were included. Hence, we assume that the results will not be affected by a larger QM region.

Stefan Haacke queried: Is it correct to think of one reaction coordinate or should we not consider many low frequency modes and their energetics for promoting isomerization?

Igor Schapiro replied: One reaction coordinate is not enough to describe the photoisomerization. This reaction is triggered by a photon which is absorbed by the retinal. Consequently, the charge distribution of the retinal changes. This leads to a rearrangement of the nuclei and an inversion of the bond length alternation; double bonds become single bonds and single bonds become double bonds in the excited state. The next step is the onset of the rotation of two double bonds according to the bicycle pedal mechanism. Therefore we need to consider multiple reaction coordinates for the description of the isomerization.

Roberto Improta asked: Do you know if any dynamical studies of retinal exploiting TD-DFT exist? What is the picture that they provide with respect to the CASPT2 results?

Igor Schapiro answered: Yes, there are a few studies which use linear response TD-DFT to describe retinal in the excited state. However, this method cannot describe the degeneracy at the conical intersection or in its close proximity. Since the excited state trajectories evolve through such regions of the potential energy surface we have not used TD-DFT. However, there are some promising new methods that use the spin-flip approach or spin-restricted ensemble-DFT. Once they become available we will be happy to look into it.

Anthony Watts queried: Is there any other retinal binding protein that has a similar restricting residue to Tyr200 in your system?

Igor Schapiro responded: I have not checked if this residue is conserved in other retinal proteins. We need to do a sequence alignment and compare it with proteorhodopsin.

Ilme Schlichting remarked: Tyrosine 120 in proteorhodopsin corresponds to tyrosine 185 in bacteriorhodopsin. They are both at a van der Waals distance from the retinal. Should we not see the same effect there?

Igor Schapiro replied: Thank you for pointing this out. We have not looked at bacteriorhodopsin in as much detail as for the green-absorbing proteorhodopsin. However, I would expect that the interaction between retinal and tyrosine 185 is also important for bacteriorhodopsin.

Anvy Kuriakose communicated: I believe you mentioned in your discussion that "You don't do any potential energy minimizations for molecular dynamics; you just run the simulation, lean back and wait for the results." I thought that energy minimization was the first step of molecular dynamics?

Igor Schapiro communicated in reply: You cited me correctly. However, this citation was part of a reply to the question of Stefan Haacke. He asked about the difference between the simulations I have presented and a minimum energy path. In this context I replied that a minimum energy path calculation is a static approach where specific points on the potential energy surface are optimized. In contrast, molecular dynamics simulations describe a propagation in time which does not necessarily describe a minimum energy path.

Furthermore, energy minimization is not the first step in a molecular dynamics simulation. The geometry that is chosen as a starting point for such a simulation can be obtained from an energy minimization, but it doesn't have to be.

Rienk van Grondelle opened a general discussion of the papers by Rajaram Swaminathan, Ravindra Venkatramani and Igor Schapiro, and addressed Igor Schapiro: I am confused about your dynamics. The fast oscillation does not seem to drive the isomerisation as this is much slower. What could drive the isomerisation? How big is the hydrogen displacement? Do you know from Raman analysis *etc.* which frequency is coupled to the optical transition? This could what drives isomerisation. If you look at the PES, there could be another parabola that drives isomerisation.

Igor Schapiro replied: This is a misunderstanding. The fast modulation of the potential energy is due to the bond stretching mode. We see this in the simulation because we calculate only one retinal chromophore at a time. In experiments you measure an ensemble, which means that this fast oscillation is averaged out or not resolved. What we know now is that the photon promotes retinal on the excited state where the stretching mode is activated. This is also called the inversion of the bond length alternation. Basically, the double bonds effectively become single bonds and *vice versa*. After that the C13=C14 bond is significantly stretched and we observe an onset of rotation around this double bond in our simulations. Therefore, my answer to your question is that the driving force of the

 This journal is © The Royal Society of Chemistry 2018

isomerization is the combination of the stretching mode and the torsion of the double bond.

Bern Kohler returned to the discussion of the paper by Ravindra Venkatramani: It is striking that your predicted CT states lie more than 1 eV below the HOMO–LUMO gaps calculated for your compounds. The much lower transition energies that you report for the anionic amino acid monomers make me wonder whether the omission of a solvent in your calculations could lead to spurious CT states. There are well-known pitfalls to predicting CT states using TDDFT calculations.[1,2] What steps did you take to avoid these difficulties, and do the energies change significantly if you include a continuum solvation model?

1 A. Lange and J. M. Herbert, *J. Chem. Theory Comput.*, 2007, **3**, 1680–1690.
2 N. T. Maitra, *J. Phys. Condens. Matter*, 2017, **29**, 423001.

Ravindra Venkatramani responded: While TDDFT with specific exchange–correlation functionals may produce spurious low energy CT transitions, the deficiencies of these functionals in terms of their distance dependence that lead to such failures are well known and noted in both Neepa Maitra's review and John Herbert's study. We have used a range-separated hybrid functional (CAM-B3LYP) which provides a reasonable description of CT states in organic molecules and peptides. Furthermore, we verified that there are no spurious CT transitions in our calculated spectra by comparing computed excitation energies with the lower bound for CT excitation energies given in terms of the ground state energy gap and the exciton binding energy (see equations 4 and 5 in our paper). In general, as I mentioned in my response to Professor Crespo-Hernández's query, range-separated hybrid functionals produce much higher CT excitation energies than those obtained from B3LYP.

The issue of solvation in our systems is a bit involved. The objective of the present study is to compare the 'intrinsic' absorption features of charged amino acids as a function of their conformation within protein folds. In that sense, the absorption spectra calculated in the present manuscript are not directly experimental observables, but rather represent the essential electronic properties of charged amino acids which form the basis of the interpretation of the UV-Vis absorption of proteins that are rich in such residues. The experimentally observable spectra[1] arise when many charged amino acids are tightly packed within a protein fold. Continuum solvation and QM/MM models are not applicable here, since they require a clear demarcation of the chromophore-solvent or QM/MM boundary which is not possible for these systems. The neighbourhood of each charged amino acid chromophore includes other charged protein sidechains and bound water, all of which can undergo changes in their charged states upon photoexcitation. Thus, the only way to assess the effect of the environment is to explicitly increase the size of the QM region and systematically sample the different charged states of the amino acids. We examined this issue in detail in our previous study[1] and found that solvation can lead to two effects that can shift the wavelengths of CT transitions in opposite directions: (1) the screening/polarization effect which tends to localize CT donor/acceptor state wavefunctions and shift the transitions to higher energies, and (2) the electronic coupling effect which tends to delocalize CT donor/acceptor state wavefunctions and shift the CT transitions to lower energies.

We would like to draw attention to the subsection "Critical discussion of modelling assumption and caveats" in the "Methods" section of the present manuscript, and our previous study[1] which discusses the above considerations in more detail along with suitable references.

1 S. Prasad, I. Mandal, S. Singh, A. Paul, B. Mandal, R. Venkatramani and R. Swaminathan, *Chem. Sci.*, 2017, **8**, 5416–5433.

Richard Cogdell returned to the discussion of the paper by Rajaram Swaminathan: You can now easily synthesise the peptides that you design. If you were going to design some peptides, what would you design in order to give you a good testbed of how applicable your spectroscopy is going to be to real life peptides?

Rajaram Swaminathan replied: We could design peptides such that the amino (A) and carboxylate (C) groups of Lys and Glu, respectively, are positioned at chosen distances from each other. In this way we can establish a correlation between the distance between these charged groups and their ProCharTS absorbance.

Additionally, this approach can similarly also be repeated with AA and CC pairs. Furthermore, this can be extended to other charged residues such as His, Arg and Asp. Once all of these were designed, we would be in a better position to establish the spectral signatures in ProCharTS of the structural features that exist among the charged side chains in proteins.

Anthony Watts returned to the discussion of the paper by Ravindra Venkatramani: When you come to other residues, they too can have different pK_a values. Do you consider those as well? Are the dielectric constants related?

Ravindra Venkatramani answered: We appreciate the fact that the pK_a values of amino acids would be sensitive to the dielectric medium in which they are embedded. While the present manuscript deals with only the charged states of various amino acids, we have considered the non-charged states of Lys/Glu in our previous study.[1] Our analysis shows that protonation of Glu or deprotonation of Lys amino acids eliminates the backbone–sidechain CT character of the lowest energy transitions. This is because the polarization of filled and empty orbitals across the chromophores (Fig. 3 in the present manuscript) is lost. A comparison of singly and doubly protonated His in the present study also confirms that the lowest energy excitations are backbone–sidechain CT transitions only for the latter.

1 S. Prasad, I. Mandal, S. Singh, A. Paul, B. Mandal, R. Venkatramani and R. Swaminathan, *Chem. Sci.*, 2017, **8**, 5416–5433.

Conflicts of interest

Gebhard F. X. Schertler is a co-founder of the leadXpro and InterAx biotechnology companies. There are no other conflicts to declare.

 This journal is © The Royal Society of Chemistry 2018

Faraday Discussions

PAPER

Adenine radicals generated in alternating AT duplexes by direct absorption of low-energy UV radiation

Akos Banyasz, [a] Tiia Ketola,[a] Lara Martínez-Fernández, [b] Roberto Improta [*ab] and Dimitra Markovitsi [*a]

Received 24th July 2017, Accepted 12th September 2017
DOI: 10.1039/c7fd00179g

There is increasing evidence that the direct absorption of photons with energies that are lower than the ionization potential of nucleobases may result in oxidative damage to DNA. The present work, which combines nanosecond transient absorption spectroscopy and quantum mechanical calculations, studies this process in alternating adenine–thymine duplexes $(AT)_n$. We show that the one-photon ionization quantum yield of $(AT)_{10}$ at 266 nm (4.66 eV) is $(1.5 \pm 0.3) \times 10^{-3}$. According to our PCM/TD-DFT calculations carried out on model duplexes composed of two base pairs, $(AT)_1$ and $(TA)_1$, simultaneous base pairing and stacking does not induce important changes in the absorption spectra of the adenine radical cation and deprotonated radical. The adenine radicals, thus identified in the time-resolved spectra, disappear with a lifetime of 2.5 ms, giving rise to a reaction product that absorbs at 350 nm. In parallel, the fingerprint of reaction intermediates other than radicals, formed directly from singlet excited states and assigned to AT/TA dimers, is detected at shorter wavelengths. PCM/TD-DFT calculations are carried out to map the pathways leading to such species and to characterize their absorption spectra; we find that, in addition to the path leading to the well-known TA* photoproduct, an AT photo-dimerization path may be operative in duplexes.

Introduction

Low-energy UV radiation may damage DNA in an indirect or direct way. According to the former mechanism, photons are absorbed by molecules that are present close to the nucleic acids, which in turn react with DNA, thus modifying its chemical structure. In the case of direct damage, UV photons are absorbed by nucleobases, with the other components of nucleic acids being transparent in the

[a]*LIDYL, CEA, CNRS, Université Paris-Saclay, F-91191 Gif-sur-Yvette, France. E-mail: robimp@unina.it; dimitra.markovitsi@cea.fr*
[b]*Istituto Biostrutture e Bioimmagini – Consiglio Nazionale delle Ricerche, Via Mezzocannone 16, I-80134 Napoli, Italy*

spectral area. While it is well documented that cyclobutane pyrimidine dimers, the major UV-induced lesions in DNA, may be formed by either mechanism,[1] it is currently considered that oxidative damage is provoked exclusively in an indirect way.[2] The latter conclusion is proposed due to the high vertical ionization potential of nucleobases (>7 eV according to the most recent studies[3]), which is expected to render photo-ionization impossible at wavelengths corresponding to the lowest absorption band which peaks at around 260 nm (Fig. 1).

Surprisingly, a recent study reported the detection of a well-known oxidation marker, 8-oxo-7,8-dihydro-2′deoxyguanosine (8-oxoG) in purified naked genomic DNA irradiated at 254 and 300 nm.[4] More importantly, it was demonstrated that 8-oxoG does not result from reactions involving singlet oxygen, as had been postulated in the past,[5,6] but rather stems from guanine radical cations. This finding made a connection with older spectroscopic studies on short oligonucleotides, which observed electron ejection following UV excitation at wavelengths ranging from 260 to 300 nm.[7–9]

A quantitative characterization of both the ejected electrons and the resulting nucleobase radicals as a function of the base sequence is an important step towards understanding one-photon ionization of DNA by low-energy UV radiation. To this end, a tool of choice is nanosecond absorption spectroscopy, which is already used to study one-photon ionization of bases, nucleosides and nucleotides at 193 nm.[10] Photo-ejected electrons are rapidly hydrated, giving rise to a broad absorption band at around 700 nm,[11] which is detectable on the sub-microsecond time-scale.[7,12] Despite the apparent facility of such measurements, a series of obstacles must be overcome in order to properly characterize these transient species. (i) The excitation intensity should be reduced so as to avoid two-photon ionization of the aqueous solvent and minimize two-photon ionization of DNA. However, the drawback of this is the production of very weak optical signals. (ii) Reactions of ejected electrons with the studied system should be prevented; hydrated electrons are known to attack nucleic acids, forming electron adducts which absorb in the same spectral domain as radicals stemming from photo-ionization.[10,12,13] (iii) The transient absorption spectra associated with photo-ionization should be disentangled from those arising from UV-induced base dimerization reactions.[14,15]

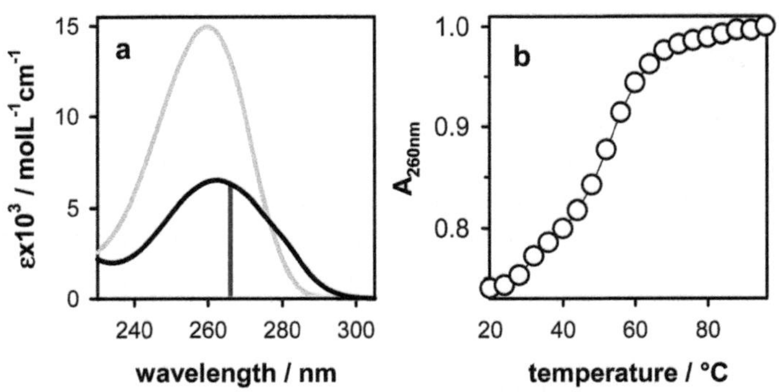

Fig. 1 (a) Absorption spectra of $(AT)_{10}$ (black) and dAMP (grey); (b) typical melting curve obtained for $(AT)_{10}$. The violet line indicates the excitation wavelength, 266 nm, that was used in transient absorption experiments.

 This journal is © The Royal Society of Chemistry 2018

In addition to these experimental difficulties, the effect of the duplex on the radical spectra has never been fully disclosed. Their assignment is based on their similarity with the corresponding mononucleotide spectra.[8,16] On the other hand, radical cations $(A)^{+\bullet}$ and deprotonated radicals $(A\text{-}H6)^{\bullet}$, from which a proton has been abstracted from the position 6 of adenine, have similar spectra,[17] thus rendering their assignment in duplexes not straightforward.

Within this general context, we have recently undertaken an investigation of one-photon ionization of model DNA multimers at 266 nm by controlling the aforementioned experimental problems and combining time-resolved absorption experiments with quantum mechanical calculations. Thus, we determined the spectral and dynamical features of adenine radicals in A-tracts[17] and identified the reaction intermediate leading to adenine dimerization.[15] Moreover, we reported the first UV-induced oxidation study of G-quadruplexes.[18]

The present work focuses on alternating adenine–thymine $(AT)_n$ double-stranded structures. Such structures correspond to the so called TATA boxes encountered in the human genome, which play a crucial role in gene transcription.[19] Experimentally, we study duplexes with 20 base pairs, abbreviated as $(AT)_{10}$, using transient absorption spectroscopy with excitation at 266 nm, near the absorption maximum (Fig. 1a). Ejected electrons are quantified with a time-resolution of 30 ns and disappear without interacting with the duplexes. Subsequently, we follow the fate of adenine radicals. In parallel, we detect the fingerprints of the dimerization reactions. Such reactions are known to occur on the singlet excited state following 254 nm excitation of AT strands,[20–23] leading to the formation of various photoproducts whose nature is still a matter of debate.[24]

Our experimental results are interpreted in the light of quantum mechanical calculations (TD-DFT). As the structural arrangement of the bases in $(AT)_n$ duplexes may be different for the AT and TA steps,[25] we studied both sequences, using the two model duplexes $(AT)_1$ and $(TA)_1$ shown in Fig. 2. We provide the absorption spectra of the base radicals within these duplexes. Moreover, following a preliminary study on single stranded dinucleotides,[26] we map, for the first time, different reaction paths leading to adenine–thymine dimerization within double-stranded structures and compute the absorption spectra of the main reaction intermediates.

Methodological details

Spectroscopic setups

Steady-state absorption spectra were recorded using a Lambda 900 (Perkin-Elmer) spectrophotometer.

Time-resolved experiments were performed using a home-made nanosecond flash photolysis setup. The excitation source was the fourth harmonic of a Nd:YAG laser (Spectra-Physics, Quanta Ray) and the excitation frequency was 0.2 Hz. The energy of the exciting pulse was measured using a NIST traceable pyroelectric sensor (OPHIR Nova2/PE25). The excited area at the surface of the sample was 0.06 cm^2, with the optical path-length on the excitation side being 0.1 cm.

The analysing beam (150 W Xe-arc lamp, Applied Photophysics) passed through the sample (optical length 1 cm) at a right angle with respect to the exciting beam, was dispersed in a SPEX 270M monochromator and detected by a Hamamatsu R928 photomultiplier. The signal from the photomultiplier was

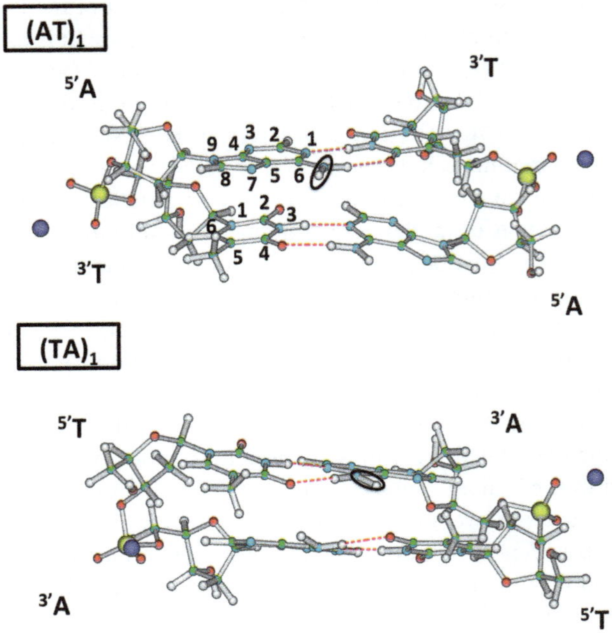

Fig. 2 Schematic drawing of the two computational models, denoted by $(AT)_1$ and $(TA)_1$, used to account for AT and TA steps in double-stranded structures. The black frames highlight the positions where deprotonation occurs.

recorded using a Lecroy Waverunner 6050 oscilloscope. Detection on the sub-microsecond time-scale was achieved by intensifying the emission intensity of the Xe-arc lamp *via* an electric discharge. Transient absorption spectra were recorded using a wavelength-by-wavelength approach. Typically, at each wavelength, a series of three successive signals, resulting from 50–150 laser shots each, were recorded; if judged to be reproducible they were averaged to increase the signal-to-noise ratio.

Fast shutters were placed in the path of both the laser and lamp beams in order to minimize sample exposure to both light sources.

Oligonucleotide handling

Oligonucleotides were purchased from Eurogentec Europe; they were purified by HPLC and their quality was tested by carrying out MALDI-TOF measurements. They were dissolved in phosphate buffer (0.15 mol L^{-1} NaH_2PO_4, 0.15 mol L^{-1} Na_2HPO_4) using ultrapure water delivered by a MILLIPORE (Milli-Q Synthesis) system. The pH of the buffer solutions was adjusted to 7 by addition of a concentrated NaOH solution. The duplexes were prepared in a dry bath (Eppendorf-ThermoStatplus); 2 mL of concentrated oligomer solution was heated to 96 °C for 5 min, then cooled to the melting point (cooling time: 1 h), where the temperature was maintained for a few minutes; subsequently, the solution was cooled to 4 °C (cooling time: 2 h) and incubated overnight. A typical melting curve is shown in Fig. 1b.

For the detection of hydrated electrons on the sub-μs scale, 2 mL of solution, contained in a 1 cm × 1 cm quartz cell, was argon saturated and mildly stirred

 This journal is © The Royal Society of Chemistry 2018

during the experiment; fresh samples were used for each excitation intensity. For recording transient signals at longer times, both quartz cells containing 3 mL of solution and a flow system allowing the circulation of 40 mL of solution were used. In all cases, the optical density on the excitation side was 0.25 ± 0.02 over 0.1 cm, corresponding to a duplex concentration of 9.6×10^{-5} mol L^{-1}. At the maximum excitation intensity (3.3 MW cm^{-2}), the concentration of ejected electrons was 1.2×10^{-6} mol L^{-1}, which was significantly lower than the duplex concentration.

During the experiments the temperature was maintained at 23 ± 0.5 °C.

Theoretical methods

The starting geometry of our model duplexes was taken from the X-ray structure 1DN9.pdb,[25] in which the AT and TA steps exhibit different stacking arrangements, with the former showing much better overlap between bases. The ground state minima of the $(AT)_1$ and $(TA)_1$ dimers, their radical species, and possible intermediates and photoproducts were optimized using Density Functional Theory (DFT). The selected functional was M052X,[27,28] due to its good performance for treating stacked systems, and was combined with the 6-31G(d) basis set. The Polarizable Continuum Model (PCM) was used to account for solvent (water) effects.[29,30] We exploited the time dependent version of DFT (TD-DFT) to compute the absorption spectra at the ground state minima and to characterize their excited state Potential Energy Surfaces (PESs). This computational approach has provided satisfactory results in previous studies both for radical spectra[17,18] and for photo-dimerization processes.[15,26,31–33] For a better comparison with the experimental absorption spectra, the computed Vertical Excitation Energies (VAE) were red-shifted by 0.6 eV and convoluted using a Gaussian with a full width at half maximum (fwhm) of 0.6 eV. The applied shift allows the VAE computed for the lowest bright $\pi\pi^*$ excited state of adenine at this level of theory to coincide with the experimental maximum of the absorption band; this accounts for the absence of vibronic and thermal effects, besides the possible inaccuracies of our electronic calculations.

Results and discussion

The photo-ionization process

Fig. 3a shows the signal of the hydrated electron obtained for an argon saturated $(AT)_{10}$ solution and the aqueous buffered solvent alone at the maximum excitation intensity. As the solvent signal is practically flat, we can conclude that the decay observed for the $(AT)_{10}$ solution corresponds to electrons stemming only from the duplex. The decay can be fitted with a single exponential function $c + \Delta A_0 \exp(-t/\tau)$ with a time constant of 0.42 μs. This value is independent of the base sequence[17,18] and is also found for the solvent alone at much higher excitation intensities, thus showing that hydrated electrons do not react with the duplex; they are scavenged by the phosphate groups[34] of the buffer at high concentration. The initial concentration of hydrated electrons $[e]_0$ is determined from the ΔA_0 value and the most recent value reported for their molar absorption coefficient (19 700 mol L^{-1} cm^{-1} at 700 nm).[11]

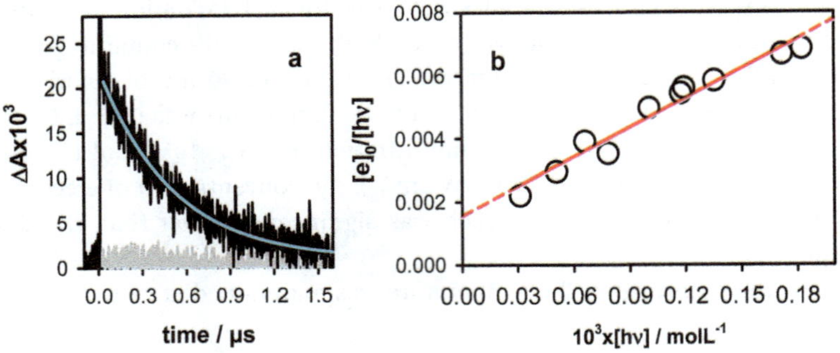

Fig. 3 (a) Transient absorption decay recorded for an argon saturated $(AT)_{10}$ solution at 700 nm (black) fitted with a mono-exponential function (cyan); in grey is the signal obtained for the solvent at the same excitation intensity of 3.3 MW cm^{-2}. (b) Ionization curve of $(AT)_{10}$ obtained by varying the excitation energy; [hν] and $[e]_0$ denote the concentrations of absorbed photons and hydrated electrons at time zero, respectively; the experimental points (circles) are fitted with a linear function (red line).

Although one-photon ionization is the biologically relevant process, laser experiments also provoke two-photon ionization of duplexes. This is probably due to absorption of a second photon during the 5 ns laser pulse by long-lived singlet excited states, charge transfer (CT) states whose lifetimes amount to tens of ps,[35,36] and high energy long-lived mixed (HELM) states surviving on the ns time-scale.[37] The electron concentration $[e]_0$ varies as a function of the concentration of absorbed photons [hν] according to the equation $[e]_0 = \phi_{1h\nu}[h\nu]_{1hn} + \phi_{2hn}[h\nu]^2$, where $\phi_{1h\nu}$ and $\phi_{2h\nu}$ represent, respectively, the yields of the one- and two-photon processes. While the quantum yield $\phi_{1h\nu}$ corresponds to an intrinsic property of the duplex, $\phi_{2h\nu}$ depends also on the experimental conditions.[38]

The so-called ionization curve is obtained by plotting $[e]_0/[h\nu]$ *versus* [hν] and fitting the experimental points with a linear function (Fig. 3b). The slope of the fitted line corresponds to two-photon ionization and the intercept on the vertical axis provides the one-photon ionization quantum yield, $(1.5 \pm 0.3) \times 10^{-3}$. The $\phi_{1h\nu}$ value determined from the plot in Fig. 3b is higher than that derived from previous measurements, such as $(1.0 \pm 0.1) \times 10^{-3}$ which was obtained with a time resolution of 200 ns;[8] we note that, for better comparison, the 2006 $\phi_{1h\nu}$ value was re-evaluated using the molar absorption coefficient (19 700 mol L^{-1} cm^{-1}) reported in 2016.[11] Following a similar re-evaluation, the $\phi_{1h\nu}$ value of the homopolymeric duplex was found to be $(1.3 \pm 0.3) \times 10^{-3}$. Consequently, we can conclude that the one-photon ionization of AT duplexes is not significantly affected by the base sequence.

We note that under the same experimental conditions, using excitation intensities of up to 3.3 MW cm^{-2}, we did not detect any hydrated electrons for neither dA nor dAMP.

Radical spectra in duplexes

The shapes of the spectra recorded for $(AT)_{10}$ between 2 μs, when hydrated electrons have already disappeared, and 0.2 ms do not exhibit any noticeable difference. The spectrum at 0.2 ms is presented in Fig. 4a. It is composed of two bands

 This journal is © The Royal Society of Chemistry 2018

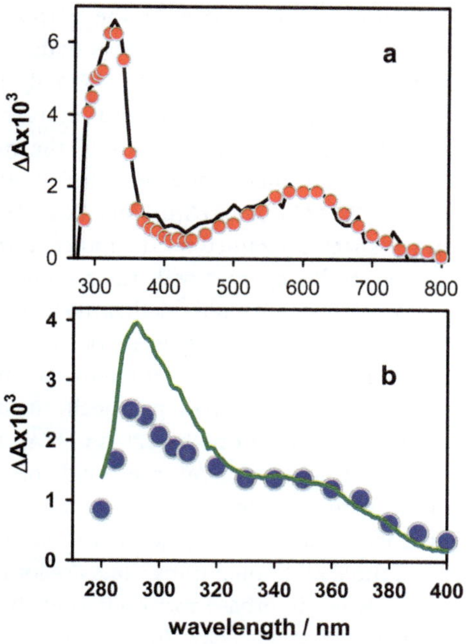

Fig. 4 Time-resolved absorption spectra obtained for an aerated solution of $(AT)_{10}$ at 0.2 ms ((a), red) and 30 ms ((b), blue). Excitation intensity: 3.3 MW cm^{-2}. The black line in (a) corresponds to the spectrum of the deprotonated radical (A-H6)$^\bullet$ determined for dAMP,[13] normalized at 600 nm with the duplex spectrum. The green line in (b) corresponds to the difference between the steady-state spectra of $(AT)_{10}$ recorded before and immediately after irradiation with 200 laser pulses, normalized with the spectrum at 30 ms at 350 nm.

that peak at around 330 and 600 nm. At 30 ms the 600 nm band has disappeared but absorption still persists below 400 nm, characterized by a peak at 300 nm and a shoulder at 350 nm (Fig. 4b). Its origin will be discussed later.

The $(AT)_{10}$ spectrum at 0.2 ms strongly resembles that obtained for dAMP at 300 ns following excitation at 193 nm,[13] also shown in Fig. 4a, which is attributed to the deprotonated adenine radical (A-H6)$^\bullet$. The latter attribution was recently confirmed by comparing the spectra of the dA radical cation and deprotonated radical in low temperature glasses, with the spectra of these transient species calculated using quantum chemistry methods.[17] This is also in agreement with pulse radiolysis studies which report that deprotonation takes place with a time constant of 50 ns.[39] However, in the case of UV-induced processes, the radical reactivity could be different from that reported in pulse radiolysis studies using two-step oxidation of bases and involving bimolecular electron transfer.[39] As a matter of fact, we recently showed[18] that the behaviour of guanine radical cations produced by photo-ionization in G-quadruplexes is not the same as that found for photo-sensitized processes in these systems.[40] Moreover, the similarity between the time-resolved differential spectra of $(AT)_{10}$ and that of the dAMP deprotonated radical observed in Fig. 4a could be fortuitous. This is particularly probable for the blue side of the transient spectrum corresponding to the difference between the radical spectrum and the ground state spectrum (Fig. 1a). As was found for adenine single strands[17] and G-quadruplexes,[18] this part of the differential spectrum is different than that of the corresponding monomer because of the modification of the ground state spectrum due to delocalization of the Frank–Condon excited states over a few bases.[37,41] Finally, the dimeric photoproduct reported by Davies (denoted by TA*),[20] absorbs in the 270–320 nm region. Therefore, we computed the electronic transitions of the radical cations and deprotonated radicals for our two model double stranded structures $(AT)_1$ and $(TA)_1$ (Fig. 2) using the PCM/TD-M052X/6-31G(d)//M052X/6-31G(d) level of theory.

In Fig. 5 we present the spectra obtained for the radical cation $(AT)_1^{+\bullet}$ and the deprotonated radical $[(AT)_1\text{-H6}]^\bullet$, together with their monomeric counterparts computed at the same level of theory. We remark that the inclusion of the adenine radical within the double-stranded structure does not induce any drastic modification in the radical spectra. The positions and relative intensities of the peaks in the UV and visible spectral domains remain practically unchanged. This result is consistent with our prediction that for both $(AT)_1$ and $(TA)_1$ the positive charge is essentially localized on a single base. Yet, the oscillator strength is slightly higher compared to that of the monomer, and a larger number of electronic transitions, albeit extremely weak, are present in the 300–900 nm range. In particular, for the cation, three additional transitions are located at wavelengths longer than 750 nm; they involve interbase ππ* excitations between the adenine bearing the positive charge and other bases, thus giving rise to a slightly more intense red tail in the spectrum. Analogously, several series of interbase nπ* dark excited states, involving transitions from the lone pairs of other bases to the positively charged adenine, fall between the 300 and 600 nm bands. As the radical localization is more pronounced for $[(AT)_1\text{-H6}]^\bullet$ compared to that of the cation, the similarity of its spectrum with that of the monomer is greater. However, some additional interbase ππ* transitions still arise in $[(AT)_1\text{-H6}]^\bullet$ at 480–520 nm, which is higher in energy in comparison with $(AT)_1^{+\bullet}$.

The features computed for our second model double-stranded structure $(TA)_1$ (Fig. 6) are quite similar to those shown in Fig. 5. As for $(AT)_1$, the relative intensities of the bands in the UV and visible regions are not altered with respect to those of the monomeric radicals. The differential ($(TA)_1^{+\bullet}-(TA)_1$ and $[(TA)_1\text{-6H}]^\bullet-(TA)_1$) absorption spectra computed for the $(TA)_1$ radicals are presented in Fig. 6.

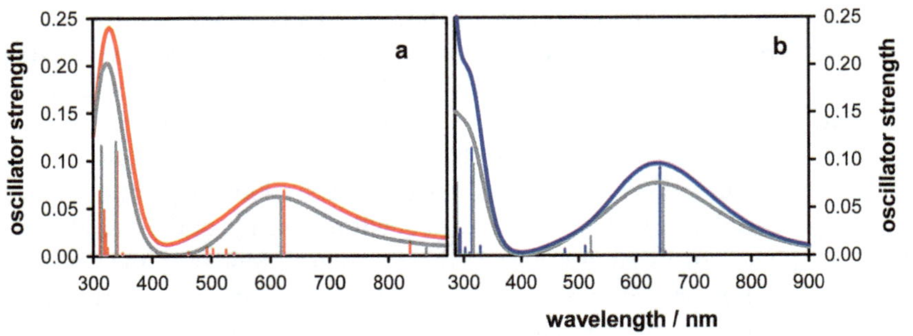

Fig. 5 Absorption spectra computed for the $(AT)_1^{+\bullet}$ radical cation ((a), red) and the $[(AT)_1\text{-6H}]^\bullet$ deprotonated radical ((b), blue). The corresponding spectra of 9-methyl-adenine are shown in grey. The sticks represent individual electronic transitions. Calculated at the PCM/TD-M052X/6-31G(d)//M052X/6-31G(d) level of theory.

 This journal is © The Royal Society of Chemistry 2018

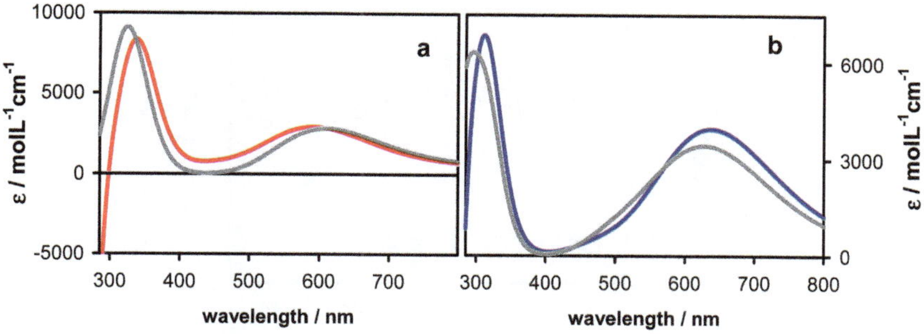

Fig. 6 Differential absorption spectra computed by subtracting the $(TA)_1$ absorption spectrum from the spectrum of the radical cation $(TA)_1^{+\bullet}$ ((a), red) and the deprotonated radical $[(TA)_1\text{-6H}]^\bullet$ ((b), blue). The corresponding spectra of 9-methyl-adenine are shown in grey. Calculated at the PCM/TD-M052X/6-31G(d)//M052X/6-31G(d) level of theory.

In the light of the computed spectra, it appears that deprotonation of the radical cation in the double stranded structures is expected to induce an increase in the transient absorbance. It is also clear that the UV peaks of both radicals in the duplexes are red-shifted compared to those of the corresponding monomeric radicals. Thus, the apparent similarity of the $(AT)_{10}$ spectrum at 0.2 ms to that of the monomeric $(A\text{-H6})^\bullet$ (Fig. 4a) indicates the presence of other absorbing species. Below we explore these aspects in more detail.

Radical dynamics

The time-dependence of the transient absorption signal at 600 nm is shown in Fig. 7. No rise is observed after 2 μs, which is the shortest time at which the radical transient spectra were recorded. This observation, together with the lack of a pronounced red tail in the 2 μs spectrum of $(AT)_{10}$ compared to that of the monomer (Fig. 4a), confirms that, at the probed time-scale, we observe deprotonated radicals. The signal decays on the ms time-scale and can be fitted with a mono-exponential function with a time constant of 2.5 ms. The same time constant is found upon reducing the incident excitation intensity so as to increase the ratio of electrons ejected *via* one-photon ionization *versus* those resulting from two-photon ionization from 0.31 to 0.71. This shows that the radical decay does not depend on the mode of their formation.

Our study on A-tracts reported that base pairing increases the survival probability of the deprotonated adenyl radical.[17] Our present work corroborates this conclusion and shows that not only base pairing but also the base sequence affects the reactivity of this radical; its half time in the alternating duplex $(AT)_{10}$ (2.0 ± 0.1 ms) is clearly shorter than that in the homopolymeric duplex with the same number of base pairs (4 ± 1 ms), but significantly longer than that found for adenine single strands (1.0 ± 0.1 ms). One possible reason for this difference could be the increased rigidity of the homopolymeric duplex, as suggested by molecular dynamics simulations.[42] Another way to reduce the amplitude of conformational motions in the duplex is to decrease the temperature of the aqueous solution. We indeed found that lowering the temperature from 23° to 5 °C lengthens the radical lifetime (Fig. 7b) and increases the initial radical concentration by about 25%. In contrast, argon saturation of the solutions has no

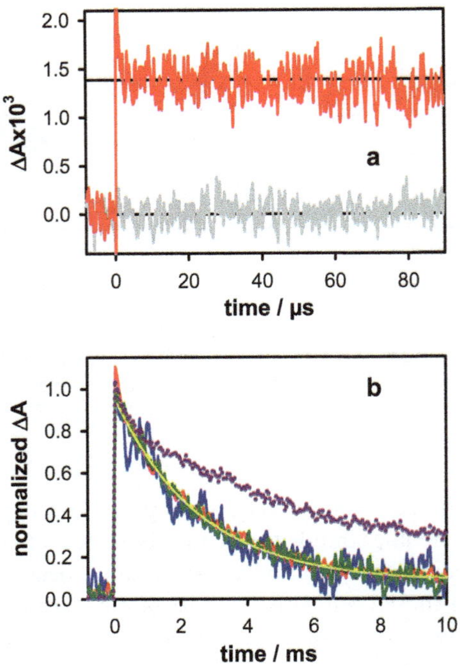

Fig. 7 Transient absorption signals recorded for $(AT)_{10}$ at 600 nm under different experimental conditions: solid lines (23 °C), dots (5 °C); the red, green and blue lines correspond, respectively, to $\phi_{1h\nu}/\phi_{2h\nu}$ ratios of 0.31, 0.63 and 0.71. (a) In grey: the signal of the buffer obtained under the same excitation intensity of 3.3 MW cm^{-2}. (b) In yellow: the fitted mono-exponential function with a time constant of 2.5 ms.

effect on the transient signals. This observation is in line with literature reports that monomeric adenyl radicals do not react with oxygen.[43]

The initial concentration of deprotonated radicals in $(AT)_{10}$ was evaluated using the differential absorbance determined at 2 μs (see an example in Fig. 7a) and the molar absorption coefficient reported for the dAMP deprotonated radical by Candeias and Steenken (1460 mol L^{-1} cm^{-1} at 600 nm). It was found to be 10% higher than the initial concentration of hydrated electrons determined for the same excitation intensity. This difference could either be due to the 10% estimated experimental error or reflect the fact the molar absorption coefficient of the deprotonated radical in double-stranded structures is higher than that of the monomeric analogue (Fig. 5 and 6).

UV-induced absorption below 400 nm

In order to discover the possible contribution of transient species other than adenine radicals to the early time spectra of $(AT)_{10}$ (Fig. 4a), we plotted in Fig. 8 typical transient absorption signals recorded at short wavelengths. All of them contain a plateau that appears at 20–30 ms and persists, at least, until 180 ms, which is the longest time that we can probe under these experimental conditions. We remark that, going from 330 nm to 280 nm, the pattern of the transient signal changes dramatically. The signals below 290 nm exhibit complex kinetics, being composed of a decay followed by a rise. An attempt to make a global fit with two

 This journal is © The Royal Society of Chemistry 2018

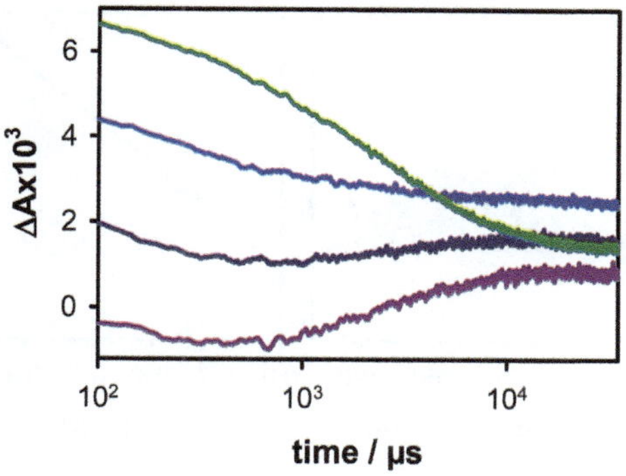

Fig. 8 Transient absorption signals recorded for $(AT)_{10}$ at 280 nm (violet), 285 nm (dark blue), 290 nm (blue) and 330 nm (green); excitation intensity 3.3 MW cm^{-2}.

exponential functions failed, thus indicating the presence of more than two species. Considering the absorption spectra computed for the duplex radicals (Fig. 5b and 6b) and the experimental transient absorption signals in Fig. 8, we can deduce that the UV part of the $(AT)_{10}$ spectra in Fig. 4 contains contributions from reaction intermediates other than radicals and/or their reaction products.

According to the literature, the TA* photoproduct, detected by Davies and identified as diazacyclooctatriene (see below), absorbs in the 280–330 nm region;[20–22,44,45] it is formed in the singlet excited state *via* a reaction intermediate, denoted hereafter as TA*–CB, because of the presence of a cyclobutane ring.[23] The relative concentration of TA*/TA*–CB *versus* that of the adenine radicals is expected to increase at low excitation intensities, for which two-photon ionization is less important (Fig. 3). We found indeed that when the excitation intensity decreases from 3.2 to 1.4 MW cm^{-2}, the intensity ratio of the 300 and 350 nm signals I_{300}/I_{350} increases from 1.7 to 2.0 at 0.2 ms and from 1.2 to 3.0 at 30 ms. This indicates that the absorption of reaction products resulting from the radicals is more intense at 350 nm compared to at 300 nm.

As a second step, we recorded steady-state absorption spectra for the $(AT)_{10}$ solutions before and after their irradiation with laser pulses. A different laser intensity was used for each sample and the number of laser pulses was adjusted so that the accumulated quantity of ejected hydrated electrons per volume to be the same for all the solutions, corresponding to a concentration of 1.7×10^{-6} mol L^{-1}, was the same for all of the solutions. The resulting steady-state differential spectra are shown in Fig. 9a. Despite the important noise due to the small amplitude of the signals, we clearly observe a peak around 295–300 nm, whose intensity increases linearly with the number of absorbed photons (Fig. 9b). In addition, a shoulder is present at around 350 nm, its intensity being roughly the same for all of the samples. Therefore, it appears that the radical reaction products are dominant at this wavelength. Interestingly, the time resolved spectrum at 30 ms also presents a shoulder at 350 nm (Fig. 4b).

From the slope of the linear regression in Fig. 9 we can deduce the equation $\Sigma\Delta\varepsilon_S \times \phi_S = 10.4$, where $\Delta\varepsilon_S$ and ϕ_S represent, respectively, the differential molar absorption coefficients and quantum yields corresponding to the various reaction products formed in the singlet excited state. Considering only the formation of

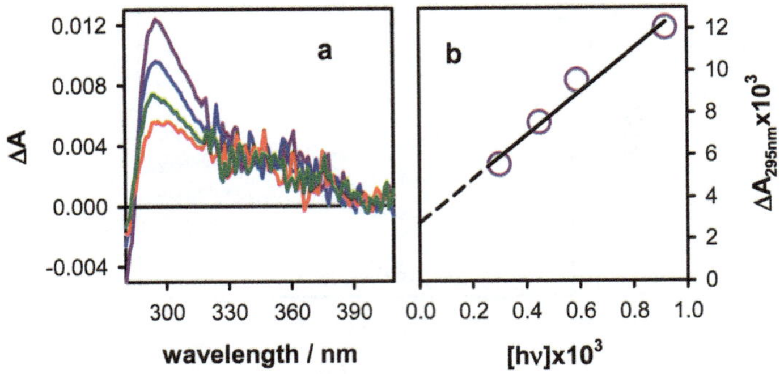

Fig. 9 (a) Steady-state differential absorption spectra obtained for different solutions of $(AT)_{10}$ irradiated by 266 nm laser pulses with intensities of 0.7 MW cm^{-2} (violet), 1.4 MW cm^{-2} (blue), 2.7 MW cm^{-2} (green) and 3.4 MW cm^{-2} (red). The total number of pulses per solution was adjusted so that the accumulated quantity of ejected electrons remained the same. (b) Variation of the maximum differential absorbance at 295 nm (a) as a function of the total concentration of absorbed photons (circles) fitted with a linear function (black line).

TA*, whose quantum yield in the $(AT)_n$ duplexes is *ca.* 10^{-4},[21] a $\Delta\varepsilon_S$ of 10^5 was found. Such a high value, encountered for strongly conjugated systems, has never been found for nucleobase derivatives. This indicates that absorption at 295 nm is due not only to TA*, but also to other photoproducts formed with a higher quantum yield, which is in line with what has been reported for adenine–thymine adducts in calf thymus DNA.[24] However, a comparison of the transient spectrum at 30 ms with the steady-state differential spectrum (Fig. 4b) shows that, for the latter spectrum, the relative intensity of I_{300}/I_{350} is higher. These observations suggest that the chemical species present in the solution evolve within the two hours that are needed to carry out the irradiation and obtain the steady-state spectrum shown in Fig. 4a, which is in line with what is reported in ref. 24.

Base dimerization in duplexes: computational approach

In this section, we first examine the reaction pathways (Fig. 10) of the singlet excited states of our two model duplexes (Fig. 2). We then compute the differential absorption spectra (Fig. 11) of the reaction intermediates, so that we can compare them with our experimental results.

Reaction paths in $(AT)_1$ duplexes. Confirming the results of previous QM studies on $(AT)_1$ model duplexes,[37] the lowest energy excited state in the ground state geometry derives from the interaction of $\pi\pi^*$ states localized on thymines (which are the dominant component) with the A → T charge transfer (CT) state. The partial CT character increases following geometry optimization of S_1, leading to a minimum, (AT_1)-S_1-min, where the two bases are very close to each other. The distance of C4A–C6T is 2.6 Å and the LUMO orbital involved in the S_1 transition exhibits a significant bonding contribution between these two atoms (Fig. 10). A relaxed scan that decreases this distance indeed shows that an almost barrierless path (barrier $\leq$ 0.1 eV) connects the $(AT)_1$-S_1-min to a degeneracy region with S_0. This region is characterized by a C4A–C6T distance of 2.00 Å. Subsequent S_0 geometry optimization leads to a stable minimum, whose energy is only 1.81 eV

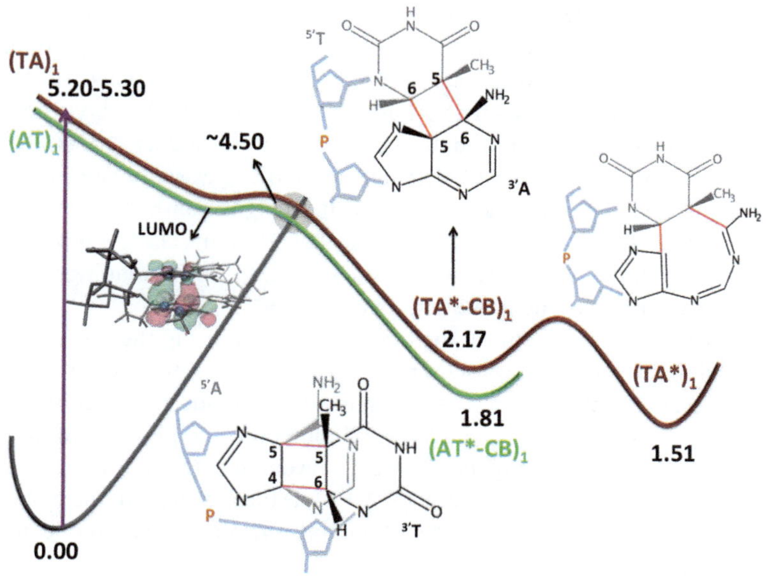

Fig. 10 Excited state pathways and schematic representation of the reaction intermediates (for simplicity we show only the strand involved in photo-dimerization) computed for $(AT)_1$ (green) and $(TA)_1$ (dark red). The energies (in eV) are given with respect to the ground state at the Franck–Condon point.

above the $(AT)_1$-S_0-min (Fig. 10). At this ground state minimum, C5A–C5T are also bonded, producing a cyclobutyl-like photo-adduct $(AT^*–CB)_1$. On the other hand, State-Specific (SS)-PCM/TD-M052X calculations (which provide a more reliable description of electronic transitions with CT character)[46] on dApT indicate that full equilibration of the solvent degrees of freedom leads to the appearance of a sizeable energy barrier (~0.5 eV) in the path connecting the $(AT)_1$-S_1-min and the crossing region.[26] Dynamical solvation effects should be less important inside the duplex. Nonetheless, we expect that our estimate of the energy barrier (≤0.1 eV) is a lower bound to the effective one in the duplex.

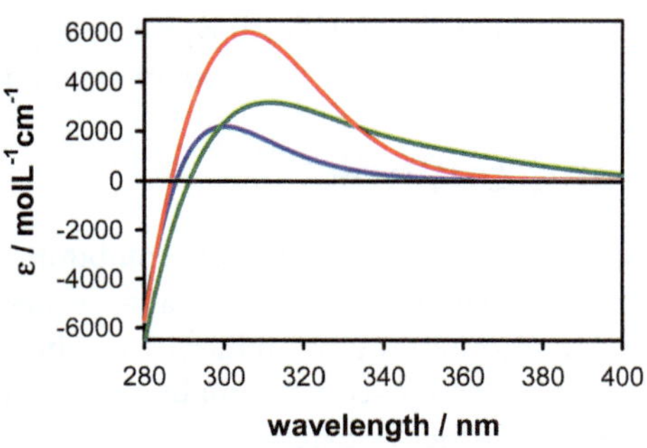

Fig. 11 Computed differential absorption spectra for $(AT^*–CB)_1$ (blue), $(TA^*–CB)_1$ (green) and $(TA^*)_1$ (red). Calculated at the PCM/TD-M052X/6-31G(d)//M052X/6-31G(d) level of theory.

Reaction paths in $(TA)_1$ duplexes. As discussed above, bases are poorly stacked in the TA steps compared to the AT steps. Consequently, S_1 geometry optimization, starting from the ground state minimum $(TA)_1$-S_0-min, predicts the localization of the excitation on a single T base. In order to study the photochemical paths, we have thus first optimized the 'Davies-like' photo-adduct, $(TA^*–CB)_1$, and then, by increasing the C6A–C5T distance, located the excited state minimum related to this $(TA)_1$-S_1-min adduct. $(TA^*–CB)_1$ is significantly less stable than $(AT^*–CB)_1$ and $(TA)_1$-S_1-min less stable than $(AT)_1$-S_1-min. On the other hand, a more stable minimum can be reached from $(TA^*–CB)_1$, corresponding to $(TA^*)_1$, the 1,3-diazacyclooctatriene product isolated for TpdA.[23]

Our calculations thus confirm the possibility of producing the Davies adduct in the duplexes. In addition, we show that, due to the geometry of the TA steps in the duplexes, this path is less effective than in single strands, thus accounting for the decreased TA* yield in the duplexes.[21,24] Moreover, the photochemical path leading to $(AT^*–CB)_1$, involving the AT steps, is favoured over that leading to $(TA^*–CB)_1$. On this ground, we propose that $(AT^*–CB)_1$ is the precursor of the hydration product found to be the most abundant "adenine–thymine" photoproduct in calf thymus DNA.[24]

Absorption spectra of reaction intermediates. We have computed the absorption spectra associated with the three species $(AT^*–CB)_1$, $(TA^*–CB)_1$ and $(TA^*)_1$ located along the potential energy surfaces in Fig. 10. The corresponding differential spectra, derived by subtracting the ground state absorption, are presented in Fig. 11. They peak, respectively, at 300, 311 and 306 nm. Considering the possible inaccuracies of our method and the use of smaller molecular models, all three spectra could be consistent with the transient spectra recorded for $(AT)_{10}$. On the other hand, it should be noted that, due to the peak position and the lack of absorption above 330 nm, the $(AT^*–CB)_1$ spectrum is the closest match to the transient spectrum at 30 ms (Fig. 4b), thus supporting our suggestion that it plays a major role in the photo-dimerization pathways.

The spectra in Fig. 11 allow us to roughly estimate the quantum yield for the formation of dimeric photoproducts. Using an average value of 2500 mol L^{-1} cm^{-1} for the differential molar absorption coefficient at 295 nm, corresponding to the three compounds in Fig. 11, and the equation $\Sigma\Delta\varepsilon_S \times \phi_S = 10.4$ derived from Fig. 9b, we find that the average quantum yield amounts to 4×10^{-3}. The latter value is of the same order of magnitude as that determined for radical formation, thus showing that the probability of base dimerization and oxidative damage is similar at 266 nm. Hopefully, future experiments will determine how these reaction paths vary with the irradiation wavelength.

Conclusions and outlook

The present study on alternating $(AT)_n$ duplexes, combining nanosecond laser photolysis and quantum mechanical calculations, presents new insights, not only into photo-ionization but also into the UV-induced base dimerization processes in these systems. Below, we summarize the main results and highlight open questions that might invite further research.

Our work shows that the direct absorption of single photons at 266 nm with energies that are 3.0 eV lower than the adenine ionization potential (7.7 eV)[3] leads to electron ejection, thus confirming the finding of an older study.[8] The one-

 This journal is © The Royal Society of Chemistry 2018

photon ionization yield ($1.5 \pm 0.3 \times 10^{-3}$), determined with the improved experimental setup used in the present work (time-resolution of 30 ns instead of 200 ns, lower excitation intensity by a factor of 2), is 50% higher than that of the first observation reported a decade ago. Under the same experimental conditions, a quantum yield of $(1.3 \pm 0.3) \times 10^{-3}$ was found for the homopolymeric analogue.[17] It appears that, within the current experimental errors, the one-photon ionization of adenine–thymine duplexes is not significantly affected by the base sequence. The mechanism of electron ejection resulting from the absorption of low-energy photons remains to be discovered. We can speculate that the CT states, which have been shown to be formed in significant yield in AT duplexes,[36,47] are the gateway for electron ejection.

Our calculations show that inclusion of the adenine radical cation and deprotonated adenine radical within double-stranded structures does not induce large modifications in their absorption spectra but does affect the differential absorption spectra because of the collective nature of the Frank–Condon excited states of the "neutral" species. In light of these theoretical predictions, our experimental results lead to the conclusion that (i) deprotonated adenine radicals are produced in equal concentration to that of the hydrated electrons, (ii) the deprotonation of radical cations occurs before 2 μs and (iii) the transient absorption spectra below 330 nm contain the fingerprint of the dimeric reaction intermediates formed in the singlet excited state.

In contrast to one-photon ionization, the survival probability of deprotonated adenine radicals in the duplexes does depend on the base sequence, with their half life being significantly longer in the homopolymeric duplex (4 ± 1 ms) than in the alternating (2.0 ± 0.1 ms) duplex. We have correlated this difference with the larger amplitude of conformational motions in the alternating structure, which could increase the probability of reaching configurations favouring radical reactivity. In order to confirm this hypothesis, both identification of the reaction products and molecular dynamics simulations are needed. However, it is also possible that the radical reaction products in the homopolymeric duplexes are different to those in the alternating duplexes.

The computed reaction pathways on the singlet excited states explained why the formation of the TA* photoproduct, isolated half a century ago, is less probable within double stranded structures.[21] More importantly, analysis of the potential energy surfaces, in combination with the computed and experimental absorption spectra, identified a reaction intermediate, which could be the precursor of the most abundant "adenine–thymine" photo-dimer formed in duplexes.[24]

Finally, our experimental spectra show that adenine radicals give rise to a reaction product characterized by an absorption band that peaks at around 350 nm, that is at a longer wavelength compared to that of other photoproducts. We hope that this spectral identification will help in the determination of its chemical structure.

Acknowledgements

Financial support from the French National Research Agency (ANR-12-BS08-0001-01, OPHID project), the CNRS-CNR PICS project (No. 6827-2015)/Bilateral CNR/

CNRS and the Université Paris-Saclay (Chaire d'Alembert No. 2016-10751) is acknowledged.

References

1 J. Cadet, A. Grand and T. Douki, in *Photoinduced Phenomena in Nucleic Acids II: DNA Fragments and Phenomenological Aspects*, ed. M. Barbatti, A. C. Borin and S. Ullrich, 2015, vol. 356, pp. 249–275.

2 J. Cadet, J. R. Wagner, V. Shafirovich and N. E. Geacintov, *Int. J. Radiat. Biol.*, 2014, **90**, 423–432.

3 C. A. Schroeder, E. Pluharova, R. Seidel, W. P. Schroeder, M. Faubel, P. Slavicek, B. Winter, P. Jungwirth and S. E. Bradforth, *J. Am. Chem. Soc.*, 2015, **137**, 201–209.

4 M. Gomez-Mendoza, A. Banyasz, T. Douki, D. Markovitsi and J. L. Ravanat, *J. Phys. Chem. Lett.*, 2016, **7**, 3945–3948.

5 J. Cadet and P. Vigny, in *Bioorganic Photochemistry*, ed. H. Morrison, John Wiley & Sons, New York, 1990, pp. 1–272.

6 T. Mohammad and H. Morrison, *J. Am. Chem. Soc.*, 1996, **118**, 1221–1222.

7 C. E. Crespo-Hernández and R. Arce, *Photochem. Photobiol.*, 2002, **76**, 259–267.

8 S. Marguet, D. Markovitsi and F. Talbot, *J. Phys. Chem. B*, 2006, **110**, 11037–11039.

9 V. Gabelica, F. Rosu, E. De Pauw, R. Antoine, T. Tabarin, M. Broyer and P. Dugourd, *J. Am. Soc. Mass Spectrom.*, 2007, **18**, 1990–2000.

10 L. P. Candeias and S. Steenken, *J. Phys. Chem.*, 1992, **96**, 937–944.

11 F. Torche and J. L. Marignier, *J. Phys. Chem. B*, 2016, **120**, 7201–7206.

12 R. Arce, *Photochem. Photobiol.*, 1987, **45**, 713–722.

13 L. P. Candeias and S. Steenken, *J. Am. Chem. Soc.*, 1992, **114**, 699–704.

14 S. Marguet and D. Markovitsi, *J. Am. Chem. Soc.*, 2005, **127**, 5780–5781.

15 A. Banyasz, L. Martinez-Fernandez, T. Ketola, A. Muñoz-Losa, L. Esposito, D. Markovitsi and R. Improta, *J. Phys. Chem. Lett.*, 2016, **7**, 2020–2023.

16 Y. Rokhlenko, J. Cadet, N. E. Geacintov and V. Shafirovich, *J. Am. Chem. Soc.*, 2014, **136**, 5956–5962.

17 A. Banyasz, T. Ketola, A. Muñoz-Losa, S. Rishi, A. Adhikary, M. D. Sevilla, L. Martinez-Fernandez, R. Improrta and D. Markovitsi, *J. Phys. Chem. Lett.*, 2016, **7**, 3949–3953.

18 A. Banyasz, *et al.*, *J. Am. Chem. Soc.*, 2017, **139**(30), 10561–10568.

19 Y. Wang, R. C. Jensen and W. E. Stumph, *Nucleic Acids Res.*, 1996, **24**, 3100–3106.

20 S. N. Bose, R. J. H. Davies, S. K. Sethi and J. A. McCloskey, *Science*, 1983, **220**, 723–725.

21 S. N. Bose, S. Kumar, R. J. H. Davies, S. K. Sethi and J. A. McCloskey, *Nucleic Acids Res.*, 1984, **12**, 7929–7947.

22 X. D. Zhao, S. Nadji, J. L. F. Kao and J. S. Taylor, *Nucleic Acids Res.*, 1996, **24**, 1554–1560.

23 R. J. H. Davies, J. F. Malone, Y. Gan, C. J. Cardin, M. P. H. Lee and S. Neidle, *Nucleic Acids Res.*, 2007, **35**, 1048–1053.

24 S. Asgatay, A. Martinez, S. Coantic-Castex, D. Harakat, C. Philippe, T. Douki and P. Clivio, *J. Am. Chem. Soc.*, 2010, **132**, 10260–10261.

 This journal is © The Royal Society of Chemistry 2018

25 C. Yoon, G. G. Prive, D. S. Goodsell and R. E. Dickerson, *Proc. Natl. Acad. Sci. U. S. A.*, 1988, **85**, 6332–6336.

26 L. Martínez-Fernández and R. Improta, *Photochem. Photobiol. Sci.*, 2017, **16**, 1277–1283.

27 Y. Zhao and D. G. Truhlar, *Acc. Chem. Res.*, 2008, **41**, 157–167.

28 Y. Zhao, N. E. Schultz and D. G. Truhlar, *J. Chem. Theory Comput.*, 2006, **2**, 364–382.

29 J. Tomasi, B. Mennucci and R. Cammi, *Chem. Rev.*, 2005, **105**, 2999–3093.

30 S. Miertus, E. Scrocco and J. Tomasi, *Chem. Phys.*, 1981, **55**, 117–129.

31 A. Banyasz, T. Douki, R. Improta, T. Gustavsson, D. Onidas, I. Vayá, M. Perron and D. Markovitsi, *J. Am. Chem. Soc.*, 2012, **134**, 14834–14845.

32 R. Improta, *J. Phys. Chem. B*, 2012, **116**, 14261–14274.

33 L. Esposito, A. Banyasz, T. Douki, M. Perron, D. Markovitsi and R. Improta, *J. Am. Chem. Soc.*, 2014, **136**, 10838–10841.

34 G. V. Buxton, C. L. Greenstock, W. P. Helman and A. B. Ross, *J. Phys. Chem. Ref. Data*, 1988, **17**, 513–886.

35 W. M. Kwok, C. S. Ma and D. L. Phillips, *J. Phys. Chem. B*, 2009, **113**, 11527–11534.

36 Y. Zhang, K. de La Harpe, A. A. Beckstead, R. Improta and B. Kohler, *J. Am. Chem. Soc.*, 2015, **137**, 7059–7062.

37 I. Vayá, J. Brazard, M. Huix-Rotllant, A. Thazhathveetil, F. Lewis, T. Gustavsson, I. Burghardt, R. Improta and D. Markovitsi, *Chem.–Eur. J.*, 2016, **22**, 4904–4914.

38 K. L. Stevenson, G. A. Papandonakis and P. R. LeBreton, *J. Photochem. Photobiol., A*, 2000, **133**, 159–167.

39 K. Kobayashi, *J. Phys. Chem. B*, 2010, **114**, 5600–5604.

40 L. D. Wu, K. H. Liu, J. L. Jie, D. Song and H. M. Su, *J. Am. Chem. Soc.*, 2015, **137**, 259–266.

41 B. Bouvier, T. Gustavsson, D. Markovitsi and P. Millié, *Chem. Phys.*, 2002, **275**, 75–92.

42 B. Bouvier, J. P. Dognon, R. Lavery, D. Markovitsi, P. Millié, D. Onidas and K. Zakrzewska, *J. Phys. Chem. B*, 2003, **107**, 13512–13522.

43 C. von Sonntag, *Free-Radical-Induced DNA damage and its Repair*, Springer-Verlag, Berlin-Heildelberg, 2006.

44 S. Kumar, N. D. Sharma, R. J. H. Davies, D. W. Phillipson and J. A. McCloskey, *Nucleic Acids Res.*, 1987, **15**, 1199–1216.

45 S. Kumar, P. C. Joshi, N. D. Sharma, S. N. Bose, R. J. H. Davies, N. Takeda and J. A. McCloskey, *Nucleic Acids Res.*, 1991, **19**, 2841–2847.

46 R. Improta, V. Barone, G. Scalmani and M. J. Frisch, *J. Chem. Phys.*, 2006, **125**, 054103.

47 D. B. Bucher, B. M. Pilles, T. Carell and W. Zinth, *Proc. Natl. Acad. Sci. U. S. A.*, 2014, **111**, 4369–4374.

Faraday Discussions

PAPER

Photoactivated proton coupled electron transfer in DNA: insights from quantum mechanical calculations†

Lara Martinez-Fernandez[a] and Roberto Improta*[ab]

Received 15th September 2017, Accepted 26th September 2017
DOI: 10.1039/c7fd00195a

The energetics of the two main proton coupled electron transfer processes that could occur in DNA are determined by means of time dependent-DFT calculations, using the M052X functional and the polarizable continuum model to include solvent effect. This approach has been applied to fragments of GC and AT alternated duplexes containing up to 4 base pairs. In GC DNA, intra-strand G → C charge transfer (CT) can trigger inter-strand Proton Transfer (PT) involving the C anion, since the reaction is exothermic and the related energy barrier is <0.3 eV. In AT DNA, the minimum resulting from intra-strand A → T CT is not predicted to undergo inter-strand PT. These conclusions are in full agreement with the indication of time-resolved IR spectra. In both systems, inter-strand CT can be coupled to inter-strand PT, and the resulting minimum, reached after overcoming a moderate energy barrier (<0.4 eV), is close to a crossing region with the ground electronic state, providing a relatively fast non-radiative deactivation route.

1. Introduction

Nucleic acids strongly absorb UltraViolet (UV) light, a process of fundamental biological interest, since it can trigger a cascade of dangerous photochemical events, potentially leading to the damage of the genetic code and to apoptosis.[1–3] For these reasons, in the last decades a huge number of experimental and theoretical papers have been devoted to characterize the photophysics and the photochemistry of DNA, often employing model oligonucleotides.[4–9] Different time-resolved spectral techniques show that the excited state dynamics of oligonucleotides is very complex, characterized by multi-exponential decay with time-constants spanning from the sub-ps up to the ns time-scale,[5,7,8,10] suggesting that several different excited states are involved, as confirmed by the available computational studies.[4,6,9] There is general agreement that the fastest

[a]Consiglio Nazionale delle Ricerche, Istituto di Biostrutture e Bioimmagini, 80136 Naples, Italy. E-mail: robimp@unina.it; Tel: +39 081 2532036

[b]LIDYL, CEA, CNRS, Université Paris-Saclay, F-91191 Gif-sur-Yvette, France

† Electronic supplementary information (ESI) available. See DOI: 10.1039/c7fd00195a

components can be mainly associated to monomer-like decay routes. Isolated nucleobases are able to dissipate the excess UV photon energy by efficient ultrafast internal conversion (IC)[5] and the subsequent picosecond vibrational energy transfer (VET) to water,[6,8] due to the existence of almost barrierless decay paths connecting the Franck–Condon (FC) region with effective Conical Intersection (CoIn) between the lowest energy bright excited state (a $\pi\pi^*$ excitation) with the ground electronic state (S_0).[4,11,12] The assignment of the long-living components (*i.e.* lasting tens to hundreds of picoseconds) has been instead the object of a very lively debate.[13–22] On the one hand, also in isolated nucleobases part of the photoexcited population, probably trapped in 'dark' $n\pi^*$ transitions,[4,23–25] can survive on a >10 ps time-scale. On the other hand, in oligonucleotides, excitations involving multiple bases can play a role.[4,18,26–31] Several studies, both experimental and computational, suggested the participation of charge-separated states,[13,21,32,33] whose weak intensity, however, makes their unambiguous determination rather difficult. Femtosecond time-resolved infrared (fs-TRIR) pump-probe experiments have therefore been extremely useful to identify the nature of these long-living excited states, since vibrational resonances are typically much narrower (and thus more distinct) compared to their electronic counterpart. Recent fs-TRIR experiments have definitively shown that in several different DNA single strands (TpT, dApT, $(dApT)_{10}$, OxoGpdA, $(dT)_{18}$) UV absorption creates, with substantial yield, charge-transfer (CT) states generated by photoinduced electron transfer (ET) between two π-stacked nucleobases.[34–42] The distinctive patterns of vibrational marker bands of these excited states nicely compare with those, predicted from Quantum Mechanical (QM) calculations or measured experimentally by two-photon ionization, of monomeric (anionic/cationic) radicals, *i.e.* not considering partial CT and the interaction within the radical ion pair. Actually, a recent study on $(dApT)_n$ shows that the difference IR spectrum between S_0 and the excited minimum of the CT state is in very good agreement with the experimental one, improving the assignments compared to ones derived from the spectra of the separate cation and anion.[40] There are thus clear indications that in single strands CT states are formed in less than several hundred fs in quantum yields that can reach unity,[37] and subsequently decay by back ET in tens to hundreds of picoseconds.

When going from the single strand to duplex the picture is more complex. On the one hand, it has been shown the CT states formed in the single strands have a much shorter lifetime in the duplex.[35] In a subsequent TRIR study, Zhang *et al.*[41] observed different long-lived excited-state vibrational spectra and lifetimes for two sequence isomers, *i.e.* $(AT)_9$ homo- and alternated duplexes, which contain identical numbers of A·T base pairs but differ in how the base pairs are stacked. This clearly indicates that both base stacking and pairing must play crucial roles in DNA photophysics. By comparing the vibrational spectra of the long-lived excited state to those predicted by ground state DFT calculations, the authors proposed a novel proton-coupled electron transfer (PCET) mechanism that is operative in UV-excited DNA duplexes.[41] As for single-stranded DNA, intra-strand ET between two π-stacked bases creates a pair of opposite charge radical ions in duplexes, and triggers inter-strand proton transfer (PT) whenever the driving force allows (Scheme 1).[41] As a result, UV-excitation generates a distonic ion pair in which the spin and charge are separated on the two complementary strands. PCET has been shown to be operative also in miniduplexes containing only two

 This journal is © The Royal Society of Chemistry 2018

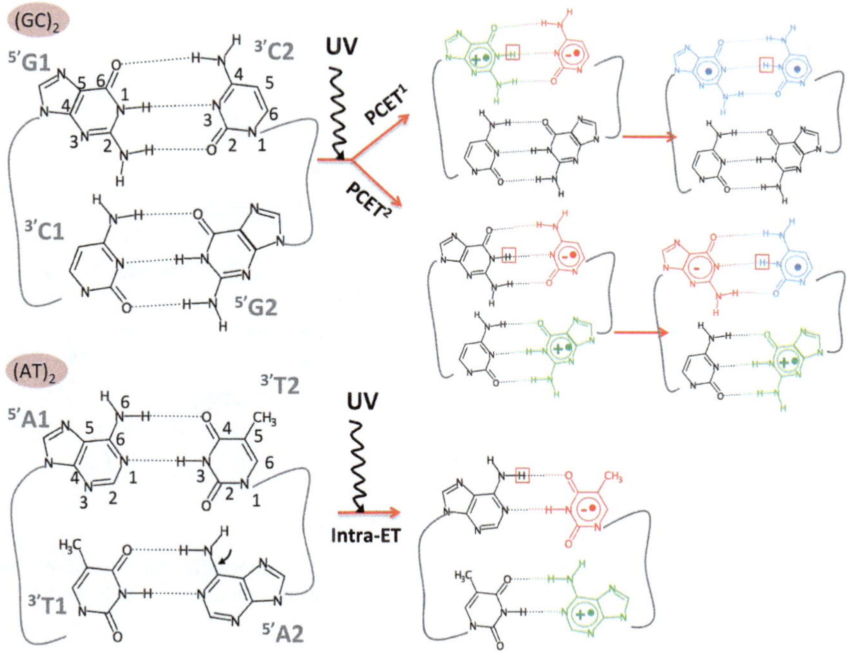

Scheme 1 General scheme of the possible intra and inter-strand ET/PT mechanisms (PCET^1 and PCET^2) triggered by UV absorption in $(GC)_2$ and $(AT)_2$ duplexes.

CG pairs,[43] indicating that the presence of a stacked tetramer is sufficient to trigger it. PCET is sequence dependent: it occurs in alternated $(A)_n(T)_n$ homoduplexes and in $(GC)_n$ duplex, both in B- and Z-DNA form,[41] but not at low pH, when C is protonated and the two strands are connected by Hoogsteen (HG) base pairing.[42] In alternated $(AT)_n$ duplexes, instead, the intra-strand CT reaction is not followed by inter-strand PT.[41]

The PCET mechanism is supported by the experimental acid dissociation constants (pK_a) for the neutral (isolated) nucleobases and their radical ions[44–48] and by the IR spectra computed on the different cation/anionic radicals within DNA (in the ground electronic state).[41] On the other hand, to the best of our knowledge, with the notable exception of a very interesting study on model GC systems not including the phospho-deoxyribose backbone,[49] no excited state study of the PCET process has been performed in a 'realistic' DNA environment. We thus report here a study of the energetics of the PCET processes in two representative model systems, *i.e.* alternated $(GC)_n$ and $(AT)_n$, chosen because the former exhibits the spectral signatures of PCET whereas, according to TR-IR spectra, for the latter formation of intra-strand $A^+ \rightarrow T^-$ CT states is not followed by PT. Our reference system (Fig. 1) will be dinucleotide (dGpdC)(dGpdC) (hereafter $(GC)_2$) and (dApT)(dApT) (hereafter $(AT)_2$), *i.e.* the minimal, though representative,[43] models to account for both base stacking and pairing effects; for $(GC)_2$ we also report the first results considering 2 additional base pairs ($(GC)_4$ tetramer, see Fig. 1), where the central dimer is considered at the PCM/M052X level and the external base pairs at the MM level. We characterize the energetic paths associated to the ET and PCET, focusing on the interplay between intra-

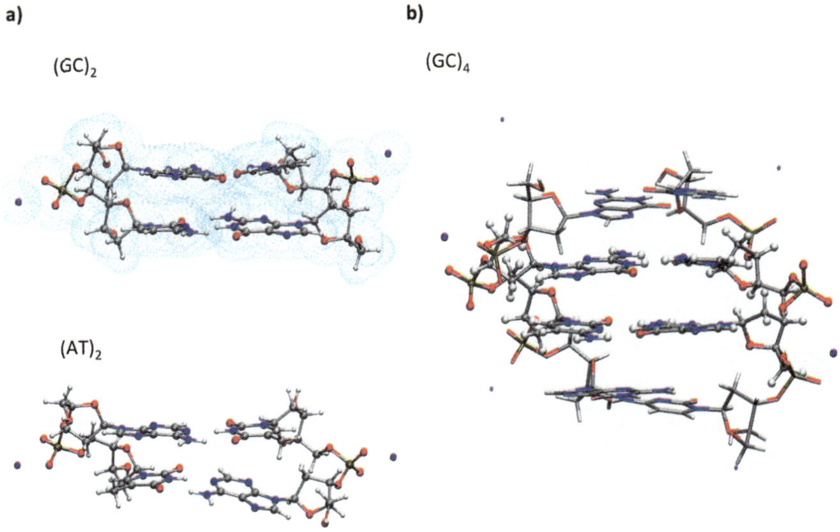

Fig. 1 (a) $(GC)_2$ (PCM solvation model in dots) and $(AT)_2$ dinucleotides considered in the PCM/QM calculations, (b) the QM/MM $(GC)_4$ system: QM part depicted in balls and MM region in sticks.

strand and inter-strand reactions, and, in particular, on the two processes described in Scheme 1: (i) the process, involving two base pairs (therefore labeled in the following $PCET^2$) where intra-strand ET is coupled with inter-strand PT (ii) the process (involving only one base pair, $PCET^1$) where inter-strand ET and PT are coupled. This latter reaction path has been shown to be operative for Watson–Crick (WC) paired nucleobases in the gas phase,[50–56] giving account of their sub-ps excited state decay; on the other hand, its importance for HB dimer in solution or within DNA is lively debated.[43,57–60] We show that, in agreement with the indication of TRIR spectra, in $(GC)_2$ the $PCET^2$ process is exothermic on the potential energy surface (PES) of the lowest energy intra-strand CT state, and a small energy barrier is associated to this process. $PCET^1$ is also feasible, though its relevance should be decreased by the lower stability of inter-strand G → C CT states with respect to the intra-strand one. In $(AT)_2$ $PCET^2$ is instead confirmed to be strongly endothermic and the reaction is predicted to 'stop' in the minimum of the intra-strand A → T CT state.

On these grounds we shall discuss some perspectives for the study of PCET in DNA, for what concerns especially the challenges involved in a dynamical characterization of these processes.

2. Computational methods

We adopted Density Functional Theory (DFT) to optimize the ground electronic state (S_0) of the different duplexes, and its Time Dependent (TD) version (TD-DFT) to characterize their excited state PES. We selected the M052X functional, due to its good performance when treating stacked systems and fairly accurate description of CT transitions,[61,62] as shown, *inter alia*, by many previous studies concerning the photochemistry and the photophysics of oligonucleotides.[26,32,63–67]

 This journal is © The Royal Society of Chemistry 2018

Due to the large size of the system (for each tetramer we consider, at the QM level, the full backbone including 2′-deoxyribose and phosphate groups together with two Na^+ counter ions, Fig. 1a) most of our analysis has been performed by using the 6-31G(d) basis set.

Solvent effects were accounted for by the polarizable continuum model (PCM),[68] exploiting its linear response (LR) version, for which analytical gradients are available, in excited state geometry optimizations.[69] For some specific cases, when considering dynamic solvation effects becomes important (ET and PT S_1 minima and barriers), we have resorted to State Specific (SS-PCM) calculations,[70,71] which provide a more accurate estimate of electronic transitions energies with significant CT character. We shall consider two limit time-regimes. In the non-equilibrium time-regime (NEQ), only fast solvation degrees of freedom are in equilibrium with the excited-state electron density, whereas the slow ones are still in equilibrium with the ground state. In the equilibrium limit (EQ) solvent is fully equilibrated with the excited state electron density. In order to mimic the possible decrease of the effective dielectric constant experienced by a base pair within a DNA duplex, we performed our analysis by considering three different dielectric constants in our PCM calculations: 78.36, that of bulk water, 8 and 4, mimicking a less polar environment.

In order to model an even more 'realistic' DNA environment we have performed QM/MM calculations in a $(GC)_2$ step extracted from the B-DNA X-ray structure 1CGC.pdb (Fig. 1b).[72] The QM part is described with the same level of theory used in the tetramer calculations (M052X/6-31G(d)) whereas the MM part is modeled using the AMBER parameter parm96.dat.[73] Both QM and MM parts are coupled using the ONIOM[74] interface as implemented in Gaussian 09.[75] Since this study involve 'charged' species and thereby either the backbone and/or counter ions could be important, we have selected the QM region as similar as possible to the tetramer calculations (see Fig. 1b). Also for this system bulk solvent effect has been included by PCM.

Formal charges present on the bases have been estimated by a simple Mulliken population analysis, considering the atoms of the purine and pyrimidine rings and the sugar C atom involved in the glycosidic bond, in terms of δq, *i.e.* the difference between the populations in the excited state of the selected stationary point and in the ground state at the FC point.

All the calculations have been performed by the Gaussian 09 program.[75]

3. Results and discussion

3.1 Franck–Condon region

3.1.1 $(GC)_2$. As reported in Table 1, the two lowest energy excited states in $(GC)_2$ exhibit a significant intra-strand G1 → C1 (G2 → C2 for S_2) CT character: $\delta q \sim +0.8$ a.u. for G1 and $\delta q \sim -0.8$ a.u. for C1 (for S_2 $\delta q \sim +0.6$ a.u. and ~ -0.6 a.u., for G2 and C2, respectively). The S_3 state corresponds to a $\pi\pi^*$ excitation localized in the G1 base (L_a excited state). These transitions are followed by two states that can be assigned to $\pi\pi^*$ excitations mainly localized over the two C bases. For more symmetric arrangements (especially for Z-DNA and for longer GC repeats) these two transitions can give rise to Frenkel excitons and, after some mixing with G → C CT transition, to the High Energy Long-Living Mixed (HELM) responsible for the long-living blue-shifted emission in $(GC)_n$ polymers.[27] S_6 can be described

Table 1 Vertical absorption energies (VAEs) and oscillator strengths (f) of the lowest excited states computed for $(GC)_2$ and $(AT)_2$ in water at the PCM/TD-M052X/6-31G(d) level of theory. The associated excited state density difference can be found in Fig. S1

	Main excitation	VAE	f
$(GC)_2$			
S_1 (intraET)	$\pi(G_1) \rightarrow \pi^*(C_1)$	4.96	0.004
S_2 (intraET)	$\pi(G_2) \rightarrow \pi^*(C_2)$	5.04	0.024
S_3	$\pi(G_1) \rightarrow \pi^*(G_1)$	5.23	0.063
S_4	$\pi(C_1) \rightarrow \pi^*(C_1)$	5.27	0.131
	$\pi(G_2) \rightarrow \pi^*(G_2)$		
S_5	$\pi(C_2) \rightarrow \pi^*(C_2)$	5.30	0.145
	$\pi(C_1) \rightarrow \pi^*(C_1)$		
S_6	$\pi(C_2) \rightarrow \pi^*(C_2)$	5.35	0.313
	$\pi(C_1) \rightarrow \pi^*(C_1)$		
	$\pi(G_2) \rightarrow \pi^*(G_2)$		
S_7 (interET)	$\pi(G_1) \rightarrow \pi^*(C_2)$	5.64	0.002
S_8	$\pi(G_1) \rightarrow \pi^*(G_1)$	5.65	0.843
$(AT)_2$			
S_1	$\pi(T_1) \rightarrow \pi^*(T_1)$	5.27	0.159
S_2	$\pi(T_1) \rightarrow \pi^*(T_1)$	5.31	0.029
S_3	$\pi(T_2) \rightarrow \pi^*(T_2)$	5.31	0.008
S_4	$\pi(T_2) \rightarrow \pi^*(T_2)$	5.36	0.242
S_5	$\pi(A_1) \rightarrow \pi^*(A_1)$	5.44	0.698
S_6	$\pi(A_2) \rightarrow \pi^*(A_2)$	5.51	0.083
S_7	$\pi(A_2) \rightarrow \pi^*(A_2)$	5.55	0.062
S_8	$\pi(A_1) \rightarrow \pi^*(A_1)$	5.57	0.015
	$\pi(A_1) \rightarrow \pi^*(T_1)$		
S_9	$\pi(A_1) \rightarrow \pi^*(T_1)$	5.63	0.067
	$\pi(A_1) \rightarrow \pi^*(A_1)$		
S_{10} (intraET)	$\pi(A_2) \rightarrow \pi^*(T_2)$	5.71	0.025
S_{11}	$\pi(A_1) \rightarrow \pi^*(A_1)$	5.73	0.001
S_{12}	$\pi(A_2) \rightarrow \pi^*(A_2)$	5.74	0.001
S_{13} (interET)	$\pi(A_1) \rightarrow \pi^*(T_2)$	6.12	0.001
S_{14} (interET)	$\pi(A_2) \rightarrow \pi^*(T_1)$	6.18	0.001

as an exciton delocalized over the two strands, deriving from the interactions between the $\pi\pi^*$ excitations of three bases. S_7 receives the largest contribution from G1 → C2 inter-strand CT transition, and a smaller one from the G2 → C2 transition. Finally, S_8 is reminiscent of the most intense $\pi\pi^*$ excitation in isolated G (L_b excited state) and S_{10} corresponds to the second inter-strand CT G2 → C1 at 5.70 eV ($f = 0.005$). Confirming the results of previous analysis, including the study of GC Watson–Crick,[27,49–52,57,76] G → C CT transitions are predicted to be close in energy to the bright excitons deriving from the interaction of the lowest energy $\pi\pi^*$ excited states of G and C. On the other hand, for what concerns duplexes, we show that in the FC region the lowest energy CT state has a predominant intra-strand CT character and is ~0.7 eV more stable than that with mainly inter-strand CT character.

3.1.2 $(AT)_2$. The two lowest energy bright excited states in the FC region (S_1 and S_4, see Table 1) correspond to localized $\pi\pi^*$ excitation on the two T. For longer $(AT)_n$ sequences the interaction between the Frenkel exciton formed by these two states with A → T CT transition can give rise to the HELM states similar

 This journal is © The Royal Society of Chemistry 2018

to those discussed above for $(GC)_2$.[28] Two dark excited states, $n\pi^*$ transitions involving the carbonyl lone pair of T, are confirmed[32,58,77] to fall close in energy (S_2 and S_3) to these bright excited states. S_5 and S_6 can be described as excitons deriving from the interaction between the bright $\pi\pi^*$ excited states of A bases (L_a). Then, at higher energies than other excited states related to the L_b state of A (S_7 and S_8), we find the most stable state with a clear CT character. It corresponds to S_{10} (VAE 5.71 eV) and has a predominant intra-strand A2 → T2 CT character. The lowest excited states with a clear inter-strand CT character are S_{13} and S_{14}, which fall at 6.12 and 6.18 eV.

In short, our calculations show that CT transitions are relatively much less stable for $(AT)_2$ than for $(GC)_2$. On the other hand, a population of CT states seem feasible also for the former systems, though mainly for intra-strand CT, is only 0.3–0.4 eV less stable than the bright excited states of $(AT)_2$ and appears vibronically coupled also to the lowest energy transitions, which have a small CT contribution.

3.2 Mapping the CT excited states paths

3.2.1 $(GC)_2$

3.2.1.1 Intra-strand ET. Excited state geometry optimizations from FC of S_1 and S_2 predict an increase of the CT and lead directly (Fig. 2) to minima in the first excited state PES, hereafter denoted as intra-ET-min at 4.06 (4.04 for S_2) eV, with a clear G1 → C1 (G2 → C2) CT character. During the geometry optimization in G1 (G2) the C2–NH_2 bond distance decreases, while in C1 (C2) the C2–O and C5–C6

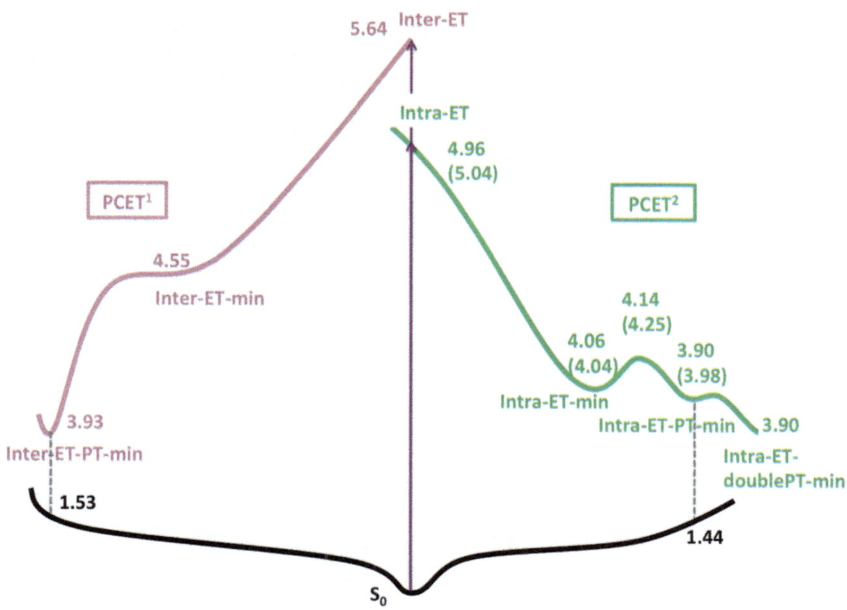

Fig. 2 Excited state pathways of $(GC)_2$ involving the inter-ET (pink) and intra-ET (green: the energies of the minima obtained by optimizing the S_1 state in the FC region are also shown, those issuing by S_2 geometry optimizations are in parenthesis) states together with the possible PT associated processes calculated at the LR-PCM/TD-M052X/6-31G(d) level of theory. Energies in eV with respect to the ground state minimum.

bond lengths increase and the two bases approach the geometry of G^+ cation and C^- anion, respectively (Fig. S2†). Indeed, Mulliken population analysis show that at intra-ET-min G1 (G2) bears a $\delta q = +0.80$ (0.76) a.u. while C1 (C2) has a negative one, δq −0.76 (0.71). The relative energies of the two intra-ET-min for the two strands are very similar (see Fig. 2), though their geometry exhibits some differences, with the two bases showing a smaller stacking distance in the minimum obtained for S_1.

Starting from intra-ET-min, we then performed two relaxed paths along the PT coordinate involving the proton of the G (H1) hydrogen bonded with the N3 nitrogen atom of the anionic C (see Scheme 1), *i.e.* G2–N1–H···N3–C1$^-$ or G1–N1–H···N3–C2$^-$.

The two paths are very similar, indicating that the PCET process does not exhibit any significant dependence on the features of the intra-strand CT minimum. After overcoming a moderate energy barrier of 0.1–0.2 eV (depending on the path considered, Fig. 2), the minimum provided by the PCET reaction (in the following intra-ET-PT-min) is found. Intra-ET-PT-min is slightly more stable than intra-ET-min, by ~0.1 eV (see Fig. 2), and the anionic C transfers its partial negative charge to the adjacent G.

We have refined the energy of the most relevant points of the path connecting the intra-ET-min and intra-ET-PT-min (minima and barrier) by performing SS-PCM/TD-M052X calculations (both in the EQ and NEQ limit, Table 2). The picture is similar to that obtained at the LR-PCM/TD-M052X level. As expected, since LR-PCM/TD-DFT is known to underestimate the stability of CT states, both

Table 2 Linear response (LR), state specific (SS) S_1 (S_0) energies at the equilibrium (EQ) and non-equilibrium (NEQ) levels for specific stationary points of the $(GC)_2$ using different values for the dielectric constant (ε) and QM/MM energies for $(GC)_4$ (in bold)

$\varepsilon = 78.36$		ET-min	Barrier maximum	ET-PT-min	ET-doublePT-min
PCET^2	LR	4.04, **3.38**	4.25	3.98, **3.27**	3.90
	SS-NEQ	4.18	4.53	4.25 (1.55)	4.29
	SS-EQ	3.13	3.45	3.08 (1.80)	3.18
PCET^1	LR	4.55	—	3.93	
	SS-NEQ	4.56	4.76	4.43 (1.76)	
	SS-EQ	2.64	3.05	2.45 (2.03)	
$\varepsilon = 8$					
PCET^2	LR	3.93	4.11	3.92	
	SS-NEQ	3.95	4.25	4.04	
	SS.EQ	3.21	3.50	3.22	
PCET^1	LR	4.43	—	3.74	
	SS-NEQ	4.25	4.42	3.95	
	SS-EQ	2.86	3.18	2.51	
$\varepsilon = 4$					
PCET^2	LR	3.82	4.00	3.86	
	SS-NEQ	3.75	4.02	3.86	
	SS-EQ	3.28	3.54	3.34	
PCET^1	LR	4.30	—	3.57	
	SS-NEQ	3.97	4.11	3.53	
	SS-EQ	3.06	3.30	2.60	

 This journal is © The Royal Society of Chemistry 2018

intra-ET-min and intra-ET-PT-min (PCET^2) are stabilized with respect to that predicted by LR-PCM/TD-M052X calculation (Table 2). On the other hand their relative energy does not significantly change, but the energy barrier increases up to ~0.3 eV (Table 2). A decrease of the dielectric constant used in the SS-PCM/TD-DFT calculations does not significantly affect the PCET^2 path: as could be expected, the absolute stability of the two minima slightly decreases, but the energy barrier separating them is almost unaffected.

Finally, PCM/TD-M052X/MM calculations on $(GC)_4$ confirm the picture obtained on $(GC)_2$, indicating that intra-ET-PT-min is more stable by 0.11 eV than intra-ET-min. Interestingly the absolute energy of these states is similar to that provided by SS-PCM/TD-M052X calculations by using $\varepsilon = 4$.

On the grounds of their experimental pK_a (see below), it could also be possible that the G cation base transfers a proton to a neutral C, leading, in our case, to a double PCET process. In order to investigate this process, we perform a new scan starting from intra-ET-PT-min, but relaxing now the $^+$G2–N1–H···N3–C1 coordinate. A new optimized S_1 minimum, intra-ET-doublePT-min (Fig. 2) is found. At the LR-PCM/TD-M052X level, this minimum is slightly more stable than intra-ET-PT-min, though by less than 0.1 eV. On the other hand, SS-PCM/TD-M052X predicts the opposite result, with PT in the cationic part being slightly (<0.1 eV) endergonic (Table 2).

We can now compare the results of our excited state calculations with the estimates based on the pK_a of the isolated bases[44–48] (and their radical cation/anion) or on ground state calculations. As anticipated above, the equilibrium constant (K) for the reaction

$$G + C^{\bullet -} \rightarrow G(-H1)^{-} + C(+H3)^{\bullet}$$

is $K = 10^{+3.4}$ suggesting that PCET reaction is strongly favored. Within DNA, ground state PCM/M052X/6-31G(d) calculation on a $(GC)_2$ anion (the negative charge is localized on a C) indicates that PT to the G leads to a minimum more stable by 0.11 eV, an estimate very close to that provided by our excited state calculations.

For what concerns the reaction

$$G^{\bullet +} + C \rightarrow G(-H1)^{\bullet} + C(+H3)^{+}$$

the pK_a of the partners suggests a substantial equilibrium, $K = 10^{+0.55}$. Within DNA, ground state PCM/M052X/6-31G(d) calculation on a $(GC)_2$ cation (the positive charge is localized on a G) indicates that PT is weakly disfavored (by 0.06 eV).

If PT in the cationic pair takes place in addition to PT in the anionic pair, then photoexcitation of GC/GC would result in the transfer of one electron and two protons, an example of a $1e^-/2H^+$ multisite PCET reaction.[78] Actually, our excited state calculations seem to suggest a substantial equilibrium, since the minima are close in energy and their relative stability results depend on the PCM model (LR-PCM *vs.* SS-PCM).

3.2.1.2 Inter-strand ET. Despite the corresponding excited state lying higher in energy at the FC region (Table 1), it has been proposed that the PCET^1 route could also play a role in DNA decay mechanisms.[52,55,56,76,79] As a consequence we studied it for $(GC)_2$. We have then explored this pathway by optimizing the S_7 state (with predominant G2 → C1 CT character) starting from the FC region. A steep path leads to an energy plateau in the S_1 adiabatic PES (Fig. 2, VAE 4.55 eV). In this pseudo-minimum (inter-ET-min*), C2 shows the structural features of a C anion, *e.g.* longer C2–O and C5–C6 bond distances (Fig. S2†), and has a significant negative partial charge $\delta q = -0.69$ a.u. Accordingly, for G1 $\delta q = +0.71$ a.u. formal positive charge and the structural features of a G^+ cation. Starting from inter-ET-min*, geometry optimization predicts a barrierless G1 → C2 PT reaction, leading to inter-ET-PT-min, with VAE 3.93 eV (Fig. 2). Therefore, inter-ET-PT-min is significantly more stable than inter-ET-min* and has a similar energy to intra-ET-PT-min.

On the other hand, in the case of inter ET, PT reaction leads to the formation of two neutral radicals, 'quenching' the strong dipole moment of inter-ET-min. As a consequence, solvation dynamics extensively affects inter-ET-PT reactions, as shown by SS-PCM/TD-M052X calculations ($PCET^1$ in Table 2), which predict that an energy barrier (0.20–0.40 eV) is associated to PT, which is barrierless at the LR level. Furthermore the driving force towards the PT reaction decreases, since inter-ET-min* and inter-ET-PT-min are much closer in energy. At the SS-PCM/TD-M052X level, inter-ET-PT-min is rather close to the ground state PES and the energy gap critically depends on the full equilibration of the solvent degrees of freedom. At the NEQ level this is ~2.7 eV, whereas at the EQ level only 0.4 eV, suggesting the proximity of a crossing with S_0.

3.2.2 $(AT)_2$

3.2.2.1 Intra-strand ET. Excited state geometry optimization of the lowest energy intra-strand CT state (S_{10}) provides a picture that, in the first part of the PES, is very similar to that found for $(GC)_2$, leading directly (see Fig. 3) to a minimum in the first excited state PES (hereafter denoted as intra-ET-min at 4.62 eV). This minimum shows a clear A2 → T2 CT character, with charges ($\delta q = +0.68$ and -0.65 a.u.) and geometries similar to that of A^+ cation and T^- anion, respectively (Fig. S4†). In order to verify the possibility of a $PCET^2$ process, starting from intra-ET-min, we have performed a relaxed path along the A1–N6–H···O4–$T2^-$ coordinate, *i.e.* involving the T anion. The energy of the intra-strand-CT state starts increasing and, after surmounting a barrier of almost 0.4 eV, hole transfer from A2 to A1 leads to the diabatic PES of the inter-strand CT, whose path is discussed below. We succeed in optimizing a minimum for the intra-ET-PT state (intra-ET-PT-min) but it is quite high in energy (5.12 eV, *i.e.* 0.5 eV less stable than intra-ET-min) and located in the S_2 PES (Fig. 3). The $PCET^2$ process appears to be strongly disfavored and intra-ET-min is predicted to be the stable minimum. This conclusion agrees with that obtained by analysing the pK_a of T anion and A; T^- is a very weak base and the equilibrium constant for the reaction $A + T^{\bullet -} \rightarrow A(-H6)^- + T(+H4)^\bullet$ can be estimated to be $K = 10^{-7.1}$. Actually, PCM/M052X ground state optimizations of the PT reaction in the $(AT)_2$ anion were unsuccessful, and decay to the minimum where the negative charge is localized on T is always observed.

3.2.2.2 Inter-strand ET. Inter-strand CT states fall quite high in energy (>6 eV) at the FC region and LR-PCM/TD-M052X optimization of this state ends up in the most stable intra-ET-min. However, as described above, PT leads to the decay of the inter-ET state to a stable inter-ET-PT-min in the S_1 surface, following a $PCET^1$ reaction. The features of inter-ET-PT-min are similar to those described above for

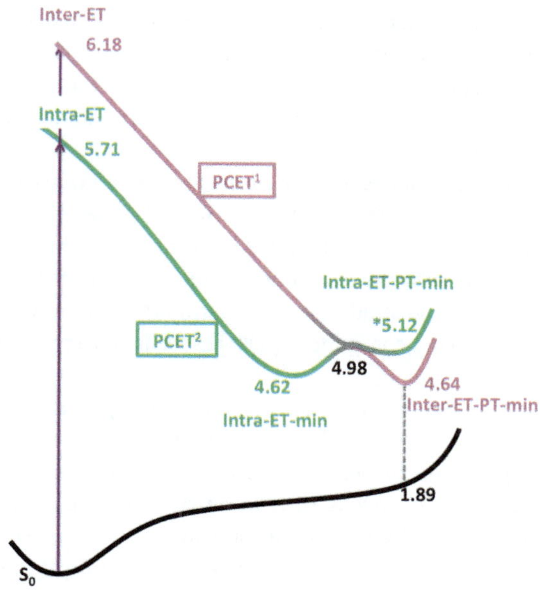

Fig. 3 Excited state pathways of $(AT)_2$ involving the inter-ET and intra-ET states together with the possible PT associated processes calculated at the PCM/TD-M052X/6-31G(d) level of theory. Energies in eV with respect to the ground state at the FC region. At the SS-PCM/M052X level the absolute energy of intra-ET min is 4.64 eV (NEQ) and 3.56 eV (EQ); the absolute energy of inter-ET-PT-min is 4.93 (NEQ) and 3.20 (EQ, $S_0 = 2.47$ eV). The energy of the barrier maximum is 5.02 eV (NEQ) and 3.88 eV (EQ).

the corresponding minimum in $(GC)_2$. At the LR-PCM/TD-M052X level it has a similar energy to stable intra-ET-min (see Fig. 3), and it is strongly stabilized at the SS-PCM/TD-M052X EQ level (numbers in Fig. 3 caption). Full equilibration of the solvent degrees of freedom leads very close to a crossing with S_0, the energy gap being smaller than 1 eV.

4. Concluding remarks

We have reported here the first quantum mechanical characterization of the excited state PES associated to the main PCET processes that could occur in DNA, labeled $PCET^2$ and $PCET^1$, for two representative sequences, *i.e.* alternated GC and AT DNA. In $PCET^2$, intra-strand ET is coupled to inter-strand PT; $PCET^1$ corresponds to that extensively discussed for isolated base pairs, and it involves both inter-strand ET and PT.

4.1 GC DNA

Our study fully confirms the results of previous studies, based on TR-IR spectra and ground state calculations, indicating that in $(GC)_9$ DNA PCET was operative and the resulting diradical species was responsible for the long-living components of the excited state decay with a time constant of 32 ± 1 ps.[41] The lowest energy excited states have a strong intra-strand G → C CT component, which increases following geometry optimizations, leading to a stable minimum (intra-

ET-min). Though we have not studied here the excited state decay of the bright excited state of GC, the existence of an effective decay route leading to a non-emissive CT minimum, together with the presence of the so-called HELM states,[22,27,28] gives account of the small contribution of 'monomer-like' emission to the fluorescence spectra.[27,28]

After the system has reached intra-ET-min our calculations predict that the $PCET^2$ reaction is slightly exergonic, by 0.05–0.1 eV depending on the level of calculation, and a small energy barrier (<0.3 eV) separates the two minima. These results (fully consistent with the experimental pK_a of the different bases[44–48] and the predictions based on ground state calculations)[41] give account of the large amplitude of the long-living component (60%) associated to the $PCET^2$ process in $(GC)_9$.[43]

Our calculations do not allow us to reach an unambiguous conclusion on the occurrence of PT also in the cationic WC pairs (those involving the cationic G), since the result depends on the adopted PCM model. The possibility of PT in the G·C base pair radical cation needs thus to be studied further, but even if this process is thermodynamically favourable, it may occur too slowly to influence excited-state dynamics. Considering also the experimental pK_a, and waiting for a purposely tailored dynamical study, our picture is consistent with that proposed for radiation-damaged DNA, *i.e.* that both $G^{\bullet+}$·C (non-distonic) and G(−H1)$^{\bullet}$·C(+H3)$^+$ (distonic) base pairs are present in equilibrium.[80]

For what concerns $PCET^1$, on balance, our calculations suggest that it should play a less relevant role than the $PCET^2$ route, mainly because of the much greater stability of the intra-strand CT state compared to the inter-strand CT states. On the other hand, this latter state could be populated either from the bright excited state localized on G and C or by the intra-ET-min, *via* hole-transfer between the two G. Once in the inter-strand CT state, $PCET^1$ reaction is predicted to be feasible, being exergonic and characterized by an energy barrier <0.4 eV. Since inter-ET-PT-min is very close to a crossing with S_0, this route should not significantly contribute to the slower (>10 ps) excited state dynamics but pertains to the fastest part (<5 ps) evidenced by TR-IR experiments. The presence of a sizeable energy barrier, which could trap the system in inter-ET-min, and the dependence of the energy gap with S_0 on the full equilibration of solvation degrees of freedom (likely on a few ps time scale) suggest, however, that this deactivation path should not be as fast as found in the gas phase (lifetime < 1 ps).[50,51] Interestingly, time resolved experiments of a GC base pair in chloroform[59] and for a miniduplex[43] containing a single GC pair show the presence of excited state (associated to the $PCET^1$ route) decaying on a time scale of 2.9 and 4.6 ps, respectively. These values are, roughly, consistent with the energetics we find for this process by our calculations.

4.2 AT DNA

Geometry optimizations of the lowest energy intra-strand A → T CT state in $(AT)_2$ lead to a stable intra-ET-min, as it happens for $(GC)_2$. On the other hand, excited state calculations, in full agreement with the prediction based on the pK_a of the bases and on ground state calculations, indicate that $PCET^2$ reaction is strongly disfavored. This result gives account of the experimental TR-IR spectra of $(AT)_9$ duplex, where the long living components exhibit the spectral signature of A^+

cation and T^- anion. According to our calculations, the relative stability of the intra-strand CT state is smaller in $(AT)_2$ than in $(GC)_2$, suggesting that the decay paths connected to this state could play a relatively less important role in the former system. This conclusion is in line with the comparison of the relative amplitude associated to the long-living component in $(AT)_9$ and in $(GC)_9$ duplexes.

The unavailability of the $PCET^2$ decay channel also affects the $PCET^1$ path. Our calculations indicate indeed that in this case, starting from intra-ET-min PT reaction involving the T anion triggers hole transfer between the two A bases, decay to the intra-strand ET state, for which the $PCET^1$ path is strongly favoured. The energy barrier associated to this transfer (~0.4 eV) is not too large and, considering also the tunneling effect, this path could provide an additional, relatively fast excited state decay route. As predicted for $(GC)_2$, the di-radical species resulting from the $PCET^1$ is rather close to a crossing region with the ground electronic state. Considering that 'monomer' like decay paths, which could in principle also give access to the $PCET^1$ path, play a more important role in $(AT)_n$ than in $(GC)_n$ duplexes, we can explain why TR-IR spectra show that the excited state decay in $(AT)_9$ is more complex than in $(GC)_9$.

4.3 Some methodological considerations

On the grounds of the results reported here, it is possible to propose some considerations on the main methodological challenges to be faced when studying the photoactivated dynamics of DNA and, more in general, of strongly coupled multichromophore systems. For what concerns the static description of the PES, it is confirmed that, when employing a suitable functional, our PCM/TD-DFT approach, based on LR-PCM/TD-DFT optimizations complemented, if necessary, by SS-PCM single point calculations, is able to provide a fairly accurate description of realistic DNA fragments. Here we have treated at the full QM level a system containing 4 bases and the phospo-deoxyribose backbone; on the one hand this is the minimal model for describing intra-strand and inter-strand interactions on the same foot. On the other hand it can capture the main physical effects ruling the photoactivated dynamics of longer sequences, as suggested by the studies on HELM states[22,27,28] and, especially, by the similarity of the results obtained on a GC miniduplex and longer sequences by TR-IR techniques.[43] Our computational approach, which has shown good performance in describing vibrational effects in absorption and emission spectra in several classes of compounds,[81,82] has already provided an accurate description of the excited state vibrational spectra in isolated bases and in nucleotide single strands.[40,83]

Considering the recent availability of TD-DFT analytical hessians,[84] a more direct comparison between experimental and computed ΔIR spectra is possible, eventually including also anharmonic effects. For the cases where excited state vibrational spectra are not easily accessible, our previous studies showed that ground state tetramer calculations can provide sufficiently reliable predictions for TRIR spectra.[36,37] Actually, as discussed in the ESI,† the geometries of the different species (*e.g.* cation, anion) – when computed within a duplex – are similar to those optimized in the excited state. The presence of a polar environment probably decreases the importance of the electrostatic interaction between oppositely charged units, which are ignored when the IR spectrum of a CT minimum is approximated as the sum of the spectra of a separated cation and anion.

Our approach obviously suffers from some limitations. Besides those related to the possible pitfalls of the adopted density functional, making the comparison with wave-function based methods always very fruitful,[66] the solvation model is based on an implicit description of the solvent, treated as an isotropic medium. Mixed implicit/explicit approaches, where a small number of solvent molecules (*e.g.* the first solvation shell) are explicitly included in the calculations, work pretty well for small molecules (such as nucleobases)[85] but they are not suitable for oligonucleotides. As a consequence, the use of explicit solvation models in combination with a dynamical simulation would likely bring important information.[86,87] The integration with dynamical approaches would be very useful to tackle what is probably the most important limit of our treatment, *i.e.* being based on a single minimum. When dealing with a macromolecule such as DNA, with a high degree of conformational freedom, it cannot be taken for granted that the results (both energy and vibrations) obtained by exploring a single conformational family can give account of the conformational variability of our system. Our previous studies have shown indeed that soft and hard vibrational degrees of freedom are significantly coupled within DNA and that, therefore, also vibrational modes that should be localized (say the stretching of a carbonyl) are coupled not only with the HB partners but also with the stacked bases.[40,41] Since the stacking geometry depends on the general conformation adopted by the duplex, it is clear that different conformations could exhibit different ΔIR spectra. Actually, the ground state IR spectra of $(AT)_n$ and $(A)_n(T)_n$ duplexes are different (though both duplexes are based on AT WC pairs)[41] and the IR spectrum of $(GC)_n$ changes when going from B-DNA to Z-DNA conformation.

Our study provides some indication also on some of the challenges to be faced in order to attain a proper dynamical description of the photoactivated dynamics of oligonucleotides, *i.e.* going beyond the simple use of static calculations to interpret time resolved experiments. In recent years quantum and semi-classical dynamical methods have known impressive advances,[88–92] making an exhaustive review impossible, as witnessed, *inter alia*, by the first studies tackling the photochemistry of dinucleotides.[93] On the other hand, as shown here, a complete dynamic study of PCET still poses severe challenges: (i) as discussed above, the minimum meaningful model is formed by 4 bases; (ii) it should be necessary to include many excited states ($\sim$10), to study the interplay between the combination of the bright excited states over the four monomers and the CT states (intra and inter-strand); (iii) the sensitivity of the energy of the CT excited states to dynamical solvation effects calls for a very accurate treatment of environmental effects, explicitly coupling intramolecular dynamics with solvent degrees of freedom. In this respect, it would be useful to enlarge as much as possible the portion of the system treated at the QM level, or, at least, to use a polarizable MM force field;[94] (iv) the PT reaction would require an accurate quantum dynamical treatment; (v) a proper sampling of slow degrees of freedom is necessary, requiring very long simulation. Actually, a previous, preliminary study on a model dinucleotide[95] shows that in a given minima the amount of CT character (ruling the excited state dipole, *inter alia*) strongly depends on soft degrees of freedom, such the interbase stacking distance or the mutual orientation of the bases. In a polar solvent, small changes in the CT character strongly affect the energetics of the excited states and, thus, CT and charge recombination rate, besides also modulating the

 This journal is © The Royal Society of Chemistry 2018

experimental IR spectrum. This requires a treatment considering simultaneously the contribution of fast and slow degrees of freedom to the CT rate.

Acknowledgements

The authors thank Dr Yuyuan Zhang, Prof. Bern Kohler and Dr Fabrizio Santoro for very useful discussion. RI thanks Program D'Alembert of the University Paris-Saclay for financial support.

References

1 T. Douki, *Photochem. Photobiol. Sci.*, 2013, **12**, 1286–1302.
2 J. S. Taylor, *Acc. Chem. Res.*, 1994, **27**, 76–82.
3 J. Cadet and P. Vigny, The Photochemistry of Nucleic Acids, in *Bioorganic Photochemistry*, ed. H. Morrison, Wiley, New York, 1990, vol. 1, pp. 1–272.
4 R. Improta, F. Santoro and L. Blancafort, *Chem. Rev.*, 2016, **116**, 3540–3593.
5 C. E. Crespo-Hernández, B. Cohen, P. M. Hare and B. Kohler, *Chem. Rev.*, 2004, **104**, 1977–2019.
6 M. Barbatti, A. C. Borin and S. Ullrich, in *Photoinduced Phenomena in Nucleic Acids I: Nucleobases in the Gas Phase and in Solvents*, ed. M. Barbatti, C. A. Borin and S. Ullrich, Springer International Publishing, Cham, 2015, pp. 1–32, DOI: 10.1007/128_2014_569.
7 W. J. Schreier, P. Gilch and W. Zinth, *Annu. Rev. Phys. Chem.*, 2015, **66**, 497–519.
8 C. T. Middleton, K. de La Harpe, C. Su, Y. K. Law, C. E. Crespo-Hernández and B. Kohler, *Annu. Rev. Phys. Chem.*, 2009, **60**, 217–239.
9 A. Giussani, J. Segarra-Martí, D. Roca-Sanjuán and M. Merchán, in *Photoinduced Phenomena in Nucleic Acids I: Nucleobases in the Gas Phase and in Solvents*, ed. M. Barbatti, C. A. Borin and S. Ullrich, Springer International Publishing, Cham, 2015, vol. 355, pp. 57–97.
10 D. Markovitsi, *Photochem. Photobiol.*, 2016, **92**, 45–51.
11 L. Serrano-Andrés and M. Merchán, *J. Photochem. Photobiol., C*, 2009, **10**, 21–32.
12 M. Barbatti, A. J. A. Aquino, J. J. Szymczak, D. Nachtigallová, P. Hobza and H. Lischka, *Proc. Natl. Acad. Sci. U. S. A.*, 2010, **107**, 21453–21458.
13 C. E. Crespo-Hernández, B. Cohen and B. Kohler, *Nature*, 2005, **436**, 1141–1144.
14 D. Markovitsi, F. Talbot, T. Gustavsson, D. Onidas, E. Lazzarotto and S. Marguet, *Nature*, 2006, **441**, E7.
15 B. Kohler, *J. Phys. Chem. Lett.*, 2010, **1**, 2047–2053.
16 D. Markovitsi, T. Gustavsson and I. Vaya, *J. Phys. Chem. Lett.*, 2010, **1**, 3271–3276.
17 D. B. Bucher, A. Schlueter, T. Carell and W. Zinth, *Angew. Chem., Int. Ed.*, 2014, **53**, 11366–11369.
18 V. A. Spata, W. Lee and S. Matsika, *J. Phys. Chem. Lett.*, 2016, **7**, 976–984.
19 G. W. Doorley, D. A. McGovern, M. W. George, M. Towrie, A. W. Parker, J. M. Kelly and S. J. Quinn, *Angew. Chem., Int. Ed.*, 2009, **48**, 123–127.
20 M. C. Stuhldreier and F. Temps, *Faraday Discuss.*, 2013, **163**, 173–188.
21 T. Takaya, C. Su, K. de La Harpe, C. E. Crespo-Hernández and B. Kohler, *Proc. Natl. Acad. Sci. U. S. A.*, 2008, **105**, 10285–10290.

22 I. Vayá, F.-A. Miannay, T. Gustavsson and D. Markovitsi, *ChemPhysChem*, 2010, **11**, 987–989.
23 C. Ma, C. C.-W. Cheng, C. T.-L. Chan, R. C.-T. Chan and W.-M. Kwok, *Phys. Chem. Chem. Phys.*, 2015, **17**, 19045–19057.
24 L. Martinez-Fernandez, A. J. Pepino, J. Segarra-Martí, J. Jovaisaite, I. Vayá, A. Nenov, R. Improta, *et al.*, *J. Am. Chem. Soc.*, 2017, **139**, 7780–7791.
25 P. M. Hare, C. E. Crespo-Hernández and B. Kohler, *Proc. Natl. Acad. Sci. U. S. A.*, 2007, **104**, 435–440.
26 R. Improta and V. Barone, *Angew. Chem., Int. Ed.*, 2011, **50**, 12016–12019.
27 M. Huix-Rotllant, J. Brazard, R. Improta, I. Burghardt and D. Markovitsi, *J. Phys. Chem. Lett.*, 2015, **6**, 2247–2251.
28 I. Vayá, J. Brazard, M. Huix-Rotllant, A. K. Thazhathveetil, F. D. Lewis, T. Gustavsson, D. Markovitsi, *et al.*, *Chem.–Eur. J.*, 2016, **22**, 4904–4914.
29 I. Conti, A. Nenov, S. Hofinger, S. Flavio Altavilla, I. Rivalta, E. Dumont, M. Garavelli, *et al.*, *Phys. Chem. Chem. Phys.*, 2015, **17**, 7291–7302.
30 G. Olaso-González, M. Merchán and L. Serrano-Andrés, *J. Am. Chem. Soc.*, 2009, **131**, 4368–4377.
31 F. Plasser, A. J. A. Aquino, H. Lischka and D. Nachtigallová, in *Photoinduced Phenomena in Nucleic Acids II: DNA Fragments and Phenomenological Aspects*, ed. M. Barbatti, A. C. Borin and S. Ullrich, Springer International Publishing, Cham, 2015, pp. 1–37, DOI: 10.1007/128_2013_517.
32 F. Santoro, V. Barone and R. Improta, *J. Am. Chem. Soc.*, 2009, **131**, 15232–15245.
33 F. Santoro, V. Barone and R. Improta, *Proc. Natl. Acad. Sci. U. S. A.*, 2007, **104**, 9931–9936.
34 G. W. Doorley, M. Wojdyla, G. W. Watson, M. Towrie, A. W. Parker, J. M. Kelly and S. J. Quinn, *J. Phys. Chem. Lett.*, 2013, **4**, 2739–2744.
35 D. B. Bucher, B. M. Pilles, T. Carell and W. Zinth, *Proc. Natl. Acad. Sci. U. S. A.*, 2014, **111**, 4369–4374.
36 Y. Zhang, J. Dood, A. A. Beckstead, X. B. Li, K. V. Nguyen, C. J. Burrows, B. Kohler, *et al.*, *Proc. Natl. Acad. Sci. U. S. A.*, 2014, **111**, 11612–11617.
37 Y. Zhang, J. Dood, A. A. Beckstead, X. B. Li, K. V. Nguyen, C. J. Burrows, B. Kohler, *et al.*, *J. Phys. Chem. B*, 2015, **119**, 7491–7502.
38 B. M. Pilles, D. B. Bucher, L. Z. Liu, P. Gilch, W. Zinth and W. J. Schreier, *Chem. Commun.*, 2014, **50**, 15623–15626.
39 P. M. Keane, M. Wojdyla, G. W. Doorley, J. M. Kelly, A. W. Parker, I. P. Clark, S. J. Quinn, *et al.*, *Chem. Commun.*, 2014, **50**, 2990–2992.
40 L. Martinez-Fernandez, Y. Zhang, K. de La Harpe, A. A. Beckstead, B. Kohler and R. Improta, *Phys. Chem. Chem. Phys.*, 2016, **18**, 21241–21245.
41 Y. Zhang, K. de La Harpe, A. A. Beckstead, R. Improta and B. Kohler, *J. Am. Chem. Soc.*, 2015, **137**, 7059–7062.
42 Y. Zhang, K. de La Harpe, A. A. Beckstead, L. Martínez-Fernández, R. Improta and B. Kohler, *J. Phys. Chem. Lett.*, 2016, **7**, 950–954.
43 Y. Zhang, X.-B. Li, A. M. Fleming, J. Dood, A. A. Beckstead, A. M. Orendt, B. Kohler, *et al.*, *J. Am. Chem. Soc.*, 2016, **138**, 7395–7401.
44 P. O. P. Ts'o, in *Basic Principles in Nucleic Acid Chemistry*, ed. P. O. P. Ts'o, Academic Press, New York, 1974, pp. 454–584.
45 S. Steenken, *Free Radical Res. Commun.*, 1992, **16**, 349–379.
46 S. Steenken, *Chem. Rev.*, 1989, **89**, 503–520.

 This journal is © The Royal Society of Chemistry 2018

47 K. Kobayashi, *J. Phys. Chem. B*, 2010, **114**, 5600–5604.
48 D. M. Close, *J. Phys. Chem. A*, 2013, **117**, 473–480.
49 C. Ko and S. Hammes-Schiffer, *J. Phys. Chem. Lett.*, 2013, **4**, 2540–2545.
50 A. L. Sobolewski, W. Domcke and C. Hättig, *Proc. Natl. Acad. Sci. U. S. A.*, 2005, **102**, 17903–17906.
51 A. L. Sobolewski and W. Domcke, *Phys. Chem. Chem. Phys.*, 2004, **6**, 2763–2771.
52 T. Schultz, E. Samoylova, W. Radloff, I. V. Hertel, A. L. Sobolewski and W. Domcke, *Science*, 2004, **306**, 1765–1768.
53 S. Perun, A. L. Sobolewski and W. Domcke, *J. Phys. Chem. A*, 2006, **110**, 9031–9038.
54 J. P. Gobbo, V. Saurí, D. Roca-Sanjuán, L. Serrano-Andrés, M. Merchán and A. C. Borin, *J. Phys. Chem. B*, 2012, **116**, 4089–4097.
55 V. Sauri, J. P. Gobbo, J. J. Serrano-Pérez, M. Lundberg, P. B. Coto, L. Serrano-Andrés, D. Roca-Sanjuán, *et al.*, *J. Chem. Theory Comput.*, 2013, **9**, 481–496.
56 A. Francés-Monerris, J. Segarra-Martí, M. Merchán and D. Roca-Sanjuán, *Theor. Chem. Acc.*, 2016, **135**, 31.
57 L. Biemann, S. A. Kovalenko, K. Kleinermanns, R. Mahrwald, M. Markert and R. Improta, *J. Am. Chem. Soc.*, 2011, **133**, 19664–19667.
58 M. Dargiewicz, M. Biczysko, R. Improta and V. Barone, *Phys. Chem. Chem. Phys.*, 2012, **14**, 8981–8989.
59 K. Röttger, H. J. B. Marroux, M. P. Grubb, P. M. Coulter, H. Böhnke, A. S. Henderson, G. M. Roberts, *et al.*, *Angew. Chem., Int. Ed.*, 2015, **54**, 14719–14722.
60 K. Röttger, H. J. B. Marroux, A. F. M. Chemin, E. Elsdon, T. A. A. Oliver, S. T. G. Street, G. M. Roberts, *et al.*, *J. Phys. Chem. B*, 2017, **121**, 4448–4455.
61 Y. Zhao and D. G. Truhlar, *Acc. Chem. Res.*, 2008, **41**, 157–167.
62 Y. Zhao, N. E. Schultz and D. G. Truhlar, *J. Chem. Theory Comput.*, 2006, **2**, 364–382.
63 L. Martinez-Fernandez and R. Improta, *Photochem. Photobiol. Sci.*, 2017, **16**, 1277–1283.
64 A. Banyasz, L. Martinez-Fernandez, T. Ketola, A. Muñoz-Losa, L. Esposito, D. Markovitsi and R. Improta, *J. Phys. Chem. Lett.*, 2016, **7**, 2020–2023.
65 A. Banyasz, L. Esposito, T. Douki, M. Perron, R. Improta and D. Markovitsi, *J. Phys. Chem. B*, 2016, **120**, 4232–4242.
66 I. Conti, L. C. Martinez Fernandez, L. Esposito, S. Hofinger, A. Nenov, M. Garavelli and R. Improta, *Chem.–Eur. J.*, 2017, **23**, 15177–15188.
67 A. Banyasz, T. Gustavsson, D. Onidas, P. Changenet-Barret, D. Markovitsi and R. Improta, *Chem.–Eur. J.*, 2013, **19**, 3762–3774.
68 J. Tomasi, B. Mennucci and R. Cammi, *Chem. Rev.*, 2005, **105**, 2999–3093.
69 G. Scalmani, M. J. Frisch, B. Mennucci, J. Tomasi, R. Cammi and V. Barone, *J. Chem. Phys.*, 2006, **124**, 094107.
70 R. Improta, V. Barone, G. Scalmani and M. J. Frisch, *J. Chem. Phys.*, 2006, **125**, 054103.
71 R. Improta, G. Scalmani, M. J. Frisch and V. Barone, *J. Chem. Phys.*, 2007, **127**, 074504.
72 U. Heinemann, C. Alings and M. Bansal, *EMBO J.*, 1992, **11**, 1931–1939.
73 W. D. Cornell, P. Cieplak, C. I. Bayly, I. R. Gould, K. M. Merz, D. M. Ferguson, P. A. Kollman, *et al.*, *J. Am. Chem. Soc.*, 1995, **117**, 5179–5197.

74 S. Dapprich, I. Komáromi, K. S. Byun, K. Morokuma and M. J. Frisch, *J. Mol. Struct.: THEOCHEM*, 1999, **461**, 1–21.
75 M. J. Frisch, *et al.*, *Gaussian 09, revision A.1*, Gaussian, Inc., Wallingford, CT, 2009.
76 G. Groenhof, L. V. Schäfer, M. Boggio-Pasqua, M. Goette, H. Grubmüller and M. A. Robb, *J. Am. Chem. Soc.*, 2007, **129**, 6812–6819.
77 F. Plasser, A. J. A. Aquino, W. L. Hase and H. Lischka, *J. Phys. Chem. A*, 2012, **116**, 11151–11160.
78 M. H. V. Huynh and T. J. Meyer, *Chem. Rev.*, 2007, **107**, 5004–5064.
79 P. R. L. Markwick, N. L. Doltsinis and J. Schlitter, *J. Chem. Phys.*, 2007, **126**, 045104.
80 A. Kumar and M. D. Sevilla, *Chem. Rev.*, 2010, **110**, 7002–7023.
81 F. J. A. Ferrer, J. Cerezo, E. Stendardo, R. Improta and F. Santoro, *J. Chem. Theory Comput.*, 2013, **9**, 2072–2082.
82 F. Santoro and D. Jacquemin, *Wiley Interdiscip. Rev.: Comput. Mol. Sci.*, 2016, **6**, 460–486.
83 Y. Zhang, R. Improta and B. Kohler, *Phys. Chem. Chem. Phys.*, 2014, **16**, 1487–1499.
84 M. J. Frisch, G. W. Trucks, H. B. Schlegel, G. E. Scuseria, M. A. Robb, J. R. Cheeseman, D. J. Fox, *et al.Gaussian 16*, Wallingford, CT, 2016.
85 R. Improta and V. Barone, in *Photoinduced Phenomena in Nucleic Acids I: Nucleobases in the Gas Phase and in Solvents*, ed. M. Barbatti, A. C. Borin and S. Ullrich, Springer International Publishing, Cham, 2015, pp. 329–357, DOI: 10.1007/128_2013_524.
86 L. Martínez-Fernández, A. J. Pepino, J. Segarra-Martí, A. Banyasz, M. Garavelli and R. Improta, *J. Chem. Theory Comput.*, 2016, **12**, 4430–4439.
87 A. Petrone, J. Cerezo, F. J. A. Ferrer, G. Donati, R. Improta, N. Rega and F. Santoro, *J. Phys. Chem. A*, 2015, **119**, 5426–5438.
88 E. Brunk and U. Rothlisberger, *Chem. Rev.*, 2015, **115**, 6217–6263.
89 M. K. Lee, P. Huo and D. F. Coker, *Annu. Rev. Phys. Chem.*, 2016, **67**, 639–668.
90 I. Burghardt, K. Giri and G. A. Worth, *J. Chem. Phys.*, 2008, **129**, 174104.
91 B. F. E. Curchod, C. Rauer, P. Marquetand, L. González and T. J. Martínez, *J. Chem. Phys.*, 2016, **144**, 101102.
92 H.-D. Meyer, F. Gatti and G. A. Worth, in *Multidimensional Quantum Dynamics*, Wiley-VCH Verlag GmbH & Co. KGaA, 2009, pp. 17–30, DOI: 10.1002/9783527627400.ch3.
93 C. Rauer, J. J. Nogueira, P. Marquetand and L. González, *J. Am. Chem. Soc.*, 2016, **138**, 15911–15916.
94 M. F. S. J. Menger, S. Caprasecca and B. Mennucci, *J. Chem. Theory Comput.*, 2017, **13**, 3778–3786.
95 J. Cerezo, L. Martínez-Fernández, R. Improta and F. Santoro, *Theor. Chem. Acc.*, 2016, **135**, 221.

 This journal is © The Royal Society of Chemistry 2018

Faraday Discussions

PAPER

Fluorescent excimers and exciplexes of the purine base derivative 8-phenylethynyl-guanine in DNA hairpins†

Kristen E. Brown, Arunoday P. N. Singh, Yi-Lin Wu, Lin Ma, Ashutosh K. Mishra, Brian T. Phelan, Ryan M. Young, * Frederick D. Lewis * and Michael R. Wasielewski *

Received 21st August 2017, Accepted 1st September 2017
DOI: 10.1039/c7fd00186j

The ground- and excited-state electronic interactions between the nucleobase analog 8-(4′-phenylethynyl)deoxyguanosine, **EG**, with natural nucleobases and 7-deazaguanine, as well as between adjacent **EG** base analogs, have been characterized using a combination of steady-state spectroscopy and time-resolved fluorescence, absorption, and stimulated Raman spectroscopies. The properties of the nucleoside **EG-H$_2$** are only weakly perturbed upon incorporation into synthetic DNA hairpins in which thymine, cytosine or adenine are the bases flanking **EG**. Incorporation of the nucleoside to be adjacent to guanine or deazaguanine results in the formation of short-lived (40–80 ps) exciplexes, the charge transfer character of which increases as the oxidation potential of the donor decreases. Hairpins possessing two or three adjacent **EG** base analogs display exciton-coupled circular dichroism in the ground state and form long-lived fluorescent excited states upon electronic excitation. Incorporation of **EG** into the helical scaffold of the DNA hairpins places it adjacent to its neighboring nucleobases or a second **EG**, thus providing the close proximity required for the formation of exciplex or excimer intermediates upon geometric relaxation of the short-lived **EG** excited state. The three time-resolved spectroscopic methods employed permit both the characterization of the several intermediates and the kinetics of their formation and decay.

1 Introduction

Fluorescent analogs and derivatives of the nucleobases found in nature have been widely investigated in diverse applications, including DNA sequencing and

Department of Chemistry, Argonne-Northwestern Solar Energy Research (ANSER) Center, Institute for Sustainability and Energy at Northwestern, Northwestern University, Evanston, Illinois 60208-3113, USA. E-mail: ryan.young@northwestern.edu; fdl@northwestern.edu; m-wasielewski@northwestern.edu; Tel: +1-2224-866-3571

† Electronic supplementary information (ESI) available: Additional experimental details for time-resolved spectroscopy and data analysis, hairpin characterization, DFT calculations, time-resolved spectroscopy measurements, and Raman mode assignments. See DOI: 10.1039/c7fd00186j

structural studies, and in photophysical studies, including electron transfer and exciplex and excimer formation.[1,2] Fluorescent bases differ from the natural bases either by minor structural modifications such as in 2-aminopurine (AP),[3–5] an isomer of adenine, extending the conjugation of the base as in pyrrolocytosine,[6,7] appending a fluorophore such as pyrene to the base,[8] or replacing the entire base with a fluorophore.[9] The fluorescent base used most extensively in photophysical studies is AP. Unfortunately, the difference in base pairing for AP *vs.* adenine (A) with thymine (T) results in perturbation of the duplex structure. This results in the observation of multiple exponential fluorescence decays for duplexes containing a single AP–T base pair, irrespective of the identity of the neighboring base pairs.[10]

A different fluorescent nucleobase, 8-phenylethynylguanine (**EG**, Chart 1), and its aryl derivatives were prepared and incorporated into nucleosides, oligodeoxynucleotides and duplexes by Firth[11] and by Saito and co-workers.[12,13] The latter group investigated the relative fluorescence intensities of duplexes in which the complementary strand contained a mismatched base opposite **EG**, and observed enhanced **EG** fluorescence when opposite T and excimer fluorescence when opposite A. We were attracted to the use of **EG** as a base analog in studies of photoinduced electron transfer based on the potential use of the oxidized triple bond stretch as a Raman-active signature for its cation radical.[14,15] To our knowledge, the time-resolved spectroscopy of **EG**-containing duplexes possessing both natural and unnatural flanking bases has not previously been reported. We report here an investigation of the excited-state properties of DNA hairpins possessing a T_3 hairpin linker and either a single **EG** base with different neighboring bases or two or three stacked sequential **EG** bases (Chart 1). A combination of steady-state and femtosecond fluorescence upconversion, transient absorption (fsTA), and stimulated Raman (FSRS) spectroscopies reveals the formation of excited states possessing locally excited, charge-transfer exciplex, or excimer characteristics, depending on the identity and oxidation potential of the adjacent base.[16]

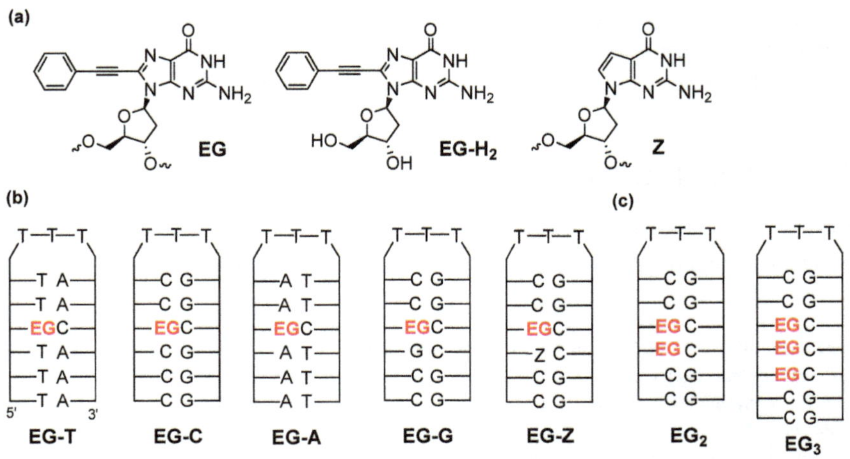

Chart 1 Structures of (a) 8-(4′-phenylethynyl)deoxyguanosine (**EG**) and its isolated form (**EG-H**$_2$), and the base analog 7-deazaguanosine, Z, (b) **EG-X** hairpins, and (c) **EG**$_n$ hairpins.

2 Materials and methods

Reagent grade chemicals were used as received. Flash column chromatography was carried out with SiO_2 (particle size 0.040–0.063 mm, 230–400 mesh) and technical solvents. ^{1}H and ^{13}C NMR spectra were obtained using a Bruker AVANCE III 500 instrument, and the chemical shifts were determined in ppm relative to the signal of $Si(CH_3)_4$. Residual solvent signals in the NMR spectra were used as an internal reference. High-resolution HR-ESI-MS spectra were obtained using an Agilent 6210 LC-TOF, with the signals reported in *m*/*z* units. The nucleoside **EG-H$_2$** was prepared by the method reported by Firth *et al.*,[11] and the respective DNA conjugates were prepared and purified by HPLC (Fig. S1, ESI†) as previously described.[17] DNA sequences were characterized using MALDI-TOF mass spectrometry, and their steady-state photophysical properties were examined using UV absorption, circular dichroism and fluorescence spectroscopies. All optical measurements of the DNA hairpins were carried out in buffered water (10 mM phosphate, 0.1 M NaCl, pH 7.2). Thermal dissociation temperatures (T_m) and *m*/*z* values are reported in the ESI (Table S1†). Fluorescence spectra were acquired using a Jasco fluorescence spectrometer, and the quantum yields (Φ_{fl}) were determined relative to the quantum yield of quinine sulfate.[18]

Time-resolved fluorescence data were collected at room temperature using a laboratory-constructed fluorescence up-conversion apparatus. The excitation laser source was a commercial direct diode-pumped 100 kHz amplifier (Spirit 1040-4, Spectra-Physics) which produced the fundamental beam of 1040 nm (350 fs, 4.5 W). The third harmonic of the output was used to pump a non-collinear optical parametric amplifier (Spirit-NOPA-3H, Spectra-Physics) which delivered 320 nm, high repetition rate pulses with pulse durations as short as sub-20 fs (10 nJ per pulse at the sample). The gate pulse was generated with a second harmonic-pumped NOPA (Spirit-NOPA-2H, Spectra-Physics) tuned to 800 nm. The samples were agitated throughout the experiment. The fluorescence was collected and focused into a β-barium borate nonlinear crystal where it was overlapped with the gate pulse. The phase-matching angle was optimized for sum-frequency mixing of the 800 nm gate and the fluorescence at 430 nm. The up-converted light was directed into a monochromator (SPEX 270M, Jobin-Yvon) and detected with a photomultiplier tube. The total instrument response was ~150 fs. Details of our fsTA and FSRS experiments have been presented elsewhere.[19,20] The ESI† contains details of the data processing and analysis methods.

3 Results and analysis

3.1. Steady-state spectroscopy

The CD spectra of the **EG-X** (X = T, C, A, G, and Z) and **EG$_n$** ($n = 2$ or 3) hairpins in aqueous buffer are shown in Fig. 1a and b. All of the **EG**-containing hairpins exhibit positive bands between 260–300 nm and negative bands between 240–260 nm, which is consistent with the formation of B-DNA hairpin structures and rules out perturbation of the DNA backbone by the modified **EG** base.[21] The **EG-T**, **-C** and **-A** hairpins exhibit weak bisignate CD curves, resulting from the weak coupling between **EG** and their neighboring bases which have different absorption maxima (Fig. 2a).[22] Stronger exciton coupling is observed for **EG-G** and **EG-Z** which exhibit more similar absorption maxima for **EG** and their adjacent purine

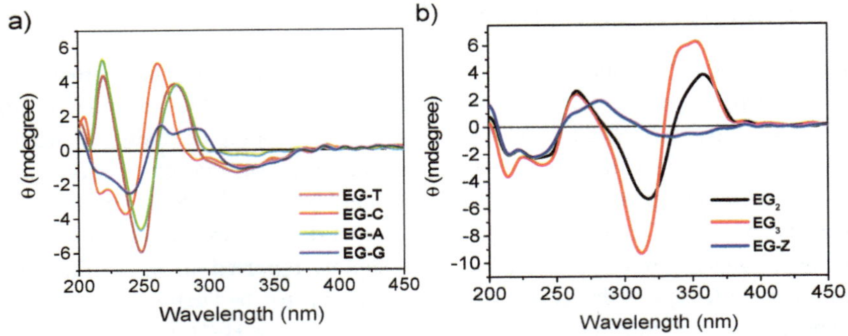

Fig. 1 Circular dichroism spectra of the **EG-X** hairpins used in this study in phosphate buffer (10 mM phosphate, 0.1 M NaCl, pH 7.2).

bases. The hairpins **EG$_2$** and **EG$_3$** exhibit strong positive bands at 358 and 353 nm, respectively, and negative bands at 317 and 312 nm, respectively, which is characteristic of exciton coupling between identical chromophores.

The UV absorption and fluorescence spectra of **EG-H$_2$** in methanol are shown in Fig. 2a and b, and its spectra in other solvents with varying dielectric constants are shown in Fig. S2 (ESI†). The **EG-H$_2$** spectrum in methanol exhibits an absorption maximum at 320 nm with a vibronic progression of *ca.* 1800 cm^{-1}. Minimal solvatochromatic response is observed in other solvents. The UV absorption spectra of the **EG-X** hairpins in phosphate buffer are also shown in Fig. 1a and b. These hairpins exhibit long-wavelength maxima between 330–335 nm (Table 1), with the weak shoulders at around 355 nm attributed to the **EG** chromophore, and the short-wavelength maxima between 261–263 nm attributed to the natural nucleobases. The UV spectra of **EG$_2$** and **EG$_3$** (Fig. 1b) exhibit long-wavelength maxima that are blue-shifted with respect to those of **EG-H$_2$** and the

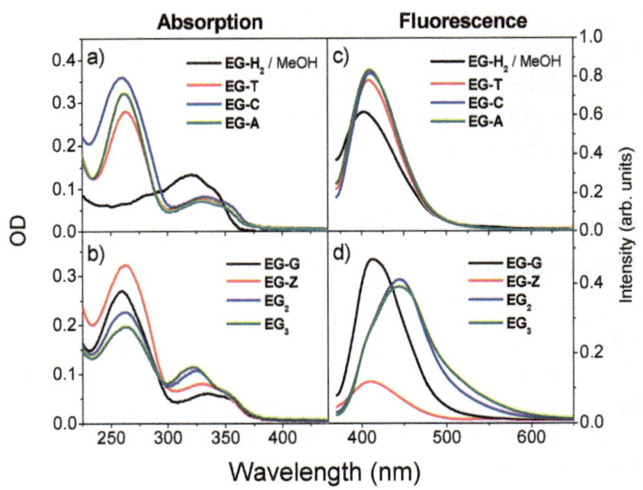

Fig. 2 UV-vis (a and b) and fluorescence (c and d) spectra of the **EG-X** hairpins (X = T, A, G and zG) and **EG$_n$** hairpins (n = 2 or 3) used in this study in phosphate buffer (10 mM phosphate, 0.1 M NaCl, pH 7.2).

 This journal is © The Royal Society of Chemistry 2018

Table 1 Neighboring purine (X) oxidation potential, absorption and fluorescence maxima, fluorescence quantum yields (%), fluorescence lifetimes and associated population fractions, as determined by fluorescence up-conversion spectroscopy, and the lifetimes of decay-associated transient absorption spectra in methanol for deoxynucleotide **EG** and in phosphate buffer (10 mM phosphate, 0.1 M NaCl, pH 7.2) for all hairpins

	E_{ox},[a] V	λ_{UV}, nm	λ_{fl},[b] nm	Φ_{fl}, %	τ_{fl},[c] ps	τ_{DAS},[d] ps
EG-H$_2$/MeOH	1.12[e]	320	390	1.3	0.6 (rise), 10.5 (decay)	4.6, 16
EG-T	—	330	410	1.4	8.0 (55%), 43 (45%)	4.4, 40
EG-C	—	332	410	1.4	0.4 (rise), 22.7 (decay)	3.8, 28
EG-A	1.36	330	414	1.5	18 (55%), 55 (45%)	6.9, 42
EG-G	1.11	336	408	0.7	3.2 (49%), 37 (51%)	2.1, 14, 73
EG-Z	0.81	331	410	0.2		~0.3, 3.3, 26, 68
EG$_2$	1.12[e]	325	444	0.8		~0.3, 3.1, 30, 490, 2900
EG$_3$		321	444	0.8		0.7, 5.1, 71, 670, 4110

[a] *V vs.* Ag/AgCl in phosphate buffer (pH 7.2) with 0.1 M NaCl from ref. 16. [b] λ_{ex} = 320 nm. [c] Determined from fluorescence up-conversion spectroscopy. The fits for **EG-Z**, **EG$_2$** and **EG$_3$** were of poor quality and are not reported (see text). [d] Determined from transient absorption spectroscopy. [e] *V vs.* SCE in DMF from ref. 15.

other **EG-X** hairpins (Table 1). Density functional theory (DFT) calculations for the **EG** chromophore were carried out at the B3LYP/6-31G(d) level (see ESI† for details).[23] The long-wavelength absorption band is assigned to a long-axis-polarized allowed HOMO–LUMO transition between delocalized π-orbitals (Fig. S3 and Tables S2 and S3, ESI†).

The fluorescence spectra of the nucleoside **EG-H$_2$** in methanol and the hairpins in aqueous buffer are shown in Fig. 2c and d. In MeOH, the spectral maximum appears at 390 nm and the fluorescence quantum yield (Φ_{fl}) is *ca.* 1.3%. The fluorescence maxima (λ_{fl}) for the **EG-X** and **EG$_n$** hairpins are red-shifted with respect to that for **EG-H$_2$**, however the fluorescence quantum yields for the hairpins **EG-X** (X = C, T, or A) are similar to that for **EG-H$_2$** in polar protic and aprotic solvents (Fig. S2c, ESI†). The value of Φ_{fl} in the non-polar aprotic solvent THF is significantly higher (Φ_{fl} = 16%), which is consistent with the report of Saito and co-workers.[12,13] Somewhat lower values of Φ_{fl} are observed for **EG-G**, **EG-Z**, **EG$_2$** and **EG$_3$**, and the fluorescence spectra of **EG$_2$** and **EG$_3$** are red-shifted and broadened. The values of Φ_{fl} and λ_{fl} are summarized in Table 1.

3.2. Time-resolved spectroscopies

3.2.1. Fluorescence up-conversion. Fluorescence decay traces for the nucleoside **EG-H$_2$** in methanol and THF and for the **EG-X** hairpins (X = T, -C, -A or -G) in aqueous buffer, determined at 430 nm following 320 nm pulsed laser excitation, are shown in Fig. S4 (ESI†). The decay of **EG-H$_2$** is best fitted as a single exponential with a decay time of 10.5 ps following a 600 fs rise, which is commensurate with the low (1%) quantum yield of emission. In THF, the fluorescence decay of **EG-H$_2$** is biexponential, with τ_1 = 13.6 ps and τ_2 = 220 ps comprising 85% and 15% of the decay profile, respectively. The **EG-X** hairpins, with the exception of **EG-C**, exhibit dual exponential decays with short decay times (3–20 ps) that are similar to the decay time of **EG-H$_2$** in methanol, followed by a longer (37–55 ps) decay. **EG-C** exhibits a single 23 ps decay following a fast rise, which is similar to that of **EG-H$_2$** in methanol. The fluorescence decays of **EG-Z**, **EG$_2$** and **EG$_3$** were of poor quality, which is plausibly due to the low fluorescence intensity at 430 nm and the presence of multiple intermediates (see below). The decay times obtained from the exponential fits of the transient data shown in Fig. S4 (ESI†) are summarized in Table 1.

3.2.2. Femtosecond transient absorption spectroscopy. The femtosecond transient absorption spectra of **EG-H$_2$** in methanol and the hairpins **EG-C** and **EG-G** acquired after 350 nm excitation are shown in Fig. 3. These spectra are dominated at short delay times by a single band between 450–600 nm. Global analysis of the transient absorption data with bi- or tri-exponential decays yields two decay-associated spectra (DAS) with similar maxima but differing in their band shapes and lifetimes. The DAS are shown as insets in Fig. 3. The decay times for **EG-H$_2$** and **EG-C** are consistent with those reported in our prior study.[15] The fast component has a maximum at 506 nm and a decay time of ~2–4 ps, accompanied by a growth in intensity near 645 nm which is clearly seen in the case of **EG-H$_2$**. The longer-lived component (tens of ps) exhibits a blue-shifted maximum at 495 nm and a tail that extends beyond 700 nm (Fig. 3a). The relaxation dynamics of **EG-H$_2$** are highly solvent dependent. The fsTA data for **EG-H$_2$** in aqueous buffer

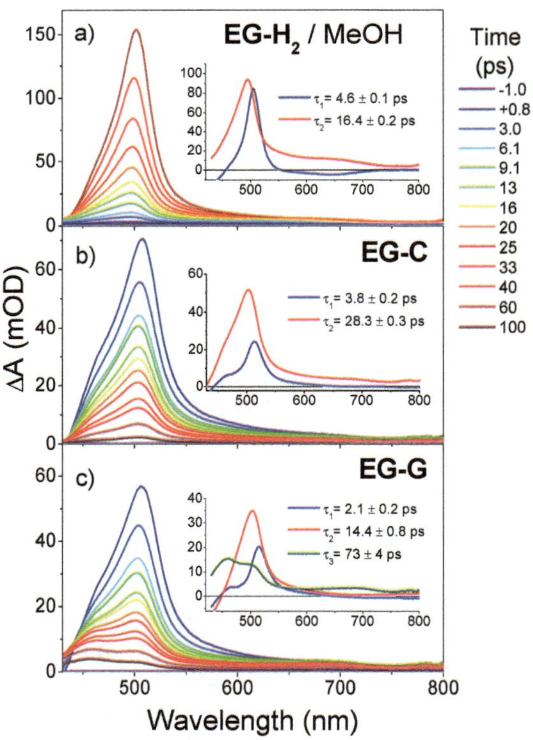

Fig. 3 Transient absorption spectra for (a) nucleoside **EG** in methanol and the hairpins (b) **EG-C** and (c) **EG-G** in phosphate buffer (10 mM phosphate, 0.1 M NaCl, pH 7.2) following excitation by a 350 nm, ~60 fs pulse (1.0 μJ per pulse). Insets: decay-associated spectra following multi-wavelength global fitting.

 This journal is © The Royal Society of Chemistry 2018

(Fig. S5, ESI†) show two short components with similar DAS to those in methanol, but with lifetimes that are almost twice as short (~2 and 9 ps, compared to ~5 and 16 ps in methanol). In THF, both the fluorescence up-conversion and the transient absorption data demonstrate more complex decays; a fast ~2 ps decay is accompanied by a shift in the absorption maximum from 510 to 497 nm. The first two DAS resemble the spectra obtained in methanol but have longer decay times. In addition there is a third broad absorption between 450–500 nm that has a decay time of 207 ps, which likely arises from the same population that gives the long-lived emission (Fig. S4, ESI†). Similar behavior is observed in acetonitrile, with faster initial components (1 and 5 ps) and a slower 240 ps decay (Fig. S5, ESI†).

The transient absorption spectra for **EG-C**, **-T** and **-A** are shown in Fig. 3b and S6 (ESI†). They exhibit DAS and short-lived decay components that are similar to that of **EG-H$_2$**, however their long decay times are longer than that of **EG-H$_2$** in methanol (Table 1). The long-lived components of the DAS are also similar to those extracted from the fluorescence up-conversion data (Fig. S4, ESI†). The fsTA spectra of **EG-G** (Fig. 3c and S7a, ESI†) exhibit two short-lived DAS components that are similar in band shape to those of **EG-C**, and a third, longer-lived component which has band maxima at 455 nm and 700 nm. The differences between the transient spectra of the hairpin **EG-Z** and those of the hairpins **EG-C**, **-T** and **-A** are even more pronounced (Fig. 4a and S7b, ESI†), with new bands appearing with maxima at around 460 and 680 nm. The initially formed 508 nm band is broader and shorter-lived than the initial transients of the other **EG-X** hairpins. The DAS for these processes in **EG-Z** exhibit both decreases at around 500 nm and rises near 460 and 680 nm, which matches the observed behavior of the spectra. The lifetimes of these first two DAS are ~0.3 and 3.3 ± 0.3 ps, respectively. Following these, the 460 and 680 nm peaks decay bi-exponentially, with the major and minor components exhibiting decay times of 26 ± 1 ps and 68 ± 4 ps, respectively.

The transient absorption spectra for the hairpins **EG$_2$** and **EG$_3$** (Fig. 4b and c and Fig. S8a and b, ESI†) exhibit even larger divergence from those of **EG-C**, **-T** and **-A**. As in the case of **EG-Z**, both **EG$_2$** and **EG$_3$** display strong bands at ~450 and 680 nm after a few ps. However, these bands exhibit longer lifetimes than those observed for **EG-Z.** Similarly, the five DAS needed to fit the spectra of **EG$_2$** and **EG$_3$** exhibit significantly longer decay times than those of **EG-Z.** In **EG$_2$**, the decay-associated fitting shows that the initially formed band with a maximum near 508 nm decays bi-exponentially, accompanied by a growth near 450 nm and a broad feature centered at 680 nm. This process occurs with ~0.3 and 3.0 ps time constants. **EG$_2$** also exhibits a slower, 30 ± 1 ps decay of the 500 nm band, with a concomitant increase of the amplitude near 430 nm, but growth near 680 nm is not observed in this component. The resulting intermediate exhibits peaks at 450 nm and 685 nm that dominate the spectra and which decay bi-exponentially, with the major and minor components exhibiting decay times of 490 ± 10 ps and 2900 ± 170 ps, respectively.

The dynamics and DAS for **EG$_3$** are quite similar to those for **EG$_2$**, but with even longer-lived states. The 508 nm band is rapidly replaced by features at 450 and 680 nm, similar to those of **EG$_2$**, but with longer time constants of 0.7 ± 0.3 and 5.1 ± 0.3 ps, respectively. Further decay of the 508 nm band occurs over 71 ± 4 ps with no associated growth of the 680 nm band. The resultant intermediate again

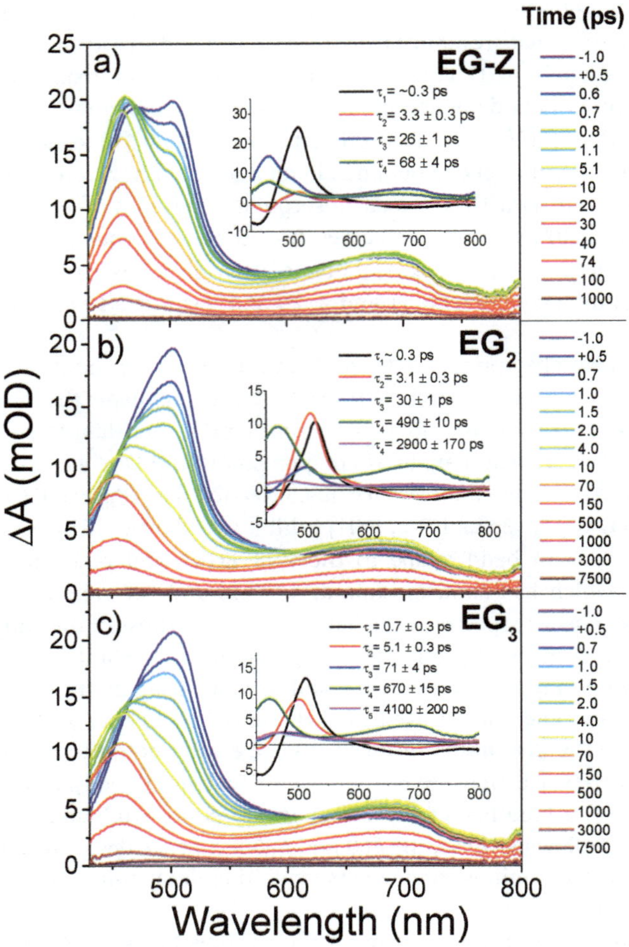

Fig. 4 Transient absorption spectra for the hairpins (a) **EG-Z**, (b) $\mathbf{EG_2}$ and (c) $\mathbf{EG_3}$ in phosphate buffer (10 mM phosphate, 0.1 M NaCl, pH 7.2) following excitation by a 350 nm, ~60 fs pulse (1.0 μJ per pulse). Insets: decay-associated spectra following multi-wavelength global fitting.

decays bi-exponentially, with decay times of 670 ± 15 and 4100 ± 200 ps. The lifetimes extracted from the decay-associated fitting are given in Table 1.

3.2.3. Femtosecond stimulated Raman spectroscopy. The FSRS spectrum of $\mathbf{EG\text{-}H_2}$ in methanol, acquired with a 480 nm Raman pump following 350 nm actinic excitation, is shown in Fig. S9a (ESI†) and exhibits bi-exponential decay with ~1 and 13 ps decay times. Mode assignments are given in Table S4 (ESI†). The decay-associated Raman spectra exhibit the same general shape, with subtle changes in the relative intensities of the transitions near 1000 cm^{-1} (Fig. S9a, ESI†). We also observe a broadening and small shift of the C≡C stretch at 2106 cm^{-1}. These differences indicate that the coupling of these vibrations to the 480 nm electronic transition changes as the state evolves following excitation. The FSRS spectra of the hairpins **EG-C, EG-G, EG-Z** and $\mathbf{EG_2}$ are shown in Fig. 5. The two highest frequency bands for **EG-C** and **EG-G** are narrower and more intense

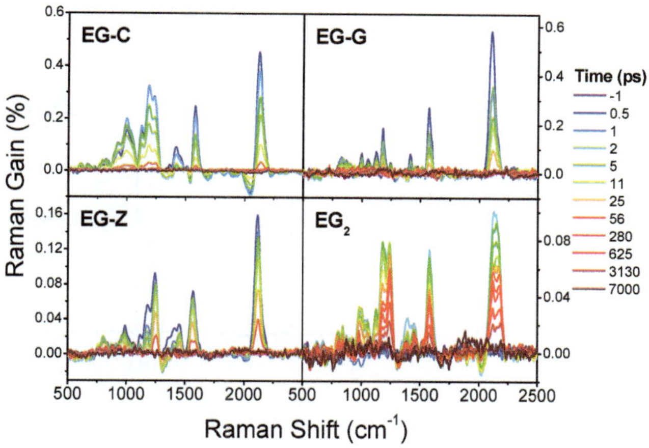

Fig. 5 Femtosecond stimulated Raman spectra of the hairpins **EG-C**, **EG-G**, **EG-Z** and **EG$_2$** following 350 nm actinic excitation and collection using a 480 nm Raman pump.

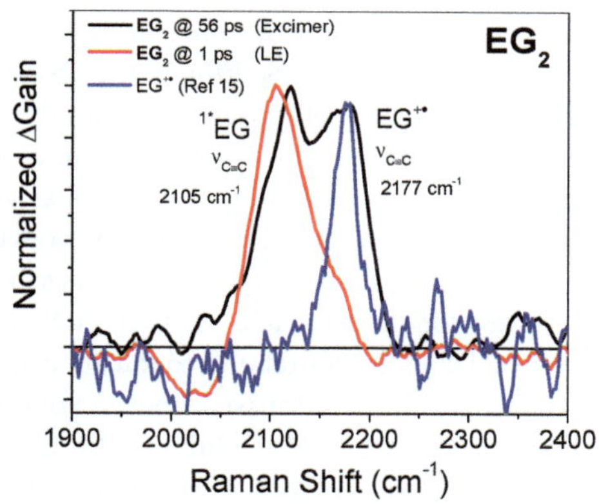

Fig. 6 Overlay of the normalized FSRS spectrum of **EG$_2$** at 56 ps (black curve) with the ^{1*}**EG** spectrum of **EG$_2$** at 1 ps (red curve) and the **EG$^{+\bullet}$** spectrum (blue curve) from ref. 15.

than those of the other hairpins. The bands of **EG-Z** are notably less intense, broader and shifted to a lower frequency. The **EG$_2$** bands are most distinct, with the highest frequency band split into two peaks near 2110 and 2180 cm^{-1} that decay over several nanoseconds (Fig. 6), in keeping with the long decays observed in the fsTA experiments (Table 1). Additional variations in peak intensities are observed in the lower frequency 750–1500 cm^{-1} region of **EG$_2$** (Fig. 5), thus demonstrating its unique behavior relative to the other hairpins. The strong coupling between nucleobases complicates the computation and assignment of the excited-state Raman bands of these hairpins.

4 Discussion

4.1. Excited-state dynamics of EG-H$_2$

The excited-state behavior of **EG-H$_2$** in MeOH can be summarized as shown in Scheme 1a. The bi-exponential evolution of the fsTA is attributed to the geometric

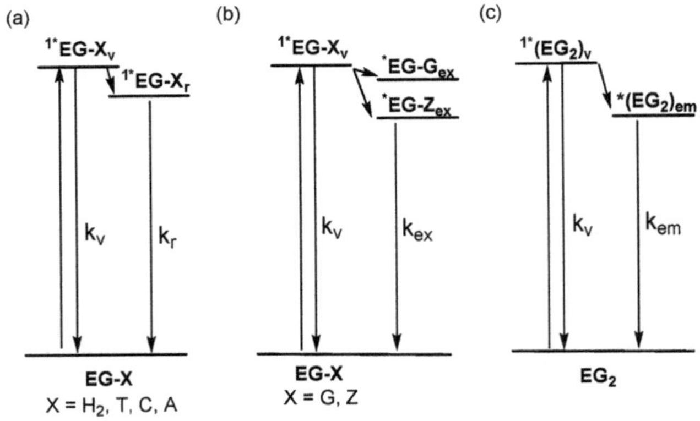

Scheme 1 Schematic for excited-state evolution showing the formation and decay of vertical and relaxed states for (a) nucleoside **EG-H$_2$** and the hairpins **EG-C**, **-T** and **-A**, (b) exciplex states for **EG-G** and **EG-Z** and (c) the excimer state for **EG$_2$**. The decay times are given in Table 1. Assignments of the fast and slow decay components are discussed in the text.

relaxation of the vertical excited state ^{1*}**EG-H$_{2v}$** to the relaxed excited state ^{1*}**EG-H$_{2r}$**. Indeed, there may be significant nuclear rearrangement from the vertically accessed state during this time, resulting in the observed bi-exponential decay. Such behavior has been recently proposed to explain the photophysics of 8-vinylguanine (VG).[24] The fluorescent vertical state of VG can cross a low barrier to access a part of the potential energy surface that more closely resembles the L_a state of unsubstituted guanine,[25] where rapid internal conversion or motion through a conical intersection (CI) can be accessed.[26] CIs have been established as the primary decay pathway for photoexcited natural nucleobases.[27] A similar twisting of the phenylethynyl substituent relative to the guanine core may explain the changes in the Raman spectrum discussed above. Notably, computational work predicts that the vertical state of VG exhibits a higher degree of charge-transfer (CT) character than the relaxed non-fluorescent state.[24] A similar relaxation mechanism for **EG-H$_2$** would explain the observed trends in the fluorescence quantum yield Φ_{fl} with solvent polarity and hydrogen bonding. In their original work, Saito and co-workers noted that the fluorescence of **EG-H$_2$** is solvent-dependent, being weaker and red-shifted in the polar solvents acetonitrile and methanol than in the less polar solvent ethyl acetate.[12] A higher, solvent-dependent barrier to accessing the L_a-like region of the potential energy surface in **EG-H$_2$** could result in a reduction of the rate of approach to the CI, thereby increasing the fluorescence quantum yield. This mechanism is also consistent with the multi-exponential decay observed in both polar and non-polar solvents, and the increased complexity of the decay dynamics with increasing solvent polarity (Fig. S4 and S5, ESI†). The observation of the ~200 ps emissive state of **EG-H$_2$** in THF might suggest that the barrier to accessing the L_a-like state is modulated *via* hydrogen bonding; the decays are relatively fast in the protic solvent MeOH but are slowed in the aprotic solvents THF and acetonitrile.

4.2. Hairpins with flanking natural nucleobases

Saito *et al.* investigated the fluorescence properties of duplexes possessing a *p*-acetyl substituted **EG** flanked by either C or T bases with matched or mismatched

 This journal is © The Royal Society of Chemistry 2018

bases in the complementary strand opposite **EG**, and found that the fluorescence intensity for the mismatch of **EG** with T was higher than that for a mismatch with A or G or base pairing with C.[13] The hairpins in our study have the **EG** base paired with C; only the identity of the flanking base pairs is varied (Chart 1). The small shifts in the UV and fluorescence maxima of **EG-C** and **EG-T** when compared to **EG-H$_2$** (Fig. 1a) and the weak negative long-wavelength bands in their CD spectra (Fig. 2a) are indicative of weak ground-state interactions of **EG** with the neighboring pyrimidine bases.[28]

The time-resolved spectroscopy data suggest that the excited-state behavior of the **EG-C**, **EG-T** and **EG-A** hairpins is similar in some respects to that of **EG-H$_2$**, thus implying minimal excited-state interaction with the neighboring nucleobases. The hairpins **EG-C** and **EG-T** have Φ_{fl} values and short-lived components of fluorescence and transient absorption decay that are similar to those for **EG-H$_2$** (Table 1). The long-lived components of the fluorescence and transient absorption decay of **EG-C** are longer than those of **EG-H$_2$** and are assigned to the decay of the relaxed excited state ^{1*}**EG-X$_r$** *via* a mechanism similar to that for **EG-H$_2$** discussed above (Scheme 1a). The phenylethynyl group extends into the minor groove adjacent to the guanosine deoxyribose, and thus decay of the relaxed excited states of **EG-T** and **-C** may be slowed relative to **EG-H$_2$** by the ordered solvent within the minor groove as well as base pairing and base stacking within the DNA scaffold. The FSRS spectra for **EG-C** and **EG-H$_2$** (Fig. 5 and S9a, ESI†) exhibit the same relative band structure and shift in the ethynyl stretch at longer times, which suggests that the relaxation process is not substantially perturbed by the base-paired structure. Although **EG-A** deploys purine bases on both sides of the **EG** nucleoside, it possesses ground- and excited-state properties that are similar to those of **EG-C** and **-T**. Based on the difference in the oxidation potentials of A and **EG** (Table 1), their HOMO energies are likely still too disparate in energy to significantly mix, thus resulting in their similar excited-state dynamics with only a small extension of the excited-state lifetime (Table 1). The excited-state behavior of the hairpins **EG-C**, **-T** and **-A** is summarized in Scheme 1a.

The CD spectrum of **EG-G** exhibits stronger exciton coupling in the 300–350 nm region than for the spectra of **EG-C**, **-T** and **-A** (Fig. 1a). More pronounced interactions also appear in the excited-state behavior of **EG-G**, even though it possesses a single flanking purine base rather than two as is the case for **EG-A**. The absorption maximum is also shifted to 336 nm, which is the longest wavelength in the series (Table 1). While the maximum of the dominant emission of **EG-G** is similar to those of **EG-C**, **-T** and **-A**, its fluorescence is more strongly quenched, with $\Phi_{fl} = 0.7\%$, which is the lowest of the sequences possessing flanking natural bases. It also exhibits weak long-wavelength fluorescence extending beyond 550 nm, which is attributed to a weakly fluorescent exciplex (Fig. 2d). Accordingly, both the vertical and relaxed excited states of **EG-G** in the fluorescence up-conversion measurements decay more rapidly than those of **EG-A**. We attribute this behavior to the quenching of the fluorescent ^{1*}**EG-G$_v$** state *via* the formation of the weakly-fluorescent exciplex (Scheme 1b).

The transient absorption spectra of **EG-G** (Fig. 3c) support this mechanism. These spectra exhibit two short-lived components of the DAS that are similar in band shape to those of **EG-C**, and a third, longer-lived component which has a band maximum at 455 nm as well as at 700 nm. This state decays with a 73 ps lifetime, which is longer than the longest time observed by time-resolved fluorescence. We assign this state to the weakly emissive exciplex ^{1*}**EG-G$_{ex}$**. Since **EG** and G have similar oxidation potentials, this intermediate is best viewed as a "heteroexcimer" with resonance contributions from both of the charge resonance states EG^{-}-G^{+} and EG^{+}-G^{-}, as well as the locally excited state ^{1*}**EG-G**.[29] The similar redox potentials of **EG** and G[15] allow for sufficient mixing to form an exciplex with moderate CT character; the extended 70 ps lifetime reflects the stabilization afforded by this mixing. The lower LUMO energy for **EG** *vs.* G makes it likely that EG^{-}-G^{+} is an important charge-resonance contributor. However, we have also demonstrated[15] that **EG**$^{+\bullet}$ absorbs at 460 nm with a spectral width similar to that shown in Fig. 3c, so this absorption may indicate a contribution from that charge-resonance state. The perturbation of the low frequency portion of the **EG-G** FSRS spectra relative to **EG-H$_2$** also indicates a minor impact of excited-state coupling between **EG** and G.

In summary, ^{1*}**EG** does not appear to strongly interact with the natural nucleobases, with the exception of guanine where the HOMO energy levels are the most favorable for exciplex formation (*vide infra*). In the case of **EG-C**, **-T** and **-A**, the solvent and nuclear relaxation of the vertical excited state yields a longer-lived relaxed singlet state, with minimal electronic mixing between the locally excited state of **EG** and the flanking bases (Fig. 1a). In the case of **EG-G**, the vertical excited state forms a weakly fluorescent exciplex that possesses a small amount of charge-transfer character (Fig. 1b). Exciplex formation with G might compete with the formation of a relaxed **EG-G** or **-C** complex, thus accounting for the several components of the fsTA. However, in the absence of direct observation of additional intermediates, only the exciplex is shown in Fig. 1b.

4.3. Hairpins with flanking deazaguanine

We next consider the excited-state behavior of ^{1*}**EG** when it is adjacent to the modified nucleobase Z. Unlike the other **EG-X** conjugates, **EG-Z** exhibits strong exciton-coupled CD bands at long wavelengths (Fig. 1b). The absorption and fluorescence maxima for **EG-Z** are similar to those for **EG-T** and **-C**, however, its value of Φ_{fl} is markedly smaller and it exhibits a weak fluorescence band near 560 nm, similar to that of **EG-G**, which is attributed to exciplex fluorescence (Fig. 2d). The values of Φ_{fl} and the short- and long-lived components of τ_{fl} for the **EG-X** hairpins with flanking purines follow the order of purine oxidation potential: Z < G < A (Table 1). These data corroborate the assertion that purines can form exciplexes with **EG**, with the extent of the CT character being determined by the purine oxidation potential. Similar behavior has been observed for the fluorescent base aminopurine in its dinucleotides, with the most efficient quenching being observed when AP is stacked with G.[30,31]

The transient absorption spectra of **EG-Z** (Fig. 4a) also differ in appearance from those of the other **EG-X** hairpins (Fig. 3b and c). This difference is attributed to the greater extent of charge-transfer character in the exciplex as a consequence of the increased contribution of the charge-resonance state EG^{-}-Z^{+} to the resonance description of the exciplex. The initially formed ^{1*}**EG-Z$_v$** state ($\lambda_{max} \sim 500$ nm) decays *via* rapid exciplex formation, leading to the exciplex ^{1*}**EG-Z$_{ex}$** with maxima near 460 and 680 nm without the formation of a relaxed singlet state (Scheme 1c). The prompt decay of the weak fluorescence and the 500 nm DAS is attributed to the decay of the vertical singlet state, whereas the slower decays of

 This journal is © The Royal Society of Chemistry 2018

the fluorescence and the 460 and 680 nm DAS are attributed to the exciplex. The dual exponential decay of the exciplex plausibly arises as a result of the solvent or nuclear relaxation of the polar exciplex within the solvated DNA duplex. Alternatively, the non-rigid DNA backbone may simply lead to a distribution of conformers with variations in the π-stacking and subsequently the exciplex decay rates.[5]

4.4. Hairpins with adjacent phenylethynylguanine

The hairpins **EG$_2$** and **EG$_3$** which possess two and three stacked **EG** bases, respectively (Chart 1), exhibit strong electronic interactions in both the ground and excited states. The CD spectra of these hairpins exhibit strong exciton coupling unlike that of the **EG-X** hairpins (Fig. 1a and b). The blue-shifted CD maxima and minima and increased CD intensity for **EG$_3$** and **EG$_2$** are similar to those for duplexes that possess helical arrays of the modified nucleobase pyrollocytosine, which has a calculated torsion angle of ~30° between the long axes of its chromophores.[7] The long-wavelength absorption maxima of **EG$_2$** and **EG$_3$** (Fig. 2b) are blue-shifted with respect to that of **EG-G**, with the shift being largest for **EG$_3$**, however, the fluorescence maxima of **EG$_2$** and **EG$_3$** exhibit the largest redshifts (Fig. 2d). Less pronounced changes in the UV band shape and similar, but weaker, EC-CD spectra have been reported for DNA duplexes possessing adjacent 2-aminopurine[32] and pyrrolocytosine[7,33] chromophores. The more intense EC-CD spectra for adjacent **EG** no doubt reflect the fully allowed nature of its lowest energy π–π* transition (Fig. S3†). The fluorescence spectra of **EG$_2$** and **EG$_3$** are broadened and red-shifted with respect to those of the natural-base **EG-X** hairpins (Fig. 1), which is consistent with their assignment to excimer or overlapping monomer and excimer fluorescence. The values of Φ_{fl} and both the short- and long-lived components of τ_{fl} are smaller than those for **EG-C** (Table 1), which is as expected for an excimer state wherein non-radiative decay dominates.[34,35] Since **EG$_2$** and **EG$_3$**, like **EG-C**, have flanking cytosine bases, the reduction in Φ_{fl} reflects the non-radiative decay of the ^{1*}**EG$_{2em}$** excimer rather than the quenching of excimer fluorescence.

The fsTA DAS of **EG$_2$** and **EG$_3$** are similar in appearance to those of **EG-Z**, thus suggesting similar relaxation dynamics (Fig. 4). However, the lifetimes of the longest states are significantly longer for **EG$_2$** than for **EG-Z** (2.9 ns *vs.* 68 ps), as is typical for aromatic hydrocarbon excimers within DNA.[35,36] The observation of two absorption maxima with a large spectral separation for the transient absorption spectrum of an excimer has previously been observed for perylenediimide dimers, and was attributed to the non-aligned geometry of their transition dipoles.[37] Alignment of the transition dipoles of adjacent bases within B-DNA would require substantial torsional motion of the duplex and is therefore unlikely. Transient absorption measurements of π-stacked dimers have shown that ultrafast excimer formation can be followed by an increase in the oscillator strength of excimer absorption on the ps timescale attributed to structural relaxation of the excimer.[38–40] The ultrafast component of the growth of the exciplex feature would thus reflect the excimer formation process, while the slower growth may represent a similar structural relaxation event. The excited-state properties of **EG$_2$** and **EG$_3$** are similar but not identical, as would be expected for excimers possessing two and three identical chromophores. However, the environment of the **EG** chromophores differs within the two hairpins, with the excimer from **EG$_2$** possessing two cytosine neighbors and the excimer from **EG$_3$** possessing either of two possible cytosines and one **EG** as neighbors.

The excited-state behavior of **EG$_2$** and **EG$_3$** can be summarized as shown in Scheme 1c. Despite the observation of strong exciton coupling in the UV and CD spectra, the transient absorption spectra of **EG$_2$** and **EG$_3$** (Fig. 4a–c) immediately following excitation are similar to those of the **EG-X** hairpins and are thus assigned to the vertical excited state of a single **EG** chromophore. Again, the multiple formation rates extracted from the decay-associated fitting of the fsTA data are consistent with bi-exponential fluorescence decay. The initially populated short-lived fluorescent state can either form the fluorescent relaxed state ^{1*}**EG–EG$_r$** or the long-lived excimer ***EG–EG$_{ex}$**.

The FSRS spectra of **EG$_2$** exhibit interesting features for the **EG** excimer. The lower intensity of the **EG$_2$** FSRS spectra compared to those of the other hairpins at similar concentrations is consistent with previously observed excited-state interactions between cofacial chromophores, which have been demonstrated to exhibit substantially weakened and broadened band intensities.[41] Uniquely, the broad **EG$_2$** ethynyl stretch, initially at 2140 cm^{-1}, bifurcates into two well resolved bands at 2105 and 2177 cm^{-1} within the first 10 ps. This change is accompanied by the loss of intensity in the 1178 cm^{-1} C–N stretching mode associated with the decay of the locally excited ^{1*}**EG** state into the **EG–EG** excimer. Importantly, the 2177 cm^{-1} band is of a similar frequency to that of **EG**$^{+\bullet}$ observed in our previous work.[15] Fig. 6 shows an overlay of the **EG$_2$** excimer FSRS spectrum and that of **EG**$^{+\bullet}$ adapted from ref. 15, along with the ^{1*}**EG** spectrum in the ethynyl stretch region. The simultaneous appearance of transitions associated with both **EG**$^{+\bullet}$ and ^{1*}**EG** is consistent with the classic picture of an excimer as the mixture of those two electronic states, and suggests a substantial contribution from the charge-resonance configuration in the excimer.[41] Overall, the high CT-character of the **EG** excimer offers an interesting probe into the electronic interactions within excimers and allows us to track their structural evolution over time.

5 Concluding remarks

The fluorescent base analog 8-phenylethynylguanine (**EG**) has proven to be a versatile probe of ground- and excited-state interactions with flanking nucleobases in fully base-paired DNA hairpins. Base pairing with C places **EG** at the edge of the minor groove of B-DNA where it does not disrupt the base-paired structure. The fluorescence quantum yields and decay times of the hairpins possessing flanking T, C and A bases in aqueous buffer are similar to those for the nucleoside in methanol, which is indicative of weak ground- and excited-state interactions between **EG** and these bases. The excited states of these hairpins decay *via* fluorescence and non-radiative pathways from both vertical and relaxed singlet states. ^{1*}**EG** forms charge transfer-stabilized exciplexes with G and Z, with the extent of charge separation increasing as the purine oxidation potential decreases. When paired with itself, ^{1*}**EG** forms an excimer, having significantly longer lifetimes than its exciplexes with G or Z. The **EG** excimer appears to have significant CT character, as evidenced by the observation of both ^{1*}**EG** and **EG**$^{+\bullet}$ bands in the FSRS spectra.

The interplay between electron transfer and exciplex formation remains an area of deep interest.[42,43] The work discussed above strongly highlights the continued need to develop new systems and techniques to probe fundamental interactions between cofacial, electron-rich molecules in functional systems.

Acknowledgements

This material is based upon work supported by the U.S. Department of Energy, Office of Science, Office of Basic Energy Sciences, Chemical Sciences, Geosciences, and Biosciences Division under the Awards DE-FG02-96ER14604 (F. D. L.) and DE-FG02-99ER14999 (M. R. W.).

References

1 J. N. Wilson and E. T. Kool, *Org. Biomol. Chem.*, 2006, **4**, 4265–4274.
2 R. W. Sinkeldam, N. J. Greco and Y. Tor, *Chem. Rev.*, 2010, **110**, 2579–2619.
3 C. Wan, T. Fiebig, O. Schiemann, J. K. Barton and A. H. Zewail, *Proc. Natl. Acad. Sci. U. S. A.*, 2000, **97**, 14052–14055.
4 T. Gelot, P. Tourón-Touceda, O. Crégut, J. Léonard and S. Haacke, *J. Phys. Chem. A*, 2012, **116**, 2819–2825.
5 V. A. Spata, W. Lee and S. Matsika, *J. Phys. Chem. Lett.*, 2016, **7**, 976–984.
6 K. C. Thompson and N. Miyake, *J. Phys. Chem. B*, 2005, **109**, 6012–6019.
7 P. P. Neelakandan, M. McCullagh, G. C. Schatz and F. D. Lewis, *J. Phys. Chem. B*, 2012, **116**, 5199–5204.
8 A. Trifonov, M. Raytchev, I. Buchvarov, M. Rist, J. Barbaric, H.-A. Wagenknecht and T. Fiebig, *J. Phys. Chem. B*, 2005, **109**, 19490–19495.
9 E. T. Kool, *Acc. Chem. Res.*, 2002, **35**, 936–943.
10 A. Dallmann, L. Dehmel, T. Peters, C. Mügge, C. Griesinger, J. Tuma and N. P. Ernsting, *Angew. Chem., Int. Ed.*, 2010, **49**, 5989–5992.
11 A. G. Firth, I. J. S. Fairlamb, K. Darley and C. G. Baumann, *Tetrahedron Lett.*, 2006, **47**, 3529–3533.
12 Y. Shinohara, K. Matsumoto, K. Kugenuma, T. Morii, Y. Saito and I. Saito, *Bioorg. Med. Chem. Lett.*, 2010, **20**, 2817–2820.
13 Y. Saito, K. Kugenuma, M. Tanaka, A. Suzuki and I. Saito, *Bioorg. Med. Chem. Lett.*, 2012, **22**, 3723–3726.
14 Y.-L. Wu, K. E. Brown, D. M. Gardner, S. M. Dyar and M. R. Wasielewski, *J. Am. Chem. Soc.*, 2015, **137**, 3981–3990.
15 K. E. Brown, A. P. N. Singh, Y.-L. Wu, A. K. Mishra, J. Zhou, F. D. Lewis, R. M. Young and M. R. Wasielewski, *J. Am. Chem. Soc.*, 2017, DOI: 10.1021/jacs.7b06998.
16 A. S. Jalilov, S. Patwardhan, A. Singh, T. Simeon, A. A. Sarjeant, G. C. Schatz and F. D. Lewis, *J. Phys. Chem. B*, 2014, **118**, 125–133.
17 F. D. Lewis, T. Wu, X. Liu, R. L. Letsinger, S. R. Greenfield, S. E. Miller and M. R. Wasielewski, *J. Am. Chem. Soc.*, 2000, **122**, 2889–2902.
18 W. H. Melhuish, *J. Phys. Chem.*, 1961, **65**, 229–235.
19 R. M. Young, S. M. Dyar, J. C. Barnes, M. Juríček, J. F. Stoddart, D. T. Co and M. R. Wasielewski, *J. Phys. Chem. A*, 2013, **117**, 12438–12448.
20 K. E. Brown, B. S. Veldkamp, D. T. Co and M. R. Wasielewski, *J. Phys. Chem. Lett.*, 2012, **3**, 2362–2366.

21 K. E. V. Holde, W. C. Johnson and P. S. Ho, *Principles of Physical Biochemistry*, Pearson/Prentice Hall, 2006.
22 N. Berova and K. Nakanishi, in *Circular Dichroism*, ed. N. Berova, K. Nakanishi and R. W. Woody, Wiley-VCH, New York, 2000.
23 M. J. Frisch, G. W. Trucks, H. B. Schlegel, G. E. S. M. A. Robb, J. R. Cheeseman, G. Scalmani, V. Barone and G. A. Petersson, *Gaussian 09*, Gaussian, Inc., Wallingford CT, 2009.
24 M. A. Kochman, M. Pola and R. J. D. Miller, *J. Phys. Chem. A*, 2016, **120**, 6200–6215.
25 A. F. El-Yazbi, A. Palech and G. R. Loppnow, *J. Phys. Chem. A*, 2011, **115**, 10445–10451.
26 C. T. Middleton, K. d. L. Harpe, C. Su, Y. K. Law, C. E. Crespo-Hernández and B. Kohler, *Annu. Rev. Phys. Chem.*, 2009, **60**, 217–239.
27 C. E. Crespo-Hernandez, B. Cohen, P. M. Hare and B. Kohler, *Chem. Rev.*, 2004, **104**, 1977–2019.
28 F. D. Lewis, L. Zhang, X. Liu, X. Zuo, D. M. Tiede, H. Long and G. S. Schatz, *J. Am. Chem. Soc.*, 2005, **127**, 14445–14453.
29 M. Ottolenghi, *Acc. Chem. Res.*, 1973, **6**, 153–160.
30 O. F. A. Larsen, I. H. M. van Stokkum, F. L. de Weerd, M. Vengris, C. T. Aravindakumar, R. van Grondelle, N. E. Geacintov and H. van Amerongen, *Phys. Chem. Chem. Phys.*, 2004, **6**, 154–160.
31 O. J. G. Somsen, L. B. Keukens, M. N. de Keijzer, A. van Hoek and H. van Amerongen, *ChemPhysChem*, 2005, **6**, 1622–1627.
32 N. P. Johnson, W. A. Baase and P. H. von Hippel, *Proc. Natl. Acad. Sci. U. S. A.*, 2004, **101**, 3426–3431.
33 N. P. Johnson, W. A. Baase and P. H. von Hippel, *Proc. Natl. Acad. Sci. U. S. A.*, 2005, **102**, 7169–7173.
34 K. E. Brown, W. A. Salamant, L. E. Shoer, R. M. Young and M. R. Wasielewski, *J. Phys. Chem. Lett.*, 2014, **5**, 2588–2593.
35 E. A. Chandross and H. T. Thomas, *J. Am. Chem. Soc.*, 1972, **94**, 2421–2424.
36 F. D. Lewis, T. Wu, E. L. Burch, D. M. Bassani, J.-S. Yang, S. Schneider, W. Jaeger and R. L. Letsinger, *J. Am. Chem. Soc.*, 1995, **117**, 8785–8792.
37 P. P. Neelakandan, T. A. Zeidan, M. McCullagh, G. C. Schatz, J. Vura-Weis, C. H. Kim, M. R. Wasielewski and F. D. Lewis, *Chem. Sci.*, 2014, **5**, 973–981.
38 J. Sung, P. Kim, B. Fimmel, F. Würthner and D. Kim, *Nat. Commun.*, 2015, **6**, 8646.
39 C. M. Mauck, R. M. Young and M. R. Wasielewski, *J. Phys. Chem. A*, 2017, **121**, 784–792.
40 E. S. S. Iyer, A. Sadybekov, O. Lioubashevski, A. I. Krylov and S. Ruhman, *J. Phys. Chem. A*, 2017, **121**, 1962–1975.
41 Y. Wu, J. Zhou, B. T. Phelan, C. M. Mauck, J. F. Stoddart, R. M. Young and M. R. Wasielewski, *J. Am. Chem. Soc.*, 2017, **139**(40), 14265–14276.
42 M. Koch, G. Licari and E. Vauthey, *J. Phys. Chem. B*, 2015, **119**, 11846–11857.
43 B. Dereka, M. Koch and E. Vauthey, *Acc. Chem. Res.*, 2017, **50**, 426–434.

 This journal is © The Royal Society of Chemistry 2018

Faraday Discussions

PAPER

Two-dimensional electronic spectroscopy as a tool for tracking molecular conformations in DNA/RNA aggregates

Javier Segarra-Martí, *[a] Vishal K. Jaiswal,[b] Ana Julieta Pepino,[b] Angelo Giussani,[c] Artur Nenov,[b] Shaul Mukamel,[d] Marco Garavelli *[b] and Ivan Rivalta *[a]

Received 30th September 2017, Accepted 17th October 2017
DOI: 10.1039/c7fd00201g

A computational strategy to simulate two-dimensional electronic spectra (2DES) is introduced, which allows us to analyse ground state dynamics and to sample and measure different conformations attained by flexible molecular systems in solution. An explicit mixed quantum mechanics/molecular mechanics (QM/MM) approach is employed for the evaluation of the necessary electronic excited state energies and transition dipole moments. The method is applied towards a study of the highly flexible water-solvated adenine–adenine monophosphate (ApA), a system featuring two interacting adenine moieties that display various intermolecular arrangements, known to deeply affect their photochemical outcome. Molecular dynamics simulations and cluster analysis have been used to select the molecular conformations, reducing the complexity of the flexible ApA conformational space. By using our sum-over-states (SOS) approach to obtain the 2DES spectra for each of these selected conformations, we can discern spectral changes and relate them to specific nuclear arrangements: close lying π-stacked bases exhibit a splitting of their respective ^1L_a signal traces; T-stacked bases exhibit the appearance of charge transfer states in the low-energy Vis probing window while displaying no ^1L_a splitting, being particularly favoured when promoting amino to 5-ring interactions; unstacked and distant adenine moieties exhibit signals similar to those of the adenine monomer, as is expected for non-interacting nucleobases. 2DES maps reveal the spectral fingerprints associated with specific molecular conformations, and are thus a promising option to enable their quantitative spectroscopic detection beyond standard 1D pump-probe techniques. This is expected to aid the understanding of how nucleobase aggregation controls and modulates the photostability and photo-damage of extended DNA/RNA systems.

[a]*Univ Lyon, Ens de Lyon, CNRS, Université Claude Bernard Lyon 1, Laboratoire de Chimie UMR 5182, F-69342, Lyon, France. E-mail: Javier.segarra-marti@ens-lyon.fr; ivan.rivalta@ens-lyon.fr*

[b]*Dipartimento di Chimica Industriale "Toso Montanari", Università di Bologna, Viale del Risorgimento, IT-40136, Italy. E-mail: marco.garavelli@unibo.it*

[c]*Department of Chemistry, University College London, 20 Gordon Street, London WC1H 0AJ, UK*

[d]*Department of Chemistry, University of California, Irvine, California 92697-2025, USA*

Introduction

The conformations attained by molecular systems in condensed (liquid or solid) phases are often surrounded by controversy given the difficulties faced by theorists and experimentalists in locating and assigning specific spectroscopic fingerprints pertaining to a given molecular motif. Various molecular arrangements, which often exhibit completely different photophysical properties due to their featured intermolecular interactions, are thus key to properly assessing and rationalising certain bulk and interfacial phenomena that are of paramount interest, such as the ultrafast and efficient electronic energy transfer processes registered in light harvesting complexes or the specific charge/transport mechanisms behind state-of-the-art solar cells.[1] This is especially true in biomolecular systems, where a specific conformation may impair or even completely alter a concrete cellular function. One particularly important case that highlights this effect would be the photostability of genetic material,[2,3] and how the photophysics of DNA/RNA sequences are severely affected in dimeric and multimeric species by the presence of additional neighbouring stacked nucleobases within a given strand in the double helix. Nucleobase stacking might indeed increase their excited state lifetime and potentially make them more prone to photochemical mutations,[4–7] which is in contrast to the efficient and ultrafast decay registered in isolated nucleobases,[8,9] which is expected to strongly promote its photoprotection mechanisms.[10] This effect, which is believed to strongly modulate the complexity of DNA photophysics,[11,12] is particularly hard to measure using standard pump-probe set-ups, which are the experimental workhorse in photochemistry. In standard 1D pump-probe spectroscopy, different signals from diverse molecular motifs overlap strongly in space and time, leading to highly congested spectral regions whose complex interpretation remains elusive.

A recently developed technique that can overcome these aforementioned limitations is two-dimensional (2D) electronic spectroscopy (2DES), which is able to disentangle signals arising from different intermolecular interactions that remain hidden in 1D pump-probe spectra. In these multi-pulse techniques, which are frequently used in the IR[13,14] and which have been recently extended to the electronic regime,[15–17] the system is excited by a sequence of three pulses with controlled delays, t_1 (coherence time) and t_2 (population/waiting time), with the third pulse at time t_3 emitting the third-order nonlinear signal S (t_1, t_2, t_3).[18–21] By carrying out Fourier transform with respect to t_1 and t_3, one can extract a sequence of 2D maps S (Ω_1, t_2, Ω_3) for different waiting times, t_2. These 2D maps contain a vast number of signals that can potentially be employed for tracking all sorts of localised, delocalised, energy and charge transfer photoinduced phenomena with unprecedented detail. However, their interpretation proves to be challenging. It is in this context that theoretical chemistry comes in handy; by simulating the spectra we can separate the different peaks into their contributing specific electronic transitions, thus providing a route map to recognise, understand and interpret the main fingerprints observed.[22,23]

In this contribution we introduce a computational strategy that combines ground-state molecular dynamics with QM/MM approaches in order to sample the conformational space of dimeric and multimeric flexible bio-chromophores and separate their different nuclear arrangements to be characterised spectroscopically by 2DES. The methodology is applied to ApA, a model system that contains both unstacked (non-interacting) and π- or T-stacked (interacting) adenine moieties, and is shown to display distinct fingerprints for these different conformations, thus proving to be a promising tool to separate them and understand precisely how they affect the underlying photophysical and photochemical pathways in such complex and extended DNA/RNA systems.

Computational details

Molecular dynamics simulations were carried out using the AMBER 11 software package.[24,25] A 12 × 12 × 12 Å cubic box was employed making use of the parmbsc0 (ref. 26) and TIP3P[27] force fields for the ApA moiety and water molecules, respectively. A 100 ns trajectory was carried out, extracting one snapshot every 2.5 ps, totalling 40 000 frames. Root mean square deviation (RMSD) cluster analysis was carried out on those 40 000 structures and was set within 1.5 Å, extracting 89 different clusters. From these, 17 were retained for further electronic structure theory evaluations, accounting for up to ∼75% of the conformational space sampled by the highly flexible ApA system. The selected snapshots, which are the closest structures to each centroid, were subsequently optimised by employing a complete active space self-consistent field (CASSCF) comprising 8 electrons in 8 orbitals, 4π occupied and 4π* unoccupied, constraining phosphate and sugar groups and allowing relaxations only on the adenine moieties in order to obtain more accurate bond length distances than those given by the force field. This has been done due to the strong dependence of the high-energy excited state manifold on the bond length distances displayed by adenine, as has been shown elsewhere.[28]

All electronic structure computations were carried out using the MOLCAS 8 package.[29] The restricted active space self-consistent field (RASSCF)[30] and its second-order perturbation theory extension (RASPT2)[31,32] were applied to the aforementioned selected structures. Cholesky decomposition was used to speed-up the two-electron integrals[33–35] and in conjunction with Atomic Natural Orbital (ANO)[36] large-type basis sets with valence double-ζ polarised contractions. Only the adenine nucleobases were included in the quantum mechanical region, with atom links connecting the QM (adenine moieties) and MM (phosphate and water molecules) layers.[37] An active space of 24 electrons in 18 orbitals was employed, comprising the whole π occupied valence space of both adenine moieties and all except the highest π* virtual orbital of each moiety. Up to four holes/electrons were included in the RAS1/RAS3 subspaces, respectively, leading to RASPT2(4,12|0,0|4,6), and these were averaged with equal weights over 100 states. RASPT2(4,6|0,0|4,3) was used for the monomer as previously benchmarked[28] and averaged over 25 states, with its active space being duplicated for the dimer. This particular active space was used considering previous benchmark studies where its adequacy to represent the energies and transition dipole moments with respect to reference RASPT2(0,0|12,10|2,12) computations on the adenine monomer was assessed.[28] An imaginary level shift[38] of 0.6 a.u. was employed for the dimer, with 0.5 a.u. employed for the monomer as previously benchmarked,[28] and with the IPEA shift being set to zero.[39,40] The restricted active space state interaction (RASSI) method[41,42] was employed to obtain the required transition dipole moments among the different electronic ground and excited states.

The resulting energy levels and transition dipole moments were then used in conjunction with the Sum-Over-States (SOS) approach[43] within the dipole approximation and a QM/MM scheme (SOS//QM/MM)[18] to simulate the 2DES as implemented in the Spectron 2.7 program.[44] Further details on the working equations have been given elsewhere.[28] Spectral lineshapes were simulated assuming that the dephasing is caused by pure dephasing in the Markovian approximation. Effectively, the spectra of the individual snapshots are subject to a constant line broadening of 1000 cm^{-1}, which has been set so that the broadenings along Ω_1 for the total ApA spectrum mimic the experimental broadening recently measured in 2DUV experiments for adenine.[17] Although several computational approaches have been devised recently to consider spectral diffusion and non-adiabatic effects on the simulated 2DES,[45–49] the size of the current system prevents the usage of more sophisticated approaches. Quasi-absorptive signals were simulated for the condition of a collinear pump pulse pair (thus simultaneously recording rephrasing and non-rephasing contributions). All reported signals used the all-parallel *xxxx* pulse polarisation configuration, were plotted on a linear scale and considered white light (infinite broadband) pulses. This allowed us to explore the variety of signals that may be present in the 2D spectra. Both ground state bleaching (GSB) and stimulated emission (SE) contributions appear as negative (blue) peaks, whereas excited-state absorptions (ESAs) appear as positive (red) peaks in the 2D spectra.

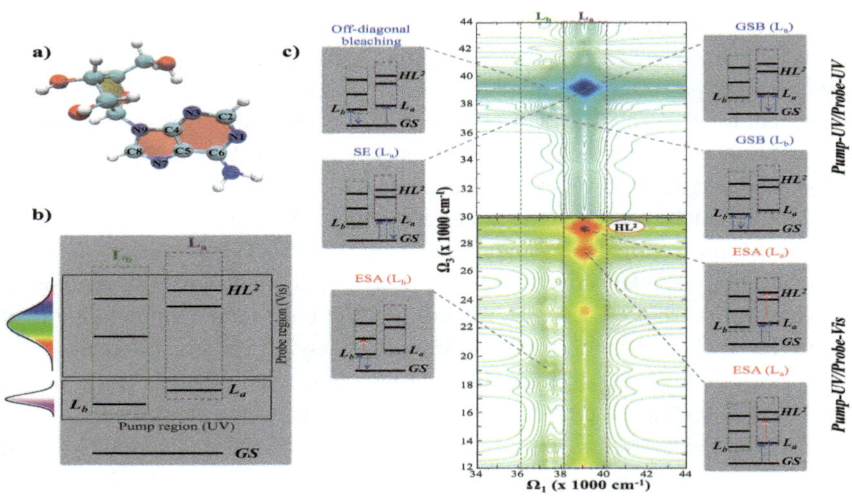

Fig. 1 2DES spectrum of adenosine monophosphate, where (a) displays the molecular system and atom labelling, (b) shows a scheme of the energy levels that are accessible with different laser pulses and (c) shows the 2DES in the Vis (lower panel) and UV (upper panel) probe windows, together with energy level schemes depicting the different signals featured in terms of the laser pulse sequence. A red colour denotes positive (ESA) signals. A blue colour denotes negative (GSB, off-diagonal bleach and SE) signals. The SE signal associated with the ^1L_b state was not included in the scheme for simplicity, being placed on top of GSB(L_b). The signal intensity in each spectrum is normalised according to the absolute value of the most intense signal.

Results

The 2DES of water-solvated adenosine is considered first (see Fig. 1) and is then used as a reference in order to interpret the signals arising in the ApA dimer. As shown in Fig. 1a, adenosine features a sugar group connected to the nucleobase through the N9 position, analogous to the adenine moieties contained in ApA, and is expected to exhibit mostly the same spectral signals as those of adenine while not exhibiting the known mix of tautomers (7*H*- and 9*H*-adenine) in the bare nucleobase embodying different excited state lifetimes[50,51] that complicate its analysis. As can be seen in Fig. 1b, by pumping (photo-exciting) in the absorbing UV region, the two main excited states of $\pi\pi^*$ character are expected to be populated, with a more stable and less intense ^1L_b state and a more intense and higher in energy ^1L_a state, even though their relative energetic state ordering is still under debate.[28,52,53] These two neighbouring excited states that are accessible from the ground state give rise to two traces along Ω_1 in the 2DES spectra, featuring ESAs in the Vis probing window and GSB and off-diagonal bleaching signals in the UV. This strong GSB contribution mostly obscures all signals from ESAs in the UV window. Fig. 1c shows the spectra together with a simple scheme depicting the pulse sequence leading to each of the signals, with the arrows representing the electronic transitions produced by the multi-pulse laser sequence. GSB arises from an excitation from the ground state to a given excited state, de-excitation from that state back to the ground state and again an excitation to the very same excited state. Off-diagonal bleaching follows the same first two excitations as GSB, with the third pulse promoting excitation to another excited state. SE and ESA signals are formed from a two-pulse coherent excitation from the ground state to a given excited state, which leads to population of the excited state, from which the third pulse either promotes excitation to higher lying states (ESA) or emission back to the ground state (SE).

Our results for water-solvated adenosine were compared against previous reference computations on gas-phase adenine performed with larger active spaces and basis sets *in vacuo*,[28] with notable differences observed along Ω_1 (linear absorption of the photo-activated ^1L_a and ^1L_b states of adenine) which is red-shifted by 0.2–0.3 eV, which is partially expected due to the addition of a polar protic solvent such as water that produces an analogous red-shifting effect in DNA/RNA nucleobases.[37,54–56] It is worth noting that more accurate approaches could be used to compute the spectrum of the monomer, as shown in previous benchmarks,[28] yet those cannot be extended to dimeric species, thus justifying our use of a cost-effective RASPT2(4,6|0,0|4,3) scheme. Beyond the red-shift observed along Ω_1, the other main difference observed with respect to the reference is the relative intensity of the ^1L_b trace with respect to that of ^1L_a, which appears to be obscured due to the usage of infinitely wide pulses that enhance the ^1L_a trace, as opposed to the tailored finite pulses employed in the reference computations to highlight this relatively dark state.[28] The absence of signals in the low-energy window as opposed to the reference computations[28] can be attributed to the use of infinitely broad pulses, which obscure them as has also been reported for the ^1L_b trace. The usage of such pulses allows us to focus on the overall most intense signals without considering particular pump and probe set-ups that could potentially be used with finite pulses to enhance specific peaks, which is for the moment out of scope of the present manuscript. Our reduced RASPT2(4,6|0,0|4,3) model is able to quantitatively reproduce the most important signal, that is, the one relating to the transition between ^1L_a and the doubly excited HOMO to LUMO state placed ~29 000 cm^{-1} along Ω_3 (HL^2), which is very intense and appears in different spectral regions for the different nucleobases,[28,57] thus being the main monomer-based fingerprint to potentially follow excited state decays on separate monomers simultaneously. An additional signal right below HL^2 can be observed in the spectrum (see Fig. 1c), which relates to a $\pi\pi^* \leftarrow {}^1L_a$ transition that appears to be less intense and that, given that additional homogeneous broadenings were included, would most likely merge with that of HL^2, with its in-depth analysis being considered not as important. In the higher energy UV-probe window all signals are obscured in solution, as opposed to what was observed in the gas phase reference computations,[28] where ESAs along both the ^1L_b and ^1L_a traces could still be observed, albeit at low intensity compared to the GSB signals.

As indicated previously, in order to obtain a 2DES spectrum that represents the whole conformational space, a reductionist approach was employed whereby cluster analysis was performed and the most populated clusters along the dynamics were used to reduce the number of QM/MM evaluations, which is the actual bottleneck of the computational procedure. In this way, by considering the first 17 clusters (from the most to the least populated), which have been labelled in alphabetical (from A to Q) order, we can account for up to ~75% of the sampled conformational space of ApA. All of the details are given in Table 1, which contains the concrete number of structures contained in each cluster, their contribution to the overall conformational space, a normalised estimate

Table 1 Clustering parameters obtained from the 100 ns molecular dynamics simulation of ApA. Relative weights are normalised and fitted to ‰ for the 2DES simulations. C5–C5 distance ranges are given (in Å) for each cluster.

Cluster	No. of snapshots contained	% over total (40 000 snapshots)	Normalised % (over 29 421 snapshots)	Weight (‰)	C5–C5 distance (Å)
A	4063	10.16	13.81	138	>6.0
B	3542	8.86	12.04	120	4.0–4.5
C	3470	8.68	11.79	118	<4.0
D	1916	4.79	6.51	65	<4.0
E	1744	4.36	5.93	59	4.5–5.0
F	1713	4.28	5.82	58	4.5–5.0
G	1689	4.22	5.74	57	4.0–4.5
H	1681	4.20	5.71	57	5.0–6.0
I	1661	4.15	5.65	57	4.5–5.0
J	1498	3.75	5.09	51	>6.0
K	1496	3.74	5.08	51	<4.0
L	1245	3.11	4.23	42	4.5–5.0
M	917	2.29	3.12	31	>6.0
N	821	2.05	2.79	28	<4.0
O	793	1.98	2.70	27	5.0–6.0
P	601	1.50	2.04	20	4.0–4.5
Q	571	1.43	1.94	19	5.0–6.0
Total	29 421	73.55%	99.99%	998	

 This journal is © The Royal Society of Chemistry 2018

(assuming sampling of the full conformational space) and the corresponding weights (in ‰) employed in the simulation of the total 2DES. As shown in Table 1, populations drop quickly in weight, with the first 8 clusters accounting for ~50% of the population while the next 9 represent ~25% of the conformational space; cluster A accounts for 7 times the contribution of cluster Q. The ~75% conformational space considered is assumed to be made into the full space for obtaining the concrete weighting factors to be used for each of the clusters, thus accounting for their different contributions to the overall 2DES of ApA. At this point it is important to note that a more accurate clustering method than the RMSD adopted here would be desirable in order to obtain less clusters yet be more representative, thus further minimising the number of QM/MM evaluations required and provide better weightings for the simulation of the full 2DES spectrum. Accurate weightings leading to a direct comparison with future experiments are beyond the scope of this work, which focuses instead on setting a viable strategy to simulate 2DES spectra and the inhomogeneous broadenings due to different molecular conformations. We have opted, indeed, for a qualitative treatment where the length of the ground state molecular dynamics simulation might not be enough to sample the full conformational space of ApA. Moreover, the force field employed was parameterised for double helix DNA chains (and not strictly for flexible systems like ApA), which may introduce an additional source of error. However, by employing the so-called "stacking coordinate" introduced by Florián and co-workers,[58] we obtain a ~45% population within a stacking coordinate of 4.2, which, assuming it is representative of stacked instances, seems to be in agreement with reported circular dichroism ApA estimates and is attributed to these interacting arrangements.[7] Considering all the caveats, we envision that this computational strategy could be systematically improved in various ways by employing, for instance, longer simulations, more suitable force fields, more accurate clustering techniques and *ab initio* simulations as well as more sophisticated treatments for the spectral line shapes.

Fig. 2 shows the molecular structure of water-solvated ApA, and both pump-UV/probe-Vis and pump-UV/probe-UV 2DES windows. The key fingerprint in the Vis probing window is the HL^2 transition that was described previously, which features as the main fingerprint of each of the adenine monomers, with a less localised monomer transition displayed right below HL^2 at ~25k cm^{-1} that is expected to form part of the same peak. The main difference from the monomer is the large broadening that is obtained by considering several structures, which is partially induced due to the interaction between the ^1L_a states of the different adenine moieties, thus leading to a splitting in their energy estimates along Ω_1, as will be analysed in detail below. Further considerations would have to be made in order to properly account for the signal line shape, which is here accounted for by homogeneous broadening due to pure dephasing and is thus symmetric in the single snapshots, with asymmetries in the total spectrum arising due to the averaging of the peaks over many different conformational arrangements. The other difference that is easily observed upon comparison with the monomer is the appearance of a signal in the low-energy Vis window (~15k cm^{-1} along Ω_3), which is attributed mainly to contributions that arise from CT states, as demonstrated by looking at their specific wave function contributions in our model. This signal is covered in some cases by localised monomer excitations that can also be observed at ~13 000 and ~19 000 cm^{-1} and which obscure the CT signals, thus

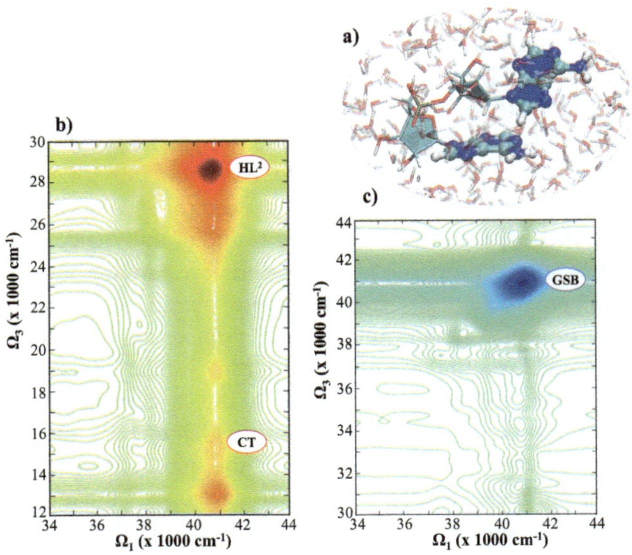

Fig. 2 2DES of water-solvated ApA considering the full conformational space (see Table 1), showing (a) the molecular structure, (b) the pump-UV/probe-Vis spectrum and (c) the pump-UV/probe-UV spectrum. The colour scheme for the 2DES spectra is set according to Fig. 1.

making their experimental characterisation challenging. These signals are not further discussed since they are associated with localised monomeric fingerprints. Moving onto the pump-UV/probe-UV window (Fig. 2c), all that can be seen is the inhomogeneous broadening of the GSB signal along the diagonal and anti-diagonal widths, which arises mainly due to the different nuclear arrangements considered and which also weakly displays the contributions of the ^1L_b state diagonal GSB located at $\Omega_1 = \Omega_3 \approx 38\,000\ \text{cm}^{-1}$. It is worth noting that the anti-diagonal width of the signal has been reported to hold information regarding the formation of aggregates,[59,60] species that are formed upon light absorption due to exciton delocalisation over several monomers that can be monitored through this particular parameter. The diverse water arrangements around the different clusters will also contribute to this broadening, with its concrete role being analysed in detail on a monomer for simplicity in an upcoming work and which is therefore out of scope of the present manuscript.

Analysis of the 2DES spectra of highly flexible systems like ApA is rather cumbersome as many of the fingerprints are specific to certain conformations. Our computational strategy allows us to dissect the full spectrum into the different clusters or representative nuclear arrangements giving rise to these signals, thus enabling us to extract the specific interactions behind a given contribution. In order to do so, we have separated the different clusters depending on their C5–C5 distance (see Fig. 1 for atom labelling), which is close to the centre of mass of the molecule and is thus a decent descriptor to group the different structures. Nevertheless, we are aware that not only the distance but also the specific orientation would play a role, which may also be said of the particular orientation of the adenine moieties with respect to the sugar groups to which they

 This journal is © The Royal Society of Chemistry 2018

are attached. Thus, we use the simplest descriptor (C5–C5 distance) to partition the clusters into groups, while discussing specific nucleobase arrangements for each group of clusters.

The first group that is analysed is composed of clusters with C5–C5 distances of up to 4 Å, which are close and thus correspond to different π-stacked arrangements with both near face-to-face and various degrees of rotated stacked instances. Due to the lack of fingerprints in the pump-UV/probe-UV window beyond the GSB and off-diagonal bleaching, in the following discussion we will only consider the signal-rich pump-UV/probe-Vis region. Fig. 3 displays clusters C, D, K and N (see Table 1), which possess different rotated and π-stacked adenine moieties. These types of structures account for ~19% of the conformational space and, importantly, do not possess fully (or perfectly) stacked moieties, which is similar to how the bases appear rotated when embedded in a double helix DNA chain.[61,62] As can be seen in Fig. 3, these bases exhibit strong contributions of the doubly excited HL^2 signal in the 26–30k cm^{-1} range, oscillating along Ω_1 and also along Ω_3. This is more obvious in cluster K, where the ^1L_a traces (consisting of two close-lying traces due to the ^1L_a states of the two adenine moieties) along Ω_1 are slightly split but strongly red-shifted with respect to the other clusters due to the presence of close-lying water molecules pointing to the C4–C5 bond that are not featured in the other conformations, which seems to strongly red-shift these excitations. Despite treating the neighbouring water molecules at the MM level, their inclusion at this lower level has recently been shown for related systems to reproduce qualitative features of the low-lying electronic excitations,[63] such as ^1L_a, and is expected to qualitatively reproduce the solvatochromic shift for the states in the high-energy window like HL^2.

The splitting of the two ^1L_a traces appears to be the main fingerprint of these nuclear arrangements, where the bases are close and strongly interact leading to

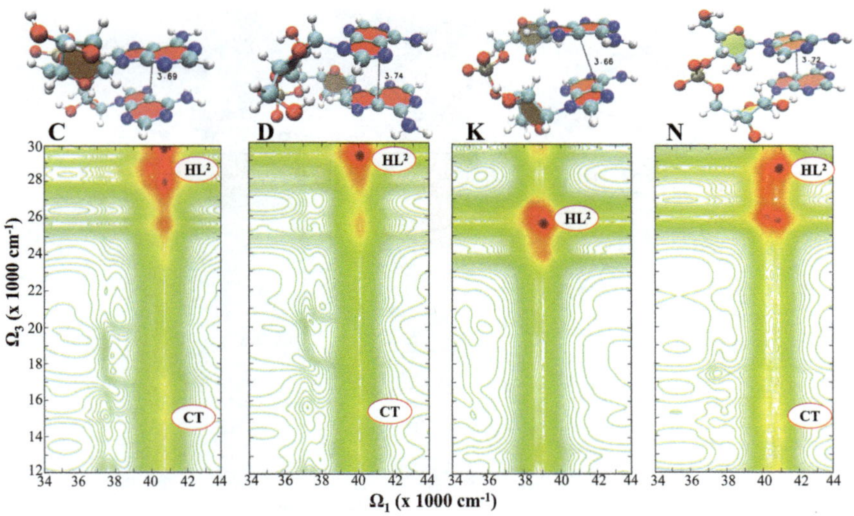

Fig. 3 2DES of the clusters with a C5–C5 distance of under 4 Å: C, D, K and N (see Table 1). The C5–C5 distances are shown on the structures. HL^2 and CT labels are shown to aid in the assignment. The colour scheme for the 2DES spectra is set according to Fig. 1.

a 500–1500 cm^{-1} splitting between these two analogous states centred at the different adenine moieties, one being red-shifted and one blue-shifted. Moreover, this fingerprint occurs along Ω_1, which is not accessible using standard pump-probe set-ups, therefore highlighting the potential of 2DES for disentangling conformations even for difficult cases such as homo-dimers. Besides the HL^2 signal, other $\pi\pi^*$ states are accessed within 2000–3000 cm^{-1} of the latter, which are expected to merge and broaden that signal further. Additional signals appear in the low-energy window, with some of them being related to localised ESAs that were obscured in the monomer but which appeared in our previous gas-phase reference computations,[28] and with some (highlighted in Fig. 3) corresponding to CT states that arise due to the proximity between adenine moieties. However, these CT states appear to be quite weak and are hard to discern from the localised ESAs on the different monomers (~13 000 cm^{-1} along Ω_3), with the more intense contributions observed for C, D and N (Fig. 3) at ~15 000 cm^{-1}.

Fig. 4 displays the clusters with C5–C5 distances of between 4 and 4.5 Å, namely clusters B, G and P (see Table 1). These account for ~17% of the overall conformational space that was sampled, and exhibit a range of π-stacked and tilted geometries, which resemble those previously shown in Fig. 3. Again, the main fingerprints observed are associated with the very intense HL^2 ESA, which is present between 25–28k cm^{-1} along Ω_3, red-shifted in comparison with the clusters possessing shorter C5–C5 distances, and which is strongly modulated by the position of the ^1L_a traces along Ω_1, which oscillates as previously discussed. A splitting of the ^1L_a traces can also be observed, particularly in cluster G, but can also be observed to a minor extent in the others. Additional $\pi\pi^*$ ESAs either slightly above or below the HL^2 signal are again exhibited, with this peak being found (and labelled as signal 11) in reference computations,[28] which refers to a HOMO-2 to LUMO+2 transition accessed through the ^1L_a state. Less intense CT

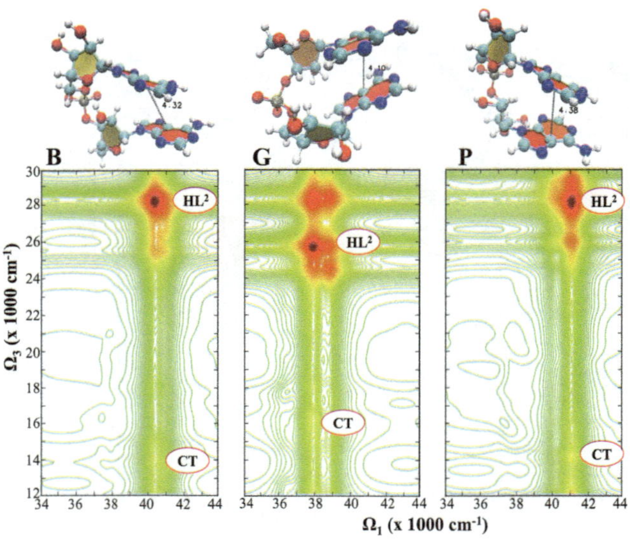

Fig. 4 2DES of the clusters with a C5–C5 distance of between 4 and 4.5 Å: B, G, and P (see Table 1). The C5–C5 distances are shown on the structures. HL^2 and CT labels are shown to aid in the assignment. The colour scheme for the 2DES spectra is set according to Fig. 1.

 This journal is © The Royal Society of Chemistry 2018

and localised ESAs appear in the low-energy region (~15 000 and ~13 000 cm^{-1}, respectively) in comparison with those observed at shorter C5–C5 distances (see Fig. 3).

At longer C5–C5 distances, between 4.5 and 5 Å, the four clusters E, F, I and L are featured, which contribute to ~16% of the full conformational space of ApA. As can be seen in Fig. 5, upon increasing the C5–C5 distance the traces of the ^1L_a states of the different adenine moieties overlap with one another, giving rise to a narrower signal along Ω_1 that is analogous to that of the monomer and producing the main intense HL^2 ESA signal at the same energies (~29 000 cm^{-1}) as that of the monomer. An analogous set of signals is produced by both adenine moieties, which overlap displaying the other relatively intense ππ* ESAs at slightly different energies and thus giving rise to what appears to be more signals, as can be seen in cluster F. Due to the larger distance, the interaction among ^1L_a traces is not observed and the additional broadening along Ω_1 is lost, thus demonstrating their non-interactive (unstacked) nature, which also exhibits no intense contributions in the low-energy Vis window (12–15k cm^{-1}) as was observed in the monomer. However, it is worth noting that slightly intense ESAs with CT character may still be observed in conformations displaying these long C5–C5 distances, as can be seen for clusters I and L in Fig. 5. The former presents a tilted base pointing towards the other with its amino group, which appears to enhance the appearance of CT states, whereas the latter, despite the long distances displayed, possesses a fully face-to-face stacked conformation that is expected to enhance the π-stacking interaction thus facilitating the appearance of these barely bright CT states in the low-energy Vis window, as reported above.

Fig. 6 shows C5–C5 distances of between 5 and 6 Å, comprising the clusters H, O and Q that amount to ~9% of the total conformational space, showing different arrangements of distant adenine moieties that do not interact strongly with one

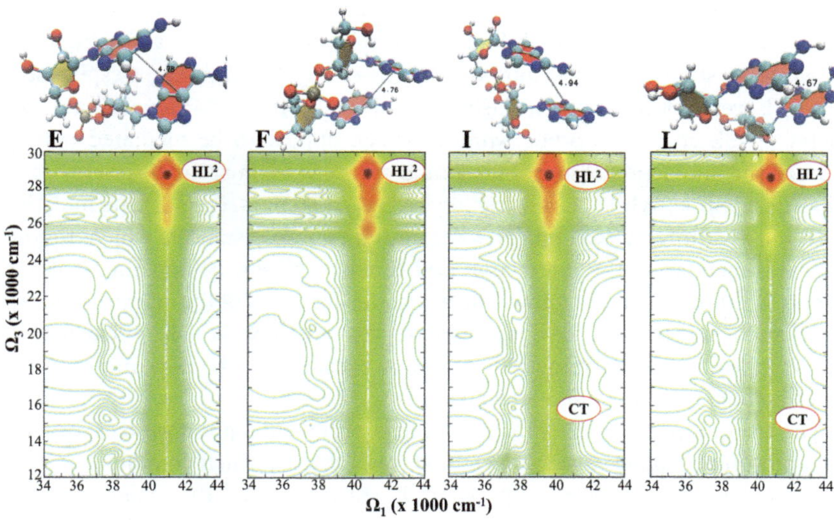

Fig. 5 2DES of the clusters with a C5–C5 distance of between 4.5 and 5 Å: E, F, I and L (see Table 1). The C5–C5 distances are shown on the structures. HL^2 and CT labels are shown to aid in the assignment. The colour scheme for the 2DES spectra is set according to Fig. 1.

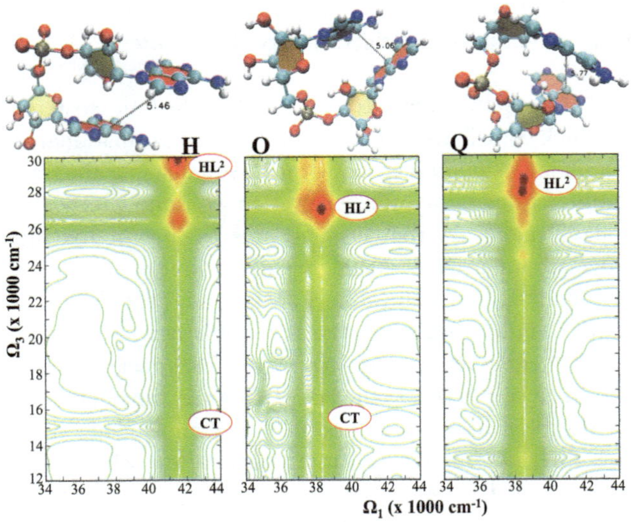

Fig. 6 2DES of the clusters with a C5–C5 distance of between 5 and 6 Å: H, O, and Q (see Table 1). The C5–C5 distances are shown on the structures. HL^2 and CT labels are shown to aid in the assignment. The colour scheme for the 2DES spectra is set according to Fig. 1.

another. As can be seen, and upon comparison with the monomer spectrum given in Fig. 1, the fingerprints displayed essentially relate to those registered from their monomeric contributions, with the main HL^2 ESA in the 26–28k cm^{-1} range, except for cluster O which experiences a strong red-shift along both Ω_1 and Ω_3 due to the presence of a close-lying water molecule coordinating directly to the C4 position. No interaction is observed for clusters H and Q, as indicated by the absence of the ^1L_a splitting that is a feature of the closer and more interacting clusters, despite the observation of slightly displaced HL^2 signals along Ω_3. Cluster O, on the other hand, displays the shortest distance within this cluster group (5.09 Å, see Fig. 6) and exhibits a slight splitting in the trace along Ω_1, while showing no strong contributions in the low-energy Vis window. Despite possessing larger C5–C5 distances, clusters H and O exhibit CT ESA signals in the low energy (~15k cm^{-1}) Vis probe window, which appear to be mediated mainly through the interaction of one of the amino groups with an aromatic group of the opposite adenine. While possessing large C5–C5 distances, the adenine moieties in H and O still remain at much closer amino-to-base distances, which are the distances facilitating the interaction and thus the reason behind the weak CT signals in the low-energy window. Therefore, the proximity of the amino groups to the aromatic moieties on the opposing base, also noted previously for cluster I in Fig. 5, appears to be an important aspect to consider for the appearance of CT states among adenine dimers like ApA.

The last group of clusters consists of those with the longest distances, beyond 6 Å, and comprises the clusters A, J and M (see Table 1), which amount to 16% of the sampled conformational space, and which contain the most repeated conformation (A) that accounts on its own for 10% of the overall population. Fig. 7 shows the different spectra, showcasing three apparently different behaviours. The most populated cluster A features joint ^1L_a traces along Ω_1, and does not

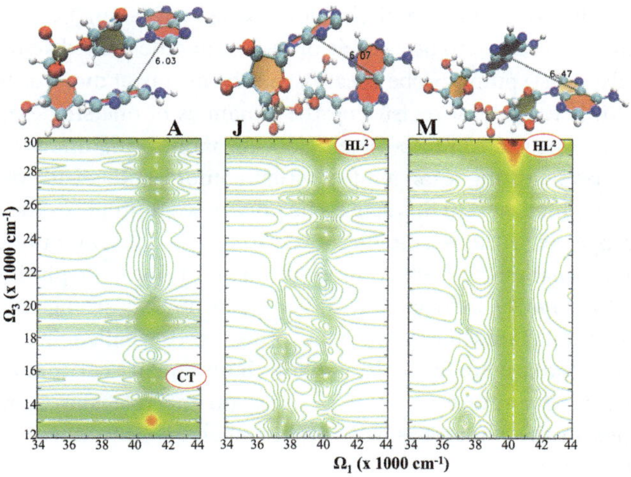

Fig. 7 2DES of the clusters with a C5–C5 distance of more than 6 Å: A, J and M (see Table 1). The C5–C5 distances are shown on the structures. HL^2 and CT labels are shown to aid in the assignment. The colour scheme for the 2DES spectra is set according to Fig. 1.

display the previously reported splitting due to strongly interacting π-stacked systems, while exhibiting a strong localised ESA at ~13 000 and a CT signal at ~15–16k cm^{-1}, with the more intense HL^2 being placed in this case beyond the Vis probe window. The relative intensity of the CT state is attributed to the proximity of one of the adenine's amino groups to the 5-membered-ring of the opposing moiety. This interaction, which appears to stabilise CT transitions, is behind most T-stacked interactions, particularly those mediated by the interaction of the amino group with the opposing nucleobase, and is thus expected to be a fingerprint for these particular conformations, coupled with the lack of broadening along Ω_1 which is characteristic of these distant nuclear arrangements. Cluster J shows a qualitatively different spectrum due to the partial appearance of the HL^2 state in the high-energy window, which obscures the rest of the localised signals that appear in the lower energy window and displays no CT contributions. Cluster M, on the other hand, exhibits monomer-like ESAs, displaying a weak signal at ~14 000 cm^{-1} that is attributed to a localised ESA. Overall, more distant bases lack the additional broadening along Ω_1, which arises due to the splitting of the two ^1L_a states of the different strongly interacting adenine moieties in ApA, while exhibiting a degree of charge transfer character in T-stacked-like conformations. This is particularly the case for those structures where the amino group points towards the opposing 5-ring moiety, thus making them fingerprints for these particular conformations coupled with narrow bands along Ω_1. It is worth noting that both π- and T-stacked CT signals appear in the same energy window, with the different expected broadenings along Ω_1 again being the only way to disentangle them.

Unfortunately, no 2DES experimental data is currently available for ApA to compare with our present findings, with the first time-resolved 2D spectrum for water-solvated adenine in the UV-pump and probe windows being recently reported in the literature,[17] together with a theoretical model rationalising the different signals exhibited and their monitoring.[64] This high-energy window employed has been shown to be capable of registering the excited state dynamics of adenine also in 1D pump-probe spectroscopy by means of monitoring the GSB recovery.[50] While capable of registering the dynamics of one single nucleobase, this window is expected to not be as efficient for monitoring dimeric[7] and multimeric[65] species featuring a range of different conformations with very distinct associated decay lifetimes. In this sense, the present work proposes the UV-pump/Vis-probe range as a more suitable spectral window for monitoring of the different conformations observed in multimeric species, displaying distinct fingerprints that may be associated with particular nuclear arrangements, albeit their experimental characterisation may prove more challenging. Upcoming experimental efforts, coupled with carefully built theoretical models, will be required in the near future in order to ascertain the role played by different conformations in the photophysics and photochemistry of dimeric and multimeric species like ApA.

Conclusions

A computational protocol is introduced that enables the simulation of 2DES of highly flexible solvated monophosphate dinucleosides and the disentangling of the spectroscopic fingerprints of various stacked/unstacked configurations of the ApA dimer case. The 2DES spectra are expected to reveal information about the molecular arrangements in dimeric and multimeric systems, thus going beyond the limitations of standard 1D pump-probe techniques for the study of photoinitiated processes. By employing ground state molecular dynamics and clustering techniques, 17 structures of the water-solvated ApA dimer have been selected in order to represent $\sim$75% of the sampled conformational space. Accurate QM/MM RASPT2 computations have been carried out on each of the selected structures to obtain the electronic energy levels and transition dipole moments in order to simulate the 2DES signals within the SOS//QM/MM approach. By weighting in the contributions of each cluster to the signal according to their relative populations, an overall 2DES spectrum is obtained, showing fingerprints relating to both unstacked (non-interacting) and either π- or T-stacked (interacting) ApA conformations. The extracted clusters can be analysed one-by-one, thus enhancing our understanding of the nuclear arrangements behind specific spectral signals and allowing us to characterise particular fingerprints that arise for each different possible intermolecular interaction. We show how for π-stacked arrangements, a splitting of the two ^1L_a traces of both adenine moieties leads to a marked broadening of the ^1L_a trace along Ω_1, which can potentially mark the appearance of strongly interacting stacked species, thus being a specific fingerprint of 2DES that cannot be observed by standard pump-probe set-ups. T-stacked conformations, on the other hand, are shown to exhibit relatively intense ESA CT signals in the low-energy Vis window through the interaction between the amino group of one adenine monomer and the opposing base, and are particularly intense for those cases promoting amino to 5-membered-ring interactions. Unstacked ApA conformations are shown to display very similar spectra to those of the monomer, as expected from these non-interacting species. Overall, 2DES is shown to be a promising alternative to standard pump-probe experiments for elucidating and studying the concrete

 This journal is © The Royal Society of Chemistry 2018

photo-initiated events in DNA multimeric systems, being potentially able to separate contributions arising from a wide range of conformations and thus providing unique insights into how photostability and photo-damage is modulated due to the presence of extended and interacting nucleobase sequences in our genomic material.

Acknowledgements

M. G. acknowledges the support from the European Research Council Advanced Grant STRATUS (ERC-2011-AdG No. 291198). I. R. gratefully acknowledges the support from the Agence National de la Recherche (FEMTO-2DNA, ANR-15-CE29-0010) and ENS-Lyon (grants 900/S81/BS81-FR14 and MI-LOURD-FR15). J. S.-M. and I. R. acknowledge the use of the HPC resources of the "Pôle Scientifique de Modélisation Numérique" (PSMN) at ENS-Lyon, France. S. M. gratefully acknowledges the support of the National Science Foundation (grant CHE-1663822), and the Chemical Sciences, Geosciences, and Biosciences division, Office of Basic Energy Sciences, Office of Science, U.S. Department of Energy, award DE-FG02-04ER15571.

Notes and references

1 G. D. Scholes, G. R. Fleming, L. X. Chen, A. Aspuru-Guzik, A. Buchleitner, D. F. Coker, G. S. Engel, R. van Grondelle, A. Ishizaki, D. M. Jonas, J. S. Lundeen, J. K. McCusker, S. Mukamel, J. P. Ogilvie, A. Olaya-Castro, M. A. Ratner, F. C. Spano, K. B. Whaley and X. Zhu, *Nature*, 2017, **543**, 647–656.

2 C. E. Crespo-Hernandez, B. Cohen, P. M. Hare and B. Kohler, *Chem. Rev.*, 2004, **104**, 1977–2019.

3 C. T. Middleton, K. de La Harpe, C. Su, Y. K. Law, C. E. Crespo-Hernandez and B. Kohler, in *Annual Review of Physical Chemistry*, Annual Reviews, Palo Alto, 2009, vol. 60, pp. 217–239.

4 D. Markovitsi, T. Gustavsson and I. Vayá, *J. Phys. Chem. Lett.*, 2010, **1**, 3271–3276.

5 W. J. Schreier, P. Gilch and W. Zinth, *Annu. Rev. Phys. Chem.*, 2015, **66**, 497–519.

6 W. J. Schreier, T. E. Schrader, F. O. Koller, P. Gilch, C. E. Crespo-Hernández, V. N. Swaminathan, T. Carell, W. Zinth and B. Kohler, *Science*, 2007, **315**, 625–629.

7 T. Takaya, C. Su, K. d. L. Harpe, C. E. Crespo-Hernández and B. Kohler, *Proc. Natl. Acad. Sci. U. S. A.*, 2008, **105**, 10285–10290.

8 A. Giussani, J. Segarra-Martí, D. Roca-Sanjuán and M. Merchán, in *Photoinduced Phenomena in Nucleic Acids I*, ed. M. Barbatti, A. C. Borin and S. Ullrich, Springer International Publishing, 2015, ch. 501, vol. 355, pp. 57–97.

9 M. Barbatti, A. J. A. Aquino, J. J. Szymczak, D. Nachtigallova, P. Hobza and H. Lischka, *Proc. Natl. Acad. Sci. U. S. A.*, 2010, **107**, 21453–21458.

10 L. Serrano-Andrés and M. Merchán, *J. Photochem. Photobiol., C*, 2009, **10**, 21–32.

11 D. B. Bucher, B. M. Pilles, T. Carell and W. Zinth, *Proc. Natl. Acad. Sci. U. S. A.*, 2014, **111**, 4369–4374.

12 D. B. Bucher, C. L. Kufner, A. Schlueter, T. Carell and W. Zinth, *J. Am. Chem. Soc.*, 2016, **138**, 186–190.

13 L. E. Buchanan, J. K. Carr, A. M. Fluitt, A. J. Hoganson, S. D. Moran, J. J. de Pablo, J. L. Skinner and M. T. Zanni, *Proc. Natl. Acad. Sci. U. S. A.*, 2014, **111**, 5796–5801.
14 A. Ghosh, J.-J. Ho, A. L. Serrano, D. R. Skoff, T. Zhang and M. T. Zanni, *Faraday Discuss.*, 2015, **177**, 493–505.
15 B. A. West, J. M. Womick and A. M. Moran, *J. Phys. Chem. A*, 2011, **115**, 8630–8637.
16 B. A. West and A. M. Moran, *J. Phys. Chem. Lett.*, 2012, **3**(18), 2575–2581.
17 V. I. Prokhorenko, A. Picchiotti, M. Pola, A. G. Dijkstra and R. J. D. Miller, *J. Phys. Chem. Lett.*, 2016, **7**, 4445–4450.
18 I. Rivalta, A. Nenov, G. Cerullo, S. Mukamel and M. Garavelli, *Int. J. Quantum Chem.*, 2014, **114**, 85–93.
19 S. Mukamel, *Principles of Nonlinear Optical Spectroscopy*, O.U.P., New York, 1995.
20 S. Mukamel, *Annu. Rev. Phys. Chem.*, 2000, **51**, 691–729.
21 D. Polli, I. Rivalta, A. Nenov, O. Weingart, M. Garavelli and G. Cerullo, *Photochem. Photobiol. Sci.*, 2015, **14**, 213–228.
22 Q. Li, A. Giussani, J. Segarra-Martí, A. Nenov, I. Rivalta, A. A. Voityuk, S. Mukamel, D. Roca-Sanjuán, M. Garavelli and L. Blancafort, *Chem.–Eur. J.*, 2016, **22**, 7497–7507.
23 A. Nenov, S. a. Beccara, I. Rivalta, G. Cerullo, S. Mukamel and M. Garavelli, *ChemPhysChem*, 2014, **15**, 3282–3290.
24 D. Case, T. A. Darden, T. E. Cheatham, C. Simmerling, J. Wang, R. Duke, R. Luo, M. Crowley, R. Walker, W. Zhang, K. M. Merz, B. Wang, S. Hayik, A. Roitberg, G. Seabra, I. Kolossváry, K. F. Wong, F. Paesani, J. Vanicek, X. Wu, S. Brozell, T. Steinbrecher, H. Gohlke, L. Yang, C. Tan, J. Mongan, V. Hornak, G. Cui, D. H. Mathews, M. G. Seetin, C. Sagui, V. Babin and P. Kollman, *Amber 11*, University of California, San Francisco, 2011.
25 D. A. Case, T. E. Cheatham, T. Darden, H. Gohlke, R. Luo, K. M. Merz, A. Onufriev, C. Simmerling, B. Wang and R. J. Woods, *J. Comput. Chem.*, 2005, **26**, 1668–1688.
26 A. Pérez, I. Marchán, D. Svozil, J. Sponer, T. E. Cheatham III, C. A. Laughton and M. Orozco, *Biophys. J.*, 2007, **92**, 3817–3829.
27 W. L. Jorgensen, J. Chandrasekhar, J. D. Madura, R. W. Impey and M. L. Klein, *J. Chem. Phys.*, 1983, **79**, 926–935.
28 A. Nenov, A. Giussani, J. Segarra-Martí, V. K. Jaiswal, I. Rivalta, G. Cerullo, S. Mukamel and M. Garavelli, *J. Chem. Phys.*, 2015, **142**, 212443.
29 F. Aquilante, J. Autschbach, R. Carlson, L. Chibotaru, M. G. Delcey, L. De Vico, I. Fernández Galvan, N. Ferré, L. M. Frutos, L. Gagliardi, M. Garavelli, A. Giussani, C. Hoyer, G. Li Manni, H. Lischka, D. Ma, P. A. Malmqvist, T. Müller, A. Nenov, M. Olivucci, T. B. Pedersen, D. Peng, F. Plasser, B. Pritchard, M. Reiher, I. Rivalta, I. Schapiro, J. Segarra-Martí, M. Stenrup, D. G. Truhlar, L. Ungur, A. Valentini, S. Vancoillie, V. Veryazov, V. Vysotskiy, O. Weingart, F. Zapata and R. Lindh, *J. Comput. Chem.*, 2016, **37**, 506–541.
30 P. A. Malmqvist, A. Rendell and B. O. Roos, *J. Phys. Chem.*, 1990, **94**, 5477–5482.
31 P. Å. Malmqvist, K. Pierloot, A. R. M. Shahi, C. J. Cramer and L. Gagliardi, *J. Chem. Phys.*, 2008, **128**, 204109.
32 V. Saurí, L. Serrano-Andrés, A. R. M. Shahi, L. Gagliardi, S. Vancoillie and K. Pierloot, *J. Chem. Theory Comput.*, 2011, **7**, 153–168.

 This journal is © The Royal Society of Chemistry 2018

33 F. Aquilante, R. Lindh and T. Bondo Pedersen, *J. Chem. Phys.*, 2007, **127**, 114107.
34 F. Aquilante, T. B. Pedersen and R. Lindh, *J. Chem. Phys.*, 2007, **126**, 194106.
35 T. Pedersen, F. Aquilante and R. Lindh, *Theor. Chem. Acc.*, 2009, **124**, 1–10.
36 P. O. Widmark, B. Joakim Persson and B. O. Roos, *Theor. Chim. Acta*, 1991, **79**, 419–432.
37 S. F. Altavilla, J. Segarra-Martí, A. Nenov, I. Conti, I. Rivalta and M. Garavelli, *Front. Chem.*, 2015, **3**, 29.
38 N. Forsberg and P. A. Malmqvist, *Chem. Phys. Lett.*, 1997, **274**, 196–204.
39 J. P. Zobel, J. J. Nogueira and L. Gonzalez, *Chem. Sci.*, 2017, **8**, 1482–1499.
40 G. Ghigo, B. O. Roos and P.-Å. Malmqvist, *Chem. Phys. Lett.*, 2004, **396**, 142–149.
41 P.-Å. Malmqvist and B. O. Roos, *Chem. Phys. Lett.*, 1989, **155**, 189–194.
42 P. Å. Malmqvist, B. O. Roos and B. Schimmelpfennig, *Chem. Phys. Lett.*, 2002, **357**, 230–240.
43 S. Mukamel, R. Oszwaldowski and D. Abramavicius, *Phys. Rev. B: Condens. Matter Mater. Phys.*, 2007, **75**, 245305.
44 D. Abramavicius, B. Palmieri, D. V. Voronine, F. Šanda and S. Mukamel, *Chem. Rev.*, 2009, **109**, 2350–2408.
45 M. Richter and B. P. Fingerhut, *J. Chem. Theory Comput.*, 2016, **12**, 3284–3294.
46 A. Nenov, A. Giussani, B. P. Fingerhut, I. Rivalta, E. Dumont, S. Mukamel and M. Garavelli, *Phys. Chem. Chem. Phys.*, 2015, **17**, 30925–30936.
47 R. Tempelaar, C. P. van der Vegte, J. Knoester and T. L. C. Jansen, *J. Chem. Phys.*, 2013, **138**, 164106.
48 C. P. van der Vegte, A. G. Dijkstra, J. Knoester and T. L. C. Jansen, *J. Phys. Chem. A*, 2013, **117**, 5970–5980.
49 M. Kowalewski, B. P. Fingerhut, K. E. Dorfman, K. Bennett and S. Mukamel, *Chem. Rev.*, 2017, **117**(19), 12165–12226.
50 B. Cohen, P. M. Hare and B. Kohler, *J. Am. Chem. Soc.*, 2003, **125**, 13594–13601.
51 T. Gustavsson, A. Sharonov, D. Onidas and D. Markovitsi, *Chem. Phys. Lett.*, 2002, **356**, 49–54.
52 F. Santoro, R. Improta, T. Fahleson, J. Kauczor, P. Norman and S. Coriani, *J. Phys. Chem. Lett.*, 2014, **5**, 1806–1811.
53 P. G. Szalay, T. Watson, A. Perera, V. F. Lotrich and R. J. Bartlett, *J. Phys. Chem. A*, 2012, **116**, 6702–6710.
54 R. Improta and V. Barone, Excited States Behavior of Nucleobases in Solution: Insights from Computational Studies, in *Photoinduced Phenomena in Nucleic Acids I. Topics in Current Chemistry*, ed. M. Barbatti, A. Borin and S. Ullrich, Springer, 2014, vol. 355, pp. 329–357.
55 A. J. Pepino, J. Segarra-Martí, A. Nenov, R. Improta and M. Garavelli, *J. Phys. Chem. Lett.*, 2017, **8**, 1777–1783.
56 L. Martinez-Fernandez, A. J. Pepino, J. Segarra-Martí, J. Jovaisaite, I. Vayá, A. Nenov, D. Markovitsi, T. Gustavsson, A. Banyasz, M. Garavelli and R. Improta, *J. Am. Chem. Soc.*, 2017, **139**, 7780–7791.
57 A. Giussani, J. Segarra-Martí, A. Nenov, I. Rivalta, A. Tolomelli, S. Mukamel and M. Garavelli, *Theor. Chem. Acc.*, 2016, **135**, 1–18.
58 S. Jafilan, L. Klein, C. Hyun and J. Florián, *J. Phys. Chem. B*, 2012, **116**, 3613–3618.

59 I. Stiopkin, T. Brixner, M. Yang and G. R. Fleming, *J. Phys. Chem. B*, 2006, **110**, 20032–20037.

60 N. S. Ginsberg, Y.-C. Cheng and G. R. Fleming, *Acc. Chem. Res.*, 2009, **42**, 1352–1363.

61 I. Conti, A. Nenov, S. Ho, S. F. Altavilla, I. Rivalta, E. Dumont, G. Orlandi and M. Garavelli, *Phys. Chem. Chem. Phys.*, 2015, **17**, 7291–7302.

62 A. Nenov, J. Segarra-Martí, A. Giussani, I. Conti, I. Rivalta, E. Dumont, V. K. Jaiswal, S. F. Altavilla, S. Mukamel and M. Garavelli, *Faraday Discuss.*, 2015, **177**, 345–362.

63 L. Martínez-Fernández, A. J. Pepino, J. Segarra-Martí, A. Banyasz, M. Garavelli and R. Improta, *J. Chem. Theory Comput.*, 2016, **12**, 4430–4439.

64 A. G. Dijkstra and V. I. Prokhorenko, *J. Chem. Phys.*, 2017, **147**, 064102.

65 C. Su, C. T. Middleton and B. Kohler, *J. Phys. Chem. B*, 2012, **116**, 10266–10274.

Faraday Discussions

PAPER

Direct observation of the external force mediated conformational dynamics of an IHF bound Holliday junction†

Subhas C. Bera,‡[a] Tapas Paul,‡[a] A. N. Sekar Iyengar[b] and Padmaja P. Mishra *[a]

Received 21st August 2017, Accepted 1st November 2017
DOI: 10.1039/c7fd00184c

We have investigated the isomerization dynamics and plausible energy landscape of 4-way Holliday junctions (4WHJs) bound to integration host factor (IHF, a DNA binding protein), considering the effect of applied external force, by single-molecule FRET methods. A slowing down of the forward as well as the backward rates of the isomerization process of the protein bound 4WHJ has been observed under the influence of an external force, which indicates an imposed restriction on the conformational switching. This has also been reflected by an increase in rigidity, as observed from the increase in the single-molecule FRET (smFRET)-anisotropy values (0.270 ± 0.012 to 0.360 ± 0.008). The application of an external force has assisted the conformational transitions to share the unstacked open structure intermediate, with different rate-limiting steps and a huge induced variation in the energy landscape. Furthermore, the associated landscape of the 4WHJ is visualized in terms of rarely interconverting states embedded into the two isoforms by using nonlinear dynamics analysis, which shows that the chaoticity of the system increases at intermediate force (0.4 to 1.6 pN). The identification of chaos in our investigation provides useful information for a comprehensive explanation of the origin of the complex behavior of the system, which effectively helps us to perceive the dynamics of IHF bound 4WHJs under the influence of external force, and also demonstrates the applicability of nonlinear dynamics analysis in the field of biology.

Introduction

Nucleic acid rearrangement during homologous recombination is one of the crucial events in cellular processes,[1–3] in which DNA four-way Holliday junctions

[a]Chemical Sciences Division, Saha Institute of Nuclear Physics, HBNI Mumbai, 1/AF Bidhannagar, Kolkata 700064, India. E-mail: padmaja.mishra@saha.ac.in; Fax: +91-33-2337-4637; Tel: +91-33-2337-5345-49

[b]Plasma Physics Division, Saha Institute of Nuclear Physics, Kolkata, 700064, India

† Electronic supplementary information (ESI) available: The structure and sequence of 4WHJ, theory of smFRET-anisotropy calculation, schematic of smFRET-optical tweezers setup, and equations for the nonlinear dynamics. See DOI: 10.1039/c7fd00184c

‡ Both authors contributed equally to this work.

(4WHJs) play an essential role in forming a central intermediate.[4–6] Since the inception of 4WHJs,[7] there have been substantial reports that have explored their conformations and dynamics, amongst which crystal structure studies have revealed the adoption of an extended open-X conformation while bound to proteins and a stacked-X conformation in the presence of ions.[8,9] The dynamic interconversion between stacked-X and extended-X conformations of 4WHJs has also been shown to be greatly influenced by ionic effects (Fig. S1, ESI†).[10,11] The former conformation further adopts two conformations, namely, parallel and anti-parallel,[12–14] with the anti-parallel conformation reported to exhibit *iso*I and *iso*II conformers[15] depending on the environmental conditions, the basic structure of DNA and its interaction with the ions present. Single-molecule FRET studies[10,16] on an immobilized junction have shown that the dynamic switching between different conformations with a bias to a particular conformer depends on the sequence and surroundings.[17,18]

Additionally, the majority of the dynamic processes inside the cell are immensely affected and sometimes induced by several architectural proteins.[19–21] On a related note, crystal structures of the 4WHJ–protein complex predominantly show an open-X conformation of the 4WHJ,[1,22] which is possibly due to the freedom of migration of the junction cross-over along the DNA strands. Dimeric proteins preferably recognize the stacked-X conformation while most of the tetrameric proteins prefer the open-X form.[1,13] It has been indicated that an increase in the number of different open-X intermediates upon binding to the proteins can be caused by slowing down the rates of isomerization.[23] However, these types of DNA–protein recognition provide structural rigidity, hence the stability of DNA during cellular processes, and they are frequently driven by molecular complexes to occur at the tertiary structural level, thus it is challenging to explore their mechanistic pathway. Additionally, during the course of the cellular processes involving 4WHJ formation and the post processing phenomena, DNA suffers tremendous mechanical strain due to the structural alternation.[24] Though the transitions between the alternative stacking conformers of 4WHJs (Fig. S1, ESI†) have received adequate attention, the dynamics during the internal conformational changes of these protein bound molecular machines have not been thoroughly explored. This could be due to the limitations of the widely used conventional techniques as well as some of the single-molecule or manipulation techniques, which mean that they are unable to capture the full complexity of the system. As the smFRET technique relies on the ratiometric detection of fluorescence from the donor and acceptor, it has been found to be more suitable than traditional ensemble techniques for studying complex biophysical processes. This method also allows for the determination of the subtle conformational fluctuation of most of the biological systems, including 4WHJs, under the influence of internal and external factors.[25]

Since an understanding of the overall scenario demands a careful investigation of the structural as well as mechanistic aspects of 4WHJs while interacting with DNA/junction binding proteins, further detail could be obtained by analyzing their kinetics/mechanisms under the influence of an externally applied force. Furthermore, a combination of smFRET with optical tweezers allows us to explore the effect of externally applied force on the conformational energy landscape of 4WHJs, thus providing clues about the role of intracellular mechanical stress. The global structure, as well as the local structure, of a 4WHJ has recently been probed

 This journal is © The Royal Society of Chemistry 2018

using a combination of magnetic tweezers and smFRET.[26] However, none of the study demonstrated the interaction of a junction binding protein, *e.g.* IHF, with 4WHJ by utilizing hybrid single-molecule tools *i.e.* smFRET-force techniques. There are reports, using smFRET only, that show the slowing of kinetics upon the binding of a junction binding protein, *viz.* architectural protein, to 4WHJs, which makes the conjugate comparatively rigid by restricting the fast conformational switching.[23] In this context, we have used integration host factor (IHF) from *E. coli*, which is one of the most studied architectural proteins that is mainly known for its sequence specific binding and for creating a sharp bend in linear dsDNA.[27–29] Additionally, integrative and excessive recombination of viral chromosome into and out of the host chromosome also requires the host-encoded IHF in addition to Int (λ integrase, a type-I topoisomerase).[30–32] 4WHJs are an important intermediate precursor in the recombination event, hence understanding the effects and nature of IHF interaction with them may reveal new insights into the mechanism of binding and the variation in the global structure during the entire course of interaction.

Here, we have used smFRET-anisotropy (Scheme 1), a relatively new endeavour, along with smFRET[33–35] to obtain mechanistic insight into the dynamics of *E. coli* IHF bound to 4WHJs. Furthermore, we combined optical tweezers to exert external force with the smFRET technique, with a view to understanding the role of mechanical stress. For proper execution, our synthetically designed 4WHJ possesses an IHF specific binding sequence (H1)[27] near the junction and the FRET donor–acceptor pair is labeled at two different arms. The third and fourth arms are either used to immobilize the system with the surface or are linked to the dielectric bead to be trapped by optical tweezers. Our conclusions are based on a combination of smFRET-force experiments followed by nonlinear dynamics analysis, which is a relatively unconventional approach in this field.

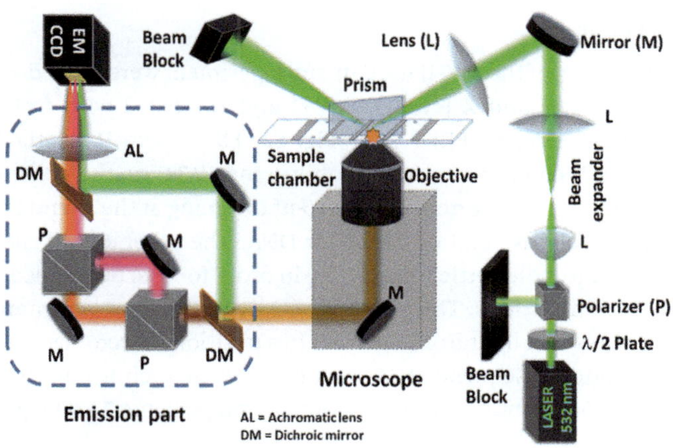

Scheme 1 Schematic of the instrumental setup used for the smFRET acceptor fluorescence polarization measurements. The donor (Cy3) is excited using a parallel polarized 532 nm laser employing the total internal reflection technique. The donor and acceptor (Cy5) fluorescence signals, collected through a 1.2 NA water-immersion objective, are then divided into two corresponding spectral regions. The Cy5 emission is further divided into two polarization components ($Cy5_{\perp}$ and $Cy5_{\parallel}$). An EMCCD camera is used to capture the time traces of the three different polarized components.

Experimental section

Chemicals

All chemicals were purchased from Sigma-Aldrich and used as received without any further purification, unless otherwise mentioned. Oligos were purchased from IDT, USA and linker-DNA was purchased from New England Biolabs (Ipswich, MA). All buffers were made in Milli-Q water and were sterilised either by 0.22 μm membrane filtration or by autoclaving, unless otherwise mentioned.

Sequences of oligo/DNA used to construct the 4WHJ (IHF-H1 sequence shown in bold)

5′-Cy5-CCGGTTTTTCG**TTAGTT**CCACGTGCTCGCCACCC-3′

5′-Cy3-CCACGTTGTT**ATTCATCAACTAA**CGAAAAAACCGG-3′

5′-Biotin-AAAAACCGTAGCAGGCGGCTTA**GATGAAT**AACAACGTGG-3′

5′-GCGCAGCGATCTCGAAGGGTGGCGAGCACGTGGTAAGCCGCCTGCTACGG-3′

The 4WHJ arms were named as B (blue), H (magenta), R (gray) and X (cyan) (the letters were not set according to the color). The 16-nt overhang is used to hybridize with the linker DNA (see Fig. S2 for the 4WHJ structure, ESI†).

4WHJ construct design

The four helical arms of the 4WHJ used in our experiment were named as B (blue), H (magenta), R (gray) and X (cyan), with H and B labeled with Cy3 and Cy5, respectively, as an efficient FRET pair (Fig. S2, ESI†). Arm R is attached with a biotin tag to its 5′-end in order to immobilize the 4WHJ on the polymer coated quartz surface, and arm X is extended by a 16-nt overhang at the 5′-end to allow for annealing with a cohesive end of the linker DNA. The other end of the tethered DNA was attached to a dielectric bead (1 μm) in order for it to be trapped and to be pulled in all three directions. The position of the trapped bead was monitored by guiding the scattered IR light from the bead to a position detector *via* a condenser lens. The IHF binding sequence (H1) was inserted in such a way that the IHF could bind itself to the middle of the junction region (Scheme 2 and Fig. S2, ESI†).

Annealing of Holliday junction and linker-DNA

Biotinylated labeled probes and non-biotinylated strands (each in a stock conc. of 100 μM) were mixed in a 1 : 1 molar ratio in a buffer containing 10 mM Tris (pH 8) and 50 mM NaCl to a final conc. of 10 μM. The mixture was heated for 4 minutes in a water bath and then gradually cooled to room temperature in the dark. In order to make linear form λ-DNA, λ-DNA (~10 nM) was heated at 87 °C for 5

 This journal is © The Royal Society of Chemistry 2018

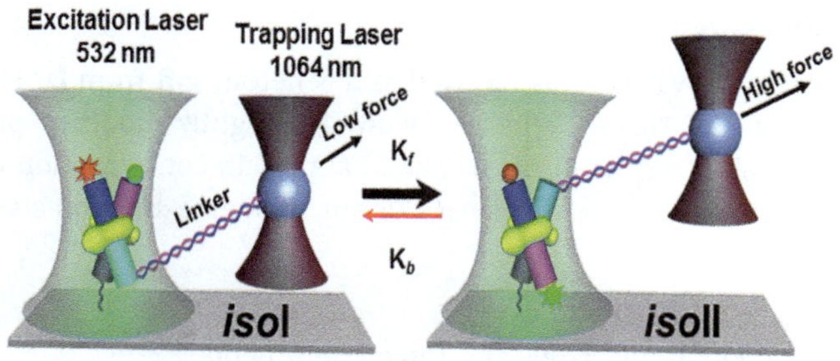

Scheme 2 Experimental scheme describing the isomerization dynamics of the IHF bound 4WHJ under an applied force. The four helical arms of the 4WHJ are named as B (blue), H (magenta), R (gray) and X (cyan). The high FRET state (*iso*I) is more stable under low force conditions, whereas the application of a higher force shifts the equilibrium towards the low FRET state (*iso*II). An external force is applied through a dielectric bead, linked to a 4WHJ-arm, trapped by single-beam optical tweezers. A PTIR-based wide field smFRET technique is used to monitor the FRET signals from the surface immobilized 4WHJ.

minutes in 0.5 M NaCl and then quickly cooled down in wet ice in order to prevent the intermolecular hybridization of complementary overhangs. Then the λ-DNA solution (40 μl, ~10 nM) and pre-annealed 4WHJ (1 μl, ~200 nM) were mixed in a 2 : 1 molar ratio with BSA (1 μl, ~20 mg ml^{-1}) and NaCl (2 μl, 1 M). The mixture was incubated at room temperature for 90 min, followed by 60 min incubation at 4 °C. The digoxigenin coated bead (see below), containing a complementary sequence to that of the λ-DNA overhang, was added to the mixture in a 10-fold molar excess (1 μl, 4 μM) over that of the λ-DNA, and incubated at 4 °C for ~2 h. Finally, the stock was stored at −20 °C in aliquots.[36]

Preparation of anti-digoxigenin coated beads[37,38]

Protein G-coated polystyrene beads (1.0 μm, polysciences) were cross-linked to sheep anti-digoxigenin (Roche Applied Science) by following Wei Cheng's (U.C. Berkeley) protocol.

(1) 0.5 ml of the protein G-coated bead was taken and transferred into a crosslinking buffer (0.1 M Na_2HPO_4, 0.1 M NaCl) by 3–4 instances of repeated centrifugation (4 minutes at 7000 rpm), followed by the removal of the supernatant and subsequent addition of 0.5 ml of crosslinking buffer.

(2) 40 μl of anti-digoxigenin (1 mg ml^{-1} dissolved in 0.019 M NaH_2PO_4, 0.081 M Na_2HPO_4, 0.14 M NaCl and 2.7 mM KCl) and 15 μl of DMP (50 mg ml^{-1} dissolved in the crosslinking buffer) was added to the tube.

(3) The reaction tube was mixed well for 2 hours at room temperature.

(4) To stop the reaction, 30 μl of Tris buffer (1 M, pH 6.8) was added and mixing was continued for another two hours.

(5) The beads were transferred into a bead storage buffer (0.039 M NaH_2PO_4, 0.061 M Na_2HPO_4, 0.14 M NaCl, 2.7 mM KCl) by 3–4 instances of repeated centrifugation (4 minutes at 7000 rpm), followed by the removal of the supernatant and subsequent addition of 0.5 ml of bead storage buffer.

(6) 5 μl of sodium azide (2%) was added to the tube and the beads were stored at 4 °C.

(7) 10–100 times diluted bead in a Tris 10 mM, pH 8.0 buffer was used for sample preparation.

Expression and purification of IHF

The *E. coli* strain of WT-IHF was received as a generous gift from Dr H. A. Nash, National Institute of Health, and we followed a slightly modified protocol as described by Nash *et al.*[39] to purify it. The final protein concentration was determined by measuring the absorbance at 276 nm, with an extinction coefficient of 5800 M^{-1} cm^{-1}.[40]

Single-molecule FRET-anisotropy and force experiments

PEG/biotin-PEG-coated predrilled quartz microscope slides were used to monitor the fluorescence signals arising from the Cy3 and Cy5 labeled DNA substrate immobilized through biotin–streptavidin interaction using the standard protocol.[41] An excess amount of IHF was delivered to the premade channels containing the DNA substrate in order to ensure maximum binding, with free IHF washed out before the measurements were carried out. A 532 nm laser was used to excite the Cy3, and the subsequent polarized fluorescence intensities from Cy3 and Cy5 were recorded simultaneously using a prism-type total internal reflection (PTIR) based inverted microscope (Scheme 1).

The smFRET setup was also combined with an optical tweezers setup, omitting the polarizers (Fig. S3, ESI†), in order to monitor the FRET time traces as a function of applied force. The FRET time traces were analyzed with a hidden Markov model (HMM) using a home-built HMM optimization code. The detailed experimental setup, calibration, and calculation procedure have been explained in detail below.

Optical trapping and force calibration

The FRET-force measuring setup is constructed on an inverted microscope by spatially separating the trapping and fluorescence excitation beams by about 15 μm. The 1064 nm trapping beam was fixed along the optical axis of the microscope, and force was applied by moving the surface-tethered 4WHJ using a piezoelectric sample scanner. The 532 nm excitation beam was programmed to follow the HJ using a piezo-controlled mirror. Bead images were taken with a 100× oil immersion objective lens with 1.49 NA (CFI Apochromat TIRF, Nikon, Japan) at 33 Hz using a CCD camera. Bright field illumination was achieved with a tungsten lamp through a condenser mounted on the upper side of the sample slide. The center position of the bead was determined in real-time using a QPD (quadrant photodiode) by an algorithm that finds the centroid of the intensity pattern of the bead image on the CCD camera device. The precision of the position determined using the method was typically within 52 nm based on a series of position determinations of a bead stuck on a glass surface. The position of the trap beam was determined with a trapped free bead in a solution by measuring the position of the trapped bead. The stiffness of the optical trap was determined using the power spectrum method and confirmed using the equipartition method. We used a quadrant diode (diode and amplifier from Current Designs, Philadelphia, PA) as the high-frequency detection device for stiffness determination. To record the positional fluctuations of a bead at a high frequency (5 kHz), we used the scattered trap beam by a trapped free bead imaged on a quadrant diode by the objective lens mounted on the upper side of the sample slide. The stiffness of the trap at 1.50 W of a 1064 nm trapping beam (before entering the microscope) was 0.15 pN nm^{-1}.

After incubating the beads with the 4WHJ–IHF complex, fluctuating beads were visually identified in the sample chamber. Once a fluctuating bead was identified, the bead was moved to the center of the optical trap by moving a piezo-electrically driven microscope stage based on a P-733.3DD stage from Physik Instrumente (Karlsruhe, Germany). The piezo-stage was controlled by home-built software written in Visual C. After precisely moving the bead to the predetermined center of the optical trap by controlling the piezo-stage with the software, the power of the trap laser beam (1064 nm 2.5 W max, COMPASS 1064–2500 mN; Coherent, Santa Clara, CA) was turned up to 1.50 W from 0.00 W (set to be at the minimum power and blocked with a shutter). The piezo-stage was moved in one direction at a rate of 100 nm s^{-1} until, shortly after a rupture event was identified, either the motion of the piezo-stage reached its maximum traveling distance in one direction (15 mm) or the bead displacement from the trap beam center reached 300 nm.

Results and discussion

Our customized 4WHJ has the IHF binding sequence (H1) located at the junction (Fig. S2, ESI†), instead of having a homologous sequence, in order to facilitate the interaction of protein at the junction to reveal the role of junction binding protein in the 4WHJ conformational dynamics. To monitor the conformational alteration, if any, we probed a FRET donor (Cy3)–acceptor (Cy5) pair at two different 4WHJ-arms (here arms H and B, respectively, Scheme 2 and Fig. S2, ESI†). We employed a wide-field fluorescence imaging technique for analysis of the surface immobilized fluorophore tagged-4WHJ in order to obtain information from a single FRET pair with millisecond time resolution.[41] Though higher order time resolution methods, which analyse FRET in diffusive molecules, are also available, they are lacking in suitability for analysis of a complex system.[42] For surface immobilization of the 4WHJ, one of the four arms (arm R here, Scheme 2 and Fig. S2, ESI†) is attached to biotin in order to conjugate with streptavidin (protocol as described in the Experimental section).

Single-molecule FRET-anisotropy

The use of fluorescence emission anisotropy is the most accepted method for the determination of the comparative orientation of fluorophores and hence, the flexibility of the system or a part of the system attached to the fluorophore. However, in order to probe the relative flexibilities of two different parts of the same system, it is more appropriate to utilize FRET acceptor emission anisotropy instead of traditional fluorescence anisotropy.[42,43] As the fluorescence emission anisotropy of a FRET acceptor provides information about the flexibility of the donor-labeled part, the acceptor-labeled part and the linker between the two fluorophores, it therefore allows for a greater opportunity to determine the relative flexibility at a higher accuracy.[42] Furthermore, the advantages of single-molecule techniques over ensemble methods for monitoring molecular dynamics have encouraged us to perform the FRET-anisotropy experiments at a single-molecule level.[41,42] The commensurate timescale of the fluorescence lifetime of the fluorophores with that of the rotational diffusion endows this method with further advantages.[42]

The relative flexibilities of the IHF-free and -bound 4WHJs, attached with donor and acceptor pair, were estimated by smFRET-anisotropy. The corresponding anisotropy values for the free 4WHJ (0.270 ± 0.012) and IHF bound 4WHJ (0.360 ± 0.008) were found to be significantly different from each other.

The errors indicate the standard deviation of the peak positions in a Gaussian fit and are not to be confused with the FWHM (Fig. 1). A detailed derivation relating to the smFRET-anisotropy calculations is given in the ESI.†

The acceptor emission fluorescence anisotropies distinctly reflect the variation of the orientational stability of the 4WHJs for the imposed restriction with its flexibility upon binding to IHF. The higher fluorescence anisotropy for the IHF bound 4WHJ could be as a result of the lower positional flexibility and more rigid nature of the IHF–4WHJ complex. However, this result is not sufficient to explain the dynamics of the junction, although the higher anisotropy value could be attributed to the imposed restriction on the switching of the 4WHJ conformation.

Effect of force on the smFRET trajectories of the IHF-bound 4WHJ

Although smFRET-anisotropy provides some clues regarding the effect of imposed constraint on the conformational alterations, it does not provide legitimate information about the conformational switching. Additionally, the total energy transfer efficiency cannot be obtained with accuracy from the anisotropy results, as the obtained signals are already split into two polarized directions. On the other hand, the implementation of smFRET experiments without polarization results in overall FRET that contains all possible FRET states *i.e.*, all of the possible conformations that contribute to the kinetics of the dynamic processes can be extracted. The smFRET experiments were carried out at an IHF concentration at which all of the 4WHJs were believed to be bound, and the free protein was washed out before data acquisition. The smFRET time trajectories reflect the inter-conversion of 4WHJ between two isolated conformations, *i.e. iso*I [high FRET state, E(fret) 0.65] and *iso*II [low FRET state, E(fret) 0.28] (Fig. 2A, no force

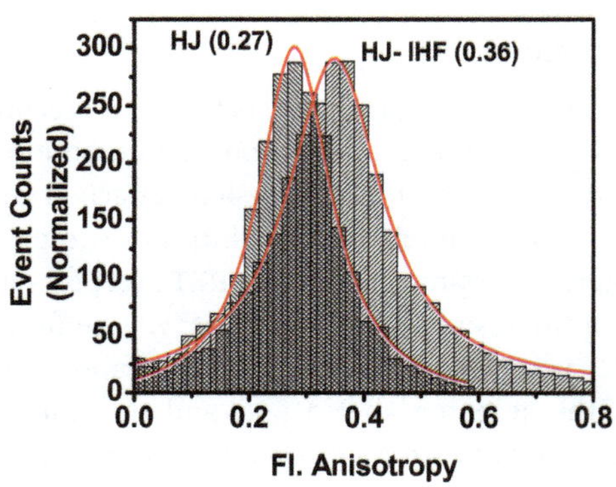

Fig. 1 smFRET anisotropy histograms obtained for the 4WHJ only and the IHF bound 4WHJ. The values in the legend are the average FRET anisotropy values obtained from the Gaussian fittings. Each histogram is an ensemble of the smFRET-anisotropy values collected from 200 single-FRET pairs.

 This journal is © The Royal Society of Chemistry 2018

applied).[14,15] The same trajectories calculated for the IHF free 4WHJ are of the order of 0.36 (low FRET state) and 0.75 (high FRET state). The corresponding distances between the donor and acceptor are 7.0 nm and 5.4 nm for IHF bound 4WHJ, and 6.6 nm and 5.0 nm for IHF free 4WHJ, considering R_0 as 6.0 nm. However, it has been found from HMM analysis that the rate constants of both the forward (*iso*I to *iso*II, K_f) and backward (*iso*II to *iso*I, K_b) processes are considerably lower (2.2 ± 0.2 s^{-1} & 3.7 ± 0.7 s^{-1}) for the IHF bound 4WHJ, compared to those of the unbound, free 4WHJ (4.2 ± 1.9 s^{-1} & 6.1 ± 1.6 s^{-1}).[26,27] This could be because of the increase in the activation energy barrier either due to a lowering of the potential energy of both conformations which results in an increase in the stability, or because of the increase in the potential energy of the transition state (TS), thus destabilizing it (Fig. 3, left panel). It can also be speculated that IHF binds to both of the conformations and restricts the conformational switching once bound (Scheme 2), thus eventually lowering the rate constant of conformational fluctuation due to imposed restriction upon IHF binding.

In order to carry out a detailed investigation of the influence of force on the dynamics and free energy landscape[24] of the restricted 4WHJ by utilizing smFRET measurements, a dielectric bead was attached with a linker to the 4WHJ, so that it could be trapped and manipulated by single-beam optical tweezers (Scheme 2). Our designed 4WHJ has been shown to bias the equilibrium towards the *iso*II (low FRET) conformation over *iso*I (high FRET) upon the application of force to the arm-XR. Fig. 2A shows the smFRET time traces (in gray) along with the idealized fits (in red) obtained from HMM analysis[44] for five different applied forces, 0.0 pN, 0.4 pN, 0.8 pN, 1.6 pN, 3.2 pN and 5.2 pN, each with a 12 ms integration time and over a 12 s duration. It has been observed that the 4WHJ is more biased towards the high FRET state (*iso*I) conformation at lower values of applied force (<0.8 pN

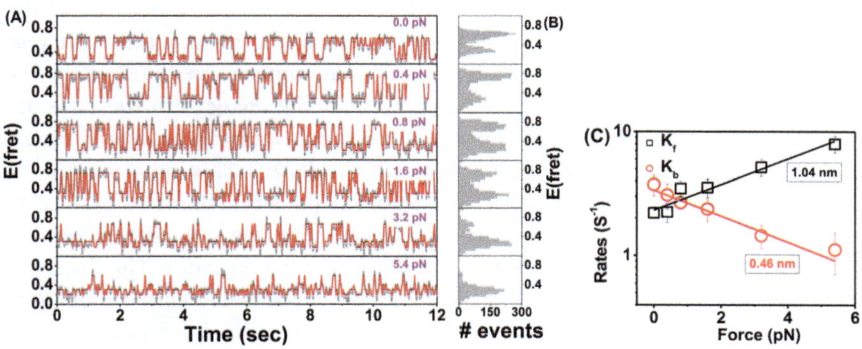

Fig. 2 (A) smFRET traces (gray lines) with applied forces of 0.0 pN, 0.4 pN, 0.8 pN, 1.6 pN, 3.2 pN and 5.2 pN, indicating an increase in the population of the low-FRET state (*iso*II) with an increase in force. The red lines represent the idealized HMM analysis of the corresponding time traces with a 12 ms integration time. (B) Histogram of the E(fret) states obtained from the smFRET time traces at different forces from an ensemble of at least 50 molecules, reflecting a switching of the population distribution from the high to low FRET state as well as a broadening of the histogram due to an increase in the possibility of the number of intermediate state(s). (C) A log–linear plot of the corresponding rate constants, K_f for *iso*I to *iso*II and K_b for *iso*II to *iso*I, indicating an increase in the rate of conformational switching for the former and a decrease in the rate for the latter. The slopes represent the relative distances from the TS.

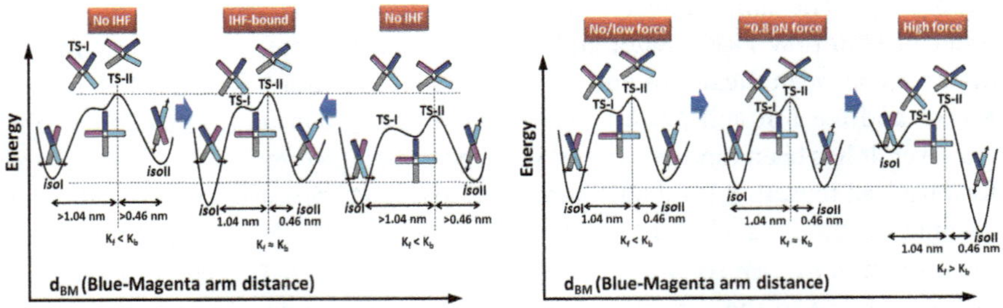

Fig. 3 Mapping of the potential energy landscapes of the 4WHJ over different applied forces. The *y*-axes represent the relative activation energy and the *x*-axes represents the separation of the blue-magenta (B-H) or cyan-gray (X-R) arms due to the application of force (indicated by the arrows). Left panel: plausible energy landscapes of the IHF free and bound 4WHJs. The higher activation energy barrier (middle) is probably due to the lowering of the potential energy of both conformations which results in an increase in the stability (left), or because of the increased potential energy of the transition state (right). Right panel: at no/low force, the *iso*I conformation is at a lower energy hence has a higher population compared to that of *iso*II. As the applied force increases, the energy landscape shifts towards the *iso*II conformer and the potential energy of *iso*II becomes almost similar to that of *iso*I at a force of ~0.8 pN, thus resulting in a nearly equal population of both conformers. However, at a higher force, the opposite phenomenon is observed.

force, Fig. 2A). Beyond a force of 0.8 pN, the low FRET state (*iso*II) conformation predominates (>0.8 pN, Fig. 2A). The two isomeric forms, however, exhibit almost equal populations at a force of ~0.8 pN (0.8 pN, Fig. 2A), which is also clear from the histogram analysis of an ensemble of at least of 50 molecules (Fig. 2B). This is also reflected by the phase space analysis carried out using the smFRET time trajectories (Fig. 4, left panel). The conformational switching is considered to occur through a single transition state (TS), thus neglecting the existence of any intermediate conformations at those forces. We observed that the rate of the forward transition rate (K_f) increases upon increasing the external force, whereas that of the backward transition rate (K_b) follows an opposite trend (Fig. 2C). The log–linear plot of K *vs.* applied forces shows that the changes in the rates are apparently linear in nature. The slope for K_f (1.04) was found to be approximately 2.3 times higher than that of the slope for K_b (0.46, Fig. 2C). This indicates that the distance of the *iso*I conformer from its TS is further (0.58 nm) than that of the *iso*II conformer (Fig. 3, right panel).[24] The distances here represent the separation between the XR-arm of the 4WHJ under different force conditions. The distances observed are shorter, compared to those in reports published earlier looking at 4WHJs only.[36,37] Although the smFRET time traces were analyzed expecting the existence of an unavoidable open-X intermediate[37] during the course of the conformational switching, it could not be traced with consistency. However, its existence has been considered when plotting the energy landscape, assuming this to be a limitation of our instrumental resolution.

The above result also demonstrates the fact that the arms come closer due to bending upon binding to IHF, thus resulting in an increase in rigidity, which is also supported by the smFRET anisotropy experiments. It is worthwhile to mention that the stability and therefore the potential energy are closely related to the activation energy of any process.[45] This indicates that once bound to IHF, the activation energy barrier of the TS increases compared to that of the free 4WHJ. At

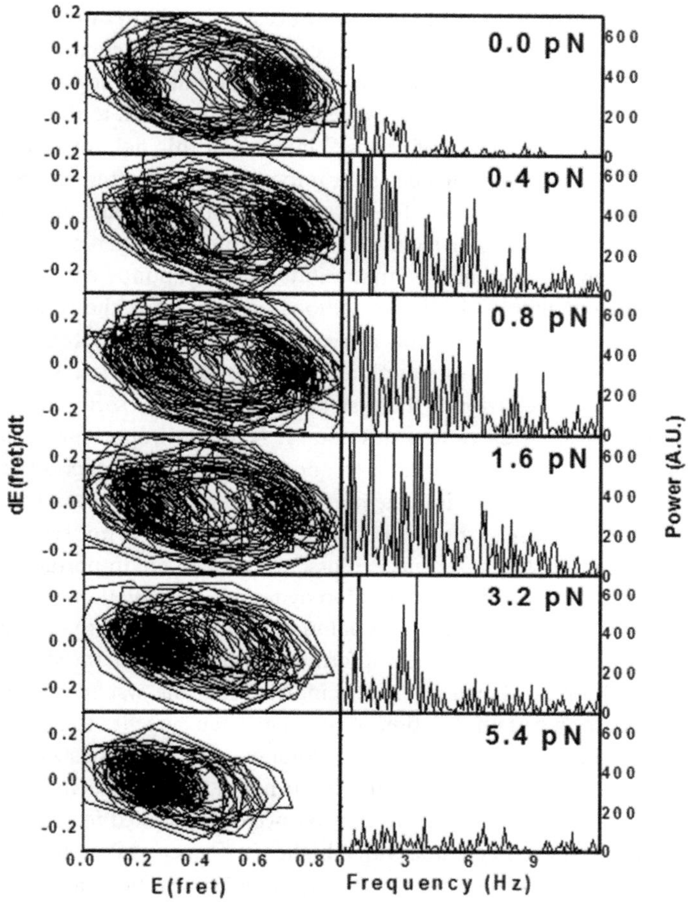

Fig. 4 Nonlinear dynamics analysis of the time traces of the conformational dynamics of the IHF bound 4WHJ at different applied forces. The phase space plots (left) show the system oscillating in one-half for a certain period of time and then switching to the other half. Power spectrum plots (right) representing the frequency domain transformation of the time series data, and showing the number of different types of fluctuations at each force.

the same time, the potential energy level of the *iso*I state becomes higher than that of the *iso*II state, as the rate of the forward process (K_f) increases and the rate of the backward process (K_b) decreases upon increasing the external force (Fig. 2C and 3, right panel).

Nonlinear dynamics analysis of conformational switching

As the majority of biological systems are nonlinear in nature, their dynamics could be best explored as nonlinear as well.[46] Nonlinearity is an essential condition for chaos, which essentially does not mean that the output is random but only seemingly random, and can be considered to be a deterministic system. It also has the merit of replacing the study of such systems within the framework of nonlinear dynamics to define the actual underlying mechanism by considering the observed behavior of coupled dynamic processes.[46] In order to interpret and corroborate the nature and behavior of the conformational fluctuations, we have followed a nonlinear analytic approach, which is certainly a new attempt and which is considered to be an alternative accession in this field, to the best of our knowledge. Since the discovery of chaos,[47] interest in this field of research has risen rapidly.[48] Chaos theory, which is an important topic within the regime of nonlinear dynamics, is a good tool for the analysis of an experimental time series even from a one-dimensional signal. Two important parameters include the phase space reconstruction and computation of the Lyapunov exponent which quantifies the sensitivity to initial conditions (see ESI† for the mathematical expression). The temporal evolution of a complex dynamical system is usually represented by a time series as a function of some control parameters. In order to further elucidate the behavior of the system, nonlinear time series analysis was carried out on the smFRET time series data. It is always important to carry out a Fourier transform analysis[49] in order to understand the frequencies that could be present in the system since it is the nonlinear interaction between the various frequencies that gives rise to the odd behavior. Here, we seek to understand chaos and its nature through phase space plots, which are an important tool for understanding the nature of the fluctuation dynamics of a 4WHJ, if chaotic at all. This is usually done by plotting the signal *vs.* the differential of the signal.[48]

The smFRET time traces of the conformational changes of the 4WHJ under different applied forces, shown in Fig. 2A, indicate that the system resides in the high FRET state for a certain duration of time and then switches to the low FRET state. It again resides in that state for some duration of time before switching back to the high FRET state. The residence time at the high FRET state seems to be larger than the time at the low FRET state in the absence of an applied force and even at low force. Additionally, it fluctuated with a higher frequency while staying in either of the two states. The residence times at the high FRET state were found to vary from 0.2 to 5 seconds. The system is always observed to be chaotic from a force of 0 to 5.4 pN. Considering the characteristics of this chaos, we have represented the phase space plots in the left panel of Fig. 4. The phase space plots of the FRET time series data (shown in Fig. 2A) demonstrate that the initial double scroll attractor observed at zero force becomes a single scroll at around 3.2 pN. This also shows that the majority of the oscillations are in the low FRET state and that the system oscillates in one state for a certain period of time before it switches back to the other state, as expected (Fig. 4, left). Looking at the time series data, it is clear that although the system is switching from one state to another it is quite chaotic in both of the equilibrium positions. Perhaps the acceptor and donor are not stationary but also oscillate around a mean position while switching from one state to the other.

The right panel of Fig. 4 shows the power spectra of the fluctuations shown in Fig. 2A. At a force of 0.0 pN, there is a band of frequencies from 2 to 7 Hz which increase when the applied force increases and which reach a maximum at 0.8 pN. A further increase in the force reduces this band of frequencies significantly, which eventually become one single dominant frequency at a force of 5.4 pN. A frequency bifurcation diagram[50] summarizes this behavior (Fig. 5A).

From analysis of the time series data, the phase space plots and the power spectrum plots, it is evident that the dominant frequencies as well as the chaotic nature increase with applied force up to 0.8 pN and then decrease significantly for higher values of applied force. This is further verified by the Largest Lyapunov

 This journal is © The Royal Society of Chemistry 2018

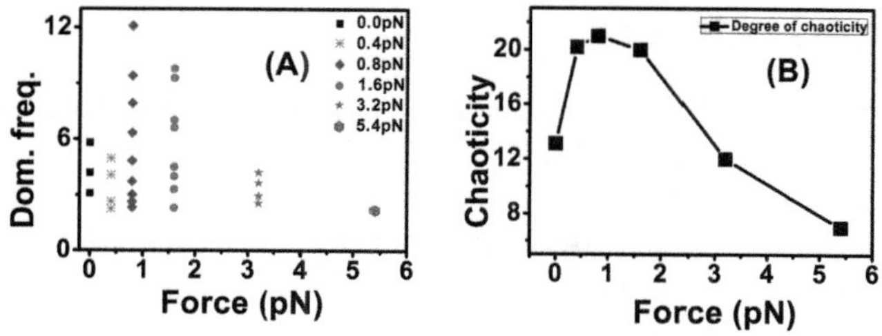

Fig. 5 (A) Frequency bifurcation diagram of the band of dominant frequencies obtained from Fourier transform analysis *vs.* different applied forces, with the maximum seen at a force of 0.8 pN. (B) A plot of Lyapunov exponents *vs.* force, representing a degree of chaoticity, thus supporting the frequency bifurcation diagram.

Exponent (LLE, details in ESI†) which shows a fall in the LLE estimates, which are a measure of chaoticity, from 3.2 pN onwards (Fig. 5B). This is more likely to be due to a possible increase in the number of intermediate states as the energy landscape is influenced by an external force (Fig. 3, right panel).[45] This could be explained in the following way; the imposed restriction on the HJ due to the binding of IHF reduces when increasing the applied force. Thus, the HJ experiences a combined effect of two competing forces. The increase in applied force reduces the tightness of the binding between IHF and the 4WHJ, and at a force of 0.8 pN the HJ experiences almost no bias towards the individual forces *i.e.* towards any particular isomer. Hence, under those conditions, it switches between the two conformations at a higher rate. This is further supported by the broadening of the histogram obtained from the smFRET time traces due to an increase in the possibility of the number of intermediate states (Fig. 3, left panel).

Conclusions

In summary, the enhanced rigidity and reduced flexibility that the 4WHJ experiences upon binding to IHF results in a slowing down of its isomerization dynamics. This is primarily shown in the smFRET-anisotropy study and is further supported by the smFRET-force experiments. HMM analysis of the time traces indicates a dramatic increase in the tightness of the 4WHJ upon binding to IHF, hence reducing the distance between the arms and changing the potential energy landscape of the TS. Nonlinear dynamics analysis was found to provide a practical framework for analysing single-molecule data in order to understand the origin of the complex behavior of the system and to decipher the energy landscape. It is worth mentioning that as we have used the Holliday junction as an example to quantitatively explore the concept of heterogeneity due to applied force, our conclusions are far reaching. This study not only opens up new aspects of 4WHJ–protein interactions, but the concept is also expected to be applied to study protein–DNA interactions and the effect of junction binding protein in recombination, irrespective of the presence of an applied force.

Acknowledgements

We thank Prof Tae-Hee Lee, Pennsylvania State University, for his generous help in providing the IDL-based data analysis codes.

Notes and references

1 P. A. Khuu, A. R. Voth, F. A. Hays and P. S. Ho, *J. Mol. Recognit.*, 2006, **19**, 234–242.

2 H. L. Klein, in *Reviews in Cell Biology and Molecular Medicine*, Wiley-VCH Verlag GmbH & Co. KGaA, 2006.

3 K. Dybvig, *Mol. Microbiol.*, 1993, **10**, 465–471.

4 R. Holliday, *Genet. Res.*, 1964, **5**, 282–304.

5 D. R. Duckett, A. Murchie and D. Lilley, *EMBO J.*, 1990, **9**, 583.

6 A. Schwacha and N. Kleckner, *Cell*, 1995, **83**, 783–791.

7 Y. Liu and S. C. West, *Nat. Rev. Mol. Cell Biol.*, 2004, **5**, 937–944.

8 W. Wang, L. M. Nocka, B. Z. Wiemann, D. M. Hinckley, I. Mukerji and F. W. Starr, *Sci. Rep.*, 2016, **6**, 22863.

9 M. A. Karymov, M. Chinnaraj, A. Bogdanov, A. R. Srinivasan, G. Zheng, W. K. Olson and Y. L. Lyubchenko, *Biophys. J.*, 2008, **95**, 4372–4383.

10 S. A. McKinney, A.-C. Déclais, D. M. Lilley and T. Ha, *Nat. Struct. Mol. Biol.*, 2003, **10**, 93–97.

11 K. Okamoto and Y. Sako, *Biophys. Chem.*, 2016, **209**, 21–27.

12 E. G. Wheatley, S. N. Pieniazek, I. Mukerji and D. Beveridge, *Biophys. J.*, 2012, **102**, 552–560.

13 H. Shinagawa and H. Iwasaki, *Adv. Biophys.*, 1995, **31**, 49–65.

14 D. M. Lilley, *Q. Rev. Biophys.*, 2000, **33**, 109–159.

15 R. J. Grainger, A. I. Murchie and D. M. Lilley, *Biochemistry*, 1998, **37**, 23–32.

16 C. Joo, S. A. McKinney, D. M. Lilley and T. Ha, *J. Mol. Biol.*, 2004, **341**, 739–751.

17 S. M. Miick, R. S. Fee, D. P. Millar and W. J. Chazin, *Proc. Natl. Acad. Sci. U. S. A.*, 1997, **94**, 9080–9084.

18 P. S. Eis and D. P. Millar, *Biochemistry*, 1993, **32**, 13852–13860.

19 M. S. Luijsterburg, M. F. White, R. van Driel and R. T. Dame, *Crit. Rev. Biochem. Mol. Biol.*, 2008, **43**, 393–418.

20 C. A. Bewley, A. M. Gronenborn and G. M. Clore, *Annu. Rev. Biophys. Biomol. Struct.*, 1998, **27**, 105–131.

21 J. Zlatanova and K. Van Holde, *FASEB J.*, 1998, **12**, 421–431.

22 A.-C. Déclais and D. M. Lilley, *Curr. Opin. Struct. Biol.*, 2008, **18**, 86–95.

23 B. Cannon, A. H. Kachroo, I. Jarmoskaite, M. Jayaram and R. Russell, *J. Biol. Chem.*, 2015, **290**, 22734–22746.

24 C. Bustamante, Y. R. Chemla, N. R. Forde and D. Izhaky, *Annu. Rev. Biochem.*, 2004, **73**, 705–748.

25 T. Ha, *Biochemistry*, 2004, **43**, 4055–4063.

26 F. E. Kemmerich, M. Swoboda, D. J. Kauert, M. S. Grieb, S. Hahn, F. W. Schwarz, R. Seidel and M. Schlierf, *Nano Lett.*, 2015, **16**, 381–386.

27 S. Wang, R. Cosstick, J. F. Gardner and R. I. Gumport, *Biochemistry*, 1995, **34**, 13082–13090.

28 K. K. Swinger and P. A. Rice, *Curr. Opin. Struct. Biol.*, 2004, **14**, 28–35.

29 D. F. Browning, D. C. Grainger and S. J. Busby, *Curr. Opin. Microbiol.*, 2010, **13**, 773–780.

30 W. Tong, D. Warren, N. E. Seah, G. Laxmikanthan, G. D. Van Duyne and A. Landy, *Proc. Natl. Acad. Sci. U. S. A.*, 2014, **111**, 12366–12371.

31 N. E. Seah, D. Warren, W. Tong, G. Laxmikanthan, G. D. Van Duyne and A. Landy, *Proc. Natl. Acad. Sci. U. S. A.*, 2014, **111**, 12372–12377.

 This journal is © The Royal Society of Chemistry 2018

32 J. L. Boldt, K. V. Kepple, G. D. Cassell and A. M. Segall, *Nucleic Acids Res.*, 2006, **35**, 716–727.

33 I. Heller, T. P. Hoekstra, G. A. King, E. J. Peterman and G. J. Wuite, *Chem. Rev.*, 2014, **114**, 3087–3119.

34 K. C. Neuman and A. Nagy, *Nat. Methods*, 2008, **5**, 491–505.

35 S. Hohng, S. Lee, J. Lee and M. H. Jo, *Chem. Soc. Rev.*, 2014, **43**, 1007–1013.

36 S. Lee and S. Hohng, *J. Am. Chem. Soc.*, 2013, **135**, 18260–18263.

37 S. Hohng, R. Zhou, M. K. Nahas, J. Yu, K. Schulten, D. M. Lilley and T. Ha, *Science*, 2007, **318**, 279–283.

38 S. Sugimura and D. M. Crothers, *Proc. Natl. Acad. Sci. U. S. A.*, 2006, **103**, 18510–18514.

39 H. Nash, C. Robertson, E. Flamm, R. Weisberg and H. Miller, *J. Bacteriol.*, 1987, **169**, 4124–4127.

40 T. W. Lynch, E. K. Read, A. N. Mattis, J. F. Gardner and P. A. Rice, *J. Mol. Biol.*, 2003, **330**, 493–502.

41 P. R. Selvin and T. Ha, *Single-molecule techniques*, Cold Spring Harbor Laboratory Press, 2008.

42 P. P. Mishra, M. T. Qureshi, W. Ren and T.-H. Lee, *Biophys. J.*, 2010, **99**, 3849–3858.

43 S. C. Blanchard, H. D. Kim, R. L. Gonzalez, J. D. Puglisi and S. Chu, *Proc. Natl. Acad. Sci. U. S. A.*, 2004, **101**, 12893–12898.

44 T.-H. Lee, *J. Phys. Chem. B*, 2009, **113**, 11535–11542.

45 S. K. Sarkar, N. M. Andoy, J. J. Benítez, P. R. Chen, J. S. Kong, C. He and P. Chen, *J. Am. Chem. Soc.*, 2007, **129**, 12461–12467.

46 A. A. Lense, *Riv. Biol.*, 2006, **99**, 467–481.

47 E. N. Lorenz, *J. Atmos. Sci.*, 1963, **20**, 130–141.

48 E. Renshaw, *Math. Med. Biol.*, 1994, **11**, 17–44.

49 J. Theiler, S. Eubank, A. Longtin, B. Galdrikian and J. D. Farmer, *Phys. D*, 1992, **58**, 77–94.

50 G. A. Leonov, D. V. Ponomarenko and V. B. Smirnova, Theory and applications, *Frequency-domain methods for nonlinear analysis*, World Scientific, 1996.

Faraday Discussions

PAPER

Excited-state dynamics of mononucleotides and DNA strands in a deep eutectic solvent†

Yuyuan Zhang,[a] Kimberly de La Harpe,[b] Mahesh Hariharan [c] and Bern Kohler *[a]

Received 19th October 2017, Accepted 6th November 2017
DOI: 10.1039/c7fd00205j

The photophysics of several mono- and oligonucleotides were investigated in a deep eutectic solvent for the first time. The solvent glyceline, prepared as a 1 : 2 mole ratio mixture of choline chloride and glycerol, was used to study excited-state deactivation in a non-aqueous solvent by the use of steady-state and time-resolved spectroscopy. DNA strands in glyceline retain the secondary structures that are present in aqueous solution to some degree, thus enabling a study of the effects of solvent properties on the excited states of stacked bases and stacked base pairs. The excited-state lifetime of the mononucleotide 5′-AMP in glyceline is 630 fs, or twice as long as in aqueous solution. Even slower relaxation is seen for 5′-TMP in glyceline, and a possible triplet state with a lifetime greater than 3 ns is observed. Circular dichroism spectra show that the single strand $(dA)_{18}$ and the duplex $d(AT)_9 \cdot d(AT)_9$ adopt similar structures in glyceline and in aqueous solution. Despite having similar conformations in both solvents, femtosecond transient absorption experiments reveal striking changes in the dynamics. Excited-state decay and vibrational cooling generally take place more slowly in glyceline than in water. Additionally, the fraction of long-lived excited states in both oligonucleotide systems is lower in glyceline than in aqueous solution. For a DNA duplex, water is suggested to favor decay pathways involving intrastrand charge separation, while the deep eutectic solvent favors interstrand deactivation channels involving neutral species. Slower solvation dynamics in the viscous deep eutectic solvent may also play a role. These results demonstrate that the dynamics of excitations in stacked bases and stacked base pairs depend not only on conformation, but are also highly sensitive to the solvent.

[a]Department of Chemistry and Biochemistry, The Ohio State University, 100 West 18th Avenue, Columbus, Ohio 43210, USA. E-mail: kohler.40@osu.edu; Tel: +1-614-688-2635

[b]Department of Physics, United State Air Force Academy, USAF Academy, Colorado 80840, USA

[c]School of Chemistry, Indian Institute of Science Education and Research, Thiruvananthapuram, Kerala, India 695 016

† Electronic supplementary information (ESI) available. See DOI: 10.1039/c7fd00205j

Introduction

Intense efforts have been devoted for many years towards understanding the ultrafast photophysics of DNA and RNA monomers and oligomers. The experimental and theoretical advances and the open questions in this very active field have been reviewed a number of times since 2000.[1–7] To date, most solution studies have been carried out in water. This is clearly appropriate given water's singular importance in biology, but experiments in non-aqueous solvents are also needed in order to better understand the role of the solvent in excited-state deactivation and to test theoretical predictions. Such studies[8–14] have been carried out on nucleobase monomers (free bases, nucleosides and nucleotides), which are straightforward to study in non-aqueous solvents, provided that they are sufficiently soluble, but it is highly challenging to study assemblies of two or more nucleobases in non-aqueous conditions that maintain the secondary structures that are present in aqueous solution.

The photophysics of DNA base pairs and larger hydrogen-bonded assemblies of nucleobases have been successfully studied in non-aqueous solvents.[12,15,16] Temps and co-workers showed that photophysical studies of hydrogen-bonded structures are possible in low polarity solvents using nucleoside derivatives in which one or more H atoms bonded to the O atoms of the (deoxy)ribose moiety are replaced by bulky *tert*-butyldimethylsilyl (TBDMS) groups.[12,15] This approach was used to study single GC base pairs in chloroform[12,17] and the extended hydrogen-bonded ribbons, sheets and quartets formed by TBDMS-protected guanosine in *n*-hexane.[15,16] Although these assemblies display fascinating photophysics, their two-dimensional structures lack the base stacking that is ubiquitous in DNA strands in aqueous solution. Recent studies point to the charge-transfer excited states formed by electron transfer between π-stacked nucleobases as a particularly important class of electronic excitations in DNA strands.[18–20] In order to study how the environment affects the excited-state dynamics, it would be of great interest to study these states in a non-aqueous environment. However, to quote the authors of a 1972 paper,[21] "stacking seems to occur in water only".

While it is correct that most solvents destabilize DNA, Hud and co-workers showed in 2010 that various DNA strands form stable secondary structures in a deep eutectic solvent (DES) made from choline chloride and urea (1 : 2 mole ratio).[22] DESs constitute an important class of ionic liquids. They are easily prepared by mixing a hydrogen bond donor molecule with a salt such as choline chloride. Like other ionic liquids, DESs can be very viscous and have a very low vapor pressure. They are also extremely attractive for many applications on account of their high biocompatibility.[23] The choline ion is a commonly encountered head group in amphiphilic compounds found in biological membranes, while a hydrogen bond donor like glycerol has low toxicity. Circular dichroism (CD) spectra, UV-melting studies and MD simulations indicate that DNA structures in a DES can be highly similar to those formed in aqueous solution, although structural differences and changes in thermal stability have also been noted.[22,24,25] Two-dimensional DNA origami structures can even fold correctly, albeit slowly, in a DES.[26]

 This journal is © The Royal Society of Chemistry 2018

The structure preserving abilities of DESs provide an opportunity to study solvent effects without altering the conformations of interest. Here, we report the first study of the excited-state dynamics of nucleic acid compounds in a non-aqueous solvent that preserves base stacking. Steady-state and femtosecond transient absorption (TA) spectroscopic techniques were used to investigate the excited-state deactivation of mononucleotides and single- and double-stranded DNA oligonucleotides in a DES made from choline chloride and glycerol. The results show that stacked bases and stacked base pairs exhibit strikingly different dynamics in a non-aqueous solvent even though they adopt structures similar to those present in aqueous solution.

Experimental methods

Sodium salts of the mononucleotides adenosine 5′-monophosphate (5′-AMP) and thymidine 5′-monophosphate (5′-TMP) were purchased from Sigma-Aldrich and used as received. The oligonucleotides $(dA)_{18}$ and $d(AT)_9$ were purchased from The Midland Certified Reagent Company as lyophilized powders and used as received. The deep eutectic solvent was prepared by heating a mixture of anhydrous solid choline chloride and glycerol combined in a 1 : 2 mole ratio at 100 °C for *ca.* 2 hours. Hereafter, we will refer to this solvent as glyceline, which is the name most commonly encountered in the literature.

Solutions of 5′-AMP and 5′-TMP were prepared by directly dissolving each compound in freshly prepared glyceline. First, between 1 and 2 mg of the sample of interest was dissolved in 100 μL of water. For the oligonucleotide solutions, 30 mg of KCl was additionally added to promote thermal stability, but no additional salt was added to the mononucleotide solutions. Next, the aqueous solution was added to 4 g of glyceline and freeze dried under vacuum for 12 hours to remove the water. The final per-nucleotide concentration was approximately 1.0 mM for all solutions. The final KCl molality in the oligonucleotide-glyceline solutions was 100 mmol kg^{-1}.

All samples studied in aqueous solution were dissolved in a 100 mM phosphate buffer that was prepared using water from an ultrapurifier. The aqueous oligonucleotide solutions additionally contained 250 mM NaCl. These solutions are referred to below as water solutions for brevity. The per-nucleotide concentrations of the buffer solutions were approximately 1.0 mM as in the DES solutions. Prior to performing spectroscopic measurements, the samples containing DNA strands were thermally annealed by maintaining the solution temperature at 90 °C for 5 minutes, followed by slowly cooling to room temperature. A total sample volume of ~2.0 mL was recirculated through a 1 mm path length sample cell (Harrick) with CaF_2 windows. Ionic liquids, including the DES glyceline, are often highly viscous. The viscosity of glyceline at 20 °C, $\eta = 376$ mPa s,[27] is nearly four times lower than that of neat glycerol and is low enough to conveniently flow the solution using a peristaltic pump. Flowing the sample solution minimizes re-excitation and photodegradation effects.

The setup for two-color pump-probe experiments has been described in detail elsewhere.[28,29] Briefly, 265 nm excitation pulses were generated by pumping a white light-seeded, two stage optical parametric amplifier (OPerA Solo, Coherent Inc) with approximately 1.2 mJ of the 800 nm fundamental (Libra-HE, Coherent Inc; 3.5 W, 1 kHz, 90 fs). The signal pulses were mixed with the fundamental to generate pulses centered at 530 nm, which were subsequently frequency-doubled to 265 nm. The probe pulses at 570 nm were generated by sum-frequency mixing of the idler and fundamental in a second OPerA Solo. The frequency of the pump beam was modulated at 333 Hz by a mechanical chopper synchronized to the fundamental laser frequency. The intensity of the probe pulses was detected by a photomultiplier tube and readout by a lock-in amplifier. The pump-probe polarization was set to the magic angle (54.7°) in order to minimize the re-orientation dynamics. The pump pulses were attenuated to 0.9 μJ per pulse and focused to 210 μm (full width at half maximum) at the front surface of the optical cuvette.

All nonlinear least squares fitting, including global fits, was performed using the Igor Pro program (Wavemetrics). Stated uncertainties are twice the standard deviation.

Results

Mononucleotides

5′-AMP. The UV absorption spectrum of 5′-AMP in glyceline (green solid curve in Fig. 1a) has a maximum at a wavelength (λ_{max}) of 261 nm, and is somewhat broader than the spectrum measured in water (gray dashed curve in Fig. 1a) for

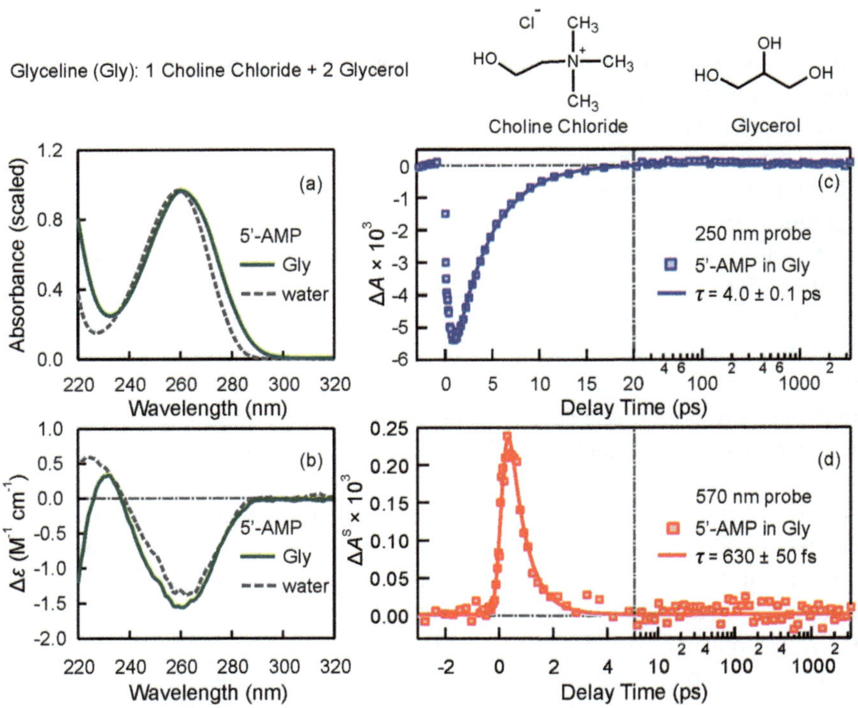

Fig. 1 UV-visible (a) and CD (b) spectra of 5′-AMP in glyceline (green solid curves) compared to those of 5′-AMP in water (gray dashed curves). Kinetic traces monitored at 250 nm (c) and 570 nm (d) after 265 nm excitation of 5′-AMP in glyceline. The solvated electron signal from the solvent has been subtracted in panel (d). The vertical dash-dotted line marks the linear-logarithmic break. The best-fit lifetimes are also shown.

which λ_{max} is 259 nm. Note that glyceline has minimal absorption in the UV (see Fig. S1 in the ESI†). Table 1 lists the λ_{max} values for 5′-AMP and the other compounds studied. In all cases, a small red shift of 1 to 3 nm is observed upon going from water to glyceline. The CD spectrum of 5′-AMP is very similar in glyceline and in water (Fig. 1b). Both spectra agree very well with a previously published spectrum of 2′-deoxyadenosine 5′-monophosphate in water.[30]

The TA signal for 5′-AMP in glyceline at 250 nm following 265 nm excitation is shown in Fig. 1c. The signal fully recovers to the baseline approximately 20 ps after the pump pulse, and is well described by a mono-exponential decay with a time constant of 4.0 ± 0.1 ps. The TA signal at a probe wavelength of 570 nm corrected for the signal from the solvent is shown in Fig. 1d. As can be seen in Fig. S2,† 265 nm excitation of the neat solvent glyceline yields a nearly constant signal from 0 to 3 ns at the probe wavelength of 570 nm (Fig. S2a†). The solvent-only signal at 570 nm was scaled to match the raw signal from 5′-AMP in glyceline seen at delay times greater than 20 ps, and then subtracted from the latter in order to yield the solvent-subtracted signal shown in Fig. 1d. A single exponential fit to the 570 nm trace yields a decay time constant of 630 ± 50 fs. For probing at 250 nm, the solvent-only signal (Fig. S2b†) is a factor of 20 weaker than the raw signal from 5′-AMP at this probe wavelength (Fig. 1c), and no correction is needed. For comparison, the raw signals recorded for 5′-AMP and for the other solutes are shown in Fig. S3.†

5′-TMP. The maximum in the absorption spectrum of 5′-TMP in glyceline (Fig. 2a) is red-shifted by 1 nm compared to the spectrum in water (Table 1). The TA signal for 5′-TMP in glyceline monitored at 570 nm after 265 nm excitation is shown in Fig. 2b after subtraction of the signal arising only from the solvent following a previously published correction procedure.[31] First, a solution of 5′-AMP in glyceline was prepared to have an identical absorbance at the pump wavelength of 265 nm to that of the 5′-TMP sample. Then, signals from both solutions were recorded sequentially. Next, the signal recorded from the solvent glyceline was scaled to best match the signal from the 5′-AMP in glyceline solution at delay times greater than 20 ps and subtracted from the 5′-TMP in glyceline transient (Fig. S3b†). This procedure is justified by the observation that the signal from 5′-AMP in glyceline at delay times greater than 20 ps is due solely to species produced by UV excitation of the solvent (see Discussion below). The solvent-corrected signal decays exponentially with a lifetime of 2.9 ± 0.5 ps to a non-decaying offset that persists to 3 ns, which is the limit of our temporal observation window.

DNA strands

Single-stranded $(dA)_{18}$. The UV absorption and CD spectra of single-stranded $(dA)_{18}$ in glyceline are shown in panels (a and b) of Fig. 3, respectively. A small

Table 1 Absorption maxima (nm) observed in glyceline and water

	5′-AMP	5′-TMP	$d(AT)_9 \cdot d(AT)_9$	dA_{18}
Glyceline	261	268	263	260
H_2O	259	267	262	257

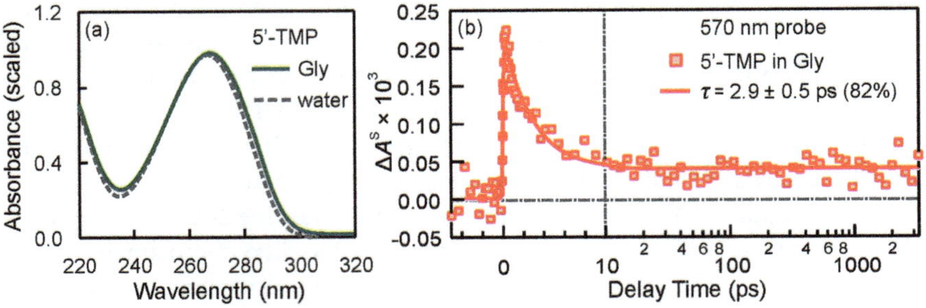

Fig. 2 (a) UV absorption spectrum of 5′-TMP in glyceline (green solid curve), compared to that in water (gray dashed curve). (b) TA signal at 570 nm after 265 nm excitation of 5′-TMP in glyceline. The solvated electron signal from the solvent has been subtracted. The vertical dash-dotted line marks the linear-logarithmic break. The best-fit lifetime is also shown.

red shift in λ_{max} is observed in the absorption spectrum in glyceline (Table 1). The CD spectrum is similar in shape to that of $(dA)_{18}$ in water; a weak positive band at 275 nm is followed by a strong negative band at 250 nm and a strong positive band at 220 nm. However, the overall intensity of the CD signal for $(dA)_{18}$ in glyceline is three times weaker than in water.

The pump-probe transient recorded for $(dA)_{18}$ in glyceline at a probe wavelength of 250 nm is shown in Fig. 3c (blue squares) and compared with the transient signal recorded under the same conditions in water (black circles). The $(dA)_{18}$ signal in glyceline shows much more extensive re-population of the ground

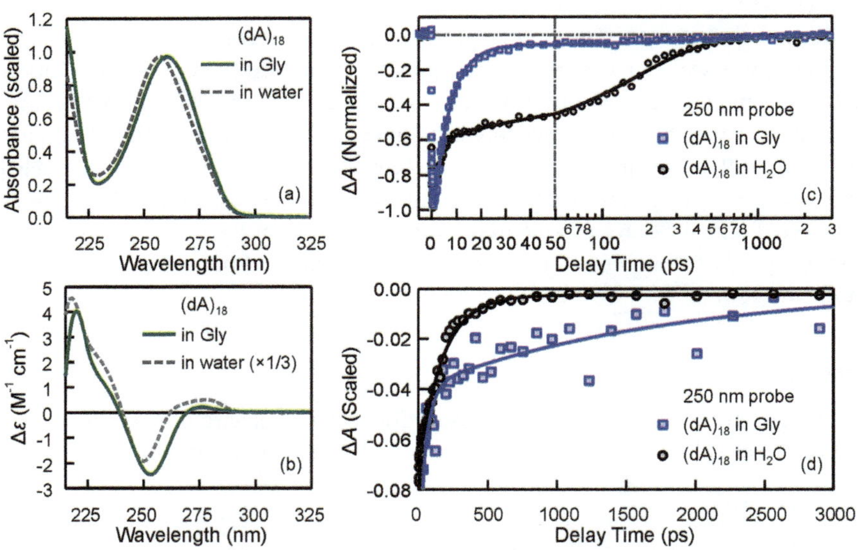

Fig. 3 UV absorption (a) and CD spectra (b) of $(dA)_{18}$ in glyceline *vs.* water. Transient absorption was monitored at 250 nm (c) after 265 nm excitation of $(dA)_{18}$ in both solvents. The vertical dash-dotted line marks the linear-logarithmic break. The signals in panel (c) are shown in panel (d) on an expanded scale to show the long-time behavior. The fits are represented by solid curves and the fitting parameters are listed in Table 2.

state within the first 100 ps after excitation than is the case for $(dA)_{18}$ in water, which has a very prominent long-lived signal component. 50 ps after excitation, 95% of the initial signal for $(dA)_{18}$ in glyceline has decayed. In contrast, the signal from $(dA)_{18}$ in water has only recovered by approximately 50% at 50 ps. In Fig. 3d, both signals are shown on an expanded vertical scale with a linear time axis. The signal in water has been scaled in order to have the same amplitude as that of the glyceline signal at 50 ps. This view shows that the weak $(dA)_{18}$ signal in glyceline clearly decays much more slowly than the signal for $(dA)_{18}$ in water. Fitting the $(dA)_{18}$ signal in glyceline to three exponentials yields the parameters shown in Table 2. The large uncertainties of the τ_1 and τ_2 parameters reflect the low signal-to-noise ratio of the weak signal at long delay times. For comparison, parameters for the fit to the signal recorded in aqueous solution are also shown.

$d(AT)_9 \cdot d(AT)_9$ duplex. The duplex $d(AT)_9 \cdot d(AT)_9$ made up of 18 A·T base pairs was studied in glyceline and water using steady-state and TA spectroscopy. The UV absorption and CD spectra for the $d(AT)_9 \cdot d(AT)_9$ duplex in glyceline change significantly with temperature (Fig. 4a and b). The intensities of the CD signals decrease, and the absorbance of the lowest-energy band in the absorption spectrum increases as the temperature is raised. The low-temperature CD spectra show the signature of exciton coupling: a positive band at 270 nm and a negative band at 250 nm. The spectral shapes and temperature-dependent changes closely resemble those observed for the same duplex in water (compare Fig. 4a and b with Fig. S1a and b† from ref. 32). The melting point, determined from the inflection point on a graph of absorbance at 260 nm *vs.* temperature, is 55 °C for $d(AT)_9 \cdot d(AT)_9$ in water, but 45 °C in glyceline (Fig. 4c). The CD melting point data recorded in glyceline (Fig. 4d) confirm a melting point that falls roughly between 40 and 45 °C.

The 265 nm pump/250 nm probe TA signal for $d(AT)_9 \cdot d(AT)_9$ in glyceline is shown in Fig. 4e. The black data points and fitted curve in Fig. 4e show the signal reported earlier[33] for the same duplex at the same wavelengths in water. Both signals decay rapidly during the first 10 to 20 ps, after which a slower decay is seen. The latter component contributes much more strongly to the signal in water than in glyceline. The solvent-subtracted TA signal at 570 nm (Fig. 4f) also shows biphasic decays, similar to those seen at 250 nm. The majority of the glyceline signal (89%) at 250 nm decays on a time scale of 8.2 ± 0.7 ps, while a small portion (10%) decays on a slower time scale of 170 ± 50 ps. A constant offset of 1% was also used to fit the data. The low amplitude of the 570 nm signal at times greater than 50 ps combined with the uncertainty introduced by the solvent correction made fitting to the long-time signal difficult. For this reason, the longest decay constant at 570 nm was linked to the same parameter for the 250 nm transient

Table 2 Best fit parameters for AMP and $(dA)_{18}$ in glyceline and water

Substrate	Solvent	λ_{probe} (nm)	% A_0	τ_0 (ps)	% A_1	τ_1 (ps)	% A_2	τ_2 (ns)
AMP	Glyceline	250	100%	4.0 ± 0.1	—	—	—	—
$(dA)_{18}$	Glyceline	250	78%	6.1 ± 0.4	13%	50 ± 50	9%	1.8 ± 1.3
$(dA)_{18}$	H_2O	250	52%	2.7 ± 0.3	47%	170 ± 50	1%	—

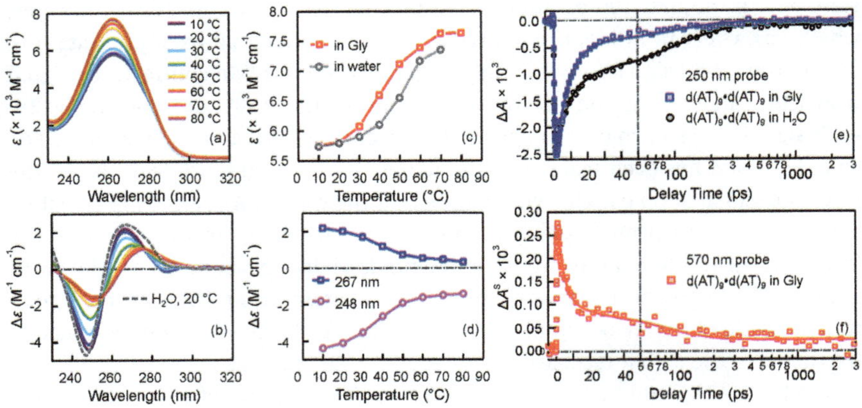

Fig. 4 Temperature-dependent UV-visible (a) and CD spectra (b) of $d(AT)_9 \cdot d(AT)_9$ in a glyceline solution. The CD spectrum of the same duplex in water at 20 °C (dashed curve) is shown in panel (b) for comparison. The absorbance at 260 nm (c) and CD signal amplitudes at 267 nm and 248 nm (d) are plotted *vs.* temperature for $d(AT)_9 \cdot d(AT)_9$ in glyceline. Panel (c) also shows the temperature-dependent absorbance at 260 nm for the same duplex in water. UV-visible and CD spectra of the duplex in water are from ref. 32. Kinetic traces monitored at 250 nm (e) and 570 nm (f) after 265 nm excitation of $d(AT)_9 \cdot d(AT)_9$ in glyceline. Panel (f) shows the solvent-subtracted signal (see the text for details).

during global fitting. The fitting parameters for the transients shown in Fig. 4e and f are summarized in Table 3.

Discussion

Mononucleotides

The wavelength of the maximum absorption (λ_{max}) is red-shifted by 1 to 2 nm in glyceline compared to in water for both mononucleotides (Table 1). These shifts are unusual as λ_{max} generally decreases for nucleobases in solvents that are less polar than water. For example, λ_{max} of thymine is blue-shifted by 1 nm in methanol and ethanol.[34] A larger blue shift of 4 nm is seen upon going from water to acetonitrile.[34] For the adenine derivative, 9-methyladenine, λ_{max} is 261 nm in both water and methanol and 258 nm in acetonitrile.[11] These shifts have been rationalized by the decreased solvent polarity of these solvents compared to water. Solvatochromic studies suggest that DESs are also somewhat less polar than water. The $E_T(30)$ value, which is an empirical measure of solvent polarity, is

Table 3 Best fit parameters for the $d(AT)_9 \cdot d(AT)_9$ duplex in glyceline and water

Solvent	λ_{probe} (nm)	% A_0	τ_0 (ps)	% A_1	τ_1 (ps)	% A_2
Glyceline	570	62%	4 ± 3	31%	170 ± 50	7%
Glyceline	250	89%	8.2 ± 0.7	10%	170 ± 50	1%
H_2O^a	250	52%	5.6 ± 1.4	45%	84 ± 14	3%

^{a} Data from ref. 33.

58.0 kcal mol^{-1} for glyceline (a similar value was measured for glycerol) compared to 63.1 kcal mol^{-1} for water.[35] The red shift of λ_{max} is not expected from the reduced polarity of glyceline, therefore possibly indicating that specific solute–solvent interactions are more important than dipolar solvation effects for determining the steady-state absorption properties.

The corrected 570 nm signal and the 250 nm bleach signal for 5′-AMP both decay to the baseline (Fig. 1). This presents strong evidence that there is no long-lived population of excited states or photoproducts. The long-lived offset seen at 570 nm in the raw signal for 5′-AMP in glyceline (Fig. S3a†) must therefore arise from the solvent. This solvent-only contribution is likely due to two-photon ionization by the UV pump pulse as seen, for example, in water[36] and methanol.[11] Solvated electrons are readily generated by multiphoton ionization of alcohol solvents.[37]

In glyceline, solvated electrons could originate from multiphoton ionization of glycerol or the choline ion, or possibly even through photodetachment from chloride ions, which are present in high concentration (>3 M). In glycerol, the solvated electron absorbs broadly throughout the visible region with a λ_{max} equal to 537 nm at 299 K.[37] Bonin *et al.*[38] found that femtosecond pulses at 263 nm produce solvated electrons in neat glycerol at 333 K by two-photon ionization. In their experiments, the TA signal at 560 nm showed little decay from 0 to 50 ps, and geminate recombination resulted in a signal loss of less than 15% at 450 ps after the pump pulse. The slow rate of geminate recombination was attributed to the high solvent viscosity of 76 mPa s for glycerol at 333 K. In fact, the viscosity of glyceline at room temperature is approximately five times higher (376 mPa s),[27] so there should be even less recombination under our experimental conditions. The solvated electrons created by the UV pump pulse in our measurements result in a constant signal offset, which is then removed from the 570 nm signals by subtraction. The lack of any decay in the solvent-only signal means that any scaling error is unlikely to introduce spurious decay components.

Solvent effects on the photophysics of monomeric nucleobases have been studied previously.[9,10,13] The femtosecond fluorescence upconversion technique was used to study the ultrafast fluorescence of thymine in water,[10] acetonitrile,[10] and in the alcohols methanol, ethanol and 1-propanol.[13] Average lifetimes were up to 30% shorter in the alcohol solvents than in aqueous solution, but the results showed little to no correlation with solvent polarity or viscosity.[13] The authors proposed that solute–solvent hydrogen bonding destabilizes the $^1n\pi^*$ state. In acetonitrile and alcohols, weaker hydrogen bonding interactions allow the initial bright $^1\pi\pi^*$ state to couple to a dark $^1n\pi^*$ state, leading to a decreased emission lifetime.

In contrast to this trend of shorter lifetimes, the 630 fs excited-state lifetime of 5′-AMP in glyceline is approximately twice the lifetime of adenosine in H_2O (290 fs).[39] The latter value was also measured by the TA technique using the same pump and probe wavelengths as those used in this study. Lifetimes of 360 fs and 430 fs were reported for 9-methyladenine in solutions of water and methanol, respectively.[11] In a previous TA study, the lifetime of thymidine in water was reported as 540 fs.[39] Deactivation is believed to take place *via* out-of-plane motions that facilitate the $S_1 \rightarrow S_0$ internal conversion (IC). It is possible that the high viscosity of glyceline impedes the required motions, thus extending the lifetimes of the lowest excited states of the mononucleotides.

In addition to changing the energies of the various excited states, solvation can also alter the distribution of tautomers present in solution. Different tautomers of the same base can exhibit very different excited state dynamics, as has been seen for the case of 9-H- *vs.* 7-H-adenine.[40] Our use of 9-substituted nucleotides eliminates certain tautomers, but we cannot exclude the possibility that other tautomers are present that contribute to the dynamics. Nevertheless, the polarity of glyceline is sufficiently similar to that of water[35] that we expect a similar distribution of tautomers in both solvents.

The shortest component obtained from fitting the single decay trace for 5′-TMP is 2.9 ± 0.5 ps. For comparison, the lowest-energy $^1\pi\pi^*$ state lifetime of aqueous thymidine is 540 fs.[39] A component that is substantially faster than our instrument response function (~200 fs) could have gone undetected in our measurements, but we tentatively assign the 2.9 ps component to the lowest-energy $^1\pi\pi^*$ state of 5′-TMP, based on the increased lifetime seen for 5′-AMP in glyceline. The 5′-TMP signal also shows a long-lived offset both before and after solvent correction (see Fig. 3b). Based on its long lifetime, this component is assigned to the lowest triplet state. The triplet quantum yields of pyrimidine bases are known to increase with decreasing solvent polarity.[8]

The pattern of a slower decay component at 250 nm than at visible probe wavelengths has been seen before for many nucleobase monomers in aqueous solution and is attributed to vibrational cooling (VC) dynamics.[2] Here, the 4.0 ps decay in the 250 nm signal for 5′-AMP is longer than the 650 fs lifetime present at 570 nm. Ground-state bleach (GSB) signals like those measured in this study at 250 nm are sensitive to the vibrational cooling (VC) of hot ground state molecules created by ultrafast internal conversion.[11] The vibrational cooling time constant of 4.0 ps observed here for 5′-AMP in glyceline is similar to the value of 4.5 ps that was observed previously for 9-methyladenine in methanol, but is slower than the decay time of 2.4 ps observed in water.[11]

Terazima proposed that the rate of thermalization of a hot solute is correlated with the number of OH groups per unit volume in the solvent when solute–solvent hydrogen bonds are present.[41] For DNA bases, high-frequency O–H stretching and medium-frequency bending modes involving the O–H group accelerate intermolecular vibrational energy transfer to the surrounding solvent molecules.[42] From the density of glyceline of 1.181 g mL^{-1} at 20 °C,[27] a density of 15 OH bonds nm^{-3} is calculated. Virtually the same value is calculated for methanol, which is consistent with the similar VC time constants measured in both solvents. Faster VC in water is explained in this model by its high OH bond density of 67 nm^{-3} at 20 °C.

To summarize our findings for the mononucleotides 5′-AMP and 5′-TMP in glyceline, deactivation of the $^1\pi\pi^*$ state of each takes place 2 to 5 times more slowly than in aqueous solution, but nevertheless occurs rapidly enough for VC dynamics to be observable in the GSB signals at 250 nm. VC takes place more slowly than in water, but at a rate that is reasonable for a solvent with a high-density network of hydrogen bonds.

DNA strands

Ultrafast spectroscopy experiments on single DNA strands and supporting theoretical work have revealed long-lived excited states that appear to be localized on

 This journal is © The Royal Society of Chemistry 2018

two bases and have significant charge transfer (CT) character.[43,44] These excimer/exciplex states exhibit vibrational modes that are characteristic of a pair of oppositely charged radical ions formed by photoinduced electron transfer (ET) between two π-stacked bases, and they decay efficiently by charge recombination (CR) over tens to hundreds of picoseconds.[18–20] Importantly, these states are only observed when two or more nucleobases are π-stacked at the time of photoexcitation, and no such states are seen when base stacking is disrupted by raising the temperature or by adding a denaturing co-solvent.[30] Monomer-like decay channels appear to dominate whenever an excitation is localized on an unstacked nucleobase. In duplex DNA structures, intrastrand ET can couple to interstrand proton transfer (PT), as experimentally observed in several duplexes.[32,45]

Although it has been reported that duplex and higher-order structures retain their secondary structure when dissolved in a DES,[22,24,46,47] it is unknown if bases in single-stranded DNA remain partially stacked in a DES like glyceline, as is the case in water. The CD spectra of $(dA)_{18}$ in glyceline and in water are similar (Fig. 3b), which suggests that base stacking is still present, but the three-fold decrease in the magnitude of the CD signal in the former solvent could indicate that a smaller percentage of all bases are stacked. Consistent with this expectation, long-lived excited states, which have been shown to be a reliable diagnostic of base stacking in aqueous solution,[30] are formed in lower yields for $(dA)_{18}$ in glyceline than in water (compare the relative amplitudes in Table 2).

The lifetime of the fast component in the GSB signal of $(dA)_{18}$ in glyceline (6.3 $\pm$ 0.3 ps) is ~60% longer than the lifetime measured for 5′-AMP in the same solvent (4.0 ps). As was explained above, GSB signals are often dominated by vibrational cooling, which could conceivably occur at a different rate for an unstacked subunit within an oligonucleotide than for a free and fully solvated mononucleotide. However, the GSB signals of $(dA)_2$ and $(dA)_{18}$ in water recorded at 80 °C, in which most of the subunits are unstacked, show nearly identical dynamics with a fast component of 2.5 to 3 ps.[48] Moreover, this lifetime is nearly identical to the VC time of 9-methyladenine in aqueous solution at room temperature,[11] thus suggesting that the backbone does not significantly alter the VC dynamics of a nucleobase after UV-excitation. Fast lifetimes of between 2.2 and 2.8 ps have also been observed when the bases in $(dA)_2$ are unstacked by the addition of 10–50% methanol (v/v) to an aqueous solution.[30]

These considerations suggest that VC of an unstacked nucleobase that is part of a DNA strand in water should take place at the same rate as that of a fully solvated monomer. On this basis, the longer 6.3 ps fast component seen in the $(dA)_{18}$ GSB signal compared to the 4.0 ps lifetime observed for 5′-AMP, both measured in glyceline, could represent an average lifetime due to somewhat faster deactivation by unstacked monomers and somewhat slower, but still rapid (<10 ps), decay by excitations associated with stacked bases. We speculate that the dielectric response of the viscous DES, which is expected to fully develop over slower time scales than is the case for water, may only be capable of stabilizing a smaller percentage of excimers. In other words, not only the decay rate of an excimer, but also the kinetics of its formation may depend sensitively on the solvent. Although the three-fold lower amplitude of the CD signal for $(dA)_{18}$ in glyceline suggests that there is less base stacking than in water, inefficient excimer formation even when bases are stacked could additionally contribute to the very low amplitudes of the long-lived signal components (τ_1 and τ_2 in Table 2).

It must also be kept in mind that the decay of $(dA)_{18}$ in glyceline is complex and that the three-exponential fit (see Table 2) is not unique. It would thus be unwise to over-interpret the individual lifetimes from these initial observations. Nonetheless, Fig. 3d shows that there is a very small population of excited states of $(dA)_{18}$ in glyceline that relax over several nanoseconds, or much slower than the slowest detectable component for this substrate in water. The precise reasons for this behavior are unclear, but could be due to a small population of excimers with strong CT character that decay more slowly in glyceline than in water due to the lower solvent reorganization energy of the former solvent.

The very similar CD spectra seen for the $d(AT)_9 \cdot d(AT)_9$ duplex in glyceline and water at room temperature suggest that this double strand adopts the same structure in both solvents. Earlier, Hud and co-workers[22] reported that the $d(AT)_{16} \cdot d(AT)_{16}$ duplex exhibits similar CD signals in water and in a DES made from a 1 : 2 mole ratio of choline chloride and urea. The $d(AT)_9 \cdot d(AT)_9$ duplex is somewhat less stable in glyceline than in water based on its slightly lower melting temperature in the former solvent. However, the melting temperature in glyceline is still significantly higher (45 °C) compared to that of $d(AT)_{16} \cdot d(AT)_{16}$ in the urea-based DES (29 °C).[22] Despite the difference in melting points, the CD signals demonstrate similar strengths at 20 °C, thus suggesting that bases are stacked (and paired) to approximately the same extent in both solvents.

The long-lived components are easier to see in the GSB signal from the duplex than in the signal from the single strand, which is possibly because the latter system is more disordered. Stabilization of AT-rich DNA is promoted by binding of the choline ion in the DNA minor groove,[49] although this favorable interaction may be partially counteracted by the denaturing activity of glycerol.[50,51] We speculate that without the possibility of stabilization through minor groove binding, single-stranded DNA sequences in glyceline may exhibit reduced base stacking than when present in a duplex structure in the same solvent.

While the steady-state spectra for the duplex in both solvents are very similar, the excited-state dynamics are remarkably different. The long-lived decay component (τ_1 in Table 3) for the $d(AT)_9 \cdot d(AT)_9$ duplex in glyceline is much less prominent compared to the long-time component seen in water, as can be seen by comparing the amplitudes (A_1 in Table 3). The 84 ± 14 ps decay component in aqueous solution was assigned to an $A^{\bullet +}T^{\bullet -}$ exciplex formed by photoinduced ET from A to T on an ultrafast time scale.[33] The contact radical ion pair-character of this state was established by time-resolved IR (TRIR) experiments on the same duplex in a D_2O solution.[32] The exciplex decays *via* back ET, which takes place in the Marcus inverted region, resulting in the nearly 90 ps lifetime. We propose that the weak component with a lifetime of 170 ± 50 ps observed for UV-excited $d(AT)_9 \cdot d(AT)_9$ in glyceline is also due to the decay of an exciplex state. The somewhat slower rate of back ET may be due in this case to a lower solvent reorganization energy of glyceline, thus leading to a decrease in the rate of back ET.

In glyceline, the long-lived decay component (τ_1) for the $d(AT)_9 \cdot d(AT)_9$ duplex is weak and contributes only a small fraction (10%) of the total decay of the excited-state population as judged from the GSB signals (Table 3). In contrast, the τ_0 component (Table 3) with a lifetime of less than 10 ps accounts for nearly 90% of the GSB signal decay (Fig. 4e and Table 3). On the other hand, in water the τ_0 component accounts for only about 50% of the total GSB decay. At first glance, the larger fast-decaying population seems to contradict the conclusion derived from the CD spectrum that $d(AT)_9 \cdot d(AT)_9$ exhibits a similar degree of stacking in both solvents. A possible explanation is that forward ET proceeds more slowly in glyceline than in water, thus making exciplex formation less competitive with internal conversion. However, we did not observe a rise in time in the 250 nm or 570 nm signals, thus suggesting that any ET must take place within the experimental time resolution ($\sim$200 fs). Furthermore, the lower reorganization energy provided by glyceline is expected to place the forward ET closer to the apex of the Marcus parabola, making its rate faster.

The possibility that rapid, monomer-like internal conversion could take place even in base-stacked sequences is supported by experiments on DNA duplexes with reduced structural disorder.[52,53] Experiments on DNA hairpins and cyclized oligonucleotides reveal that a fast component of several ps is observed, even when the fraction of unstacked nucleobases is believed to be very small. That ultrafast internal conversion can take place even in sequences of stacked bases could be due to the inability of the DES to stabilize an intrastrand excimer sufficiently rapidly as discussed above.

In the stacked base pairs present in double-stranded DNA, ultrafast internal conversion could be a consequence of the electron-driven proton transfer channel proposed by Domcke and Sobolewski,[54] in which a hydrogen atom is transferred between bases in a base pair. Evidence for this interstrand decay pathway was recently obtained in a study of single GC base pairs in a chloroform solution.[17] The authors of ref. 17 found that the biradical formed by hydrogen atom transfer decays with a lifetime of 3 ps. It is possible that the short decay component ($\sim$4 ps) we have observed for the $d(AT)_9 \cdot d(AT)_9$ duplex in glyceline arises from a similar process. If this is correct, then the weaker contribution that longer-lived excited states make to the duplex GSB signal in glyceline would indicate that interstrand decay is favored over an intrastrand mechanism. In this case, the DES would clearly have a strong effect on the electronic branching to these two channels. It is possible that the DES environment favors the transfer of a neutral hydrogen atom, creating two neutral radicals, over intrastrand ET. The latter process gives rise to oppositely charged radical ions that might be better solvated in water. Further experimental and theoretical studies are needed to investigate this possibility.

Conclusions

As this study of DNA photophysics in glyceline shows, DESs hold great promise for investigating the effects of solvent on the excited states of DNA strands. Such studies will be crucial for developing comprehensive models of photoinduced charge transfer in DNA. This study has also shown that DESs are well suited for femtosecond pump-probe measurements with UV excitation pulses. Other types of time-resolved experiments should also be possible, and a recent study[55] using two-dimensional IR spectroscopy to investigate the solvation dynamics of an ion in a DES demonstrates the feasibility of performing experiments on solutes in DESs with mid-infrared laser pulses. Time-resolved emission studies should also be possible.

Both time-resolved emission and TRIR experiments can address the questions raised by this study. Emission experiments can confirm whether excimers/

exciplexes are formed in lower quantum yields in glyceline than in water. They can also reveal how their emission wavelengths change compared to in water. Meanwhile, TRIR experiments will be extremely useful for identifying inter- *vs.* intrastrand decay pathways through the vibrational signatures of nucleobase radicals and radical ions. These experiments can also establish whether the PCET channel identified in double-stranded DNA in water[32] also operates in a DES.

The solvation response of a viscous DES like glyceline is certain to extend over many time scales from ultrafast to many nanoseconds, or even microseconds. The possibility that solvent relaxation occurs more slowly than excited-state depopulation should give rise to effects such as excitation wavelength-dependent emission (the so-called red-edge effect),[56] as has been seen for some fluorophores in ionic liquids.[57]

As discussed above, the solvent response could determine not only the decay rates, but also the formation times of excimers and other long-lived excited states in DNA. The efficiency of intramolecular excimer formation in a cyclophane was shown to change with solvent polarity,[58] so it is reasonable that solvent effects can influence excited-state decay in ways that go beyond changes to the supramolecular organization of the nucleobases. Indeed, this preliminary study provides evidence that this is so. Further experiments on DNA strands in a non-aqueous solvent such as the DES investigated here will help to clarify the poorly understood competition between the various deactivation pathways available to excited states in DNA strands. Such studies can establish that non-radiative relaxation depends not just on structure, but also on dynamical solvent properties. The growing understanding of ionic liquid properties and DNA photophysics makes this a feasible and attractive quest.

Acknowledgements

Work at The Ohio State University was supported by the U.S. National Science Foundation (CHE-1465277) and by startup funding. M. H. acknowledges Department of Science and Technology (DST), Govt. of India for the support of this work, SR/NM/NS-23/2016(C). K. d. L. H. thanks USAFA for support during a sabbatical leave.

References

1 C. E. Crespo-Hernández, B. Cohen, P. M. Hare and B. Kohler, *Chem. Rev.*, 2004, **104**, 1977–2019.
2 C. T. Middleton, K. de La Harpe, C. Su, Y. K. Law, C. E. Crespo-Hernández and B. Kohler, *Annu. Rev. Phys. Chem.*, 2009, **60**, 217–239.
3 B. Kohler, *J. Phys. Chem. Lett.*, 2010, **1**, 2047–2053.
4 D. Markovitsi, T. Gustavsson and I. Vayá, *J. Phys. Chem. Lett.*, 2010, **1**, 3271-3276.
5 D. Markovitsi, T. Gustavsson and A. Banyasz, *Mutat. Res.*, 2010, **704**, 21–28.
6 K. Kleinermanns, D. Nachtigallova and M. S. de Vries, *Int. Rev. Phys. Chem.*, 2013, **32**, 308–342.
7 R. Improta, F. Santoro and L. Blancafort, *Chem. Rev.*, 2016, **116**, 3540–3593.
8 P. M. Hare, C. E. Crespo-Hernández and B. Kohler, *J. Phys. Chem. B*, 2006, **110**, 18641–18650.

 This journal is © The Royal Society of Chemistry 2018

9 T. Gustavsson, N. Sarkar, E. Lazzarotto, D. Markovitsi, V. Barone and R. Improta, *J. Phys. Chem. B*, 2006, **110**, 12843–12847.

10 T. Gustavsson, N. Sarkar, E. Lazzarotto, D. Markovitsi and R. Improta, *Chem. Phys. Lett.*, 2006, **429**, 551–557.

11 C. T. Middleton, B. Cohen and B. Kohler, *J. Phys. Chem. A*, 2007, **111**, 10460–10467.

12 N. K. Schwalb and F. Temps, *J. Am. Chem. Soc.*, 2007, **129**, 9272–9273.

13 T. Gustavsson, A. Banyasz, N. Sarkar, D. Markovitsi and R. Improta, *Chem. Phys.*, 2008, **350**, 186–192.

14 T. Gustavsson, N. Sarkar, I. Vayá, M. C. Jiménez, D. Markovitsi and R. Improta, *Photochem. Photobiol. Sci.*, 2013, **12**, 1375–1386.

15 K. Röttger, N. K. Schwalb and F. Temps, *J. Phys. Chem. A*, 2013, **117**, 2469–2478.

16 K. Hunger, L. Buschhaus, L. Biemann, M. Braun, S. Kovalenko, R. Improta and K. Kleinermanns, *Chem.–Eur. J.*, 2013, **19**, 5425–5431.

17 K. Röttger, H. J. B. Marroux, M. P. Grubb, P. M. Coulter, H. Böhnke, A. S. Henderson, M. C. Galan, F. Temps, A. J. Orr-Ewing and G. M. Roberts, *Angew. Chem., Int. Ed.*, 2015, **54**, 14719–14722.

18 G. W. Doorley, M. Wojdyla, G. W. Watson, M. Towrie, A. W. Parker, J. M. Kelly and S. J. Quinn, *J. Phys. Chem. Lett.*, 2013, **4**, 2739–2744.

19 D. B. Bucher, B. M. Pilles, T. Carell and W. Zinth, *Proc. Natl. Acad. Sci. U. S. A.*, 2014, **111**, 4369–4374.

20 Y. Zhang, J. Dood, A. A. Beckstead, X.-B. Li, K. V. Nguyen, C. J. Burrows, R. Improta and B. Kohler, *Proc. Natl. Acad. Sci. U. S. A.*, 2014, **111**, 11612–11617.

21 D. Pörschke and F. Eggers, *Eur. J. Biochem.*, 1972, **26**, 490–498.

22 I. Mamajanov, A. E. Engelhart, H. D. Bean and N. V. Hud, *Angew. Chem., Int. Ed.*, 2010, **49**, 6310–6314.

23 A. P. Abbott, R. C. Harris, K. S. Ryder, C. D'Agostino, L. F. Gladden and M. D. Mantle, *Green Chem.*, 2011, **13**, 82–90.

24 H. Tateishi-Karimata and N. Sugimoto, *Nucleic Acids Res.*, 2014, **42**, 8831–8844.

25 H. Zhao, *J. Chem. Technol. Biotechnol.*, 2015, **90**, 19–25.

26 I. Gállego, M. A. Grover and N. V. Hud, *Angew. Chem., Int. Ed.*, 2015, **54**, 6765–6769.

27 A. P. Abbott, R. C. Harris and K. S. Ryder, *J. Phys. Chem. B*, 2007, **111**, 4910–4913.

28 J. Chen and B. Kohler, *Phys. Chem. Chem. Phys.*, 2012, **14**, 10677–10682.

29 Y. Zhang, A. A. Beckstead, Y. Hu, X. Piao, D. Bong and B. Kohler, *Molecules*, 2016, **21**, 1645.

30 J. Chen and B. Kohler, *J. Am. Chem. Soc.*, 2014, **136**, 6362–6372.

31 C. Crespo-Hernandez and B. Kohler, *J. Phys. Chem. B*, 2004, **108**, 11182–11188.

32 Y. Zhang, K. de La Harpe, A. A. Beckstead, R. Improta and B. Kohler, *J. Am. Chem. Soc.*, 2015, **137**, 7059–7062.

33 D. J. Skowron, Y. Zhang, A. A. Beckstead, J. M. Remington, M. Strawn and B. Kohler, *ChemPhysChem*, 2016, **17**, 3558–3569.

34 T. Gustavsson, N. Sarkar, A. Banyasz, D. Markovitsi and R. Improta, *Photochem. Photobiol.*, 2007, **83**, 595–599.

35 A. Pandey, R. Rai, M. Pal and S. Pandey, *Phys. Chem. Chem. Phys.*, 2014, **16**, 1559–1568.

36 J. Pecourt, J. Peon and B. Kohler, *J. Am. Chem. Soc.*, 2001, **123**, 10370–10378.

37 F. Y. Jou and G. R. Freeman, *Can. J. Chem.*, 1979, **57**, 591–597.

38 J. Bonin, I. Lampre, P. Pernot and M. Mostafavi, *J. Phys. Chem. A*, 2008, **112**, 1880–1886.
39 J.-M. L. Pecourt, J. Peon and B. Kohler, *J. Am. Chem. Soc.*, 2001, **123**, 10370–10378.
40 B. Cohen, P. M. Hare and B. Kohler, *J. Am. Chem. Soc.*, 2003, **125**, 13594–13601.
41 M. Terazima, *Chem. Phys. Lett.*, 1999, **305**, 189–196.
42 Y. Zhang, J. Chen and B. Kohler, *J. Phys. Chem. A*, 2013, **117**, 6771–6780.
43 J. Chen, Y. Zhang and B. Kohler, *Top. Curr. Chem.*, 2015, **356**, 39–88.
44 A. Banyasz, T. Gustavsson, D. Onidas, P. Changenet-Barret, D. Markovitsi and R. Improta, *Chem.–Eur. J.*, 2013, **19**, 3762–3774.
45 Y. Zhang, K. de La Harpe, A. A. Beckstead, L. Martínez-Fernández, R. Improta and B. Kohler, *J. Phys. Chem. Lett.*, 2016, **7**, 950–954.
46 D. Mondal, M. Sharma, C. Mukesh, V. Gupta and K. Prasad, *Chem. Commun.*, 2013, **49**, 9606–9608.
47 C. Q. Zhao, J. S. Ren and X. G. Qu, *Langmuir*, 2013, **29**, 1183–1191.
48 D. J. Skowron, Ph.D. dissertation, Montana State University, 2017.
49 M. Nakano, H. Tateishi-Karimata, S. Tanaka and N. Sugimoto, *J. Phys. Chem. B*, 2014, **118**, 379–389.
50 L. Levine, W. P. Jencks and J. A. Gordon, *Biochemistry*, 1963, **2**, 168–175.
51 G. Kodali, K. A. Kistler, S. Matsika and R. J. Stanley, *J. Phys. Chem. B*, 2008, **112**, 1789–1795.
52 J. Chen, A. K. Thazhathveetil, F. D. Lewis and B. Kohler, *J. Am. Chem. Soc.*, 2013, **135**, 10290–10293.
53 Y. Zhang, X.-B. Li, A. M. Fleming, J. Dood, A. A. Beckstead, A. M. Orendt, C. J. Burrows and B. Kohler, *J. Am. Chem. Soc.*, 2016, **138**, 7395–7401.
54 A. L. Sobolewski and W. Domcke, *Eur. Phys. J. D*, 2002, **20**, 369–374.
55 Y. Cui, K. D. Fulfer, J. Ma, T. K. Weldeghiorghis and D. G. Kuroda, *Phys. Chem. Chem. Phys.*, 2016, **18**, 31471–31479.
56 A. P. Demchenko, *Luminescence*, 2002, **17**, 19–42.
57 A. Samanta, *J. Phys. Chem. B*, 2006, **110**, 13704–13716.
58 Y. Tamai, H. Ohkita, J. Shimada, H. Benten, S. Ito, S. Yamanaka, K. Hisada, K. Tani, K. Kubono and T. Shinmyozu, *J. Phys. Chem. A*, 2013, **117**, 7776–7785.

 This journal is © The Royal Society of Chemistry 2018

Faraday Discussions

DISCUSSIONS

Photocrosslinking between nucleic acids and proteins: general discussion

Vinayak Bhat, Richard Cogdell, Carlos E. Crespo-Hernández, Ankona Datta, Arijit De, Stefan Haacke, John Helliwell, Roberto Improta, Joshy Joseph, Tolga Karsili, Bern Kohler, Retheesh Krishnan, Mahil L, Frederick Lewis, Imon Mandal, Dimitra Markovitsi, Padmaja P. Mishra, Sneha Paul, Ganga Periyasamy, P. I. Pradeepkumar, Priyadarshi Roy Chowdhury, Manas Sarangi, Devika Sasikumar, Igor Schapiro, Gebhard F. X. Schertler, Ilme Schlichting, Javier Segarra-Martí, Rajaram Swaminathan, Vishnu V, Rienk van Grondelle, Reji Varghese and Ravindra Venkatramani

DOI: 10.1039/C8FD90005A

Tolga Karsili opened the discussion of the paper by Dimitra Markovitsi: It is well known that acidic substituents in aromatic systems contain $\pi\sigma^*$ states that are dissociative with respect to X–H bond fission (where X = N, O or S). It is also well known that the photochemistry of simple aromatic chromophores such as phenol and pyrrole are dominated by such states. In 9*H*-adenine, $\pi\sigma^*$ states have been predicted to give rise to N–H bond fission, localized on the amine moiety. In the AT duplex, the expected stabilization of the radical formed by N–H bond fission may lead to the stabilization of $\pi\sigma^*$ states. Such states have also been predicted to contain the initial signature of the solvated electron, formed by solute-to-solvent charge-transfer. Therefore, do you expect such states to dominate the observed initial N–H photodissociation in AT duplexes?

Dimitra Markovitsi answered: The role of the $\pi\sigma^*$ states in the photochemistry of nucleobases has been a subject of debate in the literature over the last few years. We stress that the N9–H bond of adenine involved in this state is accessible to water only for the bare nucleobase, and not for nucleosides or DNA multimers. The involvement of N6–H bonds can also be ruled out because we showed that base-pairing does not have an important effect on one-photon ionization.

Bern Kohler remarked: You observe adenine radicals that persist for milliseconds. What makes them relatively unreactive, and did you check to see whether the lifetimes depend on the nature or concentrations of the pH buffer components?

Dimitra Markovitsi responded: In the case of adenine single strands, we observed that replacing the phosphate buffer with a NaCl solution of the same ionic strength did not modify the lifetime of the deprotonated radicals.[1] It is worth noting that deprotonated guanine radicals produced by photosensitizers survive for hundreds of milliseconds;[2] the authors of that study stressed that hydrogen bonding decreases the radical lifetime, which is the opposite of our observations. A better knowledge of the final reaction products and possible intermediates is necessary in order to explain the observed kinetics.

1 A. Banyasz, T.-M. Ketola, A. Muñoz-Losa, S. Rishi, A. Adhikary, M. D. Sevilla, L. Martinez-Fernandez, R. Improta and D. Markovitsi, *J. Phys. Chem. Lett.*, 2016, 7, 3949–3953.
2 Y. Rokhlenko, J. Cadet, N. E. Geacintov and V. Shafirovich, *J. Am. Chem. Soc.*, 2014, **136**, 5956–5962.

Carlos E. Crespo-Hernández asked: As you are aware, we have previously performed similar photoionization experiments for a wide range of monomers[1] and dinucleotides.[2] In those works, evidence was presented for the formation of hydrated electrons in monomers and dinucleotides through a combination of one- and two-photon absorption processes.

In the first study,[1] the observation of a one-photon ionization mechanism in the monomers at a photon energy of 4.66 eV (266 nm), which is much lower than their gas phase ionization energies (9 to 7.77 eV), was interpreted in terms of the effect that solvation has on the stabilization of the excited states of the monomers and the putative formation of hydrated electrons from either a charge-transfer-to-solvent state or *via* the photo-transfer of an electron in the excited molecule into a trap formed by favorably oriented molecular dipoles of water.[3,4] Do you think that a similar mechanism may be operative in the single- and double-stranded oligonucleotides that you have investigated in this and other recent works?

In our second study,[2] we observed that an increase in the concentration of the GpC, CpG, ApG and GpA dinucleotides in the range of 4 to 80 μM resulted in a systematic increase in the net photoionization yields (*i.e.* the combined one- and two-photon ionization yields), whereas no change in the photoionization yield was observed for the monomers within experimental uncertainty. Have you studied the dependence of the one- and two-photon ionization yields on the initial concentrations of the oligonucleotides that you have investigated? Such systematic investigations could potentially shed some light on the photoionization mechanism and the precursor state(s) leading to the one- and two-photon ionization of these systems.

1 C. E. Crespo-Hernández and R. Arce, *Photochem. Photobiol.*, 2002, **76**, 259–267.
2 C. E. Crespo-Hernández and R. Arce, *J. Phys. Chem. B*, 2003, **107**, 1062–1070.
3 A. L. Sobolewski and W. Domcke, *Chem. Phys. Lett.*, 2000, **329**, 130–137.
4 D. N. Nikogosyan, *Int. J. Radiat. Biol.*, 1990, **57**, 233–299.

Dimitra Markovitsi replied: Your papers have indeed been very stimulating for our studies. However, in our experiments, using much lower excitation intensities (<4 mW cm^{-2}) than yours (>4 mW cm^{-2}), we found that the one-photon ionization of mono-nucleosides and mono-nucleotides is lower than 3×10^{-4}, which corresponds to our detection limit. In contrast, when using similar excitation intensities to those reported in your 2002 paper we do observe ionization from the buffer. The fact that we observe one-photon ionization only for oligomers

 This journal is © The Royal Society of Chemistry 2018

exhibiting a pronounced secondary structure, in which nucleobases are, in principle, less exposed to the solvent compared to monomeric chromophores, does not corroborate the hypothesis that the interaction with water molecules is the dominant factor responsible for this phenomenon.

We did not vary the oligomer concentration because we optimized the experimental conditions in order to keep the excitation intensity as low as possible (<4 MW cm^{-2}). The best signal-to-noise ratio is obtained when the absorbance on the excitation side is ~0.24 over 0.1 cm, which is used for all of our DNA systems.

Rajaram Swaminathan queried: What are the consequences of the ionisation of DNA due to low energy UV radiation inside the living cell?

Dimitra Markovitsi answered: One-photon ionization of DNA may be caused by solar light reaching the surface of the Earth (UVB). The resulting base radicals give rise to oxidative damage which is known to induce carcinogenic mutations.

Manas Sarangi asked: Why is the radical lifetime observed for double stranded AT longer than that found for polyadenine single strands?

Dimitra Markovitsi responded: Base pairing decreases the amplitude of the conformational motions of the system and, hence, the probability that the system adopts the configurations that are required for the radical reaction. This explanation is supported by two observations regarding the radical lifetime: (i) it is longer at lower temperatures (see Fig. 7b in our Faraday Discussions paper), and (ii) it is longer in the homopolymeric AT duplexes,[2] whose structures fluctuate less than the alternating AT duplexes.[1]

1 B. Bouvier, J.-P. Dognon, R. Lavery, D. Markovitsi, P. Millié, D. Onidas and K. Zakrzewska, *J. Phys. Chem. B*, 2003, **107**, 13512–13522.
2 A. Banyasz, T. Ketola, A. Muñoz-Losa, S. Rishi, A. Adhikary, M. D. Sevilla, L. Martinez-Fernandez, R. Improta and D. Markovitsi, *J. Phys. Chem. Lett.*, 2016, 7, 3949–3953.

P. I. Pradeepkumar questioned: Since G-quadruplex (G4) can form multiple topologies depending upon the sequence and salt conditions, what effect would these topologies have on the formation of radical cations? Furthermore, do you know which guanine base in G-quadruplex preferentially undergoes radical cation formation?

Dimitra Markovitsi replied: We are currently studying how the topology of the G-quadruplex and the type of cation that is present in the solution affect the formation and dynamics of radical cations. With regards to your second question, our experimental results do not provide information about the base that is responsible for electron ejection.

Ankona Datta asked: How do you characterize base pair dimerization in DNA? Does it make any difference to the structure of the DNA when the base pairs dimerize?

Dimitra Markovitsi answered: To the best of our knowledge, covalently linked dimers resulting from a photoreaction between Watson–Crick pairs have not been characterized thus far. If such dimers are indeed formed, their effect on the conformation of the duplex should depend on their precise chemical structures.

John Helliwell communicated: In your paper, in the subsection "Reaction paths in $(AT)_1$ duplexes", distances of 2.6 Å and 2.0 Å are quoted. What are the standard uncertainties for these distances? This leads on to another question; what is the sensitivity of the authors' calculations to these distance values, *i.e.* what range of distance values must be allowed for due to error?

Dimitra Markovitsi communicated in reply: The accuracy of TD-DFT optimized geometries has been extensively benchmarked.[1] The errors in the distances compared to those obtained from wavefunction based methods depend on the type of bond and the nature of the optimized excited state, although they are typically $\sim 10^{-2}$ Å. Furthermore, we paid attention to select an appropriate functional (M052X) for the description of the Potential Energy Surface based on the character of our excited states. Thus, we are confident of the reported distances.

1 A. D. Laurent and D. Jacquemin, *Int. J. Quantum. Chem.*, 2013, **113**, 2019–2039.

Priyadarshi Roy Chowdhury communicated: Please could you discuss the conformational stability that is associated with the quadruplex formation?

Dimitra Markovitsi communicated in reply: G-quadruplexes were prepared by folding the corresponding single strand; the detailed protocol is described in our previous study.[1] Once formed, G-quadruplexes are quite stable at room temperature, as shown by their thermodynamic properties.[2] We also tested the stability of G-quadruplexes *versus* time-resolved experiments by checking their steady-state absorption spectra before and after laser irradiation.

1 A. Banyasz, L. Martínez-Fernández, C. Balty, M. Perron, T. Douki, R. Improta and D. Markovitsi, *J. Am. Chem. Soc.*, 2017, **139**, 10561–10568.
2 J.-L. Mergny, A.-T. Phan and L. Lacroix, *FEBS Lett.*, 1998, **435**, 74–78.

Tolga Karsili opened the discussion of the paper by Roberto Improta: Did you benchmark the energetics returned from the M052X functional against those returned from other Minnesota and long-range corrected functionals? Do you expect such benchmarking to reveal any significant changes in the reaction barriers for PCET?

Roberto Improta responded: We carried out several test calculations by employing both CAM-B3LYP and M06-2X functionals, for both the duplex and for an isolated Watson–Crick pair, and our conclusions are solid with respect to the choice of the functional.

Ravindra Venkatramani remarked: Your study addresses interstrand and intrastrand photoinduced CT processes in nucleobase stacks. However, the electronic states in such pi-stacks are expected to be delocalized over several nucleobases. What are your thoughts on photoinduced charge transfer and its mechanism in this context of strong electronic coupling between nucleobases?

Roberto Improta replied: I agree that the strong electronic coupling between the bases is an important factor for consideration. Actually, the decay paths involving charge transfer states, either intra- or inter-strand, are competitive with many other deactivation paths, both photophysical and photochemical. We explicitly tackled this issue in our previous work.[1] Since then, other possible decay paths have been recognized, involving both single (the excitation localizes on a single base and 'monomer like' decay is observed) and multiple bases (*e.g.* excimer minima or photodimerization paths).

1 R. Improta and V. Barone, *Angew. Chem., Int. Ed.*, 2011, **50**, 12016–12019.

Carlos E. Crespo-Hernández asked: In the absence of dynamics simulations, have you investigated the role of different base stacking conformations (such as base-stack roll, tilt, twist, rise, shift and slide) on the energy barriers that you calculated for the intra- and inter-strand proton coupled electron transfer (PCET) in the AT and GC systems that you have investigated? Do you think that different base stacking conformations can play a major role in the energetics and probabilities of participation of these relaxation pathways? Can you comment on the potential role of water molecules and counter ion interactions in modulating the rates and energetics (participation) of these different reaction pathways?

Previous experiments with AT and GC oligonucleotides and related systems (cited in your manuscript) have shown that the excited stacked complexes (often called excimers and exciplexes) in these systems overwhelmingly decay back to the ground state by charge recombination in tens of picoseconds. Have you investigated the mechanism of repopulation of the ground state in these systems after the PCET events?

Roberto Improta answered: The Charge Transfer (CT) processes, especially intrastrand CT between stacked bases, are surely affected by the stacking geometry of the bases and, therefore, by the conformational equilibria in the oligonucleotide. Such effects, which were investigated in a previous preliminary study on a GC model system,[1] are very important in a single strand. They also play a role in the duplex, although in this case our conclusion should be valid, at least for the structural family that we have investigated, which is representative of a GC repeat crystal structure. Interactions with water molecules are also likely to be less important for a duplex, whereas they should significantly modulate CT in a single strand, as discussed in a previous study.[2] In the two studies cited above we made some attempts to estimate the charge recombination rate, also including the vibrational effect,[1] based on Marcus theory, which gave promising results.

1 J. Cerezo, L. Martínez-Fernández, R. Improta and F. Santoro, *Theor. Chem. Acc.*, 2016, **135**, 221.
2 L. Martínez-Fernández, Y. Zhang, K. de La Harpe, A. A. Beckstead, B. Kohler and R. Improta, *Phys. Chem. Chem. Phys.*, 2016, **18**, 21241–21245.

Bern Kohler remarked: My question concerns the excited state pathway involving intrastrand electron transfer in the GC tetramer. Does the weak barrier that you observe between the two minima, corresponding to electron transfer with and without proton transfer, have implications for whether the PCET state is reached *via* concerted or sequential transfers? As you know, it has not yet been possible to observe the formation kinetics of this state as it is formed faster than the experimental time resolution.

Roberto Improta responded: It is clear that only a purposefully tailored quantum dynamical study could unambiguously assess this issue. However, by looking only at the energetics, it seems likely that in the case of GC DNA the PCET process is sequential, whereas in other systems, such as AT DNA homo-polymer, it is concerted.

Vinayak Bhat queried: In your paper, you consider GC and AT DNA. I wonder whether proton-coupled electron transfer would be possible in DNA containing a mixture of GC and AT. For instance, if C1 and G2 in Scheme 1 were to be replaced with T1 and A1 respectively, would the PCET[1] pathway be favorable?

Roberto Improta replied: The intrastrand charge transfer processes would surely be affected by the nature of the adjacent bases, and a comparison between the ionization potentials and the electron affinities of the different bases, as well as their relative pK_a values, could provide a useful guide to help understand the possible reactions. With regards to PCET,[1] its energetics depend mainly on the Watson–Crick pair involved. On the other hand, the possibility of populating the inter-strand CT state will depend on the relative energy of this excited state with respect to the spectroscopic states, and on vibronic coupling with these latter states. As a consequence, the PCET[1] reaction can also be modulated by the nature of the stacked bases.

Rienk van Grondelle asked: I have a question about dynamical environmental effects; could you get a better look at PCET by taking into account some specific vibrations?

Roberto Improta answered: We plan to analyse the effect of vibrations, including the vibrational modes involving solvent molecules, in forthcoming studies. We have actually already gained some insight into the importance of vibronic effects for the charge transfer and charge recombination processes in stacked bases.[1–3] These studies demonstrated the role of vibrational modes in tuning the CT process. On the other hand, it is clear that for a fluctuating system as a single strand, where several energetics and vibrational parameters depend strongly on the conformation, it is not easy to take the interplay between slow and fast degrees of freedom into proper account. This latter point is an important challenge that could be tackled in the future.

1 J. Cerezo, L. Martínez-Fernández, R. Improta and F. Santoro, *Theor. Chem. Acc.*, 2016, **135**, 221.
2 F. Santoro, R. Improta, F. Avila, M. Segado and A. Lami, *Photochem. Photobiol. Sci.*, 2013, **12**, 1527–1543.
3 R. Improta, F. Santoro, V. Barone and A. Lami, *J. Phys. Chem. A*, 2009, **113**, 15346–15354.

Priyadarshi Roy Chowdhury communicated: Please can you discuss the role played by the dynamical environmental effects?

Roberto Improta communicated in reply: As extensively discussed in the paper, dynamical solvation effects strongly modulate the CT processes in DNA. First of all, full equilibration of the solvent degrees of freedom is important in order to stabilize all of the minima of the Charge Transfer (CT) states. Furthermore, the Proton Transfer (PT) reaction that occurs between two bases involved in an intra-strand CT state leads to a strong quenching of the excited state dipole (since it leads to a diradical 'neutral' state). Consequently, in this case, equilibration of the solvent degrees of freedom in the minimum of the CT state leads to an increase in the energy barrier associated with PT, and to a decrease in the driving force for this reaction.

Priyadarshi Roy Chowdhury communicated: Please can you also discuss the boundary conditions associated with the quantum mechanical calculations, which lead to photoactivated proton coupled electron transfer in DNA?

Roberto Improta communicated in reply: There are no particular boundary conditions included in our quantum mechanical calculations.

P. I. Pradeepkumar opened the discussion of the paper by Frederick Lewis:† How do you compare the effect of 7-deazadeoxyguanosine (ZdG) with that of dG in exciplex formation in hairpin DNAs?

Frederick Lewis answered: ZdG is a much better electron donor (lower oxidation potential) than dG and thus forms an exciplex with EG that has much more charge transfer character.

Rienk van Grondelle asked: Do you get real charge separation in the end in these structures?

Frederick Lewis responded: We think that EG-Z and EG2 form an exciplex and excimer, respectively, and exhibit extensive charge transfer character. However, the charges never separate since there is no other hole trap in these hairpins. We have studied hairpins that have EG and the hole trap Sd separated by two AT or GC base pairs. There is no evidence for charge separation in these hairpins.[1] Evidently, the decay of 1*EG is faster than hole transport to Sd.

1 K. E. Brown, A. P. N. Singh, Y.-L. Wu, A. K. Mishra, J. Zhou, F. D. Lewis, R. M. Young and M. R. Wasielewski, *J. Am. Chem. Soc.*, 2017, **139**, 12084–12092.

Stefan Haacke queried: Are the fluorescence spectra you show obtained after excitation at the lower end? How do they change when you excite in the oligomer bands at 250–260 nm? Is there energy transfer going on and the formation of excimer states?

Frederick Lewis replied: The fluorescence spectra shown were obtained with excitation at 350 nm. We did not determine fluorescence quantum yields for

† Frederick Lewis' paper was presented by Mahesh Hariharan, Indian Institute of Science Education and Research (IISER), Thiruvananthapuram, India.

250–260 nm excitation. I would expect them to be lower as a consequence of the competition between the energy transfer and nonradiative decay of the natural bases which would absorb most of the incident light, but the band shapes should be similar.

Retheesh Krishnan communicated: In your paper you mentioned that the excitonic coupling for EG-G is strong, whereas that for EG-C, -T and -A is weak. However the CD spectra for all of these look very similar. Can you please comment on this?

Frederick Lewis communicated in reply: You are correct that the CD excitonic coupling for EG-G is only slightly stronger than for EG-C, -T and -A. This is indicated by the red-shifted positive CD band at 295 nm, which is similar to that observed for EG-Z which demonstrates stronger exciton coupling than for EG-G.

Vishnu V communicated: This study essentially focuses on guanosine derivatives. Could the same study be carried out using derivatives of other nucleobases and, if so, do you think that the same results would be observed?

Frederick Lewis communicated in response: This is an interesting question. We looked for similar behavior in the cytosine derivative, pyrrolocytosine, but did not observe it.[1] The observation of strong exciton coupling requires the base derivatives to have strong electronic transition dipoles. This is not the case for most purines and pyrimidines.

1 P. P. Neelakandan, M. McCullagh, G. C. Schatz and F. D. Lewis, *J. Phys. Chem. B*, 2012, **116**, 5199–5204.

Priyadarshi Roy Chowdhury communicated: In the DNA structure, is the loop stability dependent on the unpaired bases that are present within the loop?

Frederick Lewis communicated in reply: Yes, the loop stability will depend on the number and identity of the unpaired bases. The TTT loop was selected because it does form stable hairpins with the base-paired stems employed in this study and does not serve as a potential electron donor. Other hairpins could have been used for this study.

Priyadarshi Roy Chowdhury communicated: How does timescale variation affect the studies associated with the structural aspects of DNA?

Frederick Lewis communicated in response: This is an excellent question. Geometric relaxation of the vertical excited state of EG, relaxation of the local duplex geometry with an increase in EG-purine stacking, and solvent relaxation could all be occurring on the timescales of the ps fluorescence and transient absorption decays. Unfortunately, time-resolved Raman is not sensitive to these changes.

Dimitra Markovitsi communicated: Can you compare the results obtained by fluorescence upconversion for your monomeric guanine analog EG-H_2 with those obtained for the natural guanine nucleotide using the same experimental technique?[1]

1 F.-A. Miannay, T. Gustavsson, A. Banyasz and D. Markovitsi, *J. Phys. Chem. A*, 2010, **114**, 3256–3263.

Frederick Lewis communicated in reply: The excited state behavior of G has been studied in much greater detail than that of EG-H_2. Both additional experimental and computational studies similar to those conducted for G and vinyl G would have been desirable, but were not carried out prior to the end of our collaborative project. What we do know is that the lowest singlet of EG-H_2 is most likely of π,π^* character, is longer-lived, and has a higher fluorescence quantum yield than that of the lowest singlet(s) of G.

Rienk van Grondelle opened the general discussion of the papers by Dimitra Markovitsi, Roberto Improta and Frederick Lewis: What do you think is the overwhelming theme of this session?

Frederick Lewis replied: All three papers describe charge-transfer interactions between purine bases and adjacent bases. The first two papers deal with the duplexes of natural base pairs. The third paper deals with a base derivative (phenylethynylguanine) embedded in a hairpin. The unifying theme of the session is the role of these charge-transfer interactions in both the decay pathways and chemical reactions of duplex DNA.

Roberto Improta responded: I think that the importance of charge transfer processes in DNA is one of the main unifying themes of this session.

Dimitra Markovitsi answered: Although the three talks of this session reported different approaches to describe photo-induced processes in various DNA systems, the emerging common feature is the importance of charge transfer states. Such excited states could be responsible for electron ejection giving rise to base radicals that are observed on the millisecond timescale (see also my answer to Bern Kohler's question below). Quantum chemistry calculations showed how states with partial charge transfer character may decay *via* a mechanism involving proton coupled electron transfer. Finally, the charge transfer character of exciplexes in DNA hairpins containing non-natural bases was explored using ultrafast optical spectroscopy.

Rienk van Grondelle asked a general question: Why is nobody using EPR to detect radicals?

Bern Kohler replied: This is an interesting suggestion. EPR has of course been used to study radicals formed in DNA by ionizing radiation for many years, especially in low-temperature glasses where these radicals can be easily trapped. Charge recombination following UV excitation occurs efficiently on the picosecond timescale in DNA strands and is too fast to be observed by time-resolved EPR. However, it could be worth looking for any long-lived radicals that may have escaped geminate recombination.

Bern Kohler returned to the discussion of the paper by Dimitra Markovitsi: It is very interesting that you observe little to no photoionization of the monomeric bases under conditions that result in electron ejection from oligonucleotides. This would seem to point to the involvement of an excitation such as an excimer or CT excited state that is only formed in a DNA strand containing stacked bases. Your observation that the ionization yield is higher for your duplex at a lower temperature supports this notion, as a decreased temperature enhances the base stacking that is required for the formation of CT excited states. However, it is very difficult to understand how one-photon excitation could produce a CT state that then goes on to spontaneously lose an electron to the solvent, and so I am curious as to how you envision this possibility from an electronic structure standpoint? Indeed, it is difficult to imagine how an electron could be ejected from a nucleobase by photons with energies that are three eV below the gas-phase ionization potential of adenine. Furthermore, it would seem that this difficult task would be somewhat easier when adenine is fully solvated by water than when it is part of a base stack and exposed to a less polar environment, but you observe the opposite. It is much easier to imagine two-photon ionization in which a longer-lived excited state absorbs a second photon because the final energy is now greater than the gas-phase ionization potential.

Another comment concerns geminate recombination following one- *vs.* two-photon ionization. The greater final energy of the electron ejected by two-photon excitation should promote its escape from the field of its parent ion, whereas an electron released by one-photon ionization, which has little to no excess energy, should recombine efficiently. This suggests that even if one-photon ionization could take place with a high primary quantum yield, most electrons would decay by geminate recombination and would not be detected in nanosecond experiments. Can you comment on how you believe one-photon ionization to take place for bases in strands, but not for monomeric nucleobases?

Dimitra Markovitsi responded: Our working hypothesis is indeed that one-photon ionization is related to charge transfer states, which you have shown are formed in high yields upon low energy UV excitation of DNA multimers.[1] Although the great majority of these states decay rapidly, a small part of their population may undergo charge migration and charge separation.[2] These processes should be favoured in well-stacked base sequences. Conformational motions may prevent ion-pair recombination assisting electron ejection from the negatively charged base. According to this scenario, charge migration and charge separation are expected to have a more important role than the degree of exposure to solvent. It should be noted that the one-photon ionization quantum yields that we observe are extremely low ($\sim 10^{-3}$), thus making it very difficult to study the ionization and recombination of the ejected electron by ultrafast spectroscopy, as you have done for other systems.[3] Only purposefully tailored dynamical computational studies could assess this working hypothesis.

1 J. Chen, Y. Zhang and B. Kohler, *Top. Curr. Chem.*, 2015, **356**, 39–87.
2 D. B. Bucher, B. M. Pilles, T. Carell and W. Zinth, *Proc. Natl. Acad. Sci. U.S.A.*, 2014, **111**, 4369–4374.
3 J. Peon, G. C. Hess, J.-M. L. Pecourt, T. Yuzawa and B. Kohler, *J. Phys. Chem. A*, 1999, **103**, 2460–2466.

 This journal is © The Royal Society of Chemistry 2018

Rienk van Grondelle asked: If you can measure with pump–probe normal DNA, should the product not also be observable?

Dimitra Markovitsi answered: The final AT photodimers formed in genomic DNA have been characterized by analytical methods.[1] They cannot be identified by transient absorption spectroscopy because of the presence of many types of photolesions (the most important being pyrimidine dimers), which all absorb in the UV region.

1 S. Asgatay, A. Martinez, S. Coantic-Castex, D. Harakat, C. Philippe, T. Douki and P. Clivio, *J. Am. Chem. Soc.*, 2010, **132**, 10260–10261.

Carlos E. Crespo-Hernández queried : Can you comment on the potential effect that the temporal pulse width (*i.e.* ns *versus* fs) and repetition rate of the laser system may have on the one- and two-photon ionization yields for the DNA systems that you have investigated?

Dimitra Markovitsi responded: In our early study[1] we used 8 ns pulses with a repetition rate of 2 Hz. In our current experimental setup, which has an improved time-resolution (30 ns instead of 200 ns), the laser pulses are slightly shorter (5 ns) and the repetition rate is fixed to the lowest possible value (0.2 Hz), in order to avoid exciting transient species. However, the one-photon ionization quantum yields reported in our 2006 study and those presented in the current paper do not differ by more than 50%. More important differences are encountered in the two-photon ionization yields which, as highlighted also for other systems,[2] strongly depend on the experimental conditions.

1 S. Marguet, D. Markovitsi and F. Talbot, *J. Phys. Chem. B*, 2006, **110**, 11037–11039.
2 K. L. Stevenson, G. A. Papadantonakis and P. R. LeBreton, *J. Photochem. Photobiol. A*, 2000, **133**, 159–167.

Ravindra Venkatramani remarked: I wonder what the lower bound or threshold of the excitation energy would be to form a nucleobase radical. Would the absorption cross-sections at low energies matter for detecting such radicals?

Dimitra Markovitsi replied: Assuming that the one-photon ionization quantum yield for UVB excitation is the same as that determined at 266 nm, radical detection in transient absorption experiments should be possible if higher DNA oligomer concentrations (around 10^{-4} mol L^{-1}) are used.

Ganga Periyasamy returned to the discussion of the paper by Roberto Improta: The relationship between the number of base pairs and the conductivity has been discussed during the meeting. Similarly, is it possible to correlate the charge transfer/migration timescale with the number of DNA base pairs?

Roberto Improta answered: Although the CT processes that we have discussed mainly involve a pair of bases (either stacked or WC paired), charge transfer and charge migration processes are surely affected by the number of bases present in the polynucleotides, and this is critical for modulating the conformational equilibria and the charge separation processes. The latter can significantly affect the reactivity of CT states.

Dimitra Markovitsi made a general comment: While the ns transient absorption technique has been available for several decades, only recently has it begun to be used to study chemical reactions triggered in DNA by the direct absorption of low energy UV radiation. These investigations have largely been stimulated by ultrafast spectroscopy and quantum chemistry studies which have provided important insights into excited state relaxation in DNA. Surprisingly, many of the UV-triggered reactions are very slow. New technical developments, which allow for the detection of very weak signals at times longer than 50 ms, could provide important information on biologically relevant processes. In this respect, it would be interesting to obtain inspiration from the original flash photolysis apparatus built by George Porter,[1] which I saw as a museum piece when I worked at the Davy-Faraday Research Laboratory, and combine it with modern electronic and optical devices.

1 G. Porter, *Nobel Lecture*, 1967, 1–23.

Rienk van Grondelle returned to the discussion of the paper by Roberto Improta: In my view if you look at, for example, a DNA duplex, charge transfer states are going to be present. Is there anyone who does not believe that the basic mechanism of creating DNA damage/radical states is *via* charge transfer states? The modern consensus on this now is amazing.

Roberto Improta responded: I agree, but it is important to remember that until a few years ago the involvement of CT states in DNA photoactivated dynamics was a hotly debated and challenged topic.

Dimitra Markovitsi opened the discussion of the paper by Javier Segarra-Martí: Can you please comment on the agreement of your calculated transient absorption bands in the UV and visible regions with those that have been experimentally determined for adenosine and $(dA)_{20}$?[1]

1 W.-M. Kwok, C. Ma and D. L. Phillips, *J. Am. Chem. Soc.*, 2006, **128**, 11894–11905.

Javier Segarra-Martí responded: The agreement is quite remarkable with the experimental pump–probe data of Kwok *et al.*,[1] where the main fingerprint is placed in the 300–360 nm probe window, which refers to our estimates in the 26–32k cm^{-1} probing window along Ω_3 and which we assign to the main excited state absorption signal featuring a doubly excited ($HOMO^2 \rightarrow LUMO^2$) character. This comparison validates the results obtained from our simulations and points towards the correct theoretical description of these high-lying and challenging electronic excited states. Moreover, the experimental 1D pump–probe data shows the difficulties faced by these experiments in resolving the differences between monomeric, dimeric and multimeric species when their photo-absorption signals overlap. It is in such cases where we believe multidimensional spectroscopies, such as those simulated in our paper, may provide new information that may be concealed in 1D pump–probe signals.

1 W.-M. Kwok, C. Ma and D. L. Phillips, *J. Am. Chem. Soc.*, 2006, **128**, 11894–11905.

Bern Kohler asked: In your spoken remarks, you referenced the 2DES experiments carried out by Prokhorenko *et al.*[1] Those authors reported both transient absorption and 2DES signals for adenine. Surprisingly, the former signals did not show evidence of the approximately 9 ps decay that is expected for the 7H tautomer of adenine, which accounts for 22% of the population in aqueous solution.[2] Did you simulate the 2DES spectra of the 7H and 9H tautomers of adenine, and are significant differences predicted?

1 V. I. Prokhorenko, A. Picchiotti, M. Pola, A. G. Dijkstra and R. J. Dwayne Miller, *J. Phys. Chem. Lett.*, 2016, **7**, 4445–4450.
2 B. Cohen, P. M. Hare and B. Kohler, *J. Am. Chem. Soc.*, 2003, **125**, 13594–13601.

Javier Segarra-Martí responded: We were aware of the mixture of 7H and 9H tautomers in adenine, which is why we resorted to working with the adenosine nucleoside. By doing this, we only needed to work with the 9H tautomer, which is the most biologically relevant and the tautomer present in DNA/RNA. We considered simulating 7H adenine but we expect some differences given their different lifetimes,[1] which should influence the broadenings observed in the 2D maps and would require us to work with a time dependent framework in order to clearly observe the differences between 7H and 9H adenine. 2DES measurements of adenine were reported by Moran and co-workers,[2] where they used 7-methyl and 9-methyl adenine to observe potential differences between the 7H and 9H tautomers. They observed shifts along the pump (Ω_1) frequency and were able to observe their different lifetimes, and also found that these particular energy shifts could also induce changes to the probe frequency (Ω_3), shifting the energetic position of the recorded absorption signals. If the main excited state absorption signals were different enough, one would be able to potentially differentiate between them. Otherwise one could also look at the broadening of the GSB and its time dependence, which would be similar to what has already been done with the 1D pump–probe technique,[1] in order to separate these two tautomers.

1 B. Cohen, P. M. Hare and B. Kohler, *J. Am. Chem. Soc.*, 2003, **125**, 13594–13601.
2 B. A. West, J. M. Womick and A. M. Moran, *J. Phys. Chem. A*, 2011, **115**, 8630–8637.

Rienk van Grondelle remarked: If you look at these ApA dimers, the two As are externally coupled. Why don't you see dimer/excitonic states in your spectra? If you look at the time dependence you should see an oscillatory phenomenon.

Javier Segarra-Martí responded: The time dependence (delay time) was set to zero in this particular study as we focused on analysing the excited state absorption signals that are characteristic of the Franck–Condon (equilibrium) region in order to understand which energy windows may be more convenient to monitor in upcoming experiments. If we included the time dependence we would indeed observe the oscillatory phenomenon mentioned in your question and which features in many other 2DES experiments due to the formation of excitons/excimers in close-lying and interacting absorbing moieties.

Stefan Haacke queried: When you compared different stacking interactions there seemed to be very little difference between the absorption spectra. If you sum them all up there would be little expectation of cross-peaks or differences in the excitation effect. Can you comment on this? Would you expect different lifetimes?

Javier Segarra-Martí answered: There is indeed very little difference in terms of the main absorption signatures, which is partially due to it being a homodimer (two adenine moieties) that absorbs at the very same pump (Ω_1) frequencies. I would expect more marked differences were you to consider heterodimers, as we observed in a recent study on adenine-uracil monophosphate (ApU),[1] given that adenine and uracil absorb at different Ω_1 and their fingerprints along the probe (Ω_3) frequency are also quite diverse. Moreover, these differences would be more marked when looking at delay times different from zero (considering a time dependence that is neglected in the manuscript), as the geometrical distortions driven along the excited state decay result in massive changes in the accessible high-energy levels and thus strongly modulate their position along Ω_3.[1] For the present case, different lifetimes could still be potentially extracted from specific regions of the spectra, particularly long-living decays arising due to intermolecular interactions (excimer formation) along Ω_1, which would in principle be well separated from the monomeric contributions that would show a smaller broadening along this particular dimension.

1 Q. Li, A. Giussani, J. Segarra-Martí, A. Nenov, I. Rivalta, A. A. Voityuk, S. Mukamel, D. Roca-Sanjuán, M. Garavelli and L. Blancafort, *Chem.–Eur. J.*, 2016, **22**, 7497–7507.

Roberto Improta asked: How do the qualitative features of the computed spectra depend on the details and 'level' of the CASPT2 calculation?

Javier Segarra-Martí replied: That is indeed a very good point, which we addressed in a previous study[1] for gas phase adenine. In that study we systematically addressed this point by increasing the active space size until convergence was reached at very large spaces featuring up to 12 additional π^* orbitals to increase the correlation. We then employed these reference values to build gas-phase dimers, similar to the systems under study (ApA) in our paper, and observed a generalised red-shift of the signals along both the Ω_1 (pump) and Ω_3 (probe) frequencies, which we recovered by employing an imaginary level shift as a semi-empirical correction of CASPT2 when too small active spaces must be used. Reduced active space approaches are required for large systems such as dimeric species (ApA) due to computational demands, and a qualitative agreement of the results with those employing larger active spaces can already be observed, with the quantitative agreement being improved with the aforementioned use of non-standard and calibrated imaginary level shift values.

1 A. Nenov, A. Giussani, J. Segarra-Martí, V. K. Jaiswal, I. Rivalta, G. Cerullo, S. Mukamel and M. Garavelli, *J. Chem. Phys.*, 2015, **142**, 212443.

Richard Cogdell opened the discussion of the paper by Padmaja P. Mishra: Could you please provide some more detail on how you hold onto the DNA on the surface so that you can exert the force? Furthermore, when you look at individual molecules, how long does the fluorescence last? Is the effect reversible?

Padmaja P. Mishra replied: To hold the DNA on the surface and to apply force, we followed the experimental procedure that is described in the subsection "Annealing of Holliday junction and linker-DNA" in the Experimental section of our paper. In our experiments, PEG/Biotin-PEG-coated predrilled quartz microscope slides were used to monitor the fluorescence signals from the Cy3 and Cy5 labeled DNA substrate that was immobilized through biotin–streptavidin interactions.

We used a catalytic oxygen scavenging unit that consists of a mixture of protocatechuate 3,4-dioxygenase (PCD) and protocatechuic acid (PCA), which makes the fluorophores stable for 4–5 minutes. The process is not reversible, as once the fluorophores photobleach, they do not return to the original state.

Tolga Karsili remarked: In your quantum chemical calculations, a uniform field is applied to the zero-order electronic Hamiltonian. How do you choose the coordinate along which to position this field?

Padmaja P. Mishra replied: There seems to be some confusion here. There are no quantum calculations involved here. The field that is discussed here is a light induced optical force, which is applied at the Holliday junction arm in order to monitor the conformation Dynamics.

Arijit De addressed Padmaja P. Mishra and Tolga Karsili: I would like to rephrase Tolga's question. You focus the laser beam (along Z) to optically trap and hold a dielectric particle which is tethered to the single (protein) molecule by a linker, so that if you steer the laser beam focus in the focal plane (along X or Y) you pull the single molecule, therefore you apply a force to it. In which direction do you apply this force?

Padmaja P. Mishra responded: There seems to be some confusion with Tolga's Question. There are no quantum calculations involved here. The field that is discussed here is a light induced optical force, which is applied at the Holliday Junction arm in order to monitor the conformation Dynamics.

P. I. Pradeepkumar asked: What is the magnitude of the force exerted by the IHF on the 4WHJ in comparison to the applied external force? In other words, would it be possible to determine the force exerted by the IHF when it interacts with the 4WHJ?

Padmaja P. Mishra answered: Atomic force microscopy (AFM) based single-molecule force spectroscopy could be a suitable tool for quantifying the forces and binding mechanisms that lead to the formation of protein–DNA complexes. AFM and dynamic force spectroscopy are exciting tools that allow for the quantitative analysis of biomolecular interactions.

Arijit De queried: You described your results using sketches of Potential Energy Surfaces (PESs), with regard to how they are positioned relative to each other.

However, apart from estimating the force (in pN), what is also important (and perhaps more important) is the curvature of these PESs for estimating the force constant (in pN nm^{-1}) so that you know the trap stiffness. Have you measured force *vs.* distance curves in order to estimate the stiffness?

Padmaja P. Mishra replied: The position of the intermediates during branch migration (conformational fluctuation) is determined from the slope of the log-linear plot of K *vs.* applied force. For example, in our results the slope for K_f (1.04 nm, Fig. 2C) was found to be approximately 2.3 times higher than the slope for K_b (0.46 nm). This indicates that the distance of the *iso*I conformer from its TS is 0.58 nm further than that of the *iso*II conformer (Fig. 3, right panel). The distances represented here correspond to the separation between the X-R arms of the 4WHJ under different force conditions.

In our case, we have not estimated the force in these experiments. Instead, the force was calibrated using a parallel protocol, as described in the manuscript.

Rajaram Swaminathan questioned: Single molecule FRET experiments are not ensemble averaged, so can you please explain how you took care of noise related fluctuations in the data?

Padmaja P. Mishra responded: Photon trajectories from single molecule experiments usually report the structural changes and motions of biomolecules. We have used Hidden Markov Models (HMMs) that facilitate the extraction of the sequence of hidden states from noisy data through the construction of probabilistic models. Typically, the true number of states is determined by the Bayesian Information Criterion (BIC). However, constraints resulting from short data sets and Poisson-distributed photons in radiative processes such as fluorescence can limit the successful application of goodness-of-fit statistics. The HMM models used here enabled the unambiguous and unbiased separation of noise from state-to-state transitions and for a reliable analysis of noisy data, as well as enabling the examination and detection of significantly more complicated systems, including systems with multiple states, limited only by signal to noise. The molecule-by-molecule nature of the algorithm preserves one's ability to detect heterogeneities in the dynamics between molecules, which is critical in single-molecule studies. Potentially, the algorithm could also be used to discern states with the same FRET level but with different lifetimes.

Reji Varghese asked: Can you please comment on the dipole orientation of the donor and the acceptor?

Padmaja P. Mishra answered: The dipole orientation factor κ2, which describes the relative orientation of the donor and acceptor transition dipoles, is an important parameter for FRET that ranges from 0 to 4. For example, when the donor and acceptor transition dipoles are aligned with each other, κ2 = 4 and the orientation is ideal for energy transfer, whereas in a perpendicular orientation, κ2 = 0, thus preventing any energy transfer regardless of the distance. It is important to realize that κ2 matters only during the lifetime of the donor excited state when energy coupling is possible. For most fluorophores, this time is 1–10 ns. This means that if the fluorophores, their chromophore part, or the molecules

 This journal is © The Royal Society of Chemistry 2018

that the fluorophores are attached to move at frequencies that are significantly higher than 108–109 Hz, the fluorophore pair will experience a varying orientation while FRET occurs. In biological systems, proteins labelled with fluorophores can generally adopt a variety of conformations, and the covalent linkage between the fluorophore and the host protein can rotate. All of these factors lead to the randomization of the orientation, both spatially and temporally. Therefore, for most macromolecular interactions in solution, a κ2 value of 2 or 3 is often assumed.

Minor polarizations of the donor and acceptor molecules will not lead to any major variations in the determination of R_0. However, if one assumes that a range of static donor–acceptor orientations are present that do not change during the lifetime of the excited state, a κ2 value of 0.476 should be used instead. Nevertheless, under certain experimental conditions, such an assumption does not hold and so more rigorous treatments of κ2 are called for.

Mahil L communicated: Why is the orientational stability restricted due to the binding of IHF with 4WHJs? What will happen to the DNA due to the binding of IHF?

Padmaja P. Mishra communicated in reply: The orientational stability that is restricted due to the binding of IHF with 4WHJs is a result of the imposed restriction on the 4WHJ conformational switching. HMM analysis of the time traces indicated a dramatic increase in the tightness of the 4WHJ upon binding to IHF, therefore reducing the distance between the arms. As IHF is a histone line protein, it usually bends the DNA upon binding. However, it should be mentioned that IHF is a sequence specific protein.

Vishnu V communicated: Will the length of the arms and their constituent nucleotides of the 4WHJ affect the force that is applied? I believe that the Holliday junction is only a small part of the sequence. On what basis is the rest of the sequence constructed?

Padmaja P. Mishra communicated in response: The length of the arms and their constituent nucleotides of the 4WHJ certainly affect the force that is applied. The conformational fluctuations of the arms also depend on the nucleotide sequence of the junction. The particular sequences that were used in this study were designed in such a way that, once annealed to the junction structure, the protein binding sequence will remain more towards the junction area. The rest of the sequence was randomly selected, while keeping in mind the necessity to have 50% purines and pyrimidines.

Sneha Paul opened the discussion of the paper by Bern Kohler: Is there any specific reason for choosing this particular deep eutectic solvent (glyceline)? The high viscosity of the solvent might make transient absorption studies difficult, and there are other less viscous deep eutectic solvents that are known, *e.g.* ethaline, that do not perturb the nucleic acids.

Bern Kohler replied: We were initially interested in glyceline on account of its high viscosity, which is nevertheless low enough to allow the solution to be recirculated using a peristaltic pump. We have recently reported a time-resolved IR study of the $d(AT)_9 \cdot d(AT)_9$ duplex in ethaline (1 : 2 mol ratio of choline chloride and ethylene glycol).[1]

1 K. de La Harpe, F. R. Kohl, Y. Zhang and B. Kohler, *J. Phys. Chem. A*, 2018, **122**, 2437–2444.

Sneha Paul asked: Have you tried conducting these experiments in neat glycerol? If so, how different were the results?

Bern Kohler responded: We have not conducted experiments in neat glycerol. This is certainly feasible for the nucleobase monomers, but DNA strands are greatly destabilized in this solvent. Bonner and Klibanov showed that the melting temperature of a 22-mer double-stranded oligonucleotide with 52% GC content was lowered by approximately 30 °C (from 62 °C to 30 °C) upon going from a buffered aqueous solution to a glycerol solution.[1] Lower melting point AT-rich strands, such as those investigated in our study, are unlikely to retain the secondary structures of interest at room temperature in neat glycerol. The important point is that DNA strands are significantly more stable in a deep eutectic solvent made from choline chloride and an alcohol than they are in the alcohol alone.

1 G. Bonner and A. M. Klibanov, *Biotechnol. Bioeng.*, 2000, **68**, 339–344.

Carlos E. Crespo-Hernández queried: Given the estimation of a melting temperature in the range of 40–45 °C, reported in Fig. 4(c) and (d) of your manuscript for the double-stranded $d(AT)_9 \cdot d(AT)_9$ oligonucleotide in the deep eutectic solvent, what fraction of this oligonucleotide is present in a double stranded *versus* a single stranded form in the transient absorption experiments reported in Fig. 4(e) and (f) at room temperature?

Bern Kohler answered: In the two-state model of DNA melting, the absorbance *vs.* temperature curve provides a direct estimate of the fraction of folded DNA strands. From analysing the melting curve in Fig. 4(c), we estimate the fraction of intact duplexes at 20 °C to be greater than 90%.

Dimitra Markovitsi questioned: How do you subtract the signal that arises from two-photon ionization of the solvent? Do you consider that some of the exciting photons could be absorbed by DNA, therefore its intensity should be lower compared to that measured for the solvent alone?

Bern Kohler replied: Absorption by DNA (or by any absorbing solute for that matter) does reduce the total yield of solvated electrons that is formed by two-photon ionization of the solvent. For this reason, a transient absorption signal is recorded in a back-to-back measurement from a solution of adenosine 5′-monophosphate (AMP), which has an identical absorbance at the pump wavelength to that of the DNA solution of interest. This ensures that both solutions experience the same degree of two-photon excitation of the solvent. The AMP signal after 5 ps arises solely from solvated electrons and is used to predict the

 This journal is © The Royal Society of Chemistry 2018

signal contribution from these species at all delay times, as described in detail previously.[1]

1 C. E. Crespo-Hernández and B. Kohler, *J. Phys. Chem. B*, 2004, **108**, 11182–11188.

Rienk van Grondelle asked: Can you please expand on the effect of the solvent on inter-/intra-strand charge transfer? What is the timescale?

Bern Kohler responded: We have only made the first few steps towards understanding these solvent effects. Diverse measurements in aqueous solution make it very clear that CT states are only seen in strands made up of native nucleobases when the bases are closely stacked with each other. These deep eutectic solvents are virtually the only non-aqueous solvents that are known to stabilize DNA, so this is really the first opportunity to ask these kinds of questions. We have been unable to observe forward rates in any solvent thus far. Instead, the vibrational signatures of the CT states appear promptly within our time resolution. What have been amenable to study so far are the dynamics of charge recombination. Of course, even if the formation of CT states cannot be easily observed, it could be possible to reach conclusions about the effects of solvent on competing pathways such as inter- and intra-strand CT, if the yields of these channels can be estimated.

Ankona Datta questioned: Do you know the dielectric constant of the deep eutectic solvent that was used in this study?

Bern Kohler answered: Pandey *et al.*[1] estimated a value of 22 for glyceline from the measured Stokes shift of a solvatochromic probe molecule, but this was an indirect method. Their value is half that of neat glycerol, which has a static dielectric constant of 44 at 20 °C.[2]

1 A. Pandey, R. Rai, M. Pal and S. Pandey, *Phys. Chem. Chem. Phys.*, 2014, **16**, 1559–1568.
2 R. Behrends, K. Fuchs, U. Kaatze, Y. Hayashi and Y. Feldman, *J. Chem. Phys.*, 2006, **124**, 144512.

Roberto Improta remarked: I am curious about your results concerning the monomer. When you say that there is a two times slower decay, is the excited state lifetime twice as long or you are referring to the ground state recovery due to a slower vibrational cooling?

Bern Kohler replied: This statement refers to the excited state lifetimes as measured by the decay of excited-state absorption at a probe wavelength of 570 nm. Vibrational cooling also occurs more slowly in glyceline than in water, as is expected due to the former solvent's lower density of hydrogen bonds.

Joshy Joseph queried: I have a question regarding the photochemical stability of DNA in deep eutectic solvent systems. Since the generated excited states are much longer, will there be increased photo damage (*e.g.* thymine–thymine dimer formation) to DNA in these environments? What effect will the salt concentration have on the stability of the DNA duplex in deep eutectic solvent systems? Do mono- and di-cationic salts exhibit the same role in stabilizing DNA in DESs as in aqueous systems?

Bern Kohler answered: To answer your first question, the photochemistry of DNA in deep eutectic solvents has not yet been investigated to the best of my knowledge. It would clearly be very interesting to do so. Although quite long-lived excited states are formed in glyceline (Fig. 3(d)), it is not clear whether these are suitable photoproduct precursors. Thymine–thymine dimers form in DNA in aqueous solution on an ultrafast time scale,[1] at least when excited near 260 nm. To answer your second question, 100 mM NaCl is often added to aqueous DNA solutions to stabilize duplex formation, but it is unnecessary to do this in deep eutectic solvents because the ionic components are already present in such high concentrations that there is plenty of dielectric screening. Of course, replacing the choline ion with something else could alter the stability, but this has not been explored very much yet.

1 W. J. Schreier, T. E. Schrader, F. O. Koller, P. Gilch, C. E. Crespo-Hernández, V. N. Swaminathan, T. Carell, W. Zinth and B. Kohler, *Science*, 2007, **315**, 625–629.

Ilme Schlichting remarked: In the previous session it was mentioned that structural dynamics have an influence that is difficult to probe. By using glyceline you are changing the electrostatics and viscosity. Have you considered using something like cellulose to change only the viscosity in order to see what effect the structural dynamics have?

Bern Kohler responded: It would be very interesting to isolate the effect of viscosity without changing the dielectric properties too much if this could be done without disrupting the structure of the DNA strand.

Imon Mandal communicated: Do the charges of the deep eutectic solvent (choline chloride) have any role in the stabilisation (modulation) of the CT states of the DNA duplex, except to change the polarity of the medium?

Bern Kohler communicated in reply: A number of properties beyond solvent polarity must be important when it comes to correctly describing the microscopic interactions in these solvents. After all, measurements of solvatochromic probe molecules in deep eutectic solvents have shown that they have similar polarities to those of the neat, hydrogen-bond donating alcohol solvents that they are made from.[1] Yet the latter solvents denature double-stranded DNA (or stabilize it only very weakly), while the deep eutectic solvents do not. In this initial study, we have invoked relatively simple and qualitative arguments about how, for example, the reduced solvent polarity expected in a deep eutectic solvent would affect parameters such as solvent reorganization energy that influence electron transfer rates. As a next step, additional time-domain spectroscopy and theory will be very important for understanding both the static and dynamic effects of deep eutectic solvents on photoinduced charge transfer reactions.

1 A. Pandey, R. Rai, M. Pal and S. Pandey, *Phys. Chem. Chem. Phys.*, 2014, **16**, 1559–1568.

Dimitra Markovitsi opened the general discussion of the papers by Javier Segarra-Martí, Padmaja P. Mishra and Bern Kohler, and asked Javier Segarra-Martí and Bern Kohler: Do you expect two-dimensional spectroscopy to provide better insight into UV-induced processes by studying DNA in deep eutectic solvents instead of aqueous solvents?

Javier Segarra-Martí answered: I would expect to be able to observe the differences between water and deep eutectic solvents more clearly through the specific inhomogeneous broadenings that they induce in the 2D signals. Water rearranges very quickly and produces a marked inhomogeneous broadening along the diagonal of the 2D peaks, whereas deep eutectic solvents are much more dense and I would expect their dynamics to be slower and thus provide a less pronounced effect in comparison. Additionally, and as Professor Kohler has discussed in his contribution, deep eutectic solvents appear to slow down the excited state decay, which would also reduce their homogeneous broadening with respect to the faster decays observed in water. By monitoring the shape of the peaks with 2DES and their evolution over time, one could potentially weight these two different contributions and potentially assess their relative importance.

Bern Kohler replied: I expect multidimensional ultrafast spectroscopy to reveal many new insights into DNA photophysics, no matter what solvent is used. There is already some exciting work in this direction, and there will be a great deal more as UV pulses with the requisite broad bandwidths become easier to generate.

Ravindra Venkatramani returned to the discussion of the paper by Bern Kohler: In addition to reorganization energy effects, the solvent can also electronically couple charge donor–acceptor states (the prefactor term in the Marcus rate expression). Can you comment on how the electronic coupling provided by the solvent may modulate the electron transfer rates in DNA duplexes (*e.g.* the forward ET for AT duplexes in your paper)?

Bern Kohler responded: We have not considered these effects as it has thus far been possible to satisfactorily describe ET rates using a very simple Marcus model in the few cases where this has been done (all in aqueous solution). It should be noted that it has not yet been possible to observe forward ET in UV-excited DNA strands as this appears to take place faster than the experimental time resolution.

Richard Cogdell returned to the discussion of the paper by Padmaja P. Mishra: I would like to ask more about the pulling of the force in your experiments. If you are pulling on a protein where there is a little flexibility with regards to where you pull, the moment of the force might be over a certain variability around where you actually measure. How do you account for that? How much elasticity is in your protein structure? Do you model it as stiff or elastic?

Padmaja P. Mishra answered: In our experiment, we did not pull the protein, rather we pulled one arm of the Holliday junction, while the protein is bound to the holiday junction. Details of the force calibration method and the experimental procedure for the application of force to the DNA holiday junction can be found in the "Optical trapping and force calibration" subsection in the Experimental section of our paper.

P. I. Pradeepkumar returned to the discussion of the paper by Bern Kohler: Are there any crystal or NMR structure of duplex DNA or RNA reported in the presence of deep eutectic solvents? Is there any possibility of the formation of cation-pi interactions between the solvent and the nucleobases?

Bern Kohler replied: I am unaware of any experimental determinations of nucleic acid structures in deep eutectic solvents. Your second question about cation–pi interactions is very interesting. To my knowledge, this has not yet been discussed as a contributing factor towards stabilization. An atomistic simulation of the binding of the choline ion in the minor-groove of AT-DNAs showed that the hydroxy group of the choline ion interacts with the bases *via* hydrogen bonding.[1] The lower charge density of the choline cation compared with that of a monovalent or divalent metal ion also suggests that cation–pi interactions are likely to be weak. Finally, cation–pi interactions, at least those involving metal ions, have been suggested to stabilize unstacked conformations.[2] For these reasons, I do not expect cation–pi interactions to contribute significantly to the stabilization of nucleic acids in deep eutectic solvents.

1 M. Nakano, H. Tateishi-Karimata, S. Tanaka and N. Sugimoto, *J. Phys. Chem. B*, 2014, **118**, 379–389.
2 L. McFail-Isom, X. Shui and L. D. Williams, *Biochemistry*, 1998, **37**, 17105–17111.

Roberto Improta remarked: I have a couple of comments. (1) In these two sessions there has been a lot of focus on CT states, but we have not discussed the mechanisms explaining how they are populated. For example the monomer-like decay channel could be an important 'doorway' to inter-strand CT states, and if you increase the lifetime of the monomer-like decay channel (as happens in a deep eutectic solvent) this is going to affect all of the other channels.

(2) A significant part of the solvation effect is instantaneous, but there is an important contribution from a slow component, which in DNA could also coincide with a conformational rearrangement of the backbone structure, which could be made more difficult by the deep eutectic solvent. This suggests that the solvent dielectric constant is not the only parameter affecting the charge transfer pathways.

Bern Kohler replied: I couldn't agree more about the need to understand how CT states (both intra- and inter-strand) form in DNA strands. DNA is a complex, multichromophoric system and experiments have shown that UV excitation yields several distinct classes of excited states, which can differ in their spectroscopic signatures and dynamics. Many of the ultrafast laser experiments reported thus far for DNA strands are 'blind' to the dynamics that occur during the first few hundred femtoseconds, and it is in this brief window that the branching to the various decay channels likely takes place. It is also important to mention that CT states are not populated to any significant extent by direct photoexcitation at UV wavelengths where the DNA bases absorb strongly, such as at 260 nm. Consequently, the initially excited electronic states or excitons must rapidly evolve to CT

 This journal is © The Royal Society of Chemistry 2018

states, and this is perhaps influenced by the solvent. As we observe in our paper, the slower dynamical response of the deep eutectic solvent compared to that of water may affect how quickly CT states are formed, and may influence the competition between inter- and intra-strand channels, especially when these differ in charge transfer character. The ability to study the photophysics of DNA strands in non-aqueous environments makes it possible to systematically study effects such as these.

Rienk van Grondelle made a general comment: With multidimensional spectroscopy you are going to see every channel. If you could combine it with time resolved X-ray crystallography too, this could be a great tool for the future.

Gebhard F. X. Schertler responded: I think that further advances in multidimensional spectroscopy are very promising, and they will further aid in the interpretation of FEL measurements that are aimed towards a mechanistic explanation of quantum yields and stereoselectivity.

Devika Sasikumar returned to the discussion of the paper by Bern Kohler: This is interesting work Bern. Can you tell me whether the triplet state that is observed in 5′-TMP is merely due to the slow structural dynamics or due to any perturbation in the electronic states of the nucleobase? If I wished to populate the triplet state of an organic molecule, could a DES restrict the singlet modes that are responsible for fluorescence decay and cause intersystem crossing?

Bern Kohler responded: Nonradiative decay from an excited electronic state, whether by internal conversion or by intersystem crossing, always involves structural (nuclear) dynamics, so it may be difficult to disentangle the effects you mention. Triplet quantum yields for the pyrimidine nucleobase monomers increase as the solvent polarity decreases, even in solvents that are much less viscous than glyceline, and Crespo-Hernández and co-workers have made similar observations for 2-aminopurine.[1] These trends are explained by changes in the energies of nearby excited electronic states, but this of course results in changes in the nuclear dynamics.

1 C. Reichardt, C. Wen, R. A. Vogt and C. E. Crespo-Hernández, *Photochem. Photobiol. Sci.*, 2013, **12**, 1341–1350.

Rajaram Swaminathan asked: Is it possible to observe DNA–protein interactions and enzyme catalysis in deep eutectic solvents? Have such studies been reported?

Bern Kohler answered: A variety of enzymes have been shown to retain activity in deep eutectic solvents and catalyze synthetically valuable reactions.[1–6]

1 J. T. Gorke, F. Srienc and R. J. Kazlauskas, *Chem. Commun.*, 2008, **0**, 1235–1237.
2 D. Lindberg, M. de la Fuente Revenga and M. Widersten, *J. Biotechnol.*, 2010, **147**, 169–171.
3 P. Domínguez de María and Z. Maugeri, *Curr. Opin. Chem. Biol.*, 2011, **15**, 220–225.
4 V. Stepankova, S. Bidmanova, T. Koudelakova, Z. Prokop, R. Chaloupkova and J. Damborsky, *ACS Catal.*, 2013, **3**, 2823–2836.
5 Z.-L. Huang, B.-P. Wu, Q. Wen, T.-X. Yang and Z. Yang, *J. Chem. Technol. Biotechnol.*, 2014, **89**, 1975–1981.

6 H. Zhao, C. Zhang and T. D. Crittle, *J. Mol. Catal. B: Enzym.*, 2013, **85–86**, 243–247.

Reji Varghese asked: Is the cationic component of the eutectic mixture bound to the phosphate backbone of DNA? Could this be a possible way to dissolve DNA in organic solvents?

Bern Kohler responded: One study has suggested that choline ions have a high affinity for the minor groove of AT-rich DNAs where they form hydrogen bonds.[1] In my opinion, much remains to be learned about the behavior of DNA in deep eutectic solvents and its ultimate applications.

1 M. Nakano, H. Tateishi-Karimata, S. Tanaka and N. Sugimoto, *J. Phys. Chem. B*, 2014, **118**, 379–389.

Conflicts of interest

Gebhard F. X. Schertler is a co-founder of the leadXpro and InterAx biotechnology companies. There are no other conflicts to declare.

 This journal is © The Royal Society of Chemistry 2018

Faraday Discussions

PAPER

Engineered photoproteins that give rise to photosynthetically-incompetent bacteria are effective as photovoltaic materials for biohybrid photoelectrochemical cells

Juntai Liu,[a] Vincent M. Friebe,[b] David J. K. Swainsbury,†[a] Lucy I. Crouch,‡[a] David A. Szabo,[a] Raoul N. Frese[b] and Michael R. Jones*[a]

Received 24th August 2017, Accepted 4th September 2017
DOI: 10.1039/c7fd00190h

Reaction centre/light harvesting proteins such as the RCLH1X complex from *Rhodobacter sphaeroides* carry out highly quantum-efficient conversion of solar energy through ultrafast energy transfer and charge separation, and these pigment-proteins have been incorporated into biohybrid photoelectrochemical cells for a variety of applications. In this work we demonstrate that, despite not being able to support normal photosynthetic growth of *Rhodobacter sphaeroides*, an engineered variant of this RCLH1X complex lacking the PufX protein and with an enlarged light harvesting antenna is unimpaired in its capacity for photocurrent generation in two types of bio-photoelectrochemical cells. Removal of PufX also did not impair the ability of the RCLH1 complex to act as an acceptor of energy from synthetic light harvesting quantum dots. Unexpectedly, the removal of PufX led to a marked improvement in the overall stability of the RCLH1 complex under heat stress. We conclude that PufX-deficient RCLH1 complexes are fully functional in solar energy conversion in a device setting and that their enhanced structural stability could make them a preferred choice over their native PufX-containing counterpart. Our findings on the competence of RCLH1 complexes for light energy conversion *in vitro* are discussed with reference to the reason why these PufX-deficient proteins are not capable of light energy conversion *in vivo*.

[a]*School of Biochemistry, University of Bristol, Medical Sciences Building, University Walk, Bristol BS8 1TD, UK. E-mail: m.r.jones@bristol.ac.uk*

[b]*Department of Physics and Astronomy, LaserLaB Amsterdam, VU University Amsterdam, De Boelelaan 1081, 1081 HV, Amsterdam, The Netherlands*

† Present address: Department of Molecular Biology and Biotechnology, University of Sheffield, Sheffield S10 2TN, UK.

‡ Present address: Institute for Cell and Molecular Biosciences, The Medical School Newcastle University, Newcastle upon Tyne NE2 4HH, UK.

1. Introduction

In the quest to develop new sustainable materials for solar energy conversion, there has been increasing interest in combining pigment-proteins from photosynthetic organisms with man-made materials in biohybrid device architectures.[1–9] The primary events in light harvesting (LH) and reaction center (RC) proteins transduce solar energy with a near to unity quantum efficiency (charges separated per photon absorbed).[10–15] In layman's terms, natural photosystems act as solar batteries, using sunlight to create a potential difference across the photosynthetic membrane. Connection of additional electron transfer "wires" to the positive and negative "terminals" of these solar batteries powers subsequent biological processes. A variety of RC or combined RC/LH proteins have been incorporated into device architectures in order to harness charge separation for photocurrent generation and other applications such as sensing.[1–9] Much of this work has employed the RC from *Rhodobacter* (*Rba.*) *sphaeroides*, a pigment-protein that has informed much of our understanding of solar energy conversion in photosynthetic RCs.[16–20]

Purple bacterial RCs are typically enclosed within a hollow cylinder of the LH1 pigment-protein, forming the so-called RC–LH1 core complex (Fig. 1(a)).[21–23] In many species the light harvesting capacity of the photosystem is augmented by a peripheral LH2 antenna.[24] The RC, LH1 and LH2 pigment-proteins can assemble independently of one another in *Rba. sphaeroides*, and because the photosystem components are assembled during respiratory growth at moderate oxygen levels in the dark in this bacterium, they can be expressed in any combination.[25] This modularity is very useful if the bacterium is being used as a "factory" for the production of tailored proteins, enabling tuning of the relative contributions of light harvesting and charge separation to suit a particular device setting or application. The pigmentation of the photosystem can also be varied by selection of spontaneously-arising strains that have a dysfunction in one or more of the enzymes of the carotenoid synthesis pathway. The best known of these are "green strains" where the native red/brown carotenoids, spheroidene and spheroidenone, are replaced by their precursor neurosporene.[26,27] Red and green variants of the RC–LH1 complex were recently used in a comparison of photocurrent generation by bio-photoelectrochemical cells in mixed or tandem configurations, exploiting natural variation.[28] The carotenoids incorporated into the *Rba. sphaeroides* photosystem can also be diversified by heterologous expression of non-native genes and pathways.[29,30]

The *Rba. sphaeroides* RC–LH1 complex includes a PufX polypeptide that breaks the continuity of the LH1 cylinder (see ref. 31 for a review), and its location was revealed in a 7.8 Å resolution X-ray crystal structure of the "RCLH1X complex" (Fig. 1(a), yellow ribbon).[22] Each LH1 comprises 14 pairs of α and β membrane-spanning polypeptides that scaffold 28 carotenoids and 28 BChls as LH pigments, and when viewed perpendicular to the plane of the membrane forms a C-shaped antenna with a gap held open by PufX (Fig. 1(b), left). In addition, these RCLH1X proteins assemble into dimers around a two-fold symmetry axis, such that two RCs are surrounded by a continuous S-shaped LH1 antenna.[21,22,32–34] If the gene encoding PufX is deleted, the resulting "RCLH1 complexes" are uniformly monomeric,[35] with a single RC encased in a complete LH1

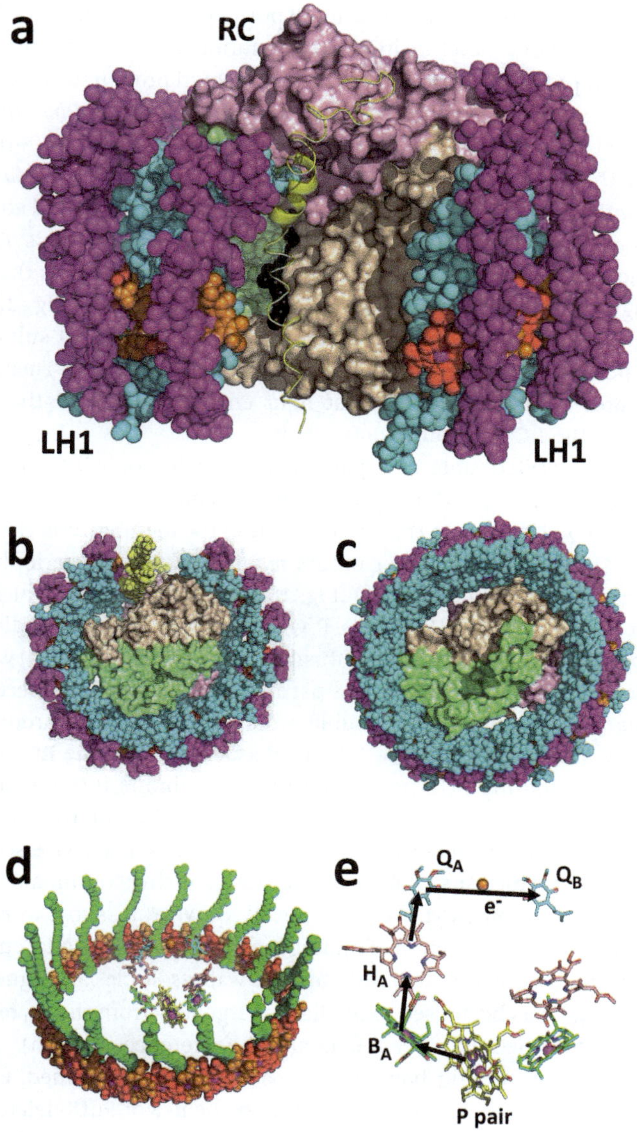

Fig. 1 Structure and mechanism of the *Rba. sphaeroides* RCLH1X and RCLH1 complexes. (a) A view approximately in the plane of the photosynthetic membrane of a RCLH1X monomer. The colour/representation coding is: pink surface – RC H-polypeptide, lime-green surface – RC L-polypeptide, beige surface – RC M-polypeptide, yellow ribbon – PufX, cyan spheres – LH1 α-polypeptides, magenta spheres – LH1 β-polypeptides, red/orange spheres – alternating LH1 BChls with the central Mg shown in magenta, black spheres – side chain of the RC Q_B ubiquinone-10. (b) A view from the periplasmic side of the membrane of the *Rba. sphaeroides* RCLH1X complex. (c) The same view of the *T. tepidum* RCLH1 complex. (d) The 32 BChls and 16 carotenoids of *T. tepidum* LH1 (spheres) arranged around the cofactors of the central RC (sticks). The colour coding for the LH1 cofactors is: red/orange spheres – alternating LH1 BChls with the central Mg shown in magenta, green spheres – carotenoids. The colour coding for the RC cofactors is: yellow carbons – P BChls, green carbons – monomeric BChls, pink carbons – bacteriopheophytins, cyan carbons – ubiquinones, brown spheres – iron atoms, magenta spheres – BChl Mg atoms. (e) Cofactors of the *Rba. sphaeroides* RC and the route of

cylinder[33,36,37] comprising 16 pairs of α and β polypeptides, 32 carotenoids and 32 BChls. Removal of PufX therefore increases the number of LH pigments servicing each RC and simplifies the composition of the prepared protein to one of uniform monomers. The X-ray crystal structure of the PufX-deficient *Rba. sphaeroides* complex is not available, but Fig. 1(c) shows the 3.0 Å resolution X-ray crystal structure of the RCLH1 complex from *Thermochromatium* (*T.*) *tepidum*.[23] This lacks PufX and has a closed 16-member LH1 ring around the RC, and so serves as a useful model for the RCLH1 complex from *Rba. sphaeroides*. This *T. tepidum* structure includes one carotenoid per LH1 subunit, and so enables the arrangement of pigments around the RC cofactors to be visualised (Fig. 1(d)). *Rba. sphaeroides* RCLH1(X) complexes contain two carotenoids per LH1 subunit,[38] but these were not resolved in the available X-ray crystal structure.[22] These rings of BChl and carotenoid pigments harvest light energy and "feed" the resulting excited state to the RC electron transfer chain.

Arrival of the excited state at a pair of bacteriochlorophyll (BChl) electron transfer cofactors (*P*) in the RC triggers a four-step charge separation to a mobile ubiquinone-10 (Q_{10}) located on the opposite side of the membrane at the so-called Q_B site (Fig. 1(e)).[16–20] This transfer occurs *via* a BChl (B_A), bacteriopheophytin (BPhe–H_A) and an immobile Q_{10} (Q_A) (Fig. 1(e)), with the system evolving along the sequence $P^* \rightarrow P^+B_A^- \rightarrow P^+H_A^- \rightarrow P^+Q_A^- \rightarrow P^+Q_B^-$. The final radical pair is stabilised by the reduction of P^+ by a diffusible c-type cytochrome (cyt) whose role is to shuttle electrons to the RC from a partner cyt bc_1 complex. A second light-powered charge separation elicits double reduction and double protonation of Q_B to produce a ubiquinol ($Q_{10}H_2$).[39,40] This dissociates from the RC and passes through the surrounding LH1 protein into the membrane interior in order to supply electrons to the cyt bc_1 complex. Several studies of the photovoltaic capacity of the *Rba. sphaeroides* RC or RCLH1X complexes in a device setting have recapitulated the interactions of the oxidising and reducing "terminals" of the RC with cyt *c* and ubiquinone. Cyt *c* has been used to "wire" RCs to a working electrode, thus enabling the reduction of photogenerated P^+ by a cathodic current (see ref. 41 and the references therein), and a water-soluble analogue of Q_{10}, ubiquinone-0 (Q_0), has been used to mediate charge flow from the Q_B terminal of the RC to a counter electrode (see ref. 42 and the references therein).

Despite its enhanced light harvesting capacity and its simplified, exclusively monomeric architecture, a point of concern over the use of PufX-deleted RCLH1 complexes for device applications is that this modification renders *Rba. sphaeroides* incapable of growth under standard anoxic, illuminated conditions.[43–47] This impairment is often portrayed in terms of the enlarged LH1 ring blocking the escape of $Q_{10}H_2$ from the RCLH1 complex and blocking its replacement by oxidised Q_{10} from the intramembrane quinone pool, diffusional processes that are normally facilitated by PufX keeping the LH1 ring open (see ref. 38 for a recent discussion). In the representation of an RCLH1X monomer in Fig. 1(a) the atoms coloured in black, visible behind PufX (yellow), are part of the hydrocarbon side chain of the Q_B ubiquinone; the quinone head-group is above these, buried in a binding pocket in the interior of the RC. It is easy to conceptualise that the

electron transfer. The cofactor representation is the same as for (d). For clarity, cofactor hydrocarbon side chains are not shown.

replacement of PufX by an extra section of LH1 pigment-protein could fill the gap maintained by PufX and prevent quinone diffusion. Regardless of whether this "blockage mechanism" is correct (see Section 3.7), the fact that strains of *Rba. sphaeroides* with PufX-deficient RCLH1 complexes are incapable of photosynthetic growth raises obvious concerns over their suitability as a material for photocurrent generation. The fragmentary experimental data published to date supports such concerns, as bio-photoelectrochemical cells fabricated using RCLH1 complexes have been reported to generate markedly lower steady-state photocurrent densities, in the range of 0.15 to 8.6 μA cm^{-2},[28,48–50] than is the case for PufX-containing RCLH1X complexes produced from equivalent strains and by equivalent purification procedures, where current densities of up to 166 μA cm^{-2} have been described.[51]

In this report, we profile the types of RCLH1X complex that can be isolated from strains of *Rba. sphaeroides* with different types of carotenoid, examine how the removal of PufX affects these profiles, and compare the abilities of the RCLH1X and RCLH1 complexes to support a photocurrent. We also compare the abilities of the two complexes to interact with synthetic antenna nanocrystals, and their structural stabilities under stress conditions. We discuss why the removal of PufX produces a photosynthesis-minus phenotype, and also discuss the validity of using the resulting RCLH1 complexes for photocurrent generation in biohybrid devices.

2. Experimental

2.1 Biological materials

Rba. sphaeroides strains with either red or green carotenoid expressing RCLH1X and RCLH1 complexes were constructed as described previously[28,51] and were named as RCLH1Xr, RCLH1r, RCLH1Xg or RCLH1g. Bacteria were grown either under dark/semi-aerobic conditions in an unilluminated orbital incubator operating at 34 °C and 180 rpm,[52] or in completely filled 500 mL medical flat bottles in a glass circulating water bath at 34 °C that was illuminated with four 100 Watt incandescent light bulbs. The RCLH1(X) and RC complexes were purified using a His$_{10}$-tag on the PufM polypeptide of the RC component as described elsewhere.[51,53]

2.2 Protein profiling on sucrose density gradients

Intracytoplasmic membranes prepared using a French pressure cell[52] were suspended in 20 mM HEPES (pH 8) to a concentration equivalent to an absorbance of 60 at the maximum of the LH1 Q_y absorbance band, and DDM was added to 4% final concentration (w/v). After incubation on ice for 30 min in the dark, membrane debris was removed through 1 hour of centrifugation in a TLA 100 rotor at 78 100*g* and 4 °C. Sucrose density gradients comprising 2 mL steps of 20, 21.25, 22.5, 23.75 and 25% (w/v) sucrose in 20 mM HEPES (pH 8)/0.04% DDM were prepared in transparent ultracentrifuge tubes. Each gradient was loaded with a 150 μL aliquot of solubilised protein at an absorbance of 25 at 875 nm. The loaded gradients were centrifuged for 20 hours in a Sorvall TH-641 swing-out rotor at 180 000*g* and 4 °C.

2.3 Photochronoamperometry of biohybrid protein/silver electrodes

Nanostructured silver electrodes were prepared from 3 mm diameter planar silver electrodes (Metrohm Autolab) as described previously.[51] RCLH1X, RCLH1 or RC proteins at a concentration of 30 μM were drop-casted onto the prepared electrodes in the dark at 4 °C for 15 minutes and unbound protein was removed by repeated mechanically-controlled dipping in 20 mM Tris (pH 8) at 4 °C. Protein-coated electrodes were mounted in a photoelectrochemical cell fitted with a platinum counter electrode and a Ag/AgCl/3 M KCl reference electrode, and immersed in an electrolyte solution comprising 20 mM Tris (pH 8) supplemented with 200 μM cyt *c* and 1.0 mM ubiquinone-0 (Q_0).[51] Photocurrents were measured at room temperature and a bias potential of −100 mV *vs.* Ag/AgCl under the control of a PGSTAT128N potentiostat (Metrohm Autolab). Illumination was supplied by an 870 nm LED (Roithner Lasertechnik) with an irradiance of 46 mW cm^{-2} at the electrode surface.

2.4 Photochronoamperometry of biohybrid protein/gold electrodes

Planar gold electrodes with a 2 mm diameter (CHI Instruments) were prepared as described previously.[53] RCLH1X or RCLH1 proteins at a concentration of 30 μM were drop-casted onto cleaned electrodes in the dark at 4 °C for 60 minutes and unbound protein was removed by repeated mechanically-controlled dipping in 20 mM Tris (pH 8) at 4 °C. Protein-coated planar gold electrodes were mounted in a photoelectrochemical cell as described above for the protein/silver electrodes, but in an electrolyte solution comprising 20 mM Tris (pH 8)/20 μM cyt *c*/100 μM UQ_0.[42] Photocurrents were measured as for the protein/silver electrodes.

2.5 Titrations of quantum dot emission

Water-soluble 6.5 nm-diameter CdTe QDs were purchased from PlasmaChem GmbH (Berlin, Germany). Solutions of QDs and RCLH1Xg or RCLH1g complexes in 20 mM Tris (pH 8)/0.04% DDM were mixed to give protein : QD ratios that varied between 0 and 10 : 1. The emission in response to 515 nm or 645 nm excitation was recorded using a Cary Eclipse fluorescence spectrophotometer (Agilent).

2.6 Sucrose pull-down assays

Protein/QD conjugates were separated from free proteins by centrifugation on sucrose density gradients comprising 5 mL 60% sucrose and 5 mL 30% sucrose, both (w/v) in 20 mM Tris (pH 8)/0.04% DDM. Protein-only or QD/protein mixed samples were added in volumes of 400 μL and overlaid with 1 mL of 20 mM Tris (pH 8.0)/0.04% DDM. The protein concentration was fixed at 250 nM and the QD concentration was varied in order to achieve the indicated QD/protein ratios. Loaded tubes were centrifuged for 16 h at 247 104*g* and 4 °C in a TH-641 swinging bucket rotor. The gradients were deconstructed by piercing the bottom of the tube and collecting 11 × 1 mL fractions dropwise, and the pigment-protein content of each fraction was determined by absorbance spectroscopy.

2.7 Assays of thermal stability

Assays were carried out using solutions of RCLH1(X) complexes in completely filled, stoppered quartz cuvettes incubated at the desired temperature in a water-thermostatted multi-cell holder linked to a circulating water bath and mounted in a Cary 60 scanning spectrophotometer (Agilent). The temperature in the cuvette was monitored using a Fluke 51 thermometer.

 This journal is © The Royal Society of Chemistry 2018

2.8 Molecular models

Images of the protein structure were produced using the PyMOL molecular graphics system (Schrödinger, LLC) using the Protein Data Bank entries 4V9G for the *Rba. sphaeroides* RCLH1X complex,[22] 3WMM for the *T. tepidum* RCLH1 complex[23] and 1HRC for the cyt *c* from *Equus caballus*.[54]

3. Results and discussion

3.1 The RC–LH1 complexes obtainable from *Rba. sphaeroides*

The pigment-proteins that can be extracted from LH2-deficient strains of *Rba. sphaeroides* are profiled in Fig. 2(a). For strain RCLH1Xr, fractionation of DDM-extracted proteins on a five-step sucrose density gradient produced three coloured bands (Fig. 2(a), left) corresponding to free carotenoid (arrow Crt), RCLH1X monomers (arrow Mon) and RCLH1X dimers (arrow Dim). The colour of the protein bands was determined by the type of carotenoid present, this being spheroidenone when strain RCLH1Xr is grown in the presence of oxygen (Fig. 2(a), top) and spheroidene when strain RCLH1Xr is grown in the absence of oxygen (Fig. 2(a), bottom). The absorbance of spheroidene in these proteins largely cuts off before the yellow/orange region of the visible spectrum (Fig. 2(b), brown trace), hence the colour change from the red colour conferred by spheroidenone (Fig. 2(a), top left) to the brown colour conferred by spheroidene (Fig. 2(a), bottom left).

The same pattern was seen for strain RCLH1Xg (Fig. 2(a), column 2) which synthesises neurosporene and its hydroxy and methoxy derivatives.[25] The absorbance of the neurosporene family carotenoids largely cuts off before the green region of the visible spectrum (Fig. 2(b), green spectra), and the absorbance line shape is not variable with the presence/absence of oxygen.

The architectures of the isolated complexes depended on the type of carotenoid present. With strain RCLH1Xg, and strain RCLH1Xr grown in the absence of oxygen, the percentage of RCLH1X complexes in the dimer form (Fig. 2(a)) was much greater than that in the monomeric form. In contrast, when strain RCLH1Xr was grown in the dark in the presence of oxygen, causing the incorporation of spheroidenone, most of the extracted RCLH1X complexes were monomeric (Fig. 2(a), top left). As expected, the removal of PufX to produce a pair of otherwise equivalent strains, denoted as RCLH1r and RCLH1g, resulted in the complete loss of the dimer form in cells grown under dark/semiaerobic conditions (Fig. 2(a), top right). These variations in the relative amounts of the monomeric and dimeric complexes, dependent on the growth conditions, carotenoid type and presence of PufX, were also seen in the strains expressing LH2 (data not shown).

3.2 Absorbance changes caused by the removal of PufX

RCLH1(X) complexes were purified from strains expressing red or green carotenoids grown under dark/semiaerobic conditions (see Experimental). Analysis using sucrose density gradient ultracentrifugation showed that the purified RCLH1r and RCLH1g complexes were exclusively monomeric, as were the RCLH1Xr complexes (data not shown). A minority of the RCLH1Xg complexes

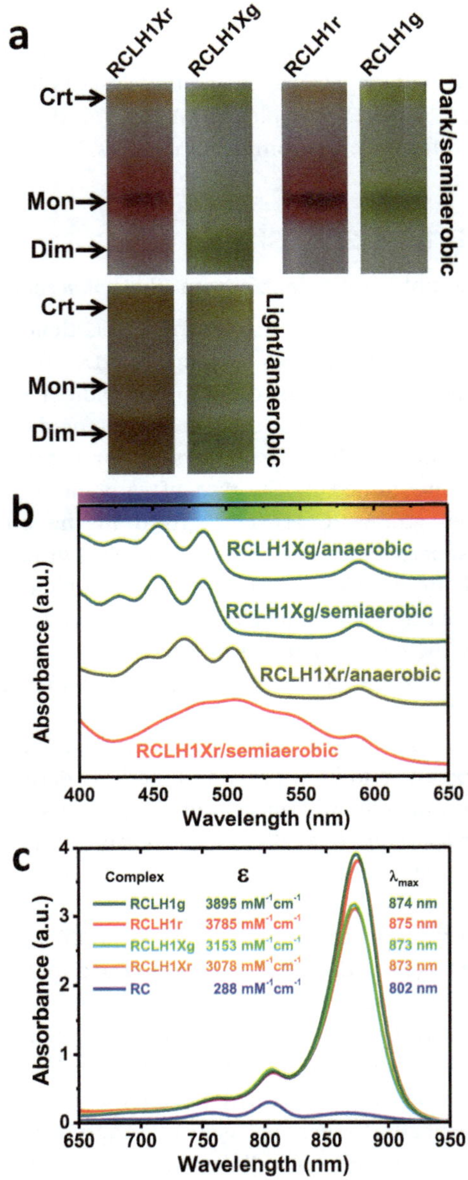

Fig. 2 A variety of RCLH1 complexes from *Rba. sphaeroides*. (a) Bands formed by pigment-proteins separated on sucrose density gradients, (b) visible region absorbance spectra for the RCLH1Xr and RCLH1Xg complexes, (c) molar absorption coefficients for the purified RCLH1(X) or RC complexes and near-IR absorbance spectra for 1 μM solutions.

were dimeric, but were not separated from the major monomeric population before further analysis.

Molar absorption coefficients for the four variants were estimated by recording absorbance spectra for aliquots of the concentrated protein diluted in 20 mM Tris (pH 8.0)/0.04% DDM and also recording absorbance spectra for aliquots diluted in 7 : 2 acetone : methanol. The concentration of extracted BChl in the latter was

 This journal is © The Royal Society of Chemistry 2018

determined using a molar absorption coefficient of 65.3 mM^{-1} cm^{-1} for BChl *a* in this solvent.[55] A molar absorption coefficient for each protein was then calculated at the maximum of the LH1 Q_y absorbance band between 873 nm and 875 nm based on the assumption that a RCLH1X complex has 32 BChl *a* and a RCLH1 complex has 36 BChl *a*.

The determined values are shown in Fig. 2(c), along with the calculated absorbance spectra for each complex at a concentration of 1 μM and purified RCs at the same concentration. For each carotenoid type, PufX removal caused an increase in intensity and a 1–2 nm red-shift of the LH1 Q_y absorbance band, as well as an increase in the carotenoid region (not shown), which is in accordance with the expectation that the PufX-deficient version will contain four extra BChl and four extra carotenoids per RC. The estimated molar absorption coefficient was 1.23-fold larger for the RCLH1r complexes compared to the RCLH1Xr complexes, and 1.24-fold larger for the green counterparts.

3.3 Photocurrents from the RCLH1X, RCLH1 and RC complexes

The capacity for photocurrent generation was tested using purified RCLH1Xr or RCLH1r complexes drop-casted onto a nanostructured silver working electrode. This was immersed in a buffer containing 1 mM Q_0 and 200 μM cyt *c*.[51] Illumination of the working electrode produced a cathodic photocurrent, with the underlying mechanism investigated in-depth in cells with RC complexes[41,42] which is summarised in Fig. 3(a). Cyt *c* facilitates current flow from the working electrode to the protein,[41] with Q_0 carrying electrons to the counter electrode in a process that requires full cycles of two-electron/two-proton chemistry at the Q_B site.[42] An additional feature of the RCLH1(X) complexes used in the present work is the expected presence of up to ~25 molecules of endogenous Q_{10} per monomer.[56] The role these play in enabling electron flux from the RC into solution is the subject of ongoing investigation, but it is clearly established that Q_0H_2 is the actual charge carrier to the counter electrode, as Q_{10} in solution has been shown to be unable to support a photocurrent.[42]

All transients showed an initial spike of cathodic photocurrent following light-on that stabilised over a minute or so to a lower steady-state level (Fig. 4(b)). This initial decline has been attributed to a limitation of current output due to diffusion-limited mass transport of the Q_0 electrolyte.[51] When the light was switched off the recombination of the accumulated Q_0H_2 product with the working electrode produced a transient spike of anodic current.

For the working electrodes with adhered RCLH1Xr complexes, the average steady-state photocurrent density determined over the last 80 s of the illumination period was approximately −69 μA cm^{-2} (Fig. 3(b) and (c)). This was much larger than the average density of −16 μA cm^{-2} determined for the RC complexes lacking the LH1 antenna adhered to the same nanostructured silver electrodes in an otherwise identical measuring system (Fig. 3(b) and (c)), thus demonstrating the benefit of using a combined LH/RC protein for photocurrent generation rather than a "naked" RC.

Despite being isolated from a strain of *Rba. sphaeroides* that is not capable of photosynthetic growth, PufX-deficient RCLH1r complexes generated a photocurrent that was not significantly different to that seen for native RCLH1Xr complexes (Fig. 3(b) and (c)). A similar result was obtained by equivalent

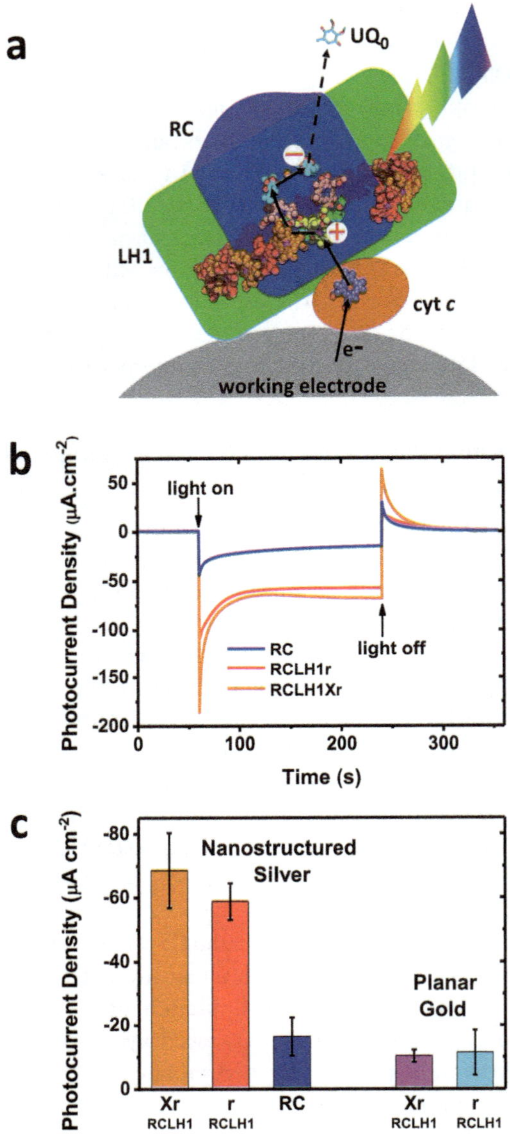

Fig. 3 Photocurrents from the RCLH1Xr, RCLH1r and RC complexes interfaced with metal electrodes. (a) Photoexcitation of the BChls (alternating red/orange) and carotenoids (not shown) of the LH1 antenna (green domain) produces charge separation in the RC (blue domain). The RC is re-reduced by electrons from the working electrode in a process mediated by cyt *c* (orange protein). The transfer of electrons to the counter electrode is mediated by 1 mM Q_0. For a representation of the RCLH1 components, see Fig. 1. The haem of cyt *c* is shown as slate-blue spheres with a brown Fe sphere. (b) Photocurrent transients in response to 180 seconds of illumination of the RCLH1Xr, RCLH1r or RC complexes adhered to nanostructured silver working electrodes. (c) A comparison of the steady state photocurrent densities for proteins adhered to nanostructured silver or planar gold working electrodes.

 This journal is © The Royal Society of Chemistry 2018

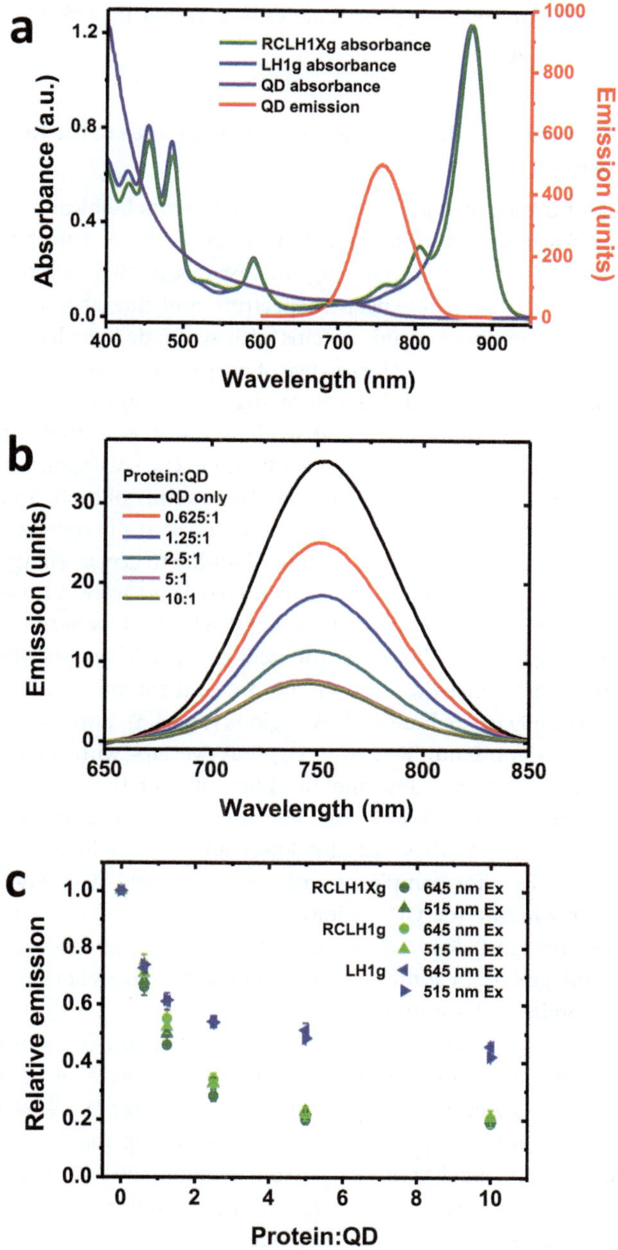

Fig. 4 Quenching of QD emission by the RCLH1Xg or RCLH1g complexes. (a) Absorbance spectra of the RCLH1Xg and LH1g complexes, and absorbance and emission spectra of 6.5 nm-diameter water soluble CdTe QDs, (b) quenching of emission of 50 nM QDs by 31.25–500 nM RCLH1Xg complexes, with excitation at 515 nm, (c) decay of relative QD emission as a function of protein : QD ratio, using data collected at two excitation wavelengths. The data points are means with standard deviation ($n = 3$).

measurements using planar gold as the working electrode (Fig. 3(c), right). The absolute photocurrent densities were lower with the planar gold electrodes than with nanostructured silver, which is in accordance with the lower surface area for protein adhesion and a lack of plasmonic enhancement provided by the nanostructured silver surface.[51]

3.4 Energy transfer from the synthetic antenna structures to the RCLH1X and RCLH1 complexes

One drawback of using natural proteins in a photoelectrochemical cell is that low currents are produced at excitation wavelengths where their LH pigments have weak absorbance. As a result, there is growing interest in the augmentation of native absorbance across the UV/visible/near-IR spectrum and the fabrication of hybrid photosystems that combine natural proteins with synthetic light harvesting materials. One option is to modify the absorbance of purple bacterial RCs with quantum dots (QDs), nanocrystalline materials whose absorbance and emission properties are dependent on their composition and size.[57] In addition to their broad visible region absorbance, these exhibit strong absorbance in the UV region which may be of interest with regard to photoprotection and the stability of biohybrid systems.

As RCLH1(X) complexes appear to be more effective photovoltaic materials than naked RCs, in this work we looked at their ability to accept energy from QDs and any effects of the removal of PufX. The QDs used were 6.5 nm-diameter nanostructures comprising cadmium telluride (CdTe) that were rendered water-soluble by coating with carboxyl terminated groups. These exhibit a broad absorbance that commences at around 800 nm and extends across the visible region, rising strongly in the blue and UV regions (Fig. 4(a), purple). They possess a symmetrical emission band (Fig. 4(a), red) that overlaps with the RC Q_y absorbance bands at 760 and 800 nm and the blue edge of the dominant LH1 Q_y absorbance band, centred at 873 nm. For this work, complexes with green carotenoids were used, as these exhibit lower absorbance in the 500–650 nm region than is the case for complexes with red carotenoids (see Fig. 2(b)). This creates a better window for QD excitation at 515 nm, in-between the strong carotenoid absorbance below 505 nm and the BChl Q_x band between 565 and 625 nm (Fig. 4(a), green), in addition to excitation at 645 nm where absorbance of the pigment-protein is at a minimum.

To look for energy transfer a fixed concentration of 50 nM QDs was mixed with up to a 10 : 1 molar ratio of RCLH1Xg or RCLH1g protein. A concentration-dependent drop in QD emission occurred that levelled off at ~80% quenching above a molar ratio of 5 : 1 (Fig. 4(b)). The QD emission maximum also blue-shifted by several nm, likely reflecting stronger quenching on the red side of the emission band due to spectral overlap with the protein being much larger on the red side of the band than on the blue side.

Repeat titrations showed that there was no dependence of the extent of quenching on the excitation wavelength (compare the circles and triangles in Fig. 4(c)). As the absorbance of the protein was around four-fold stronger at 515 nm than at 645 nm, this suggested that the drop in QD emission upon increasing the protein concentration was not due to shading by the pigments in the added protein (the protein and QD concentrations were kept low to avoid this, with the protein absorbance always below 0.05 absorbance units at 645 nm and 0.2 at 515 nm). In addition, the spectra of the most strongly quenched samples did not show any evidence of the appearance of inverse band structures that would indicate that the drop in QD emission was being affected by the

 This journal is © The Royal Society of Chemistry 2018

reabsorption of light emitted by the QDs by BChl (such artefacts, negative bands at 760 nm and to a lesser extent at 800 nm, were apparent in titrations carried out at fifty-fold higher QD and protein concentrations). Our conclusion, therefore, was that the observed quenching was most likely due to energy transfer from the QDs to the protein by Förster resonance energy transfer (FRET), which would imply binding of the protein to the QDs to form a conjugate between the two.

No significant differences were seen between the quenching brought about by the RCLH1g or RCLH1Xg complexes (Fig. 4(c)), thus indicating that energy transfer was not affected by the structural or optical differences between the two. A lesser extent of quenching (maximum of ~55%) was achieved using LH1g complexes purified from a green strain lacking RCs, PufX and LH2 (Fig. 4(c), blue), thus showing that the quenching of QD emission was not solely dependent on the presence of RCs.

3.5 Binding interactions between QDs and RCLH1(X) complexes

The nature of the conjugates formed between the 6.5 nm diameter CdTe QDs and the RCLH1(X) complexes is speculated upon in Fig. 5(a) and (b). The periplasmic and cytoplasmic surfaces of the RCLH1X complex do not possess a cluster of mostly basic residues that could provide an interaction site for a QD coated with carboxyl groups (Fig. 5(a)). Thus, if binding between RCLH1(X) complexes and QDs is electrostatic, as has been proposed for RCs,[57] it is not obvious that this would occur at a unique site on the protein surface. Regarding the stoichiometry of protein binding to the QDs, based on the available X-ray crystal structure the diameter of an RCLH1X complex is found to be around 12 nm, which is a little under twice that of a 6.5 nm diameter QD. As a result, as illustrated in Fig. 5(b), it seems unlikely that more than three RCLH1(X) complexes could fit around a single QD (bearing in mind that the true diameter of the former is further extended by a DDM micelle), or two if the binding site is on a recessed area of the protein surface. However, this estimate seems at odds with the finding that between five and ten proteins were needed to maximise the quenching of QD emission (Fig. 4(c)).

To explore this further, "pull-down" assays were carried out using three-step sucrose density gradients in which QDs were titrated against a fixed concentration of protein. The basis of this assay is that RCLH1(X) complexes band at a 0%/30% sucrose interface but QDs band at a lower 30%/60% sucrose interface, with any RCLH1(X) complexes bound to QDs also being pulled down to the lower interface. The upper pigmented band observed after centrifugation therefore comprises free RCLH1(X) complexes (Fig. 5(c), green arrow) whereas the lower band comprises RCLH1(X)/QD conjugates and free QDs (Fig. 5(c), red arrow).

The absorbance spectra of fractions from sets of gradients loaded with RCLH1Xg and RCLH1g complexes showed that, despite the presence of an equivalence or molar excess of QDs in three of the six gradients, the amount of protein binding to the QDs levelled off at around 40% of the total (Fig. 5(d)). The most likely interpretation of this is that the RCLH1(X) complexes exist in two populations, and one of which, the major part, is not able to bind to the QDs. Regardless of the cause, this observation clarified why between five and ten proteins were needed to saturate the quenching of QD emission (Fig. 4(c)), as it implied that the ratio of "competent" proteins was actually between two and four per QD, which is in better accordance with the expected nature of a protein–QD

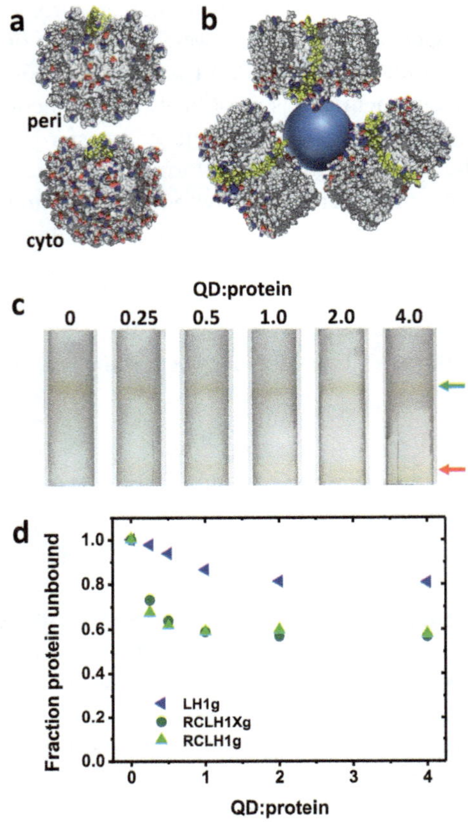

Fig. 5 Binding of RCLH1(X) complexes to 6.5 nm CdTe QDs. (a) Distribution of acidic and basic residues on the surface of the *Rba. sphaeroides* RCLH1X complex exposed at the periplasmic (top) and cytoplasmic (bottom) side of the membrane. Glu and Asp residues are shown in red and Lys and Arg residues are shown in blue. PufX is shown in yellow. (b) Schematic of three 12 nm diameter RCLH1X complexes (view parallel to the membrane) positioned around a 6.5 nm diameter QD (blue sphere). (c) Separation of the unbound RCLH1Xg protein (upper band, green arrow) from the RCLH1Xg protein bound to the QDs (lower band, red arrow) on three step sucrose density gradients. (d) Variation with QD : protein ratio of unbound protein as a fraction of the total.

conjugate given the sizes of the two components (Fig. 5(b)). The same analysis with the LH1g complexes revealed weaker binding, levelling off at around 20% of the total protein. By the same rationale this would equate to one or two LH1 per QD, which is lower than for the RCLH1(X)g complexes but consistent with the observed weaker quenching. LH1g titration confirmed the inference from the emission quenching experiments that the binding of RCLH1(X) complexes to QDs was not solely mediated by the RC component.

3.6 How does the removal of PufX affect protein stability?

The possible effects of removal of PufX on the stability of the RCLH1 complex were addressed by repeated absorbance scanning of protein solutions incubated at 60 °C (Fig. 6). For both the RCLH1Xr and RCLH1r complexes, heating produced

 This journal is © The Royal Society of Chemistry 2018

a decrease of the main absorbance band due to the unfolding of LH1 and release of BChl; the sample spectra in Fig. 6(a) show how the absorbance changes over the first six hours. There was also a drop in the smaller absorbance band at 805 nm, indicating the simultaneous unfolding of the RC, and the appearance of absorbance at 770 nm due to free BChl. The extent of this loss of the native spectrum was markedly greater for the native RCLH1Xr complex (Fig. 6(b), orange to cyan) than for the engineered RCLH1r complex (Fig. 6(a), red to blue). The time dependence of this diagnostic decrease in the LH1 Q_y band during incubation at 60 °C (Fig. 6(b)) revealed that the engineering of a closed ring of LH1 protein was associated with a significant increase in the stability of the pigment-protein. Although the absolute kinetics varied, this enhanced stability was seen over the range of 30 °C to 70 °C (data not shown), and for equivalent complexes with green carotenoids (data not shown).

As can be seen from Fig. 6(b) the decay of native LH1 absorbance was biphasic over the period monitored, with a satisfactory fit requiring a two-exponential function with time constants of <10 minutes and >100 minutes. The significance of this is not yet clear, but could again indicate two populations of RCLH1(X) complexes, in this case a minor population that is very sensitive to unfolding in response to heat stress and a major population that is more resistant, and whose resistance is markedly enhanced through the removal of PufX.

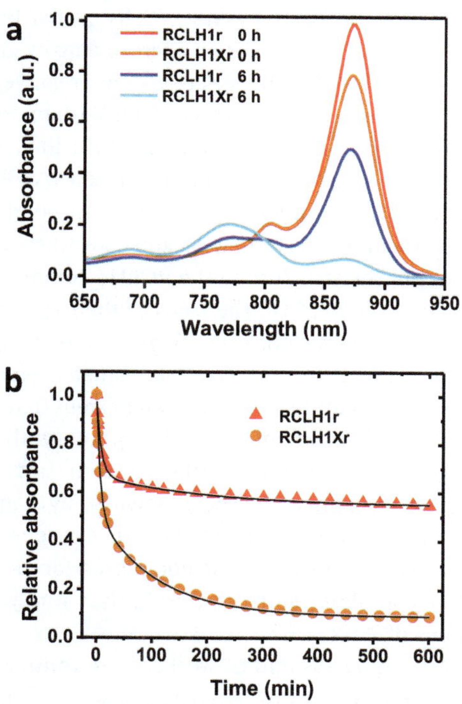

Fig. 6 Thermal stabilities of the RCLH1Xr and RCLH1r proteins. (a) Near-IR absorbance spectra of the pigment-protein solutions before and after heating at 60 °C for six hours, (b) decay of the amplitude of the Q_y absorbance band of the LH1 BChls over a 10 hour incubation at 60 °C. The data were fitted with two exponential terms and an offset to illustrate their biphasic nature.

3.7 Why are PufX-minus mutants non-photosynthetic?

The recent description of the location and gross conformation of PufX[22] supports the conclusions arrived at previously through research on PufX-deficient *Rba. sphaeroides* and *Rba. capsulatus* mutants that one role of this protein is to prevent closure of the ring of LH1 that surrounds the RC (see ref. 31 for a review). A commonly-held view is that, in doing this, PufX enables the exchange of reduced and oxidised quinone between the RC Q_B site and the intra-membrane quinone pool, and that the removal of PufX blocks this exchange, leading to a photosynthesis-minus phenotype (see ref. 38 for a recent discussion). However, two lines of evidence argue against this interpretation. First, a detailed examination of electron transfer involving RCLH1X and RCLH1 complexes in photosynthetic membranes by Comayras and co-workers showed that, although it is slowed by a factor of two, transfer of $Q_{10}H_2$ from the RC to the cyt bc_1 complex still takes place in the absence of PufX.[58] The slowing was attributed to a change in membrane organisation, with the removal of PufX actually delaying the release of $Q_{10}H_2$ from the Q_B site by only ~1 ms.[58] Second, anaerobic/photosynthetic growth can be restored to a PufX-deficient strain of *Rba. sphaeroides* by supplementation of the growth medium with dimethyl sulphoxide (DMSO), which acts as an oxidant for the quinone pool through the action of DMSO reductase.[59–61] To validate this second (often overlooked) observation using the strains that provided the proteins for our photocurrent measurements, in the present work we established that strains RCLH1r and RCLH1g would grow under photosynthetic conditions in the presence of 20 mM DMSO. Cells grown in this way yielded exclusively monomeric RCLH1 complexes on sucrose density gradients (Fig. 7(a)) that exhibited the elevated LH1 absorbance at ~873 nm relative to RC absorbance at 805 nm, which is diagnostic of the absence of PufX (data not shown).

If these two lines of evidence imply that quinone diffusion between the RC and intramembrane pool is not actually blocked in the absence of PufX, why do PufX-minus mutants not grow under standard photosynthetic conditions? From their study of PufX-deficient membranes, Comayras and co-workers concluded that the likely cause is a change to the structure of the RCLH1 complex that modifies the properties of the Q_B ubiquinone.[58] On the basis of their spectroscopic data they proposed that this change, as-yet not identified, stabilises Q_B^- and lowers the equilibrium constant for the second electron transfer reaction $Q_A^-Q_B^- \rightarrow Q_AQ_BH_2$. As a result, Q_A^- is accumulated to a greater extent than normal under the reducing conditions prevalent in an anaerobic photosynthetic culture, thus "closing" the RCs and shutting down cyclic electron transfer. Experimental evidence supporting this mechanism has been provided by Stahl and co-workers[62] who used ultrafast infrared spectroscopy to show that, under identical reducing conditions, charge separation in RCLH1Xr complexes forms $P^+Q_A^-$ but charge separation in RCLH1r complexes stops at $P^+H_A^-$, thus implying that Q_A is pre-reduced in the absence of PufX.

In their 2005 report, Comayras and co-workers determined the percentage of Q_A quinones that are reduced at any given redox state of the intramembrane quinone pool in the presence and absence of PufX (see Fig. 9 in ref. 58). Data from that plot are incorporated into the schemes in Fig. 7(b) and (c), which illustrate the mechanism Comayras and co-workers proposed to account for the consequences of removal of PufX. The schemes show the locked-in Q_A and mobile Q_B

 This journal is © The Royal Society of Chemistry 2018

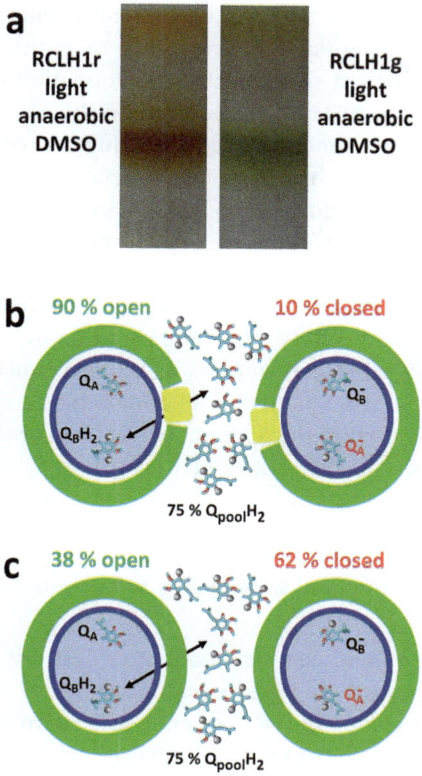

Fig. 7 Proteins from cells grown in DMSO and the cause of the phenotype of PufX-minus strains. (a) Monomeric RCLH1 complexes from cells grown under light/anaerobic conditions in the presence of DMSO. (b) In native RCLH1X complexes, a 75% reduction of the intramembrane quinone pool results in 10% of the RCs being closed due to Q_A being reduced (the electrons are represented as grey spheres). (c) In engineered RCLH1 complexes, 62% of the RCs are closed under the same conditions, but quinone exchange with the intramembrane pool is not prevented.

quinone in the RC (blue ellipse), and the intramembrane pool with which the quinone at Q_B can exchange by passing through the surrounding LH1/PufX ring (green/yellow). When 75% of the quinones in the intramembrane pool are reduced (electrons represented as grey spheres) only 10% of the RCs in the RCLH1X complexes are closed (Fig. 7(b)) because the reaction $Q_A^-Q_B^- \rightarrow Q_AQ_BH_2$ favours the product. In contrast, due to the destabilisation of Q_B^- and a shift in the equilibrium for this reaction, 62% of the RCs are closed in the RCLH1(X) complexes under the same conditions (Fig. 7(c)). This mechanism provides an explanation for the inability of PufX-deficient strains to grow under standard photosynthetic conditions despite the fact that quinone diffusion between the Q_B site and the (largely reduced) intramembrane pool is not prevented by the removal of PufX. The effect of DMSO can be explained by redox poising,[63,64] with the oxidant draining electrons from the quinone pool and hence the Q_B and Q_A sites. This lowers the percentage of RCs with a reduced Q_A and so opens them for charge separation. It is known that such redox poising by an auxiliary oxidant is required for anaerobic/photoheterotrophic growth of purple bacteria on strongly reducing carbon sources such as butyrate,[63] and this phenomenon has also been demonstrated *in vitro* using flash spectroscopy.[64] In addition, ultrafast infrared spectroscopy has been used to show that the formation of $P^+Q_A^-$ by charge separation is restored in RCLH1 complexes upon the addition of DMSO, thus demonstrating its role in opening RCs by alleviating pre-reduction of Q_A.[62]

This explanation for the inability of PufX-deficient strains to grow under standard photosynthetic conditions also explains why the RCLH1Xr and RCLH1r complexes compared in this report were equally able to support a photocurrent. The dysfunction associated with the absence of PufX should only manifest under strongly reducing conditions, and so has no impact if such conditions are avoided in a photoelectrochemical cell. We therefore conclude that the lower photocurrents published to date for RCLH1 complexes in comparison to those for RCLH1X complexes are due to the particular electrode materials, electrolytes and measuring conditions used in the different studies, rather than being due to an inherent problem with the ability of RCLH1 complexes to support a continuous flux of electrons.

3.8 Enhancement of structural stability in a sub-population of RCLH1 complexes

Given that mutated protein complexes are often less structurally-stable than their native counterparts, an unexpected outcome of our comparison of RCLH1X and RCLH1 complexes was the enhanced thermal stability displayed by the latter. The origins of this are as yet unclear, but presumably relate to the assembly of a closed ring of LH1 pigment-protein around the RC rather than the open structure dictated by PufX. Therefore, in addition to a higher light harvesting capacity, another potential benefit of using PufX-deficient RCLH1 complexes for research on bio-photoelectrochemical cells could be the enhanced stability.

Based on the kinetics of LH1 unfolding shown in Fig. 6(b) it seems likely that there is structural heterogeneity within the populations of the purified RCLH1 and RCLH1X complexes. In both cases, loss of the native absorbance spectrum due to thermal unfolding was biphasic, markedly so in the case of the RCLH1 complex, with a sub-population that is apparently unable to tolerate high temperatures for more than a few minutes and a sub-population that is substantially more tolerant. This difference was observed across a range of temperatures and with either red or green carotenoids present. An investigation of the quenching of QD emission by the RCLH1X and RCLH1 complexes also demonstrated evidence of structural heterogeneity, with only around 40% of either the RCLH1X or RCLH1 complexes being able to bind to an excess of QDs. The basis of this structural heterogeneity is the subject of ongoing investigations, alongside an exploration of methods for physically separating the apparently less stable sub-population of RCLH1 complexes from the sub-population that seems to exhibit a strongly enhanced stability compared to that of its RCLH1X counterpart. An important consideration in the future development of biohybrid photoelectrochemical cells is the stability of their output, and it seems feasible that substantial improvements in stability could be achieved through the sort of protein engineering described here or through the use of protein/material combinations that enhance the robustness of the protein in a device setting.[51,65]

 This journal is © The Royal Society of Chemistry 2018

4. Conclusions

The purpose of this work was to address concerns that PufX-deficient RCLH1 complexes from photosynthetically-incompetent bacteria may not be suitable as photovoltaic materials for incorporation into biohybrid photoelectrochemical cells because they exhibit a dysfunction that limits the photocurrent densities achievable. The finding that RCLH1 and RCLH1X proteins are equally effective in supporting photocurrents in otherwise identical photoelectrochemical cells provides reassurance that this is not the case. Alongside increasing the light harvesting capacity of the RCLH1 complex, we also indicate that the removal of PufX and enlargement of the LH1 pigment-protein ring confers enhanced structural stability upon the RCLH1 complex that may be beneficial for the future use of these proteins in technological applications.

Conflicts of interest

There are no conflicts of interest to declare.

Acknowledgements

DJKS and MRJ acknowledge the support from the Biotechnology and Biological Sciences Research Council of the UK (project BB/I022570/1). JL and MRJ acknowledge the funding from the Engineering and Physical Sciences Research Council of the UK and the BBSRC Synthetic Biology Centre for Doctoral Training (EP/L016494/1). LIC acknowledges the studentship funding from the BBSRC. RNF acknowledges the support from the Dutch science foundation NWO for a vidi grant. MRJ, RNF, DJKS and VMF acknowledge the support from the EU COST Action TD1102 – Photosynthetic proteins for technological applications: biosensors and biochips (PHOTOTECH).

References

1 D. A. LaVan and J. N. Cha, *Proc. Natl. Acad. Sci. U. S. A.*, 2006, **103**, 5251–5255.
2 A. Badura, T. Kothe, W. Schuhmann and M. Rogner, *Energy Environ. Sci.*, 2011, **4**, 3263–3274.
3 A. A. Boghossian, M.-H. Ham, J. H. Choi and M. S. Strano, *Energy Environ. Sci.*, 2011, **4**, 3834–3843.
4 F. Wang, X. Liu and I. Willner, *Adv. Mater.*, 2013, **25**, 349–377.
5 Y. Kim, S. A. Shin, J. Lee, K. D. Yang and K. T. Nam, *Nanotechnology*, 2014, **25**, 342001.
6 K. Nguyen and B. D. Bruce, *Biochim. Biophys. Acta, Bioenerg.*, 2014, **1837**, 1553–1566.
7 O. Yehezkeli, R. Tel-Vered, D. Michaeli, I. Willner and R. Nechushtai, *Photosynth. Res.*, 2014, **120**, 71–85.
8 A. Operamolla, R. Ragni, F. Milano, R. R. Tangorra, A. Antonucci, A. Agostiano, M. Trotta and G. Farinola, *J. Mater. Chem. C*, 2015, **3**, 6471–6478.
9 S. K. Ravi and S. C. Tan, *Energy Environ. Sci.*, 2015, **8**, 2551–2573.
10 R. van Grondelle, J. P. Dekker, T. Gillbro and V. Sundstrom, *Biochim. Biophys. Acta, Bioenerg.*, 1994, **1187**, 1–65.

11 R. E. Blankenship, *Molecular Mechanisms of Photosynthesis*, Blackwell Science, 2002.
12 Y. C. Cheng and G. R. Fleming, *Annu. Rev. Phys. Chem.*, 2009, **60**, 241–262.
13 G. D. Scholes, G. R. Fleming, A. Olaya-Castro and R. van Grondelle, *Nat. Chem.*, 2011, **3**, 763–774.
14 T. Mirkovic, E. E. Ostroumov, J. M. Anna, R. van Grondelle, Govindjee and G. D. Scholes, *Chem. Rev.*, 2017, **117**, 249–293.
15 E. Romero, V. I. Novoderezhkin and R. van Grondelle, *Nature*, 2017, **543**, 355–365.
16 G. Feher, J. P. Allen, M. Okamura and D. Rees, *Nature*, 1989, **339**, 111–116.
17 A. J. Hoff and J. Deisenhofer, *Phys. Rep.*, 1997, **287**, 2–247.
18 W. Zinth and J. Wachtveitl, *ChemPhysChem*, 2005, **6**, 871–880.
19 M. R. Jones, *Biochem. Soc. Trans.*, 2009, **37**, 400–407.
20 P. Heathcote and M. R. Jones, *Comprehensive Biophysics, volume 8, Bioenergetics*, Academic Press, 2012, pp. 115–144.
21 S. Bahatyrova, R. N. Frese, C. A. Siebert, J. D. Olsen, K. O. van der Werf, R. van Grondelle, R. A. Niederman, P. A. Bullough, C. Otto and C. N. Hunter, *Nature*, 2004, **430**, 1058–1062.
22 P. Qian, M. Z. Papiz, P. J. Jackson, A. A. Brindley, I. W. Ng, J. D. Olsen, M. J. Dickman, P. A. Bullough and C. N. Hunter, *Biochemistry*, 2013, **52**, 7575–7585.
23 S. Niwa, L. J. Yu, K. Takeda, Y. Hirano, T. Kawakami, Z. Y. Wang-Otomo and K. Miki, *Nature*, 2014, **508**, 228–232.
24 R. J. Cogdell, A. Gall and J. Kohler, *Q. Rev. Biophys.*, 2006, **39**, 227–324.
25 M. R. Jones, G. J. S. Fowler, L. C. D. Gibson, G. G. Grief, J. D. Olsen, W. Crielaard and C. N. Hunter, *Mol. Microbiol.*, 1992, **6**, 1173–1184.
26 J. B. Crounse, R. P. Feldman and R. K. Clayton, *Nature*, 1963, **198**, 1227–1228.
27 P. A. Scolnik, M. A. Walker and B. L. Marrs, *J. Biol. Chem.*, 1980, **255**, 2427–2432.
28 S. K. Ravi, Y. Zhimeng, D. J. K. Swainsbury, J. Ouyang, M. R. Jones and S. C. Tan, *Adv. Energy Mater.*, 2017, **7**, 1601821.
29 G. Garcia-Asua, R. J. Cogdell and C. N. Hunter, *Mol. Microbiol.*, 2002, **44**, 233–244.
30 S. C. Chi, D. J. Mothersole, P. Dilbeck, D. M. Niedzwiedzki, H. Zhang, P. Qian, C. Vasilev, K. J. Grayson, P. J. Jackson, E. C. Martin, Y. Li, D. Holten and C. N. Hunter, *Biochim. Biophys. Acta, Bioenerg.*, 2015, **1847**, 189–201.
31 K. Holden-Dye, L. I. Crouch and M. R. Jones, *Biochim. Biophys. Acta, Bioenerg.*, 2008, **1777**, 613–630.
32 C. Jungas, J. L. Ranck, J. L. Rigaud, P. Joliot and A. Verméglio, *EMBO J.*, 1999, **18**, 534–542.
33 S. Scheuring, F. Francia, J. Busselez, B. A. Melandri, J. L. Rigaud and D. Lévy, *J. Biol. Chem.*, 2004, **279**, 3620–3626.
34 P. Qian, C. N. Hunter and P. A. Bullough, *J. Mol. Biol.*, 2005, **349**, 948–960.
35 F. Francia, J. Wang, G. Venturoli, B. A. Melandri, W. P. Barz and D. Oesterhelt, *Biochemistry*, 1999, **38**, 6834–6845.
36 T. Walz, S. J. Jamieson, C. M. Bowers, P. A. Bullough and C. N. Hunter, *J. Mol. Biol.*, 1998, **282**, 833–845.
37 C. A. Siebert, P. Qian, D. Fotiadis, A. Engel, C. N. Hunter and P. A. Bullough, *EMBO J.*, 2004, **23**, 690–700.
38 J. D. Olsen, E. C. Martin and C. N. Hunter, *FEBS Lett.*, 2017, **591**, 573–580.

 This journal is © The Royal Society of Chemistry 2018

39 M. Y. Okamura, M. L. Paddock, M. S. Graige and G. Feher, *Biochim. Biophys. Acta*, 2000, **1458**, 148–163.
40 C. A. Wraight, *Front. Biosci.*, 2004, **9**, 309–337.
41 V. M. Friebe, D. Millo, D. J. K. Swainsbury, M. R. Jones and R. N. Frese, *ACS Appl. Mater. Interfaces*, 2017, **9**, 23379–23388.
42 V. M. Friebe, D. J. K. Swainsbury, P. K. Fyfe, W. van der Heijden, M. R. Jones and R. N. Frese, *Biochim. Biophys. Acta, Bioenerg.*, 2016, **1857**, 1925–1934.
43 G. Klug and S. N. Cohen, *J. Bacteriol.*, 1988, **170**, 5814–5821.
44 J. W. Farchaus, H. Gruenberg and D. Oesterhelt, *J. Bacteriol.*, 1990, **172**, 977–985.
45 T. G. Lilburn, C. E. Haith, R. C. Prince and J. T. Beatty, *Biochim. Biophys. Acta*, 1992, **1100**, 160–170.
46 J. W. Farchaus, W. P. Barz, H. Gruenberg and D. Oesterhelt, *EMBO J.*, 1992, **11**, 2779–2788.
47 T. K. Fulcher, J. T. Beatty and M. R. Jones, *J. Bacteriol.*, 1998, **180**, 642–646.
48 S. C. Tan, L. I. Crouch, M. R. Jones and M. E. Welland, *Angew. Chem., Int. Ed.*, 2012, **51**, 6667–6671.
49 S. C. Tan, L. I. Crouch, S. Mahajan, M. R. Jones and M. E. Welland, *ACS Nano*, 2012, **6**, 9103–9109.
50 S. C. Tan, F. Yan, L. I. Crouch, J. Robertson, M. R. Jones and M. E. Welland, *Adv. Funct. Mater.*, 2013, **23**, 5556–5563.
51 V. M. Friebe, J. D. Delgado, D. J. K. Swainsbury, J. M. Gruber, A. Chanaewa, R. van Grondelle, E. L. von Hauff, D. Millo, M. R. Jones and R. N. Frese, *Adv. Funct. Mater.*, 2016, **26**, 285–292.
52 M. R. Jones, M. Heer-Dawson, T. A. Mattioli, C. N. Hunter and B. Robert, *FEBS Lett.*, 1994, **339**, 18–24.
53 D. J. Swainsbury, V. M. Friebe, R. N. Frese and M. R. Jones, *Biosens. Bioelectron.*, 2014, **58**, 172–178.
54 G. W. Bushnell, G. V. Louie and G. D. Brayer, *J. Mol. Biol.*, 1990, **214**, 585–595.
55 M. van der Rest and G. Gingras, *J. Biol. Chem.*, 1974, **249**, 6446–6453.
56 M. Dezi, F. Francia, A. Mallardi, G. Colafemmina, G. Palazzo and G. Venturoli, *Biochim. Biophys. Acta*, 2007, **1767**, 1041–1056.
57 I. Nabiev, A. Rakovich, A. Sukhanova, E. Lukashev, V. Zagidullin, V. Pachenko, Y. P. Rakovich, J. F. Donegan, A. B. Rubin and A. O. Govorov, *Angew. Chem., Int. Ed.*, 2010, **49**, 7217–7221.
58 R. Comayras, C. Jungas and J. Lavergne, *J. Biol. Chem.*, 2005, **280**, 11214–11223.
59 W. P. Barz, F. Francia, G. Venturoli, B. A. Melandri, A. Vermeglio and D. Oesterhelt, *Biochemistry*, 1995, **34**, 15235–15247.
60 W. P. Barz, A. Verméglio, F. Francia, G. Venturoli, B. A. Melandri and D. Oesterhelt, *Biochemistry*, 1995, **34**, 15248–15258.
61 A. Verméglio and P. Joliot, *Biochim. Biophys. Acta, Bioenerg.*, 2002, **1555**, 60–64.
62 A. D. Stahl, L. I. Crouch, M. R. Jones, I. van Stokkum, R. van Grondelle and M. L. Groot, *J. Phys. Chem. B*, 2012, **116**, 434–444.
63 D. J. Richardson, G. F. King, D. J. Kelly, A. G. McEwan, S. J. Ferguson and J. B. Jackson, *Arch. Microbiol.*, 1988, **150**, 131–137.
64 M. R. Jones, D. J. Richardson, A. G. McEwan, S. J. Ferguson and J. B. Jackson, *Biochim. Biophys. Acta*, 1990, **1017**, 209–216.
65 D. J. K. Swainsbury, S. Scheidelaar, R. van Grondelle, J. A. Killian and M. R. Jones, *Angew. Chem., Int. Ed.*, 2014, **53**, 11803–11807.

Faraday Discussions

PAPER

Mechanistic insights into photoinduced damage of DNA and RNA nucleobases in the gas phase and in bulk solution†

Pratip Chakraborty,[a] Tolga N. V. Karsili, *[a] Barbara Marchetti[b] and Spiridoula Matsika *[a]

Received 21st August 2017, Accepted 18th September 2017
DOI: 10.1039/c7fd00188f

DNA/RNA photohydrates represent a class of well-known biomolecular lesions formed by the absorption of near- to mid-UV light. They are formed *via* a photoinduced nucleophilic hydrolysis reaction in which water is split (*via* nucleobase sensitisation) into H + OH radicals. These nascent radicals can then add across C5=C6, saturating the preexisting double bond. If unrepaired, such lesions can lead to mutagenic carcinogenesis, which is responsible for several forms of cancer. Using high-level electronic structure theory (CASPT2), we map the key excited-state reaction paths associated with the reactivity of DNA (guanine and thymine) and RNA (uracil) nucleobases with water. At the outset, we consider the intrinsic reactivity in the isolated gas phase – in which the water (cluster) + chromophore complex is free from environmental perturbations. We then extrapolate the thymine nucleobase to the bulk DNA environment in aqueous solution in order to ascertain the relative importance of hydrate formation in a more complex biological environment. In this latter study we use high-level mixed quantum/classical (QM/MM: CASPT2/AMBER) methods.

1 Introduction

The photophysics and photochemistry of DNA/RNA nucleobases have been the focus of extensive experimental and theoretical studies in the past two decades.[1–6] Such nucleobases are π-conjugated systems that absorb light in the near-UV region. Upon UV-irradiation, the nascent excited electronic state can create harmful photoproducts that may result in photocarcinogenesis. Such processes are usually minimised by nature, ensuring nucleobase photostability. In the present context, the well-known photostability of canonical nucleobases is defined by their inherently short-lived excited states, characterised by low

[a]Department of Chemistry, Temple University, Philadelphia, PA 19122, USA. E-mail: tolga.karsili@temple.edu; smatsika@temple.edu; Tel: +1 215 204 2102

[b]Department of Chemistry, University of Pennsylvania, Philadelphia, PA 19104, USA

† Electronic supplementary information (ESI) available. See DOI: 10.1039/c7fd00188f

fluorescence quantum yields.[1,2,6] Such short excited-state lifetimes are manifestations of non-radiative relaxation *via* internal conversion (IC) to the ground electronic state. The excess vibrational energy is thereby dissipated as heat – which in the cellular environment is transferred to the bulk. Such ultrafast processes are now well-understood to involve conical intersections (CIs) between excited/excited and excited/ground states.[3,7–9]

In addition to non-radiative decay, CIs are now recognised as mediators of many of the other ultrafast processes observed in molecular photochemistry. Photophysical investigations have found a variety of mechanisms through which chemical evolution has ensured photostability. These include (but are not limited to) out-of-plane ring deformations about C=C and C=N bonds, bond stretches along $\pi\sigma^*$ states and electron-driven H-atom transfer in nucleotides and base-pairs. In many of these cases, the photoexcited nuclear wavepacket on the excited state surface evolves towards a CI characterised by one of the aforementioned mechanisms, to revert to its ground electronic state in an ultrafast manner. The excess energy is usually partitioned as heat, which, in the bulk, is dissipated to the surrounding environment.[3,9,10]

Notwithstanding, prolonged exposure to harsh mid- to deep-UV irradiation can lead to deleterious photoproducts – which are commonly referred to as photo-lesions. The associated quantum yields of such photo-lesions are typically <1%,[11,12] but they can nonetheless have profound consequences, such as cancers, if formed. The most common form of photo-lesion is the cyclobutane pyrimidine dimer (CPD). CPDs form *via* a photoinduced [2 + 2] cycloaddition between the C5=C6 bonds of adjacent pyrimidine bases. The subsequent adducts can alter the structure of bulk DNA/RNA and are known to interfere with the functions of polymerases. Fortunately such adducts are usually repaired by nucleotide excision repair, but rare cases can cause mutations which lead to photocarcinogenesis.[5] In addition to UV-light, analogous lesions can be formed by the oxidative addition of radicals to DNA/RNA. The well-known lesion 8-oxoguanine (8-OG) is formed *via* the oxidation of guanine, which contains a lower redox potential than its canonical counterparts, enabling it to form base-pairs with adenine that can lead to harmful mutations. Recent experiments by Ravanat and co-workers have evidenced the photoinduced formation of 8-OG following near-UV excitation of guanosine in aqueous solution.[13] The authors speculate that photoexcitation of the guanine nucleobase leads to photoionisation, and the nascent cation reacts with the surrounding water solution to form 8-OG.

Another class of lesions involves the formation of photohydrate adducts of the aforementioned DNA/RNA nucleobases in an aqueous environment.[14–18] UV-induced photohydrate adducts form when a water molecule adds across the C=C/C=N double bond of the pyrimidine/purine ring in a so-called nucleophilic hydrolysis reaction. Remsen *et al.* have shown that uridine photohydrates are formed following UV-irradiation of single-stranded R17-RNA.[16] Synonymous chemistry occurs in aqueous solutions of uracil and thymine. In both cases, the nascent hydrate adducts are stable at 37 °C in neutral aqueous solution.[17] In comparison, cytosine hydrate is highly unstable and reverts to cytosine in neutral aqueous solution.[18] The formation of photohydrates has been speculated to be attributable to ultrafast IC and the subsequent reactivity of the vibrationally excited ground state nucleobase, but this assumption has not been verified, and our work will show that it is not necessary.[19]

 This journal is © The Royal Society of Chemistry 2018

In the cellular environment, the formation of such hydrate adducts represents nucleic acid damage.[18] If unrepaired, nascent lesions can lead to deleterious mutations and interfere with transcription. Thymine hydrate also partially inhibits the activity of DNA polymerase I, which could lead to lethal mutations.[20] Though the lesions are clearly important, a detailed mechanistic study of the formation of such harmful photohydrates is still lacking. In this work, we present high-level single- and multi-reference electronic structure calculations that detail a given nucleobase + water photoreaction in order to ascertain the extent to which such lesion processes compete with the intrinsic nonradiative decay of a given nucleobase. We start by outlining the reaction of guanine in a small model cluster of water molecules in an attempt to understand the first-order effects of forming 8-oxoguanine *via* UV-irradiation. We then extend such water reactivity to uracil and thymine in order to characterise the experimentally observed uracil and thymine hydrates. Finally, we extend the thymine + water photoreaction to the bulk DNA environment in order to ascertain the extent to which UV-induced photohydrate lesions persist in the bulk.

2 Results and discussion

2.1 The gas-phase guanine + water photoreaction

This section will highlight the details of the mechanism associated with the guanine (henceforth G; structure shown in Fig. 1(a)) + water photoreaction to devise a plausible path for forming 8-oxoguanine (henceforth 8-OG; structure shown in Fig. 1(c)). In so doing, a full quantum chemical description of solute–solvent interactions is required. To this end, a reduced solute–solvent model, comprising the G chromophore complexed with five proximal water molecules (Fig. 1(a)), was constructed in order to reduce the computational expense whilst ensuring an adequate description of the dominant solute interactions with the proximal solvent water molecules. Our solute–solvent model cluster size and orientation was motivated by the elementary first step in forming 8-OG, which is widely accepted to proceed *via* a molecular rearrangement of the guanine photohydrate (8-hydroxy-7*H*-guanine; henceforth 8-HG). The optimised geometry of 8-HG is displayed in Fig. 1(b), which shows an OH moiety bound to the C8

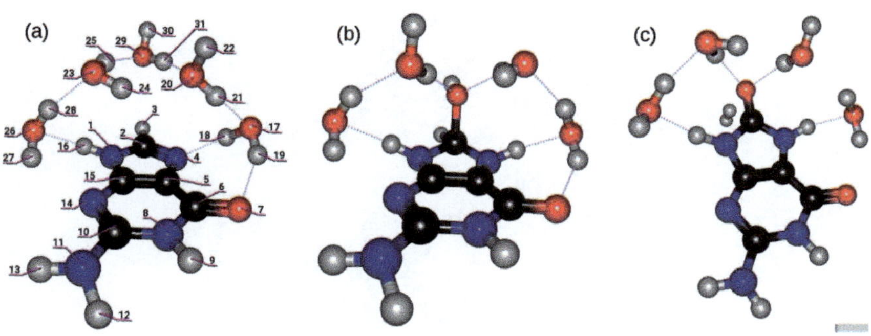

Fig. 1 Molecular structures associated with (a) the parent guanine + water complex, (b) the photohydrate adduct intermediate and (c) the 8-oxoguanine product. The numbers in (a) represent the assigned atomic numbering that we use when describing the atomic positions at which the reactivity is localised.

Table 1 Vertical excitation energies (with oscillator strengths in parentheses) and state characters associated with the lowest three excited states of the G + water complex

Transition	Excitation energy/eV	Character
S_1–S_0	5.00 (0.1406)	ππ*
S_2–S_0	5.52 (0.0030)	nπ*
S_3–S_0	5.65 (0.3232)	ππ*

position at an instinctive perpendicular angle with respect to the plane of the G molecule. It is therefore intuitive that at least one of the cluster water molecules be bound dispersively above the plane defined by G (G–1H_2O). Through previous studies on analogous systems,[21–24] five solvent water molecules were deemed sufficient to qualitatively describe all the significant orbital interactions – ensuring that all the solvent molecules are optimally aligned for an accurate description of the ensuing G + H_2O reaction, whilst maintaining a reasonable computational expense. The forthcoming description is therefore steered towards the excited-state chemistry of G in a cluster of five proximal solvent water molecules (henceforth simply G–H_2O), the ground-state minimum energy geometry of which is displayed in Fig. 1(a). The assigned atomic numberings will be used

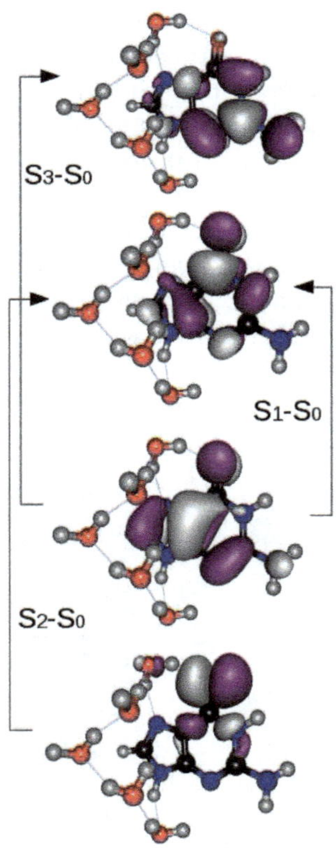

Fig. 2 Orbitals and orbital promotions associated with the electronic transitions to the lowest three excited states of the G + water complex.

henceforth when describing the relevant atom motions. Fig. 1 shows that the optimised geometry of G–H_2O is composed of two terminal water molecules, each of which contains hydrogen bonding to either N4 or N1–H. The HO17H–N4 and H_2O26–H–N1 hydrogen bond distances are returned as 2.2 Å and 2.5 Å, respectively, and are typical of such chromophores – reinforcing our cluster size choice. In the present optimised configuration a water-wire extends from H_2O26 to H_2O17 and positions H_2O29 in an optimal C2 position, so as to form a dispersive interaction with the π-system of G.

Table 1 lists the Vertical Excitation Energies (VEEs) and oscillator strengths accompanying the first three excited states of the G–H_2O complex. The orbital promotions associated with the preparation of the various excited states are given in Fig. 2, which shows that the lowest three vertically excited singlet states involve orbitals localised exclusively on the G-chromophore. States arising from such chromophore-localised orbitals are commonly (and will henceforth be) referred to as locally excited (LE) states. The S_1 and S_3 states are both of ππ* character, the participating orbitals of which contain substantially large spatial overlap, manifest in a large oscillator strength. In contrast, the S_2 state is best characterised as an $^1n\pi^*$ state; its dominant orbital transition involves a $\pi^* \leftarrow n(O\ 2p_y)$ electron promotion. The characteristically weak oscillator strength of the S_2 ($^1n\pi^*$) state is a manifestation of the poor spatial overlap between the participating n and π* orbitals. The returned near-UV electronic configurations and energetic orderings listed in Table 1 are analogous to those of isolated G.[3]

The ensuing reactivity for forming 8-OG *via* the G + H_2O photoreaction is presented in a two-step sequential process (summarized in Scheme 1), in which the initial step leads to the formation of the 8-HG intermediate, followed by H_2 elimination from 8-HG to yield 8-OG. The potential energy (PE) profiles associated with each step are presented in Fig. 3(a) and (b), respectively. The first step is predicted to involve a two-step sub-path in which a solvent-to-chromophore electron-driven proton-transfer (EDPT – middle panel in Fig. 3(a)) initiates the photoreaction from the O17–H18 donor to the N4 acceptor. We note that as well as N4, there are other hydrogen-bond acceptors (*e.g.* N and CO) and donors (*e.g.* NH and NH_2) localised around the G chromophore. Our focus on EDPT at N4 is

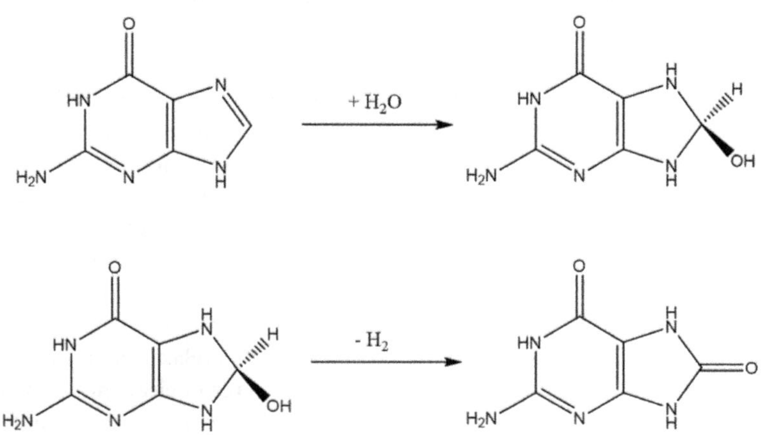

Scheme 1

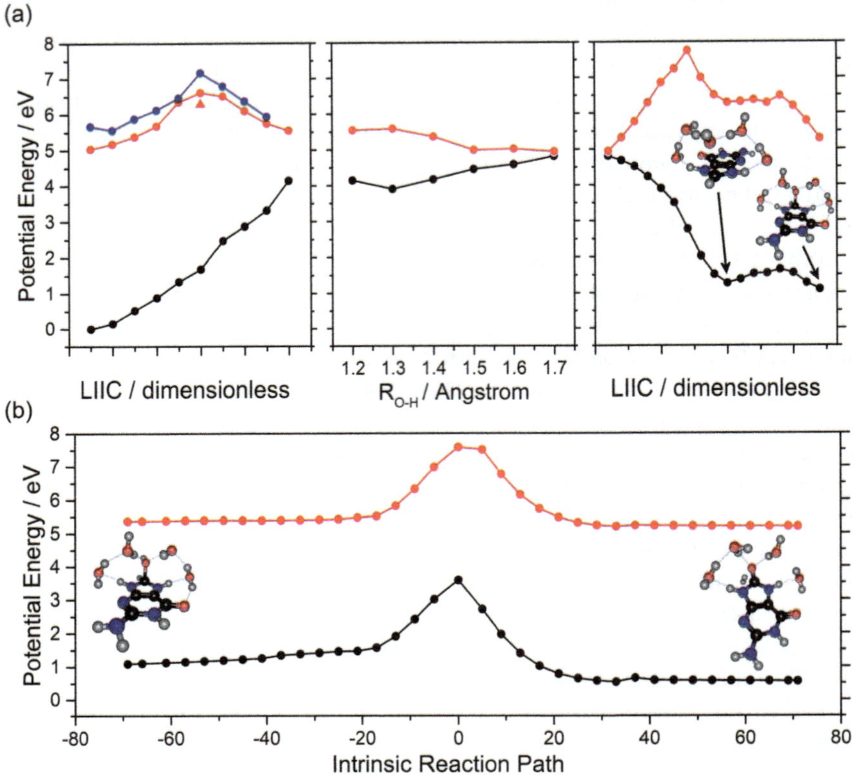

Fig. 3 CASPT2 potential energy profiles of the lowest three singlet states associated with the G + water photoreaction. The reader is directed to the main text for a fuller description of the details of the various profiles. The S_0, S_1 and S_2 states are the black, red and blue curves, respectively.

however motivated by the 8-OG end-product, which shows a hydrogenated N4 position. The inherent EDPT process is mediated by water–chromophore hydrogen-bonding and proceeds *via* an initial excitation to either the S_1 or S_2 LE state and is often referred to as proton-coupled electron transfer (PCET). Such LE states are then likely to couple to near-lying charge-transfer (CT) states – so characterised by electronic configurations in which an electron is promoted from a $2p_y$ oxygen lone pair local to a proximal solvent water to a π^* orbital of the G chromophore. The S_1 relaxed PE profile along $R_{O17-H18}$ (middle panel in Fig. 3(a)) captures one such electronic configuration. A linear interpolation in the internal coordinates (LIIC – left-most panel in Fig. 3(a)) shows the way in which the S_1(LE) configuration connects adiabatically to the S_1(CT) configuration. The returned LIIC shows a barrier height of 1.5 eV from the Franck–Condon (FC) geometry, which necessarily represents an upper limit. The 'true' barrier height connecting the S_1(LE) to S_1(CT) was optimised (red triangle on the left-most panel of Fig. 3(a)) and shown to reduce the barrier height by 0.3 eV. Such a relaxation in the barrier height, coupled with the recognition that the dominant nuclear motion is a H-atom migration, may lead to tunnelling beneath this barrier. Once populated, Fig. 3(a) reveals that the S_1(CT) state is reactive with respect to O17–H18 bond elongation, representing proton-transfer from the O17–H18 hydrogen-bond donor to the N4 hydrogen-bond acceptor. However, the equivalent coordinate on S_0 is unfavourable and shows a gradual increase in PE as a function of O17–H18 bond elongation. This favourable reaction on the S_1(CT) can be understood by recognising that the initial solvent-to-chromophore electron transfer creates a charge-separation between the donor water molecule and the acceptor G chromophore (*i.e.* creating $G^- + H_2O^+$). Migration of H^+ from H_2O^+ to G^- leads to a neutralisation of this charge separation, manifest in a stabilisation of S_1(CT) as a function of $R_{O17-H18}$. In contrast, the closed-shell stable configuration of S_0 leads to instability of the system upon O17–H18 bond elongation, manifest in a rise in PE as a function of $R_{O17-H18}$. This subsequent rise and decline of PE in the S_0 and S_1(CT) states, respectively, inevitably leads to an energy crossing at $R_{O17-H18} \sim 1.7$ Å. Such crossings between states of a common spin-symmetry comprise CIs and are therefore likely to mediate nonadiabatic transitions of the excited-state population back to the S_0 PE surface *via* IC. If IC prevails at this long-range $R_{O17-H18}$ crossing, the system retains an excess energy of ~4 eV, which is nominally above many reaction barriers to various isomeric products in the ground state. We focus on two such reactions that may lead to 8-OG, as outlined in reactions (1) and (2) in Scheme 1. The first is the initial formation of 8-HG (reaction (1) in Scheme 1), which is predicted to contain a barrierless PE profile upon linear interpolation from the crossing geometry at $R_{O17-H18} \sim 1.7$ Å to the hydrate product, as displayed in the right-hand panel of Fig. 3(a). The subsequent formation of 8-OG from 8-HG requires an elimination of H_2, as outlined in reaction (2) in Scheme 1. Fig. 3(b) shows the minimum-energy PE profile connecting 8-HG to 8-OG through an optimised transition state. It is noteworthy that this transition state is energetically below the crossing at $R_{O17-H18} \sim 1.7$ Å, as well as in the near-UV excitation wavelength range.

In the present contribution, we show that it is indeed possible to form 8-OG *via* a one-photon (near-UV excitation) induced neutral photochemical path without the necessity of preparing a cationic form of G.

2.2 The gas-phase uracil/thymine + water photoreaction

In this section, we extend the above chromophore–water reactivity to that of uracil (U) and thymine (T) in an attempt to understand the mechanisms underlying the UV-induced formation of U- and T-hydrate products measured experimentally. Our initial structures comprise U and T chromophores complexed with a single water (H_2O) molecule, giving rise to dispersively-bound uracil–water (U–H_2O) and thymine–water (T–H_2O) complexes. Our choice of a single water molecule is motivated by the need to describe the first-order water–chromophore reactivity without the bulk cluster effect. More specifically, the hydration reaction occurs *via* a nucleophilic hydrolysis reaction of a single water molecule, in which the nascent hydroxyl (–OH) and hydrogen (H) radical intermediates add across the C5=C6 double bond of the pyrimidine ring to form the 5-hydroxy-6*H*-uracil (henceforth 5-HU), 6-hydroxy-5*H*-uracil (henceforth 6-HU), 5-hydroxy-6*H*-thymine (henceforth 5-HT) or 6-hydroxy-5*H*-thymine (henceforth 6-HT) adduct, as shown in Fig. 4 (structures A, B, C and D, respectively). In order to ascertain the most stable hydrate adducts, the ground-state geometries of structures A, B, C and D were optimised – the outcome of which returned 6-HU and 6-HT as lower-energy

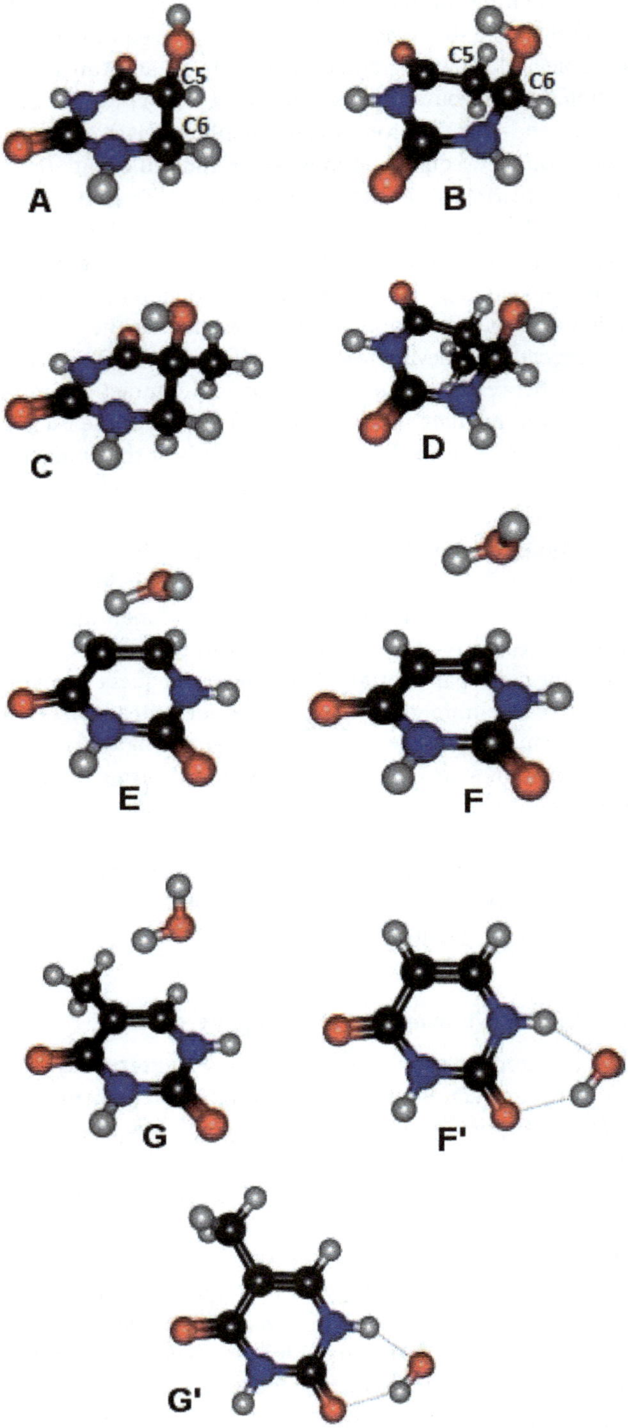

Fig. 4 Molecular structures associated with the various isomers and products of the uracil/thymine + water complex. The top two molecules depict the isolated uracil and thymine nucleobases; the assigned atomic numberings are used in the main text.

adducts when compared to their C5-substituted analogues. This can be understood by considering the step-wise addition of H^+ and OH^-, in which H^+ initially adds to either C5 or C6 *via* π electron-pair donation localised along C5=C6. The addition of H^+ to C6 delocalises the partial positive charge to C5, rendering it electrophilic. The adjacent CO acceptor is a net π-withdrawing group and thus destabilises the nascent C5-centred carbocation intermediate. In contrast, the addition of H^+ at C5 renders C6 electrophilic, with a partial positive charge. The adjacent NH donor moiety serves to stabilise this electrophilic site *via* resonance, thus stabilising the C6-centred carbocation intermediate. The more stable C6-centred carbocation intermediate thus contains an enhanced rate and Boltzmann population. Charge recombination then leads to OH^- adding to the C6-centre. This agrees with experimental results which show that 6-HU/6-HT are the observed products.[16,19,25] With these stabilising effects and energetics in mind, the forthcoming text therefore focuses on the U/T–H_2O reactivity, in which the nascent hydrate products are 6-HU and 6-HT. Working back from these products, we then located a transition state (TS) on the electronic ground state, connecting 6-HU/6-HT to the corresponding dispersively-bound chromophore–water reactant (see structure E in Fig. 4). The eigenvectors associated with the single imaginary normal mode eigenvalue (see Fig. S1 in the ESI†) confirmed that the optimised TS involves the hydrolysis and subsequent addition of water across C5=C6 of U/T. The minimum energy TS was then used as the starting configuration for the construction of an IRC path, revealing the minimum energy reaction path connecting the possible reactants to the hydrate products. In so doing, the IRC path naturally reveals the dispersively-bound reactant structure with the lowest possible energy from which we can tentatively start our studies on the excited-state chemistry associated with the U/T + H_2O photoreaction. Such initial reactant configurations are given in structures F and G in Fig. 4. Whilst these do not represent the global minimum energy geometry returned from the IRC, they are the lowest-energy dispersively-bound structures that contain the optimal configurations for an ideal description of the dominant orbital interactions required for an excited-state nucleophilic hydrolysis reaction. However, the global minimum energy structures displayed alongside structures F and G in Fig. 4 (F′ and G′) are ~0.2 eV more stable than F and G and represent a configuration in which the lone water molecule is hydrogen-bonded to the N–H donor and CO acceptor sites. Though intuitively lower in energy, in solution such hydrogen-bonded configurations will co-exist with the dispersively-bound configurations. The small energy difference between the minimum-energy geometries and the orbital-favoured geometries also serves to reinforce our choice to initiate our studies from structures F and G, rather than from F′ and G′.

Table 2 lists the VEEs to, and oscillator strengths associated with, the first two electronically excited singlet states of the U–H_2O and T–H_2O complexes. These agree well with previous studies on complexes of uracil with water.[26] The VEEs of more states, in addition to more levels of theory, can be found in the ESI.† The orbital promotions associated with the preparation of the lowest two excited singlet states of U–H_2O are shown in Fig. 5. The CASSCF (and subsequent CASPT2) energies reveal that the lowest two excited states absorb in the near-UV region and are analogous to those of isolated U. The S_1 and S_2 states are of $^1n\pi^*$ and $^1\pi\pi^*$ character, respectively. The $^1n\pi^*$ state involves an orbital promotion from the $2p_y$ carbonyl O-centred lone pair to a ring-centred π^* orbital, and thus

Table 2 Vertical excitation energies (with oscillator strengths in parentheses) and state characters associated with the lowest two excited states of the T/U + water complex at the CASPT2(20,14)/6-31G(d) level

Transition	Excitation energy/eV	Character
Uracil		
S_1–S_0	4.96 (0.0007)	$n\pi^*$
S_2–S_0	5.20 (0.2309)	$\pi\pi^*$
Thymine		
S_1–S_0	5.00 (0.0007)	$n\pi^*$
S_2–S_0	5.12 (0.2241)	$\pi\pi^*$

contains a characteristically weak oscillator strength on account of the poor spatial overlap of the participating orbitals. In contrast, the S_2 ($^1\pi\pi^*$) state contains a characteristically large oscillator strength manifesting from the appreciable spatial overlap between the participating ring-centred π and π^* orbitals. This strongly absorbing state is predicted to be the initially populated state leading to the photohydrate.

Armed with knowledge of the electronic state characters and associated energetics, we now describe the ensuing excited-state reaction path for forming 6-HU from U–H_2O (henceforth $Q_{hydrate}$). Fig. 6 depicts the CASPT2 PE profiles associated with the lowest three singlet states, as a function of $Q_{hydrate}$. Equivalent PE profiles calculated at the TDDFT levels of theory are displayed in Fig. S2 in the

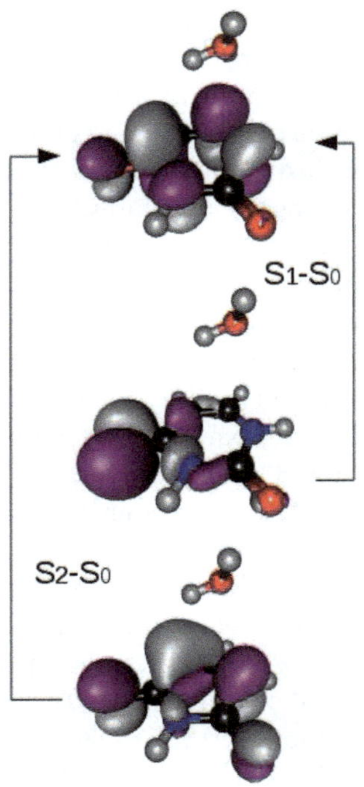

Fig. 5 Orbitals and orbital promotions associated with the lowest two excited electronic states of the uracil–water complex.

ESI† and are qualitatively similar to those obtained with CASPT2. For conciseness, we divide the description of the PE profiles along $Q_{hydrate}$ into two parts, such that part (1) represents an initial LIIC path connecting the U–H_2O reactants to a minimum energy CI at $Q_{hydrate} \geq 0.8$ (henceforth LIIC1) and part (2) represents a second LIIC path connecting CI to the 6-HU hydrate adduct (henceforth LIIC2). The dominant nuclear motions associated with LIIC1 involve the migration of the water-centred H-atom to C5, whilst those of LIIC2 comprise the subsequent addition of the OH radical intermediate to C6. The profiles associated with LIIC1 reveal that at small displacements from the U–H_2O reactant (*i.e.* $Q_{hydrate} = 0$–0.4), all states are bound and thus show no reactivity with respect to $Q_{hydrate}$. At intermediate to long-range geometries (corresponding to $0.4 \leq Q_{hydrate} \leq 0.8$), the S_1 and S_2 states become reactive with respect to $Q_{hydrate}$, which is implied by the decrease in PE as a function of $Q_{hydrate}$. The initial rise and subsequent decline in PE leads to a local maximum on S_1. S_2 at $Q_{hydrate} = 0.4$ represents a change of electronic configuration from LE to CT character and is reminiscent of the aforementioned G + H_2O reaction. The S_1 and S_2 CT states are analogous in character, such that both contain an electronic configuration in which an electron is promoted to a common ring-centred π orbital from either the O $2p_y$ or O $2p_x$ orbital of water, respectively. Such an electronic configuration represents an inherent long-range charge-transfer, which is manifest in weak oscillator strengths for the S_1(CT) and S_2(CT) states. Notwithstanding this, non-adiabatic coupling to the near-lying symmetry-equivalent LE states can promote internal conversion from LE to CT. At near-threshold excitations to S_1, such couplings would need to surmount the aforementioned energy barrier at $Q_{hydrate} = 0.4$, which at first glance appears restrictive but merely represents an upper boundary of the true energy at which such a transition would occur. Test optimisations of this local maximum were performed using TDDFT/B3LYP/6-31G(d), by fixing the O–H distance of water at the value returned by the LIIC point at $Q_{hydrate} = 0.4$ and allowing the remaining components of the nuclear framework to relax to their respective minima on S_1. A LIIC path connecting the ground-state minimum energy geometry to the returned relaxed geometry at $Q_{hydrate} = 0.4$ (see Fig. S3 in

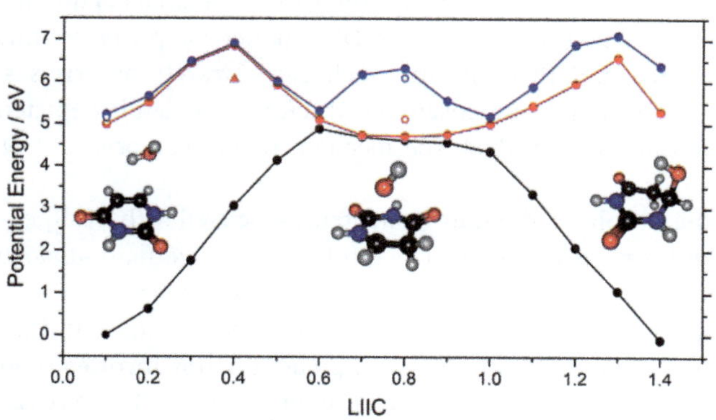

Fig. 6 CASPT2 PE profiles associated with the uracil/thymine + water photoreaction. The filled coloured circles represent the *ab initio* points associated with the uracil + water reaction, whilst the open circles represent analogous points for various optimised points for the thymine + water reaction. The S_0, S_1 and S_2 states are the black, red and blue curves, respectively.

the ESI†) reveals a 0.8 eV relaxation of this barrier (which now has a value of about 0.9 eV). Such an energy relaxation, coupled with the knowledge that LIIC1 predominantly represents PT motion, may indicate that this barrier may be surmounted by H-atom tunnelling. Any nascent population of S_1(CT) will then evolve towards the minimum energy CI and will most likely undergo internal conversion to S_0 *via* nonadiabatic coupling at this CI. The profile associated with LIIC2 shows that states beyond (and including) S_1 show a rise in PE as a function of $Q_{hydrate}$. In contrast, the S_0 state continues to decline in PE in the range of $0.8 < Q_{hydrate} < 1.4$, the final geometry of which represents the 6-HU hydrate adduct (the actual 6-HU product is point $Q_{hydrate} = 1.5$, but this could not be converged with CASPT2. See the ESI† for pathways including that point). Therefore, following internal conversion at the CI and in the limit of the 1-dimensional PE profile given in Fig. 6, the evolving population may bifurcate and form nascent 6-HU products or reform the ground state molecule – but we recognise that in reality an excess of 5 eV following IC at this CI may surmount reaction barriers associated with orthogonal nuclear motions to alternative reactions.

Equivalent CASPT2 PE profiles of the ground and various excited states associated with the T + water reaction were also constructed. The returned profiles are analogous in topography to those of the U + water reaction and are therefore displayed in Fig. S4 in the ESI.† The energies associated with the FC geometry, as well as the energy of the equivalent S_1/S_0 CI, are displayed as open points in Fig. 6. As is evident in Fig. 6, the FC transitions are largely analogous in energy to those of U + water, whereas the CI is 0.3 eV less stable in T + water when compared to U + water. Such a destabilisation in the CI associated with the T + water reaction can be understood by considering that the initial PT reaction yields an electrophilic C6 position. The +I inductive effect associated with the methyl group at C5 serves to destabilise the biradical intermediate when compared to the analogous null inductive effect of H at the C5 position in the CI associated with U + water.

2.3 The thymine + water photoreaction in bulk DNA solution

In this section we extend the above isolated gas-phase study to a somewhat complex biological environment. In so doing, we use a model of double helix DNA solvated in bulk aqueous solution. The DNA model comprises strands of deoxyribose – adenine and thymine, in which each strand comprises a common nucleobase – the equilibrium ensemble structure of which is depicted in Fig. 7. Since we are interested in DNA, the mechanism will comprise T + H_2O in this section.

As with the gas phase, the state characters associated with the S_1, S_2 and S_3 are $n\pi^*$, $\pi\pi^*$ and $\pi\pi^*$, respectively. The returned states contain analogous orbital promotions to those of the gas-phase U/T + water complex.

The above studies on the T/U + water photoreaction show an initial path to a low energy CI – motion towards which predominantly involves H-atom migration from water to C5–U/T. The ensuing reaction from the CI involves the addition of the nascent OH radical to C6–T/U, forming the photohydrate adduct. Given this two-step reaction, two PE profiles were computed following ground-state DFT/AMBER optimisations, which are outlined in more detail in the methodology. Briefly, the first comprises a relaxed scan in which the O–H stretch coordinate of water (henceforth R_{OH}) is varied, fixed at various values and the remaining

 This journal is © The Royal Society of Chemistry 2018

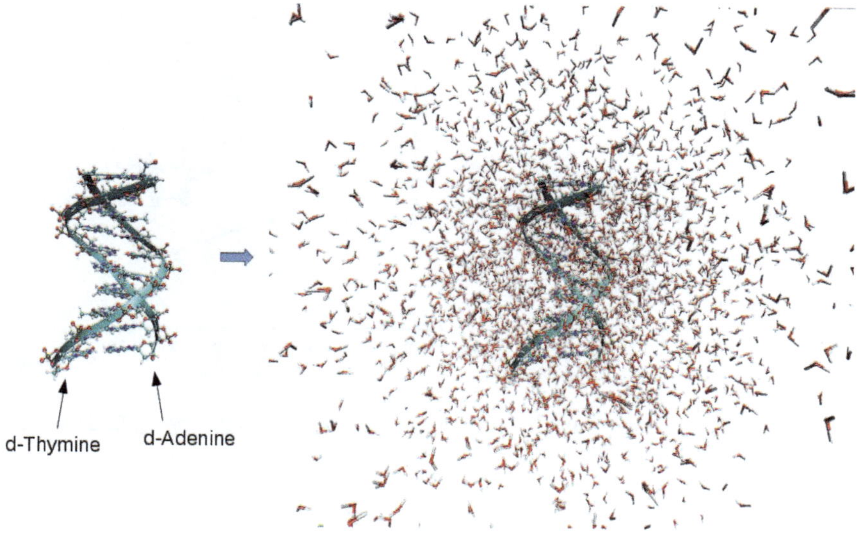

Fig. 7 Equilibrated starting structure of bulk DNA.

components of the nuclear framework are allowed to relax (at the (TD)DFT/AMBER level – see methodology) to their respective minima in S_0 and S_1. This coordinate represents proton-transfer from the water molecule to T. The second step is modelled by varying the C6–OH stretch distance (henceforth R_{C6-OH}), again fixing at various values and allowing the remaining nuclear degrees-of-freedom to relax (at the (TD)DFT/AMBER level – see methodology ((TD)DFT denotes DFT or TDDFT)) to their respective minima. The returned PE profiles associated with R_{OH} and R_{C6-OH} are presented in Fig. 8. As shown, the profiles in the bulk biological environment bear certain resemblances (in topography and electronic state character) to those obtained in the gas phase (Fig. 6, above), reinforcing the idea that the above description of gas-phase photoreaction is active in the bulk. As before, the ground state shows a sharp rise in PE as a function of R_{OH}, indicating a lack of driving force for PT from water to T. In the range of $1.0 < R_{OH} < 1.3$ Å, the S_1, S_2 and S_3 states also show a rise in energy as a function of R_{OH} but become progressively reactive with respect to proton transfer at $R_{OH} > 1.3$ Å. This initial rise and then decline in PE, at $R_{OH} < 1.3$ Å and $R_{OH} > 1.3$ Å, respectively, is again a manifestation of the change in electronic-state character from LE to CT. As before, the LE state contains orbital promotions that are intrinsic to the T chromophore, whereas the CT state arises *via* an electron promotion from one of two O 2p orbitals localised on the water molecule to a π^* orbital local to T. The returned barrier top of $R_{OH} = 1.3$ Å (which again necessarily represents an upper limit) is 1 eV above that of the S_1 FC geometry. As with the aforementioned gas-phase reaction, we note that S_1 is a dark $n\pi^*$ state, whilst the vertical S_2 state is a bright $\pi\pi^*$ state. The return maximum along the S_1 adiabatic path is approximately 0.2 eV above that of the vertical geometry of the bright S_2 state, showing a significantly smaller energy gap compared with that of the gas phase. Additionally, given that this inherent motion is an H-atom motion, the long-range CT state may be populated by tunnelling from the LE state at near-threshold

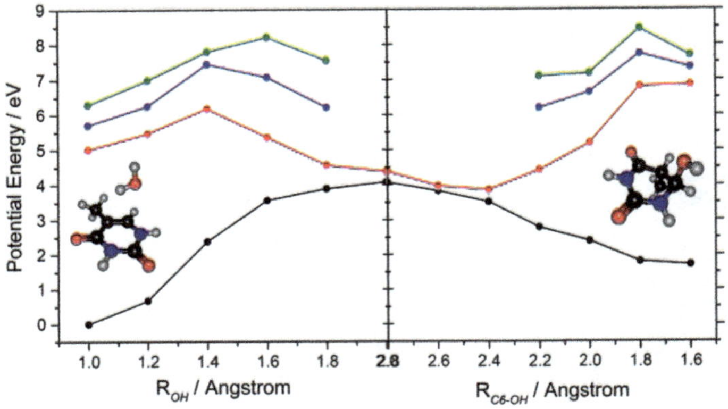

Fig. 8 CASPT2 PE profile for the T + water reaction in bulk solution. The S_0, S_1, S_2 and S_3 states are the black, red, blue and green curves, respectively.

excitations. As before, the subsequent decline and rise in PE of the S_1 and S_0 states, respectively, is inevitably manifest in a low-energy CI. From this low-energy crossing point, the remainder of the PE profile (right panel in Fig. 8) shows a gradual decline in PE as a function of the progressive shortening of C6–OH, whereas S_1 shows a net rise in PE. The ground-state optimised structure of the photohydrate (right-most point in the right panel in Fig. 8) is 1.9 eV less stable than the dispersively-bound T + water complex. This is in contrast to the gas-phase reactivity outlined in Section 2.2, which shows that the relative stabilities of the photohydrate and the dispersively-bound T/U + water complex are almost isoenergetic. This destabilisation in bulk solution may be a manifestation of the greater steric hindrance arising *via* the puckered ring geometry in the double-helix DNA environment.

Having considered the reactivity of the T + water photoreaction in bulk DNA, future studies will compare and contrast the above described bulk T + H_2O photoreaction with those of the photocycloaddition reactions between adjacent T nucleobases. This is the sole reason for choosing DNA strands that contain common nucleobases on a single strand. In a future study we also expect the above QM/MM studies to be extrapolated to U + H_2O.

3 General discussion and conclusions

Using high-level electronic structure calculations we have detailed the reaction between a given DNA/RNA nucleobase and proximal water molecules, in an attempt to understand the ways in which photoinduced nucleophilic hydrolysis reactions lead to harmful hydrate adducts. Within this class of photoreaction, the chosen nucleobases – G and T/U – represent well-known experimental cases in which G forms 8-OG and U/T form hydrate adducts. In the case of the G + water photoreaction we have shown that the well-known 8-OG adducts can be formed *via* an initial photoinduced splitting of water, initiated *via* an EDPT from water to the N4 acceptor of G. The proton-transfer reaction leads to a protonated guanine + OH biradical, which represents a low-energy crossing between the S_1 and S_0 states. IC at this crossing, followed by the subsequent population of S_0, is predicted to

 This journal is © The Royal Society of Chemistry 2018

give rise to a vibrationally-hot ground-state biradical, with an energy in excess of 5 eV. We have shown that such an energy is in excess of the reaction barrier associated with the ultimate formation of 8-OG, which is predicted to proceed *via* an 8-HG intermediate formed *via* the addition of the nascent OH radical to the C2 position. As mentioned above, our theoretical studies complement the work of Ravanat and co-workers, who have shown experimental evidence for the formation of 8-OG *via* a photoreaction of G with water.[13] Such a photoreaction also shows the capability of G to behave as a sensitiser for the initial step of water splitting in the near- to mid-UV region. In a prebiotic, ozone-free earth, an abundance of hard tropospheric UV radiation may have initiated the important first steps to photosynthesis *via* DNA/RNA nucleobase sensitisation. The well-known photorepair properties of 8-OG (coupled with the recently-described labile photoreaction for forming 8-OG in the mid-UV region) may have also been important in preserving canonical DNA/RNA nucleobases in such a prebiotic environment.

Such photoreactions are also extrapolated to U and T – both of which show a diverse range of differences and similarities to the case of G. An important difference is the addition of H and OH across adjacent C-atoms, in which the initial protonation occurs at C5. In the equivalent G + water reaction, the initial protonation step is predicted to occur *via* proton-transfer to an N-acceptor local to G. Despite the differences in the nature of the acceptor atom, the potential energy landscape and the S_1/S_0 CI and state characters associated with the initial EDPT reaction are analogous to G + water. In all cases, the nascent protonated GH/UH/TH + OH biradical exhibits a 'downhill' ground-state energy profile as a function of the addition of OH to C6/N4. In the case of 8-HG, we have shown that H_2 elimination leads to the formation of 8-OG, the profile of which displays an energy barrier of approximately 3 eV. An equivalent H_2 elimination reaction is also expected, following the formation of the UOH and TOH hydrate adducts. Such reactions have not been detailed experimentally and are thus beyond the scope of this theoretical study.

In comparing the exited-state chemistry of U + water with T + water, we note that the S_1/S_0 CI, associated with the TH + OH biradical, is less stable than that of the UH + OH biradical. This suggests a less stable CT state, which is likely manifested in the experimentally-measured lower yield of thymine hydrate adducts when compared to uracil hydrate adducts.[15] Instead, it is well-known that photoinduced [2 + 2] cycloaddition reactions between adjacent stacked thymine nucleobases lead to the formation of thymine–thymine CPD lesion adducts. The equivalent photoreaction in stacked uracil nucleobases occurs with a significantly lower yield.[12] This observation may be reinforced by the recently-obtained energetics, which show a more favourable driving force for hydrate formation in U as compared to T. The expected larger quantum yield for uracil hydrate adducts may quench the formation of U–U cycloadducts. In a fuller future study, we aim to compare and contrast the competition between cyclo- and hydrate-adduct formations in bulk DNA and RNA. Our present model DNA system is ideal for such a comparison, since the thymine nucleobases are common to a single strand and are ideally positioned for a given [2 + 2]-cycloaddition study. In the present study, we use this bulk DNA model to compare the T + water reaction in a bulk environment to that in the gas phase. It should be noted that the quantum calculations in the gas phase and bulk environment are not equivalent (as will be discussed in the Methods section, the active space in the bulk is smaller for computational reasons), so an exact quantitative comparison is not appropriate. However, we can make qualitative comparisons. In comparing the gas- and condensed-phase reactivity of the T + water reaction, we note that the PE profiles of the two phases are qualitatively similar. Both show an adiabatic S_1 PE profile that contains a net reactivity with respect to proton-transfer, whilst that of S_0 shows no net reactivity. The S_1/S_0 CI at long-range R_{OH} is also structurally similar in both the gas phase and in bulk solution. In contrast, two key energetic differences are noted. The first is the relative stability of the hydrate adduct, which in the bulk environment is predicted to be approximately 1.9 eV less stable than in the gas phase. The second is the noteworthy energy difference between the bright S_2 state and the local maximum in the S_1 adiabatic profile, which is significantly smaller in the bulk DNA environment compared with the gas phase. In bulk solution, this smaller energy difference leads to an enhancement in the competition between the photohydrate-forming reaction and the non-radiative (nucleobase-preserving) decay paths that are intrinsic to the T nucleobase.

With the above discussion in mind, detailed studies of this class of water–chromophore reaction in the gas phase – in which the intrinsic excited-state reactivity of the unperturbed chromophore complex is qualified – are an important stepping stone for extrapolation to the bulk. In so doing, our present study shows that despite the minor energetic differences, the excited-state reaction is equally as active in a complex bulk environment as in the gas phase, leaving the door open for possible experiments with nucleobase water clusters in the gas phase. Such experiments have already shown promise for related chromophore–water clusters.[27] In conclusion, we have shown that such hydrate adducts can compete with other well-known lesion-forming adducts and, more importantly, with photostabilising non-radiative decay paths.

4 Methods

4.1 The guanine + water reaction

The 9*H*-guanine molecule was clustered in an ensemble of five solvent water molecules. The five cluster molecules of water were selected in order to optimally describe the significant dispersive and hydrogen-bonding effects required for the ultimate formation of a hydrate adduct in which the OH adds to C2. Such a cluster is also motivated by previous studies on analogous systems.[23,24,28] Five water molecules represent the limit at which we can obtain a detailed qualitative description of the excited-state energetics intrinsic to this specific class of water–chromophore reaction, whilst maintaining a decent computational expense. Relaxed geometries, as a function of the R_{OH} driving coordinate of a H_2O–N7-centred hydrogen bond, were obtained by fixing R_{OH} at various values and optimising the remaining internal degrees-of-freedom using RI-ADC(2)[29]/cc-pVDZ[30] and RI-MP2 (ref. 31)/cc-pVDZ[30] for the S_1 and S_0 states, respectively. The final energies associated with the returned relaxed MP2/ADC(2) profiles were computed using the complete active space with second-order perturbation theory[32,33] (CASPT2) based on a state-averaged complete active space self-consistent field reference wavefunction[34] (SA4-CASSCF). The active space was carefully chosen to strike a balance between an adequate description of the excited-state chemistry and a reasonable computational expense. After careful testing, an

 This journal is © The Royal Society of Chemistry 2018

optimal active space of ten electrons in ten orbitals (3π, $3\pi^*$, 2O (2p lone-pair, $2\sigma^*$ orbitals)) was used. An imaginary level shift of $0.5E_H$ was used to aid convergence and to mitigate the involvement of intruder states. All ADC(2) and MP2 calculations were undertaken in Turbomole v2.2,[35] whilst all CASPT2/CASSCF calculations were undertaken in Molpro 2015.1.[36]

4.2 The uracil/thymine + water reaction in the gas phase

The ground-state minimum-energy geometries of the uracil–water complex, thymine–water complex and the uracil and thymine hydrate adducts were optimised using the Becke 3-parameter Lee–Yang–Parr functional[37–40] of Density Functional Theory, coupled to the 6-31G(d) Pople basis set[41] (DFT/B3LYP/6-31G(d)). The subsequent real 3N-6 normal-mode wavenumbers confirmed that all the optimised structures were at true minima. Transition states, connecting the uracil/thymine + water complexes to the hydrate adducts, were optimised using the Synchronous Transit-Guided Quasi-Newton method[42] at the DFT/B3LYP/6-31G(d) level of theory. The returned transition state was used to compute an intrinsic reaction path (IRC) in order to obtain the minimum energy path connecting the reactions to the products. CIs between the S_0 and S_1 states were located for both uracil/water and thymine/water, using the SA3-CASSCF method coupled to a 6-31G(d) Pople basis set. The CIs thus derived were computed using an active space of six electrons in six orbitals. The PE profiles connecting the ground-state minimum-energy geometries to the minimum-energy CIs were obtained by means of LIIC and thus necessarily represent an upper boundary to the true energetics. The energies along the LIIC were obtained with CASPT2, based on an SA6-CASSCF reference wavefunction and a full-valence active space of twenty electrons in fourteen orbitals. As with the guanine + water reaction, an imaginary level shift of $0.5E_H$ was used in all CASPT2 calculations. Energies along the LIIC were also computed using Time-Dependent Density Functional Theory[43–45] (TDDFT/B3LYP/6-31G(d)) in order to qualitatively compare the topography of the PE profile to that returned by CASPT2/CASSCF(20,14)/6-31G(d). The reactant and product excitation energies were also calculated using the Equation-of-motion Coupled Cluster Singles and Double method[46,47] (EOM-EE-CCSD), coupled to a 6-311G(d) Pople basis set. The (TD)DFT and EOM-EE-CCSD calculations were performed in Q-Chem.[48] The CIs were located using Columbus 7.0 [49–53] and the CASPT2/CASSCF(20,14) single-point calculations were performed using Molpro 2015.1.[36] The transition state optimisations and the subsequent IRC paths were obtained using Gaussian 09.[54]

4.3 QM/MM methods for the T + water photoreaction

Using AmberTools v16,[55] a double-helix structure of B-DNA, comprising ten deoxyriboseadenine and deoxyribosethymine (in which each strand contained a common nucleoside), was constructed and solvated with a TIP3P ensemble of solvent water molecules in a truncated octahedral box. Na^+ counter-ions were used to neutralise the net negative charge provided by the phosphate moieties. The AMBER ff99 parameter set[56] (in combination with ff99bsc0 corrections[57]) was adopted in the present study and has been shown to work well in analogous systems.[58,59] Using the Sander module in Amber 16,[60] the subsequent structure was minimised for 2500 steps in order to mitigate any steric clashes, then gradually annealed from 0 K to 298 K for 20 000 steps with a time step of 1 fs. Positional restraints with a force constant of 25 kcal mol^{-1} Å^{-2} were applied to the solute during heating. At 298 K, the applied restraint was gradually removed in 5 increments. A Langevin thermostat, with a collisional frequency of 5 ps^{-1}, was used in order to maintain a constant temperature. A constant pressure of 1 bar was maintained by applying a weak-coupling algorithm with a relaxation time of 1 ps. A particle mesh Ewald summation method was used for calculating long-range electrostatic interactions. The SHAKE algorithm was used to constrain all the degrees-of-freedom associated with hydrogen-atom motions. The system was equilibrated for 1 ns prior to the QM/MM calculations.

The mixed quantum and classical (QM/MM) calculations were performed using our in-house QutELa (Quantum Electronic structure with Layered molecular mechanics) programme, in which Gaussian or Molpro (for the quantum QM region) and Amber 16 (for the classical MM region) were coupled in order to obtain QM/MM minima and associated energetics. The QM region comprised a single thymine molecule at the end of the strand and the 'nearest-neighbour' water molecule returned from the equilibration. The remaining residues and water molecules made up the classical MM region. A standard additive QM/MM Hamiltonian, as displayed in eqn (1), is solved in order to determine the total energy of the bulk system. The terms in eqn (1) represent the isolated QM part (H_{QM}), describing the QM atoms, the isolated MM part (H_{MM}), describing the MM atoms, and a coupling term ($H_{QM/MM}$), describing the intrinsic coupling between the QM and MM regions. In the present case, $H_{QM/MM}$ is introduced by augmenting the zero-order electronic Hamiltonian H_{QM} of the QM region with an additional potential energy term, as given in eqn (2). The first term in eqn (2) represents the electrostatic effect that an atom M with charge Q_M at R_M (in the MM region) has on the electronic coordinates r_i of the solute. The second term represents the electrostatic interaction between an atom N with charge Z_N at R_N (in the QM region) with a point charge Q_M at R_M. Polarisability and van der Waals terms are ignored in $H_{QM/MM}$, as the former has been shown to introduce minor effects on the returned energetics in related systems, whilst the latter is independent of the electronic coordinates of the solute represented by the QM region.[61,62] A linking hydrogen atom, connecting the QM and MM regions, was introduced in the QM region in order to maintain valency. The point charge of the MM atom associated with the link-atom was set to zero in order to prevent overpolarisation.

$$H_{\mathrm{Total}} = H_{\mathrm{QM}} + H_{\mathrm{MM}} + H_{\mathrm{QM/MM}} \tag{1}$$

$$H_{\mathrm{QM/MM}} = \sum_{\mathrm{MN}} \frac{Z_{\mathrm{N}} Q_{\mathrm{M}}}{R_{\mathrm{M}} - R_{\mathrm{N}}} - \sum_{\mathrm{iM}} \frac{Q_{\mathrm{M}}}{r_{\mathrm{i}} - R_{\mathrm{M}}} \tag{2}$$

For the initial proton transfer step, relaxed geometries along the S_0 and S_1 states were obtained by using the R_{OH} coordinate of the proximal water molecule (of the thymine + water complex) as the driving coordinate, in which R_{OH} was fixed at various values and the remainder of the nuclear framework associated with the QM and MM regions was allowed to relax at the (TD)DFT/ωB97XD/6-31G(d):AMBER level of theory (coupling Gaussian 09 with Amber v16.0). This relaxation was performed by the well-known micro-optimisation method for the MM region. The

 This journal is © The Royal Society of Chemistry 2018

ωB97XD[63] functional was chosen since the present system requires optimal descriptions of short- and long-range correlations, as well as dispersion interactions. This same procedure was repeated for the second reaction step, in which the nascent OH radical adds to C6, but this time using $R_{C\text{-}OH}$ as the driving coordinate. Single-point energies and the associated returned relaxed geometries along R_{OH} and $R_{C\text{-}OH}$ were then recomputed using CASPT2. The CASPT2 calculations were performed based on an SA4-CASSCF reference wavefunction and comprised an active space of ten electrons in ten orbitals (3π, 3π*, 2σ* and 2O (2p$_y$)). An imaginary level shift of $0.5E_H$ was used.

Acknowledgements

The authors thank Carina Tse for driving earlier studies on the U/T + water photoreaction. P. C., T. N. V. K. and S. M. gratefully acknowledge the National Science Foundation (NSF, grant number: CHE-1465138) for funding. B. M. gratefully acknowledges the European Union Horizon 2020 research and innovation programme for the award of a Marie Skłodowska-Curie fellowship (PhARRAO initiative; grant number: 746593).

References

1 C. E. Crespo-Hernandez, B. Cohen, P. M. Hare and B. Kohler, *Chem. Rev.*, 2004, **104**, 1977–2019.
2 C. T. Middleton, K. de La Harpe, C. Su, Y. K. Law, C. E. Crespo-Hernandez and B. Kohler, *Annu. Rev. Phys. Chem.*, 2009, **60**, 217–239.
3 R. Improta, F. Santoro and L. Blancafort, *Chem. Rev.*, 2016, **116**, 3540–3593.
4 *Topics in Current Chemistry – Photoinduced Phenomena in Nucleic Acids I*, ed. S. Ullrich, A. C. Borin and M. Barbatti, Springer, Berlin – Heidelberg, 2015, vol. 355.
5 *Topics in Current Chemistry – Photoinduced Phenomena in Nucleic Acids II*, ed. S. Ullrich, A. C. Borin and M. Barbatti, Springer, Berlin – Heidelberg, 2015, vol. 356.
6 T. Gustavsson, R. Improta and D. Markovitsi, *J. Phys. Chem. Lett.*, 2010, **1**, 2025–2030.
7 S. Mai, M. Richter, P. Marquetand and L. González, *Top. Curr. Chem.*, 2015, **356**, 99–153.
8 W. Domcke, D. R. Yarkony and H. Köppel, *Conical Intersections*, World Scientific, Singapore, 2004.
9 S. Matsika and P. Krause, *Annu. Rev. Phys. Chem.*, 2011, **62**, 621–643.
10 A. L. Sobolewski and W. Domcke, *Phys. Chem. Chem. Phys.*, 2010, **12**, 4897.
11 *CRC Handbook of Organic Photochemistry and Photobiology*, CRC Press, New York, 1995.
12 C. L. Greenstock, I. H. Brown, J. W. Hunt and H. E. Johns, *Biochem. Biophys. Res. Commun.*, 1967, **27**, 431–436.
13 M. Gomez-Mendoza, A. Banyasz, T. Douki, D. Markovitsi and J.-L. Ravanat, *J. Phys. Chem. Lett.*, 2016, **7**, 3945–3948.
14 J. G. Burr, E. H. Park and A. Chan, *J. Am. Chem. Soc.*, 1972, **94**, 5866.
15 H. Görner, *J. Photochem. Photobiol., B*, 1991, **10**, 91–110.

16 J. Remsen, M. Mattern, N. Miller and P. Cerutti, *Biochemistry*, 1971, **10**, 524–529.
17 T. Ganguly and N. J. Duker, *Nucleic Acids Res.*, 1991, **19**, 3319–3323.
18 R. J. Boorstein, T. P. Hilbert, J. Cadet, R. P. Cunningham and G. W. Teebor, *Biochemistry*, 1989, **28**, 6164–6170.
19 S. Franzen, B. Skalski, L. Bartolotti and B. Delley, *Phys. Chem. Chem. Phys.*, 2014, **16**, 20164–20174.
20 T. Ganguly and N. J. Duker, *Mutat. Res., DNA Repair*, 1992, **293**, 71–77.
21 R. Szabla, R. W. Gora, M. Janicki and J. Sponer, *Faraday Discuss.*, 2016, **195**, 237–251.
22 R. Szabla, D. Tuna, R. W. Góra, A. L. S. J. Sponer and W. Domcke, *J. Phys. Chem. Lett.*, 2013, **4**, 2785–2788.
23 X. Wu, T. N. V. Karsili and W. Domcke, *ChemPhysChem*, 2016, **17**, 1298–1304.
24 M. Barbatti, *J. Am. Chem. Soc.*, 2014, **136**, 10246–10249.
25 G. Fisher and H. Johns, in *Pyrimidine Photohydrates*, ed. S.-Y. Wang, 1976, vol. 1, pp. 169–224.
26 A. Yoshikawa and S. Matsika, *Chem. Phys.*, 2008, **347**, 393–404.
27 N. Esteves-Lopez, S. Coussan, C. Dedonder-Lardeux and C. Jouvet, *Phys. Chem. Chem. Phys.*, 2016, **18**, 25637–25644.
28 R. Szabla, J. Sponer and R. W. Góra, *J. Phys. Chem. Lett.*, 2015, **6**, 1467–1471.
29 A. B. Trofimov and J. Schirmer, *J. Phys. B: At., Mol. Opt. Phys.*, 1995, **28**, 2299–2324.
30 T. H. Dunning Jr, *J. Chem. Phys.*, 1989, **90**, 1007–1023.
31 M. Feyereisen, G. Fitzgerald and A. Komornicki, *Chem. Phys. Lett.*, 1993, **208**, 359–363.
32 B. O. Roos and K. Andersson, *Chem. Phys. Lett.*, 1995, **245**, 215–223.
33 K. Andersson, P. A. Malmqvist, B. O. Roos, A. J. Sadlej and K. Wolinski, *J. Phys. Chem.*, 1990, **94**, 5483–5488.
34 B. O. Roos, P. R. Taylor and P. E. Sigbahn, *Chem. Phys.*, 1980, **48**, 157–173.
35 R. Ahlrichs, M. Bär, M. Häser, H. Horn and C. Kölmel, *Chem. Phys. Lett.*, 1989, **162**, 165–169.
36 H.-J. Werner, P. J. Knowles, G. Knizia, F. R. Manby, M. Schütz, P. Celani, W. Györffy, D. Kats, T. Korona, R. Lindh, A. Mitrushenkov, G. Rauhut, K. R. Shamasundar, T. B. Adler, R. D. Amos, A. Bernhardsson, A. Berning, D. L. Cooper, M. J. O. Deegan, A. J. Dobbyn, F. Eckert, E. Goll, C. Hampel, A. Hesselmann, G. Hetzer, T. Hrenar, G. Jansen, C. Köppl, Y. Liu, A. W. Lloyd, R. A. Mata, A. J. May, S. J. McNicholas, W. Meyer, M. E. Mura, A. Nicklass, D. P. O'Neill, P. Palmieri, D. Peng, K. Pflüger, R. Pitzer, M. Reiher, T. Shiozaki, H. Stoll, A. J. Stone, R. Tarroni, T. Thorsteinsson, and M. Wang, *MOLPRO, version 2015.1, a package of ab initio programs*, 2015.
37 A. D. Becke, *Phys. Rev. A*, 1988, **38**, 3098–3100.
38 C. Lee, W. Yang and R. G. Parr, *Phys. Rev. B: Condens. Matter Mater. Phys.*, 1988, **37**, 785–789.
39 S. H. Vosko, L. Wilk and M. Nusair, *Can. J. Phys.*, 1980, **58**, 1200–1211.
40 A. D. Becke, *J. Chem. Phys.*, 1993, **98**, 5648–5652.
41 R. Ditchfield, W. J. Hehre and J. A. Pople, *J. Chem. Phys.*, 1971, **54**, 724–728.
42 C. Peng and H. B. Schlegel, *Isr. J. Chem.*, 1993, **33**, 449–454.
43 E. Runge and E. K. U. Gross, *Phys. Rev. Lett.*, 1984, **52**, 997–1000.
44 R. Bauernschmitt and R. Ahlrichs, *Chem. Phys. Lett.*, 1996, **256**, 454–464.

 This journal is © The Royal Society of Chemistry 2018

45 F. Furche and R. Ahlrichs, *J. Chem. Phys.*, 2002, **117**, 7433–7447.
46 J. F. Stanton and R. J. Bartlett, *J. Chem. Phys.*, 1993, **98**, 7029–7039.
47 A. I. Krylov, *Annu. Rev. Phys. Chem.*, 2008, **59**, 433–462.
48 Y. Shao, Z. Gan, E. Epifanovsky, A. T. B. Gilbert, M. Wormit, J. Kussmann, A. W. Lange, A. Behn, J. Deng, X. Feng, D. Ghosh, M. Goldey, P. R. Horn, L. D. Jacobson, I. Kaliman, R. Z. Khaliullin, T. Kus, A. Landau, J. Liu, E. I. Proynov, Y. M. Rhee, R. M. Richard, M. A. Rohrdanz, R. P. Steele, E. J. Sundstrom, H. L. Woodcock III, P. M. Zimmerman, D. Zuev, B. Albrecht, E. Alguire, B. Austin, G. J. O. Beran, Y. A. Bernard, E. Berquist, K. Brandhorst, K. B. Bravaya, S. T. Brown, D. Casanova, C.-M. Chang, Y. Chen, S. H. Chien, K. D. Closser, D. L. Crittenden, M. Diedenhofen, R. A. DiStasio Jr, H. Do, A. D. Dutoi, R. G. Edgar, S. Fatehi, L. Fusti-Molnar, A. Ghysels, A. Golubeva-Zadorozhnaya, J. Gomes, M. W. D. Hanson-Heine, P. H. P. Harbach, A. W. Hauser, E. G. Hohenstein, Z. C. Holden, T.-C. Jagau, H. Ji, B. Kaduk, K. Khistyaev, J. Kim, J. Kim, R. A. King, P. Klunzinger, D. Kosenkov, T. Kowalczyk, C. M. Krauter, K. U. Lao, A. D. Laurent, K. V. Lawler, S. V. Levchenko, C. Y. Lin, F. Liu, E. Livshits, R. C. Lochan, A. Luenser, P. Manohar, S. F. Manzer, S.-P. Mao, N. Mardirossian, A. V. Marenich, S. A. Maurer, N. J. Mayhall, E. Neuscamman, C. M. Oana, R. Olivares-Amaya, D. P. O'Neill, J. A. Parkhill, T. M. Perrine, R. Peverati, A. Prociuk, D. R. Rehn, E. Rosta, N. J. Russ, S. M. Sharada, S. Sharma, D. W. Small, A. Sodt, T. Stein, D. Stueck, Y.-C. Su, A. J. W. Thom, T. Tsuchimochi, V. Vanovschi, L. Vogt, O. Vydrov, T. Wang, M. A. Watson, J. Wenzel, A. White, C. F. Williams, J. Yang, S. Yeganeh, S. R. Yost, Z.-Q. You, I. Y. Zhang, X. Zhang, Y. Zhao, B. R. Brooks, G. K. L. Chan, D. M. Chipman, C. J. Cramer, W. A. Goddard III, M. S. Gordon, W. J. Hehre, A. Klamt, H. F. Schaefer III, M. W. Schmidt, C. D. Sherrill, D. G. Truhlar, A. Warshel, X. Xu, A. Aspuru-Guzik, R. Baer, A. T. Bell, N. A. Besley, J.-D. Chai, A. Dreuw, B. D. Dunietz, T. R. Furlani, S. R. Gwaltney, C.-P. Hsu, Y. Jung, J. Kong, D. S. Lambrecht, W. Liang, C. Ochsenfeld, V. A. Rassolov, L. V. Slipchenko, J. E. Subotnik, T. Van Voorhis, J. M. Herbert, A. I. Krylov, P. M. W. Gill and M. Head-Gordon, *Mol. Phys.*, 2015, **113**, 184–215.
49 H. Lischka, T. Müller, P. G. Szalay, I. Shavitt, R. M. Pitzer and R. Shepard, *Wiley Interdiscip. Rev.: Comput. Mol. Sci.*, 2011, **1**, 191–199.
50 H. Lischka, R. Shepard, R. M. Pitzer, I. Shavitt, M. Dallos, T. Muller, P. G. Szalay, M. Seth, G. S. Kedziora, S. Yabushita and Z. Zhang, *Phys. Chem. Chem. Phys.*, 2001, **3**, 664–673.
51 H. Lischka, R. Shepard, I. Shavitt, R. M. Pitzer, M. Dallos, T. Müller, P. G. Szalay, F. B. Brown, R. Ahlrichs, H. J. Böhm, A. Chang, D. C. Comeau, R. Gdanitz, H. Dachsel, C. Ehrhardt, M. Ernzerhof, P. Höchtl, S. Irle, G. Kedziora, T. Kovar, V. Parasuk, M. J. M. Pepper, P. Scharf, H. Schiffer, M. Schindler, M. Schüler, M. Seth, E. A. Stahlberg, J.-G. Zhao, S. Yabushita, Z. Zhang, M. Barbatti, S. Matsika, M. Schuurmann, D. R. Yarkony, S. R. Brozell, E. V. Beck, J.-P. Blaudeau, M. Ruckenbauer, B. Sellner, F. Plasser, and J. J. Szymczak, *COLUMBUS, an ab initio electronic structure program, release 7.0*, 2012.
52 H. Lischka, M. Dallos, P. G. Szalay, D. R. Yarkony and R. Shepard, *J. Chem. Phys.*, 2004, **120**, 7322–7329.

53 M. Dallos, H. Lischka, R. Shepard, D. R. Yarkony and P. G. Szalay, *J. Chem. Phys.*, 2004, **120**, 7330–7339.
54 M. J. Frisch, G. W. Trucks, H. B. Schlegel, G. E. Scuseria, M. A. Robb, J. R. Cheeseman, G. Scalmani, V. Barone, B. Mennucci, G. A. Petersson, H. Nakatsuji, M. Caricato, X. Li, H. P. Hratchian, A. F. Izmaylov, J. Bloino, G. Zheng, J. L. Sonnenberg, M. Hada, M. Ehara, K. Toyota, R. Fukuda, J. Hasegawa, M. Ishida, T. Nakajima, Y. Honda, O. Kitao, H. Nakai, T. Vreven, J. A. Montgomery, J. E. Peralta, F. Ogliaro, M. Bearpark, J. J. Heyd, E. Brothers, K. N. Kudin, V. N. Staroverov, R. Kobayashi, J. Normand, K. Raghavachari, A. Rendell, J. C. Burant, S. S. Iyengar, J. Tomasi, M. Cossi, N. Rega, J. M. Millam, M. Klene, J. E. Knox, J. B. Cross, V. Bakken, C. Adamo, J. Jaramillo, R. Gomperts, R. E. Stratmann, O. Yazyev, A. J. Austin, R. Cammi, C. Pomelli, J. W. Ochterski, R. L. Martin, K. Morokuma, V. G. Zakrzewski, G. A. Voth, P. Salvador, J. J. Dannenberg, S. Dapprich, A. D. Daniels, O. Farkas, J. B. Foresman, J. V. Ortiz, J. Cioslowski, and D. J. Fox, *Gaussian 09, Revision B.01*, 2009.
55 D. Case, D. Cerutti, T. Cheatham III, T. Darden, R. Duke, T. Giese, H. Gohlke, A. Goetz, D. Greene, N. Homeyer, S. Izadi, A. Kovalenko, T. Lee, S. LeGrand, P. Li, C. Lin, J. Liu, T. Luchko, R. Luo, D. Mermelstein, K. Merz, G. Monard, H. Nguyen, I. Omelyan, A. Onufriev, F. Pan, R. Qi, D. Roe, A. Roitberg, C. Sagui, C. Simmerling, W. Botello-Smith, J. Swails, R. Walker, J. Wang, R. Wolf, X. Wu, L. Xiao, D. York, and P. Kollman, *Amber 17*, 2017.
56 J. Wang, P. Cieplak and P. A. Kollman, *J. Comput. Chem.*, 2000, **21**, 1049–1074.
57 A. Perez, I. Marchan, D. Svozil, J. Sponer, T. E. Cheatham, C. A. Laughton and M. Orozco, *Biophys. J.*, 2007, **92**, 3817–3829.
58 W. Lee and S. Matsika, *Phys. Chem. Chem. Phys.*, 2017, **19**, 3325–3336.
59 W. Lee and S. Matsika, *Phys. Chem. Chem. Phys.*, 2015, **17**, 9927–9935.
60 D. Case, T. Darden, T. Cheatham III, C. Simmerling, J. Wang, R. Duke, R. Luo, R. Walker, W. Zhang, K. Merz, B. Roberts, S. Hayik, A. Roitberg, G. Seabra, J. Swails, A. Götz, I. Kolossváry, K. Wong, F. Paesani, J. Vanicek, R. Wolf, J. Liu, X. Wu, S. Brozell, T. Steinbrecher, H. Gohlke, Q. Cai, X. Ye, J. Wang, M.-J. Hsieh, G. Cui, D. Roe, D. Mathews, M. Seetin, R. Salomon-Ferrer, C. Sagui, V. Babin, T. Luchko, S. Gusarov, A. Kovalenko, and P. Kollman, *AMBER 12*, University of California, San Francisco, 2012.
61 Z. Xu and S. Matsika, *J. Phys. Chem. A*, 2006, **110**, 12035–12043.
62 J. Kongsted, A. Osted, K. V. Mikkelsen, P.-O. Åstrand and O. Christiansen, *J. Chem. Phys.*, 2004, **121**, 8435–8445.
63 J.-D. Chai and M. Head-Gordon, *Phys. Chem. Chem. Phys.*, 2008, **10**, 6615–6620.

 This journal is © The Royal Society of Chemistry 2018

Faraday Discussions

PAPER

Photochemical relaxation pathways of S^6-methylthioinosine and O^6-methylguanosine in solution†

Brennan Ashwood, ‡ Luis A. Ortiz-Rodríguez and Carlos E. Crespo-Hernández *

Received 1st September 2017, Accepted 7th November 2017
DOI: 10.1039/c7fd00193b

S^6-Methylthioinosine and O^6-methylguanosine are byproducts resulting from the enzymatic reactions of sulfur-substituted prodrugs in cells and from the interaction of alkylating agents with cellular DNA, respectively. Their photochemistry has not been investigated, and it is currently unknown whether light absorption by these byproducts may pose any threat to the cell. In this contribution, their photoinduced processes upon absorption of UVB radiation are reported using broadband transient absorption spectroscopy. Plausible electronic relaxation mechanisms are proposed for both biological molecules, which are supported by steady-state absorption and emission measurements, and by singlet and triplet vertical excitation energies performed on a large subset of ground-state optimized conformational isomers in solution. The results are compared to the body of knowledge gathered in the scientific literature about the light-induced processes in the sulfur-substituted and canonical purine monomers. In particular, it is shown that S^6-methylation decreases the rate to populate the lowest-energy triplet state and blueshifts the ground-state absorption spectrum compared to those for the sulfur-substituted prodrugs and for the 6-thioguanosine metabolite. Similarly, O^6-methylation decreases the rate of internal conversion to the ground state observed in the guanine monomers by more than 10-fold in acetonitrile and 40-fold in aqueous solution, while it redshifts the ground-state absorption spectrum. Collectively, this investigation provides relevant new insights about the relationship between structural modifications of the purine chromophore and the electronic relaxation mechanisms in this important group of biological molecules.

Department of Chemistry, Center for Chemical Dynamics, Case Western Reserve University, 10900 Euclid Avenue, Cleveland, Ohio, 44106, USA. E-mail: cxc302@case.edu

† Electronic supplementary information (ESI) available: Steady-state emission and excitation spectra of 6MeTIno and 6MeGuo, optimized geometries, vertical excitation energies for each conformation of 6MeTIno 6TIno, 6MeGuo, and 6EnolGuo. See DOI: 10.1039/c7fd00193b

‡ Present address: Department of Chemistry, Institute for Biophysical Dynamics, and James Franck Institute, The University of Chicago, Chicago, Illinois, 60637.

Introduction

Sulfur-substituted purine derivatives have been prescribed for decades as effective prodrugs to treat acute lymphoblastic leukemia, immunosuppression, rheumatoid arthritis, and numerous autoimmune disorders.[1–5] 6-Thioguanine (6TGua), 6-mercaptopurine (6MP), and azathioprine (Aza) are the most common thiopurine prodrugs and metabolize to 6-thioguanosine (6TGuo) in the cell.[6–9] Unfortunately, these prodrugs have photocytotoxic consequences triggered by the exposure of patients to sunlight. Most notable in this regard is the 10- and 65- to 250-fold increased incidence of basal and squamous cell carcinomas, respectively, in patients with prolonged usage of thiopurines.[9–11] This increased risk of photodamage is thought to originate from the interaction of UVA radiation with the 6TGuo metabolite, which has led to many studies focusing on the photochemistry of this nucleoside derivative.[12–19,22–25]

Importantly, the 6-thioinosine (6TIno) nucleotide is also generated from 6MP through phosphorylation by hypoxanthine phosphoribosyltransferase.[20,21] The 6TIno nucleotide in turn is converted into the S^6-methylthioinosine (6MeTIno) nucleotide *via* the enzyme *S*-methyl transferase,[20,21] and is generated in significantly higher concentrations than the 6TGuo nucleotides in cells.[22,23] 6MeTIno can further inhibit purine *de novo* synthesis *in vitro*,[24–26] and is linked to an increased risk of hepatotoxicity.[27–29] Although the 6MeTIno nucleotide is not metabolized into DNA or RNA, it may still be capable of inflicting photochemical damage through light-induced reactions within the intracellular pool, as recently shown for Aza and 6TGua,[9] and also suggested for other thionated DNA derivatives.[30–32] Similarly, while the thione tautomer of 6TGuo predominates in solution, it may easily undergo *in situ* non-enzymatic methylation to generate S^6-methylthioguanosine (6MeTGuo) when incorporated into DNA.[8,33,34] 6MeTGuo is misread as an adenosine base by DNA polymerase and in turn induces DNA mismatch repair, which can produce DNA lesions. Methylation of 6TGuo is largely responsible for the cytotoxic effects observed from thiopurine prodrugs, specifically Aza in the absence of UVA light, which have been correlated with an increased risk in leukemia[34] and enhanced mutation frequency in circulating lymphocytes.[35] However, the photoreactivity of 6MeTIno and 6MeTGuo have yet to be investigated and it remains unknown whether these S^6-methylated derivatives of 6TIno and 6TGuo contribute to the phototoxicity of thiopurine prodrugs.

O^6-Methylation also plays a significant role in the cytotoxicity resulting from alkylating agents to cellular DNA.[36,37] Methylation of the guanine (Gua) base at the O^6 position in DNA by exogenous or endogenous alkylating agents leads to the mutagenic and carcinogenic O^6-methylguanosine (6MeGuo) adduct.[36,37] This structural modification forces an N1–C6 double bond, which effectively deprotonates the N1 position and alters the hydrogen bond properties of the base. These different hydrogen bond properties inhibit the Watson–Crick base-pairing configuration and subsequently lead to mispairing by DNA polymerase.[38,39] Unless 6MeGuo is removed by enzymatic repair,[40–42] it accumulates in DNA and can be misread as an adenine base and subsequently mispaired with the thymine base,[36,38,42] sparking the generation of DNA mutations.

In addition to their biological relevance, 6MeTIno and 6MeGuo serve as important nucleic acid derivatives for revealing the photophysical effects of S^6-

 This journal is © The Royal Society of Chemistry 2018

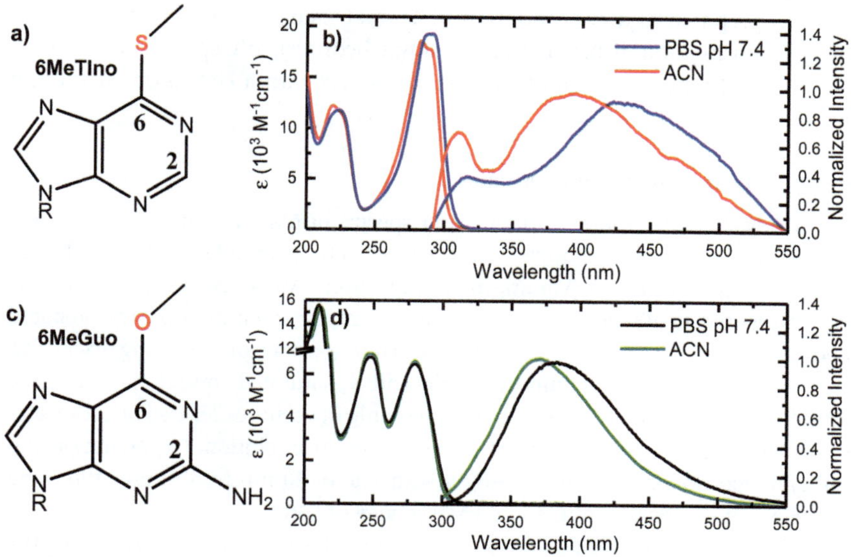

Fig. 1 (a) Molecular structure of 6MeTIno (R = ribose) and (b) its steady-state absorptivity and emission spectra in PBS at pH 7.4 and in acetonitrile upon 260 nm excitation. The emission spectrum does not change upon 290 nm excitation, but the Raman emission from the solvent strongly overlaps with the high-energy side of the spectrum. (c) Molecular structure of 6MeGuo (R = ribose) and (d) absorptivity and emission spectra in PBS at pH 7.4 and in acetonitrile upon 290 nm excitation. The absorption and emission spectra of 6MeGuo in PBS are reproduced from ref. 44. The emission spectra were corrected for the Raman band of the solvent, which affects the accuracy of the high-energy side of these spectra.

and O^6-methylation in thio- and canonical purine base derivatives, respectively. They also serve as model compounds for 6MeTGuo, 6-thiolguanine (6ThiolGua) and 6-enolguanine (6EnolGua), and allow for investigation into the impacts of solvent environment on their respective excited-state dynamics, which have been recently reported for 6ThioGua and 6EnolGua in the gas phase.[43]

In this contribution, the steady-state and ultrafast time-resolved photophysical properties of 6MeTIno (Fig. 1a) in phosphate buffered saline (PBS) solution at pH 7.4 and in acetonitrile are reported. Similar investigations are done for 6MeGuo (Fig. 1c) in acetonitrile to gain further insight into the relaxation mechanism of 6MeGuo recently reported in PBS,[44] and as a model to compare the photophysical effects of O^6- and S^6-methylation in solution. The experimental measurements are complemented with time-dependent density functional (TD-DFT) calculations, and with previous theoretical and experimental studies for the Gua monomers, in order to provide a state-specific relaxation model for each compound.

Methods

Chemicals

S^6-Methylthioinosine (6MeTIno, 99% purity) and O^6-methylguanosine (6MeGuo, 97.7% purity) were obtained from Sigma-Aldrich and used as received. Solutions were prepared in phosphate-buffered saline (PBS) at pH 7.4 or in acetonitrile. The

PBS was prepared at a concentration of 16 mM using 0.480 g of sodium dihydrogen phosphate and 0.354 g of disodium hydrogen phosphate dissolved in 400 mL of ultrapure water and adjusted to pH 7.4 with a concentrated sodium hydroxide solution.

Steady-state absorption and emission

The steady-state absorption and emission spectra of 6MeTIno and 6MeGuo were measured at room temperature using a Cary 100 spectrophotometer and Cary Eclipse spectrofluorimeter (Varian, Inc.), respectively. Molar extinction coefficients were measured in PBS and in acetonitrile by making serial dilutions of prepared stock solutions. Emission spectra of 6MeTIno were recorded using excitation wavelengths of 260 and 290 nm under air- and N_2-saturated conditions, where the emission spectrum of 6MeGuo was obtained only in acetonitrile. Excitation spectra were collected along the emission band of the respective solution. The excitation and emission spectra were collected using a scan rate of 20 nm min^{-1}, excitation and emission slit widths of 5 nm, and a PMT voltage of 800 V.

Emission quantum yields were determined in PBS and in acetonitrile using the method described by Lackowicz.[45] Tryptophan in pH 7.4 PBS ($\Phi_{Fl} = 0.13 \pm 0.01$)[46] was used as a fluorescent standard. The same excitation wavelength of 290 nm was used for the standard, 6MeTIno, and 6MeGuo. The same quantum yield was obtained when exciting tryptophan at 280 nm and using an identical ground-state absorbance at the excitation wavelength.

Quantum chemical calculations

Quantum-chemical calculations were performed using the Gaussian 09 suite of programs.[47] Ground-state optimizations were performed at the PBE0/IEFPCM/6-311++G(d,p) level of theory for 6MeTIno, 6-thiol thioinosine (6ThiolIno), 6-thione thioinosine (6TIno), 6MeGuo, 6-enol guanosine (6EnolGuo), and 6MeTGuo.[48–50] Calculations were performed for both rotamers of 6MeTIno, 6ThiolIno, 6MeGuo, and 6EnolGuo, in which the S^6/O^6 group was oriented towards the N1 or N7 atoms (see Fig. 1 for atomic numbering). For all the compounds investigated in this study, both the *syn* and *anti*-sugar conformations were investigated. Vertical excitation energies were calculated at the TD-PBE0/IEFPCM/6-311++G(d,p) level of theory.[51] Bulk solvent dielectric effects on the ground-state geometries as well as excited-state vertical energies were modeled by including self-consistent reaction field (SCRF) calculations using the polarizable continuum model (PCM)[52] with the integral equation formalism (SCRF = IEFPCM).[53]

The adiabatic triplet state energy of 6MeTIno was estimated as the energy difference between ground- and triplet-state optimizations performed at the restricted and unrestricted PBE0/IEFPCM/6-311++G(d,p) levels of theory, respectively, in water or in acetonitrile, as has been reported previously.[54]

Broadband transient absorption spectroscopy

Broadband transient absorption spectroscopy was employed to study the excited-state dynamics of 6MeTIno and 6MeGuo. The laser and spectrometer setup are described in significant detail in previous publications.[55,56] Briefly, the output of a regenerative amplifier (Coherent Libra-HE, 800 nm, 1 kHz, 100 fs) was split into two paths. One of the paths pumps an optical parametric amplifier (OPA, TOPAS

 This journal is © The Royal Society of Chemistry 2018

Quantronix/Light Conversion) that is used to generate 290 nm light for exciting 6MeTIno and 6MeGuo. The pump pulses were attenuated to an intensity of ≤1 μJ at the sample using a neutral density optical filter in order to minimize cross-phase modulation effects and sample degradation.[57] The second path corresponds to a small fraction of the fundamental beam that is focused through a constantly-translating 2 mm CaF_2 crystal to generate broadband probe pulses (320–710 nm). Prior to white-light generation, the second path travels through a mechanical delay stage that adjusts the time it takes the probe pulse to reach the sample up to a maximum delay of 3 ns. The broadband probe pulses are split into two fractions. One beam passes through the sample (2 mm path length) to probe the transient signals, while the other serves to correct for changes in the white light continuum throughout the experiment. The pulses are focused into optical fibers leading to CMOS detection units. Generation of the reference signal comes *via* a chopper wheel, which blocks every other pump pulse, giving an alternating sequence of spectra with and without sample excitation, and thus providing ΔA transient absorption data. During the experiments, solutions were continuously stirred using a Teflon stir bar. Solutions of 6MeTIno were prepared at concentrations of 0.4 and 0.2 mM in PBS and in acetonitrile, respectively. Experiments were also performed in acetonitrile at a concentration of 0.4 mM for 6MeTIno, but no differences in the transient absorption kinetics were observed. Solutions of 6MeGuo were prepared at a concentration of 0.5 mM in acetonitrile.

A homemade software (Labview, National Instruments, Inc.) was used for data collection. Prior to analysis, the data was corrected for group velocity dispersion of the probe pulse. Approximately 50 kinetic traces were extracted from the multidimensional transient absorption data of each sample, which were used for global and target analysis (Igor Pro 6.36). Traces for 6MeTIno and 6MeGuo were each globally fit to a two-component sequential model using target analysis.[58,59] Evolution associated difference spectra or EADS were extracted from these target and global analyses.[59] A constant offset was also needed in the global fit analysis of 6MeTIno because a long-lived transient species is populated, which does not decay within the maximum 3 ns time window of the experimental setup used in this work.

Results

Steady-state properties of 6MeTIno

Fig. 1a shows the molecular structure and standard atomic numbering of 6MeTIno, while Fig. 1b shows its absorptivity and emission spectra in PBS at pH 7.4 and in acetonitrile. 6MeTIno exhibits two broad absorption maxima centered at 291 nm (19 280 M^{-1} cm^{-1}) and 223 nm (11 750 M^{-1} cm^{-1}) in PBS. In acetonitrile, the UVB band exhibits more vibronic structure, with two maxima at 283 nm (18 530 M^{-1} cm^{-1}) and 289 nm (17 720 M^{-1} cm^{-1}) along with a maximum in the UVC region at 219 nm (12 120 M^{-1} cm^{-1}).

Following excitation at 260 or 290 nm, 6MeTIno produces a low-intensity emission band centered at 425 and 400 nm in PBS and in acetonitrile, respectively, which increases in magnitude by 1.4-fold in PBS and 1.3-fold in acetonitrile under N_2-saturated conditions (Fig. S1†). The excitation spectra taken at an emission wavelength of 400 nm in both solvents are very similar to the ground-state absorption spectrum (Fig. S2†), demonstrating that the band is due to emission from 6MeTIno. An additional emission maximum is observed at 315 and 310 nm in PBS and in acetonitrile, respectively, which is independent of O_2 concentration. The excitation spectrum taken at an emission wavelength of 315 nm in PBS overlaps the ground-state absorption (Fig. S2a†), evidencing that this band is also due to emission from 6MeTIno.

Due to its quenching by O_2 and significant Stokes shift from the lowest-energy ground-state absorption maximum (PBS: 10 323 cm^{-1}, 1.28 eV; ACN: 9920 cm^{-1}, 1.23 eV), the emission band centered at 425 nm in PBS and 400 nm in acetonitrile is assigned to phosphorescence from the lowest-energy triplet state. This assignment is supported by the calculation of an adiabatic triplet energy of 3.0 eV (410 nm) in both PBS and in acetonitrile at the UPBE0/IEFPCM/6-311++G(d,p) level of theory (see the Methods section for details). Phosphorescence quantum yields of 1.1×10^{-4} and 1.3×10^{-4} in PBS and in acetonitrile, respectively, were determined. They are of the same order of magnitude as those reported for other thionated nucleobases at room temperature.[54,60–63] The higher energy emission band in each solvent is assigned to fluorescence with estimated quantum yields on the order of 10^{-5}.

Steady-state absorption and emission properties of 6MeGuo in acetonitrile

Fig. 1c depicts the molecular structures and standard atomic numbering of 6MeGuo, while Fig. 1d shows its steady-state absorptivity and emission spectra in acetonitrile and in PBS.[44] 6MeGuo exhibits absorption maxima at 280 nm (6600 M^{-1} cm^{-1}), 247 nm (6970 M^{-1} cm^{-1}), and 211 nm (15 370 M^{-1} cm^{-1}) in acetonitrile, which are very similar to those in PBS. Based on the calculations presented below in acetonitrile, or in ref. 44 in PBS, the 280 and 247 nm bands correspond to the L_a and L_b transitions, respectively. Following 290 nm excitation, a broad emission band is formed with maximum at 370 nm in acetonitrile (382 nm in PBS[44]) and a tail extending to ~550 nm in both solvents. The emission band is assigned to fluorescence from the L_a state with a quantum yield of 0.005 ± 0.001 in acetonitrile and 0.0040 ± 0.0005 (ref. 44) in PBS. The point of overlap between the excitation and emission spectra was used to determine the $E_{0,0}$ energy of 31 600 cm^{-1} (3.92 eV, Fig. S3†), which is similar to the value of 3.91 eV previously reported in PBS.[44]

Quantum chemical calculations for 6MeTIno

The ground-state geometries of 6MeTIno were optimized for two different rotamers in which the S^6-methyl group was oriented towards the N1 or N7 atom, and for both *syn* and *anti*-sugar conformations at the PBE0/IEFPCM/6-311++G(d,p) level of theory in vacuum, water, and in acetonitrile environments (see Tables S1 to S3†). Similar ground-state optimizations were performed for 6TIno for comparison (Tables S4 to S6†). Vertical excitation energies were then computed for each geometry at the TD-PBE0/IEFPCM/6-311++G(d,p) level of theory in vacuum, water, and in acetonitrile environments. The relevant vertical transition energies (*i.e.*, those at or below 290 nm excitation) for the most stable conformations of 6MeTIno (*syn* with S^6 methyl pointing towards N1) and of 6TIno

 This journal is © The Royal Society of Chemistry 2018

(*syn*) are listed in Table 1. All other vertical excitation energies are reported in the ESI (Tables S7 to S24†).

The lowest-energy singlet state of 6MeTIno has $\pi\pi^*$ character and corresponds to the strong L_a absorption band in the UVB region (Fig. 1), where Platt's nomenclature is used,[64] consistent with the blueshift of the UVB absorption band in going from PBS to acetonitrile. It also increases in energy by 0.32/0.31 eV relative to the lowest-energy $^1\pi\pi^*$ of 6TIno in water/acetonitrile. As expected, S^6-methylation of 6TIno destabilizes the lowest-energy $^1n\pi^*$ state by 1.24/1.25 eV in water/acetonitrile, while still exhibiting a negligible oscillator strength (Table 1). Hence, excitation at 290 nm is not expected to populate the $^1n\pi^*$ state directly in the Franck–Condon region. The lowest-energy $^3n\pi^*$ state of 6TIno also increases in energy upon S^6-methylation, becoming the third lowest-energy triplet state in 6MeTIno in both solvents. The predicted changes in the order of the singlet and triplet electronic states upon S^6-methylation of 6TIno in the Franck–Condon region could potentially alter the rates and mechanism of intersystem crossing (ISC) between these molecules to a significant extent.

Table 1 also compares the vertical excitation energies calculated in this work for 6MeTIno with those reported recently for 6ThiolGua in vacuum.[43] Relevant for the discussion below, the order and character of singlet and triplet electronic states are identical for each molecule, even though the exact magnitude of their vertical energies is different. Thus, the amino group at the C2 position is predicted to play a minor role in the steady-state photophysics and initial dynamics of 6ThiolGua. Keeping in mind that these vertical excitation energies were obtained at different levels of theory, and that we are comparing a nucleobase (6ThiolGua) with a nucleoside (6TIno) in Table 1, the similar character and order of these vertical excitations suggest that the excited-state dynamics of 6MeTIno may be comparable to that of 6ThiolGua. Hence, at least to a first approximation, 6MeTIno may be used as a model compound to investigate how the solvent affects the excited-state dynamics reported for 6ThiolGua in the gas phase.[43]

Table 1 Vertical excitation energies for the *syn*-sugar conformation of 6MeTIno and 6TIno calculated at the TD-PBE0/IEFPCM/6-311++G(d,p) level of theory. Oscillator strengths are shown in parentheses

	Water		Acetonitrile		Vacuum	
State	6MeTIno[a] (eV)	6TIno (eV)	6MeTIno[a] (eV)	6TIno (eV)	6MeTIno[a] (eV)	6ThiolGua[b,43] (eV)
$^1\pi\pi^*$, L_a	4.51 (0.437)	4.19 (0.483) (S_2)	4.50 (0.439)	4.19 (0.483) (S_2)	4.54 (0.358)	4.66 (0.193)
$^1n\pi^*$	4.69 (0.003)	3.45 (0.000) (S_1)	4.69 (0.003)	3.44 (0.000) (S_1)	4.63 (0.004)	5.17 (0.004)
$^1\pi\pi^*$	5.07 (0.016)	4.30 (0.124)	5.017 (0.017)	4.29 (0.133)	5.04 (0.026)	—
$^3\pi\pi^*$	3.21	2.76	3.21	2.75	3.18	3.59
$^3\pi\pi^*$	4.36	3.44 (T_3)	4.36	3.43 (T_3)	4.32	4.69
$^3n\pi^*$	4.38	3.18 (T_2)	4.38	3.17 (T_2)	4.33	4.90

[a] S^6-Methyl pointing towards N1. [b] Taken from ref. 43 with S^6-hydrogen pointing towards N1 and calculated at the EOM-CCSD/6-31+G(2d,p) level of theory.

Quantum chemical calculations for 6MeGuo

As for 6MeTIno, we performed optimizations of the ground-state geometries on two rotamers of 6MeGuo and 6EnolGuo in which the O^6-methyl group was oriented towards the N1 or N7 atom for both *syn* and *anti*-sugar conformations at the PBE0/IEFPCM/6-311++G(d,p) level of theory. These calculations were performed in vacuum and in acetonitrile (see Tables S25 to S27†). Vertical excitation energies were computed at the TD-PBE0/IEFPCM/6-311++G(d,p) level of theory in vacuum and in acetonitrile (see Tables S28 to S39†). Similar ground-state optimization and vertical excitation energies calculations were previously performed for 6MeGuo, 6EnolGuo, and Guo in water.[44] The vertical transition energies of the most stable conformation of 6MeGuo (*syn* with O^6-methyl pointed towards N1) and 6EnolGuo (*syn* with O^6-hydrogen pointed towards N1), relevant for excitation at 290 nm (4.27 eV), are listed in Table 2.

According to these calculations, excitation of 6MeGuo or 6EnolGuo with 290 nm radiation is expected to populate the $^1\pi\pi^*(L_a)$ state, where the Platt's nomenclature was used.[64] The lowest-energy $^1n\pi^*$ state lies 0.5 to 0.6 eV above the $^1\pi\pi^*(L_a)$ state and has negligible oscillator strength, so it is not expected to be directly populated upon 290 nm excitation. The calculations predict that there are three triplet states below the $^1\pi\pi^*(L_a)$ state in the Franck–Condon region, all with $\pi\pi^*$ character in acetonitrile. Hence, no El-Sayed allowed singlet-to-triplet transition[65] is predicted to occur from the Franck–Condon region of the $^1\pi\pi^*(L_a)$ potential energy surface (PES). Also shown in Table 2 are recent calculations performed for 6EnoGua in vacuum.[43] These calculations are in reasonable agreement with those presented in this work for 6MeTIno and 6EnolGuo, keeping in mind that they were performed for the enol tautomer of the nucleobase and at a different level of theory.

Ultrafast transient absorption of 6MeTIno

The excited-state dynamics of 6MeTIno were studied using broadband transient absorption spectroscopy in PBS and in acetonitrile exciting at 290 nm

Table 2 Vertical excitation energies for the *syn*-sugar conformation of 6MeGuo and 6EnolGuo calculated at the TD-PBE0/IEFPCM/6-311++G(d,p) level of theory. Oscillator strengths are shown in parentheses

State	Acetonitrile 6MeGuo[a] (eV)	Vacuum 6MeGuo[b] (eV)	Vacuum 6EnolGuo[c] (eV)	Vacuum 6EnolGua[d,43] (eV)
$^1\pi\pi^*$, L_a	4.68 (0.173)	4.79 (0.198)	4.79 (0.145)	5.12 (0.107)
$^1n\pi^*$	5.30 (0.005)	5.20 (0.003)	5.29 (0.003)	5.48 (0.003)
$^3\pi\pi^*$	3.45	3.45	3.46	4.01
$^3\pi\pi^*$	4.30	4.28 ($^3n\pi^*$)	4.28	4.82
$^3\pi\pi^*$	4.77	4.82	4.86	5.39

[a] O^6-Methyl pointing towards N7. The T_2 state has mostly $n\pi^*$ character when the O^6-methyl is pointing toward N1 (see ESI). [b] O^6-Methyl pointing towards N1. The T_2 state has mostly $\pi\pi^*$ character when the O^6-methyl is pointing towards N7 (see ESI). [c] O^6-Hydrogen pointing towards N1. The O^6-hydrogen rotamer pointing towards N7 is only 0.4 kcal mol^{-1} higher in energy. See ESI for details. [d] Taken from ref. 43 with O^6-hydrogen pointing towards N1 and calculated at the EOM-CCSD/6-31+G(2d,p) level of theory.

 This journal is © The Royal Society of Chemistry 2018

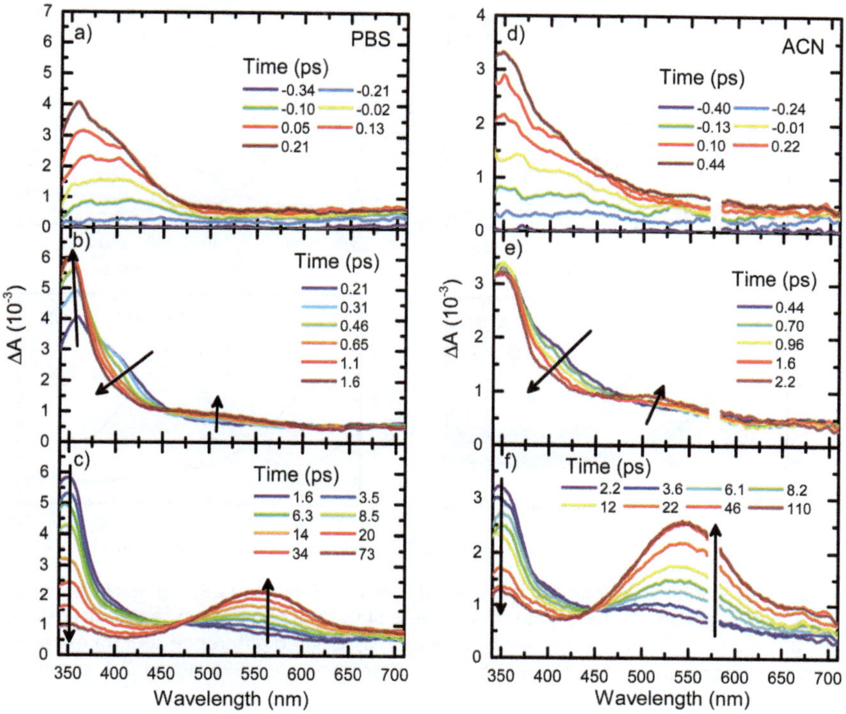

Fig. 2 Transient absorption spectra of 6MeTIno in PBS pH 7.4 (a to c) and acetonitrile (d to f) following 290 nm excitation. The spectral region from *ca.* 570 to 580 nm is omitted in acetonitrile due to Rayleigh scattering from the overtone of the pump beam.

(Fig. 2). The relaxation dynamics were globally fit to a two-component sequential model plus a constant offset in PBS and in acetonitrile (Fig. 3), and the associated lifetimes are given in Table 3. A broad transient absorption band in the UVA region, which has a maximum between 375 to 425 nm and a long tail in the complete probe window, is observed within the cross-correlation of the pump and probe beams in PBS (Fig. 2a). From *ca.* 0.2 to 1.6 ps time delay (Fig. 2b), the red-side of the UVA band blueshifts, while the amplitude of its blue-side continues to increase and blueshift to 350 nm. Occurring simultaneously with this process is the rise of a band centered at 515 nm. The initial dynamics are probe-wavelength dependent across the UVA band (Fig. S4†). After *ca.* 1.6 ps, the transient absorption spectrum decays, with a simultaneous observation of a transient species with absorption maxima at 550 nm and at 350 nm (Fig. 2c). An apparent isosbestic point at 472 nm is also observed. After *ca.* 35 ps, the transient species does not decay appreciably during the maximum time window used in this work of 3 ns (Fig. 3a).

Excitation of 6MeTIno at 290 nm in acetonitrile reveals very similar dynamics to those observed in PBS, as shown in Fig. 2. The major difference observed is that the initial broad bands below 450 nm overlap more strongly in acetonitrile, while similar blueshift and simultaneous rise of the band with maximum around 350 nm is observed during the first 2.2 ps (Fig. 2d and e). The transient species at 2.2 ps time delay starts decaying, with the simultaneous observation of a long-

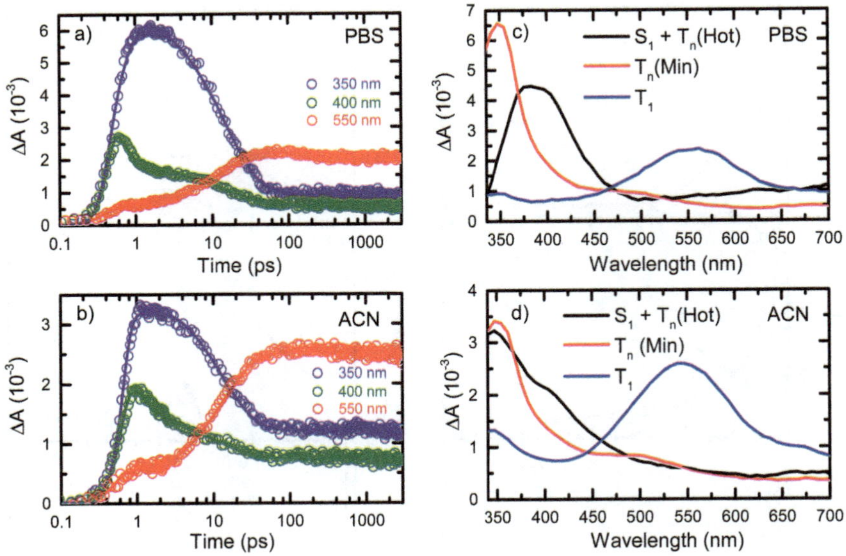

Fig. 3 Representative kinetic decay traces of 6MeTIno following 290 nm excitation in (a) PBS at pH 7.4 and in (b) acetonitrile. Traces were fit to two-component sequential models plus a constant offset in both solvents. Evolution associated difference spectra extracted from global analysis of kinetic TAS data for 6MeTIno in (c) PBS at pH 7.4 and in (d) acetonitrile.

lived transient species with absorption maxima at 540 and 350 nm (Fig. 2f). An apparent isosbestic point is observed at 448 nm in acetonitrile. After 25 ps, the transient species does not decay appreciably during the maximum time window used in this work of 3 ns (Fig. 3b). Decay-associated spectra were extracted from global analyses in both solvents and match the transient absorption spectra (Fig. 3c and d).

Ultrafast transient absorption of 6MeGuo

The ultrafast electronic relaxation dynamics of 6MeGuo were monitored in acetonitrile (Fig. 4). Upon excitation with 290 nm radiation, two transient absorption bands are observed within the cross-correlation of the pump and probe beams (Fig. 4a); a broad band with maximum around 600 nm and another band with maximum below 350 nm. Both bands rise during the initial 0.5 ps time

Table 3 Decay lifetimes obtained from global and target analyses of the transient data for 6MeTIno

Solvent	$\tau_1{}^a$ (ps)	$\tau_2{}^b$ (ps)
Acetonitrile	1.01 ± 0.35	12.3 ± 0.9
PBS, pH 7.4	0.29 ± 0.24	15.3 ± 0.1

[a] Assigned to intersystem crossing from the $^1\pi\pi^*(L_a)$ state to a receiver triplet state (T_n) plus IVR in the T_n state. Reported lifetime corresponds to the average (global value) lifetime in the probe region from 365 to 450 nm (Fig. S4). [b] Assigned to internal conversion from the T_n minimum to the lowest-energy triplet state, $T_1(\pi\pi^*)$. See the Discussion section for details.

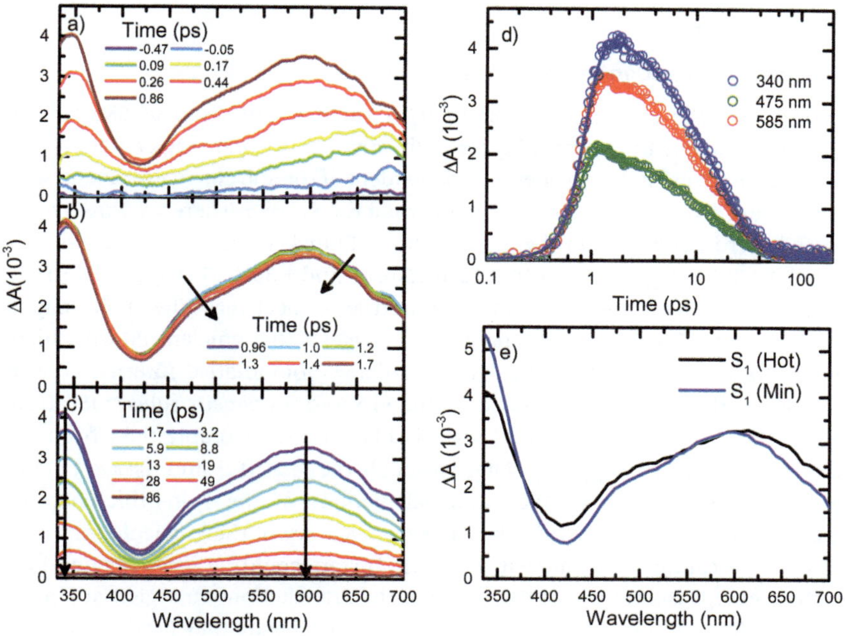

Fig. 4 (a–c) Transient absorption spectra of 6MeGuo in acetonitrile following 290 nm excitation. (d) Representative kinetic decay traces of 6MeGuo at select probe wavelengths. Traces were fit to a two-component sequential model. Best-fit curves are also shown as solid lines. (e) Evolution associated difference spectra extracted from the global analysis.

delay, but the UV band continues to rise until 1.2 ps (Fig. 4b), while the absorption band with maximum around 600 nm narrows for up to 1.6 ps. This process is then followed by a uniform decay across the entire probe window until no signal is detected after 86 ps (Fig. 4c). The transient absorption data were globally fit using a two-component sequential model, and the extracted lifetimes are provided in Table 4. Representative decay traces and best global-fit curves are shown in Fig. 4d. Evolution associated difference spectra were extracted from the global analysis across the entire probe window and follow the dynamics observed in the transient absorption spectra (Fig. 4e).

Table 4 Decay lifetimes obtained from global and target analyses of the transient data for 6MeGuo

Solvent	τ_1[a] (ps)	τ_2[b] (ps)	τ_3[c] (ps)
Acetonitrile	0.30 ± 0.10	15.4 ± 0.5	nd[d]
PBS, pH 7.4	0.95 ± 0.07	39 ± 1	2.4 ± 1.6[e]

[a] Assigned to conformational relaxation and internal conversion through the C2 reaction coordinate. [b] Assigned to conformational relaxation from the $^1\pi\pi^*(L_a, FC)$ to $S_1(L_a, min)$ along the C6 reaction coordinate. [c] Assigned to vibrational cooling in the ground-state. [d] nd = not detected. [e] Average (global) value obtained for the probe-wavelength dependent lifetimes across the UV band (Fig. S4). The error represents the standard deviation for the range of the lifetimes.

Discussion

Excited-state relaxation mechanism for 6MeTIno

According to the calculations presented in Table 1, excitation of 6MeTIno at 290 nm populates the $S_1(\pi\pi^*, L_a)$ state directly. The second singlet state is predicted to be 0.2 eV higher in energy, independent of the solvent environment, with an oscillator strength *ca.* 145-fold smaller than the S_1 state. There are three triplet states lower in energy than the S_1 state in the Franck–Condon region, with predicted $(S_1–T_n)$ energy gaps of 0.10/0.12, 0.15/0.14, and 1.30/1.29 eV for the $T_3(n\pi^*)$, $T_2(\pi\pi^*)$, and $T_1(\pi\pi^*)$ states in PBS/acetonitrile, respectively. The small magnitude of the $(S_1–T_2)$ and $(S_1–T_3)$ energy gaps suggests that efficient ISC to a high-energy triplet state may compete with radiative/nonradiative internal conversion (IC) to the ground state. Based on the El-Sayed propensity rules,[65] ISC from the $^1\pi\pi^*(L_a)$ state to the $T_3(n\pi^*)$ state is likely more favorable than from the $^1\pi\pi^*(L_a)$ state to the $T_2(\pi\pi^*)$ state, although the $^1\pi\pi^*(L_a)$ and $T_2(\pi\pi^*)$ states could also be strongly coupled and calculations of the spin-orbit coupling elements are necessary. Within the expected accuracy of these calculations of $\pm$ 0.2 eV,[66] we cannot rule out that the $S_2(n\pi^*)$ state could be isoenergetic to (or lower in energy than) the $^1\pi\pi^*(L_a)$ state in the Franck–Condon region at a more advanced level of theory. In such a case, the $S_2(n\pi^*)$ state may participate in IC and/or ISC to the singlet and triplet manifolds, respectively.

The femtosecond transient absorption data shown in Fig. 2 and 3 reveal that photoexcited 6MeTIno relaxes through a combination of at least two decay pathways. The relaxation process during τ_1 corresponds primarily to a narrowing and blueshifting of a transient absorption band with a maximum in the UVA region (Fig. 2). The magnitude of this lifetime decreases in going from 365 to 430 nm probe wavelengths (Fig. S4†), and the average (global) lifetime is significantly longer in acetonitrile than in PBS (Table 3).

Collectively, the spectral features, together with the wavelength- and solvent-dependence, suggest that τ_1 is associated with an intra- or intermolecular vibrational relaxation (IVR) process occurring in the state responsible for the primary absorption in the UVA probe region. According to the calculations discussed in the previous paragraph, the first lifetime (τ_1) is most likely assigned to ultrafast ISC from the $^1\pi\pi^*(L_a)$ state to the $T_3(n\pi^*)$ state. However, ultrafast IC from the initially populated $^1\pi\pi^*$ state to the $^1n\pi^*$ state that occurs outside of the Franck–Condon region cannot be completely ruled out. A transient species in the UVA probe region has been observed at early time delays in previous transient absorption experiments with other sulfur-substituted nucleic acid derivatives,[67–69] which was assigned to a lower-energy $^1n\pi^*$ state.

During τ_2, the state responsible for the absorption in the UVA probe region decays with the simultaneous rise of a transient species that shows a broad band with a maximum in the visible region. The observation that this transient species does not decay within a time delay of 3 ns in Fig. 3a and b suggests that it should be assigned to the $T_1(\pi\pi^*)$ state, as observed in many other thionated nucleobases.[12,17,60,61,70–75] The room-temperature phosphorescence reported in Fig. 1 for 6MeTIno lends further support to this assignment. Previous transient absorption investigations have shown that several thiopurine monomers populate a triplet state on an ultrafast time scale and in nearly unity yield.[13,17,73]

Based on the discussion above, two plausible relaxation mechanisms can be postulated to explain the experimental observations for 6MeTIno upon excitation at 290 nm (Scheme 1): (a) $^1\pi\pi^*(L_a) \rightarrow {}^3n\pi^*/{}^3\pi\pi^* \rightarrow {}^3\pi\pi^*$ or (b) $^1\pi\pi^*(L_a) \rightarrow {}^1n\pi^* \rightarrow {}^3\pi\pi^*$. In the first mechanism (Scheme 1a), ultrafast intersystem crossing (ISC) to a high-energy triplet state (T_n, *i.e.*, T_3 or T_2 state, according to the predictions in Table 1) is followed by slow internal conversion (IC) to the $T_1(\pi\pi^*)$ state.

The receiver T_n state experiences significant intra- or intermolecular vibrational relaxation during τ_1, which together with the large energy gap between the receiver T_n state and the T_1 state (1.2 eV) may result in poor Franck–Condon factors for IC.[76] Isoenergetic states usually have large coupling matrix elements, whereas electronic states with large energy mismatch are generally less efficient for depopulation. The relatively slow rate of IC may thus be a manifestation of the energy gap law for nonradiative transitions.[77,78] We note that nonradiative IC from upper-energy triplet states to the T_1 state on a picosecond time scale has been reported for other aromatic compounds.[55,57,79–82]

In the second mechanism, depicted in Scheme 1b, ultrafast IC from the $^1\pi\pi^*(L_a)$ state to the $^1n\pi^*$ state is followed by, or occurs simultaneously with, intra- or intermolecular vibrational relaxation in the $^1n\pi^*$ state within τ_1, followed by ISC to the $T_1(\pi\pi^*)$ state in 12 to 15 ps, depending on the solvent used. In this mechanism, ISC to an upper-energy triplet state of $\pi\pi^*$ character cannot be ruled out by the calculations. Here again, the large energy gap between the $^1n\pi^*$ state and the $T_1(\pi\pi^*)$ state of *ca.* 1.4 eV (in the Franck–Condon region) may explain the relatively slow ISC compared to other thionated nucleobases.[12,17,60,61,70–75] Given the experimental and computational results currently on hand, we propose that Scheme 1a most likely explains the relaxation mechanism of 6MeTIno in solution. However, higher-level static and dynamics calculations are necessary to provide a more thorough prediction of the electronic relaxation mechanism leading to the population of the $T_1(\pi\pi^*)$ state in this molecule.

Further support that ultrafast ISC is followed by slow IC in the relaxation mechanism of 6MeTIno comes from the experimental and computational results

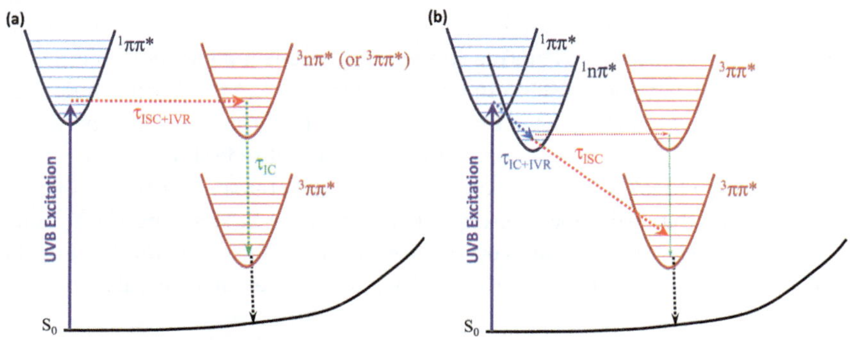

Scheme 1 Plausible electronic-energy relaxation mechanisms for 6MeTIno in aqueous and in acetonitrile solutions upon excitation at 290 nm. The lifetimes correspond to those reported in Table 3. The thickness of the line in (b) represents the anticipated likelihood of a given relaxation pathway, with thicker lines representing a higher probability. Mechanism (a) is suggested to be more probable than (b) based on the experimental and computational results available to this date for 6MeTIno and 6ThiolGua (see text for details).

recently reported for 6ThiolGua in the gas phase.[43] In that work, de Vries and co-workers investigated the excited-state dynamics of 6ThiolGua with picosecond time resolution and nanosecond temporal window. The authors observed complex dynamics depending on the amount of excess excitation energy used, which they argued could be due, at least to some extent, to the participation of other tautomers/rotamers in their experiments. In general, three lifetimes were extracted from the pico- to nanosecond excited-state dynamics of 6ThiolGua in the gas phase.[43] The first lifetime varies from 90 ps to 4.3 ns; the second from *ca.* 1 to 24 ns; and the third from *ca.* 0.46 to 3.0 μs; all depending on the excess excitation energy. Vertical excitation energies calculated at the EOM-CCSD/6-31+G(2d,p)//CCSD/6-31+G(2d,p) level of theory in vacuum suggest that the S_1 state has $\pi\pi^*$ state character with a large oscillator strength, followed by a $S_2(n\pi^*)$ state with negligible oscillator strength, as is the case for 6MeTIno reported in Table 1. Three triplet states were found lower or close in energy to the Franck–Condon region of the $S_1(\pi\pi^*)$ state,[43] with $T_1(\pi\pi^*)$, $T_2(\pi\pi^*)$, and $T_3(n\pi^*)$ character, which is in good agreement with the calculations reported in Table 1 for 6MeTIno. Siouri *et al.*[43] also perform nonadiabatic molecular dynamics simulations at the CASSCF level of theory using the SHARC program developed by González and co-workers.[83] With the caveats that only 10 trajectories were computed in their simulation,[43] and that CASSCF is known to underestimate energy barriers due to the lack of dynamical electron-electron correlation,[84] the authors identified one conical intersection (CI) between the S_1 and the ground state (S_0) and two ISC geometries at the $S_1(\pi\pi^*)/T_3(n\pi^*)$ and the $S_1(\pi\pi^*)/T_2(\pi\pi^*)$ crossing regions. Access to the S_1/S_0 CI involves the loss of the S^6-hydrogen, which can be accessed after reaching a minimum in the S_1 PES at the CASSCF level of theory. Access to the $S_1(\pi\pi^*)/T_3(n\pi^*)$ ISC pathway involves a distortion of the pyrimidine heterocycle, whereas access to the $S_1(\pi\pi^*)/T_2(\pi\pi^*)$ ISC pathway corresponds to the rotation of the thiol group out of the molecular plane of the main chromophore. Both ISC pathways can be accessed in a barrierless fashion from the Franck–Condon region of the S_1 PES. The picosecond pump-probe data reveal the initial population of a receiver state from which a long-lived dark state is populated. No evidence of IC to the S_0 state was obtained. Thus, the authors concluded that excitation of 6ThiolGua to the $S_1(\pi\pi^*)$ state leads to ultrafast ISC to a receiver T_2 or T_3 state, from which very slow (*ca.* 1 to 24 ns) IC to the long-lived $T_1(\pi\pi^*)$ state occurs. The relaxation mechanism proposed for 6ThioGua in the gas phase is analogous to the mechanism we proposed for 6MeTIno in solution (Scheme 1a). We further note that access to a $(S_1/S_0)_{CI}$, analogous to the one identified for 6ThioGua,[43] is expected to require a larger amount of excess energy in the case of 6MeTIno because of the requirement to break a S^6-methyl chemical bond. Of course, we cannot rule out the possibility that other lower-energy CIs between the S_1 state and the S_0 state may exist, which could enable competitive pathways for IC to the S_0 state in addition to the ultrafast ISC to the triplet manifold revealed in this work for 6MeTIno.

Interestingly, 6MeTIno in solution undergoes ISC to the triplet manifold more than an order of magnitude faster than 6ThiolGua in the gas phase.[43] It is currently unknown whether this difference originates from the different environments used in each study (*i.e.*, solution *versus* gas phase), lack of adequate time resolution and/or tautomer selectivity in the gas phase experiments, or the structural differences between 6MeTIno and 6ThioGua.

 This journal is © The Royal Society of Chemistry 2018

Excited-state relaxation mechanism of 6MeGuo in solution

Excited-state relaxation in 6MeGuo occurs within a time window of *ca.* 86 ps *via* a two-component mechanism. The first lifetime ($\tau_1 = 0.30$ ps) corresponds to a rise of the UV band and narrowing of the visible band, which is followed by a uniform decay across the probe window with a lifetime of 16 ps (τ_2) in acetonitrile. These spectral features are consistent with those previously reported for 6MeGuo in PBS at pH 7.4,[44] suggesting that they correspond to a similar relaxation process (Scheme 2a). In PBS, it was proposed that 6MeGuo undergoes parallel relaxation pathways from the Franck–Condon region of the $^1\pi\pi^*(L_a)$ PES upon 290 nm excitation, in which ultrafast IC through a CI along the C2-puckered reaction coordinate competes with conformational relaxation to a $^1\pi\pi^*(L_a)$ minimum along the C6-puckered coordinate. The population that internally converts through the C2-puckered CI undergoes vibrational cooling in the ground state, as observed for the canonical guanine monomers and other nucleobases in solution.[85–92] The population that reaches the $^1\pi\pi^*(L_a)$ minimum either fluoresces or decays nonradiatively to the ground state in *ca.* 40 ps after surmounting a small energy barrier towards a conical intersection along the C6 coordinate.

Interestingly, in contrast with the transient absorption experiments done in PBS,[44] the data shown in Fig. 4 do not show clear evidence of vibrational cooling in the ground state when excitation of 6MeGuo is performed at 290 nm in acetonitrile. The UVA absorption band was globally fit independently from the visible transient band, as it was done in the previous study,[44] in an attempt to extract a vibrational cooling lifetime. However, no clear evidence was found for the need to use more than two lifetimes, with equal values as those for the visible band within experimental uncertainties, in acetonitrile. Additionally, the first lifetime across the UVA band does not show clear evidence of probe-wavelength-dependency when the experiments are performed in acetonitrile, in contrast to what was observed in PBS.[44] This finding suggests that the relaxation pathway leading to vibrational cooling in the ground state of 6MeGuo in PBS cannot be accessed (or participates to a lesser extent) when acetonitrile is used as the solvent. Thus, according to this hypothesis, excitation of 6MeGuo at 290 nm in

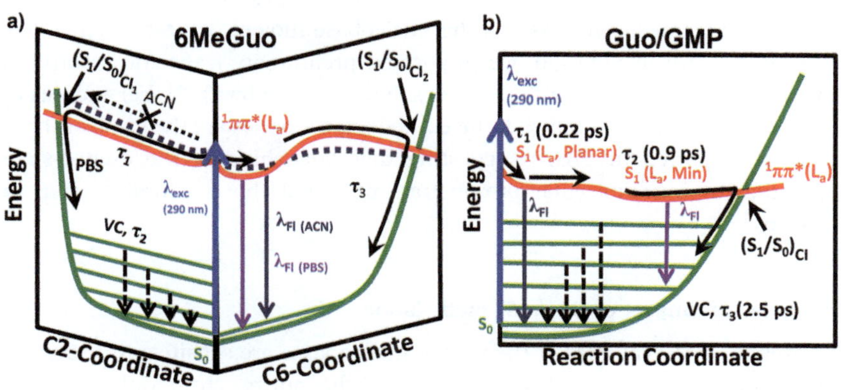

Scheme 2 (a) Proposed relaxation mechanism for 6MeGuo in PBS at pH 7.4 (solid line) and acetonitrile (dashed line). (b) Relaxation mechanism for GMP reported by Karunakaran *et al.*[91]

acetonitrile leads the majority of population in the Franck–Condon region towards the C6-puckered minimum along the $^1\pi\pi^*(L_a)$ PES in 0.30 ps (Scheme 2a). This relaxation process is *ca.* 650 fs faster than that in PBS, which suggests that conformational relaxation along the C6 coordinate is more favorable in acetonitrile than along the C2 coordinate. If so, this may explain why vibrational cooling in the ground state is not observed. The low fluorescence quantum yield in acetonitrile shows that most of the excited-state population in 6MeGuo relaxes to the ground state nonradiatively, which we argue here and elsewhere[44] is through a CI along the C6 coordinate. The 2.4-fold faster IC in acetonitrile than in PBS (*i.e.*, τ_2 in Table 4) suggests that a lower energy barrier needs to be surmounted to reach the $(S_1/S_0)_{CI}$ along the C6 coordinate in acetonitrile than in water. It is also conceivable that a higher energy barrier may need to be surmounted to access the $(S_1/S_0)_{CI}$ along the C2 coordinate in acetonitrile compared to in aqueous solution. If these propositions are correct, conformational relaxation to reach the $(S_1/S_0)_{CI}$ along the C6 coordinate would be expected to be more favorable in acetonitrile than in aqueous solution, possibly because acetonitrile has a *ca.* 3-fold lower viscosity ($\eta = 0.344$ mPa s)[93] than water ($\eta = 0.890$ mPa s).[94] The lifetime for conformational relaxation in Guo and GMP has also been proposed to be dependent on solvent viscosity.[90,95] Hence, the results presented in this study suggest that a solvent-dependent energy barrier must be surmounted to access the S_1/S_0 crossing regions along both the C2 and C6 coordinates. Clearly, however, high-level quantum-chemical calculations that include solvent effects are necessary to scrutinize the proposed relaxation mechanism.

It should also be noted that recent computational studies have suggested that adenosine[96] and 8-oxo-7,8,dihydro-2′-deoxyguanosine (8-oxodGuo),[97] when in the *syn* conformation, may relax *via* intramolecular proton transfer from the 5′-OH to N3 positions. This charge-transfer state is predicted to become inaccessible in the anionic form of 8-oxodGuo (deprotonation at the O6 position),[97] leading to a *ca.* 40-fold longer excited-state lifetime relative to the neutral species.[98] Thus, it may seem reasonable to suggest that a similar mechanism may be at play in Guo and 6MeGuo. Under this model, GMP is expected to undergo different electronic decay dynamics than Guo due to capping of the 5′-OH group with a phosphate. However, a previous study showed that GMP and Guo undergo identical electronic relaxation kinetics in aqueous solution,[95] suggesting that their primary decay mechanisms are intrinsic to the nucleobase moiety. To test whether this pathway is active in 6MeGuo, we performed preliminary transient absorption experiments in phosphate-buffered D_2O solution (not shown). Global analysis of the time-domain data reveals that the excited-state kinetics are the same of those previously reported in PBS,[44] suggesting that quenching of the excited-state population by intramolecular proton transfer is not the rate-limiting step in 6MeGuo.

Photophysical impact of O^6- or S^6-methylation

The excited-state dynamics of the purine nucleobases are sensitive to structural modifications.[70,84,99] It has been shown that the purine chromophore is not responsible for the ultrafast IC to the ground state observed in the adenine and guanine monomers.[84] Instead, the functionalization of the methine group at the C6 position of the purine chromophore with an oxo or an amino group to form the

 This journal is © The Royal Society of Chemistry 2018

guanine or adenine monomer, respectively, has been shown to facilitate the ultrafast decay of the excess electronic energy to the ground state.[84] The present study demonstrates that O^6- or S^6-methylation also modulates the electronic relaxation mechanism of the purine or thiopurine monomers, respectively.

O^6-Methylation lengthens the overall $^1\pi\pi^*(L_a)$ decay lifetime of Guo by more than 10-fold in acetonitrile and 40-fold in PBS (Table 2 and Scheme 2). It is proposed that this longer lifetime corresponds to relaxation through the C6 reaction coordinate, which occurs in parallel with ultrafast relaxation through the C2 coordinate (Scheme 2). Thus, it is argued that O^6-methylation effectively increases the probability of relaxation through the C6 coordinate. Surface hopping simulations of Gua including solvent effects indicate that the probability of undergoing IC through a C6-puckered conical intersection increases significantly in solution relative to in the gas phase,[100] supporting the hypothesis that relaxation through the C6 coordinate in the guanine monomers may be competitive with relaxation along the C2 coordinate in solution. The single bond type of the C6–OMe and C6–SMe should also facilitate relaxation through the C6 coordinate compared to the Gua monomers. In addition, the relative population of the C2 and C6 coordinates appear to be sensitive to the specific solvent environment, as suggested by the minor role that the C2 coordinate plays in acetonitrile for 6MeGuo.

This study, as well as previous computational and experimental investigations, indicates that an increase in the singlet lifetime in the Gua derivatives may be related to a change in the C6–O bond order of the Gua chromophore,[43,44,98,101,102] which may have a significant impact on the topology of the $^1\pi\pi^*(L_a)$ PES. In particular, 6EnolGua has been shown to relax to a long-lived dark state in the gas phase,[43] and TD-DFT[101] and CASPT2 (ref. 102) calculations reveal that the C6 reaction coordinate is favored in 6EnolGua. Additionally, the S_1 lifetime of the oxidation product of guanosine, 8-oxodGuo, increases from 0.9 ± 0.1 ps in its neutral form to 43 ± 3 ps in its anionic form in aqueous solution, in which an enolate has been proposed to be formed at the O^6 position.[98] Anionic 8-oxodGuo also exhibits a time constant of 3.0 ± 0.8 ps, which was assigned to vibrational cooling in the ground state following ultrafast IC-suggesting relaxation through parallel pathways. The authors initially proposed that deprotonation of 8-oxodGuo allows a fraction of the $^1\pi\pi^*(L_a)$ state population to access an $^1n\pi^*$ state that is not accessible in the neutral form.[98] However, further calculations by the same groups suggested that the anionic form of 8-oxodGuo may populate parallel reaction coordinates along the $^1\pi\pi^*(L_a)$ PES in vacuum.[103] Two key reaction coordinates were identified: (1) folding about the C4–C5 bond and (2) out-of-plane distortion of the amino group at the C2 position, together with distortion of the pyrimidine ring. Both reaction coordinates were predicted to have sizable energy barriers to reach their corresponding CI, thus potentially explaining the *ca.* 40-fold longer $^1\pi\pi^*(L_a)$ lifetime compare to the neutral species. The relaxation mechanism put forward by Lu *et al.*,[103] which was proposed using the 8-oxodGuo nucleobase (8-oxoGua) as a model compound, was recently supported with calculations for the anionic form of 8-oxodGuo.[97] While the calculations performed by Lu *et al.*[103] and Tuna *et al.*[97] for the O^6 deprotonated form of 8-oxodGuo in vacuum suggest a different relaxation mechanism than that proposed in this study for O^6-methylation of Guo along the $^1\pi\pi^*(L_a)$ PES in solution, it is noticeable that electronic relaxation in both molecules occurs with nearly identical lifetimes in aqueous solution. Regardless, the excited-state dynamics of 6MeGuo, 6EnolGua, and the anionic form of 8-oxodGuo lend further support to the idea that modifications of the functional group at the C6 position of the purine chromophore significantly modulates the topology of the $^1\pi\pi^*(L_a)$ PES and the relaxation dynamics of the purine derivatives in solution.[84]

Similarly, S^6-methylation is expected to modify the topology of the $^1\pi\pi^*(L_a)$ PES of 6TIno. Unfortunately, the ultrafast excited-state dynamics of 6TIno have yet to be investigated. What is apparent from the calculations presented in Table 1 is that S^6-methylation destabilizes the $^1n\pi^*$ state of 6TIno, such that it is predicted to be inaccessible upon excitation of 6MeTIno at 290 nm. 6MeTIno is also predicted to have larger $(S_1–T_n)$ and $(S_1–T_1)$ energy gaps compared to 6TIno. Still, ultrafast ISC dynamics to the triplet manifold is operative in 6MeTIno, as has been reported for other thiopurine derivatives such as 6MP, 6TGua, and 6TGuo.[13,17,73] Interestingly, the $^1n\pi^*$ state acts as a doorway to the triplet manifold in those thiopurine derivatives.[17–19,73] Regardless, S^6-methylation leads to a nearly 100-fold longer lifetime to populate the T_1 state in 6MeTIno compared to 6MP or 6TGuo,[13,73] whether it be through slow IC from a T_n state or through a direct ISC from the $^1n\pi^*$ state (Scheme 1). As discussed above, the results recently reported for 6ThiolGua in the gas phase,[43] where it was proposed that IC from a receiver triplet state (T_3 or T_2) to the T_1 state occurs on a nanosecond time scale, are supportive of the relaxation mechanism proposed for 6MeTIno in Scheme 1a, in which ISC occurs faster than IC in the triplet manifold.

The results presented in this work, and in previous studies,[44,48,49,84,101–103] suggest that the functional group at the C6 position plays an important role in regulating the excited-state dynamics of the purine monomers. However, further discussion is warranted to generalize the photophysical effects of O^6- or S^6-methylation to other purine monomers that do not or do have an amino group at the C2 position, respectively. Fluorescence up-conversion and transient absorption measurements revealed that hypoxanthine and inosine repopulate the ground state with a lifetime that is *ca.* 5-fold shorter than Guo.[104,105] The authors proposed that the absence of the amino group likely steepens the gradient of the $^1\pi\pi^*(L_a)$ PES toward the CI along the C2 coordinate relative to the "flat" PES proposed for guanine monomers.[104,105] If the absence of the amino group at the C2 position of Gua increases the rate of relaxation along the C2 coordinate, one may expect that O^6-methylation of inosine will also increase the probability of relaxation through the C2 coordinate compared to 6MeGuo. If that is the case, the observation of vibrational dynamics in the ground state of O^6-methyl inosine upon 290 nm excitation, particularly in acetonitrile, could afford further support to the relaxation mechanism proposed in Scheme 2a for 6MeGuo.

It is also currently unknown whether the relaxation mechanism proposed for 6MeTIno is operative in 6MeTGuo. Direct comparison of the ultrafast dynamics in 6MP and 6TGua indicates that absence of the amino at the C2 position reduces the magnitude of the ISC lifetime nearly 3-fold.[17,73] In addition, a minor IC pathway reported for 6TGua and 6TGuo from the S_1 state to the ground state was not observed in 6MP,[13,17] suggesting that access to the CI with the ground state is inhibited (or is less competitive) in 6MP. On the other hand, vertical excitation energies for the *syn*-sugar conformation of 6MeTGuo in water, where the S^6-methyl group is pointing towards N1 (Table S40†), predict that the amino group does not significantly alter the ordering of electronic states below the $^1\pi\pi^*(L_a)$

 This journal is © The Royal Society of Chemistry 2018

state or either of the (S_1–T_n) and (T_n–T_1) energy gaps compared to the same conformer for 6MeTIno in water (Table 1). Hence, these calculations suggest that the lifetime for ISC from the $^1\pi\pi^*(L_a)$ state to T_n state, and for IC from the T_n state to the T_1 state, may be similar to those reported herein for 6MeTIno in aqueous solution. Evidently, however, additional experimental and computational work is needed to further clarify the electronic relaxation mechanisms in these important biological molecules.

Conclusions

This study reports on the excited-state dynamics of 6MeTIno in PBS at pH 7.4 and in acetonitrile, and those of 6MeGuo in acetonitrile following excitation at 290 nm. The time-resolved results are complemented with steady-state measurements, and with a comprehensive set of ground- and excited-state calculations at the DFT and TD-DFT levels of theory. It is shown that S^6- and O^6-methylation significantly modulate the rates and relaxation pathways previously reported for the guanine and thiopurine monomers in solution. Specifically, it is proposed that excitation of 6MeTIno at 290 nm populates the $^1\pi\pi^*(L_a)$ state. ISC to a receiver T_n state is strongly coupled with intra- or intermolecular vibrational relaxation in the T_n state, both processes occurring within *ca.* 0.3 and 1 ps in PBS and in acetonitrile, respectively, according to the TD-DFT predictions and the broadband transient absorption measurements. The relaxed T_n state then internally converts to the $T_1(\pi\pi^*)$ state in 15 ps in PBS, or in 12 ps in acetonitrile.

Excitation of 6MeGuo at 290 nm is also shown to populate the $^1\pi\pi^*(L_a)$ state in acetonitrile, as reported recently in PBS.[44] The $^1\pi\pi^*(L_a)$ state population is proposed to relax in 0.95 ps along the C2- and C6-puckering reaction coordinates in PBS,[44] whereas it is suggested to primarily relax in 0.30 ps along the C6-puckering reaction coordinate in acetonitrile. The population reaching the $^1\pi\pi^*(L_a)$ minimum along the C6 coordinate decays radiatively to the ground state with a *ca.* 10^{-3} fluorescence yield while the majority of the population decays nonradiatively in 16 ps. In contrast to what has been observed in PBS, ground-state vibrational cooling is not observed in acetonitrile, suggesting that ultrafast relaxation through the C2 coordinate is inaccessible in this solvent. Hence it is proposed that the topology of the $^1\pi\pi^*(L_a)$ PES along the C2- and C6-puckering reaction coordinates is a key element in controlling the excited-state dynamics in 6MeGuo. Furthermore, the results suggest that solvent-dependent energy barriers must be surmounted in order to access the S_1/S_0 crossing regions along the C2 and C6 reaction coordinates.

Collectively, the results presented in this study, together with those from a previous investigation,[44] demonstrate that O^6-methylation significantly decreases the rate of IC to the ground state observed in the guanine monomers. Similarly, S^6-methylation significantly decreases the rate of the population of the T_1 state, while simultaneously blueshifting the ground-state absorption spectrum compared to those for the 6TIno, 6TGua, and 6TGuo monomers. The observation that the T_1 state of 6MeTIno is populated suggests that UVB light absorption by 6MeTIno could potentially lead to oxidatively-generated cellular damage, as has been shown for 6TGuo and other thiopurine prodrugs.[8,11,16,33,106–108] This study provides relevant new insights into the relationship between structural modifications of the purine chromophore (methylation of the carbonyl group at the C6 position in particular) and the excited-state dynamics of this important group of biological molecules.

Conflicts of interest

The authors declare no conflicts of interest.

Acknowledgements

The authors acknowledge funding from the National Science Foundation (Grant # CHE-1255084 and CHE-1539808). The authors also thank Mr Matthew Brister for acquiring some of the transient absorption data for 6MeGuo in acetonitrile. This work made use of the High Performance Computing Resource in the Core Facility for Advanced Research Computing at CWRU.

Notes and references

1 M. Presta, M. Belleri, A. Vacca and D. Ribatti, *Leukemia*, 2002, **16**, 1490–1499.
2 J. F. Colombel, W. J. Sandborn, W. Reinisch, G. J. Mantzaris, A. Kornbluth, D. Rachmilewitz, S. Lichtiger, G. D'haens, R. H. Diamond and D. L. Broussard, *N. Engl. J. Med.*, 2010, **362**, 1383–1395.
3 D. C. Pearson, G. R. May, G. H. Fick and L. R. Sutherland, *Ann. Intern. Med.*, 1995, **123**, 132–142.
4 M. C. Dubinsky, E. J. Feldman, M. T. Abreu, S. R. Targan and E. A. Vasiliauskas, *Am. J. Gastroenterol.*, 2003, **98**, 1058–1063.
5 A. Massey, Y.-Z. Xu and P. Karran, *DNA Repair*, 2002, **1**, 275–286.
6 D. J. Warren, A. Andersen and L. Slørdal, *Cancer Res.*, 1995, **55**, 1670–1674.
7 C. Cuffari, E. Seidman, S. Latour and Y. Theoret, *Can. J. Physiol. Pharmacol.*, 1996, **74**, 580–585.
8 R. Brem, M. Guven and P. Karran, *Free Radical Biol. Med.*, 2017, **107**, 101–109.
9 M. S. Cooke, T. L. Duarte, D. Cooper, J. Chen, S. Nandagopal and M. D. Evans, *DNA Repair*, 2008, **7**, 1982–1989.
10 N. R. Attard and P. Karran, *Photochem. Photobiol. Sci.*, 2012, **11**, 62–68.
11 R. Brem and P. Karran, *Photochem. Photobiol.*, 2012, **88**, 5–13.
12 M. Pollum, L. A. Ortiz-Rodríguez, S. Jockusch and C. E. Crespo-Hernández, *Photochem. Photobiol.*, 2016, **92**, 286–292.
13 C. Reichardt, C. Guo and C. E. Crespo-Hernández, *J. Phys. Chem. B*, 2011, **115**, 3263–3270.
14 X. Zou, H. Zhao, Y. Yu and H. Su, *J. Am. Chem. Soc.*, 2013, **135**, 4509–4515.
15 Y. Zhang, A. N. Barnes, X. Zhu, N. F. Campbell and R. Gao, *J. Photochem. Photobiol., A*, 2011, **224**, 16–24.
16 X. Zhang, G. Jeffs, X. Ren, P. O'Donovan, B. Montaner, C. M. Perrett, P. Karran and Y.-Z. Xu, *DNA Repair*, 2007, **6**, 344–354.
17 B. Ashwood, S. Jockusch and C. E. Crespo-Hernández, *Molecules*, 2017, **22**, 379.
18 L. Martínez-Fernández, I. Corral, G. Granucci and M. Persico, *Chem. Sci.*, 2014, **5**, 1336–1347.
19 L. Martínez-Fernández, L. González and I. Corral, *Chem. Commun.*, 2012, **48**, 2134–2136.

 This journal is © The Royal Society of Chemistry 2018

20 R. Weinshilboum, *Drug Metab. Dispos.*, 2001, **29**, 601–605.
21 K. G. Van Scoik, C. A. Johnson and W. R. Porter, *Drug Metab. Rev.*, 1985, **16**, 157–174.
22 N. Erb, D. O. Harms and G. Janka-Schaub, *Cancer Chemother. Pharmacol.*, 1998, **42**, 266–272.
23 U. Hindorf, C. Peterson and S. Almer, *Ther. Drug Monit.*, 2004, **26**, 673–678.
24 T. Dervieux, J. G. Blanco, E. Y. Krynetski, E. F. Vanin, M. F. Roussel and M. V. Relling, *Cancer Res.*, 2001, **61**, 5810–5816.
25 T. Dervieux, T. L. Brenner, Y. Y. Hon, Y. Zhou, M. L. Hancock, J. T. Sandlund, G. K. Rivera, R. C. Ribeiro, J. M. Boyett and C.-H. Pui, *Blood*, 2002, **100**, 1240–1247.
26 S. A. Coulthard, L. A. Hogarth, M. Little, E. C. Matheson, C. P. Redfern, L. Minto and A. G. Hall, *Mol. Pharmacol.*, 2002, **62**, 102–109.
27 U. Nygaard, N. Toft and K. Schmiegelow, *Clin. Pharmacol. Ther.*, 2004, **75**, 274–281.
28 M. C. Dubinsky, S. Lamothe, H. Y. Yang, S. R. Targan, D. Sinnett, Y. Théorêt and E. G. Seidman, *Gastroenterology*, 2000, **118**, 705–713.
29 M. C. Dubinsky, H. Yang, P. V. Hassard, E. G. Seidman, L. Y. Kam, M. T. Abreu, S. R. Targan and E. A. Vasiliauskas, *Gastroenterology*, 2002, **122**, 904–915.
30 M. Pollum, B. Ashwood, S. Jockusch, M. Lam and C. E. Crespo-Hernández, *J. Am. Chem. Soc.*, 2016, **138**, 11457–11460.
31 M. Pollum, L. Minh, S. Jockusch and C. Crespo-Hernández, in preparation.
32 M. Pollum, L. Guan, S. Ahsanuddin, E. Baron, M. Lam and C. Crespo-Hernández, *J. Invest. Dermatol.*, 2016, **136**, S105.
33 P. F. Swann, T. R. Waters, D. C. Moulton and Y.-Z. Xu, *Science*, 1996, **273**, 1109–1111.
34 J. Offman, G. Opelz, B. Doehler, D. Cummins, O. Halil, N. R. Banner, M. M. Burke, D. Sullivan, P. Macpherson and P. Karran, *Blood*, 2004, **104**, 822–828.
35 T. Nguyen, P. M. Vacek, P. O'Neill, R. B. Colletti and B. A. Finette, *Cancer Res.*, 2009, **69**, 7004–7012.
36 M. J. Horsfall, A. J. Gordon, P. A. Burns, M. Zielenska, G. M. van der Vliet and B. W. Glickman, *Environ. Mol. Mutagen.*, 1990, **15**, 107–122.
37 G. P. Margison, M. F. Santibáñez Koref and A. C. Povey, *Mutagenesis*, 2002, **17**, 483–487.
38 E. L. Loechler, C. L. Green and J. M. Essigmann, *Proc. Natl. Acad. Sci. U. S. A.*, 1984, **81**, 6271–6275.
39 J. J. Warren, L. J. Forsberg and L. S. Beese, *Proc. Natl. Acad. Sci. U. S. A.*, 2006, **103**, 19701–19706.
40 M. Esteller, S. R. Hamilton, P. C. Burger, S. B. Baylin and J. G. Herman, *Cancer Res.*, 1999, **59**, 793–797.
41 M. Esteller, M. Toyota, M. Sanchez-Cespedes, G. Capella, M. A. Peinado, D. N. Watkins, J.-P. J. Issa, D. Sidransky, S. B. Baylin and J. G. Herman, *Cancer Res.*, 2000, **60**, 2368–2371.
42 M. Esteller, R.-A. Risques, M. Toyota, G. Capella, V. Moreno, M. A. Peinado, S. B. Baylin and J. G. Herman, *Cancer Res.*, 2001, **61**, 4689–4692.
43 F. Siouri, S. Boldissar, J. Berenbeim and M. de Vries, *J. Phys. Chem. A*, 2017, **121**, 5257–5266.

44 B. Ashwood, L. A. Ortiz-Rodríguez and C. E. Crespo-Hernández, *J. Phys. Chem. Lett.*, 2017, **8**, 4380–4385.
45 J. R. Lackowicz, *Priniciples of Fluorescence Spectroscopy*, Springer, US, New York, 3rd edn, 2006.
46 R. F. Chen, *Anal. Lett.*, 1967, **1**, 35–42.
47 M. J. Frisch, G. W. Trucks, H. B. Schlegel, G. E. Scuseria, M. A. Robb, J. R. Cheeseman, G. Scalmani, V. Barone, B. Mennucci, G. A. Petersson, H. Nakatsuji, M. Caricato, X. Li, H. P. Hratchian, A. F. Izmaylov, J. Bloino, G. Zheng, J. L. Sonnenberg, M. Hada, M. Ehara, K. Toyota, R. Fukuda, J. Hasegawa, M. Ishida, T. Nakajima, Y. Honda, O. Kitao, H. Nakai, T. Vreven, J. A. Montgomery Jr, J. E. Peralta, F. Ogliaro, M. Bearpark, J. J. Heyd, E. Brothers, K. N. Kudin, V. N. Staroverov, R. Kobayashi, J. Normand, K. Raghavachari, A. Rendell, J. M. Millam, M. Klene, J. E. Knox, J. B. Cross, V. Bakken, C. Adamo, J. Jaramillo, R. Gomperts, R. E. Stratmann, O. Yazyev, A. J. Austin, R. Cammi, C. Pomelli, J. Ochterski, R. L. Martin, K. Morokuma, V. G. Zakrzewski, G. A. Voth, P. Salvador, J. J. Dannenberg, S. Dappich, A. D. Daniels, O. Farkas, J. B. Foresman, J. V. Ortiz, J. Cioslowski and D. J. Fox, *Gaussian 09 (Revision A.01)*, Gaussian, Inc., Wallingford, CT, 2009.
48 C. Lee, W. Yang and R. G. Parr, *Phys. Rev. B: Condens. Matter Mater. Phys.*, 1988, **37**, 785–789.
49 A. D. Becke, *J. Chem. Phys.*, 1993, **98**, 1372–1377.
50 A. D. Becke, *J. Chem. Phys.*, 1993, **98**, 5648–5652.
51 C. Adamo and V. Barone, *J. Chem. Phys.*, 1999, **110**, 6158–6170.
52 V. Barone, M. Cossi and J. Tomasi, *J. Chem. Phys.*, 1997, **107**, 3210–3221.
53 E. Cances, B. Mennucci and J. Tomasi, *J. Chem. Phys.*, 1997, **107**, 3032–3041.
54 C. Reichardt and C. E. Crespo-Hernández, *J. Phys. Chem. Lett.*, 2010, **1**, 2239–2243.
55 C. Reichardt, R. A. Vogt and C. E. Crespo-Hernández, *J. Chem. Phys.*, 2009, **131**, 224518.
56 C. Reichardt, C. Wen, R. A. Vogt and C. E. Crespo-Hernández, *Photochem. Photobiol. Sci.*, 2013, **12**, 1341–1350.
57 R. A. Vogt, T. G. Gray and C. E. Crespo-Hernández, *J. Am. Chem. Soc.*, 2012, **134**, 14808–14817.
58 C. Capellos and B. H. J. Bielski, *Kinetic Systems*, Wiley Interscience, New York, 1972.
59 I. H. van Stokkum, D. S. Larsen and R. van Grondelle, *Biochim. Biophys. Acta*, 2004, **1657**, 82–104.
60 G. Wenska, J. Koput, G. Burdziński, K. Taras-Goslinska and A. Maciejewski, *J. Photochem. Photobiol., A*, 2009, **206**, 93–101.
61 K. Taras-Goślińska, G. Burdziński and G. Wenska, *J. Photochem. Photobiol., A*, 2014, **275**, 89–95.
62 A. Favre, in *Bioorganic Photochemistry*, ed. H. Morrison, J. Wiley & Sons, New York, 1990, pp. 379–425.
63 B. Ashwood, M. Pollum and C. E. Crespo-Hernández, *Chem.–Eur. J.*, 2016, **22**, 16648–16656.
64 J. R. Platt, *J. Chem. Phys.*, 1949, **17**, 484–495.
65 M. El-Sayed, *J. Chem. Phys.*, 1963, **38**, 2834–2838.
66 C. Adamo, G. E. Scuseria and V. Barone, *J. Chem. Phys.*, 1999, **111**, 2889–2899.

 This journal is © The Royal Society of Chemistry 2018

67 J. A. Sánchez-Rodríguez, A. Mohamadzade, S. Mai, B. Ashwood, M. Pollum, P. Marquetand, L. González, C. Crespo-Hernández and S. Ullrich, *Phys. Chem. Chem. Phys.*, 2017, **19**, 19756–19766.
68 M. Pollum and C. E. Crespo-Hernández, *J. Chem. Phys.*, 2014, **140**, 071101.
69 L. Martínez-Fernández, G. Granucci, M. Pollum, C. Crespo-Hernández, M. Persico and I. Corral, *Chem.–Eur. J.*, 2017, **23**, 2619–2627.
70 M. Pollum, L. Martínez-Fernández and C. E. Crespo-Hernández, *Photoinduced Phenomena in Nucleic Acids I*, Springer, Berlin Heidelberg, 2015, vol. 355, pp. 245–327.
71 M. Pollum, S. Jockusch and C. E. Crespo-Hernández, *Phys. Chem. Chem. Phys.*, 2015, **17**, 27851–27861.
72 M. Pollum, S. Jockusch and C. E. Crespo-Hernández, *J. Am. Chem. Soc.*, 2014, **136**, 17930–17933.
73 G. Ortiz-Zayas, M. Pollum, L. A. Ortiz-Rodríguez, S. Jockusch and C. Crespo-Hernández, in preparation.
74 S. Mai, M. Pollum, L. Martínez-Fernández, N. Dunn, P. Marquetand, I. Corral, C. E. Crespo-Hernández and L. González, *Nat. Commun.*, 2016, **7**, 13077.
75 Y. Harada, C. Okabe, T. Kobayashi, T. Suzuki, T. Ichimura, N. Nishi and Y.-Z. Xu, *J. Phys. Chem. Lett.*, 2009, **1**, 480–484.
76 N. J. Turro, V. Ramamurthy and J. C. Scaiano, *Principles of Molecular Photochemistry: An Introduction*, University Science Books, Sausalito, CA, 2009.
77 R. Englman and J. Jortner, *Mol. Phys.*, 1970, **18**, 145–164.
78 K. F. Freed and J. Jortner, *J. Chem. Phys.*, 1970, **52**, 6272–6291.
79 M. Sakamoto, X. Cai, M. Hara, M. Fujitsuka and T. Majima, *J. Phys. Chem. A*, 2005, **109**, 4657–4661.
80 M. Sakamoto, X. Cai, M. Hara, M. Fujitsuka and T. Majima, *J. Am. Chem. Soc.*, 2004, **126**, 9709–9714.
81 J. S. Zugazagoitia, E. Collado-Fregoso, E. F. Plaza-Medina and J. Peon, *J. Phys. Chem. A*, 2009, **113**, 805–810.
82 R. Vogt and C. E. Crespo-Hernández, *J. Phys. Chem. A*, 2013, **117**, 14100–14108.
83 M. Richter, P. Marquetand, J. González-Vázquez, I. Sola and L. González, *J. Chem. Theory Comput.*, 2011, **7**, 1253–1258.
84 C. E. Crespo-Hernández, L. Martínez-Fernández, C. Rauer, C. Reichardt, S. Mai, M. Pollum, P. Marquetand, L. González and I. s. Corral, *J. Am. Chem. Soc.*, 2015, **137**, 4368–4381.
85 J.-M. L. Pecourt, J. Peon and B. Kohler, *J. Am. Chem. Soc.*, 2001, **123**, 10370–10378.
86 C. T. Middleton, K. de La Harpe, C. Su, Y. K. Law, C. E. Crespo-Hernández and B. Kohler, *Annu. Rev. Phys. Chem.*, 2009, **60**, 217–239.
87 C. E. Crespo-Hernández, B. Cohen, P. M. Hare and B. Kohler, *Chem. Rev.*, 2004, **104**, 1977–2020.
88 P. M. Hare, C. E. Crespo-Hernández and B. Kohler, *Proc. Natl. Acad. Sci. U. S. A.*, 2007, **104**, 435–440.
89 J.-M. L. Pecourt, J. Peon and B. Kohler, *J. Am. Chem. Soc.*, 2000, **122**, 9348–9349.
90 J. Lee, J. R. Challa and D. W. McCamant, *J. Phys. Chem. B*, 2017, **121**, 4722–4732.

91 V. Karunakaran, K. Kleinermanns, R. Improta and S. Kovalenko, *J. Am. Chem. Soc.*, 2009, **131**, 5839–5850.
92 F.-A. Miannay, T. Gustavsson, A. Banyasz and D. Markovitsi, *J. Phys. Chem. A*, 2010, **114**, 3256–3263.
93 G. Moumouzias, D. K. Panopoulos and G. Ritzoulis, *J. Chem. Eng. Data*, 1991, **36**, 20–23.
94 L. Korson, W. Drost-Hansen and F. J. Millero, *J. Phys. Chem.*, 1969, **73**, 34–39.
95 C. C.-W. Cheng, C. Ma, C. T.-L. Chan, K. Y.-F. Ho and W.-M. Kwok, *Photochem. Photobiol. Sci.*, 2013, **12**, 1351–1365.
96 D. Tuna, A. L. Sobolewski and W. Domcke, *J. Phys. Chem. A*, 2013, **118**, 122–127.
97 D. Tuna and W. Domcke, *Phys. Chem. Chem. Phys.*, 2016, **18**, 947–955.
98 Y. Zhang, J. Dood, A. Beckstead, J. Chen, X.-B. Li, C. J. Burrows, Z. Lu, S. Matsika and B. Kohler, *J. Phys. Chem. A*, 2013, **117**, 12851–12857.
99 S. Matsika, *Photoinduced Phenomena in Nucleic Acids I*, Springer, Berlin Heidelberg, 2015, vol. 355, pp. 209–243.
100 B. Heggen, Z. Lan and W. Thiel, *Phys. Chem. Chem. Phys.*, 2012, **14**, 8137–8146.
101 C. M. Marian, *J. Phys. Chem. A*, 2007, **111**, 1545–1553.
102 H. Chen and S. Li, *J. Phys. Chem. A*, 2006, **110**, 12360–12362.
103 Z. Lu, A. A. Beckstead, B. Kohler and S. Matsika, *J. Phys. Chem. B*, 2015, **119**, 8293–8301.
104 K. Röttger, R. Siewertsen and F. Temps, *Chem. Phys. Lett.*, 2012, **536**, 140–146.
105 J. P. Villabona-Monsalve, R. Noria, S. Matsika and J. Peón, *J. Am. Chem. Soc.*, 2012, **134**, 7820–7829.
106 R. Brem and P. Karran, *Cancer Res.*, 2012, **72**, 4787–4795.
107 P. O'Donovan, C. M. Perrett, X. Zhang, B. Montaner, Y.-Z. Xu, C. A. Harwood, J. M. McGregor, S. L. Walker, F. Hanaoka and P. Karran, *Science*, 2005, **309**, 1871–1874.
108 Q. Gueranger, F. Li, M. Peacock, A. Larnicol-Fery, R. Brem, P. Macpherson, J.-M. Egly and P. Karran, *J. Invest. Dermatol.*, 2014, **134**, 1408–1417.

 This journal is © The Royal Society of Chemistry 2018

Faraday Discussions

PAPER

Photoinduced formation mechanism of the thymine–thymine (6–4) adduct in DNA; a QM(CASPT2//CASSCF):MM(AMBER) study†

Angelo Giussani, ‡*[a] Irene Conti,[ab] Artur Nenov[ab] and Marco Garavelli *[ab]

Received 30th September 2017, Accepted 24th October 2017
DOI: 10.1039/c7fd00202e

The UVB-induced photomechanism leading the carbonyl group of a thymine nucleobase to react with the carbon–carbon double bond of a consecutive thymine nucleobase in a DNA strand to form the thymine–thymine (6–4) photodamage adduct remains poorly understood. Key questions remain unanswered, concerning both the intrinsic features of the photoreaction (such as the contribution (or not) of triplet states, the nature of the involved states and the time-scale of the photoprocess) and the role played by the non-reactive surroundings of the two reactive pyrimidine nucleobases (such as the nature of the flanked nucleobases and the flexibility of the whole DNA molecule). A small number of theoretical studies have been carried out on the title photoreaction, most of which have used reduced model systems of DNA, consequently neglecting potential key parameters for the photoreaction such as the constraints due to the double strain structure and the presence of paired and stacked nucleobases. In the present contribution the photoactivation step of the title reaction has been studied in a DNA system, and in particular for a specific DNA hairpin for which the quantum yield of photodamage formation has been recently experimentally measured. The reaction has been characterized by carrying out high-level QM/MM computations, combining the CASPT2//CASSCF approach for the study of the reactive part (*i.e.* the two thymine molecules) with an MM-Amber treatment of the surrounding environment. The possibility of a reaction path along both the singlet and triplet manifolds has been characterized, the nature of the reactive states has been analyzed, and the role played by the flexibility of the whole system, which in turn determines the initial accessible geometrical conformations, has been evaluated, thus substantially contributing towards

[a]*Dipartimento di Chimica "G. Ciamician", Università di Bologna, Via F. Selmi 2, 40126 Bologna, Italy. E-mail: angelo.giussani@uv.es*

[b]*Dipartimento di Chimica Industriale "Toso Montanari", Università di Bologna, Viale del Risorgimento, 40136 Bologna, Italy*

† Electronic supplementary information (ESI) available. See DOI: 10.1039/c7fd00202e

‡ Current address: Department of Chemistry, University College London, 20 Gordon Street, London WC1H 0AJ, UK.

the elucidation of the photoreaction mechanism. On the basis of the obtained results, it can be observed that a charge-transfer state can decay from a pro-reactive initial structure towards a region of energy degeneracy with the ground state, from which the subsequent decay along the ground state hypersurface can lead to the photoreaction.

Introduction

Two major classes of UV-induced damage adducts can form between a pair of consecutive pyrimidine nucleobases in a DNA strand: cyclobutane pyrimidine dimers, and pyrimidine-pyrimidone (6–4) adducts.[1,2] The latter can further evolve into their Dewar isomers through a subsequent photoreaction.[3] While a large number of theoretical[4–12] and experimental[13–20] works have been published on the process leading to cyclobutane pyrimidine dimers, only a few studies have been devoted to elucidating the mechanism that operates in the production of pyrimidine–pyrimidone (6–4) adducts.[5,11,18–24] The latter reaction is believed to take place through the initial formation of an oxetane or azedine ring, which constitutes the actual photoproduct, followed by thermal ring opening into the pyrimidine–pyrimidone (6–4) adduct.[1] Experimentally it is known that the reaction is inefficient (the quantum yield is of the order of 10^{-4}),[25–27] it occurs under UVB (280–315 nm) radiation but not under UVA (315–400 nm) exposure,[28,29] and the actual efficiency of the process depends on the nature of the involved pyrimidine nucleobases and the flanked moieties.[26,30] The timescale for the formation of thymine–thymine (6–4) adducts has been determined, through transient absorption experiments carried out by Marguet and Markovitsi, to be in the ms regime.[20] It must be noted that such a timescale refers to the production of the damage (*i.e.* the pyrimidine–pyrimidone (6–4) adduct) and not to the photoactivation part of the process (*i.e.* oxetane formation) whose timescale remains to be determined. Regarding the mechanism of formation, most of the efforts aimed at its elucidation have studied the reaction of two thymine molecules leading to the thymine–thymine (6–4) adduct.[5,11,19,21–23] In such a case, the photoactivation part of the process leads to an oxetane ring between the carbonyl group of the 3-ended nucleobase and the carbon–carbon double bond of the second moiety (Fig. 1). The photochemical addition of carbonyl compounds to alkenes to form an oxetane ring is known as the Paternò–Büchi reaction,[31] whose mechanism, although normally accepted to proceed *via* the triplet $^3(n\pi^*)$ excited state of the

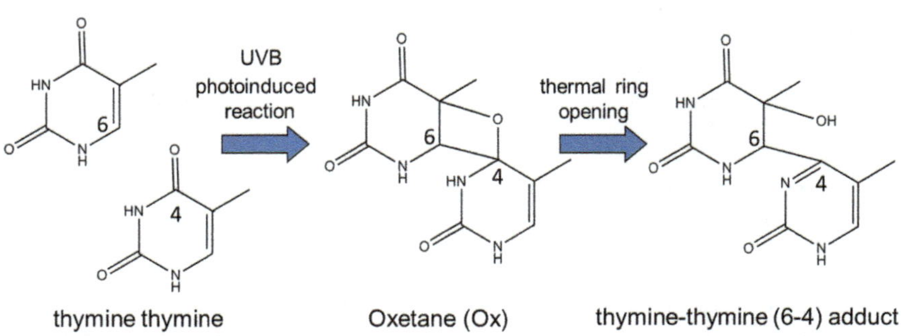

Fig. 1 Scheme of the reaction leading to the thymine–thymine (6–4) adduct.

 This journal is © The Royal Society of Chemistry 2018

carbonyl compound,[32] still constitutes a matter of debate in the scientific community.[33–35]

Two main explanations have been given in order to rationalize the photoinduced mechanism of formation of the thymine–thymine (6–4) adduct. On the basis of TD-M052X computations on dithymidine monophosphate in water and steady-state and time-resolved optical spectroscopy experiments on single strand $(dT)_{20}$, Markovitsi, Improta and co-workers concluded that the reaction proceeds on the potential energy surface of a charge-transfer state and requires the surmounting of an energy barrier, which can explain the increase in the photoreaction's quantum yield upon increasing the incident radiation energy.[11,19] A related mechanism involving the population of a charge-transfer state has also been recently documented, through performing QM/MM computations on a double strand B-fragment.[24] By performing *ab initio* CASPT2//CASSCF quantum chemical calculations on a model system composed of two thymine molecules in a vacuum, Giussani *et al.* have characterized a mechanism involving the initial decay into the $n\pi^*$ state that, leading to an elongation of the reactive carbonyl group, can subsequently react along the triplet manifold with the carbon–carbon double bond of the second nucleobase.[23]

In both works, a key geometry, previously characterized by Blancafort *et al.*,[5] in which the new carbon–oxygen bond present in the final oxetane ring is almost formed, has been indicated as the gate between the excited state of the reactant and the ground state of the photoproduct, acting as a conical intersection and/or intersystem crossing region with the ground state.

While both mechanisms provide a plausible explanation of the photoprocess, the mechanism of the photoreaction in DNA systems, where constraints due to the double strain structure and the presence of paired and stacked nucleobases can significantly alter the process, remains to be better characterized. In the present contribution the title reaction has been studied in a DNA system, in particular for a specific DNA hairpin for which the quantum yields of formation of both classes of photodamage adducts have been experimentally measured.[27] The mechanism of the reaction has been characterized by carrying out high-level QM/MM computations, combining the well-known CASPT2//CASSCF approach for the study of the reactive part (*i.e.* the two involved thymine molecules) with an MM-Amber treatment of the surrounding environment.

On the basis of the obtained results, from a pro-reactive ground state geometry characterized by a reduced distance between the $C_{4'}O$ atoms, a high-lying charge-transfer state can evolve towards a region of energy degeneracy with the ground state, from which the system can evolve on the ground state potential energy hypersurface (PEH) to either separated thymine molecules (non-reactive path), or towards the formation of an oxetane ring (reactive path), depending on the geometrical parameters of the crossing structure or the direction and velocity through which the ground state is repopulated. On the other hand, the possibility of a reaction along the triplet manifold is not confirmed here.

Computational details

The system under study here is the H_42 hairpin depicted in Fig. 2, formed by adenine and thymine nucleobases. Each of the two strands that compose the system include four nucleobases. As shown in Fig. 2, the two strands are

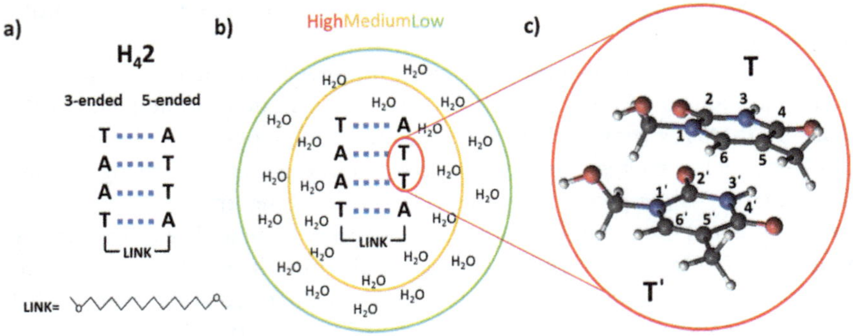

Fig. 2 (a) Structure of the H_42 hairpin; (b) high–medium–low subdivision adopted in the QM/MM computations; (c) atom and nucleobase labeling.

covalently linked through a dodecane (C12) linker, the presence of which prevents the breakdown of the double helix structure. The two reactive thymine molecules will be labelled as shown in Fig. 2c; the 5-ended thymine molecule, which contains the reactive carbon–carbon double bond, is labelled as T, while the 3-ended thymine nucleobase, where the reactive carbonyl group is located, is labeled as T′. The atoms of each moiety are enumerated as normally done for thymine, adding a ′ mark for the atoms of the T′ nucleobase.

The H_42 system has been chosen in the present study for the following main reasons; experimental data for the quantum yield of formation of both the thymine–thymine (6–4) adduct and the cyclobutane thymine dimer have been provided; the H_42 hairpin is characterized by an unusual high quantum yield of formation for the thymine–thymine (6–4) adduct (of the order of 10^{-3}, while most of the available data reports values of 10^{-4});[25–27] and despite the shortness of the oligonucleotide, the reactive dimer is inside a DNA double strand (*i.e.* both thymine molecules are paired with the two adenine nucleobases of the second strand, and each thymine is flanked with an adenine molecule).[27]

A classical 500 ns long molecular dynamics simulation (using NTP conditions at 383 K, 1 bar, and setting a 2 fs integration timestep) was performed, using the AMBER12 program and ff12SB + gaff force fields.[36,37] Explicit solvation by TIP3P water molecules (4664 waters) with eighteen Na^+ ions for neutralization has been considered. The simulation was performed with periodic boundary conditions and an Ewald sum cutoff of 10 Å.

The QM/MM computations were performed using the COBRAM interface developed by our group, whose details have been presented elsewhere.[38] The system (*i.e.* H_42 immersed in a box of water molecules) has been divided into three regions (see Fig. 2b): the high region, composed of the two reactive thymine molecules, including both the sugar ring carbon directly linked to the nucleobase and the adjacent oxygen atom, the medium region, formed of the remaining parts of the H_42 system and by water molecules whose H-atoms are within a distance of 2.5 Å from a heteroatom of the high region or whose O-atoms are within a distance of 2.5 Å from a H-atom of the high layer, covalently bound to a heteroatom; and the low region, composed of the water molecules not included in the high and medium parts. The high region is described at the QM level by means of CASPT2//CASSCF calculations,[39–43] while both the medium and low

 This journal is © The Royal Society of Chemistry 2018

regions are treated at the MM level using Amber 12.[36,37] At the QM MM interface, a hydrogen link-atom approach has been adopted.

Ground and excited state geometry optimizations were performed using HF and CASSCF,[39] (with Gaussian 03 and Molcas8, respectively)[44,45] and Amber 12 levels of theory for the high and medium regions, respectively, employing the optimization algorithm implemented in Gaussian 03.[45] During the optimizations, both the high and medium regions were free to move, while the low region was kept fixed in its initial conformation. Different CASSCF active spaces were employed according to the specific electronic state whose PEH was explored, a strategy that is normally employed in CASPT2//CASSCF photochemical studies,[46] while all the obtained critical points (*i.e.* minima, conical intersections, *etc.*) were characterized by means of CASPT2 computations, including in the active space 12 active electrons distributed in 10 active orbitals (two lone pairs, four π orbitals and 4 π^* orbitals, see Fig. S1†). In the CASSCF and CASPT2 calculations, basis sets of the L-type atomic natural orbital (ANO) corresponding to C,N,O[4s,3p,1d]/H[2s1p] were employed,[47,48] while the 6-31G* basis set was used for ground state HF optimization. Within the CASPT2 scheme, an imaginary level-shift of 0.2 au was used in order to avoid the presence of possible intruder states. The CASPT2 standard zeroth-order Hamiltonian was employed as originally implemented.[41] The core orbitals were frozen in the CASPT2 calculations. Cholesky decomposition was used to speed up the evaluation of the two-electron integrals.[49] CASPT2 computations were performed using the Molcas 8 software.[44] The method employed for the calculation of the spin–orbit coupling (SOC) is described elsewhere.[23]

Results

In order to sample the ground state accessible conformations of the system, a 500 ns MM dynamics simulation of the H_42 hairpin in water at 383 K was performed. Since the main goal of the present contribution is to elucidate the photomechanism of the title reaction, we have conducted our study starting from the MM snapshot whose structure displays the shortest internucleobase distance between the $O_{4'}$ and the C_5 atoms, assuming that such a geometrical feature (*i.e.* the proximity of the two aforementioned atoms) will strongly favor the formation of the thymine–thymine (6–4) adduct. The main reason behind such an assumption is that in different previous theoretical works a conical intersection and/or intersystem crossing region characterized by the formation of a new bond between the $O_{4'}$ and the C_5 atoms have been indicated to be the key geometry connecting the excited states of two thymine molecules with the oxetane photoproduct.[5,23] A model has been proposed in which the photoreactivity towards thymine–thymine (6–4) adduct formation is inferred by the $O_{4'}$–C_5 distance in the initial geometry under irradiation.[27]

For the selected snapshot (hereafter, snapCO) the $O_{4'}$–C_5 distance is equal to 2.810 Å (see Fig. S7†), and the double-strand structure of the H_42 hairpin is essentially preserved, but the interstrand adenine–thymine pair next to the T′ thymine is no longer forming the usual interstrand hydrogen bonds (see Fig. S6†). This structural feature can compromise the extrapolation of the following results for longer double-strand DNA structures, although it has been proposed that thymine–thymine (6–4) adduct formation indeed passes through a disruption of the double-strand structure and that the high quantum yield of formation of the thymine–thymine (6–4) adduct experimentally measured for the H_42 hairpin could be due to its flexibility and consequent ability to adopt double-strand-broken conformations.[27]

The geometrical parameters describing the reactive thymine pair in the snapCO structure are the result of an MM description. In order to use QM parameters instead, a refinement of the initial structure has been carried out by means of a ground state HF QM/MM optimization. In the latter optimization the $O_{4'}$–C_5 distance was kept frozen, so as to retain the main geometrical feature through which the selection of the snapshot was operated. The resulting optimized geometry (hereafter $^1(gs)_{min}$) displays bond lengths that are comparable with those obtained from a CASSCF optimization of the isolated nucleobase thymine (see Fig. S7†).[23]

From the $^1(gs)_{min}$ structure the accessible excited states of the system were characterized by performing CASPT2(12,10):AMBER state-averaged computations over ten singlet and ten triplet states. As discussed in the Computational details section, the (12,10) active space comprises two π, one n, and two π* orbitals on each thymine (see Fig. S1†). The resulting energies, oscillator strengths, and dipole moments are given in Table 1, where the states are reported in order of increasing energy and the natures of the different states are specified when analysis of the corresponding CASSCF(12,10) wave-functions allowed a clear identification of them. As expected, among the lowest excited states are recognizable two ππ* states localized on the C_6C_5 double bond of each thymine (states $^1\pi\pi^*$–T′ and $^1\pi\pi^*$–T, see Fig. S2†) and two nπ* states localized on the C_4O carbonyl of each nucleobase (states $^1n\pi^*$–T′ and $^1n\pi^*$–T, see Fig. S3†). The

Table 1 Calculated CASPT2(12,10):AMBER vertical excitation energies of the $^1(gs)_{min}$ geometry (E_{VA}, eV) for the lowest valence singlet and spin forbidden triplet excited states[a]

State	E_{VA} (eV)	f	$\mu(D)$
S0 (gs)	0.00	—	12.58
T1 ($^3\pi\pi^*$–T′)	3.93	—	12.73
T2 ($^3\pi\pi^*$–T)	3.97	—	12.58
S1 ($^1\pi\pi^*$–T′)	4.90	0.18487	13.81
S2 ($^1\pi\pi^*$–T)	5.21	0.93961	14.61
T3 ($^3n\pi^*$–T)	5.21	—	9.95
T4 ($^3n\pi^*$–T′)	5.23	—	10.93
S3 ($^1n\pi^*$–T)	5.25	0.00075	9.64
S4 ($^1n\pi^*$–T′)	5.27	0.00023	10.73
T5 ($^3\pi_O\pi^*$–T′)	6.18	—	10.68
S5 ($2\pi 2\pi^*$)	6.24	0.32562	13.22
T6 ($^3\pi_O\pi^*$–T)	6.30	—	10.38
S6 (ct–ππ*–TT′)	7.05	0.00690	9.46
S7 ($^1\pi_O\pi^*$–T′)	7.17	0.01275	10.92
S8	7.59	0.01781	9.25
T7	7.76	—	12.55
S9 ($^1\pi_O\pi^*$–T)	7.99	0.00512	10.76
T8	8.28	—	9.95
T9	8.33	—	10.96
T10	8.99	—	10.03

[a] The computed dipole moments (μ, D) and oscillator strengths (f) for the singlet–singlet transitions are also included.

 This journal is © The Royal Society of Chemistry 2018

corresponding triplet states (*i.e.* states $^3\pi\pi^*$–T′, $^3\pi\pi^*$–T, $^3n\pi^*$–T′ and $^3n\pi^*$–T) are also present. Additionally, ππ* states describing an excitation from the occupied π orbital mainly localized on the C_4O carbonyl (orbitals π_O–T′ and π_O–T, see Fig. S1†), a surprisingly low singlet double excited state (state S5 in Table 1, hereafter state 2π2π*, involving the π–T′, π–T, π1*–T′, and π1*–T orbitals, see Fig. S5†), and a singlet ππ* charge-transfer state from T to T′ (hereafter ct–ππ*–TT′ state, see Fig. S4†), are among the computed roots. On the basis of analysis of the obtained oscillator strengths (see Table 1), the $^1\pi\pi^*$–T′, $^1\pi\pi^*$–T, and 2π2π* states are the most populated after light interaction, although others, and in particular the ct–ππ*–TT′ state, also demonstrate non-negligible probability to be populated. The two ππ* states localized on the carbon–carbon double bond of each thymine (*i.e.* the $^1\pi\pi^*$–T′ and $^1\pi\pi^*$–T states) are separated by 0.31 eV, reflecting a significant splitting in the otherwise degenerate ππ* states, thus indicating a strong interaction among them in the $^1(gs)_{min}$ structure.[52]

As discussed in the Introduction section, two main mechanisms have been proposed regarding the photoinduced formation mechanism of the thymine–thymine (6–4) adduct: the first passing through the initial population of nπ* states localized on the reactive carbonyl, characterized as the $^1n\pi^*$–T′ and $^3n\pi^*$–T′ states, the second involving the population of a charge-transfer state from the T to T′ moieties, here characterized as the ct–ππ*–TT′ state. For that reason, in the present contribution we have focused on the optimized points reached by these three states ($^1n\pi^*$–T′, $^3n\pi^*$–T′ and ct–ππ*–TT′) as well as on the evolution of the $^1\pi\pi^*$–T′ state, with the latter being the lowest bright excited state of the system.

Regarding its initial population after light absorption, the bright and low-lying $^1\pi\pi^*$–T′ state is indeed populated. The $^1n\pi^*$–T′ and $^3n\pi^*$–T′ states could become populated due to their energy proximity with the bright $^1\pi\pi^*$–T state, with the three states placed in a 0.06 eV energy range (see Table 1). The ct–ππ*–TT′ state possesses a non-zero oscillator strength (equal to 0.00690), therefore indicating its possible direct population; however, its energetic position, 7.05 eV above the ground state, points to the need for high-energy radiation in order to reach it. A possible implication of the ct–ππ*–TT′ state in the photomechanism can explain the reported increase in the quantum yield of the photoreaction at increased excitation energies and the absence of photoproduction under UVA radiation on the basis of the high energy position of such a charge transfer state.[11,19,28,29] It would also be interesting to verify whether in the present system population of the ct–ππ*–TT′ state can occur from a minimum of the $^1\pi\pi^*$–T state though oscillations stabilizing the former state, as was recently documented for a double-strand B-fragment.[24] Further works in this respect are currently in progress.

The study of the aforementioned states was carried out by performing, unless otherwise specified, CASSCF(8,6):AMBER optimizations, using the reduced (8,6) active space as described in the Computational details section.

Optimization of the $^1n\pi^*$–T′ and $^3n\pi^*$–T′ states led in both cases to geometries that are characterized by a significant elongation of the $C_{4'}O$ bond, which increased by 0.158 and 0.172 Å with respect to the $^1(gs)_{min}$ structure, respectively (see Fig. S7†). At the minimum of the $^3n\pi^*$–T′ state, the $C_{4'}O$ bond has also undergone a sizeable bending out of the plane of the ring and towards the T thymine molecule, as reflected by the $C_{6'}C_{5'}C_{4'}O$ dihedral angle, which is equal to 171.46° for the $^1(gs)_{min}$ structure, but equal to 148.24° for the $^3n\pi^*$–T′ minimum. The electronic states for both the $^1n\pi^*$–T′ and $^3n\pi^*$–T′ minima were characterized by performing CASPT2(12,10):AMBER single point computations, with state averaging over six singlet and six triplet states. As reported in Table S2,† at the ^{1}nπ*–T′ minimum the ^{1}nπ*–T′ and ^{3}nπ*–T′ states are almost degenerate in energy (the energy gap is equal to 0.09 eV) while no other singlet state is close in energy to the ^{1}nπ*–T′ state. This makes us suppose that, despite the predictable small SOC between the ^{1}nπ*–T′ and ^{3}nπ*–T′ states due to their common nature, part of the ^{1}nπ*–T′ population will be able to decay into the ^{3}nπ*–T′ PEH, from which it can reach the corresponding minimum. In such a geometry, the ^{3}nπ*–T′ state is the lowest triplet state, T1, separated from the T2 ^{3}ππ*–T′ state by a small energy gap of 0.32 eV (see Table S3†). These facts suggest the possibility of the presence, not verified here, of a nearby conical intersection between the ^{3}nπ*–T′ and ^{3}ππ*–T′ states, as was documented for the isolated thymine molecule.[23,50] According to the first suggested mechanism described above that passes through the nπ* population, from the ^{3}nπ*–T′ minimum the population should decay to a triplet ππ* state that can evolve towards a reactive singlet–triplet crossing region with the ground state, along a path characterized by the shortening of the $O_{4'}$–C_5 distance and the gradual change in the nature of the triplet state from a localized ππ* to a charge-transfer TT′ state, with the latter being the one crossing with the ground state. Although at the ^{3}nπ*–T′ minimum the ^{3}nπ*–T′ and ^{3}ππ*–T′ states are close in energy (separated by 0.32 eV), a possible decay towards the ^{3}ππ*–T′ state and in particular the possibility that the system will then evolve in the direction of $O_{4'}$–C_5, starting from the ^{3}nπ*–T′ minimum where the $O_{4'}$–C_5 distance has the significant value of 3.141 Å, seems unfavorable, as a preliminary optimization of the ^{3}ππ*–T′ state, which kept the two molecules separated, appears to indicate. In order to evaluate the possibility of the population of a charge transfer state from the ^{3}nπ*–T′ minimum, the latter geometry was further characterized by performing a CASPT2(12,10):AMBER state-averaged computation over 15 triplet roots. The computation showed that the localization of a triplet ct–ππ*–TT′ state (hereafter ^{3}ct–ππ*–TT′) is placed at 6.57 eV with respect to the ground state at the $^1(gs)_{min}$ minimum, and consequently 2.11 eV above the ^{3}nπ*–T′ state at the ^{3}nπ*–T′ minimum. The possibility of a decay from the ^{3}nπ*–T′ state towards the ^{3}ct–ππ*–TT′ state, which in turn would be expected to favorably decay towards a reactive singlet–triplet crossing region, seems consequently remote.

On the basis of previous theoretical results for thymine and related systems,[23,51] optimization of the ^{1}ππ*–T′ state was expected to provide a minimum whose main deformation would have been a localized and significant elongation of the $C_{5'}C_{6'}$ double bond. Surprisingly, the obtained structure displays only a modest elongation of the aforementioned carbon–carbon double bond (which increases by 0.029 Å with respect to the $^1(gs)_{min}$ structure), together with a non-negligible elongation of the $C_{4'}$O bond (+0.036 Å with respect to the $^1(gs)_{min}$) and a bending of the $C_{4'}$O carbonyl group that closely resembles that which characterizes the ^{3}nπ*–T′ minimum ($C_{6'}C_{5'}C_{4'}$O dihedral angle equal to 144.87°, see Fig. S7†). It must be noted that such a minimum is the result of a QM/MM optimization in which the reduced CASSCF(4,4) active space (including the π–T′, π–T, π1*–T′, and π1*–T orbitals) has been employed, since the analogous computation with the larger CASSCF(8,6) active space did not converge. The CASPT2(12,10):AMBER vertical computation of the optimized ^{1}ππ*–T′ structure reveals that the ^{1}ππ*–T′ state is stabilized by 0.21 eV with respect to its energy at the $^1(gs)_{min}$ geometry (see Table S4†). Interestingly, at the obtained ^{1}ππ*–T′ minimum, the ^{1}ππ*–T′ state is

 This journal is © The Royal Society of Chemistry 2018

degenerate in energy with the $^{3}n\pi^{*}$–T′ state (the energy gap is equal to 0.01) and consequently part of the $^{1}\pi\pi^{*}$–T′ population is in principle able to populate the $^{3}n\pi^{*}$–T′ state. The SOC among the latter states is equal to 15.97 cm^{-1}.

For the $^{1}(gs)_{min}$ structure the decay of the ct–$\pi\pi^{*}$–TT′ state was studied by performing the corresponding CASSCF(8,6):AMBER optimization. The latter optimization did not reach a minimum, since the ct–$\pi\pi^{*}$–TT′ state evolves toward a region of the potential energy hypersurface where it is energetically close to the ground state. In order to characterize such a degenerate region, different points along the performed optimization were extracted and analyzed. In particular, three significant structures were observed. For the first structure, hereafter referred to as $(ct/\pi\pi^{*})_{CI}$, the ct–$\pi\pi^{*}$–TT′ state is degenerate in energy with the $^{1}\pi\pi^{*}$–T state, displaying an energy gap of only 0.02 eV (see Table S5†). However the energy of the ct–$\pi\pi^{*}$–TT′ state is here actually above its energy at the $^{1}(gs)_{min}$ of around 0.5 eV, thus indicating that the system will not favorably reach this region from the initial geometry. The second important structure characterized from the ct–$\pi\pi^{*}$–TT′ optimization is a conical intersection between the latter state and the ground state, hereafter referred to as $(ct/gs)_{CI\text{-}noreact}$. The two states are in fact separated by 0.06 eV; furthermore the T1 state, of ^{3}ct–$\pi\pi^{*}$–TT′ nature, is actually only 0.04 and 0.10 eV below the ct–$\pi\pi^{*}$–TT′ state and the ground state, respectively, (see Table S6†) consequently also making $(ct/gs)_{CI\text{-}noreact}$ a singlet–triplet crossing region. The $(ct/gs)_{CI\text{-}noreact}$ structure closely resembles the conical intersection that was previously reported to be the key geometry leading to oxetane formation, being in fact characterized by the formation of a new carbon–oxygen bond between the $O_{4'}$ and C_5 atoms.[5,23] Despite its geometry, the optimization of the ground state from the $(ct/gs)_{CI\text{-}noreact}$ point evolves back to the two separated thymine molecules. As indicated by the characterization of the last structure extracted from the ct–$\pi\pi^{*}$–TT′ optimization, hereafter referred to as $(ct/gs)_{CI\text{-}react}$, the formation of a new carbon–oxygen bond between the $O_{4'}$ and C_5 atoms seems to be a required but not sufficient condition for the pro-reactivity of a conical intersection. The $(ct/gs)_{CI\text{-}react}$ geometry is characterized by both the formation of the aforementioned new bond and by a sizeable puckering of the $C_{4'}$ atom toward the T nucleobase, as reflected by the value of the $N_{1'}C_{6'}C_{5'}C_{4'}$ dihedral angle, which is equal to 3.51° and 12.66° at the $(ct/gs)_{CI\text{-}noreact}$ and $(ct/gs)_{CI\text{-}react}$ points, respectively (see Fig. S7†). This additional geometrical feature appears to be key for a reactive evolution of the ground state, as proven by the ground state optimization for the $(ct/gs)_{CI\text{-}react}$ geometry, which indeed evolves to an oxetane structure, hereafter referred to as oxe (see Fig. S7†). It should be noted that $(ct/gs)_{CI\text{-}react}$ is not strictly speaking a conical intersection, as the ct–$\pi\pi^{*}$–TT′ and ground states are separated by a 0.30 eV energy gap (see Table S7†), while the degeneracy between the T1 ^{3}ct–$\pi\pi^{*}$–TT′ state and ground state, of only 0.05 eV, is sufficient to consider the structure as a proper singlet–triplet crossing region. It is plausible to believe that a similar pro-reactive conical intersection should be located in the nearby regions, although in the present contribution it has not been identified. Another possibility, which is in agreement with the aforementioned fact that only $(ct/gs)_{CI\text{-}noreact}$ is strictly speaking a conical intersection, is that actually both the $(ct/gs)_{CI\text{-}noreact}$ and $(ct/gs)_{CI\text{-}react}$ structures belong to the same seam of intersection and that the $(ct/gs)_{CI\text{-}react}$ geometry represents the direction along which the decay through the $(ct/gs)_{CI\text{-}noreact}$ conical intersection will evolve towards oxetane formation. This in turn means that a reactive or non-reactive decay will be

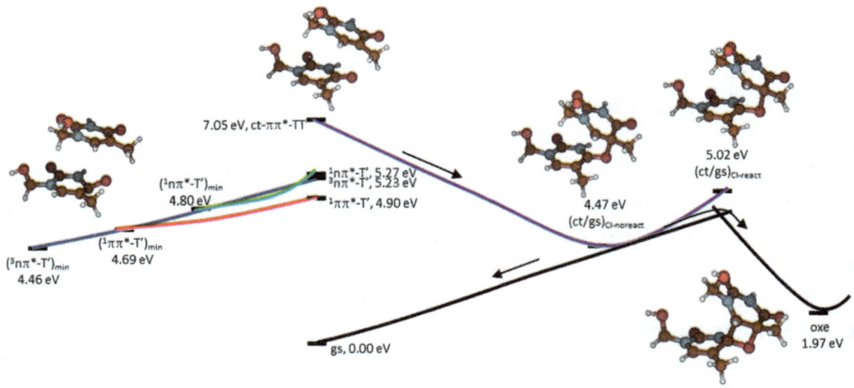

Fig. 3 Schematic representation of the characterized paths. The red, blue, green, purple and black solid lines indicate the $^1\pi\pi^*$–T′, $^3n\pi^*$–T′, $^1n\pi^*$–T′, ct–$\pi\pi^*$–TT′ and ground states, respectively.

determined by the direction and velocity with which the system passes through the $(ct/gs)_{CI\text{-}noreact}$ conical intersection: being reactive if evolving towards the $(ct/gs)_{CI\text{-}react}$ geometry, and being non-reactive in the other case.

On the basis of the described results, the charge-transfer ct–$\pi\pi^*$–TT′ state can evolve from the $^1(gs)_{min}$ state, characterized by a relatively short $O_{4'}$–C_5 distance of 2.810 Å, towards a region of energy degeneracy with the ground state where both pro-reactive structures (such as $(ct/gs)_{CI\text{-}react}$) and non-reactive structures (such as $(ct/gs)_{CI\text{-}noreact}$) can be reached.

It is difficult to predict the percentage of the ct–$\pi\pi^*$–TT′ population that will follow the reactive or the non-reactive path, since a dynamics treatment of the reaction will be required in order to answer such a question. Nevertheless, what is possible to notice is that the non-reactive $(ct/gs)_{CI\text{-}noreact}$ conical intersection is placed around 0.5 eV below the reactive $(ct/gs)_{CI\text{-}react}$ structure (see Table S6 and S7†). It should however be kept in mind that the title photoreaction is an inefficient photoreaction, and in fact for the particular system here under study, characterized by displaying a relatively high quantum yield of the photoreaction, the latter is equal to the low value of 0.75×10^{-3}.[27]

Table 2 Calculated CASPT2(12,10):AMBER energies (eV) of the main characterized states and various optimized structures[a]

	State						
Geometry	gs	$^1n\pi^*$–T′	$^3n\pi^*$–T′	$^1\pi\pi^*$–T′	$^3\pi\pi^*$–T′	ct–$\pi\pi^*$–TT′	^{3}ct–$\pi\pi^*$–TT′
$^1(gs)_{min}$	0.00	5.27	5.23	4.90	3.93	7.05	—
$^1n\pi^*$–T′ minimum	1.59	4.80	4.71	5.87	4.25	—	—
$^3n\pi^*$–T′ minimum	1.93	4.73	4.46	5.87	4.78	—	6.57
$^1\pi\pi^*$–T′ minimum	1.83	5.33	4.68	4.69	4.98	—	—
$(ct/gs)_{CI\text{-}noreact}$	4.53	—	—	—	—	4.47	4.43
$(ct/gs)_{CI\text{-}react}$	5.32	—	—	—	—	5.02	5.27

[a] All values are reported with reference to the ground state energy of the $^1(gs)_{min}$ geometry.

 This journal is © The Royal Society of Chemistry 2018

The obtained main results and energies are reported schematically in Fig. 3 and in Table 2.

Conclusions

In the present contribution QM(CASPT2//CASSCF):MM(AMBER) calculations have been performed in order to elucidate the photoinduced formation mechanism of the thymine–thymine (6–4) adduct in the specific hairpin H_42 for which the experimental quantum yield of the photoreaction has been measured.[27] Starting from a presumably pro-reactive ground state minimum, characterized by a relatively short $O_{4'}$–C_5 distance (2.810 Å), various excited states have been optimized and the corresponding minima analyzed.

The main conclusion obtained here is that from the characterized starting geometry a high-lying charge transfer ct–ππ*–TT′ state can decay towards a region of energy degeneracy with the ground state, where both reactive and non-reactive geometries have been documented. Two types of region have in fact been described along the ct–ππ*–TT′ PEH. The first, exemplified by the $(ct/gs)_{CI\text{-}noreact}$ structure, is characterized by the formation of a new carbon–oxygen $O_{4'}$–C_5 bond, and from such a geometry the ground state decays back to the two separated thymine molecules. The second, exemplified by the $(ct/gs)_{CI\text{-}react}$ structure, is again characterized by the formation of a new carbon–oxygen $O_{4'}$–C_5 bond, but it also displays a significant puckering of the $C_{4'}$ atoms towards the C_6 atom, which are the atoms among which a new carbon–carbon bond needs to be formed in order to create the oxetane ring. This latter geometrical feature seems to be fundamental for the pro-reactivity, since, as proven by the corresponding optimization, for the $(ct/gs)_{CI\text{-}react}$ structure the ground state evolves towards oxetane formation. The involvement of the here characterized ct–ππ*–TT′ charge-transfer state is in agreement with the recent work of Conti at al.[24] The identification of a reactive path from the characterized starting ground state geometry, $^1(gs)_{min}$, consequently points to the importance of a short $O_{4'}C_5$ distance for the formation of the thymine–thymine (6–4) photodamage adduct.

The possibility of a reaction mechanism passing through the initial population of nπ* states has also been partially explored here. It appears that the first part of the proposed mechanism, leading to the population of the minima of singlet and triplet nπ* states localized on the reactive carbonyl group (here referred to as the ^{1}nπ*–T′ and ^{3}nπ*–T′ minima) is indeed possible, although it is not clear what the driving force of the second part of the process will be, leading from a minimum in which the reactive $O_{4'}$–C_5 atoms are separated by more than 3 Å (as in the ^{1}nπ*–T′ and ^{3}nπ*–T′ minima, see Fig. S7†) towards a reactive singlet–triplet crossing region with the ground state where a new carbon–oxygen $O_{4'}$–C_5 bond is formed. Further computational studies are needed in order to better clarify this point.

Acknowledgements

M. G acknowledges the support from the European Research Council Advanced Grant STRATUS (ERC-2011-AdG No. 291198). A. G. acknowledges the support form the European Union's Horizon 2020 research and innovation programme under the Marie Sklodowska-Curie Grant Agreement No. 658173.

References

1 J. Cadet and P. Vigny, The Photochemistry of Nucleic Acids, in *Bioorganic Photochemistry*, ed. H. Morrison, John Wiley & Sons, Inc., New York, 1990, vol. 1, pp. 1–272.
2 R. P. Sinha and D. P. Hader, *Photochem. Photobiol. Sci.*, 2002, **1**, 225–236.
3 J. S. Taylor and M. P. Cohrs, *J. Am. Chem. Soc.*, 1987, **109**, 2834–2835.
4 M. Boggio-Pasqua, G. Groenhof, L. V. Schäfer, H. Grubmüller and M. Robb, *J. Am. Chem. Soc.*, 2007, **129**, 10996–10997.
5 L. Blancafort and A. Migani, *J. Am. Chem. Soc.*, 2007, **129**, 14540–14541.
6 D. Roca-Sanjuán, G. Olaso-González, I. González-Ramírez, L. Serrano-Andrés and M. Merchán, *J. Am. Chem. Soc.*, 2008, **130**, 10768–10779.
7 J. J. Serrano-Pérez, I. González-Ramírez, P. B. Coto, M. Merchán and L. Serrano-Andrés, *J. Phys. Chem. B*, 2008, **112**, 14096–14098.
8 T. Climent, I. González-Ramírez, R. González-Luque, M. Merchán and L. Serrano-Andrés, *J. Phys. Chem. Lett.*, 2010, **1**, 2072–2076.
9 F. Manson, T. Laino, I. Tavernelli, U. Rothlisberger and J. Hutter, *J. Am. Chem. Soc.*, 2008, **130**, 3443–3450.
10 M. Barbatti, *ChemPhysChem*, 2014, **15**, 3342–3354.
11 R. Improta, *J. Phys. Chem. B*, 2012, **116**, 14261–14274.
12 V. B. Delcheva and W. Domcke, *J. Photochem. Photobiol., A*, 2013, **271**, 1–7.
13 S. Mouret, C. Baudouin, M. Charveron, A. Favier, J. Cadet and T. Douki, *Proc. Natl. Acad. Sci. U. S. A.*, 2006, **103**, 13765–13776.
14 W. J. Schreier, T. E. Schrader, F. O. Koller, P. Gilch, C. E. Crespo-Hernández, V. N. Swaminathan, T. Carell, W. Zinth and B. Kohler, *Science*, 2007, **315**, 625–629.
15 W. M. Kwok, C. Ma and D. L. Phillips, *J. Am. Chem. Soc.*, 2008, **130**, 5131–5139.
16 T. Douki, I. Bérard, A. Wack and S. Andrä, *Chem.–Eur. J.*, 2014, **19**, 5787–5794.
17 R. R. Ramazanov, D. A. Maksimov and A. I. Kononov, *J. Am. Chem. Soc.*, 2015, **137**, 11656–11665.
18 M. C. Cuquerella, V. Lhiaubet-Vallet, F. Bosca and M. A. Miranda, *Chem. Sci.*, 2011, **2**, 1219–1232.
19 A. Banyasz, T. Douki, R. Improta, T. Gustavsson, D. Onidas, I. Vayá, M. Perron and D. Markovitsi, *J. Am. Chem. Soc.*, 2012, **134**, 14834–14845.
20 S. Marguet and D. Markovitsi, *J. Am. Chem. Soc.*, 2005, **127**, 5780–5781.
21 Z. B. Yang, R. B. Zhang and L. A. Eriksson, *Phys. Chem. Chem. Phys.*, 2011, **13**, 8961–8966.
22 Z. B. Yang, L. A. Eriksson and R. B. Zhang, *J. Phys. Chem. B*, 2011, **115**, 9681–9686.
23 A. Giussani, L. Serrano-Andrés, M. Merchán, D. Roca-Sanjuán and M. Garavelli, *J. Phys. Chem. B*, 2013, **117**, 1999–2004.
24 I. Conti, L. Martínez-Fernández, L. Esposito, S. Hofinger, A. Nenov, M. Garavelli and R. Improta, *Chem.–Eur. J.*, 2017, **23**, 1–13.
25 T. Douki, *Photochem. Photobiol. Sci.*, 2013, **12**, 1286.
26 Z. Pan, M. McCullagh, G. C. Schatz and F. D. Lewis, *J. Phys. Chem. Lett.*, 2011, **2**, 1432–1438.
27 M. Hariharan, K. Siegmund, C. Saurel, M. McCullagh, G. C. Schatz and F. D. Lewis, *Photochem. Photobiol. Sci.*, 2014, **13**, 266.

 This journal is © The Royal Society of Chemistry 2018

28 J. Cadet, E. Sageb and T. Douki, *Mutat. Res.*, 2005, **571**, 3–17.
29 J. Cadet, S. Mouret, J.-L. Ravanat and T. Douki, *Photochem. Photobiol.*, 2012, **88**, 1048–1065.
30 T. Douki and J. Cadet, *Biochemistry*, 2001, **40**, 2495.
31 E. Paterno and G. Chiefi, *Gazz. Chim. Ital.*, 1909, **39**, 431.
32 I. J. Palmer, I. N. Ragazos, F. Bernardi, M. Olivucci and M. A. Robb, *J. Am. Chem. Soc.*, 1994, **116**, 2121–2132.
33 B. S. Maeda, T. Taketsugu and K. Morokuma, *Z. Phys. Chem.*, 2013, **227**, 1421–1433.
34 S. J. Harris, D. Murdock, M. P. Grubb, I. P. Clark, G. M. Greetham, M. Towrie and M. N. R. Ashfold, *J. Phys. Chem. A*, 2014, **118**, 10240–10245.
35 R. Y. Brogaard, O. Schalk, A. E. Boguslavskiy, G. D. Enright, H. Hopf, V. Raev, E. Tarcoveanu, T. I. Sølling and A. Stolow, *Phys. Chem. Chem. Phys.*, 2012, **14**, 8572–8580.
36 D. Case, *AMBER 12*, University of California, San Francisco, 2012.
37 R. Salomon-Ferrer, D. A. Case and R. C. Walker, *Wiley Interdiscip. Rev.: Comput. Mol. Sci.*, 2013, **3**, 198–210.
38 P. Altoè, M. Stenta, A. Bottoni and M. Garavelli, *Theor. Chem. Acc.*, 2007, **118**, 219.
39 B. O. Roos, P. R. Taylor and P. E. M. Siegbahn, *Chem. Phys.*, 1980, **48**, 157.
40 K. Andersson, P.-Å. Malmqvist, B. O. Roos, A. J. Sadlej and K. Wolinski, *J. Phys. Chem.*, 1990, **94**, 5483.
41 K. Andersson, P.-Å. Malmqvist and B. O. Roos, *J. Chem. Phys.*, 1992, **96**, 1218.
42 B. O. Roos, K. Andersson, M. P. Fülscher, P.-Å. Malmqvist, L. Serrano-Andrés, K. Pierloot and M. Merchán, *Adv. Chem. Phys.*, 1996, **93**, 219.
43 R. Montero, A. Longarte, Á. Peralta Conde, C. Redondo, F. Castaño, I. Gonzalez-Ramírez, A. Giussani, L. Serrano-Andrés and M. Merchán, *J. Phys. Chem. A*, 2009, **113**, 13509.
44 F. Aquilante, J. Autschbach, R. K. Carlson, L. F. Chibotaru, M. G. Delcey, L. De Vico, I. Fdez Galván, N. Ferré, L. M. Frutos, L. Gagliardi, *et al.*, *J. Comput. Chem.*, 2016, **37**, 506.
45 M. L. Frisch, *et al.*, *Gaussian 03, Revision C.02*, Gaussian Inc, Wallingford CT, 2004.
46 A. Giussani, *J. Chem. Theory Comput.*, 2014, **10**, 3987–3995.
47 P.-O. Widmark, P.-Å. Malmqvist and B. O. Roos, *Theor. Chem. Acc.*, 1990, **77**, 291.
48 K. Pierloot, B. Dumez, P.-O. Widmark and B. O. Roos, *Theor. Chim. Acta*, 1995, **90**, 87.
49 F. Aquilante, T. B. Pedersen and R. Lindh, *Theor. Chem. Acc.*, 2009, **124**, 1.
50 J. J. Serrano-Pérez, R. González-Luque, M. Merchán and L. Serrano-Andrés, *J. Phys. Chem. B*, 2007, **111**, 11880–11883.
51 A. Giussani, J. Segarra-Martí, D. Roca-Sanjuán and M. Merchán, *Top. Curr. Chem.*, 2015, **355**, 57–97.
52 C. R. Kozak, K. A. Kistler, Z. Lu and S. Matsika, *J. Phys. Chem. B*, 2010, **114**, 1674–1683.

Faraday Discussions

DISCUSSIONS

Light induced damage and repair in nucleic acids and proteins: general discussion

Amitava Chandra, Richard Cogdell, Carlos E. Crespo-Hernández, Ankona Datta, Angelo Giussani, Stefan Haacke, John Helliwell, Roberto Improta, Ramapurath S. Jayasree, Mike Jones, Tolga Karsili, Bern Kohler, Mahil L, Imon Mandal, Dimitra Markovitsi, Himani Medhi, Padmaja P. Mishra, P. I. Pradeepkumar, Priyadarshi Roy Chowdhury, Manas Sarangi, Igor Schapiro, Ilme Schlichting, Javier Segarra-Martí, Amit Sharma, Vishnu V, Rienk van Grondelle and Anthony Watts

DOI: 10.1039/C8FD90006J

Bern Kohler opened discussion of the paper by Mike Jones: What about charge transfer as a possible explanation for the emission quenching you observe with the cadmium telluride quantum dots?

Mike Jones answered: So far the data we have are consistent with a FRET mechanism, but we cannot rule out the involvement of charge transfer states. Professor Rienk van Grondelle has suggested we use Stark spectroscopy to explore this possibility.

Anthony Watts asked: Do the quantum dots (QDs) cause any spectral shift in your engineered photoproteins, as seen with QD coupling to bacteriorhodopsin?

Mike Jones replied: No, we haven't seen any influence of these QDs on the absorbance spectra of the various wild type and mutant RCs and RCLH1 complexes we have attached to them.

Richard Cogdell said: We see the quenching of the QD, but do you see the arrival of the energy at the reaction centre?

Mike Jones replied: Yes. We see enhanced P photobleaching in the presence of QDs and we see enhanced emission from a mutant RC that is engineered to lack the primary electron donor.

Anthony Watts asked: How do you know you've got that asymmetrical arrangement shown in your solar cell figure? Could controlling orientation be worth investigating?

Mike Jones replied: In the case of electrode systems where the metal electrode is coated with a cytochrome c layer, an asymmetry of binding could be attributed to the cytochrome c binding site of the RC which is on the periplasmic side of the membrane – however the diagram is based on a rationalisation of the observed output and its likely mechanism rather than any independent evidence on the orientation of the protein at the electrode surface. There are certainly benefits to be had from controlling orientation and multiple studies have been carried out on this topic (mostly by other research groups in the field). We are currently working on strategies for engineering photoproteins for directed self assembly with electrodes and other materials, our preference being to obtain control through genetic coding (as in nature) rather than functionalisation of electrodes or complex fabrication processes.

Rienk van Grondelle asked: What is the cause of the difference between the original 8 $\mu A\ cm^{-2}$ and the now ~120 $\mu A\ cm^{-2}$ photocurrent?

Mike Jones answered: The biggest influences on current are protein concentration and the choices of electrolyte, electrode materials and cell architecture.

Anthony Watts queried: What are the absolute currents that you generate, and if you want to take this into the field for an application, how large a surface would you need to produce power outputs necessary to drive a small device?

Mike Jones answered: We are some way from this at present because the high voltage devices we've been able to construct produce low currents and *vice versa*.

Stefan Haacke asked: Is this sketch real in the sense that you only have a monolayer of proteins on the electrode layer? In addition, what is the optimal thickness in terms of photocurrent density generated and what is the value you aim for?

Mike Jones responded: In some devices you can impose a monolayer on the electrode surface, or it is likely that this is what you achieve because proteins not adhered directly to the electrode are rinsed away during device fabrication (we can check loadings from the pigment content of the adhered proteins). In other devices we deliberately aim for multilayers of protein. It is difficult at this stage to say what an optimal thickness of protein might be, other than to say that what we are ultimately aiming for is a photocurrent output of the same order as that achieved in other types of solar cell (a few tens of $mA\ cm^{-2}$).

Amit Sharma asked: How many photons are you putting into the system per protein molecule?

Mike Jones replied: We have not done that calculation.

Ilme Schlichting commented: Photosystem I has been used to photoreduce cytochrome P450, starting a catalytic reaction that is otherwise very difficult to trigger without using the natural component. Have you thought about using your system for catalysis applications?

 This journal is © The Royal Society of Chemistry 2018

Mike Jones answered: Yes, this is something we are starting to explore alongside our work on photovoltaics.

John Helliwell communicated: Would crystal structure analysis of one or more of the authors' hybrids, such as depicted in Fig. 5b in the paper, be worth a try to assist the authors' rational design towards optimisation of their nanodevices? Have they tried to crystallise one or more of their hybrids?[1]

1 As a guidebook see: N. E. Chayen, J. R. Helliwell and E. H. Snell, *Macromolecular Crystallization and Crystal Perfection*, International Union of Crystallography Monographs on Crystallography, Oxford University Press, 2010, ISBN: 9780199213252.

Mike Jones communicated in reply: We haven't attempted crystallisation of protein: QD conjugates but this is something we could consider, particularly for reaction centres where we have some experience of crystallisation.

Priyadarshi Roy Chowdhury communicated: What are the factors that are taken into consideration during the engineering of photoproteins?

Mike Jones communicated in response: Stability and expression level are really important for a mutation to be useful, so we need to avoid changes/additions to structure that interfere with assembly or destabilise the structure.

Priyadarshi Roy Chowdhury communicated: Why are PufX-deficient proteins incapable of conversion of solar energy *in vivo*?

Mike Jones communicated in reply: This is explained in our paper and summarised in Fig. 7. RCs in PufX minus RCLH1 complexes are unusually prone to "closing" under photosynthetic (*i.e.* reducing) growth conditions. By "closing" we mean the Q_A quinone becomes reduced. This prevents charge separation forming a metastable radical pair state. Instead what happens is the antenna harvests light in the usual way, energy is passed to the RC in the usual way, the first two steps of charge separation happen in the usual way, but then because Q_A is pre-reduced the $P^+H_A^-$ radical pair recombines (in nanoseconds) to the ground state and harvested energy is wasted as heat.

Dimitra Markovitsi opened discussion of the paper by Tolga Karsili: You mentioned that the 8-oxo-7,8-dihydro-2′-deoxyguanosine (8-oxoG) detected following near-UV irradiation of genomic DNA could stem from a photochemical path without involvement of the guanine radical cation. Do you expect the reaction intermediate identified in your calculations to react with spermine, which was shown to hinder 8-oxoG formation?[1]

1 M. Gomez-Mendoza, A. Banyasz, T. Douki, D. Markovitsi and J. L. Ravanat, *J. Phys. Chem. Lett.*, 2016, 7, 3945–3948.

Tolga Karsili responded: Our model does not preclude a quenching of 8-oxoG when in the presence of spermine. Instead of reacting with water the photo-basic guanine will instead react with a more potent spermine molecule. In so doing, as with water, the initial step is likely to proceed *via* a photo-induced spermine-to-chromophore electron-driven proton-transfer reaction from NH of spermine to the N7-acceptor site of G, followed by a subsequent self-reaction of the nascent G(+H) + spermine(−H) radicals at the C8-site of G.

Dimitra Markovitsi asked: Have you computed the absorption spectra of the reaction intermediates leading to 8-oxoG and pyrimidine hydrates? Such spectra could help identify these transient species by time-resolved spectroscopy, as was the case for the reaction intermediate leading to adenine dimerization.[1]

1 A. Banyasz, L. Martinez-Fernandez, T. Ketola, A. Muñoz-Losa, L. Esposito, D. Markovitsi and R. Improta, *J. Phys. Chem. Lett.*, 2016, 7, 2020–2023.

Tolga Karsili replied: We are presently developing methods to compute detailed absorption spectra in bulk solution. This form of photodamage is an avenue we plan to apply our methods to in future. That said, your spectra are unlikely to reveal the transient intermediates that lead to the formation of the hydrate adducts since these are high energy intermediates that we predict to have lifetimes of the order of tens-of-fs. The possibility of computing the stable hydrate adducts themselves would however provide great insight into the formation mechanism for 8-oxoG.

Dimitra Markovitsi commented: According to your calculations the approach of a water molecule to the guanine is necessary for 8-oxoG formation. Do you expect this approach to be hindered in double strands or G-quadruplexes? Note that 8-oxoG was detected following irradiation of G-quadruplexes with a continuous light source at 266 nm.[1]

1 A. Banyasz, L. Martinez-Fernandez, C. Balty, M. Perron, T. Douki, R. Improta and D. Markovitsi, *J. Am. Chem. Soc.*, 2017, **139**, 10561–10568.

Tolga Karsili answered: Absolutely, the same photoreaction is unlikely to hold (at least to the same facile extent) in double-helix strands or G-quadruplexes. We note however that the formation of radical cations represents a chromophore-to-solvent charge-transfer rather than the mechanism outlined in our study which shows a solvent-to-chromophore charge-transfer. The nascent solvated-electron is likely to represent a very potent nucleophile which may abstract the H-atom at the C8-site – thereby encouraging a reaction with water.

Anthony Watts said: Coming back to the question on absorption spectrum calculation, you mentioned in ref. 27 of your manuscript that some experimental data is available on this system – have you gone back and compared your data with experimental data?

Tolga Karsili replied: As highlighted in our manuscript, ref. 27 refers to a series of experiments performed on pyridine, seeded in gas phase water clusters.[1] Using high-resolution spectroscopy, the work demonstrates the possibility for studying UV-induced water-to-chromophore intermolecular proton-transfer, in a gas-phase cluster, providing the possibility for testing our theoretical predictions for the type of proton-transfer that we presently predict as the initiator for the ensuing reaction path for forming lesion products. In order to draw synergies with the

 This journal is © The Royal Society of Chemistry 2018

types of experiments performed by Prof. Markovitsi, we would need to compute the detailed absorption spectra of these hydrate intermediates in order to compare with the available data in the long-time resolved pump–probe experiments in the bulk. We are currently in the process of computing the details of such spectra at present.

1 N. Esteves-Lopez, S. Coussan, C. Dedonder-Lardeux and C. Jouvet, *Phys. Chem. Chem. Phys.*, 2016, **18**, 25637–25644.

Anthony Watts asked: If there was an ideal experiment to do to test out your model, what would it be?

Tolga Karsili responded: As I mentioned in my answer to your previous question, experiments to probe such water-to-chromophore proton-transfer studies are indeed available in the gas-phase[1] for simple heterocyclic chromophores such as pyridine. The difficulty with the DNA/RNA nucleobases is their inherently low vapour pressures that preclude effective sample concentrations in gas phase. The gas-phase however provides a first-hand, detailed and high resolution probe for such photoreactions that are free from environmental perturbations. Analogous experiments to that of jet-cooled pyridine, seeded in water clusters, are possible with the DNA/RNA nucleobases and nucleosides (and associated oligomers) but would require experimental developments of methods that would increase the sample concentrations of a low vapour pressure system in the gas-phase. One such avenue would be the development of laser-desorption methods for desorbing DNA/RNA nucleobases that are bound to a rotating graphite disk. In solution, the ideal experiment that would test our present prediction would be time-resolved transient absorption spectroscopy. This would however require further theoretical modelling of the expected absorption profiles of the reaction intermediates in order to compare with the experimentally measured spectra.

1 N. Esteves-Lopez, S. Coussan, C. Dedonder-Lardeux and C. Jouvet, *Phys. Chem. Chem. Phys.*, 2016, **18**, 25637–25644.

Igor Schapiro asked: Are the reaction pathways that you have presented generated by interpolating geometries between the Frank Condon point, the conical intersection and the product geometry? If so, what type of conical intersection optimization have you done: minimum energy or minimal distance?

Tolga Karsili replied: In part. The potential energy (PE) profiles associated with the uracil/thymine + water photo-reactions were computed by means of a linear interpolation (in internal coordinates) between the Franck–Condon geometry and the geometry of a given minimum energy conical intersection. In the case of the thymine + water photo-reaction, in bulk double-helix DNA in aqueous solution, the ground and excited state PE profiles were computed by means of an excited state relaxed scan, using the O–H stretch coordinate of the proximal water as the driving coordinate. Such relaxed scans were also used for computing the guanine + water photo-reaction in the gas phase.

Bern Kohler noted: Photohydrate yields are reported to be greater for pyrimidine nucleotides than for the free bases.[1] Can your model explain this observation? For example, could ribose substitution at N1 stabilize the carbocation?

1 J. Cadet and P. Vigny, The Photochemistry of Nucleic Acids, in *Bioorganic Photochemistry*, ed. H. Morrison, Wiley, New York, 1990, vol. 1, pp. 1–272.

Tolga Karsili answered: We would certainly expect the energetics of the various electronically excited states of the nucleotides to change relative to those of the isolated nucleobases. First however, we should clarify the nature of the reactive states as these do not lead to carbocations. In our model computations the charge-transfer (CT) state involves an electron promotion from a proximal water molecule to a π* orbital of the nucleobase. The CT state therefore contains a partial negative charge centered on the nucleobase. As with related solvent-chromophore systems, we predict that the CT state is populated by internal conversion from the initially excited locally-excited state which contains a nucleobase centered ππ*-character. The expected σ-donating nature of the ribose is therefore likely to destabilize the CT state but is likely to stabilize the final photohydrate *via* enhanced electron-density of the reduced π-system upon nucleophilic addition across C5–C6. This may lead to a greater yield of hydrate adducts in the nucleotides when compared to the nucleobases.

Javier Segarra-Martí asked: Have you compared your work to what has been done with OH-substituted complexes? While the extra hydrogen you have with respect to OH-substituted complexes will affect the decay paths, do you expect the absorption spectra to also change significantly?

Tolga Karsili replied: This is not a fair comparison. Hydroxyl-substituted uracil retains the same number of π-electrons as bare uracil but displays an understandable bathochromic shift in the absorption as a result of a π-donor substituent. C5–C6 hydrated uracil, in contrast, leads to a replacement of two of the six π-electrons of uracil into covalent bonding electrons – reducing the number of π-electrons to four. This inherent reduction of the π-electron density, upon going from uracil to 6-hydroxy-5*H*-uracil, is expected (and predicted in our manuscript) to vastly blue-shift the strong absorption bands in the hydrate adduct (*cf.* uracil).

Anthony Watts commented: You said there are two parts to the PE profile generation to the hydrate, but you were talking earlier about radicals, where do radicals come in? Is there a difference between the gas phase and solution with respect to radical generation?

Tolga Karsili responded: The two step mechanism involves an initial electron-driven water-to-nucleobase proton-transfer and then a subsequent addition of the nascent OH radical to the odd-electron radical site of the nucleobase. The initial electronic excitation leads to the formation of local-excited (LE) ππ* states – in which both the π and π* orbitals are localized on the nucleobase. These LE states then couple to a near-lying charge-transfer (CT) state which involves an electron promotion from an oxygen 2p non-bonding orbital to the π* orbital of the nucleobase. This clearly represents a charge-separation in which the water

 This journal is © The Royal Society of Chemistry 2018

contains a partial positive charge and the nucleobase contains a partial negative charge. The charge-separation is neutralized by water-to-chromophore proton (H^+) transfer (*i.e.* the first step in the reaction process). This leaves a protonated chromophore + OH radical-pair intermediate – each of which contain an odd-electron from the initial excitation step. These nascent radicals will also persist in bulk aqueous solution since the formation of the nascent radical-pairs are dependent upon electronic excitation. The corresponding ion-pair states can only be formed by heterolytic bond fission of the ground state electronic configuration or an electronic excitation which involves an electron promotion from a π orbital local to the chromophore and a π* orbital local to water. This will then follow hydrogen-atom (not proton) transfer from water to the chromophore to reveal the chromophore$^+$ + OH^- ion-pair. Although these ion-pair states are stabilized in solution, our QM/MM computations on bulk DNA in aqueous solution reveal these states as higher energy states when compared to the water-to-chromophore CT states. This can be understood since DNA/RNA nucleobases contain enhanced basicities upon electronic excitation.

Mike Jones asked: How much of a problem is water for systems? I worry about water. The problem with protein technology is trying to keep the protein stable. Should we be working in an aqueous environment or trying to find a system that is less damaging such as a deep eutectic solvent?

Tolga Karsili responded: In certain environments, we and others have shown that photoreactions of water with solute molecules are common upon continuous light irradiation. Water is also problematic for time-resolved transient absorption experiments that exploit broadband IR wavelengths as probes, since the vibrational bands of water usually obstruct the important vibrational bands of the solute. As Prof. Bern Kohler showed, deep eutectic solvents offer an exciting alternative for overcoming such experimental problems but given their ionic nature, the extent to which they are photoreactive with a photoredox-active solute is unknown.

Himani Medhi communicated: The isolated gas phase of DNA/RNA units is free from environmental perturbations such as the water (cluster) and chromophore complex which are readily found in living systems. Thus, conventional photo-induction can damage DNA/RNA units in the gas phase more drastically than in bulk solution. Has this study found the same result as conventional ones or not?

Tolga Karsili communicated in reply: It is well-known that photohydrates are formed in DNA/RNA nucleobases in aqueous solutions,[1] although their formation mechanisms are poorly understood. When comparing the gas phase potential energy profiles (Fig. 3 in our paper) with those returned in bulk aqueous double-helix DNA (Fig. 8) it is evident that the excited-state nucleophilic hydrolysis reaction is in fact enhanced in the bulk. As Fig. 8 shows, the vertically excited bright S2(1ππ*) state contains a far greater energetic driving force for coupling to the S1(CT) state when compared to the restrictive barrier height in the gas phase (Fig. 3). This is understandable since charge-transfer states are known to stabilize upon dispersive π–π interactions formed by intra-strand nucleobase stacking and inter-strand hydrogen-bonding. Our bulk DNA model features the nucleophilic hydrolysis reaction at the end-terminated nucleobase since we recognize that the interstitial regions of double helix DNA are water deficient. Such photoreactions with water are therefore unlikely to occur in chromosomal DNA but are likely to become important in short chain RNA and DNA strands as well as when bulk double-helix DNA unwinds for transcription.

1 J. Cadet and P. Vigny, The Photochemistry of Nucleic Acids, in *Bioorganic Photochemistry*, ed. H. Morrison, Wiley, New York, 1990, vol. 1, pp. 1–272.

Imon Mandal communicated: First, are 5 water molecules sufficient for the G + H_2O system to describe the environment? How are the orientation of those waters going to affect the transitions and PES? Second, why is only one water taken for T and U but 5 waters are taken for the G system for the calculations?

Tolga Karsili communicated in reply: In order to describe the photo-reactive role of proximal water molecules, a quantum mechanical description of a subset of water molecules was required. Five is on the limit of the available resources that we can include with our choice of high-level multi-reference computational methods. As described in the manuscript, five water molecules are more than sufficient in describing all dominant first-order donor–acceptor effects as well as the secondary dispersive interactions. As outlined in the manuscript, our choice of five water molecules was successful in describing the chromophore–water interactions in related systems (both in our lab and by others – see ref. 21, 23 and 24 in our present manuscript). Since the orientation of the water molecules largely represent dispersive interactions and do not represent any first-order bonding interactions, the relative chromophore–water motions will show very minor changes in potential energy as a function of chromophore–water motions. So we conclude that small changes in the chromophore–water orientations will show very minor changes in the energetics associated with the reaction barriers, minima and conical intersections.

For the second question, this is a simple entropy argument – a single water molecule in the G + H_2O photoreaction would simply preclude a kinetically facile reaction since the dominant orbitals involved in the photoreaction are simply not optimally aligned. In contrast, the dispersively bound chromophore–water interaction in the U/T + water photo-reaction leads to optimal alignment of the dominant orbitals that partake in the nucleophilic hydrolysis photoreaction.

Priyadarshi Roy Chowdhury communicated: What are the factors that lead to the photoinduced damage of DNA nucleobases?

Tolga Karsili communicated in reply: The obvious radiative factors are continuous exposure to near-UV radiation – such as those that lead to skin-cancer. Environmental factors are also important. Specifically, pH is an important factor since a reduced cellular pH is likely to enhance the probability of nucleophilic addition reactions to DNA. Other factors may include imbalances of ions/molecules that provide optimal conditions within a cellular environment.

Tolga Karsili addressed Mike Jones: How does the photocurrent compare with that measured in industrial photovoltaics?

 This journal is © The Royal Society of Chemistry 2018

Mike Jones responded: Dye sensitised solar cells typically produce an open circuit voltage of around 0.7 V and a short circuit current density of a few tens of microamps per cm^2. The best photovoltages we have achieved are also around 0.7 V, but the best photocurrents we have achieved are a little under 1 microamp per cm^2. So we have some way to go as regards to current, and as I mentioned previously we are yet to produce both high currents and high voltages in the same device.

Mahil L asked: What led to the stability of the engineered protein *vs.* the natural protein?

Mike Jones replied: Our hypothesis is that closure of the LH1 ring increases stabilising protein–protein contacts between the central reaction centre and the inner ring of LH1 alpha-polypeptides, producing an overall stabilisation of the structure. Also the complete LH1 ring does not have any free ends, which may also assist in stabilisation. At this stage this is only a hypothesis. What is needed is a clearer picture of the process through which the native structure is lost, during thermal stress and also during photo-stress.

Richard Cogdell commented: You know there are thermophilic purple bacteria. Would it be possible to get further stabilisation by making your proteins more like the thermophilic ones? You could think of doing a more random approach as was done for GPCRs.

Mike Jones replied: Possibly. The random mutagenesis route is attractive but the challenge is always to have an effective screen.

Pradeepkumar P. I. asked Tolga Karsili: While showing the mechanism of addition of water to the C5–C6 bond of dC, you have shown a subsequent deamination reaction of dC to form dU. Since deamination of dC is a common DNA damage which leads to GC to AU mutation, is there any chance that deamination happens prior to the addition of water to dC?

Tolga Karsili responded: We have not explored this subsequent reaction in our manuscript but it is an interesting query. Deamination is indeed common in DNA damage although such reactions are unlikely to proceed *via* initial C–NH_2 bond fission in the electronically excited state. This is because the only electronic state that forms a long-range dissociative interaction, with respect to C–N bond elongation, is the $^3\sigma\sigma^*$ state. Population of this state at the Franck–Condon region represents a spin-forbidden inter-system crossing from the initially excited singlet state. It is therefore more likely that the initial addition of water occurs across the C5–C6 bond of cytosine and then the nascent hydrate reacts with subsequent water molecules to eliminate NH_2 – thus forming uracil hydrate + ammonia. This has already been shown experimentally by Boorstein *et al.*[1]

1 R. J. Boorstein, T. P. Hilbert, R. P. Cunningham and G. W. Teebor, *Biochemistry*, 1990, **29**, 10455–10460.

Padmaja P. Mishra asked: Which nucleobases are more vulnerable to photodamage, DNA or RNA?

Tolga Karsili answered: This is a rather general question, since there are several examples of photodamage of DNA/RNA nucleobases, involving different photoreactions. We have shown (and it has been experimentally confirmed) that uracil forms an enhanced yield for photo-induced nucleophilic addition of water across the C5–C6 of uracil. Prof. Markovitsi showed that adenine-thymine duplexes formed deprotonated radical-cation products and cycloadducts. Guanine is known to form an 8-oxoG lesion (as shown in our manuscript and others). It is well-known that stacked thymine dimers have an enhanced rate of [2 + 2]-cycloaddition. So, all nucleobases are vulnerable to photodamage, given optimal conditions and appropriate radiative environments.

Richard Cogdell opened a general discussion of the paper by Carlos E. Crespo-Hernández: If you think how DNA evolved, it evolved in deep ocean organisms that did not have to worry about UV damage. Do you think it was fortune that oceanic organisms evolved photostable DNA that let them colonise land, or did they need to evolve this over again once they got there?

Carlos E. Crespo-Hernández responded: Understanding the events that constitute the molecular origins of life is a major challenge in prebiotic chemistry. A period of chemical evolution may have preceded and shaped the progression from simple to complex molecules capable of performing biological functions.[1,2] One hypothesis is that the chemistry that led to life began in the dark, or in highly UV-protected environments, such as in cracks and pores of solid surfaces,[3] in clay minerals,[4] beneath sacrificial molecular layers,[5] or below the surfaces of oceans.[6–8] Although it is reasonable to hypothesize that a living organism capable of actively avoiding the surface of the ocean would not have to worry about UV damage, such a UV-shielding mechanism cannot protect prebiotic molecules that should have resurfaced periodically because of diffusion and mixing.[9] As de Vries, Kohler, and co-workers eloquently proposed: "If life originated in fully UV-protected environments, UV selection pressure vanishes and less UV-hardy building blocks could just as easily have been precursors of the first self-replicating polymers. Moving prebiotic chemical evolution into the dark eliminates photochemical damage, but begs the question of how and when a transition to light-exposed environments took place. In contrast, starting life in the light is an attractive possibility that leads to UV-hardy molecules that avoid photodamage without losing the ability to support chemistry driven by longer wavelength photons."[9] Evidently, multiple selection pressures should have operated during early chemical and biological evolution as the driving force for the selection of the contemporary nucleobases.[2] Extreme selection pressure for protection against intense UV radiation and photostability was likely a decisive criterion,[6,10–12] which should have given some heterocycle analogues a selective advantage for their incorporation into the first informational polymers.[2] Encoding such early polymers with highly photostable molecules is an elegant solution to the threat of photochemical damage. The intrinsic photostability to otherwise deleterious UV radiation observed in all five RNA and DNA nucleobases, provides strong support to the idea that UV radiation should have played a key role in the natural selection of the 'building blocks of life' during prebiotic Earth.

 This journal is © The Royal Society of Chemistry 2018

1 W. Gilbert, *Nature*, 1986, **319**, 618.
2 A. C. Rios and Y. Tor, *Isr. J. Chem.*, 2013, **53**, 469–483.
3 C. C. Bryce, G. Horneck, E. Rabbow and H. G. M. Edwards, *Int. J. Astrobiol.*, 2015, **14**, 115–122.
4 J. D. Bernal, *The physical basis of life*, Routledge and Paul, London, 1951.
5 *The Ecology of Cyanobacteria*, ed. B. A. Whitton and M. Potts, Springer, 2000, pp. 591–611.
6 C. Sagan, *J. Theor. Biol.*, 1973, **39**, 195–200.
7 H. J. Cleaves and S. L. Miller, *Proc. Natl. Acad. Sci. U. S. A.*, 1998, **95**, 7260–7263.
8 M. J. Russell, A. J. Hall, J. Boyce and A. E. Fallick, *Econ. Geol.*, 2005, **100**, 419–438.
9 A. A. Beckstead, Y. Zhang, M. S. de Vries and B. Kohler, *Phys. Chem. Chem. Phys.*, 2016, **18**, 24228–24238.
10 C. E. Crespo-Hernández, B. Cohen, P. M. Hare and B. Kohler, *Chem. Rev.*, 2004, **104**, 1977–2019.
11 A. L. Sobolewski and W. Domcke, *Europhys. News*, 2006, **37**, 20–23.
12 S. Ranjan and D. D. Sasselov, *Astrobiology*, 2016, **16**, 68–88.

Ankona Datta asked: In your paper you mention that the reaction with prodrugs is going to create methylated products. What diseases are these targeted at? Do you want these molecules inserted into DNA?

Carlos E. Crespo-Hernández responded: The azathioprine, 6-mercaptopurine, and 6-thioguanine are effective anticancer, anti-inflammatory, and immunosuppressant drugs.[1] These prodrugs undergo enzymatic conversion that ultimately culminates in the formation of 6-thioinosine and 6-thioguanosine nucleotides, which are required for their clinical effectiveness. As described in our article, the 6-thioinosine nucleotide is in turn converted into the *S*6-methylthioinosine nucleotide *via* the enzyme *S*-methyl transferase.[2,3] *S*6-methylthioinosine can further inhibit purine *de novo* synthesis *in vitro*,[4–6] and is linked to an increased risk of hepatotoxicity.[7–9] The 6-thioguanosine nucleotide is metabolized into DNA, and the biological effects are partly dependent on the incorporation of 6-thioguanosine nucleotide into DNA. Once 6-thioguanosine is incorporated into DNA, it may undergo *in situ* non-enzymatic methylation to form *S*6-methylthioguanosine, which can induce lethal processing by DNA mismatch repair.[10,11] As described in our article, methylation of 6-thioguanosine in DNA is largely responsible for the cytotoxic effects observed from these prodrugs in the absence of ultraviolet light, which have been correlated with an increased risk in leukemia[11] and enhanced mutation frequency in circulating lymphocytes.[12] Hence, methylation of 6-thioinosine and 6-thioguanosine to form *S*6-methylthioinosine and *S*6-methylthioguanosine are naturally occurring events, which are at least partially responsible for the cytotoxic side effects associated with this treatment. One of the primary objectives of our study was to investigate the photochemical properties of *S*6-methylthioinosine in order to evaluate if it also has the potential to act as a phototoxic agent in the cell. Absorption of sunlight radiation by *S*6-methylthioinosine could lead to photochemical damage through light-induced reactions within the intracellular pool, as has been proposed for azathioprine and 6-thioguanine,[13] and also suggested for other thionated DNA derivatives.[14,15]

1 P. Karran and N. Attard, *Nat. Rev. Cancer*, 2008, **8**, 24–36.
2 K. G. Van Scoik, C. A. Johnson and W. R. Porter, *Drug Metab. Rev.*, 1985, **16**, 157–174.
3 R. Weinshilboum, *Drug Metab. Dispos.*, 2001, **29**, 601–605.
4 T. Dervieux, J. G. Blanco, E. Y. Krynetski, E. F. Vanin, M. F. Roussel and M. V. Rellin, *Cancer Res.*, 2001, **61**, 5810–5816.
5 T. Dervieux, T. L. Brenner, Y. Y. Hon, Y. Zhou, M. L. Hancock, J. T. Sandlund, G. K. Rivera, R. C. Ribeiro, J. M. Boyett, C.-H. Pui, M. V. Relling and W. E. Evans, *Blood*, 2002, **100**, 1240–1247.

6 S. A. Coulthard, L. A. Hogarth, M. Little, E. C. Matheson, C. P. F. Redfern, L. Minto and A. G. Hall, *Mol. Pharmacol.*, 2002, **62**, 102–109.
7 M. C. Dubinsky, S. Lamothe, H. Y. Yang, S. R. Targan, D. Sinnett, Y. Théorêt and E. G. Seidman, *Gastroenterology*, 2000, **118**, 705–713.
8 M. C. Dubinsky, H. Yang, P. V. Hassard, E. G. Seidman, L. Y. Kam, M. T. Abreu, S. R. Targan and E. A. Vasiliauskas, *Gastroenterology*, 2002, **122**, 904–915.
9 U. Nygaard, N. Toft and K. Schmiegelow, *Clin. Pharmacol. Ther.*, 2004, **75**, 274–281.
10 P. F. Swann, T. R. Waters, D. C. Moulton, Y.-Z. Xu, Q. Zheng, M. Edwards and R. Mace, *Science*, 1996, **273**, 1109–1111.
11 J. Offman, G. Opelz, B. Doehler, D. Cummins, O. Halil, N. R. Banner, M. M. Burke, D. Sullivan, P. Macpherson and P. Karran, *Blood*, 2004, **104**, 822–828.
12 T. Nguyen, P. M. Vacek, P. O'Neill, R. B. Colletti and B. A. Finette, *Cancer Res.*, 2009, **69**, 7004–7012.
13 M. S. Cooke, T. L. Duarte, D. Cooper, J. Chen, S. Nandagopal and M. D. Evans, *DNA Repair*, 2008, **7**, 1982–1989.
14 M. Pollum, B. Ashwood, S. Jockusch, M. Lam and C. E. Crespo-Hernández, *J. Am. Chem. Soc.*, 2016, **138**, 11457–11460.
15 M. Pollum, L. Guan, S. Ahsanuddin, E. Baron, M. Lam and C. E. Crespo-Hernández, *J. Invest. Dermatol.*, 2016, **136**, S105.

Manas Sarangi asked: How is the conical intersection of the excited DNA base affected if you substitute the 2 and 6 position of the bases with methyl groups?

Carlos E. Crespo-Hernández answered: A concrete answer to this question requires multiconfigurational quantum-chemical calculations to be performed for those systems, which to my knowledge have yet to be done. We have shown that the fluorescence yield and the excited singlet state lifetime of *O*6-methylguanosine increase considerably compared to those of the guanine nucleoside or nucleotide (see our paper in this issue, DOI: 10.1039/c7fd00193b and Ashwood *et al.*[1]). Similarly, methylation of the amino group at the C6 position in adenine significantly changes the photophysical properties of *N*6,*N*6-dimethyladenine compared to adenine.[2–4] These photophysical changes suggest significant modifications to the order of the electronic states, to the topology of the excited-state potential energy surfaces, and/or to the access of key conical intersections upon methylation of the guanine and adenine monomers.

1 B. Ashwood, L. A. Ortiz-Rodríguez and C. E. Crespo-Hernández, *J. Phys. Chem. Lett.*, 2017, **8**, 4380–4385.
2 B. Albinsson, *J. Am. Chem. Soc.*, 1997, **119**, 6369.
3 A. B. J. Parusel, W. Rettig and K. Rotkiewicz, *J. Phys. Chem. A*, 2002, **106**, 2293–2299.
4 N. K. Schwalb and F. Temps, *J. Phys. Chem. A*, 2009, **113**, 13113–13123.

Amitava Chandra said: I'm curious to know about possible applications of controlling the photochemistry by modifying the functional groups in nucleobases?

Carlos E. Crespo-Hernández responded: We have recently presented experimental and computational evidence showing that modification of the functional groups at the C2 and C6 positions of the purine nucleobases (or at the C2 and C4 position of the pyrimidines) regulates the rates of nonradiative and radiative decay.[1] Modifications at C6 or C4 of the purine or pyrimidine chromophore, respectively, seems to play a particularly important role for a selected group of modified nucleobases. For instance, this knowledge can be used to identify prospective modified nucleobases for phototherapeutic[2–4] and structural biology

This journal is © The Royal Society of Chemistry 2018

applications.[5,6] It could also be used to predict the photoreactivity of a modified nucleobase, which can then be investigated to caution scientists about potential unintended consequences of modifying the genetic code.[3]

1 C. E. Crespo-Hernández, L. Martínez-Fernández, C. Rauer, C. Reichardt, S. Mai, M. Pollum, P. Marquetand, L. González and I. Corral, *J. Am. Chem. Soc.*, 2015, **137**, 4368–4381.
2 M. Pollum, S. Jockusch and C. E. Crespo-Hernández, *J. Am. Chem. Soc.*, 2014, **136**, 17930–17933.
3 M. Pollum, B. Ashwood, S. Jockusch, M. Lam and C. E. Crespo-Hernández, *J. Am. Chem. Soc.*, 2016, **138**, 11457–11460.
4 M. Pollum, L. Guan, S. Ahsanuddin, E. Baron, M. Lam and C. E. Crespo-Hernández, *J. Invest. Dermatol.*, 2016, **136**, S105.
5 A. Favre, C. Saintomé, J.-L. Fourrey, P. Clivio and P. Laugâa, *J. Photochem. Photobiol. B*, 1998, **42**, 109–124.
6 M. Pollum, S. Jockusch and C. E. Crespo-Hernández, *Phys. Chem. Chem. Phys.*, 2015, **17**, 27851–27861.

Ramapurath S. Jayasree asked: Have you looked into the singlet oxygen quantum yield? Is it comparable to or better than the FDA approved protoporphyrin IX or 5-ALA?

Carlos E. Crespo-Hernández replied: Yes, we have reported the yield of singlet oxygen generation for a large number of modified bases,[1–6] with yields as high as 50%. We have also shown that some of these modified bases are effective at diminishing the proliferation of skin cancer cells.[4,7,12] Importantly, because these modified bases are often metabolized into the DNA of the skin cancer cells at a much faster rate than healthy cells, and they can be selectively excited by ultraviolet-A radiation,[1,4,7,12] they do not necessarily require the presence of molecular oxygen to inflict damage to DNA,[12] but can also damage DNA by direct photosensitized reactions.[2,8–10] This is important because solid tumors are often deprived of molecular oxygen.[11]

1 M. Pollum, S. Jockusch and C. E. Crespo-Hernández, *J. Am. Chem. Soc.*, 2014, **136**, 17930–17933.
2 M. Pollum, S. Jockusch and C. E. Crespo-Hernández, *Phys. Chem. Chem. Phys.*, 2015, **17**, 27851–27861.
3 M. Pollum, L. A. Ortiz-Rodríguez, S. Jockusch and C. E. Crespo-Hernández, *Photochem. Photobiol.*, 2016, **92**, 286–292.
4 M. Pollum, B. Ashwood, S. Jockusch, M. Lam and C. E. Crespo-Hernández, *J. Am. Chem. Soc.*, 2016, **138**, 11457–11460.
5 B. Ashwood, S. Jockusch and C. E. Crespo-Hernández, *Molecules*, 2017, **22**, 379.
6 B. Ashwood, S. Jockusch and C. E. Crespo-Hernández, *J. Phys. Chem. Lett.*, 2017, **8**, 2387–2392.
7 M. Pollum, L. Guan, S. Ahsanuddin, E. Baron, M. Lam and C. E. Crespo-Hernández, *J. Invest. Dermatol.*, 2016, **136**, S105.
8 P. Clivio, J. L. Fourrey, J. Gasche and A. Favre, *J. Am. Chem. Soc.*, 1991, **113**, 5481–5483.
9 A. Favre, C. Saintomé, J.-L. Fourrey, P. Clivio and P. Laugâa, *J. Photochem. Photobiol. B*, 1998, **42**, 109–124.
10 O. Reelfs, P. Macpherson, X. Ren, Y.-Z. Xu, P. Karran and A. R. Young, *Nucleic Acids Res.* 2011, **39**, 9620–9632.
11 M. Höckel and P. Vaupel, *J. Nat. Cancer Inst.*, 2001, **93**, 266–276.
12 M. Pollum, L. Minh, S. Jockusch, C. E. Crespo-Hernández, *ChemMedChem*, 2018, DOI: 10.1002/cmdc.201800148.

Vishnu V communicated: In the calculations performed, the solvent taken is acetonitrile. Won't the cellular environment affect the photochemical properties of the byproducts under study and what is (or can be) done to take into account such environmental effects?

Carlos E. Crespo-Hernández communicated in reply: The calculations reported in our article have been performed in vacuum, acetonitrile, and water environment for *S*6-methylthioinosine, and in vacuum and in acetonitrile for *O*6-methylguanosine. Analogous calculations for *O*6-methylguanosine in a water environment have been reported recently in a separate contribution.[1] The photochemistry of these byproducts in the cellular environment is currently unknown, but the experimental and computational results currently available (in DOI: 10.1039/c7fd00193b and in Ashwood *et al.*[1]) suggest that such investigations are necessary. Calculations that explicitly include the interaction of these byproducts with water molecules, amino acids, and/or nucleic acids could be performed at the quantum-mechanics/molecular-mechanics (QM/MM) level of theory, for instance, as have been widely discussed in this *Faraday Discussions* meeting for other systems.

1 B. Ashwood, L. A. Ortiz-Rodríguez and C. E. Crespo-Hernández, *J. Phys. Chem. Lett.*, 2017, **8**, 4380–4385.

Priyadarshi Roy Chowdhury communicated: How does the structural modification of the purine chromophore affect the electronic relaxation mechanisms?

Carlos E. Crespo-Hernández communicated in response: We have recently reported that structural modifications of the purine chromophore can significantly affect the electronic relaxation mechanism among other factors by altering the topology of the potential energy surfaces, their spin–orbit coupling interactions, and the access to key conical intersections.[1,2] For instance, while excitation of the purine free base (*i.e.*, the purine chromophore) efficiently populates a long-lived triplet state that decays in microseconds in aqueous solution,[4] direct excitation of adenine (6-aminopurine) and guanine (2-amino-6-oxopurine) monomers at 266 nm primarily results in the ultrafast decay (hundreds of femtoseconds) of the excited-state population by internal conversion to the ground state. Similarly, *O*6-methylation of the guanine monomers results in up to a 40-fold increase in the excited singlet state lifetime of *O*6-methylguanosine compared to that of the guanine nucleoside or nucleotide (see DOI: 10.1039/c7fd00193b and Ashwood *et al.*[3]).

1 C. E. Crespo-Hernández, L. Martínez-Fernández, C. Rauer, C. Reichardt, S. Mai, M. Pollum, P. Marquetand, L. González and I. Corral, *J. Am. Chem. Soc.*, 2015, **137**, 4368–4381.
2 S. Mai, M. Pollum, L. Martínez-Fernández, N. Dunn, P. Marquetand, I. Corral, C. E. Crespo-Hernández and L. González, *Nat. Commun.*, 2016, 7, 13077.
3 B. Ashwood, L. A. Ortiz-Rodríguez and C. E. Crespo-Hernández, *J. Phys. Chem. Lett.*, 2017, **8**, 4380–4385.
4 E. Quiñones and R. Arce, *J. Am. Chem. Soc.*, 1989, **111**, 8218–8223.

Dimitra Markovitsi opened a discussion of the paper by Angelo Giussani: Is the mechanism you propose for formation of thymine–thymine (6–4) adducts consistent with the excitation wavelength dependence of their quantum yield determined experimentally?[1] Experimental studies on thymine dimerization

 This journal is © The Royal Society of Chemistry 2018

using triplet–triplet energy transfer detected only CPDs but not (6–4) adducts, see for example Sauvaigo *et al.*[2]

1 A. Banyasz, T. Douki, R. Improta, T. Gustavsson, D. Onidas, I. Vayá, M. Perron and D. Markovitsi, *J. Am. Chem. Soc.*, 2012, **134**, 14834–14845.
2 S. Sauvaigo, T. Douki, F. Odin, S. Caillat, J. L. Ravanat and J. Cadet, *Photochem. Photobiol.*, 2001, **73**, 230–237.

Angelo Giussani answered: Yes, the mechanism here proposed is consistent with the experimentally determined increase of the quantum yield of formation with the increase of the excitation energy, since also in the here proposed mechanism an energy barrier needs to be surmounted in order for the reaction to proceed. In the here proposed mechanism the barrier is placed on the gs PES. Thanks for the reference regarding thymine photoproducts using triplet–triplet energy transfer.

Igor Schapiro remarked: I want to ask about these two conical intersections that you have presented. They are so much below the Franck–Condon point that I think if you did dynamics simulations you would be able to overcome the barrier and access them. Is it feasible to run non adiabatic dynamics for this system?

Angelo Giussani answered: The two characterized conical intersections are indeed well below the vertical excitation energy of the CT state at the FC region, so the system excited in the CT should indeed be able to populate such degenerated regions. It must be noticed that the reported vertical excitation energy of the CT at the FC point is quite above the experimental excitation energy normally used and for which the 6–4 TT photodamage has been detected. Regarding possible dynamics simulations, I am aware that the group of Prof. Leticia Gonzalez has recently published dynamics simulations on the formation of the DNA photodamage cyclobutane pyrimidine dimer.[1]

1 C. Rauer, J. J. Nogueira, P. Marquetand and L. González, *J. Am. Chem. Soc.*, 2016, **138**, 15911–15916.

Carlos E. Crespo-Hernández commented: Interestingly, the participation of a triplet state with significant charge transfer character (*i.e.*, a ^{3}ct-ππ*-TT′ state, according to the nomenclature used in the manuscript) in the formation of the (6–4) photodimer is probably more likely in the thymine–thymine dinucleotide analogue thymidylyl(3′-5′)-4-thiothymidine, which has been shown to also form a (6–4) thietane photodimer upon selective irradiation of the 4-thiothymine moiety at 360 nm.[1,2] This is so because recent transient absorption results for 4-thiothymidine monomers[3–5] have shown that intersystem crossing to the triplet manifold in this thymine analogue occurs in hundreds of femtoseconds and with nearly unity triplet yield.

1 P. Clivio, J. L. Fourrey, J. Gasche and A. Favre, *J. Am. Chem. Soc.*, 1991, **113**, 5481–5483.
2 P. Clivio, J.-L. Fourrey, J. Gasche and A. Favre, *Tetrahedron Lett.*, 1992, **33**, 1615–1618.
3 Y. Harada, C. Okabe, T. Kobayashi, T. Suzuki, T. Ichimura, N. Nishi and Y.-Z. Xu, *J. Phys. Chem. Lett.*, 2010, **1**, 480–484.
4 C. Reichardt and C. E. Crespo-Hernández, *J. Phys. Chem. Lett.*, 2010, **1**, 2239–2243.
5 M. Pollum, S. Jockusch and C. E. Crespo-Hernández, *J. Am. Chem. Soc.*, 2014, **136**, 17930–17933.

Tolga Karsili said: As a follow up to the comment by Prof. Crespo-Hernández, it is well-known that the efficient intersystem crossing displayed by simple carbonyl containing molecules (such as acetone) arises *via* the large changes in the C–O bond distance of the lowest singlet and triplet state minima. This large difference in C–O bond distance leads to a singlet/triplet crossing close to the minimum of the lowest singlet state, enhancing the spin–orbit coupling near the Franck–Condon geometry. The near unity quantum yield for intercrossing crossing in the Paterno–Büchi reaction of simple carbonyls is a direct consequence of the aforementioned energetics. In the case of a [2 + 2] cycloaddition of a thymine–thymine dimer to form an oxetane, the intersystem crossing yield may be quenched (when compared to a simple Paterno–Büchi system) since the aromatic system will most likely restrict sufficient elongation of the C–O bonds, manifesting in similar C–O bond distances for the first/second singlet and triplet minima. Are the authors able to comment on whether their computations return significant changes in C–O bond distance when comparing the singlet and triplet minima of common electronic state configurations?

Angelo Giussani responded: For the system here under study (*i.e.* a thymine dimer) the low-lying singlet and triplet states at the selected Franck–Condon geometry have $\pi\pi^*$ and $n\pi^*$ nature. The $\pi\pi^*$ states (states S1, S2, T1, and T2, see Table 1 in our article) are mainly localized on the two C5–C6 carbon–carbon double bonds of the two thymine molecules, while the $n\pi^*$ states (states S3, S4, T3 and T4, see Table 1), are mainly localized on the two $C_{4'}O$ carbonyl groups of the two thymine molecules.

For the studied system we have optimized the S1 $\pi\pi^*$ state, and the corresponding minima (see Fig. S7† in our article) display a slight elongation of the C5–C6 bond and a non-negligible elongation of the nearby carbonyl group (+0.036 Å). However this was the result of an optimization with a reduced active space (cas4,4) and in contrast with what was expected, that was a much pronounced elongation of the C5–C6 bond. The optimization of the singlet $n\pi^*$ S4 and triplet $n\pi^*$ T4 states have resulted in minima in which the $C_{4'}O$ bond displays a remarkable elongation, of plus 0.158 and 0.172 Å, respectively.

So we can conclude that for the present system, the optimization of the $\pi\pi^*$ S1, S2, T1, and T2 states probably leads to a slight elongation of the carbonyl group (we say probably because just the S1 state has been optimized here), while the optimization of the $n\pi^*$ S3, S4, T3, and T4 leads to a much more substantial elongation of the C–O bond. A similar degree of elongation is seen for the mentioned $n\pi^*$ minima, independently if singlet or triplet states (the $C_{4'}O$ bond is equal to 1.370 and 1.384 Å for the $1n\pi^*$ and $3n\pi^*$ minimum, respectively, see Fig. S7†), and again a similar degree of elongation is presumably forecast for the mentioned $\pi\pi^*$ states, independently if singlet or triplet states.

In comparing thymine with smaller systems such as acetone, we think that it is important to notice that while for acetone all excited states will be localized on the carbonyl group, in thymine the lowest excited states are instead mainly localized on the C5–C6 carbon–carbon double bond, which is of course not present in acetone. Probably an important difference in the Paterno–Büchi reaction between smaller molecules (as acetone and ethylene) and two thymine molecules, could in fact be caused by the fact that in thymine the main absorbing state is a $\pi\pi^*$ state

 This journal is © The Royal Society of Chemistry 2018

on the C5–C6 bond, while in acetone the main absorbing state is already localized on the reactive carbonyl group.

We give thanks for this useful comment.

Roberto Improta asked: What happens when optimizing the geometry of the two lowest energy bright excited state states? What is their nature? Do you observe a significant mixing with CT states?

Angelo Giussani replied: For the present study we have optimized just one localized $\pi\pi^*$ state, the $1\pi\pi^*$-T′ state, this being the S1 state at the selected Franck–Condon geometry. As highlighted in our article, the performed optimization gave convergence problems, and for that reason it was performed with the reduced active space cas4,4 instead of the otherwise used cas8,6 active space. Along the performed optimization of the $1\pi\pi^*$-T′ state, the state preserves its nature, although a mixing with the $1n\pi^*$-T′ state is observable, which is probably causing the final structure to present geometrical modifications similar to the ones observed along the optimization of the $1n\pi^*$-T′ state. Along the performed optimization of the $1\pi\pi^*$-T′ state a significant mixing with CT states was not observed.

Dimitra Markovitsi opened a general discussion of the papers by Carlos E. Crespo-Hernández and Angelo Giussani: Could you comment on how non-canonical bases may damage DNA?

Carlos E. Crespo-Hernández replied: This is obviously a broad and important question, but I will simply provide a generic answer herein. Non-canonical bases can potentially induce damage to DNA by both direct and indirect photosensitized reaction pathways. The specific nature of the damage will depend, among other factors (such as change in the DNA secondary or tertiary structure, base pairing and stacking interactions, *etc.*), on the electronic structure and the photophysical and physicochemical properties (such as redox potentials, fluorescence and triplet yields, *etc.*) of the precise non-canonical base. This is in part the reason for the necessity of investigating the photophysical and photodynamic properties of biologically relevant modified nucleobases thoroughly, as was one of the primary objectives of our study with *S*6-methylthioinosine and *O*6-methylguanosine.[1]

1 B. Ashwood, L. A. Ortiz-Rodríguez and C. E. Crespo-Hernández, *Faraday Discuss.*, 2018, DOI: 10.1039/c7fd00193b; B. Ashwood, L. A. Ortiz-Rodríguez and C. E. Crespo-Hernández, *J. Phys. Chem. Lett.*, 2017, **8**, 4380–4385.

Angelo Giussani remarked: Regarding the formation mechanism of the 6-4 Thy-Thy adduct, it is important to remember that the time scale of the photoactivated part of the reaction, leading to the oxetane intermediate, is still unknown, so we don't know if the reaction happens (or not) in an ultrafast fashion as with the cyclobutane pyrimidine dimer formation process.

Tolga Karsili said: I have a comment and a question. The comment is in regard to the previous question. At near-threshold excitation wavelengths, it is well-known that DNA/RNA nucleobases and related analogues decay by internal conversion *via* conical intersections that contain out-of-plane ring deformation geometries, which in the purine nucleobases (and analogues) involved out-of-plane motions of either C6 or C2. Such conical intersections are either accessed directly or indirectly *via* non-adiabatic coupling – but are usually barrierless (or at best weakly barriered) in such biological systems. Motions along N–H bond elongation coordinates do indeed contain conical intersections but these are manifestations of $\pi\sigma^*$ states. As Prof. Crespo-Hernández mentioned, $\pi\sigma^*$ states are energetically inaccessible at the Franck–Condon geometry. At such geometries, access to $\pi\sigma^*$ states (at the applicable wavelengths reported by Crespo-Hernández *et al.*) requires non-adiabatic H-atom tunneling from the initially prepared bright state. Tunneling along the N–H coordinate is however restricted in the nucleobases (and analogues) by a large potential barrier connecting the initially prepared bright state to the $\pi\sigma^*$ state.

My question is to Prof. Crespo-Hernández: In the oligonucleotides, in which out-of-plane ring deformations are quenched, what role do you expect substitution at the C6 and C2 positions to have on the relevant excited states of an oligonucleotide (*e.g.* charge-transfer states)?

Carlos E. Crespo-Hernández answered: I would like to first add to Dr Karsili's comment by directing readers to a recent work (and references therein) summarizing the evidence available regarding the participation of $\pi\sigma^*$ states in the excited-state dynamics of the nucleic acid bases.[1]

Concerning the consequences of substitution at the C6 and C2 positions of the purine chromophore on the excited state dynamics in oligonucleotides, the effect will depend on the specific modification done to the purine chromophore. The addition, removal, or substitution of a functional group from the C6 or C2 position of the purine chromophore in adenine or guanine is expected to modify the order of the electronic states and the topology of the excited-state potential energy surfaces of the modified base,[2] and their redox properties.[3] Together with the potential consequences of base stacking and base pairing interactions, photoexcitation of the modified base within a single- or double-stranded oligonucleotide can be expected to change the competition among different electronic relaxation pathways available in the oligonucleotide. A classic example is the simple substitution of adenine (6-aminopurine) by its structural isomer 2-aminopurine in oligonucleotides, which has been shown to significantly modify the excited-state dynamics in DNA.[4,5]

1 G. M. Roberts and V. G. Stavros, *Chem. Sci.*, 2014, **5**, 1698–1722.
2 C. E. Crespo-Hernández, L. Martínez-Fernández, C. Rauer, C. Reichardt, S. Mai, M. Pollum, P. Marquetand, L. González and I. Corral, *J. Am. Chem. Soc.*, 2015, **137**, 4368–4381.
3 C. E. Crespo-Hernández, D. M. Close, L. Gorb and J. Leszczynski, *J. Phys. Chem. B*, 2007, **111**, 5386–5395.
4 S. O. Kelley and J. K. Barton, *Science*, 1999, **283**, 375–381.
5 C. Wan, T. Fiebig, O. Schiemann, J. K. Barton and A. H. Zewail, *Proc. Natl. Acad. Sci. U. S. A.*, 2000, **97**, 14052–14055.

Igor Schapiro said: Both papers have pointed out the importance of intersystem crossing. Is there a possibility either experimentally or computationally to use magnetic fields to explore this further? For example can you increase or

 This journal is © The Royal Society of Chemistry 2018

decrease the intersystem crossing rate by applying a magnetic field in the nucleobases?

Carlos E. Crespo-Hernández replied: Early investigations have presented evidence that an applied magnetic field can modulate the fluorescence lifetimes and yields of pyrimidine in vacuum[1–3] or the photoionization yield of 6-methylpurine in aqueous solutions.[4] These experiments suggest mixing of singlet and triplet states with applied magnetic field for those two nucleobase derivatives. However, investigations focusing on the effect that an applied magnetic field has on the intersystem crossing rate of the canonical nucleobases have yet to be reported, perhaps because of their small triplet yields. Theoretical calculations are certainly possible.

1 N. Ohta, M. Fujita and H. Baba, *Chem. Phys. Lett.*, 1987, **135**, 330–334.
2 N. Ohta, T. Takemura, M. Fujita and H. Baba, *J. Chem. Phys.*, 1988, **88**, 4197–4203.
3 N. Ohta and T. Takemura, *Chem. Phys.*, 1992, **162**, 15–28.
4 C. Crespo-Hernández, R. Arce and E. Quiñones, *Chem. Phys. Lett.*, 2003, **382**, 661–664.

Dimitra Markovitsi asked: Could non-canonical bases behave as traps for the excitation energy in DNA as the minor natural base 5-methyl cytosine?[1]

1 A. Banyasz, L. Esposito, T. Douki, M. Perron, R. Improta and D. Markovitsi, *J. Phys. Chem. B*, 2016, **120**, 4232–4242.

Carlos E. Crespo-Hernández replied: In this article and in our previous investigation,[1] we have shown that photoexcitation with ultraviolet B (UVB) radiation of both *S*6-methylthioinosine and *O*6-methylguanosine can increase the probability of photochemical damage compared to the guanine nucleoside and nucleotide. However, it remains to be determined whether *S*6-/*O*6-methylation of guanine chromophore increases DNA photodamage when these molecules are incorporated into DNA. We have shown that both *S*6-methylthioinosine and *O*6-methylguanosine absorb radiation at lower energies than the guanosine nucleotide with larger absorptivity coefficients in the UVB spectral region, which increases the light penetration depth in cell tissue and thus probability of sunlight absorption. Furthermore, UVB excitation traps a significant fraction of the excited-state population in the excited singlet of *O*6-methylguanosine and in the triplet state of *S*6-methylthioinosine. For *O*6-methylguanosine, a 40-fold increase in the excited singlet state lifetime compared to that of the guanine nucleoside or nucleotide is observed, which could increase the probability that *O*6-methylguanosine may react with adjacent base-paired or base-stacked nucleobases when present in DNA. For *S*6-methylthioinosine, the observation that a long-lived triplet state is populated upon UVB excitation suggests that photoexcitation of this molecule could potentially lead to oxidatively-generated cellular damage when present in the intracellular pool, as has been proposed for 6-thioguasonine and other thiopurine prodrugs.[2–5] Our study shows that further research is essential to investigate the possible photo-induced damage by these biological molecules to DNA and to other biological components within the cell.

1 B. Ashwood, L. A. Ortiz-Rodríguez and C. E. Crespo-Hernández, *J. Phys. Chem. Lett.*, 2017, **8**, 4380–4385.

2 P. F. Swann, T. R. Waters, D. C. Moulton, Y.-Z. Xu, Q. Zheng, M. Edwards and R. Mace, *Science*, 1996, **273**, 1109–1111.
3 P. O'Donovan, C. M. Perrett, X. Zhang, B. Montaner, Y.-Z. Xu, C. A. Harwood, J. M. McGregor, S. L. Walker, F. Hanaoka and P. Karran, *Science*, 2005, **309**, 1871–1874.
4 Q. Gueranger, F. Li, M. Peacock, A. Larnicol-Fery, R. Brem, P. Macpherson, J. M. Egly and P. Karran, *J. Invest. Dermatol.*, 2014, **134**, 1408–1417.
5 R. Brem, M. Guven and P. Karran, *Free Radical. Biol. Med.*, 2017, **107**, 101–109.

Conflicts of interest

There are no conflicts to declare.

This journal is © The Royal Society of Chemistry 2018

Faraday Discussions

PAPER

Exploring oligomeric state of the serotonin$_{1A}$ receptor utilizing photobleaching image correlation spectroscopy: implications for receptor function

Hirak Chakraborty,[ab] Md. Jafurulla,[a] Andrew H. A. Clayton*[c] and Amitabha Chattopadhyay*[a]

Received 31st August 2017, Accepted 6th October 2017
DOI: 10.1039/c7fd00192d

The oligomerization of G protein-coupled receptors (GPCRs) represents an important process in GPCR function and drug discovery. We have addressed cholesterol-dependent oligomerization state of the serotonin$_{1A}$ receptor, a representative GPCR and an important drug target, utilizing photobleaching image correlation spectroscopy (pbICS). pbICS allows determination of oligomeric state of membrane receptors since change in cluster density upon photobleaching is dependent on the oligomeric state. Our results show that oligomeric state of the serotonin$_{1A}$ receptor is modulated by cell membrane cholesterol and a trimeric population of the receptor prevails in control (normal) cholesterol conditions. Interestingly, upon lowering membrane cholesterol, the predominant oligomeric population of the receptor changes to dimers. This is associated with an increase in specific ligand binding activity of the receptor, thereby implying a crucial role of receptor dimers in ligand binding activity. Upon cholesterol replenishment, the distribution of receptor oligomers is further changed such that the trimers become the major population, with a concomitant restoration of ligand binding activity to the control level. These results demonstrate the utility of pbICS in monitoring oligomeric states of membrane receptors in general, and the cholesterol-dependent oligomeric state of the serotonin$_{1A}$ receptor in particular. We envision that functional correlates of oligomeric states of GPCRs could provide better understanding of GPCR function in health and disease, and help design better therapeutic strategies.

[a]*CSIR-Centre for Cellular and Molecular Biology, Hyderabad 500 007, India. E-mail: amit@ccmb.res.in*
[b]*School of Chemistry, Sambalpur University, Jyoti Vihar, Burla, Odisha 768 019, India*
[c]*Centre for Microphotonics, Faculty of Science, Engineering and Technology, Swinburne University of Technology, Hawthorn, Victoria 3122, Australia. E-mail: aclayton@swin.edu.au*

Introduction

G protein-coupled receptors (GPCRs) are a family of seven transmembrane domain proteins that act as signaling hubs and transfer information from outside the cell to the cellular interior.[1–3] GPCRs facilitate a large number of physiological responses to a variety of stimuli, thereby mediating multiple cellular processes which include neurotransmission, cellular metabolism, inflammatory and immune responses, secretion, cellular differentiation and growth. As a result, GPCRs serve as major (~50%) drug targets in all clinical areas.[4,5] Work from a number of laboratories has shown that membrane lipids, particularly cholesterol and sphingolipids, have a crucial role in the function of GPCRs.[6–11] An active and exciting area in GPCR biology is the oligomerization of GPCRs and its role in receptor function.[12–16] GPCR oligomerization is believed to increase the cross-talk between receptors and their downstream signaling, which results in efficient and controlled signal transduction.[17–19] Interestingly, membrane lipids have been shown to modulate oligomerization of GPCRs.[20–27]

GPCR oligomerization has been studied using a variety of approaches. Yet, monitoring receptor oligomerization under varying conditions has proved to be challenging.[19,28–30] Due to the challenges involved in monitoring GPCR oligomers in complex and dynamic cell membranes, a number of biochemical, biophysical and simulation-based approaches have been used.[30,31] Image correlation spectroscopy (ICS) provides a useful approach for exploring the oligomerization state of membrane receptors.[32–36] ICS offers an advantage over conventional fluorescence correlation spectroscopy (FCS) as it measures spatial correlation instead of temporal correlation, which is measured in FCS. Due to the slow diffusion of membrane proteins, FCS is not suitable for analyzing receptor distribution in cell membranes.[37] ICS measures the fluctuation of fluorescence intensity in space and the spatial autocorrelation function provides the cluster density. The cluster density changes upon photobleaching of the fluorophore. In photobleaching ICS (pbICS), the cluster density is measured as a function of remaining fluorescence and the relation between cluster density and remaining fluorescence depends on the oligomeric state of the protein.[35] The average cluster density is directly proportional to remaining fluorescence (subsequent to photobleaching) for monomers, but is a nonlinear function for higher order oligomers (dimers and above). pbICS has therefore proved to be a convenient approach for determining aggregation states of membrane receptors in intact cells.[35]

The serotonin$_{1A}$ receptor is an important neurotransmitter receptor and belongs to the GPCR family. It is implicated in behavior, learning, development and cognition,[38,39] and serves as a crucial drug target for neuropsychiatric disorders and cancer.[40,41] We have previously shown that membrane cholesterol and sphingolipids play an important role in the function[6,7,9,10] and oligomerization[20–23] of the serotonin$_{1A}$ receptor. In this work, we have explored the cholesterol-dependent oligomerization state of the serotonin$_{1A}$ receptor by performing pbICS measurements on Chinese Hamster Ovary (CHO) cells stably expressing the enhanced yellow fluorescent protein (EYFP)-tagged serotonin$_{1A}$ receptor (5-HT$_{1A}$R-EYFP). Our results show that serotonin$_{1A}$ receptors exhibit a heterogeneous distribution of monomers, dimers and trimers in control cells. Upon depletion of membrane cholesterol, the distribution is changed such that the population of receptor dimers is increased. Further correlation of ligand binding

 This journal is © The Royal Society of Chemistry 2018

activity with the oligomeric state of the receptor reveals that the dimeric receptor could be most active in terms of ligand binding.

Experimental

Materials

$MgCl_2$, $CaCl_2$, cholesterol, penicillin, streptomycin, gentamicin sulfate, serotonin and methyl-β-cyclodextrin (MβCD) were obtained from Sigma Chemical Co. (St. Louis, MO). [^{3}H]8-OH-DPAT (specific activity 141.4 Ci mmol^{-1}) was purchased from MP Biomedicals (Santa Ana, CA). Dulbecco's modified Eagle's medium (DMEM)/F-12 (nutrient mixture F-12, Ham; 1 : 1), fetal calf serum, and geneticin (G418) were obtained from Invitrogen Life Technologies (Carlsbad, CA). Bicinchoninic acid (BCA) assay reagent was purchased from Pierce (Rockford, IL). GF/B glass microfiber filters were from Whatman International (Kent, UK). All other chemicals used were of highest purity. Water was purified through a Millipore (Bedford, MA) Milli-Q system and used throughout.

Cells and cell culture

CHO-K1 cells stably expressing the serotonin$_{1A}$ receptor tagged to EYFP (~10^5 receptors per cell; abbreviated as CHO-5-HT$_{1A}$R-EYFP) were used.[42] Cells were grown in DMEM/F-12 (1 : 1) supplemented with 2.4 g l^{-1} of sodium bicarbonate, 10% fetal calf serum, 60 μg ml^{-1} penicillin, 50 μg ml^{-1} streptomycin, and 50 μg ml^{-1} gentamicin sulfate in a humidified atmosphere with 5% CO_2 at 37 °C. CHO-5-HT$_{1A}$R-EYFP cells were maintained in presence of 300 μg ml^{-1} geneticin.

Treatment of cells

Cells plated at a density of 2×10^4 on cover slips or 5×10^5 in 100 mm Petri dishes were grown for 3 days followed by incubation in serum-free D-MEM/F-12 (1 : 1) medium for 3 h. Cholesterol depletion was carried out by treating cells with 10 mM MβCD in serum-free medium at 37 °C for 30 min. For replenishment of cellular cholesterol, cholesterol-depleted cells were incubated with 1 mM cholesterol as a pre-formed MβCD-cholesterol (10 : 1, mol/mol) complex (prepared as described earlier[43]) at 37 °C for 10 min. Cells were then washed with PBS (containing 1 mM $CaCl_2$ and 0.5 mM $MgCl_2$) and fixed using 4% (v/v) formaldehyde solution and used for microscopy. For measuring agonist binding activity of the serotonin$_{1A}$ receptor, cell membranes were prepared as described earlier.[44] Protein concentration in isolated membranes was assayed using BCA reagent.[45] All experiments were performed at room temperature (~23 °C).

Estimation of cholesterol

Cholesterol content in cell membranes was estimated using Amplex Red cholesterol assay kit.[46]

Radioligand binding assay

Receptor binding assays with the radiolabeled agonist 8-hydroxy-2(di-*N*-propylamino)tetralin ([^{3}H]8-OH-DPAT), at a final concentration of 1 nM, were carried out as described previously.[44]

pbICS experiments

pbICS measurements of control and cholesterol-modulated CHO-5-HT_{1A}R-EYFP cells were carried out with an inverted Zeiss LSM 510 Meta confocal laser scanning microsope (Jena, Germany) with a 63×, 1.4 NA oil-immersion objective using the 514 nm line of an argon laser as the excitation source. Fluorescence emission was collected using a Zeiss LSM Meta detector over an emission range of 520–555 nm. Typically, 128 × 128 pixel regions of interest (ROIs) on the upper cell surfaces were selected for ICS analysis. ROIs were auto-correlated using the Fourier Transformed Math option in ImageJ (National Institutes of Health, Bethesda), normalized by the number of pixels and average intensity squared, and corrected for background intensity. Data were fitted to a Gaussian-plus-offset function.[36]

Results

Analysis of pbICS results in terms of receptor oligomeric state

The amplitude of the autocorrelation function $g(0)$ and the full width at half maximum (r) are related to the cluster density (CD, number of clusters per square micron) by the equation:[36]

$$\mathrm{CD} = 1/[g(0)\pi r^2] \tag{1}$$

The above method was used to calculate autocorrelation parameters in frames in a stack that was collected as a function of photobleaching. For each image in the acquired bleaching time series, the fractional remaining fluorescence after bleaching ($p(t)$) was computed as

$$p(t) = [I(t) - I_\mathrm{b}(t)]/[I(t=0) - I_\mathrm{b}(t=0)] \tag{2}$$

where I and I_b are intensities of the ROI and background, respectively. The plot of cluster density as a function of fractional remaining fluorescence (p) can be fitted for homogenous distribution of monomers, dimers, trimers and j-mers. The expression of CD as a function of p is given by:[35]

$$\text{Monomer: } \mathrm{CD}(p) = c_1 p \tag{3}$$

$$\text{Dimer: } \mathrm{CD}(p) = \frac{c_2}{\left(\frac{1}{2p}\right) + \left(\frac{1}{2}\right)} \tag{4}$$

$$\text{Trimer: } \mathrm{CD}(p) = \frac{c_3}{\left(\frac{1}{3p}\right) + \left(\frac{2}{3}\right)} \tag{5}$$

$$j\text{-mer} = \frac{c_j}{\left(\frac{1}{jp}\right) + \left(\frac{j-1}{j}\right)} \tag{6}$$

where c_j is the mean concentration of the jth species. However, the plot of experimentally determined cluster density as a function of fractional remaining fluorescence does not fit to any homogeneous model. We therefore analyzed all

 This journal is © The Royal Society of Chemistry 2018

data in a monomer–dimer–trimer model, where we varied the fractional abundance (α_i) of monomer, dimer and trimer using the following equation:

$$g(0,p) = \frac{\alpha_1}{\mathrm{CD}(p)_{\mathrm{monomer}}} + \frac{\alpha_2}{\mathrm{CD}(p)_{\mathrm{dimer}}} + \frac{\alpha_3}{\mathrm{CD}(p)_{\mathrm{trimer}}} \tag{7}$$

where $g(0,p)$ is the autocorrelation function at zero lag as a function of photobleaching and $\alpha_1 + \alpha_2 + \alpha_3 = 1$. The fractional components are proportional to the square of the intensity contributions, *e.g.*, ($\alpha_1 = \langle i_1 \rangle^2 / \langle i_T \rangle^2$), where $\langle i_1 \rangle$ is the average intensity from monomer population and $\langle i_T \rangle$ is the sum of the intensity contributions from monomer, dimer and trimer. The fitting of the plot of cluster density as a function of fractional remaining fluorescence in a monomer–dimer–trimer model yielded the fractional components of monomer, dimer and trimer population of receptors in the cell membrane.

Oligomerization state of the serotonin$_{1A}$ receptor by pbICS

We previously showed that the heterologously expressed EYFP-tagged serotonin$_{1A}$ receptor in CHO cells is functionally similar to the native receptor.[42] Knowledge of the oligomerization status of the receptor is important since receptor function and signaling could depend on its oligomeric state.[20,47] In this work, our objective was to exploit the dependence of cluster density on photobleaching in ICS to explore the oligomeric status of the serotonin$_{1A}$ receptor and its modulation with membrane cholesterol.

ICS measures cluster density of aggregates in a given region[32,33] and cluster density is inversely proportional to the amplitude of the spatial correlation function. The cluster density reduces upon photobleaching (increases with remaining fluorescence) and the correlation between cluster density and remaining fluorescence intensity depends on the oligomeric state of the fluorophore-tagged receptor. A representative confocal microscopic image of serotonin$_{1A}$-EYFP receptors on the top membrane surface of CHO cells showing predominantly membrane localization of the receptor is shown in Fig. 1a. We selected a 128 × 128 pixel by pixel region of interest (ROI) from the confocal image and subjected it to image correlation analysis (see Fig. 1b). The spatial autocorrelation of the ROI was calculated using ImageJ software and Fig. 1c shows a representative spatial autocorrelation function. The cluster density was calculated from the intercept on the ordinate (y-axis) of the spatial autocorrelation function. Similarly, the cluster density was calculated for stacks of images that were collected upon progressive photobleaching. The cluster density was plotted as a function of fractional remaining fluorescence (Fig. 2a) and the plot was analyzed using eqn (7) to estimate the fraction of various oligomeric states.

A schematic representation of photobleaching probabilities of fluorescing populations of various oligomeric states is shown in Fig. 2b. The dependence of apparent cluster density on remaining fluorescence is shown in Fig. 2c. For monomeric receptors, the amount of fluorescence remaining would be proportional to the number of fluorescent monomers (the remaining fraction being photobleached and thereby not contributing to the observed fluorescence). The apparent cluster density in this case would be a linear function of the fractional remaining fluorescence (see Fig. 2c). For higher order oligomers, a greater number of monomer units of the oligomer need to be photobleached to obtain

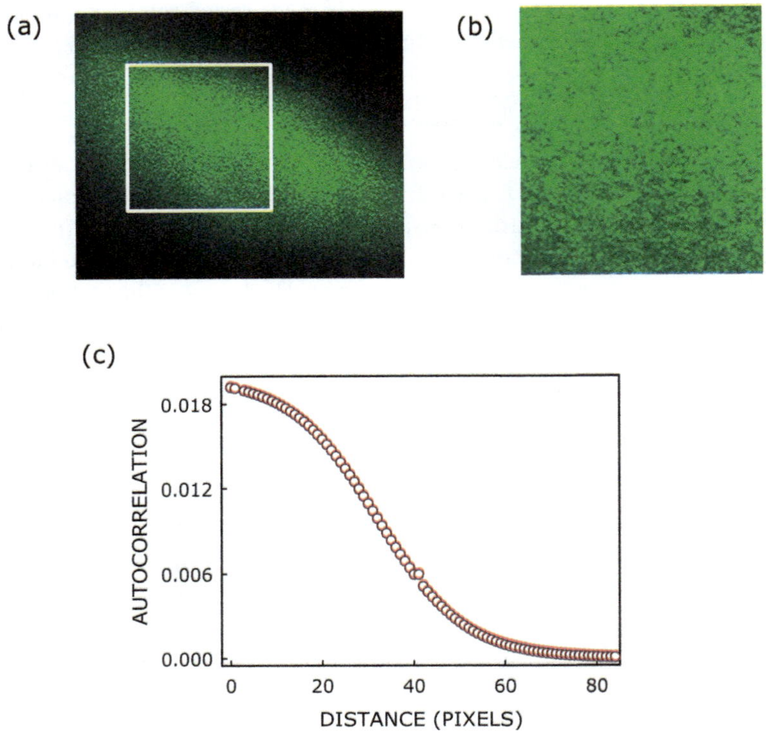

Fig. 1 (a) A representative confocal microscopic image of serotonin$_{1A}$-EYFP receptors on the top membrane surface of CHO cells. (b) A 128 × 128 region of interest (ROI, shown as a box in (a)) was cropped from the image (a) and was subjected to pbICS analysis. (c) Representative spatial autocorrelation for image (b).

complete photobleaching (Fig. 2b). The apparent cluster density, therefore, would exhibit a nonlinear dependence on remaining fluorescence for higher order oligomers (Fig. 2c). The extent of nonlinearity would increase with the increase in size of the oligomeric state.

The cluster density of serotonin$_{1A}$ receptors measured as a function of fractional remaining fluorescence upon photobleaching is shown in Fig. 3a. Data was fitted to eqn (7) to obtain the fraction of various oligomeric states of the receptor. This analysis showed that the serotonin$_{1A}$ receptor exists predominantly as trimers (~60%) with equal contributions from monomeric (~20%) and dimeric (~20%) receptors in control conditions (see Fig. 3b).

Cholesterol-dependent oligomerization state of the serotonin$_{1A}$ receptor

We have previously shown that membrane cholesterol plays a crucial role in the function and organization of the serotonin$_{1A}$ receptor.[43,48–51] Specific agonist binding to the serotonin$_{1A}$ receptor, measured using [^{3}H]8-OH-DPAT, upon modulation of membrane cholesterol, is shown in Fig. 4a. The figure shows that the specific [^{3}H]8-OH-DPAT binding exhibits an increase of ~57% upon depletion of membrane cholesterol using MβCD. MβCD is a water-soluble polymer of methylated-glucose, and has been extensively used to selectively and efficiently

 This journal is © The Royal Society of Chemistry 2018

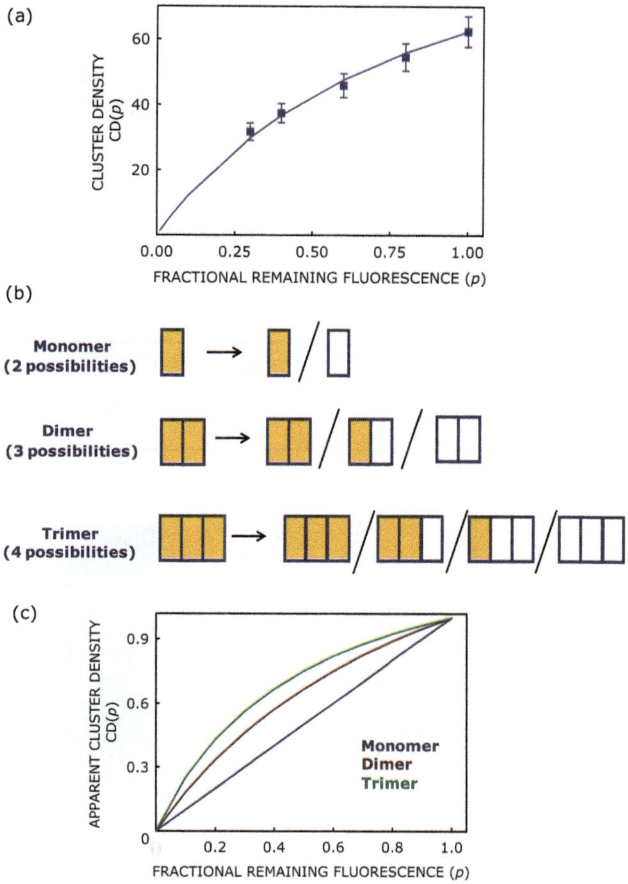

Fig. 2 (a) A representative cluster density plot as a function of remaining fluorescence upon photobleaching of serotonin$_{1A}$-EYFP receptors in CHO cells. Data shown are means ± S.E. of at least six independent measurements. (b) A schematic model illustrating various possibilities of photobleaching of oligomeric states. The arrows indicate photobleaching. (c) Theoretical plots of cluster density as a function of fraction of remaining fluorescence for monomer (blue), dimer (maroon) and trimer (green) states generated using eqn (3)–(5).

extract cholesterol from membranes by incorporating it in a central nonpolar cavity.[52] Interestingly, upon replenishment of membrane cholesterol using MβCD-cholesterol complex, the specific [^{3}H]8-OH-DPAT binding was restored to control level (see Fig. 4a). With an overall goal to address any change in oligomerization pattern accompanying the change in receptor activity due to cholesterol modulation, we carried out pbICS analysis under these conditions (see Fig. 4b).

As mentioned above (Fig. 3b), the serotonin$_{1A}$ receptor exists predominantly as trimers (~60%) with ~20% contribution from both monomeric and dimeric populations under control (normal) membrane cholesterol. The oligomeric distribution of receptors displayed a marked shift upon depletion of membrane cholesterol (see Fig. 4b). Fitting of pbICS data obtained under these conditions to eqn (7) yielded the fraction of various oligomeric states of the receptor. These

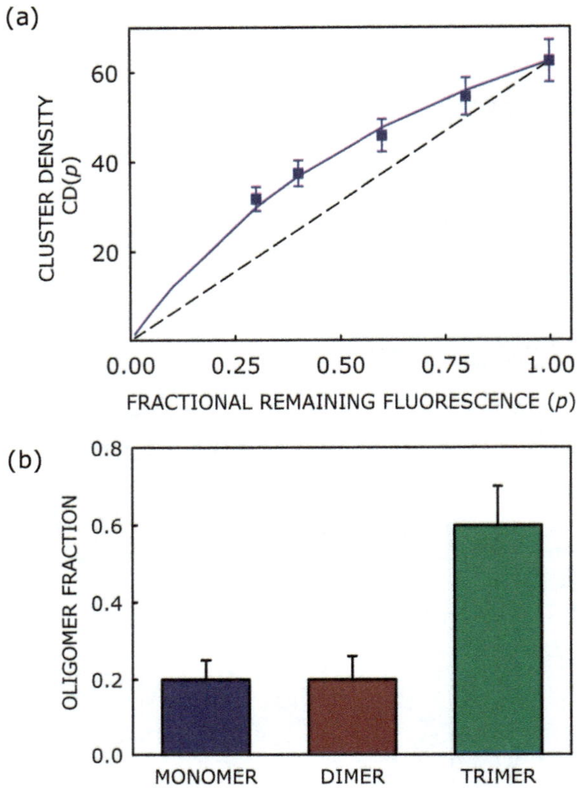

Fig. 3 (a) Cluster density as a function of remaining fractional fluorescence upon photobleaching for serotonin$_{1A}$-EYFP receptors in CHO cells. Data shown are means ± S.E. of at least six independent measurements. Experimental points were fitted to a monomer–dimer–trimer distribution of cluster sizes using eqn (7). A dashed linear line corresponding to the fit to a monomeric distribution of cluster size is shown for comparison. See the Experimental section for more details. (b) Fraction of monomer, dimer and trimer components obtained by fitting the data to a monomer–dimer–trimer distribution model. Data shown are means ± S.E. of at least six independent measurements.

results show that the serotonin$_{1A}$ receptor exists predominantly in the dimeric state (~95%) under cholesterol-depleted conditions. The population of the momomeric receptor in this condition is ~5% while the contribution from the trimeric population is negligible (<0.5%). The increase in dimeric receptor population upon cholesterol depletion is in overall agreement with our previous results on receptor oligomerization using photobleaching homo-FRET[20] and coarse-grain molecular dynamics simulations.[23] Upon comparison of Fig. 4a and b, it appears that the dimeric form of the serotonin$_{1A}$ receptor contributes most to ligand binding activity, relative to the monomeric and trimeric forms.

Fig. 4a shows that upon replenishment of membrane cholesterol in cholesterol-depleted cells using MβCD-cholesterol complex, the specific [^{3}H]8-OH-DPAT binding could be restored to control level. To monitor whether the recovery of ligand binding activity was due to change in distribution of oligomeric states of the receptor, we carried out pbICS analysis under this condition. Fig. 4b shows that there is a change in oligomeric distribution of receptors upon

 This journal is © The Royal Society of Chemistry 2018

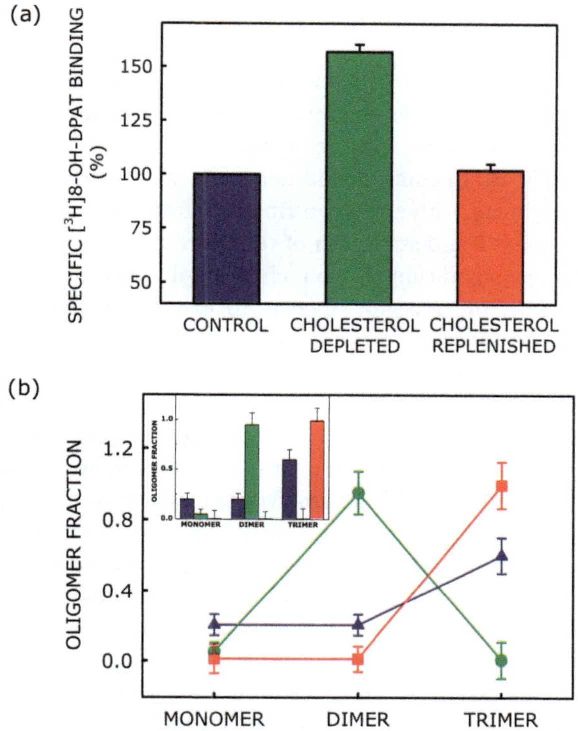

Fig. 4 (a) Specific agonist ([^{3}H]8-OH-DPAT) binding to serotonin$_{1A}$-EYFP receptors in membranes isolated from control, cholesterol-depleted and cholesterol-replenished cells. The results shown are means ± S.E. of at least four independent measurements. (b) Variations of monomer, dimer and trimer components of the serotonin$_{1A}$ receptor under conditions of varying membrane cholesterol content. The fraction of oligomeric components of the receptor in control (▲), cholesterol-depleted (●) and cholesterol-replenished (■) cells is shown. The fractions of different oligomeric states were estimated by fitting cluster density as a function of fractional remaining fluorescence using eqn (7). The results shown are means ± S.E. of at least six independent measurements. Lines joining data points are provided as viewing guides. The inset shows the same data as a bar plot (same color coding: blue: control, green: cholesterol-depleted, red: cholesterol-replenished). See the Experimental section for more details.

cholesterol replenishment. The figure shows that the predominant oligomeric state of the receptor under cholesterol-replenished conditions is trimeric (~99%). The contribution from the other oligomeric states (monomer and dimer) is negligible. The restoration of ligand binding activity to control levels upon replenishment of membrane cholesterol could therefore be attributed to the change in oligomeric state of the receptor. It should be mentioned here that our results show that replenishment of cholesterol using MβCD-cholesterol complex resulted in a cholesterol level which was somewhat higher relative to the control level. We found it experimentally difficult to manipulate conditions in such a way so as to bring back the cholesterol level exactly to control levels (prior to depletion). This could be a possible reason for not obtaining the same proportion of various oligomers in control and cholesterol-replenished conditions, although there is an overall agreement (increase in trimer population).

Discussion

Taken together, we show here that pbICS represents a convenient approach to monitor the oligomeric state of membrane receptors. Our results show that the oligomeric state of the serotonin$_{1A}$ receptor is controlled by cell membrane cholesterol (see Fig. 5). In control conditions, the majority of the receptor population exists as trimers, with equal contributions from monomeric and dimeric receptor populations. The distribution of receptors in various oligomeric states appears to change significantly upon cholesterol depletion. In cholesterol-depleted conditions, the receptor appears to exist predominantly as dimers with minor contributions from the other oligomeric forms. These results, along with the observation that the specific ligand binding activity of the serotonin$_{1A}$ receptor exhibits a considerable increase, suggest that the dimeric population of the receptor could contribute toward ligand binding activity in a major way. These results are supported upon replenishment of cell membrane cholesterol, which results in restoration of ligand binding activity to the control level and a change in the distribution of oligomeric states leading to an increase in trimeric population and a concominant reduction in the dimeric population of receptors.

GPCR oligomerization assumes greater relevance in designing better therapeutic strategies. Importantly, earlier reports have demonstrated the increased specificity of multivalent drugs,[53] as well as ligand sensitivity of the various GPCR oligomer interfaces.[47] We have recently shown that GPCR dimer organization is controlled by membrane cholesterol, and this could have potential implications in cellular physiology and drug discovery.[22,23] Since cellular cholesterol is developmentally regulated and increases with aging,[54,55] oligomeric states of GPCRs could be age-dependent. Correlation of receptor activity with oligomeric states of

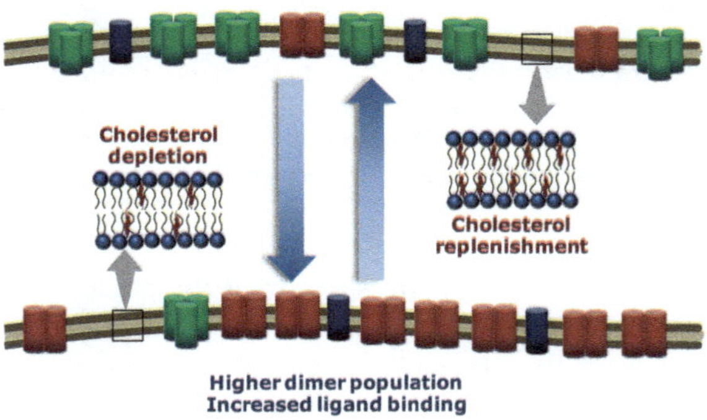

Fig. 5 A schematic model of the oligomeric status of serotonin$_{1A}$ receptors with varying cholesterol content. The model shows an initial heterogeneous oligomeric state of the serotonin$_{1A}$ receptor in control cells. Depletion of membrane cholesterol appears to favor the dimeric state of the serotonin$_{1A}$ receptors and the specific agonist binding to the receptors is higher under these conditions. Upon cholesterol replenishment, the oligomeric state of the receptor changes in such a way that the dimers disappear giving rise predominantly to a trimer population (it is to be noted here that the exact distribution of oligomers is different in control and upon cholesterol replenishment (Fig. 4b); this figure is a schematic and mainly shows the increase in trimer population upon replenishment). These results imply that the receptor dimer is primarily responsible for enhanced agonist binding.

the receptor could therefore lead to a better understanding of GPCR function in health and disease.

Conflict of interest

The authors declare that there is no conflict of interest.

Acknowledgements

This work was supported by the Science and Engineering Research Board (Govt. of India) project (EMR/2016/002294) to A. C. A. H. A. C. acknowledges ARC funding through DP110100164 and DP130101475 grants. A. C. gratefully acknowledges support from J. C. Bose Fellowship (Department of Science and Technology, Govt. of India). H. C. thanks the Council of Scientific and Industrial Research for the award of a Senior Research Associateship and the University Grants Commission for UGC-Assistant Professor position and UGC-Start-Up Grant (F.4-5(138-FRP)/2014(BSR)). A. C. is an Adjunct Professor of Tata Institute of Fundamental Research (Mumbai), Indian Institute of Technology (Kanpur), and Swinburne University of Technology (Melbourne, Australia). We thank G. Aditya Kumar for help in making figures and members of the Chattopadhyay laboratory for their comments and discussions.

References

1 K. L. Pierce, R. T. Premont and R. J. Lefkowitz, *Nat. Rev. Mol. Cell Biol.*, 2002, **3**, 639.

2 D. M. Rosenbaum, S. G. F. Rasmussen and B. K. Kobilka, *Nature*, 2009, **459**, 356.

3 A. Chattopadhyay, *Adv. Biol.*, 2014, **2014**, 143023.

4 K. A. Jacobson, *Biochem. Pharmacol.*, 2015, **98**, 541.

5 R. M. Cooke, A. J. H. Brown, F. H. Marshall and J. S. Mason, *Drug Discovery Today*, 2015, **20**, 1355.

6 T. J. Pucadyil and A. Chattopadhyay, *Prog. Lipid Res.*, 2006, **45**, 295.

7 Y. D. Paila and A. Chattopadhyay, *Subcell. Biochem.*, 2010, **51**, 439.

8 J. Oates and A. Watts, *Curr. Opin. Struct. Biol.*, 2011, **21**, 802.

9 M. Jafurulla and A. Chattopadhyay, *Curr. Med. Chem.*, 2013, **20**, 47.

10 M. Jafurulla and A. Chattopadhyay, *Eur. J. Pharmacol.*, 2015, **763**, 241.

11 G. Gimpl, *Chem. Phys. Lipids*, 2016, **199**, 61.

12 K. Shanti and A. Chattopadhyay, *Curr. Sci.*, 2000, **79**, 402.

13 M. J. Lohse, *Curr. Opin. Pharmacol.*, 2010, **10**, 53.

14 K. Palczewski, *Trends Biochem. Sci.*, 2010, **35**, 595.

15 G. Milligan, *Curr. Opin. Pharmacol.*, 2010, **10**, 23.

16 K. Herrick-Davis, E. Grinde, A. Cowan and J. E. Mazurkiewicz, *Mol. Pharmacol.*, 2013, **84**, 630.

17 S. Terrillon and M. Bouvier, *EMBO Rep.*, 2004, **5**, 30.

18 M. Bellot, S. Galandrin, C. Boularan, H. J. Matthies, F. Despas, C. Denis, J. Javitch, S. Mazères, S. J. Sanni, V. Pons, M.-H. Seguelas, J. L. Hansen, A. Pathak, A. Galli, J.-M. Sénard and C. Galés, *Nat. Chem. Biol.*, 2015, **11**, 271.

19 H. Chakraborty and A. Chattopadhyay, *ACS Chem. Neurosci.*, 2015, **6**, 199.

20 S. Ganguly, A. H. A. Clayton and A. Chattopadhyay, *Biophys. J.*, 2011, **100**, 361.
21 Y. D. Paila, M. Kombrabail, G. Krishnamoorthy and A. Chattopadhyay, *J. Phys. Chem. B*, 2011, **115**, 11439.
22 X. Prasanna, A. Chattopadhyay and D. Sengupta, *Biophys. J.*, 2014, **106**, 1290.
23 X. Prasanna, D. Sengupta and A. Chattopadhyay, *Sci. Rep.*, 2016, **6**, 31858.
24 J. Shan, G. Khelashvili, S. Mondal, E. L. Mehler and H. Weinstein, *PLoS Comput. Biol.*, 2012, **8**, e1002473.
25 S. Mondal, J. M. Johnston, H. Wang, G. Khelashvili, M. Filizola and H. Weinstein, *Sci. Rep.*, 2013, **3**, 2909.
26 K. A. Marino, D. Prada-Gracia, D. Provasi and M. Filizola, *PLoS Comput. Biol.*, 2016, **12**, e1005240.
27 S. Gahbauer and R. A. Böckmann, *Front. Physiol.*, 2016, **7**, 494.
28 A. H. A. Clayton and A. Chattopadhyay, *Biophys. J.*, 2014, **106**, 1227.
29 S. Ferré, V. Casadó, L. A. Devi, M. Filizola, R. Jockers, M. J. Lohse, G. Milligan, J.-P. Pin and X. Guitart, *Pharmacol. Rev.*, 2014, **66**, 413.
30 M. Scarselli, P. Annibale, P. J. McCormick, S. Kolachalam, S. Aringhieri, A. Radenovic, G. U. Corsini and R. Maggio, *FEBS J.*, 2016, **283**, 1197.
31 H. Guo, S. An, R. Ward, Y. Yang, Y. Liu, X.-X. Guo, Q. Hao and T.-R. Xu, *Biosci. Rep.*, 2017, **37**, BSR20160547.
32 N. O. Petersen, P. L. Höddelius, P. W. Wiseman, O. Seger and K.-E. Magnusson, *Biophys. J.*, 1993, **65**, 1135.
33 N. O. Petersen, C. Brown, A. Kaminski, J. Rocheleau, M. Srivastava and P. W. Wiseman, *Faraday Discuss.*, 1998, **111**, 289.
34 P. W. Wiseman and N. O. Petersen, *Biophys. J.*, 1999, **76**, 963.
35 G. D. Ciccotosto, N. Kozer, T. T. Y. Chow, J. W. M. Chon and A. H. A. Clayton, *Biophys. J.*, 2013, **104**, 1056.
36 N. Kozer, D. Barua, S. Orchard, E. C. Nice, A. W. Burgess, W. S. Hlavacek and A. H. A. Clayton, *Mol. BioSyst.*, 2013, **9**, 1849.
37 S. Ganguly and A. Chattopadhyay, *Biophys. J.*, 2010, **99**, 1397.
38 T. J. Pucadyil, S. Kalipatnapu and A. Chattopadhyay, *Cell. Mol. Neurobiol.*, 2005, **25**, 553.
39 C. P. Müller, R. J. Carey, J. P. Huston and M. A. De Souza Silva, *Prog. Neurobiol.*, 2007, **81**, 133.
40 E. Lacivita, M. Leopoldo, F. Berardi and R. Perrone, *Curr. Top. Med. Chem.*, 2008, **8**, 1024.
41 F. Fiorino, B. Severino, E. Magli, A. Ciano, G. Caliendo, V. Santagada, F. Frecentese and E. Perissutti, *J. Med. Chem.*, 2014, **57**, 4407.
42 T. J. Pucadyil, S. Kalipatnapu, K. G. Harikumar, N. Rangaraj, S. S. Karnik and A. Chattopadhyay, *Biochemistry*, 2004, **43**, 15852.
43 T. J. Pucadyil and A. Chattopadhyay, *Biochim. Biophys. Acta*, 2007, **1768**, 655.
44 S. Kalipatnapu, T. J. Pucadyil, K. G. Harikumar and A. Chattopadhyay, *Biosci. Rep.*, 2004, **24**, 101.
45 P. K. Smith, R. I. Krohn, G. T. Hermanson, A. K. Mallia, F. H. Gartner, M. D. Provenzano, E. K. Fujimoto, N. M. Goeke, B. J. Olson and D. C. Klenk, *Anal. Biochem.*, 1985, **150**, 76.
46 D. M. Amundson and M. Zhou, *J. Biochem. Biophys. Methods*, 1999, **38**, 43.
47 F. Mancia, Z. Assur, A. G. Herman, R. Siegel and W. A. Hendrickson, *EMBO Rep.*, 2008, **9**, 363.
48 T. J. Pucadyil and A. Chattopadhyay, *Biochim. Biophys. Acta*, 2004, **1663**, 188.

 This journal is © The Royal Society of Chemistry 2018

49 Y. D. Paila, M. R. V. S. Murty, M. Vairamani and A. Chattopadhyay, *Biochim. Biophys. Acta*, 2008, **1778**, 1508.

50 S. Shrivastava, T. J. Pucadyil, Y. D. Paila, S. Ganguly and A. Chattopadhyay, *Biochemistry*, 2010, **49**, 5426.

51 R. Saxena and A. Chattopadhyay, *J. Neurochem.*, 2011, **116**, 726.

52 R. Zidovetzki and I. Levitan, *Biochim. Biophys. Acta*, 2007, **1768**, 1311.

53 A. V. Dix, S. M. Moss, K. Phan, T. Hoppe, S. Paoletta, E. Kozma, Z.-G. Gao, S. R. Durell, K. A. Jacobson and D. H. Appella, *J. Am. Chem. Soc.*, 2014, **136**, 12296.

54 M. Martin, C. G. Dotti and M. D. Ledesma, *Biochim. Biophys. Acta*, 2010, **1801**, 934.

55 K. Smiljanic, T. Vanmierlo, A. M. Djordjevic, M. Perovic, N. Loncarevic-Vasiljkovic, V. Tesic, L. Rakic, S. Ruzdijic, D. Lutjohann and S. Kanazir, *Lipids*, 2013, **48**, 1069.

Faraday Discussions

PAPER

Gold nanorods decorated with a cancer drug for multimodal imaging and therapy†

Resmi V. Nair,‡ Hema Santhakumar‡ and Ramapurath S. Jayasree *

Received 21st August 2017, Accepted 12th September 2017
DOI: 10.1039/c7fd00185a

Cancer, a condition with uncontrolled cell division, is the second leading cause of death worldwide. The currently available techniques for the imaging and treatment of cancer have their own limitations and hence a combination of more than one modality is expected to increase the efficacy of both diagnosis and treatment. In the present study, we have developed a multimodal imaging and therapeutic system by incorporating a chemotherapeutic drug, mitoxantrone (MTX) onto PEG coated gold nanorods (GNR). Strong absorption in the near-infrared (NIR) and visible regions qualifies GNR as an efficient photothermal (PTT) agent upon irradiation with either a NIR or visible laser. Additionally, the enhanced electric field of GNR makes it a suitable substrate for surface enhanced Raman scattering (SERS). Modification of GNR with amino PEG offers biocompatibility without affecting its optical property. In order to achieve tumor specificity, GNR–PEG was conjugated with tumor specific marker that can target cancer cells, leaving the normal cells unaffected. The incorporation of fluorescent chemotherapeutic drug mitoxantrone onto GNR–PEG facilitates chemotherapy as well as fluorescence imaging. The therapeutic efficacy of the developed GNR based system is tracked using fluorescence imaging and Raman imaging. The careful design of the system also facilitates the controlled release of the drug by photothermal triggering. Likewise, the imaging modality could be chosen as either Raman or fluorescence to monitor drug release in accordance with irradiation. The physico-chemical properties, and drug release profiles under different physiological conditions have been well studied. Finally, the developed system was tested for its therapeutic efficacy using cancer cells, *in vitro*. The receptor mediated cell uptake was more effective in folate receptor over-expressing cancer cells than in the normal and low-expressing cells. Accordingly the percentage of cell death was higher in folate receptor over-expressing cancer cells, which was further enhanced due to the effect of the dual therapeutic approach. The cell uptake and treatment efficacy was monitored using fluorescence microscopy and SERS. In conclusion, the developed GNR–PEG–MTX system is found to

Division of Biophotonics and Imaging, Sree Chitra Tirunal Institute for Medical Sciences and Technology, Biomedical Technology Wing, Trivandrum 695012, India. E-mail: jayasree@sctimst.ac.in

† Electronic supplementary information (ESI) available. See DOI: 10.1039/c7fd00185a

‡ First and second authors contributed equally.

be an efficient multimodal therapeutic agent against cancer which could be tracked using two different techniques.

1. Introduction

Gold based nanoconstructs are gaining wide acceptance in the biomedical field due to their favourable electrical, optical and surface properties.[1–4] Various gold nanostructures like gold spheres, gold rods (GNR), gold stars, gold nanoclusters *etc.* have been extensively studied.[5–12] Among them gold nanoclusters exhibit unique NIR fluorescence which has been demonstrated for use in bioimaging.[6–8] However, the therapeutic efficiency of this structure is limited. The most studied gold anisotropic system is gold nanorods.[13] They have strong NIR absorption and scattering which allows the photons to penetrate biological tissue and makes them a suitable candidate for therapeutic applications.[14] However, the fluorescence imaging capacity is compromised here, except for the two photon imaging efficacy.[15–17] On the other hand, the heat generation capacity of GNR has been established for the photothermal therapy (PTT) of cancer.[19–24] GNR, when irradiated with lasers of suitable wavelength, can absorb light and convert it to heat by electron–electron or electron–phonon interactions.

Cancer is one of the leading causes of death, and researchers are in a continuous effort to improve the efficiency of cancer diagnosis and treatment by developing newer drugs and advanced techniques including nanotechnology. Among the various anticancer drugs, mitoxantrone (MTX) is commonly used for the treatment of certain types of cancers and is reported to be less cardiotoxic compared to doxorubicin.[25] It is an anthraquinone drug which is structurally related to DNA-anthracycline antibiotics, which intercalates into and cross links with DNA, followed by inhibition of DNA and RNA replication. It also binds to topoisomerase II, resulting in DNA strand breakage and inhibition of DNA repair. These properties, in addition to its emission in the NIR region for imaging applications, forms the basis of choosing MTX as the drug of this study.

Fluorescence imaging and Raman microspectroscopy are two well-established methods providing functional insight into the living cells and tissues. However, the strong fluorescence signal very often masks the weak Raman signal, posing difficulties in utilizing the combined effect of both techniques. Thus strategies to enhance the Raman signals and to reduce the fluorescence signals are quite often adopted. Likewise, surface enhanced Raman scattering (SERS) is one of the most effective techniques capable of increasing the Raman signals by many fold. Gold nanorods are reported as one of the most efficient Raman substrates for SERS.[26–30] In the present study, we used GNR as a carrier for the MTX drug, a substrate for SERS and also as a photothermal agent. The GNR is functionalized with amine polyethylene glycol to reduce its toxicity[31] and folic acid is used as a cancer targeting moiety. The cancer targeting efficacy of the developed system is studied in different cell lines with varying degrees of folate receptor expression, in order to prove its specificity. The drug delivery to the cells is controlled by the photothermal effect and is monitored through fluorescence imaging and SERS. Thus the GNR construct works as an on/off switch during SERS and fluorescence imaging, for better cancer diagnosis.

2. Experimental

2.1 Materials

Chloroauric acid ($HAuCl_4$), thio amine polyethylene glycol (thio-amino PEG) and cetyl trimethyl ammonium bromide (CTAB) were purchased from Sigma Aldrich. Other chemicals for construct synthesis were purchased from Merck. All cell culture media and reagents were purchased from HiMedia.

2.2 Methods

2.2.1 Synthesis and characterisation

Preparation of gold nanorods (GNR). Gold nanorods were synthesised using a seed mediated growth procedure. The seed solution was prepared by treating 0.2 mM $HAuCl_4$ with sodium borohydride ($NaBH_4$) as a reducing agent and CTAB as a stabilizing agent. The solution was kept at 20 °C for 3 h. The growth solution was prepared by treating 0.2 mM $HAuCl_4$ with CTAB, $AgNO_3$ and ascorbic acid followed by 200 μl of seed solution. The solution was kept overnight for growth at room temperature. On completion of growth, the solution is kept at 4 °C for 1 h, and then at room temperature to remove the excess CTAB as sediment. The supernatant collected was centrifuged at 1200 rpm for 20 minutes and the centrifugation was repeated three times to purify the GNR.

Surface modification of GNR with thio-amino polyethylene glycol (PGNR). GNR were stirred with 1 mg of thio-amino PEG for two hours. The synthesised material was recovered using centrifugation at 22 000 rpm for 15 minutes and washed twice to get PGNR.

Preparation of cancer targeting system (FPGNR). PGNR was treated with half the concentration of folic acid through EDC/NHS coupling. 3 mg/5 μl of EDC was treated with folic acid at pH 4 and the solution was stirred for 3 h. Then 3 mg/5 μl of NHS was added to the solution and kept stirring for 3 h. After 15 minutes, PGNR was added to the EDC/NHS mixture and the solution was stirred for another 12 h at pH above 9 and then centrifuged at 22 000 rpm for 15 minutes. The supernatant was washed twice to remove the unreacted folic acid. The construct was named as FPGNR.

Encapsulation of mitoxantrone drug to FPGNR (MTX@FPGNR). FPGNR was sonicated with 0.5 ml of 1.5 mg ml^{-1} MTX solution for 15 minutes. Then the solution was stirred for 12 h at pH 4. The products were washed by centrifugation with distilled water for 15 minutes at 22 000 rpm to remove the unreacted MTX. The construct was named as MTX@FPGNR.

Characterisation and instrumentation. GNR, PGNR, FPGNR and MTX@FPGNR were characterized with a Fourier transform infrared spectrometer (Carry 600, Agilent Technologies) to confirm the functionalization and chemical changes after functionalization of GNR. IR spectra were recorded in the range of 400 to 3600 cm^{-1} in transmission mode using a KBr pellet method spanning over 32 scans. Zeta potential at every step of the synthesis was investigated using a Malvern Zetasizer NanoZS 90 to investigate the change in charge after functionalization. Surface morphology was confirmed using TEM (100kv JEM-2010 HiTachi-JEOL, Tokyo, Japan) analysis. UV-Visible spectroscopy was done using a Shimadzu UV spectrometer-UV 2600 to confirm the functionalization of GNR. The emission properties were studied using a Varian Carl Eclipse fluorescence spectrophotometer at different excitation wavelengths. The heat generation potential of the material was studied by irradiating with an 808 nm laser (MDL III). Raman imaging of the cells and surface enhanced Raman scattering were investigated with the spectral imaging mode of a confocal Raman Microscope (alpha300R, WITec Inc. Germany). A spectrum at every pixel was taken (scan range: $40 \times 40\ \mu M^2$, 100×100 pixel, 10 000 spectra) using a 20× Nikon objective. The sample was excited with a 10 mW power, 633 nm frequency doubled Nd:YAG laser. Using the integrated video camera, a suitable cell was focused and scanned with an integration time of 0.5 s. All the cell imaging experiments were performed using an Olympus IX83 Inverted Fluorescence microscope (Tokyo, Japan) equipped with a cooled CCD camera (XM10, monochrome, Olympus), a metal halide lamp (X-Cite, series 120 PC Q), objective lens (LUCPLFLN 40X PH/0.6, including UPLFLN 10X/0.3, Olympus), excitation filters (360–370, 470–495, 540–585), dichroic mirrors (DM410, DM505, DM595), and emission filters (420–460, 510–550, 600IF), controlled with CellSens Imaging software.

2.2.2 Drug release. *In vitro* drug release study was done using fluorescence spectroscopy. 1 mg ml^{-1} of MTX@FPGNR was weighed separately and divided into four batches. One batch was dissolved in normal distilled water (pH 7) and another batch in DI water (pH 4). Similar batches were chosen for laser irradiation. The fluorescence of each sample was pre-recorded and all the samples were diluted to 5 ml with solvent as mentioned above. At different time intervals of 1, 2, 4, 8, 12 and 24 h, one batch of each sample was irradiated with an 808 nm laser for 1.30 s with 1 W power. All four samples were centrifuged at 15 000 rpm for 10 minutes. 1 ml of supernatant was collected and then made up to 5 ml with fresh solvent. The fluorescence intensity of each collection was recorded with 600 nm excitation. The ratio of intensity at 668 and 711 nm was recorded, which are the characteristic emissions of the MTX drug. This ratio is directly proportional to the concentration of the MTX drug.

2.3 Cell culture

Two cancerous cell lines over-expressing different levels of folate receptors were chosen for this study: HeLa (human cervical cancer) and C6 glioma (rat glial tumor). Normal fibroblast cell line L929 (mouse alveolar and adipose connective tissue) was used as a folate receptor-negative control group. L929 and HeLa cells were cultured in Dulbecco's modified Eagle's medium supplemented with 10% (v/v) fetal bovine serum (FBS), 100 μg ml^{-1} penicillin, 100 μg ml^{-1} streptomycin and 2.5 μg ml^{-1} amphotericin B. The C6 glioma cell line was cultured in Dulbecco's modified Eagle's medium/nutrient mixture F12 Ham (DMEM/F12, 1 : 1) supplemented with 1 mM sodium pyruvate, 10% (v/v) fetal bovine serum (FBS), 100 μg ml^{-1} penicillin, 100 μg ml^{-1} streptomycin and 2.5 μg ml^{-1} amphotericin B. All the cell lines were maintained in a humidified atmosphere with 5% CO_2 at 37 °C. All experiments were performed using the following protocol: cells grown to reach 85–95% confluency were harvested with trypsin–ethylenediaminetetraacetic acid and sub-cultured in culture dishes at a seeding density of 0.25×10^6 cells per millimetre.

2.3.1 *In vitro* cellular cytotoxicity. The cellular cytotoxicity of all constructs was tested in all three cell lines, L929, C6 and HeLa. Each cell line was seeded onto a 96-well plate with cell density of 1×10^4 cells per well. After 24 h of incubation, the cells were further incubated with fresh media (100 μl per well) containing all constructs in concentrations of 2, 1, 0.5, 0.25 and 0.125 mg ml^{-1} for 24 h, respectively. Then the constructs-containing media was removed and replaced with 100 μl of fresh DMEM containing 3-(4,5-dimethylthiazol-2-yl)-2,5-diphenyltetrazolium bromide (MTT-10 μl, 5 mg ml^{-1}). Following 4 h of incubation, the insoluble MTT formazan crystals formed were dissolved with DMSO (100 μl). Absorbance at 570 nm was measured using a microplate reader (Synergy H1 hybrid multi-mode microplate reader, Bio-Tek). The relative cell viability (%) for each construct related to the control cells was finally calculated.

2.3.2 Cancer cell targeting potential of MTX@FPGNR. HeLa, C6 and L929 cell lines were evaluated for the uptake of MTX@FPGNR using fluorescence microscopy. All the cells were trypsinized and seeded in 4 well plates (10^4 cells per well) for further investigation. Then the cells were incubated with 1 mg ml^{-1} of MTX@FPGNR for 6 h. After incubation, the cell lines were washed with PBS three times in order to remove the excess material in the media and then stained with nucleus specific Hoechst stain for 10 min. Excess stain was removed by PBS washing and the cells were imaged under the inverted fluorescence microscope.

2.3.3 Evaluation of drug release in cancer cells

Fluorescence. HeLa and C6 cell lines were evaluated for drug release using fluorescence microscopy. All the cells were trypsinized and seeded in 4 well plates (10^4 cells per well) for further investigation. Then the cells were incubated with 1 mg ml^{-1} of MTX@FPGNR for 6 h. After incubation, the cell lines were washed with PBS three times in order to remove the excess material in the media. The cells were divided into two batches of each cell line and were irradiated with an 808 nm laser with 1 W power for 1.30 s. Then cells were stained with nucleus specific Hoechst stain for 10 min. Excess stain was removed by PBS washing and the cells were imaged under the inverted fluorescence microscope.

Raman. Both HeLa and C6 cancer cell lines were cultured in Lab-Tek 4 well chamber slides. After the cells attained 50% confluency, the final construct MTX@FPGNR was incubated at a concentration of 1 mg ml^{-1} in both the cell lines for 6 h to get maximum uptake. One set of construct treated cells were irradiated with an 808 nm laser and another set was kept without laser irradiation. These construct treated cells with and without laser irradiation were mapped by Raman microscopy to confirm the drug release, along with untreated HeLa and C6 cells as controls to observe the changes in the Raman signatures.

2.3.4 Validation of photothermal and chemotherapies using MTX@FPGNR

Quantitative. For the quantitative determination of photothermal therapy, HeLa and C6 cells were used. Both cell lines were seeded in 96 well plates at a density of 10 000 cells per well and incubated for 24 h. Different concentrations of GNR, PGNR, FPGNR and MTX@FPGNR were prepared using cell culture medium. The existing medium in the wells was replaced with 100 μl of fresh medium containing the materials and incubated for different time periods. The wells were irradiated with an 808 nm laser for 1.30 s with a laser power of 1 W. Then the MTT assay was carried out. Similarly for chemotherapy, MTX alone was incubated with different cell lines for 6 h.

Qualitative. For the qualitative microscopy evaluation of photothermal therapy, both the cancer cell lines were treated with 1 mg ml^{-1} of MTX@FPGNR for 6 h and the cells were irradiated with a laser. A live/dead assay was done using acridine orange/ethidium bromide staining. 1 μl of each dye was added to 1 ml of PBS and

100 μl of solution was added to each well, and the cells were imaged under a fluorescence microscope.

3. Results and discussion

3.1 Synthesis and characterisation

Gold nanorods (GNR) were synthesised by a seed mediated growth method as described elsewhere, using CTAB as a capping agent.[32,33] Uniform sized GNR were obtained after removing excess CTAB by centrifugation. The absorption spectrum shows the typical absorption bands of GNR due to longitudinal oscillation of electrons at 786 nm and transverse plasmon absorption at 534 nm (Fig. 1A). The longitudinal absorption band in the NIR region is the main attraction to use GNR as an excellent candidate for photothermal therapy.

The surface of GNR was then modified with thio-amino PEG to reduce its toxicity.[34] For this, GNR was stirred with 1 mg of thio-amino PEG for 2 h to get PGNR. A slightly blue shifted absorption band at 770 nm compared to 786 nm of GNR is an indication of the presence of PEG over GNR. No shift in the transverse absorption peak was observed (Fig. 1A).

PGNR was further functionalized with folic acid using EDC/NHS coupling. The resultant system, FPGNR has absorption bands at 371, 283 and 255 nm along with

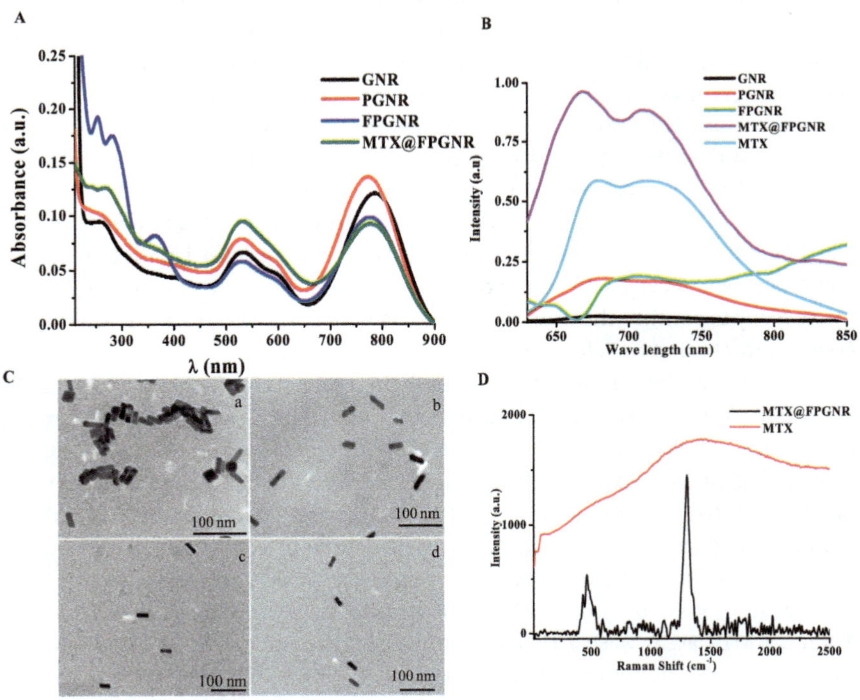

Fig. 1 (A) UV-Visible spectra of GNR, PGNR, FPGNR and MTX@FPGNR. (B) Emission spectra of GNR, PGNR, FPGNR and MTX@FPGNR. (C) Transmission electron microscopy images of (a) GNR, (b) PGNR, (c) FPGNR, (d) MTX@FPGNR. (D) Raman spectra of MTX and MTX@FPGNR.

 This journal is © The Royal Society of Chemistry 2018

the longitudinal and transverse plasmonic absorption bands of GNR indicating the presence of folic acid in FPGNR (Fig. 1A).

Mitoxantrone (MTX) was encapsulated in FPGNR for simultaneous chemo and photothermal therapy and also for fluorescence imaging. MTX was incorporated in FPGNR by stirring it with a specified amount of drug at an acidic pH (4–6) for 24 h. The absorption spectrum of the final system, MTX@FPGNR, also shows both the absorption bands of GNR (Fig. 1A), with a slight shift in the transverse oscillation band, due to the presence of MTX. The longitudinal and transverse absorption bands of MTX@FPGNR and GNR have almost equal intensity. Moreover, the shoulder peak around 594 nm observed in all other constructs vanishes in the case of MTX@FPGNR, which might be due to overlapping of the absorption frequency of MTX with that of FPGNR. This further confirms the encapsulation of the drug. The peaks at 283 and 255 nm also indicate the presence of both folic acid and MTX in the final system. The shoulder peak at 371 nm can be due to the presence of folic acid (Fig. 1A) in MTX@FPGNR. The synthesis of MTX@FPGNR is shown schematically in the ESI (Scheme S1†).

Modifications of GNR at different stages were also confirmed using FTIR spectroscopy (Fig. S1†). The initial system, GNR, showed the characteristic vibrations of the trimethyl ammonium ion of CTAB around 929, 1000, 1061 cm^{-1} (Fig. S1†). Upon surface modification with PEG, additional peaks corresponding to primary amino groups are observed in the range 3269–3196 cm^{-1}. The 2892 cm^{-1} peak is assigned to CH stretching vibrations. Vibrational bands of amino groups are also visible at 1597, 1122 and 667 cm^{-1}. Peaks corresponding to CTAB are absent in PGNR which proves the successful covering of amino PEG over the GNR. The peak at 2598 cm^{-1}, corresponding to the thiol group is absent in PGNR indicating the bonding of the same to the GNR. The FTIR spectrum of FPGNR shows peaks corresponding to amino groups and amide groups at 3391 and 653 cm^{-1}, and 1606 and 1119 cm^{-1}, respectively. The 891 cm^{-1} peak indicates the presence of aromatic groups. The peaks corresponding to mitoxantrone are prominent in MTX@FPGNR. The peak at 3268 cm^{-1} is due to the vibrations of the phenolic OH group, whereas the 2880 and 2326 cm^{-1} peaks are due to CH and C=C vibrations. The peaks at 1739 and 1373 cm^{-1} correspond to the vibrations of ester bonds formed between MTX and FPGNR. Peaks at 1655 and 860 cm^{-1} are assigned to the vibrations of the amide group and aromatic ring, respectively. The peak at 1556 cm^{-1} is assigned to the vibrations of the amine group.

The fluorescence spectroscopic study shows weak emission profiles for PGNR and FPGNR in the NIR region at 600 nm excitation. The weak emission is attributed to the fluorescence of gold nanorods, which is enhanced in the presence of PEG and folic acid. Two strong emission peaks at 668 and 711 nm are available for the final GNR construct, MTX@FPGNR. The drug MTX has two prominent emission peaks at 679 and 714 nm (Fig. 1B). In this case, the fluorescence of the drug is retained even in the presence of GNR, unlike in the case of many other studies where the fluorescence quenching of the fluorophore is in the proximity of GNR.[35] The reason behind the retaining of fluorescence in this study is due to the presence of the polymer covering over the GNR. The strong NIR emission of the final system makes it suitable for imaging.

Electron microscopic analysis reveals the formation of anisotropic GNR with an average aspect ratio of 3.2 ± 0.18. Further functionalization did not alter the aspect ratio considerably (Fig. 1C) and hence the heat generation properties. This is confirmed by monitoring the temperature increase of the final system in aqueous solution with an NIR laser irradiation. Heat generation of more than 50 °C was achieved with MTX@FPGNR within 5 minutes of laser irradiation while GNR alone generated heat of the order of 60 °C (Fig. S2†). While the temperature above 50 °C is sufficient for effective photothermal therapy, the observed decrease in the heat generation is attributed to the effect of surface functionalization.

The zeta potential at every stage of functionalization was monitored to know more about the behaviour of particles. The first system, GNR, had a zeta potential of +14.6 mV, while the zeta potential of PGNR is decreased to +7.19 mV, which could be due to the presence of NH_2 in the amino PEG. Further, conjugation with folic acid decreased the zeta potential to −5.86 mV, due to the presence of more –COOH and NH_2 groups. Finally, MTX@FPGNR at normal pH gives a zeta potential of −1.06, which in acidic pH changes to 4.93 mV (Fig. S3†).

3.1.1 SERS activity of MTX@FPGNR. The Raman spectrum of the drug MTX in aqueous medium did not show any prominent peaks, due to the high fluorescence of the drug. However, the Raman peaks are well defined in the case of MTX@FPGNR because the gold construct acts as a SERS substrate. The Raman spectrum of MTX@FPGNR shows a prominent peak at 1300 cm^{-1} which is assigned to the ring stretching mode coupled with C–O bond vibrations (Fig. 1D). The prominent peak at 464 cm^{-1} indicates the presence of C–N in the aromatic ring and C–C–O bonds present in folic acid and MTX.[36] The results are supportive of the fact that MTX@FPGNR can be used as a SERS substrate.

3.1.2 Drug release study. A controlled release of the drug followed by an increase in the fluorescence from the drug is expected due to the unique design of the final system. Increase in the fluorescence intensity following drug release was studied at different time periods with and without laser irradiation (Fig. S4†). The drug release study revealed that at all pH, the material is stable for up to 8 h. A small amount of drug release was observed at acidic pH or with laser irradiation. But the combined acidic environment along with laser irradiation induces spontaneous drug release. This finding could be used for cancer treatment where the pH is acidic with controlled laser irradiation. At normal pH, no drug release was observed even after 24 h indicating the stability of material without laser irradiation. The drug release mechanism is explained as follows: when the alcoholic side chains of the MTX molecule react with the acid groups of folic acid, ester bonds are formed. This is confirmed in the FTIR spectrum. Acidic pH and high temperature from laser irradiation induces ester hydrolysis leading to drug release followed by an increase in fluorescence intensity. Thus the system works on the principle of laser induced drug release. The targeting efficacy of MTX@FPGNR is retained by using half the moles of MTX compared to the number of moles of folic acid.

3.2 *In vitro* cell studies

3.2.1 *In vitro* cytotoxicity. To evaluate biocompatibility, the different gold constructs were tested for their long term cellular cytotoxicity in the normal L929 fibroblast cell line, using the standard MTT assay (Fig. S5†).

At 24 h of incubation, GNR showed less cellular viability which may be due to the presence of toxic CTAB. The functionalization of GNR with PEG and folic acid reduced the toxicity, showing a cell viability of more than 95% even at a higher concentration of 2 mg ml^{-1}, proving its suitability to use safely in a cellular environment. On encapsulation of MTX drug in FPGNR, the cell compatibility was retained in L929 fibroblast cells. This result indicates that the drug is not released under normal conditions without laser stimulation and is stable inside the cell. Overall, the results suggest that MTX@FPGNR can serve as a biocompatible material for the combined photothermal-chemo therapy for cancer, without affecting the neighbouring normal cells.

3.2.2 Cancer cell targeting potential of MTX@FPGNR. In this study, the targeting of the GNR construct to the cancer cells is achieved by folic acid conjugation. Different cell lines with varying folate receptor expression were chosen to prove this. It is reported that the normal L929 fibroblast cells have very low folate receptor expression compared to cancerous cell lines.[37] Between the two cancer cell lines considered, rat brain glioma cells (C6) have moderate folate receptor expression compared with human cervical cancer cells (HeLa).[38] The cell uptake study was performed using these three cell lines with an incubation period of 6 h (Fig. 2). The results show that receptor mediated uptake of MTX@FPGNR is maximum in HeLa cells where the folate receptor expression is highest, followed by C6 glioma cells, and least uptake is observed in L929 cells. From the red fluorescence of the material and the nucleus specific Hoechst staining, it is clear that the construct is concentrated mostly in the cytoplasm rather than in the nucleus. The result of the cell uptake study gives a clear indication of the folate receptor over-expressing cancer cell targeting efficiency of MTX@FPGNR.

3.2.3 Evaluation of drug release in cancer cells. On achieving excellent cancer targeting ability, MTX@FPGNR was further tested for its drug release efficacy in both the cancer cell lines, with photothermal triggering. The observed increase of red fluorescence is a clear indication of the drug release upon laser irradiation compared to non-irradiated cells (Fig. 3a–d). Both HeLa and C6 cells showed an increase in fluorescence during drug release in accordance with the material uptake efficiency, which reflected the difference in folate receptor expression

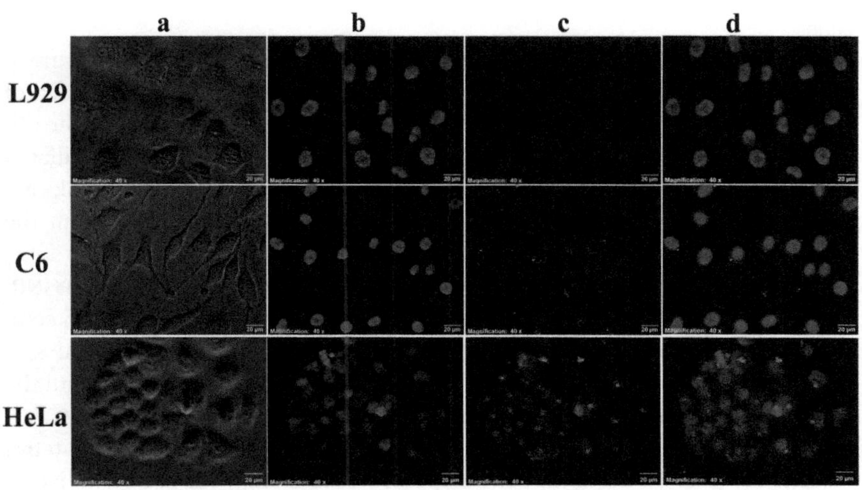

Fig. 2 Cellular uptake of MTX@FPGNR. (a) Bright field, (b) Hoechst, (c) MTX@FPGNR and (d) overlay images.

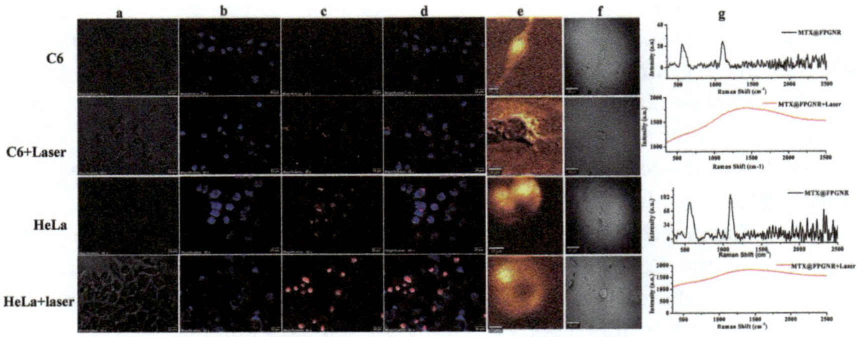

Fig. 3 Drug release evaluation using fluorescence microscopy and Raman imaging. (a) Bright field, (b) Hoechst, (c) MTX@FPGNR and (d) overlay images, (e) Raman image, (f) Raman map, (g) Raman spectra of corresponding cells.

level. The drug release study was also confirmed with Raman spectral analysis where the presence of GNR contributed to SERS activity and thus was instrumental in getting Raman signals even in the presence of fluorescent MTX (Fig. 3e–g). For this, cells incubated with the final system for 6 h, with and without laser irradiation were monitored. Due to the intense fluorescence of the released drug, no Raman peaks could be identified from the laser irradiated cells whereas the material incubated cells without laser irradiation showed distinct Raman peaks (Fig. 3g). The Raman peaks of the GNR construct in the cells were assigned in accordance with those in aqueous solution. The most prominent peak observed at 1300 cm^{-1} in the Raman spectrum of MTX@FPGNR in solution is absent in cells. This is because in the acidic environment of cancer cells, the ring stretching modes of MTX@FPGNR are restricted. HeLa cells show a Raman peak at 1104 cm^{-1} due to the C–O and C–N vibrations. Peaks at 556 and 994 cm^{-1} indicate the ring deformation and C–C vibrations of the construct. The 463 cm^{-1} peak shows C–N vibration of the drug, MTX is also visible. Less intense Raman peaks of the cells were also observed in the spectrum. Similar trends were also seen in C6 cells.

However, in HeLa cells the less intense peaks corresponding to the signature of cells are more prominent than in the C6 cells. This may be due to the lack of a sufficient amount of GNR substrate inside the cells to support SERS, as the cell uptake efficiency was less in the case of C6. The Raman spectra of laser irradiated cells appear to be similar to that of the drug MTX in solution where no peaks are identified. This is due to the switching 'on' of strong fluorescence from the released drug upon laser irradiation.

3.2.4 Validation of photothermal and chemotherapies using MTX@FPGNR. Owing to its strong NIR absorbance and high heat generation efficiency in solution, MTX@FPGNR was further studied for its combined photothermal and chemotherapy potential in cancer cells. A live/dead assay was performed with the cancerous cells, HeLa and C6 glioma, after 808 nm laser irradiation using AO/ER staining for a qualitative study and the MTT assay was used for quantitative assessment (Fig. 4). MTX@FPGNR incubated cells along with laser irradiation showed considerable cell death in comparison with non irradiated control cells (Fig. 4A and B). The results showed that MTX@FPGNR induced a more significant

 This journal is © The Royal Society of Chemistry 2018

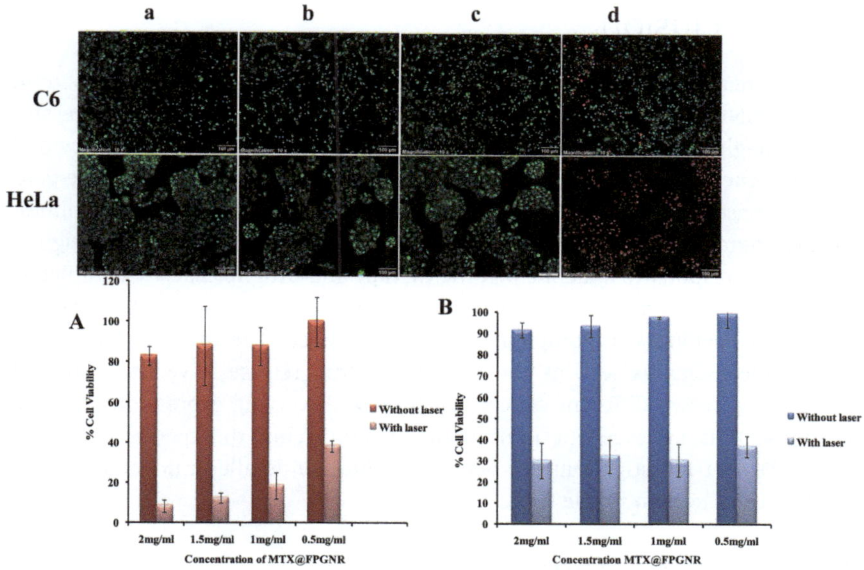

Fig. 4 Live/dead assay of MTX@FPGNR in HeLa and C6 cells, (a) control, (b) control + laser, (c) MTX@FPGNR and (d) MTX@FPGNR + laser. Plots show quantitative evaluation of therapy, with and without laser irradiation, of (A) HeLa and (B) C6 cells incubated with MTX@FPGNR.

therapeutic effect in HeLa cells than in the C6 cells, because of the high expression of folate receptors in HeLa cells. This is in accordance with the qualitative results of the live/dead assay (Fig. 4a–d). The efficacy of combined therapy was evaluated *versus* chemotherapy of MTX alone. More extensive cell death was observed in GNR construct incubated cells with laser irradiation, than in the cells treated with MTX alone (Fig. S6†). The treatment efficacy of MTX@FPGNR was further compared with GNR, PGNR and FPGNR. MTT assays were carried out using C6 and HeLa cells after 6 h of incubation with different GNR systems at different concentrations, after laser irradiation (Fig. S7†). From this, it is clear that the treatment efficacy is higher in HeLa cells compared to C6 cells. This is in accordance with all other results. The long term stability of the final construct in cells without laser irradiation was tested in HeLa cells by MTT assay (Fig. S8†). The results show that 95% of the cells are viable even after 24 h of incubation. From this it is clear that the drug remains stable inside the cells without release, until stimulated by laser. This confirms that both photothermal and chemotherapies are under control and will be activated only after laser irradiation. These results suggest that the combined photothermal and chemotherapy induced by MTX@FPGNR led to irreversible and serious damage to HeLa cells which have a high expression of folate receptors. More detailed study is necessary in order to examine the mechanism of cell death, *e.g.*, apoptosis.

To summarise, the construct MTX@FPGNR possesses several unique properties and functionalities, such as long term stability, controlled drug release with laser irradiation, excellent heat generation capacity, cancer-targeting properties and biocompatibility, which make it an ideal photothermal cum chemotherapy agent for folate receptor over-expressing cancer cells.

4. Conclusion

A gold nanorod based multimodal imaging and therapeutic system, MTX@FPGNR with cancer targeting properties was developed for successful cancer imaging and therapy. The chemotherapeutic drug mitoxantrone provided a fluorescence imaging potential along with the therapeutic efficiency. The gold nanorod based system also facilitates SERS based Raman imaging and photothermal therapy. As the gold nanorod based drug conjugated system is designed to be photo-responsive at acidic pH, the therapy and drug release are completely controlled. The perfect control over the system allows an 'ON/OFF' Raman imaging behaviour with complementary fluorescence. The imaging and therapeutic efficiencies as well as the controlled drug release have been studied successfully in two different cancer cell lines with varying expression of folate receptors. It is an excellent multimodal imaging cum therapeutic agent for cancer. Deeper investigations are needed to obtain a detailed understanding of the developed system inside cells.

Acknowledgements

ICMR and DBT, Government of India are acknowledged for the financial support. The authors also acknowledge the staff of MIT and DPMD, BMT Wing, SCTIMST for providing the plate reader and fluorescence spectrophotometer support, and Dr K. K. Maiti and team, CSIR-NIIST, Trivandrum for the Raman facility.

Notes and references

1 N. Gong, S. Chen, S. Jin, J. Zhang, P. C. Wang and X. Liang, *Regener. Biomater.*, 2015, 273–280.

2 M. Vinod, R. S. Jayasree and K. G. Gopchandran, *Curr. Appl. Phys.*, 2017, **17**, 1430–1438.

3 S. Eustis, M. A. El-sayed and M. Kasha, *Chem. Soc. Rev.*, 2006, **35**, 209–217.

4 X. Huang and M. A. El-sayed, *Int. J. Adv. Res.*, 2010, 13–28.

5 C. L. Nehl, H. Liao and J. H. Hafner, *Nano Lett.*, 2006, **6**, 683–688.

6 L. V. Nair, S. S. Nazeer, R. S. Jayasree and A. Ajayaghosh, *ACS Nano*, 2015, **9**, 5825–5832.

7 L. V. Nair, D. S. Philips and R. S. Jayasree, *Small*, 2013, **9**, 2673–2677.

8 L. V. Nair, R. V. Nair and R. S. Jayasree, *J. Chem. Soc., Dalton Trans.*, 2016, **45**, 11286–11291.

9 M. K. Popp, I. Oubou, C. Shepherd, Z. Nager, C. Anderson and L. Pagliaro, *J. Nanomater.*, 2014, 1–8.

10 C. G. Khoury and T. Vo-dinh, *J. Phys. Chem. C*, 2008, **112**, 18849–18859.

11 P. K. Jain, X. Huang, I. H. El-Sayed and M. A. El-Sayed, *Acc. Chem. Res*, 2008, **41**, 1578–1586.

12 H. Liao and J. H. Hafner, *Chem. Mater.*, 2005, **17**, 4636–4641.

13 X. Huang, S. Neretina and M. A. El-Sayed, *Adv. Mater.*, 2009, **21**, 4880–4910.

14 M. A. MacKey, M. R. K. Ali, L. A. Austin, R. D. Near and M. A. El-Sayed, *J. Phys. Chem. B*, 2014, **118**, 1319–1326.

15 M. B. Mohamed, V. Volkov, S. Link and M. A. El-sayed, *Chem. Phys. Lett.*, 2000, 517–523.

 This journal is © The Royal Society of Chemistry 2018

16 T. Wang, D. Halaney, D. Ho, M. D. Feldman and T. E. Milner, *Biomed. Opt. Express*, 2013, **4**, 584.
17 C. Molinaro, Y. El Harfouch, E. Palleau, F. Eloi, S. Marguet, L. Douillard, F. Charra and C. Fiorini-Debuisschert, *J. Phys. Chem. C*, 2016, **120**, 23136–23143.
18 N. Verellen, D. Denkova, B. De Clercq, A. V. Silhanek, M. Ameloot, P. Van Dorpe and V. V. Moshchalkov, *ACS Photonics*, 2015, **2**, 410–416.
19 X. Huang and M. A. El-sayed, *Alexandria Journal of Medicine*, 2011, **47**, 1–9.
20 S. Shen, H. Tang, X. Zhang, J. Ren, Z. Pang, D. Wang, H. Gao, Y. Qian, X. Jiang and W. Yang, *Biomaterials*, 2013, **34**, 3150–3158.
21 T. Niidome, *J. Phys. Conf. Ser*, 2010, **232**, 1–8.
22 D. Pissuwan, S. M. Valenzuela and M. B. Cortie, *Biotechnol. Genet. Eng. Rev.*, 2008, **25**, 93–112.
23 X. Huang, I. H. El-sayed, W. Qian and M. A. El-sayed, *J. Am. Chem. Soc.*, 2006, 2115–2120.
24 J. Wang, G. Zhu, M. You, E. Song, M. I. Shukoor, K. Zhang, M. B. Altman, Y. Chen, Z. Zhu, C. Z. Huang and W. Tan, *ACS Nano*, 2012, **6**, 5070–5077.
25 S. Agarwal, D. K. Jangir and R. Mehrotra, *J. Photochem. Photobiol., B*, 2013, **120**, 177–182.
26 K. Lin, J. Yi, S. Hu, B. Liu, J. Liu, X. Wang and B. Ren, *J. Phys. Chem. C*, 2016, **120**, 20806–20813.
27 M. Y. Tsvetkov, B. N. Khlebtsov, V. A. Khanadeev, V. N. Bagratashvili, P. S. Timashev, M. I. Samoylovich and N. G. Khlebtsov, *Nanoscale Res. Lett.*, 2013, **8**, 1.
28 J. V. Jokerst, A. J. Cole, D. Van De Sompel and S. S. Gambhir, 2012, 10366–10377.
29 X. Guo, *J. Biophotonics*, 2012, **5**, 483–501.
30 T. Vo-Dinh, H. N. Wang and J. Scaffidi, *J. Biophotonics*, 2010, **3**, 89–102.
31 X. Liu, N. Huang, H. Li, H. Wang, Q. Jin and J. Ji, *ACS Appl. Mater. Interfaces*, 2014, **6**, 5657–5668.
32 B. Nikoobakht and M. A. El-Sayed, *Chem. Mater.*, 2003, **15**, 1957–1962.
33 C. V. Durgadas, V. N. Lakshmi, C. P. Sharma and K. Sreenivasan, *Sens. Actuators, B*, 2011, **156**, 791–797.
34 Y. V. Kaneti, C. Chen, M. Liu, X. Wang, J. L. Yang, R. A. Taylor, X. Jiang and A. Yu, *ACS Appl. Mater. Interfaces*, 2015, **7**, 25658–25668.
35 B. Jang, J. Park, C. Tung, I. Kim and Y. Choi, *ACS Nano*, 2011, **5**, 1086–1094.
36 I. Nabiev, A. Baranov, I. Chourpa, A. Beljebbar, G. D. Sockalingum and M. Manfait, *J. Phys. Chem.*, 1995, **99**, 1608–1613.
37 D. Bhattacharya, M. Das, D. Mishra, I. Banerjee, S. K. Sahu, T. K. Maiti and P. Pramanik, *Nanoscale*, 2011, **3**, 1653–1662.
38 J. M. Saul, A. Annapragada, J. V. Natarajan and R. V. Bellamkonda, *J. Controlled Release*, 2003, **92**, 49–67.

Faraday Discussions

PAPER

Optically sensing phospholipid induced coil–helix transitions in the phosphoinositide-binding motif of gelsolin†

Samsuzzoha Mondal,‡ Amitava Chandra,‡ Ravindra Venkatramani * and Ankona Datta *

Received 22nd September 2017, Accepted 1st November 2017
DOI: 10.1039/c7fd00197e

We present a systematic experimental and computational study of phospholipid induced peptide coil–helix transitions which are relevant in the context of proteins mediating cytoskeletal rearrangement *via* membrane binding. We developed a sensitive Förster resonance energy transfer (FRET) based assay to address whether coil–helix transitions in phospholipid binding motifs of actin-binding proteins can be induced by physiologically-relevant concentrations (1–20 μM) of phosphatidylinositol-4,5-bisphosphate (PI(4,5)P2) phospholipids. Based on inter-residue distance constraints obtained from Molecular Dynamics (MD) simulations of a 20 residue peptide (Gel 150–169) from the actin-severing protein gelsolin, we synthetized and labeled the peptide with a tryptophan donor and IAEDANS acceptor pair. Upon addition of PI(4,5)P2 micelles and mixed vesicles containing PI(4,5)P2 and phosphatidylcholine to the peptide, we observed a decrease in the tryptophan emission intensity with increasing concentrations of PI(4,5)P2. The IAEDANS emission spectra showed a more complex profile exhibiting a blue shift of the emission peak and non-monotonic changes in the intensity profile with increasing concentrations of PI(4,5)P2. We showed that the IAEDANS acceptor emission response is a result of both intrinsic polarity sensitivity of the acceptor in the vicinity of the membrane surface and fluorescence energy transfer from the donor. Importantly, the fluorescence lifetime of the donor (tryptophan) showed a monotonous decrease with increasing mol% of PI(4,5)P2 from 1.13 ± 0.10 ns in the absence of phospholipids to 0.25 ± 0.03 ns in the presence of 100% PI(4,5)P2 micelles. We also showed a concomitant increase in FRET efficiency with increasing PI(4,5)P2 levels indicating a PI(4,5)P2 concentration dependent coil–helix transition. Our studies demonstrate that membrane PI(4,5)P2 concentrations as low as 2.5–5 μM can

Department of Chemical Sciences, Tata Institute of Fundamental Research, Mumbai, 400005, India. E-mail: ravi.venkatramani@tifr.res.in; ankona@tifr.res.in

† Electronic supplementary information (ESI) available: Supplementary figures, peptide purity and characterization, and SUV characterization data. See DOI: 10.1039/c7fd00197e

‡ Equal contribution.

trigger helix–coil conformational changes in gelsolin relevant for triggering regulatory processes in the cell.

Introduction

Protein-phospholipid interactions are central to the regulation of membrane mediated cellular processes.[1–5] A prevailing hypothesis in phospholipid mediated cellular function is that binding to phospholipids induces conformational changes in proteins which are relevant for the regulation of biological processes.[4,6,7] While phospholipids that induce protein conformational changes are often constitutively present in cell membranes, their concentrations and locations are subject to dynamical changes.[5] These dynamical changes act as triggers for phospholipid-mediated regulation of cell function. A key question related to phospholipid-mediated cellular regulation is: what are the triggering phospholipid levels that lead to conformational changes in phospholipid-binding proteins? Here, we address this question in the context of actin-binding proteins (ABPs) that modulate cytoskeletal rearrangements by binding to phosphoinositides, a key class of signal-mediating phospholipids.[7–17]

Circular dichroism (CD) studies on multiple ABPs indicate a coil to helix transition in their phospholipid binding domains in the presence of high concentrations (100–400 μM) of phosphoinositide micelles.[10,15,16] Here, we address whether the coil–helix transition can occur at lower physiologically relevant levels of phosphoinositides at which the ABPs regulate cytoskeletal rearrangements by developing a sensitive fluorescence-based assay that can report on coil–helix conformational changes.

For our proof-of-concept study we chose a phosphoinositide binding peptide segment from the ABP gelsolin, which participates in cytoskeletal rearrangements by promoting actin depolymerisation. Phosphoinositide binding releases gelsolin from the barbed ends of actin filaments allowing actin polymerization.[11,14,18–20] Previous studies by Janmey and co-workers have shown that phospholipid vesicles containing greater than 10 mol% of a key signal-mediating phosphoinositide, phosphatidylinositol-4,5-bisphosphate (PI(4,5)P2), can dissociate the gelsolin–actin complex.[11,20] Janmey, Braunlin and co-workers have also reported a distinct coil–helix transition in the phosphoinositide binding N-terminal sequence of gelsolin, Gel 150–169, in the presence of 400 μM PI(4,5)P2 micelles using CD studies.[15] Using computational studies, the same group predicted that the coil–helix transitions of Gel 150–169 could potentially destabilize gelsolin–actin binding interactions.[11] Coil–helix transitions in other ABPs like villin and profilin have also been reported only in the presence of PI(4,5)P2 micelles.[10,16] The reason for studying the coil–helix transitions in ABPs including gelsolin at high concentrations of PI(4,5)P2 micelles was due to the low sensitivity of CD. Moreover, micelles were chosen for the CD studies to minimize light-scattering.[10,16]

In order to associate the coil–helix transition induced by phosphoinositide-binding in gelsolin to its biological function it is important to assess if the structural transition in Gel 150–169 can occur at physiologically relevant phospholipid levels. The abundance of PI(4,5)P2 in cell membranes is around 1 mol% of total cellular phospholipids and related lipid concentrations are between 10–30 μM.[21] More importantly, since phosphoinositides are present in the plasma

 This journal is © The Royal Society of Chemistry 2018

membrane, it is necessary to investigate the effect of the presence of other major plasma membrane phospholipids like phosphatidylcholine (PC) on the coil–helix transitions of Gel 150–169. Finally, it is also important to investigate whether the coil–helix conformational change can occur within lipid vesicles along with micelles. This is because vesicles mimic the bilayer feature of the plasma-membrane and also provide the opportunity to systematically vary the mol% of phosphoinositides to investigate the levels of phosphoinositide necessary to induce the coil–helix conformational change.

To investigate the phosphoinositide levels that can trigger coil–helix transitions a peptide-conformation sensitive assay is required. A coil–helix transition in a protein-segment would be associated with a distance change between amino acid residues within the domain undergoing the transition. Fluorescence energy transfer based techniques utilizing an appropriate donor and acceptor fluorophore pair can afford information on intra-molecular distances in bio-macromolecules.[22–26] We therefore set out to develop a fluorescence energy transfer based assay to study the coil–helix transition associated with the Gel 150–169 phosphoinositide binding fragment of gelsolin. The major advantage of a fluorescence based assay is high sensitivity which enables us to investigate the effect of lower phosphoinositide concentrations on the peptide coil–helix transitions.

In order to achieve a sensitive Förster resonance energy transfer (FRET) based assay, especially in a short protein segment like Gel 150–169, the donor–acceptor pair should be strategically positioned in the peptide to undergo maximum distance change, thereby leading to the maximum contrast in donor–acceptor energy transfer efficiencies, upon the coil–helix transition.[26] While, structural information is available on the helical form of Gel 150–169, the conformational space accessed by both the helix and the intrinsically disordered coil form is required to make an informed decision on the positioning of the fluorophores. We used molecular dynamics (MD) simulations to sample the conformational space of the helix and the coil form of Gel 150–169 to create inter-residue distance maps and identified residue pairs that undergo maximum distance changes during the peptide coil–helix transition. Based on the distance estimates obtained from our computational studies, we identified an appropriate donor–acceptor fluorophore pair, tryptophan and IAEDANS,[27] for probing the coil–helix transition in the phosphoinositide-binding peptide. Our FRET studies using the labeled peptides indicate that the coil–helix transitions in Gel 150–169 can occur in the presence of PI(4,5)P2 micelles and more importantly also in the presence of mixed vesicles. Our FRET assay allowed us to investigate peptide coil–helix transitions at physiologically relevant (1–20 μM) levels of phosphoinositides and indicated that such conformation changes can occur in the Gel 150–169 peptide at PI(4,5)P2 concentrations as low as 2.5–5 μM.

Results and discussions

Molecular dynamics simulations predict the distance changes during the coil-to-helix transition in Gel 150–169 peptide

In order to choose suitable donor–acceptor fluorophores for a FRET based assay which can probe coil–helix transitions in Gel 150–169 we carried out atomistically detailed MD simulations of the peptide. The studies were aimed at capturing residue-pair distances which showed maximum change between the helical and coil forms of Gel 150–169. Braunlin and co-workers have previously reported a solution NMR based structure (PDB ID: 1SOL) for the helical form of Gel 150–169 in trifluoroethanol.[15] However, structural information on the random coil form is not available. Using the NMR structure of Braunlin and co-workers as the starting conformation we employed room temperature (300 K) and high temperature (500 K) atomistic MD simulations to computationally generate both helical and coil form conformational ensembles (Fig. 1) of the Gel 150–169 peptide as described below.

Helix form conformational ensemble of Gel 150–169. The starting NMR structure (PDB ID: 1SOL) of Gel 150–169 was equilibrated in explicit water and simulated under canonical NPT (normal pressure and temperature) conditions for 100 ns (see Methods). The peptide conformation was monitored in terms of both root mean square deviation (RMSD) of the peptide backbone atoms and percentage helical character over the time course of the 100 ns trajectory. We did not observe any significant conformation change within the 100 ns time scale of the simulation. The helical content remained within 50–60% throughout the trajectory (Fig. 1a). The RMSD of the peptide backbone also remained within 4 Å indicating minimal conformation changes (Fig. 1a). Previous circular dichroism experiments show that the Gel 150–169 peptide exists in a random coil form in aqueous buffer at pH 7.4.[15] However, the results from the 100 ns simulation on the helical form of Gel 150–169 indicated that the activation barrier for the helix to coil transition was high enough to disallow the transition within 100 ns at 300

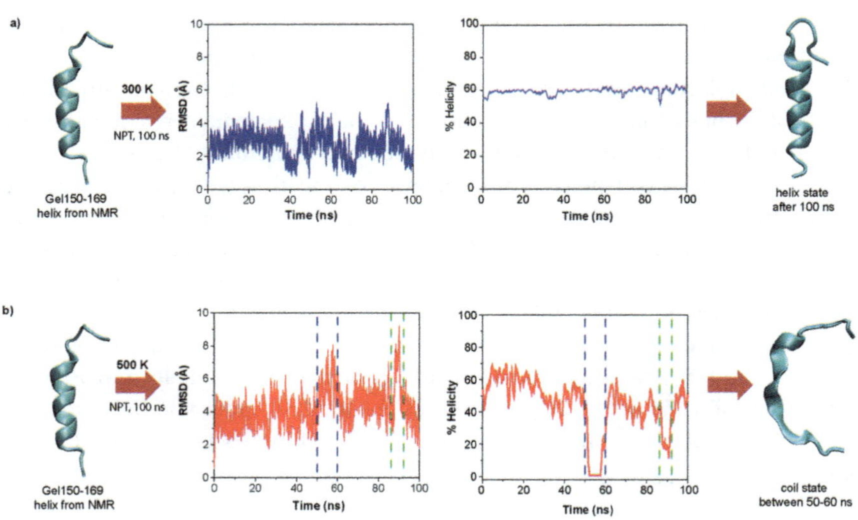

Fig. 1 MD simulations at (a) 300 K and (b) 500 K to obtain the structures of Gel 150–169 peptide in the helix state and coil state, respectively. The RMSD *vs.* time plots show the deviations in the peptide backbone from the structure at the '0' time point along the trajectories. Percentage helicity has been plotted against time to monitor the change in secondary structure over time. When simulated at 500 K, the peptide was found to be completely unfolded (0% helicity) between 50–60 ns (region between the blue dotted lines) whereas between 87–92 ns (region between the green dotted lines) the peptide conformation was more compact than the helix state. Representative images of the helix state and coil state are shown on the right hand side.

 This journal is © The Royal Society of Chemistry 2018

K. Our simulation timescale is at least 10 times shorter than the anticipated time scales for conformational transitions in proteins which occur in μs to ms time ranges.[28] Further, the CHARMM27 force field employed in our solution phase peptide study tends to over-stabilize helices[29] unfolding helical peptides at much higher temperatures than experimentally estimated.[30] We therefore employed high temperature simulations to explore the coil state of the peptide.

Generation of coil form of Gel 150–169. In order to generate a coiled state we equilibrated the NMR helix structure at an elevated temperature of 500 K and carried out 100 ns production runs. At elevated temperatures, the higher kinetic energy of the protein and solvent atoms enables molecules to cross activation barriers faster. The RMSD and percentage helicity *versus* time plots (Fig. 1b) indicated increased thermal fluctuations at the elevated temperature compared to the simulations performed at 300 K. We observed two helix–coil transitions during our 100 ns runs at 500 K that generated two sub-populations of conformations (Fig. 1b). In addition to the helix state we observed peptide coil state structures with RMSD values >6 Å relative to the starting helix conformation and with substantially lower percentage helicity (0–20%). These coil state sub-populations were observed between 50–60 ns of the trajectory as indicated by blue dashed lines and between 87–92 ns as indicated by green dashed lines in Fig. 1b. While, Gel 150–169 structures between 50–60 ns were completely unfolded exhibiting no helicity as shown in Fig. 1b, peptide structures near 90 ns were not fully unfolded and retained about 20% helicity. We therefore considered structures within the 50–60 ns time frame of the 500 K trajectories as representative conformers of the Gel 150–169 coil form in the analysis of inter-residue distances discussed in the next subsection.

Inter-residue distances in helical and coil forms of Gel 150–169 peptide. In order to identify specific residue pairs which afford maximal distance change upon coil to helix transition, we compared inter-residue distances between all residue pairs in the Gel 150–169 coil form with the corresponding distances in the helical form. Inter-residue distances were calculated in two ways as: (1) distance between alpha-carbon (C_α) atoms of amino acid residues and (2) distance between the center-of-mass (COM) of amino acid sidechain residues.

First, we averaged the inter-residue distances for all the 5000 peptide conformers between 50–60 ns of the 500 K simulation trajectory to generate average distances between residues in the coil form. The average distances were represented in the form of a 20 × 20 data matrix C. Each element in the matrix was the distance r_{ij} between the ith and j^{th} residues of the 20 amino acid peptide. Corresponding average distances for the helical form were calculated from 5000 peptide structures within the first 10 ns window during the 100 ns simulation trajectory performed at 300 K. The inter-residue distances in the helical form were represented as a 20 × 20 distance matrix H.

Next, we generated a distance difference matrix by subtracting the matrix elements of matrix H from matrix C. Two different distance difference matrices were obtained by either considering the distances between the C_α atoms of the residues or by considering distances between the COM of the individual sidechains. The matrices were color coded to generate distance difference maps. The maps afforded visual identification of the residue pairs showing maximal distance change during the coil–helix transition in Gel 150–169 (Fig. 2a and b).

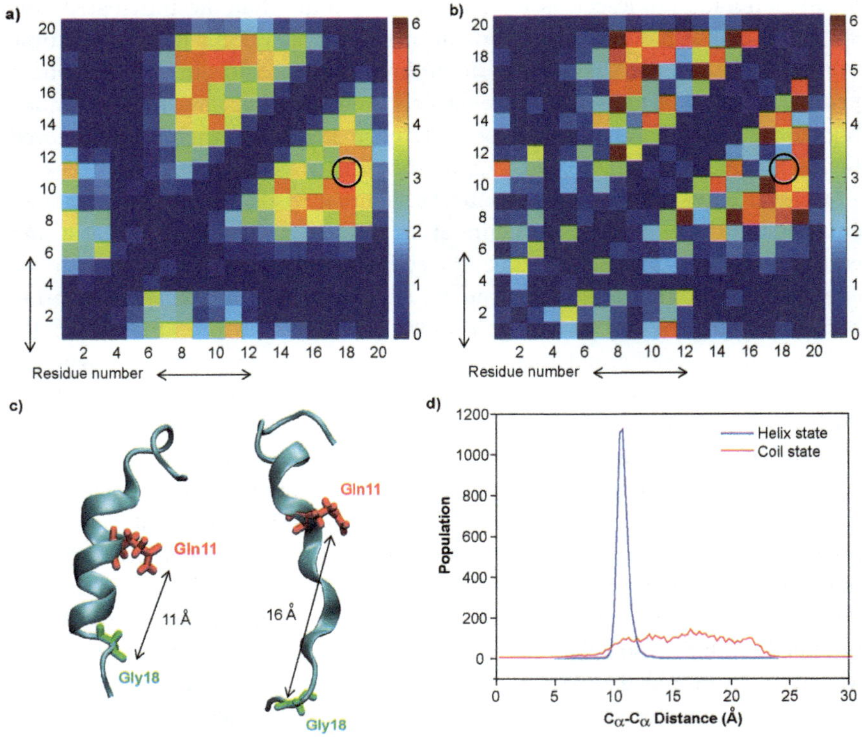

Fig. 2 Calculation of inter-residue distance changes upon coil–helix transition from the MD simulations performed on the gelsolin peptide structure (1SOL) obtained from NMR. Distance difference maps representing the inter-residue distance changes (a) between the C_α atoms and (b) between the COM of the sidechains during the coil–helix transition. The color scales represent the change in average distances between 0–6 Å. The residue pair that shows the maximum distance change in the C_α atom distance map has been highlighted with a black circle in both a and b. (c) Representative structures of the peptide in helix state and coil state. The residue pair Gln160 (# 11, red) and Gly167 (# 18, green) which shows the maximum distance change upon coil to helix transition has been highlighted. (d) Distance distribution between the C_α atoms of Gln160 (# 11) and Gly167 (# 18) residues in helix state and coil state.

We identified two residue pairs from the C_α distance difference map that provided a distance change of 5 Å or greater during the coil–helix transition. These positions were: Gln160–Gly167 (residue # 11–18) and Val159–Gly167 (residue # 10–18). The distance difference map generated using the COM of amino acid sidechains also provided several residue pairs that afforded a similar distance change. When the two distance maps were compared there were certain residue pairs like Val157–Arg161 (residue # 8–12) which showed distance changes greater than 6 Å in the sidechain COM distance difference map but a corresponding distance change of only 3 Å in the C_α distance map. Such residue pairs were not considered suitable for the placement of donor–acceptor fluorophores since the sidechain COM distances were more susceptible to change upon incorporation of donor and acceptor fluorophores and the C_α distance changes during the coil–helix transition were modest. In contrast, Gln160–Gly167 (11–18 residue pair) exhibited the maximum C_α distance change of 5 Å and

 This journal is © The Royal Society of Chemistry 2018

simultaneously >5 Å distance change in the sidechain COM distance. The residue pair Val159–Gly167 also exhibited similar distance changes in both maps. Placement of appropriate donor–acceptor fluorophore pairs at either of these residue pair positions would afford FRET response provided the inter-residue distances (r_{ij}) in the extended form and the folded form are within 1.5 to 0.5 times that of the Förster radius (R_0) for the corresponding donor–acceptor pair. The FRET efficiency from the donor to the acceptor fluorophore is 50% at a separation of R_0. We estimate r_{ij} from the distance distribution for individual residue pairs from our peptide simulations in coil and helix forms. Fig. 2d contrasts the C_α distance distributions for the Gln160–Gly167 pair. While, the C_α distance distribution for Gln160–Gly167 in the helical conformation was narrow (9.75–12.5 Å) and peaked at 11 Å, the more flexible coil conformations showed an extremely broad (8.75–23 Å) distance distribution profile with an average Gln160–Gly167 C_α distance separation of 16 Å. While a small population of the peptide coil conformations had an overlapping distance distribution (9.75–12.5 Å) with that of the helical form, the population of helical conformers in the overlapping distance range was significantly larger (4901 helical conformers *versus* 775 coiled conformers).

In addition to the backbone separations within the peptide, relative orientation and mobility of sidechains is another factor that might affect the FRET donor–acceptor separation and dynamics. Hence, we monitored the relative orientations of the Gln160 and Gly167 residues in both the helix and coil states (Fig. 3) over the same 10 ns time intervals for which the backbone distance analysis (Fig. 1 and 2) was carried out. Since Gly lacks a side-chain, we defined the relative orientation of Gln160 and Gly167 to be the angle between vectors normal to the plane containing the C_α and backbone carbonyl (C and O) atoms for these residues (Fig. 3a). Interestingly, the relative orientations of the residue pair were random for the coil state, but in a major population of the helical structures the relative orientation of the residue pair peaked around 30° (Fig. 3b and c). This indicated that the helix state would have an inherently favorable sidechain orientation for the fluorophores that might further contribute to increased FRET efficiency in the helical form over the coil form.

Since the helical conformation was predicted to form when the Gel 150–169 peptide would bind to PI(4,5)P2 present in a membrane it was necessary to check if the peptide would adopt a helical conformation in the presence of a lipid bilayer. Further if the peptide adopted a helical conformation would the distance and angle predictions still remain valid? To address this question we performed a simulation with the Gel 150–169 peptide in the presence of a fully solvated lipid bilayer containing PI(4,5)P2 (Fig. 4a). In the first step of the simulations, a lipid bilayer model composed of total 270 molecules of POPC (1-palmitoyl-2-oleoyl-*sn*-glycero-3-phosphocholine) and a single PI(4,5)P2 molecule was equilibrated in a model water box containing 120 mM NaCl. A pre-equilibrated helix form of the Gel 150–169 residue was placed within 6 Å of the PI(4,5)P2 headgroup in the next step and the whole system was further equilibrated and simulated for 100 ns. In the simulations, the C-terminal arginine residues of the peptide were found to form stable salt bridge interactions with the phosphates of the PI(4,5)P2 headgroup (Fig. 4b). Further, the Gel 150–169 peptide remained in the helical conformation throughout the 100 ns trajectory (Fig. S1†) and the distance and angle distributions mirrored those obtained from the simulations performed on

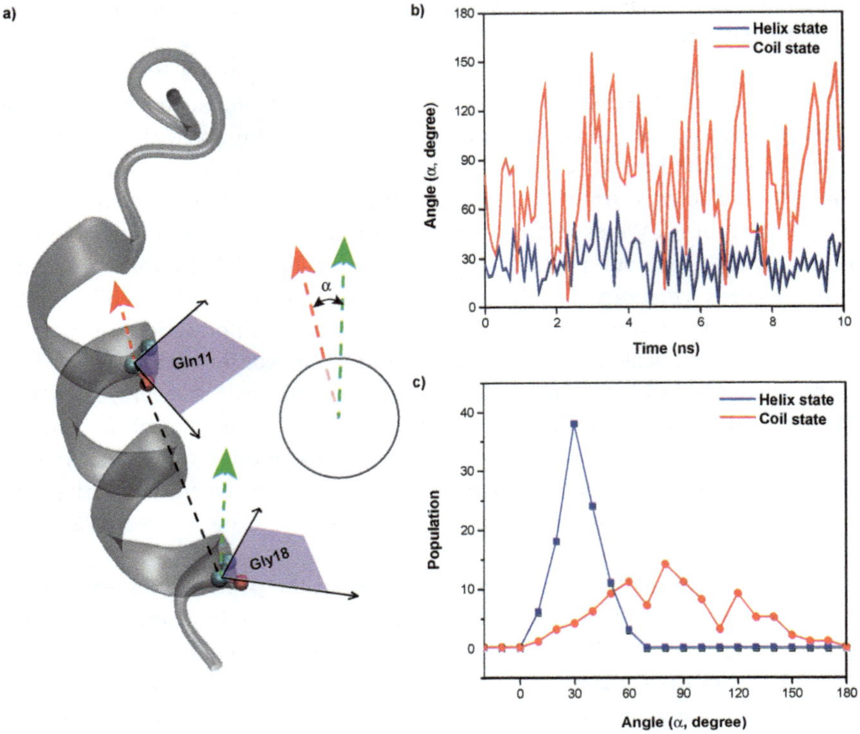

Fig. 3 Relative orientation of Gln160 and Gly167 residues in coil state and helix state. (a) A reference plane containing three backbone atoms (C_α, C and O) has been considered within each of these two residues. The angle (α) between the two normal vectors to these planes (red and green dotted arrows) represents relative orientation of the two residues within the peptide chain. (b) Variation of the angle α with time in the helix state (0–10 ns, 300 K simulation) and coil state (50–60 ns, 500 K simulation). (c) Distribution of the angle α in the helix state and the coil state.

the helical state in solution (Fig. 4c). Thus, MD simulations of the peptide performed both in solution and in the presence of a lipid bilayer indicated that the attachment of a suitable FRET pair based on residue pair distance and orientation information would afford a sensitive optical strategy for monitoring coil–helix transitions in the Gel 150–169 peptide. We thus used the computational predictions of inter-residue distance and sidechain orientation changes to develop a distance sensitive FRET assay for monitoring the coil–helix transition in Gel 150–169.

Choice of donor and acceptor fluorophores

The next step was to identify appropriate donor and acceptor fluorophores based on the computed inter-residue distances in the coiled and helical forms. For both Gln160–Gly167 and Val159–Gly167 residue pairs identified from the distance maps (Fig. 2), the average distances in the coil form were less than 20 Å. Tryptophan and 5-((((2-iodoacetyl)amino)-ethyl)amino)naphthalene-1-sulfonic acid (IAEDANS) constitute a donor–acceptor fluorophore pair with an R_0 value of 22 Å.[27] Since, most other known FRET donor–acceptor pairs have larger R_0 values, we

 This journal is © The Royal Society of Chemistry 2018

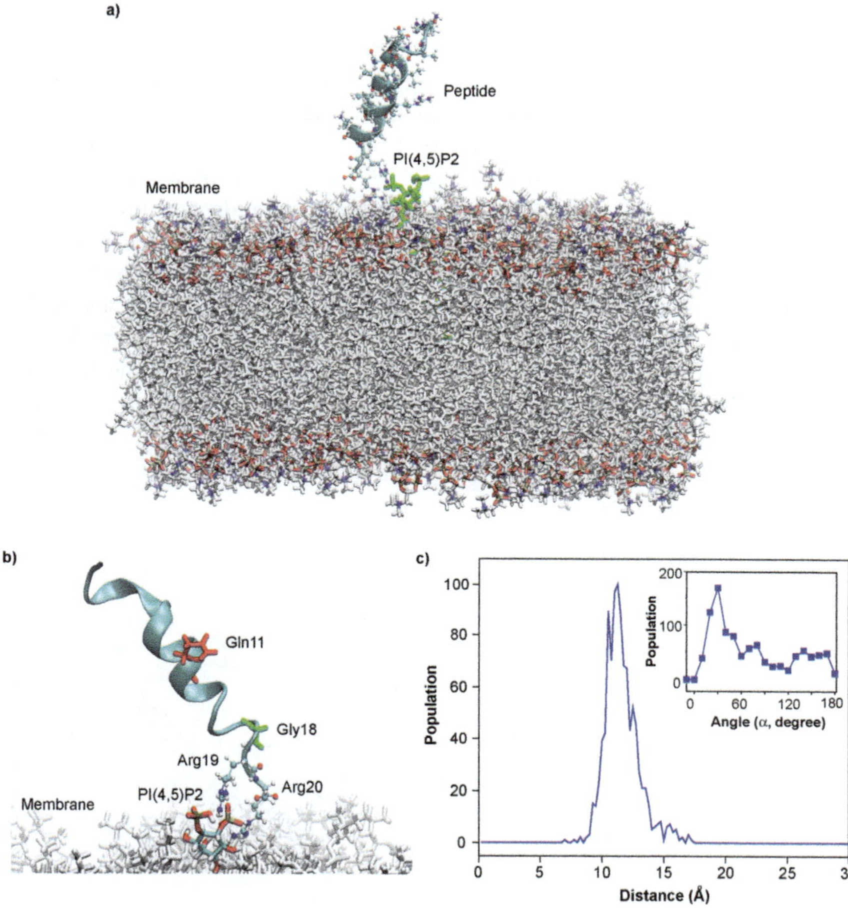

Fig. 4 MD simulations on the helical form of the Gel 150–169 peptide in the presence of a PI(4,5)P2 containing membrane. The peptide conformation has been monitored throughout a 100 ns trajectory. (a) Gel 150–169 peptide bound to a single PI(4,5)P2 molecule in a lipid bilayer composed of POPC. (b) Enlarged image to show the interaction of the peptide with the PI(4,5)P2 headgroup in the lipid bilayer. The C-terminus of the peptide with two arginine residues (Arg19 and Arg20) is observed to form electrostatic interactions with the PI(4,5)P2 phosphate groups. (c) A histogram representing distribution of distances between Gln11 (red) and Gly18 (green) from the 100 ns MD trajectory. Inset, relative orientation between Gln11 and Gly18 has been represented by the distribution of the angle α defined in Fig. 3a.

decided to use the tryptophan–IAEDANS pair as our distance sensing fluorophore pair.[26,31] Hence, for a tryptophan–IAEDANS donor–acceptor labeled Gel 150–169 peptide, the energy transfer efficiency would be greater than 50% in the coil state ($r_{ij} \sim 20$ Å) and would increase further in the helical state.

The Gln160–Gly167 (residue # 11–18) pair was the most suitable position for placement of the fluorophores when the distance changes in both distance maps were considered. The Gln160–Gly167 C_α average distance varies from 16 Å in the coil state to 11 Å in the helix state as shown in Fig. 2d. Therefore, tryptophan was incorporated in the place of Gln160 and Gly167 was replaced with cysteine to generate Gel-Q11W-G18C. Further, the cysteine residue was conjugated to IAEDANS to generate Gel-Q11W-G18C-IAEDANS (Fig. 5a).

Both the synthesized peptides had an amide group at the C-terminus. The MD simulations reported in the previous sections correspond to a C-terminus $-COO^-$ peptide derived from the NMR structure of Gel 150–169. However, we also performed MD simulations by modifying the $-COO^-$ terminus peptide with an amide terminus. The C_α distance difference map for the amide terminus peptide (Fig. S2†) indicated that the maximum distance changes were between either i^{th} and $(i+4)^{th}$ or i^{th} and $(i+7)^{th}$ pairs. Since the ith and $(i+4)^{th}$ residue pairs would be within less than 10 Å distances in both the coil and helical forms of the peptide, the ith and $(i+7)^{th}$ residue pairs would be more appropriate for the placement of the donor and acceptor fluorophores. Once again the Gln160–Gly167 residue pair afforded the maximal distance change amongst all i^{th} and $(i +$

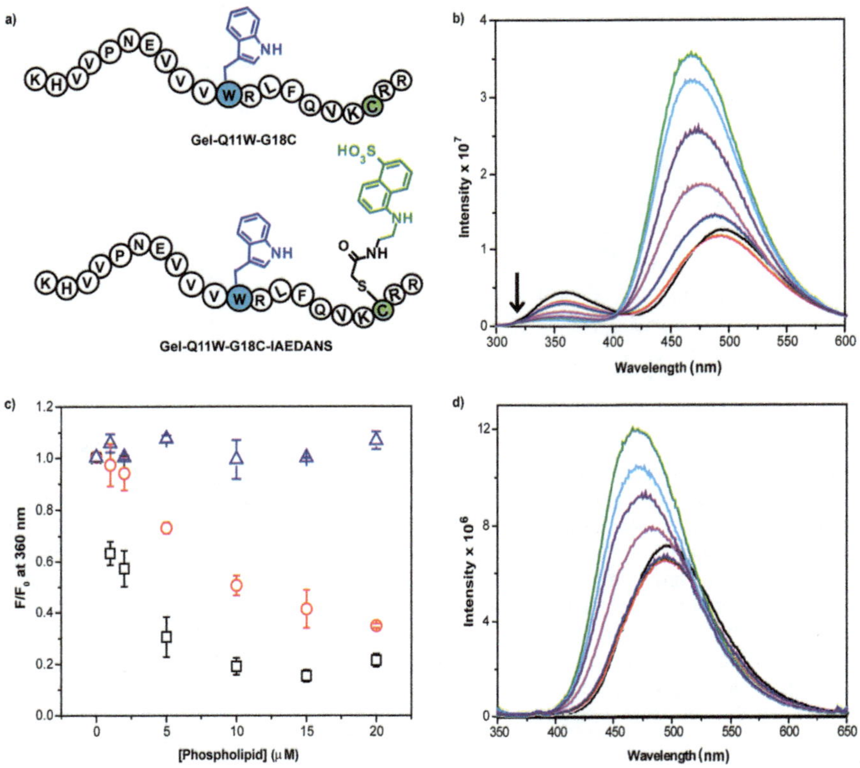

Fig. 5 (a) Peptide sequence for Gel-Q11W-G18C highlighting positions of donor tryptophan and acceptor IAEDANS. (b) Steady-state fluorescence response of the Gel-Q11W-G18C-IAEDANS (10 μM) upon addition of PI(4,5)P2 micelle solutions of concentrations: 0 μM (black), 1 μM (red), 2 μM (blue), 5 μM (pink), 10 μM (green), 15 μM (cyan), and 20 μM (purple). λ_{ex} 287 nm. (c) Relative fluorescence intensity (F/F_0) of tryptophan at 360 nm (λ_{ex} 287 nm) in Gel-Q11W-G18C-IAEDANS with varying concentrations (0–20 μM) of PC vesicles (blue triangles), 50% PI(4,5)P2–PC vesicles (red circles), and PI(4,5)P2 micelles (black squares). (d) Emission spectra of IAEDANS in Gel-Q11W-G18C-IAEDANS (10 μM) upon direct excitation at 340 nm in the presence of PI(4,5)P2 micelles. PI(4,5)P2 concentration: 0 μM (black), 1 μM (red), 2 μM (blue), 5 μM (pink), 10 μM (green), 15 μM (cyan), and 20 μM (purple).

 This journal is © The Royal Society of Chemistry 2018

7)th residue pairs further validating our choice of positions for attachment of donor and acceptor fluorophores (Fig. S2†).

Steady-state fluorescence response of the Gel 150–169 peptide with PI(4,5)P2

Previously reported CD experiments showed that the Gel 150–169 peptide sequence undergoes a coil-to-helix transition in the presence of high concentrations of PI(4,5)P2 micelles.[15] In order to test if the tryptophan–IAEDANS donor–acceptor pair, incorporated at the 11th and 18th amino acid positions respectively, could capture the distance changes in our labeled peptide, we first analyzed the steady-state fluorescence spectra of Gel-Q11W-G18C-IAEDANS in the presence of increasing concentrations of PI(4,5)P2 micelles (Fig. 5b). Upon excitation of the donor tryptophan at 287 nm, in the absence of PI(4,5)P2 micelles, characteristic emission bands from both tryptophan and IAEDANS were observed as shown in Fig. 5b. The donor excitation was performed at 287 nm since the IAEDANS absorption spectrum shows a minimum around 274–288 nm (Fig. S3†). The spectrum indicates that tryptophan and IAEDANS are already within the energy transfer distance in the absence of PI(4,5)P2. Upon addition of increasing concentrations of PI(4,5)P2 micelles (1–15 μM) to the labeled peptide the emission intensity of tryptophan monotonically decreased indicating that energy transfer from the donor was increasing with PI(4,5)P2 addition. The intensity decrease levelled at 15 μM concentrations of PI(4,5)P2 indicating binding saturation with the Gel 150–169 peptide. This result indicated that the distance between the tryptophan donor and IAEDANS acceptor decreased in the presence of PI(4,5)P2 micelles. CD experiments were also performed to monitor the coil to helix transition (Fig. S4†). The data indicated that the transitions occurred at sub-stoichiometric concentrations of PI(4,5)P2 and hence the increase in energy transfer between the donor and acceptor at sub-stoichiometric concentrations of PI(4,5)P2 micelles was indicative of structural changes occurring during the coil to helix transition.

A blue-shift and subsequent increase in the IAEDANS acceptor emission was observed upon addition of up to 1 equivalent of PI(4,5)P2 micelles with respect to the Gel-Q11W-G18C-IAEDANS peptide (Fig. 5b). However upon further increasing the concentration of micelles, the emission intensity of IAEDANS decreased. We hypothesized that the emission response of IAEDANS with increasing PI(4,5)P2 concentrations indicates a polarity-sensitive behavior as a result of the proximity of the fluorescent probe to micelles when the peptide binds to PI(4,5)P2.[32,33] To validate our hypothesis we directly excited the labeled peptide at 340 nm. A similar emission response, initial intensity increase followed by a decrease accompanied by a blue shift of the peak, was observed with increasing PI(4,5)P2 micelle concentrations as shown in Fig. 5d indicating that IAEDANS emission features were affected by the proximity of the fluorophore to lipid micelles.[23,32,33] Comparing the normalized emission intensities for the acceptor at 287 nm excitation *versus* that at 340 nm excitation, we observed an increased emission for excitation at 287 nm for all PI(4,5)P2 concentrations relative to that for excitation at 340 nm at the same concentrations indicating energy transfer from the tryptophan donor (Fig. S5†). Hence, the IAEDANS acceptor emission response is a result of both, intrinsic polarity sensitivity of the acceptor and FRET from the donor.

Next, we performed steady-state fluorescence measurements in the presence of mixed vesicles (130–140 nm diameter) containing PI(4,5)P2 and PC. A steady decrease in tryptophan donor emission in Gel-Q11W-G18C-IAEDANS was observed in the case of 50% PI(4,5)P2–PC mixed vesicles in the concentration range of 1–10 μM PI(4,5)P2 (Fig. 5c). When we further reduced the mol% of PI(4,5)P2 to 20% we also observed a decrease in tryptophan emission (Fig. S6†). The decrease in tryptophan emission was maximum for 100% PI(4,5)P2 micelles, followed by 50% PI(4,5)P2 vesicles and then 20% PI(4,5)P2 vesicles (Fig. S6†). Experiments were also performed with 100% PC as a control since previous reports indicated that PC does not bind to the Gel 150–169 peptide.[34] The PC vesicles did not affect either tryptophan or IAEDANS emission indicating that the energy transfer between the labels was selective to PI(4,5)P2 over PC (Fig. 5c and S7†). These data show that the Gel-Q11W-G18C-IAEDANS labeled peptide can indeed report on distance dependent changes in conformations of the unbound and PI(4,5)P2 bound Gel 150–169 peptide. To further confirm that the decrease in tryptophan emission was due to energy transfer to the IAEDANS acceptor, time resolved fluorescence measurements were performed.

Fluorescence lifetime measurements of labeled Gel 150–169 with PI(4,5)P2

Gel-Q11W-G18C with only the donor tryptophan fluorophore was used as a control peptide for the lifetime measurements. The fluorescence lifetimes for the Gel-Q11W-G18C peptide were measured by exciting the fluorophore at 287 nm. The average lifetime (τ_m) for the tryptophan fluorescence was obtained *via* a three exponential discrete analysis. The tryptophan fluorescence showed a τ_m of 2.56 $\pm$ 0.02 ns (Table 1) in the absence of phospholipids and did not change significantly upon addition of either PI(4,5)P2 micelles or PC vesicles (Fig. S8,† Table 1). This result clearly indicated that binding of the control peptide to PI(4,5)P2 did not affect the tryptophan lifetime. Hence, the tryptophan lifetime data was used to study the FRET response of the donor–acceptor labeled Gel-Q11W-G18C-IAEDANS peptide as described in the following paragraphs.

Next we measured the fluorescence lifetime of the tryptophan residue in Gel-Q11W-G18C-IAEDANS peptide. The fluorescence intensity decay curve for the donor–acceptor labeled peptide indicated a faster decay in the presence of PI(4,5)P2 micelles when compared to the decay in the absence of phospholipids (Fig. 6a). The average lifetime for the tryptophan fluorescence was obtained *via* either a three exponential or a four exponential discrete analysis. The tryptophan lifetime in the Gel-Q11W-G18C-IAEDANS peptide in absence of any phospholipids was 1.13 $\pm$ 0.10 ns (Table 1) with a corresponding 56% energy transfer efficiency (Fig. 6c). With increasing concentration of PI(4,5)P2 micelles (1–15 μM) the tryptophan fluorescence lifetime in the Gel-Q11W-G18C-IAEDANS peptide showed a steady decrease with a concomitant increase in the energy transfer efficiency (Fig. 6b and c). This data indicated a reduction in the distance between the donor and acceptor fluorophores, indicating that the coil to helix transition was taking place in the presence of PI(4,5)P2 micelles. In the presence of 20 μM PI(4,5)P2 micelles we observed a slight increase in tryptophan lifetime for the Gel-Q11W-G18C-IAEDANS peptide. This data might indicate a slight opening-up of the helical structure at higher PI(4,5)P2 concentrations.

 This journal is © The Royal Society of Chemistry 2018

Table 1 Time resolved fluorescence data for Gel-Q11W-G18C and Gel-Q11W-G18C-IAEDANS peptides in the presence and absence of phospholipids

Peptide (10 μM)	Phospholipid (10 μM)	α_1 (%)	τ_1 (ns)	α_2 (%)	τ_2 (ns)	α_3 (%)	τ_3 (ns)	α_4 (%)	τ_4 (ns)	Mean lifetime (ns)
Gel-Q11W-G18C	No PL	30 ± 1	0.59 ± 0.04	32 ± 1	2.22 ± 0.03	38 ± 1	4.45 ± 0.02	n/a	n/a	2.56 ± 0.02
	PC	32 ± 5	0.53 ± 0.11	33 ± 1	1.92 ± 0.49	36 ± 3	4.24 ± 0.43	n/a	n/a	2.32 ± 0.09
	PI(4,5)P2	27 ± 4	0.69 ± 0.01	46 ± 1	2.18 ± 0.08	28 ± 2	4.4 ± 0.06	n/a	n/a	2.39 ± 0.18
Gel-Q11W-G18C-IAEDANS	No PL	52 ± 3	0.36 ± 0.10	38 ± 2	1.43 ± 0.20	10 ± 3	4.10 ± 0.60	n/a	n/a	1.13 ± 0.10
	PC	51 ± 4	0.34 ± 0.04	39 ± 2	1.27 ± 0.20	11 ± 3	3.47 ± 0.60	n/a	n/a	1.02 ± 0.10
	20% PI(4,5)P2–PC	40 ± 8	0.17 ± 0.08	36 ± 2	0.65 ± 0.21	22 ± 8	1.73 ± 0.43	3 ± 1	4.22 ± 1.14	0.74 ± 0.10
	50% PI(4,5)P2–PC	48 ± 10	0.10 ± 0.02	28 ± 4	0.47 ± 0.10	21 ± 3	1.51 ± 0.20	3 ± 1	4.46 ± 0.9	0.64 ± 0.10
	PI(4,5)P2	76 ± 3	0.07 ± 0.01	18 ± 3	0.29 ± 0.10	4 ± 1	1.47 ± 0.40	2 ± 1	5.79 ± 1.64	0.25 ± 0.03

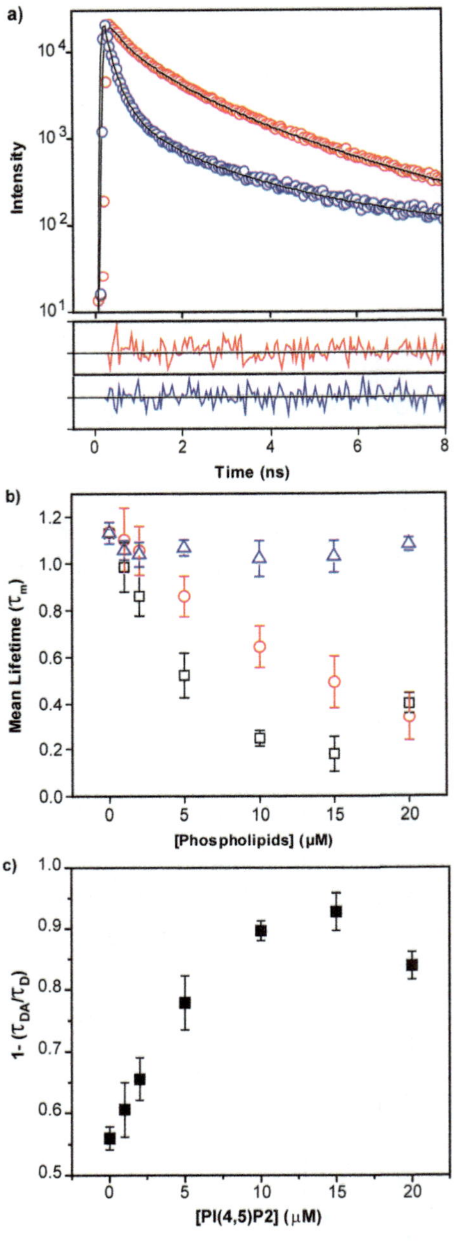

Fig. 6 Monitoring the conformation change upon coil-to-helix transition *via* time resolved fluorescence. (a) Experimental (scatter) and fitted (line) fluorescence intensity decay curves for tryptophan in Gel-Q11W-G18C-IAEDANS (10 μM) in the absence (red) and presence of PI(4,5)P2 micelles (10 μM, blue). Residuals from both the fits have been plotted at the bottom. (b) Mean lifetime (τ_m) values for tryptophan in Gel-Q11W-G18C-IAEDANS (10 μM), plotted against varying concentrations of PI(4,5)P2 micelles (black squares), 50% PI(4,5)P2–PC vesicles (red circles) and PC vesicles (blue triangles). (c) Energy transfer efficiencies ($1 - \tau_{DA}/\tau_D$) plotted against concentration of PI(4,5)P2 micelles, where τ_{DA} and τ_D are the mean lifetimes of tryptophan in Gel-Q11W-G18C-IAEDANS and Gel-Q11W-G18C respectively, at the corresponding PI(4,5)P2 concentrations.

The tryptophan fluorescence lifetime in the Gel-Q11W-G18C-IAEDANS peptide in the presence of PC vesicles was similar to that in the absence of any phospholipids indicating a PI(4,5)P2 sensitive peptide conformation change (Fig. 6b and Table 1). We therefore measured the lifetimes for both 50% PI(4,5)P2–PC (Fig. 6b and Table 1) and 20% PI(4,5)P2–PC (Fig. S9a† and Table 1) as a function of increasing vesicle concentration. The average tryptophan lifetime values showed a steady decrease with increasing concentrations of 50% PI(4,5)P2–PC vesicles. A similar trend was observed for 20% PI(4,5)P2–PC vesicles (Fig. S9a†). With increasing PI(4,5)P2 content the tryptophan lifetime progressively decreased from 20% PI(4,5)P2 vesicles to 50% PI(4,5)P2 vesicles followed by 100% PI(4,5)P2 micelles confirming a distance dependent energy transfer (Table 1 and Fig. S9b†). Importantly, the tryptophan lifetime in Gel-Q11W-G18C-IAEDANS decreased by around 20% in the presence of 2.5 μM PI(4,5)P2 (5 μM, 50% PI(4,5)P2–PC vesicles) from the lifetime value in the presence of PC vesicles. This data indicates that the probe can sense distance changes in the peptide arising from coil–helix transitions at physiologically relevant PI(4,5)P2 concentrations.

The results from the lifetime experiments indicate that energy-transfer occurs between the tryptophan and the IAEDANS fluorophores and that the energy transfer efficiency increases with increasing levels of PI(4,5)P2. Further, the labeled probe Gel-Q11W-G18C-IAEDANS is able to report on phosphoinositide induced coil–helix conformation changes in the Gel 150–169 peptide at physiologically relevant PI(4,5)P2 concentrations. Importantly, combined with the computational studies, the fluorescence steady state and lifetime experiments distinctly prove that coil–helix transitions in Gel 150–169 can occur not only in the presence of 100% PI(4,5)P2 micelles but also at 20 mol% and 50 mol% of PI(4,5)P2.

Conclusions

Janmey and co-workers had previously reported that the binding of gelsolin to actin can be disrupted at greater than 10 mol% PI(4,5)P2.[20] However, the phosphoinositide-binding induced coil–helix transition in the N-terminus segment of gelsolin had been reported only in the presence of large excess of 100% PI(4,5)P2 micelles.[15] Hence, there was no direct experimental link between the phosphoinositide induced coil–helix transition in gelsolin and the biological function of gelsolin *i.e.* dissociation from actin filaments leading to actin polymerization. We have developed a sensitive FRET based assay for monitoring the coil–helix transition in peptides and proteins. Using the assay we demonstrate that a coil–helix conformation change in the Gel 150–169 peptide can occur at low mol% and physiologically relevant concentrations of PI(4,5)P2. Hence, our data now provides a link between the phosphoinositide-induced coil to helix transition in gelsolin to its dissociation from actin.

Importantly, the systematic framework established by our study, which includes computational predictions of inter-residue distance and side-chain orientation changes during coil–helix transitions and the development of a sensitive fluorescent assay to investigate phospholipid levels that trigger protein conformational changes, can be used to probe other phospholipid–protein interactions. Phospholipid-induced coil–helix transitions are relevant not only to the entire class of actin-binding proteins that regulate cyto-skeletal rearrangements, but have also been reported for other important proteins, such as α-synuclein.[35–37] In the future, the development of sensitive fluorescence energy transfer based methods can be extended toward understanding protein–phospholipid interactions in the context of gaining insights into the mechanisms underlying phospholipid mediated signaling processes.

Experimental and computational methods

Molecular dynamics simulations of gelsolin peptide in solution

The peptide model was chosen from the helical form of Gel 150–169 peptide determined by NMR studies in a trifluoroethanol–water environment (PDB: 1SOL).[15] Two models of the peptide were considered: (1) with a carboxylate (COO^-) terminated C-terminus, and (2) with an amidated ($CONH_2$) C-terminus. The peptide was incorporated into a rectangular box of water molecules. The box size was maintained such that the distance between the peptide and its periodic image is greater than 28 Å in all directions. The solvent box was neutralized by adding the requisite number of Na^+ and Cl^- ions and total NaCl concentration was maintained to be 120 mM (100 mM NaCl and 20 mM HEPES was used in the experiments).

MD simulations were performed using the NAMD simulation package and CHARMM27 force field.[38] The system was subjected to three minimization steps: (1) keeping the non-hydrogen heavy atoms in the protein fixed; (2) with harmonic constraints of 25 kcal mol^{-1} $Å^{-2}$ on the peptide heavy atoms; (3) with harmonic constraints of 12 kcal mol^{-1} $Å^{-2}$. The minimization step was followed by heating the system from 0 K to 300 K in two steps: (1) steady increment of temperature in steps of 6 K per 1 ps from 0 K to 300 K; (2) equilibration at 300 K for 50 ps keeping the protein heavy atoms fixed; (3) repeating the heating and equilibrium steps keeping the heavy atoms under harmonic constraints of 25 kcal mol^{-1} $Å^{-2}$. The harmonic constraints were gradually released in four NPT (1 bar pressure and 300 K temperature) equilibration steps of 1 ns each. Following the equilibration step, NPT production runs were carried out either at 300 K or at 500 K for 100 ns timescales. All simulations employed periodic boundary conditions with the Particle Mesh Ewald (PME) algorithm describing long range electrostatics. The van der Waals forces were calculated with the use of a switching function that had 10 Å as the switching distance and 12 Å as the cutoff. MD trajectories were analyzed using VMD[39] and distance maps were generated using MATLAB software version 2015b.

Molecular dynamics simulations of gelsolin peptide on lipid membrane

Generation of the lipid bilayer. The lipid bilayer model was generated in VMD using the 'Membrane Builder' plug-in. We first created a 100 Å × 100 Å membrane bilayer composed of 270 POPC molecules and then inserted a single PI(4,5)P2 molecule into one leaflet of the POPC membrane. The coordinate files for 1-stearoyl-2-arachidonoyl-*sn*-glycero-3-phosphoinositol-4,5-bisphosphate with the 4-phosphate singly protonated (SAPI24) for PI(4,5)P2 molecule were downloaded from the "CHARMM-GUI Archive – Individual Lipid Molecule Library".[40] The resultant membrane model contained 1 phosphoinositide per 135 POPC molecules which corresponds to 0.74 mol% of PI(4,5)P2 on that leaflet. The model lipid bilayer was then solvated in a box of explicit water molecules followed by neutralization and ionization with Na^+ and Cl^- ions with a NaCl concentration of 120 mM.

Equilibration of the lipid bilayer. MD equilibration of the lipid membrane bilayer was carried out using the NAMD simulation package and CHARMM36 force field.[41,42] The system was subjected to two rounds of minimization and NPT (pressure set to 1 bar and temperature 300 K) equilibration steps: (1) 10 000 steps of minimization followed by 0.5 ns NPT equilibration at 300 K keeping every atom except the lipid tails fixed; (2) repeating the minimization and NPT equilibration steps keeping all the lipid non-hydrogen atoms fixed. Next, the whole system was subjected to a free 1 ns NPT equilibration run setting all atoms free. We then continued the NPT simulation for a further 10 ns to check the stability of the lipid bilayer.

Insertion of the peptide onto the lipid bilayer and simulation. A helical pre-equilibrated peptide model from our solution phase simulations (previous subsection) was aligned with the equilibrated membrane such that the C-terminus of the peptide faces the PI(4,5)P2 headgroup (Fig. 4a). The peptide–membrane system was re-solvated and ionized with 120 mM NaCl. The whole system was subjected to two minimization steps: (1) keeping the non-hydrogen atoms of the peptide and lipid molecules fixed, and (2) applying harmonic constrains of 50 kcal mol^{-1} $Å^{-2}$ on the peptide non-hydrogen atoms. The minimization step was followed by a heating step to 300 K and equilibration at the same temperature for 1 ns keeping the peptide heavy atoms constrained. The harmonic constraints were gradually released in five NPT (1 bar pressure and 300 K temperature) equilibration steps of 1 ns each. After the equilibration, NPT production runs for the gelsolin peptide bound to the lipid membrane were carried out for 100 ns.

The simulations of peptide on membrane were carried out using the NAMD simulation package and CHARMM36 force field.[41,42] All simulations employed periodic boundary conditions with the Particle Mesh Ewald (PME) algorithm describing long range electrostatic interactions. The van der Waals forces were calculated by using a switching function that had 10 Å as the switching distance and 12 Å as the cutoff.

Peptide synthesis and purification

Gel 150–169 peptide was synthesized *via* a solid state peptide synthesis strategy on Fmoc protected Rink amide MBHA resin LL (100–200 mesh, loading 0.38 mmol g^{-1} resin) in an automated peptide synthesizer (PS3, Protein Technologies Inc., USA). IAEDANS was obtained from Molecular Probes (Thermo Fisher Scientific Inc.). For the labeling reaction, peptide (0.67 μmol) was dissolved in 400 μL of aqueous buffer (Tris–HCl, pH 8.2). Tris(2-carboxyethyl)phosphine hydrochloride (TCEP, 4 equivalents) was added to the solution and the solution was kept at room temperature for 15 min to avoid cysteine oxidation. IAEDANS (5 equivalents) was dissolved in DMSO (50 μL) and added to the peptide solution in several aliquots. The reaction mixture was protected from light and left on a Rotospin at room temperature for 5 h. Peptide labeling was checked by MALDI-TOF-MS (Bruker UltrafleXtreme, Bruker Daltonics). A small fraction of the reaction mixture was desalted by using Ziptip (Thermo Fisher Scientific Inc.) and spotted along with the matrix (α-cyano-4-hydroxycinnamic acid) and MALDI mass spectra were recorded.

The labeled peptide was purified on a Shimadzu Prominence UFLC system (Shimadzu Corp.) equipped with a Phenomenex Luna® C18 (250 × 10 mm) semi-preparative column. Acetonitrile with 0.1% trifluoroacetic acid (solvent B) and water with 0.1% trifluoroacetic acid (solvent A) were used as the mobile phase and absorption at 220 nm, 280 nm, and 340 nm were monitored for product elution. The following gradient was used for the separation: 0–5 min: 5% solvent B; 5–65 min: 5–100% solvent B; 70 min: 100% solvent B; 75 min: 5% solvent B; 80 min: 5% solvent B in solvent A. Purity of the fractions were determined by LC-ESI MS and pure fractions were evaporated under vacuum.

Preparation of PI(4,5)P2 micelles

PI(4,5)P2 (0.2 mg) was suspended in aqueous buffer (400 μL, 20 mM HEPES, 100 mM NaCl, pH 7.4) the suspension was sonicated for 15 min at room temperature to form micelles. Micelle solutions were stored at −80 °C by flash freezing with liquid N_2. Stored micelles were thawed and sonicated for 15 min before use.

Preparation of small unilamellar vesicles (SUVs)

PI(4,5)P2 (L-α-phosphatidylinositol-4,5-bisphosphate, ammonium salt (Brain, Porcine)) and PC (1-palmitoyl-2-oleoyl-*sn*-glycero-3-phosphocholine) were purchased from Avanti Polar Lipids. The lipids were taken in glass vials in the desired molar ratio and mixed either in chloroform or in a mixture of chloroform : methanol : water (20 : 9 : 1). After mixing, lipid films were formed by evaporating the solvents from the vials under N_2 gas and the films were kept in vacuum overnight for drying. The dried films were hydrated in buffer (20 mM Na-HEPES, 100 mM NaCl, pH 7.4), followed by 5 cycles of freeze (liquid N_2) and thaw (37 °C). SUVs were prepared by passing the aqueous suspensions through polycarbonate membranes of 100 nm pore diameter (Whatman, GE Healthcare) at 25 °C. The formation of the vesicles of diameter around 130–140 nm (Fig. S10†) was checked using Dynamic Light Scattering (DynaPro, S9 Protein Solutions, Wyatt Technology Corp., USA). The concentrations of phospholipids in the SUV suspensions were determined by phosphate estimation following previously established protocol.[43]

Steady-state fluorescence studies

For fluorescence studies, the IAEDANS labeled peptide, Gel-Q11W-G18C-IAEDANS, was dissolved in water and concentrations of the stock solutions were measured in a NanoDrop 1000 spectrophotometer (Thermo Fisher Scientific, Inc.) by monitoring the IAEDANS absorbance at 337 nm (ε for IAEDANS at 337 nm is 4500 M^{-1} cm^{-1}). Constant volume fluorescence experiments were performed using 10 μM of the sensor and increasing concentrations (from 0 μM to 20 μM) of micelles or SUVs of lipids in aqueous buffer (20 mM HEPES, 100 mM NaCl, pH 7.4) at 25 °C with final volume 200 μL in each case. For measurements, the mixtures were taken in a quartz cuvette (10 mm × 2 mm). Fluorescence spectra were recorded in a FluoroLog-3 spectrofluorimeter (Horiba Jobin Yvon, Inc.). For each sample, spectra were recorded at 287 nm and 340 nm excitation wavelengths separately. Fluorescence emission was monitored from either 300 nm to 650 nm or 350 nm to 650 nm for 280 nm and 340 nm excitations respectively. A glass plate was used as a UV filter, to block scattered excitation light from reaching to the emission detector. Fluorescence spectra were plotted and analysed in Origin 9.1 software. Error bars for fluorescence intensity plots represent standard deviations from at least two measurements.

Time-resolved fluorescence studies

Time-resolved fluorescence intensity decay curves were measured using a time correlated single photon counting setup.[44] For fluorescence lifetime measurements, the solution was excited at 287 nm by using a dye-laser pulse at 575 nm which was frequency-doubled to 287 nm. Emission was collected (at 90° to the excitation beam) at 360 nm using a 325 nm low pass filter. The instrument response function (IRF) was taken at 287 nm and was found to be 90 ± 10 ps (FWHM). For all the fluorescence lifetime measurements, peak counts of 20 000 were collected with the emission polarizer oriented at the magic angle (54.7°) with respect to the excitation polarizer. The decay was de-convoluted with respect to IRF and analyzed using either a three exponential or four exponential model (eqn (1)).

$$I(t) = \sum \alpha_i \exp(-t/\tau_i) \qquad i = 1\text{–}4 \tag{1}$$

$I(t)$ is the fluorescence intensity at time t and α_i is the amplitude of the ith lifetime τ_i such that $\sum \alpha_i = 1$. The goodness of fit was determined by monitoring the χ^2 values and residuals.

Average lifetime (τ_m) values were calculated from the equation:

$$\tau_\text{m} = \sum \alpha_i \tau_i \tag{2}$$

All lifetime data for Gel-Q11W-G18C-IAEDANS were recorded in at least triplicates. Lifetime data for Gel-Q11W-G18C were recorded in at least duplicates. Error bars represent standard deviations.

Conflicts of interest

There are no conflicts of interest to declare.

Acknowledgements

AD and RV acknowledge TIFR intra-mural funding. We acknowledge the MALDI facility TIFR, Mumbai, India for mass analysis and Mr Srinath Ranya for generating the Python scripts for computing the distance matrices.

References

1 T. Balla, Phosphoinositides: tiny lipids with giant impact on cell regulation, *Physiol. Rev.*, 2013, **93**, 1019–1137.

2 A. Shevchenko and K. Simons, Lipidomics: coming to grips with lipid diversity, *Nat. Rev. Mol. Cell Biol.*, 2010, **11**, 593–598.

3 C. Schultz, Challenges in studying phospholipid signaling, *Nat. Chem. Biol.*, 2010, **6**, 473–475.

4 T. Balla, Z. Szentpetery and Y. J. Kim, Phosphoinositide Signaling: New Tools and Insights, *Physiology*, 2009, **24**, 231–244.

5 G. Di Paolo and P. De Camilli, Phosphoinositides in cell regulation and membrane dynamics, *Nature*, 2006, **443**, 651–657.

6 M. A. Lemmon, Membrane recognition by phospholipid-binding domains, *Nat. Rev. Mol. Cell Biol.*, 2008, **9**, 99–111.

7 P. A. Janmey and U. Lindberg, Cytoskeletal regulation: rich in lipids, *Nat. Rev. Mol. Cell Biol.*, 2004, **5**, 658–666.

8 S. Nag, M. Larsson, R. C. Robinson and L. D. Burtnick, Gelsolin: the tail of a molecular gymnast, *Cytoskeleton*, 2013, **70**, 360–384.

9 P. D. Moens and L. A. Bagatolli, Profilin binding to sub-micellar concentrations of phosphatidylinositol (4,5) bisphosphate and phosphatidylinositol (3,4,5) trisphosphate, *Biochim. Biophys. Acta*, 2007, **1768**, 439–449.

10 N. Kumar, P. Zhao, A. Tomar, C. A. Galea and S. Khurana, Association of villin with phosphatidylinositol 4,5-bisphosphate regulates the actin cytoskeleton, *J. Biol. Chem.*, 2004, **279**, 3096–3110.

11 W. Xian and P. A. Janmey, Dissecting the gelsolin-polyphosphoinositide interaction and engineering of a polyphosphoinositide-sensitive gelsolin C-terminal half protein, *J. Mol. Biol.*, 2002, **322**, 755–771.

12 D. Raucher, T. Stauffer, W. Chen, K. Shen, S. L. Guo, J. D. York, M. P. Sheetz and T. Meyer, Phosphatidylinositol 4,5-bisphosphate functions as a second messenger that regulates cytoskeleton-plasma membrane adhesion, *Cell*, 2000, **100**, 221–228.

13 P. A. Janmey, W. Xian and L. A. Flanagan, Controlling cytoskeleton structure by phosphoinositide-protein interactions: phosphoinositide binding protein domains and effects of lipid packing, *Chem. Phys. Lipids*, 1999, **101**, 93–107.

14 P. A. Janmey, T. P. Stossel and P. G. Allen, Deconstructing gelsolin: identifying sites that mimic or alter binding to actin and phosphoinositides, *Chem. Biol.*, 1998, **5**, R81–R85.

15 W. J. Xian, R. Vegners, P. A. Janmey and W. H. Braunlin, Spectroscopic studies of a phosphoinositide-binding peptide from gelsolin: Behavior in solutions of mixed solvent and anionic micelles, *Biophys. J.*, 1995, **69**, 2695–2702.

16 V. Raghunathan, P. Mowery, M. Rozycki, U. Lindberg and C. Schutt, Structural changes in profilin accompany its binding to phosphatidylinositol 4,5-bisphosphate, *FEBS Lett.*, 1992, **297**, 46–50.

17 P. J. Goldschmidt-Clermont, L. M. Machesky, J. J. Baldassare and T. D. Pollard, The actin-binding protein profilin binds to PIP2 and inhibits its hydrolysis by phospholipase C, *Science*, 1990, **247**, 1575–1578.

18 E. K. Tuominen, J. M. Holopainen, J. Chen, G. D. Prestwich, P. R. Bachiller, P. K. Kinnunen and P. A. Janmey, Fluorescent phosphoinositide derivatives reveal specific binding of gelsolin and other actin regulatory proteins to mixed lipid bilayers, *Eur. J. Biochem.*, 1999, **263**, 85–92.

19 P. A. Janmey and T. P. Stossel, Modulation of gelsolin function by phosphatidylinositol 4,5-bisphosphate, *Nature*, 1987, **325**, 362–364.

20 P. A. Janmey, K. Iida, H. L. Yin and T. P. Stossel, Polyphosphoinositide micelles and polyphosphoinositide-containing vesicles dissociate endogenous

 This journal is © The Royal Society of Chemistry 2018

gelsolin-actin complexes and promote actin assembly from the fast-growing end of actin filaments blocked by gelsolin, *J. Biol. Chem.*, 1987, **262**, 12228–12236.

21 C. Nasuhoglu, S. Feng, J. Mao, M. Yamamoto, H. L. Yin, S. Earnest, B. Barylko, J. P. Albanesi and D. W. Hilgemann, Nonradioactive analysis of phosphatidylinositides and other anionic phospholipids by anion-exchange high-performance liquid chromatography with suppressed conductivity detection, *Anal. Biochem.*, 2002, **301**, 243–254.

22 G. Cicchetti, M. Biernacki, J. Farquharson and P. G. Allen, A ratiometric expressible FRET sensor for phosphoinositides displays a signal change in highly dynamic membrane structures in fibroblasts, *Biochemistry*, 2004, **43**, 1939–1949.

23 P. Hammarstrom, R. Owenius, L. G. Martensson, U. Carlsson and M. Lindgren, High-resolution probing of local conformational changes in proteins by the use of multiple labeling: unfolding and self-assembly of human carbonic anhydrase II monitored by spin, fluorescent, and chemical reactivity probes, *Biophys. J.*, 2001, **80**, 2867–2885.

24 D. S. Talaga, W. L. Lau, H. Roder, J. Tang, Y. Jia, W. F. DeGrado and R. M. Hochstrasser, Dynamics and folding of single two-stranded coiled-coil peptides studied by fluorescent energy transfer confocal microscopy, *Proc. Natl. Acad. Sci. U. S. A.*, 2000, **97**, 13021–13026.

25 M. R. Eftink, The use of fluorescence methods to monitor unfolding transitions in proteins, *Biophys. J.*, 1994, **66**, 482–501.

26 L. Stryer, Fluorescence energy transfer as a spectroscopic ruler, *Annu. Rev. Biochem.*, 1978, **47**, 819–846.

27 J. R. Lakowicz, *Principles of fluorescence spectroscopy*, Springer Science & Business Media, 2013.

28 P. L. Freddolino, C. B. Harrison, Y. Liu and K. Schulten, Challenges in protein folding simulations: Timescale, representation, and analysis, *Nat. Phys.*, 2010, **6**, 751–758.

29 R. B. Best, N.-V. Buchete and G. Hummer, Are current molecular dynamics force fields too helical?, *Biophys. J.*, 2008, **95**, L07–L09.

30 K. Lindorff-Larsen, P. Maragakis, S. Piana, M. P. Eastwood, R. O. Dror and D. E. Shaw, Systematic validation of protein force fields against experimental data, *PLoS One*, 2012, **7**, e32131.

31 J. W. Taraska, M. C. Puljung and W. N. Zagotta, Short-distance probes for protein backbone structure based on energy transfer between bimane and transition metal ions, *Proc. Natl. Acad. Sci. U. S. A.*, 2009, **106**, 16227–16232.

32 T. Hiratsuka, ATP-induced opposite changes in the local environments around Cys(697) (SH2) and Cys(707) (SH1) of the myosin motor domain revealed by the prodan fluorescence, *J. Biol. Chem.*, 1999, **274**, 29156–29163.

33 M. Palmer, P. Saweljew, I. Vulicevic, A. Valeva, M. Kehoe and S. Bhakdi, Membrane-penetrating domain of streptolysin O identified by cysteine scanning mutagenesis, *J. Biol. Chem.*, 1996, **271**, 26664–26667.

34 S. Mondal, A. Rakshit, S. Pal and A. Datta, Cell Permeable Ratiometric Fluorescent Sensors for Imaging Phosphoinositides, *ACS Chem. Biol.*, 2016, **11**, 1834–1843.

35 S. B. Lokappa, J. E. Suk, A. Balasubramanian, S. Samanta, A. J. Situ and T. S. Ulmer, Sequence and membrane determinants of the random coil–helix transition of alpha-synuclein, *J. Mol. Biol.*, 2014, **426**, 2130–2144.

36 T. Bartels, N. C. Kim, E. S. Luth and D. J. Selkoe, N-alpha-acetylation of alpha-synuclein increases its helical folding propensity, GM1 binding specificity and resistance to aggregation, *PLoS One*, 2014, **9**, e103727.

37 O. A. Andreev, A. G. Karabadzhak, D. Weerakkody, G. O. Andreev, D. M. Engelman and Y. K. Reshetnyak, pH (low) insertion peptide (pHLIP) inserts across a lipid bilayer as a helix and exits by a different path, *Proc. Natl. Acad. Sci. U. S. A.*, 2010, **107**, 4081–4086.

38 A. D. MacKerell Jr, D. Bashford, M. Bellott, R. L. Dunbrack Jr, J. D. Evanseck, M. J. Field, S. Fischer, J. Gao, H. Guo and S. Ha, All-atom empirical potential for molecular modeling and dynamics studies of proteins, *J. Phys. Chem. B*, 1998, **102**, 3586–3616.

39 W. Humphrey, A. Dalke and K. Schulten, VMD: visual molecular dynamics, *J. Mol. Graphics*, 1996, **14**, 33–38.

40 E. L. Wu, X. Cheng, S. Jo, H. Rui, K. C. Song, E. M. Dávila-Contreras, Y. Qi, J. Lee, V. Monje-Galvan and R. M. Venable, CHARMM-GUI Membrane Builder toward realistic biological membrane simulations, *J. Comput. Chem.*, 2014, **35**, 1997–2004.

41 J. C. Phillips, R. Braun, W. Wang, J. Gumbart, E. Tajkhorshid, E. Villa, C. Chipot, R. D. Skeel, L. Kale and K. Schulten, Scalable molecular dynamics with NAMD, *J. Comput. Chem.*, 2005, **26**, 1781–1802.

42 R. B. Best, X. Zhu, J. Shim, P. E. Lopes, J. Mittal, M. Feig and A. D. MacKerell Jr, Optimization of the additive CHARMM all-atom protein force field targeting improved sampling of the backbone ϕ, ψ and side-chain χ^1 and χ^2 dihedral angles, *J. Chem. Theory Comput.*, 2012, **8**, 3257–3273.

43 S. Gupta, S. Mondal, A. Mhamane and A. Datta, Smart "Lanthano" Proteins for Phospholipid Sensing, *Inorg. Chem.*, 2013, **52**, 12314–12316.

44 S. Narayan, M. H. Kombrabail, S. Das, H. Singh, K. V. R. Chary, B. J. Rao and G. Krishnamoorthy, Site-specific fluorescence dynamics in an RNA 'thermometer' reveals the role of ribosome binding in its temperature-sensitive switch function, *Nucleic Acids Res.*, 2015, **43**, 493–503.

 This journal is © The Royal Society of Chemistry 2018

Faraday Discussions

PAPER

Chiral self-assembly of fullerene clusters on CT-DNA templates†

Sandeepa Kulala Vittala[ab] and Joshy Joseph *[ab]

Received 18th September 2017, Accepted 5th October 2017
DOI: 10.1039/c7fd00196g

Herein we discuss the differential interaction of three monosubstituted fullerene derivatives possessing pyridinium, aniline or phenothiazine end groups (**F-Py**, **F-An** and **F-PTz**, respectively) with calf thymus DNA (CT-DNA), probed *via* spectroscopic and imaging techniques. The pyridinium derivative, **F-Py** becomes molecularly dissolved in 10% DMSO–PBS and interacts with CT-DNA *via* groove binding and electrostatic interactions, leading to the initial condensation of CT-DNA into micrometer sized aggregates and subsequent precipitation. On the other hand, the aniline derivative **F-An**, which is reported to form nanoclusters of 3–5 nm size, interacts with DNA through ordered, chiral assemblies on the CT-DNA template, thus perturbing the highly networked structure of CT-DNA to form nanonetworks, which eventually transform into condensed aggregates. The binding interactions between CT-DNA and **F-An** nanoclusters were established *via* UV-Vis, AFM and TEM analysis, and the chiral nature of the fullerene nanocluster assemblies on CT-DNA was confirmed by the presence of induced circular dichroism that was exhibited around the 250–370 nm region, corresponding to **F-An** nanocluster absorption. In contrast, the phenothiazine derivative **F-PTz**, which forms larger nanoclusters of ~70 nm size in 10% DMSO–PBS, exhibited only weak interactions with CT-DNA without affecting its network structure. These results demonstrate the role of the hydrophobic–hydrophilic balance in the design of DNA interacting fullerene derivatives by controlling their cluster size and interactions with CT-DNA, and are significant in applications such as DNA condensation, gene delivery and dimension controlled nanomaterial fabrication.

Introduction

DNA condensation involves the compaction of long, highly charged genomic DNA into compact structures, usually with the aid of suitable compaction agents, in

[a]*Photosciences and Photonics Section, CSIR-National Institute for Interdisciplinary Science and Technology, Thiruvananthapuram 695019, India. E-mail: joshyja@gmail.com; joshy@niist.res.in*

[b]*Academy of Scientific and Innovative Research (AcSIR), CSIR-NIIST Campus, Thiruvananthapuram 695019, Kerala, India*

† Electronic supplementary information (ESI) available: Experimental details and figures referenced throughout the text. See DOI: 10.1039/c7fd00196g

order to accommodate the genomic material inside the cell nuclei.[1–5] Several classes of compaction agents such as proteins, synthetic peptides, lipids, polyamines, linear/branched polycations, amphiphilic small molecules *etc.* are known to aid the DNA condensation process under *in vivo* and *in vitro* conditions, and the mechanistic aspects of DNA compaction are fairly well understood.[6–10] The synthetic design of novel multivalent polycations and amphiphilic molecules, which induce DNA condensation through multiple interactions with DNA *via* electrostatic, intercalative and/or groove binding, has attracted widespread attention due to the potential applications in gene transfection and DNA nanotechnology.[11] While synthetic peptides and multivalent polycations which mimic biological proteins and polyamines remain as the most studied classes of DNA compaction agents, carbon-based amphiphilic molecules such as fullerene derivatives have also attracted considerable attention due to their unique structure and binding with DNA through groove and π–π interactions.[12–17] For example, Nakamura and co-workers have demonstrated a series of fullerene derivatives for DNA condensation and subsequent applications including DNA photocleavage and fullerene-mediated *in vivo* gene delivery.[18]

Various DNA structures such as duplex, three-way and higher order cross-over junctions, DNA origami based structures *etc.* have been widely explored as templates and scaffolds for the assembly of nanomaterials *via* non-covalent and covalent strategies.[19–21] Similarly, well-defined nanostructures of compact DNA such as toroids, rods *etc.* have also been used as templates for the construction of nanomaterials with controlled shapes and compositions.[22–24] Typically, DNA condensation with the aid of compaction agents proceeds *via* initial non-specific, non-covalent interactions of the multivalent molecules with DNA, which may give rise to isolable intermediate morphologies (*e.g.* bead-on-a-string types of nanostructures)[25] that are also interesting templates for nanomaterial fabrication. In the case of water soluble fullerene-based amphiphilic molecules, the DNA compaction properties and the obtained morphologies strongly depend on the initial fullerene nanostructures formed under aqueous/buffer conditions, which in turn offer a level of structural control *via* fullerene functionalization.[26–28] Moreover, the controlled self-assembly of DNA and fullerene derivatives into regular DNA based nanostructures incorporating fullerene molecules or clusters holds great promise in the context of DNA nanotechnology.

With a view to understanding the structure–property relationships governing the interaction of fullerene derivatives with long genomic DNA and their subsequent applications in DNA condensation and the construction of ordered nanostructures, we have studied the interactions of three mono-substituted fullerene derivatives possessing pyridinium, aniline or phenothiazine end groups (Chart 1, **F-Py**, **F-An** and **F-PTz**, respectively) with calf thymus DNA (CT-DNA). As previously reported by our group,[29] these derivatives exhibit different solvation behaviour in a 10% DMSO–PBS solvent mixture (10 mM phosphate, 2 mM NaCl, pH = 7.4), with **F-Py** remaining molecularly dissolved, while **F-An** and **F-PTz** form nanoclusters with sizes of 3–5 nm and ~70 nm, respectively, reflecting the unique hydrophilic–hydrophobic balance in their molecular structures. We have also demonstrated the mutually assisted self-assembly of 3–5 nm sized nanoclusters of the aniline substituted fullerene derivative **F-An** (Chart 1) and short DNA duplexes (20 bp, ~6.4 nm duplex length) into micrometer long semiconducting nanowires.[29] Since we used shorter DNA strands for the previous study, the interactions

 This journal is © The Royal Society of Chemistry 2018

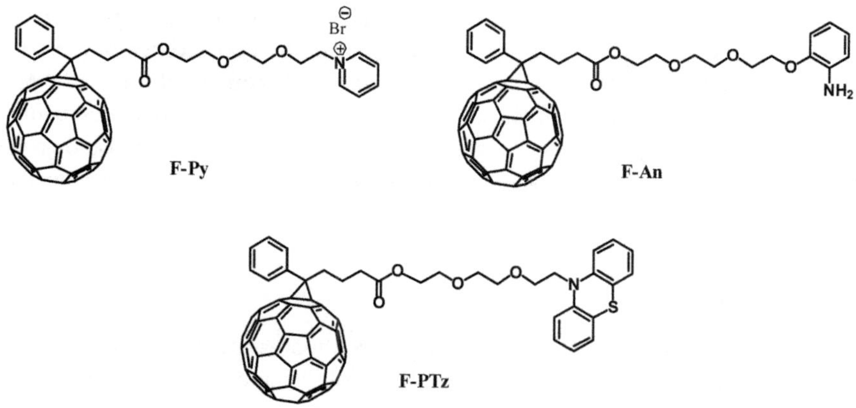

Chart 1 Chemical structures of the fullerene derivatives under study.

leading to the self-assembly invoked the inter-strand condensation of oligonucleotide duplexes to linear nanostructures.

Herein, we report the differential interactions of these derivatives with CT-DNA, probed *via* UV-Vis, fluorescence, circular dichroism (CD), atomic force microscopy (AFM), transmission electron microscopy (TEM) and dynamic light scattering (DLS) techniques. The molecularly dissolved pyridinium derivative **F-Py** interacts with CT-DNA *via* groove binding and electrostatic interactions, thus leading to the initial condensation of CT-DNA into micrometer sized aggregates and subsequent precipitation. In contrast, the phenothiazine derivative **F-PTz**, which forms larger nanoclusters with sizes of ~70 nm, exhibited only weak interactions with CT-DNA without affecting its network structure. Interestingly, the nanoclusters of **F-An** interact with DNA through ordered, chiral assemblies on the CT-DNA template, perturbing the highly networked structure of CT-DNA to form nanonetworks, which eventually transform into condensed aggregates.

Experimental section

Materials and methods

The detailed synthesis and characterization of the fullerene derivatives **F-Py**, **F-An** and **F-PTz** was reported previously.[29] Double stranded calf-thymus DNA (CT-DNA) and ethidium bromide (EB) were purchased from Sigma Aldrich. Spectroscopic grade (99.9%) DMSO solvent was purchased from Merck. All of the chemicals and reagents were used as received without further purification. Milli Q deionized water (18.2 MΩ cm) was used throughout the study.

All of the experiments were carried out in a 10% DMSO–PBS solvent mixture. The fullerene derivatives were initially dissolved in DMSO and then diluted with deionized water to obtain appropriate concentrations in the 10% DMSO–PBS solvent mixture. UV-Visible absorption spectra were recorded using a Shimadzu UV-2600 Spectrophotometer. For these studies, the concentration of fullerene derivative (3 μM) was kept constant and DNA solution (2.5 mM) was added in small aliquots. All optical measurements were carried out using 1 cm cuvettes at room temperature (25 ± 1 °C), unless otherwise mentioned. Fluorescence emission spectra were recorded using a Spex Fluorolog-3 (FL3-221) spectrofluorimeter. The wavelength of excitation was 515 nm. The fluorescence of the CT-DNA/EB complex (30 μM/30 μM) was measured by subsequent addition of **F-An** in small portions. Circular dichroism measurements were carried out using a Jasco J-810 spectropolarimeter possessing a Peltier controlled thermostatic cell holder, and the sensitivity, integration time and scan rate were chosen appropriately.

CT-DNA binding studies

A solution of calf thymus DNA was dissolved in Millipore water for 24 h in order to minimize any complexities arising from DNA flexibility and then filtered through a 0.45 μm syringe filter. The concentrations of the DNA solutions were determined by using the average value of 6600 M^{-1} cm^{-1} for the extinction coefficient of a single nucleotide at 260 nm.[30] A stock solution of 2.5 mM CT-DNA in 10 mM phosphate buffer containing 2 mM NaCl was used for all further studies. The intrinsic binding constant of **F-An** with CT-DNA was determined using the absorbance at the respective maxima (336 nm) recorded after each addition of CT-DNA. The intrinsic binding constant K_{DNA} was determined from the half reciprocal plot of $D/\Delta\varepsilon_{app}$, *vs.* D, using the equation

$$D/\Delta\varepsilon_{app} = D/\Delta\varepsilon + 1/(\Delta\varepsilon K_{DNA}) \quad (1)$$

where D is the base pair concentration of CT-DNA, $\Delta\varepsilon_{app} = [\varepsilon_a - \varepsilon_F]$ and $\Delta\varepsilon = [\varepsilon_b - \varepsilon_F]$.[31,32] The apparent extinction coefficient, ε_a, is obtained by calculating $A_{[obsd]}$/[**F-An**]. ε_b and ε_F correspond to the extinction coefficient of the bound form of **F-An** and the extinction coefficient of the free **F-An** derivatives, respectively. The data were fitted to eqn (1), with a slope equal to $1/\Delta\varepsilon_{app}$ and a y-intercept equal to $(1/K_{DNA})$. ε_b was determined from $\Delta\varepsilon$ and K_{DNA} was obtained from the ratio of the slope to the y-intercept.

AFM/TEM sample preparation

Samples for AFM analysis were prepared by drop-casting a 10 μL solution of a CT-DNA or CT-DNA/fullerene derivative (1 : 1) complex onto a freshly cleaved mica sheet under ambient conditions, followed by drying under air. A Bruker Multimode AFM operating with a tapping mode regime was used to record AFM images under ambient conditions. Micro fabricated TiN cantilever tips (NT-MDT-NSG series) with a resonance frequency of 299 kHz and a spring constant of 20 to 80 N m^{-1} were used. AFM section analysis was done offline. TEM samples were prepared by drop casting 10 μL of solution of the sample on the top of the carbon-coated copper grid (400 mesh). The samples were allowed to adsorb on the grid overnight under ambient conditions and Millipore water (2× 10 μL) was added to remove the buffer salts. TEM analysis was carried out using a JEOL 100 kV high-resolution transmission electron microscope. The accelerating voltage of the TEM was 100 kV and the beam current was 65 A. The samples were imaged using a Hamamatsu ORCA CCD camera. The formation of 3–5 nm sized **F-An** nanoclusters was confirmed through TEM analysis, as shown in Fig. S1.† DLS experiments were carried out in quartz cuvettes using the same solutions that were used

 This journal is © The Royal Society of Chemistry 2018

for the microscopic studies. The hydrodynamic diameters of the nanoclusters were determined using a Malvern Zeta Nano-ZS system.

Results and discussion

Fullerene–CT-DNA interaction studies

Initially, the interactions of the fullerene derivatives **F-Py**, **F-An** and **F-PTz** with CT-DNA were studied using UV-Visible absorption spectroscopy, and the results are summarized in Fig. 1. Upon the successive addition of CT-DNA to cationic **F-Py**, the absorption bands around 260 nm and 334 nm decreased consistently (Fig. 1a), thus indicating the initial aggregation and subsequent precipitation of the DNA–**F-Py** complex. The decrease in absorbance around 260 nm, where DNA bases have substantial absorption, even after successive addition of CT-DNA clearly indicated the **F-Py** assisted precipitation of the added CT-DNA. The amphiphilic **F-Py** derivative can interact with DNA through electrostatic interactions of the cationic, pyridinium group with the anionic DNA backbone and through groove interactions of the fullerene moiety.[13] The observed aggregation behaviour of the DNA–**F-Py** complex could be attributed to the synergistic effect of charge neutralization by the cationic pyridinium group and to the increased hydrophobicity due to the groove binding of hydrophobic fullerene moieties.

On the other hand, the sequential addition of CT-DNA to the **F-An** solution resulted in a regular increase in the absorption around 260 nm, corresponding to the absorption of the DNA bases, with a concomitant decrease in the **F-An** nanocluster absorption around 336 nm (Fig. 1b). The maximum hypochromicity observed around 336 nm was approximately 40%, with an isosbestic point at 296 nm (Fig. 1b; inset). The change in the absorption around 260 nm includes a contribution from the decreased absorption (hypochromicity) of the **F-An** nanoclusters and nucleobases due to **F-An**–DNA interactions, and an enhancement due to the increase in DNA concentration. These opposing effects result in an apparent, but diminished increase in the absorption around 260 nm, in comparison with a blank CT-DNA titration experiment without the **F-An** nanoclusters (Fig. S2†). The observed changes could be due to the binding of **F-An** nanoclusters with CT-DNA through the groove binding interactions of the fullerene and aniline moieties. The intrinsic binding constant (K_{DNA}) for the binding of the **F-An** nanoclusters was calculated using half reciprocal plots and was found to be $1.87 \times 10^4\ M^{-1}$ (Fig. S3†). A similar titration experiment with CT-

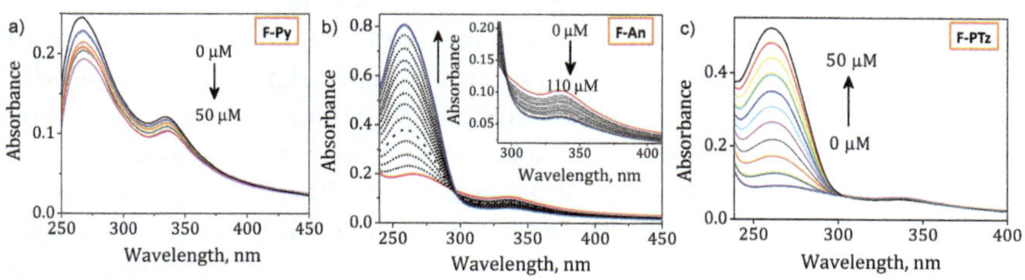

Fig. 1 Changes in the absorption spectra of (a) **F-Py**, (b) **F-An** and (c) **F-PTz** with sequential addition of CT-DNA in 10% DMSO–PBS; the inset in (b) shows a magnified portion between 290 and 400 nm.

DNA and **F-PTz** demonstrated negligible changes corresponding to the **F-PTz** cluster absorption around 340 nm (Fig. 1c), which clearly indicates that larger nanoclusters of **F-PTz** exhibit negligible interactions with DNA. Both **F-An** and **F-PTz** remain as neutral clusters under the physiological conditions used, hence any electrostatic interaction between these fullerene derivatives and DNA can be ruled out. Thus, the UV-Vis titration experiments indicate differential interactions of these fullerene derivatives with CT-DNA due to the differences in their solubility and cluster formation behaviour in 10% DMSO–PBS. Individually solvated molecules of **F-Py** and 3–5 nm sized nanoclusters of **F-An** interact efficiently through electrostatic/groove interactions, while larger nanoclusters of **F-PTz** fail to establish such interactions due to possible steric constraints.

Fluorescence displacement assay

An ethidium bromide based fluorescence displacement assay is highly useful for understanding the mode of binding of small intercalator and groove binding molecules, where competitive displacement of the intercalated ethidium bromide leads to substantial quenching of the ethidium fluorescence.[33,34] The aniline moieties in the **F-An** nanoclusters, due to their small size and planar aromatic nature, can interact with DNA through groove binding and intercalative interactions. In order to understand the role of aniline intercalation in the CT-DNA–**F-An** nanocluster interactions, we have carried out an ethidium bromide based fluorescence displacement assay. A highly fluorescent ethidium bromide–CT-DNA complex was prepared *via* titration of ethidium bromide against CT-DNA, as shown in Fig. S4.† Furthermore, **F-An** nanoclusters were titrated against this fluorescent ethidium bromide–CT-DNA complex, which exhibited only small changes in the fluorescence upon the addition of up to 30 μM of **F-An** (Fig. S5†), thus indicating the negligible displacement of the intercalated ethidium bromide. Therefore, the interaction of the **F-An** nanoclusters with CT-DNA could be attributed mostly to the groove interactions of the fullerene clusters and aniline moieties.

Circular dichroism studies

In order to understand the possible interactions between the CT-DNA/**F-An** nanoclusters and to verify the role of CT-DNA as a template in the self-assembly of nanoclusters, we have carried out circular dichroism (CD) studies of CT-DNA (100 μM) in the presence of varying amounts of **F-An** nanoclusters as shown in Fig. 2. CT-DNA alone exhibited the characteristic CD spectrum corresponding to the B-form DNA with a positive peak at 275 nm and a negative peak at 245 nm (Fig. 2), as reported previously.[35,36] Upon increasing the concentration of **F-An** nanoclusters from 1 : 0 to 1 : 1, the CD spectra exhibited significant changes with the appearance of strong induced CD around 250–370 nm, which is the region corresponding to **F-An** nanocluster absorption. Solutions of the **F-An** nanocluster alone at different concentrations did not exhibit any CD signal (Fig. S6†), hence the observed positive induced CD signal could be attributed to the chiral organization of the **F-An** nanoclusters on the CT-DNA template, facilitated by the groove binding interactions and possible π–π interactions between adjacently bound **F-An** nanoclusters. The control experiments, such as **F-Py** or **F-PTz** with and without CT-DNA, exhibited no ICD, thus confirming that only **F-An** with

 This journal is © The Royal Society of Chemistry 2018

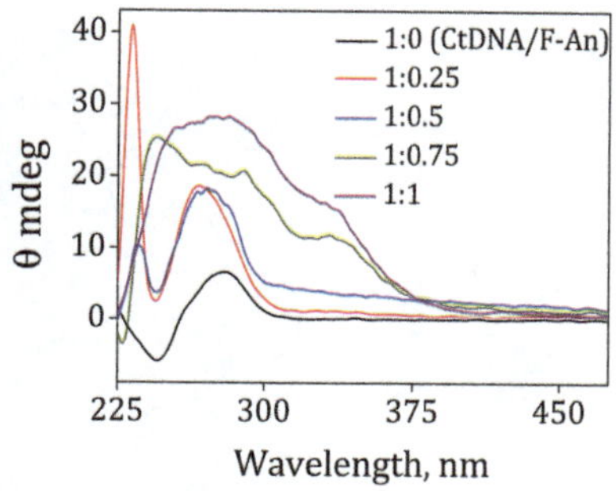

Fig. 2 Changes in the CD spectra of CT-DNA with increasing addition of **F-An**.

a unique cluster size can undergo templated, chiral self-assembly with CT-DNA. Furthermore, temperature dependent CD spectra were studied in order to understand the stability of the CT-DNA/**F-An** self-assembled system in solution. The decrease in the CD intensity upon increasing the temperature from 20 °C to 80 °C demonstrates the disassembly of the nanoclusters from the CT-DNA template (Fig. S7†). The lack of complete disappearance of the CD signal could be attributed to the higher melting temperature of CT-DNA, preventing the complete disassembly of nanoclusters from its surface.

Morphological studies

The morphology changes accompanying the interaction of the fullerene derivatives with CT-DNA were studied using AFM, TEM and DLS techniques. AFM experiments were carried out by drop casting 10 μL of solution containing CT-DNA and CT-DNA/**F-An** complex separately on a freshly cleaved mica surface. The concentrations of the samples were chosen based on the CD studies. CT-DNA

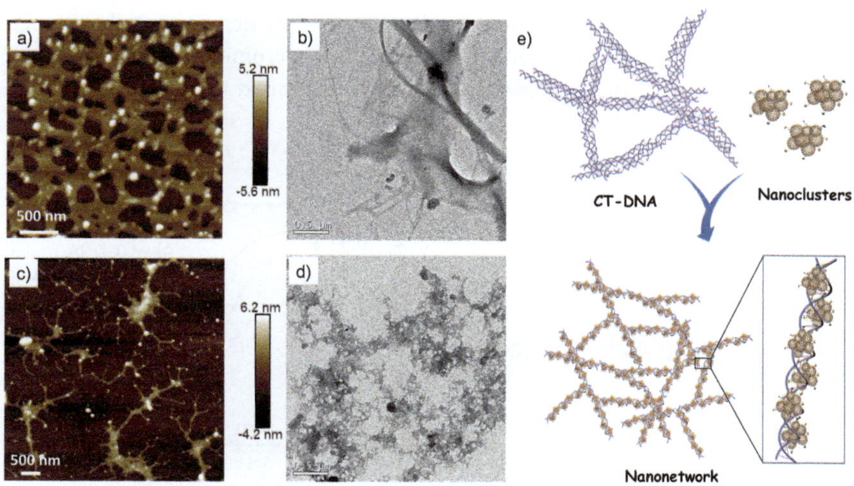

Fig. 3 AFM and TEM images of the (a and b) CT-DNA and (c and d) CT-DNA/**F-An** complex in 10% DMSO–PBS, respectively. (e) Pictorial representation of the CT-DNA templated nanonetwork formation of **F-An** nanoclusters in a 10% DMSO–PBS mixture.

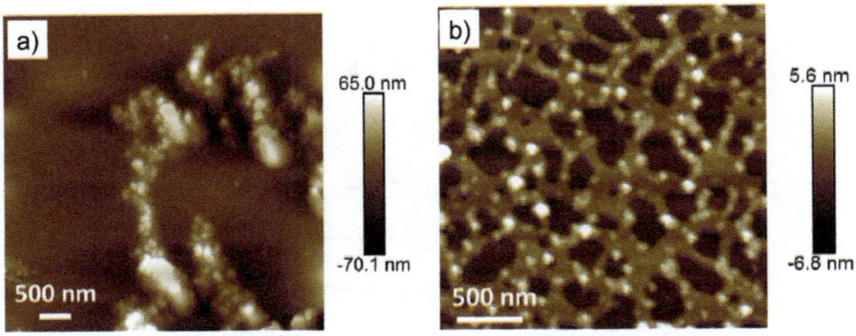

Fig. 4 AFM height image of the (a) CT-DNA/**F-Py** and (b) CT-DNA/**F-PTz** (1 : 1) complex.

alone exhibited a highly networked structure with an average width of ~250 nm, due to the strong DNA–DNA interactions, which agrees well with previous reports (Fig. 3a).[37–39] On the other hand, the CT-DNA/**F-An** complex exhibited a significantly different nanofibrous structure, where the highly networked structure of CT-DNA was broken down into smaller nanofibers with sizes of 50–100 nm and subsequent condensates (Fig. 3b and S8a†). The change in morphology of CT-DNA upon **F-An** nanocluster binding could be attributed to the reduction in the DNA–DNA interactions subsequent to the groove binding of the **F-An** nanoclusters. TEM analysis of the CT-DNA alone and the CT-DNA/**F-An** complex over a carbon coated copper grid further confirmed the disruption of the CT-DNA network structure and DNA templated **F-An** nanocluster self-assembly (Fig. 3c and d). Moreover, EDAX analysis over the obtained nanostructures indicated the occurrence of phosphate atoms, which confirmed the presence of DNA throughout the obtained nanostructures (Fig. S9†). These results, coupled with the CD analysis, further support the chiral self-assembly of the **F-An** nanoclusters along the CT-DNA template, facilitated by the groove binding and possible π–π interactions between the clusters, as shown schematically in Fig. 3e. AFM analysis of the CT-DNA/**F-Py** complex, under similar conditions, demonstrated evidence for **F-Py** assisted CT-DNA condensation leading to micrometer sized aggregates (Fig. 4a). In contrast, the CT-DNA/**F-PTz** complex exhibited no visible disruption

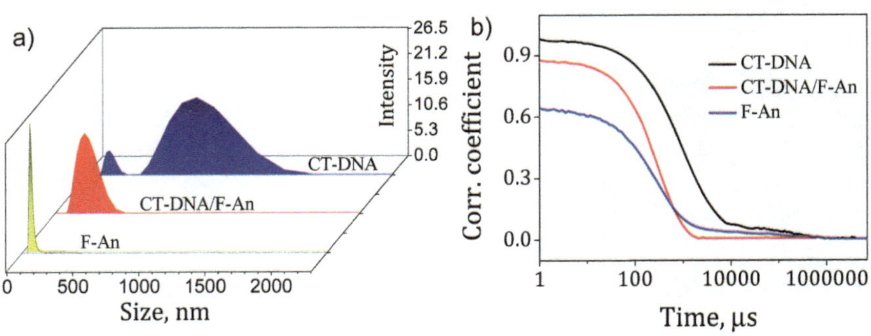

Fig. 5 (a) Size distribution of **F-An**, CT-DNA/**F-An** (1 : 1) and CT-DNA obtained from DLS measurements. (b) Corresponding correlogram.

 This journal is © The Royal Society of Chemistry 2018

of the CT-DNA fibrous network structures, as shown in Fig. 4b. In order to ascertain that the formation of the CT-DNA/**F-An** self-assembled nanostructures was not a surface dependent phenomenon, we carried out dynamic light scattering (DLS) studies under similar experimental conditions. The CT-DNA/**F-An** complex exhibited observable changes in the hydrodynamic radius compared to that of CT-DNA alone, thus indicating the breakdown of the highly networked structure of CT-DNA upon **F-An** nanocluster binding (Fig. 5).

Conclusions

In summary, we have reported the differential interaction of three monosubstituted fullerene derivatives, **F-Py**, **F-An** and **F-PTz** possessing pyridinium, aniline or phenothiazine end groups, with CT-DNA. The nature of the end group and subsequent changes in the hydrophobic–hydrophilic balance in these derivatives resulted in significant differences in their solvation and cluster formation behaviour in 10% DMSO–PBS and in their interactions with CT-DNA. For example, the cationic fullerene derivative **F-Py** becomes molecularly dissolved in 10% DMSO–PBS and interacts with CT DNA *via* electrostatic and groove binding interactions, leading to the formation of micrometer sized condensed aggregates. On the other hand, the phenothiazine derivative **F-PTz** forms larger nanoclusters and fails to interact with CT-DNA. In contrast, the aniline derivative **F-An**, which forms 3–5 nm sized nanoclusters, efficiently interacts with CT-DNA and induces perturbations in the highly networked structure of the CT-DNA. CD and morphological studies of the CT-DNA/**F-An** complex confirmed the chiral self-assembly of **F-An** nanoclusters along the CT-DNA template and the disruption of the CT-DNA network structure to form smaller, nanofibrous structures with subsequent condensation.

Acknowledgements

The financial support from the Council of Scientific and Industrial Research (CSIR 12 FYP M2D-CSC-0134) and the Department of Science and Technology, Government of India (Ramanujan Fellowship Grant RJN-19/2012) is greatly acknowledged. S. K. V. thanks the University Grant Commission for the fellowship.

References

1 A. Estevez-Torres and D. Baigl, *Soft Matter*, 2011, **7**, 6746–6756.
2 J. Bednar, R. A. Horowitz, S. A. Grigoryev, L. M. Carruthers, J. C. Hansen, A. J. Koster and C. L. Woodcock, *Proc. Natl. Acad. Sci. U. S. A.*, 1998, **95**, 14173–14178.
3 P. Carrivain, A. Cournac, C. Lavelle, A. Lesne, J. Mozziconacci, F. Paillusson, L. Signon, J.-M. Victor and M. Barbi, *Soft Matter*, 2012, **8**, 9285–9301.
4 T. Zhou, A. Llizo, C. Wang, G. Xu and Y. Yang, *Nanoscale*, 2013, **5**, 8288–8306.
5 D. Barbara, H. Zahra and B. Jean-Paul, *Curr. Gene Ther.*, 2004, **4**, 445–455.
6 A. Gonzalez-Perez and R. S. Dias, *Front. Biosci.*, 2009, **1**, 228–241.
7 V. A. Bloomfield, *Curr. Opin. Struct. Biol.*, 1996, **6**, 334–341.

8 P. Midoux, C. Pichon, J.-J. Yaouanc and P.-A. Jaffrès, *Br. J. Pharmacol.*, 2009, **157**, 166–178.

9 K. Liu, L. Zheng, C. Ma, R. Gostl and A. Herrmann, *Chem. Soc. Rev.*, 2017, **46**, 5147–5172.

10 M. Hariharan, E. Kuruvilla and D. Ramaiah, *J. Phys. Chem. Lett.*, 2010, **1**, 834–838.

11 V. A. Bloomfield, *Biopolymers*, 1997, **44**, 269–282.

12 A. M. Cassell, W. A. Scrivens and J. M. Tour, *Angew. Chem., Int. Ed. Engl.*, 1998, **37**, 1528–1531.

13 M. B. Patel, U. Harikrishnan, N. N. Valand, D. S. Mehta, K. V. Joshi, S. P. Kumar, K. H. Chikhalia, L. B. George, Y. T. Jasrai and S. K. Menon, *RSC Adv.*, 2013, **3**, 8734–8746.

14 H. Nitta, K. Minami, K. Harano and E. Nakamura, *Chem. Lett.*, 2014, **44**, 378–380.

15 K. Minami, K. Okamoto, K. Doi, K. Harano, E. Noiri and E. Nakamura, *Sci. Rep.*, 2014, **4**, 4916.

16 C. Klumpp, L. Lacerda, O. Chaloin, T. D. Ros, K. Kostarelos, M. Prato and A. Bianco, *Chem. Commun.*, 2007, 3762–3764, DOI: 10.1039/b708435h.

17 H. Isobe, S. Sugiyama, K.-i. Fukui, Y. Iwasawa and E. Nakamura, *Angew. Chem., Int. Ed. Engl.*, 2001, **40**, 3364–3367.

18 E. Nakamura and H. Isobe, *Chem. Rec.*, 2010, **10**, 260–270.

19 V. L. Malinovskii, D. Wenger and R. Häner, *Chem. Soc. Rev.*, 2010, **39**, 410–422.

20 F. A. Aldaye, A. L. Palmer and H. F. Sleiman, *Science*, 2008, **321**, 1795–1799.

21 A. L. Benvin, Y. Creeger, G. W. Fisher, B. Ballou, A. S. Waggoner and B. A. Armitage, *J. Am. Chem. Soc.*, 2007, **129**, 2025–2034.

22 N. V. Hud and K. H. Downing, *Proc. Natl. Acad. Sci. U. S. A.*, 2001, **98**, 14925–14930.

23 M. R. Shen, K. H. Downing, R. Balhorn and N. V. Hud, *J. Am. Chem. Soc.*, 2000, **122**, 4833–4834.

24 K. Osada, H. Oshima, D. Kobayashi, M. Doi, M. Enoki, Y. Yamasaki and K. Kataoka, *J. Am. Chem. Soc.*, 2010, **132**, 12343–12348.

25 Y. Fan, H. Wang, C. He, F. Qiao, S. Wang and Y. Wang, *ACS Appl. Mater. Interfaces*, 2017, **9**, 23333–23341.

26 I. Donskyi, K. Achazi, V. Wycisk, C. Bottcher and M. Adeli, *Chem. Commun.*, 2016, **52**, 4373–4376.

27 T. Homma, K. Harano, H. Isobe and E. Nakamura, *Angew. Chem., Int. Ed. Engl.*, 2010, **49**, 1665–1668.

28 T. Homma, K. Harano, H. Isobe and E. Nakamura, *J. Am. Chem. Soc.*, 2011, **133**, 6364–6370.

29 S. K. Vittala, S. K. Saraswathi and J. Joseph, *Chem.–Eur. J.*, 2017, **23**, 15759–15765.

30 B. C. Baguley and E. M. Falkenhaug, *Nucleic Acids Res.*, 1978, **5**, 161–171.

31 J. Joseph, E. Kuruvilla, A. T. Achuthan, D. Ramaiah and G. B. Schuster, *Bioconjugate Chem.*, 2004, **15**, 1230–1235.

32 E. Kuruvilla, J. Joseph and D. Ramaiah, *J. Phys. Chem. B*, 2005, **109**, 21997–22002.

33 D. L. Boger, B. E. Fink, S. R. Brunette, W. C. Tse and M. P. Hedrick, *J. Am. Chem. Soc.*, 2001, **123**, 5878–5891.

 This journal is © The Royal Society of Chemistry 2018

34 N. C. Garbett, N. B. Hammond and D. E. Graves, *Biophys. J.*, 2004, **87**, 3974–3981.
35 H. Yang, P. Tang, B. Tang, Y. Huang, X. Xiong and H. Li, *RSC Adv.*, 2017, **7**, 10242–10251.
36 B. H. Shankar, D. T. Jayaram and D. Ramaiah, *Chem.–Eur. J.*, 2015, **21**, 17657–17663.
37 H. Mahesh, K. Elizabeth and R. Danaboyina, *J. Phys. Chem. Lett.*, 2010, **1**(5), 834–838.
38 A. M. Oliveira Brett and A.-M. Chiorcea, *Langmuir*, 2003, **19**, 3830–3839.
39 A. Terawaki, Y. Otsuka, H. Lee, T. Matsumoto, H. Tanaka and T. Kawai, *Appl. Phys. Lett.*, 2005, **86**, 113901.

Faraday Discussions

PAPER

Adaptation of *Rhodopseudomonas acidophila* strain 7050 to growth at different light intensities: what are the benefits to changing the type of LH2?

A. T. Gardiner, *[a] D. M. Niedzwiedzki [b] and R. J. Cogdell[a]

Received 26th August 2017, Accepted 13th October 2017
DOI: 10.1039/c7fd00191f

Typical purple bacterial photosynthetic units consist of light harvesting one/reaction centre 'core' complexes surrounded by light harvesting two complexes. Factors such as the number and size of photosynthetic units per cell, as well as the type of light harvesting two complex that is produced, are controlled by environmental factors. In this paper, the change in the type of LH2 present in the *Rhodopsuedomonas acidophila* strain 7050 is described when cells are grown at a range of different light intensities. This species contains multiple *puc*BA genes that encode the apoproteins that form light-harvesting complex two, and a more complex mixture of spectroscopic forms of this complex has been found than was previously thought to be the case. Femtosecond time resolved absorption has been used to investigate how the energy transfer properties in the membranes of high-light and low-light adapted cells change as the composition of the LH2 complexes varies.

Introduction

The primary reactions of purple bacterial photosynthesis involve light harvesting (LH) complexes and reaction centres (RC) that are located in intracytoplasmic membranes (ICM).[1] Light energy absorbed by the light harvesting system is transferred to the RCs where it is trapped by a series of transmembrane electron transfer reactions. The cell's capacity for LH is flexible and can be adjusted in response to environmental factors, such as the incident light intensity.[2,3] Under low light (LL) conditions there is an extensive system of ICM in the cell, whereas under high light (HL) conditions the amount of ICM is a lot less. Typically, there are two types of LH complexes present, called LH1 and LH2. The LH1 complexes surround the RC to form the so called 'core' complex, with the LH2 complexes arranged more peripherally around the core complexes.[4–6] The incident light

[a]*Institute of Molecular, Cellular and Systems Biology, University of Glasgow, University Avenue, Glasgow, G12 8QQ, UK. E-mail: alastair.gardiner@glasgow.ac.uk*
[b]*Photosynthetic Antenna Research Center, Washington University in St Louis, MO 63130, USA*

intensity also regulates the ratio of the LH2 complexes to core complexes; this ratio is higher under low intensity illumination. These adaptations allow the purple bacteria to optimise their LH apparatus in response to changing environmental conditions.[7]

The LH1 complex has a single major bacteriochlorophyll a (Bchl a) Q_y absorption band in the near infra-red (NIR) region at 884 nm, whereas LH2 has two major Bchl a Q_y absorption bands around 800 and 850 nm. Chemically it is the same Bchl a molecules that give rise to these absorption bands, with the different positions being due to specific pigment–pigment and pigment–protein interactions within the various LH complexes.[8,9] Importantly, these wavelength differences establish an energy gradient to ensure that the energy from the harvested photons is directed downhill from LH2 to LH1 and then on to the RC.[10]

In purple bacteria, changes in the amount and composition of the LH system, in response to environmental stimuli, serve to enhance their ability to harvest sufficient solar energy to allow photosynthetic growth under the changing conditions experienced in different ecological niches. Some species, however, exhibit an additional response to growth at different light intensities. Within the genome, these species contain a multigene family of pucBA genes that encode the alpha and beta apoproteins that form LH2 complexes.[11] The expression of different pucBA genes results in LH2 complexes that exhibit markedly different absorption spectra in the NIR spectral range.[12]

Rhodopseudomonas (Rps.) acidophila has a form of LH2 that has its main Bchl a Q_y bands at 800 and 850 nm when the cells are grown under HL conditions, and this switches to a form that has its Bchl a Q_y bands at 800 and 820 nm under LL conditions. The question then arises as to what extra benefits this ability gives species such as *Rps. acidophila* at LL intensities? A start towards answering this question forms the major focus of this paper. Previously the energy transfer properties of the isolated B800–820 and B800–850 LH2 complexes have been investigated using ultrafast laser flash photolysis.[18–20] The time constant for 800 to 820 nm excitation energy transfer (EET) is ~0.75 ps at room temperature (RT) and 0.9 ps at 77 K. This compares with 0.8 ps at RT and 1.3 ps at 77 K for 800-to-850 nm energy transfer, *i.e.*, the EET rates for both these types of LH2 are rather similar. The rates of energy transfer from both these types of LH2 to LH1 in photosynthetic membranes are also rather similar; B820 to LH1 is just a bit slower than B850 to LH1. However, when one considers that in the absence of LH1 the lifetime of the excited singlet state of both B850 and B820 is about 1 ns, the small differences in the LH2 to LH1 EET rates do not significantly change the overall high efficiency of this energy transfer step. Therefore, the question is worth repeating; what does the organism gain by this change in the type of LH2 upon going from growth at high to low light intensities?

A partial answer has been suggested from experiments on membranes using high intensity laser flashes that induce exciton annihilation.[21] This process requires singlet–singlet fusion. Two excited singlet states must be able to migrate, 'collide' and then de-excite with the excess energy released as heat. This approach has been used to measure the size of the domain in which the excited singlet states can move around in and 'find' each other to annihilate.[22] The presence of B800–820 complexes in the LL membranes confined the excited singlet states into smaller domains around the LH1 core complexes compared with HL grown membranes containing B800–850 complexes.[21] It was, therefore, suggested that

 This journal is © The Royal Society of Chemistry 2018

the presence of B800–820 complexes slowed down the back reaction from LH1 to LH2 compared to when B800–850 was present. It was assumed that this could then give more time for the trapping of the excited state by the RC, which is more important under light limiting conditions.

Previously, we investigated how the light harvesting systems in three strains of *Rps. acidophila* (10050, 7750 and 7050) adapt to growth at different light intensities.[23,24] These studies were mainly carried out in order to learn how to grow these strains under defined, standardised conditions, where the type of LH2 complex made is consistent.

It has been known for some time that the genome of *Rps. acidophila* contains a large multi-gene family of the pucBA genes.[25] Recently, however, the genome of *Rps. acidophila* 10050 was sequenced which revealed that in this strain there are twelve copies of the pucBA genes.[26] This suggests that a full and complete understanding of LH2 complexes and their role is much more complicated than imagined. As the technology underpinning laser flash photolysis has significantly improved, compared to that used in our previous study in 1991,[21] it was decided to reinvestigate how this species of purple bacteria adapts to growth at different light intensities and the possible benefits that these adaptations confer.

Results and discussion

Fig. 1 shows the absorption spectra of the isolated, purified core and LH2 complexes from the purple bacterium *Rps. acidophila*. Fig. 2 illustrates some examples of the variety of spectra that these different forms of LH2 can exhibit.

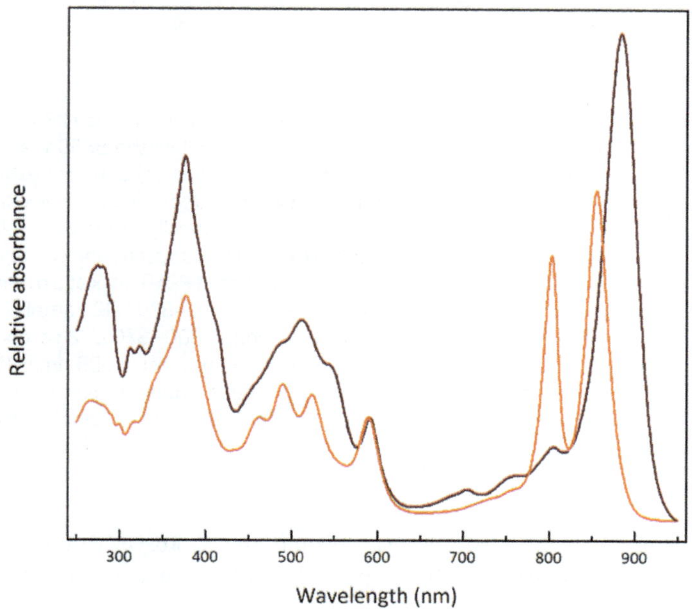

Fig. 1 Normalised absorption spectra of the LH2 (orange) and LH1–RC 'core' (brown) complexes from *Rps. acidophila* 7050. This high light intensity LH2 complex exhibits two peaks in the NIR region at 800 and 850 nm. The LH1 complex exhibits a single NIR peak at 884 nm.

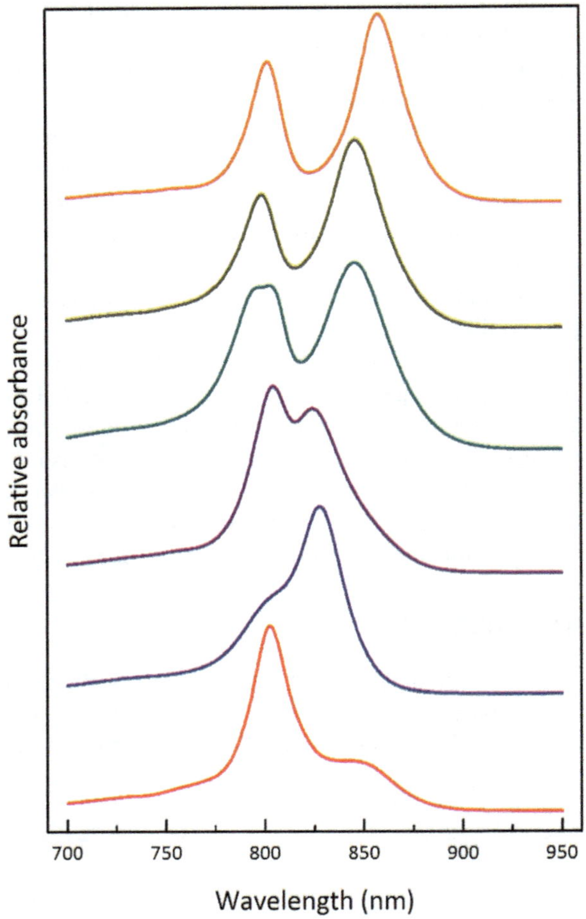

Fig. 2 Selected LH2 complexes to illustrate the spectral variability in the NIR region. The archetypal *Rps. acidophila* 10050 LH2 (orange) complex exhibits the B850 maximum at 858 nm.[13] The octameric *Phaeospirillum molischianum* LH2 (olive) complex exhibits the B850 maximum at 846 nm.[14] The B850 maximum of the *Allochromatium vinosum* LH2 (turquoise) complex is at 845 nm, but this complex additionally contains a split B800 band.[15] The most discernible low-light LH2 complex from *Rhodopseudomonas acidophila* 7050 (purple), the subject of this paper, exhibits a blue-shifted B850 peak at a maximum of 824 nm.[16] For this reason, the complex is referred to as a B800–820 LH2 complex and the old nomenclature of LH3 is no longer used. The octameric B800–830 LH2 complex from *Marichromatium purpuratum* (navy blue) exhibits its NIR maximum at 828 nm.[17] The low-light form of LH2 from *Rhodopseudomonas palustris* (red) exhibits a much reduced B850 band.[12] All of the LH2 complex spectra shown in this figure were produced by the Cogdell laboratory in Glasgow.

Fig. 3 shows how the absorption spectrum of the photosynthetic membranes from *Rps. acidophila* strain 7050 changes depending upon the incident light intensity at which the cells were grown.

The light intensity was varied by growing the cells at different distances from an incandescent bulb. This strategy ensures that the spectral composition of the incident light does not vary as these bacteria also contain

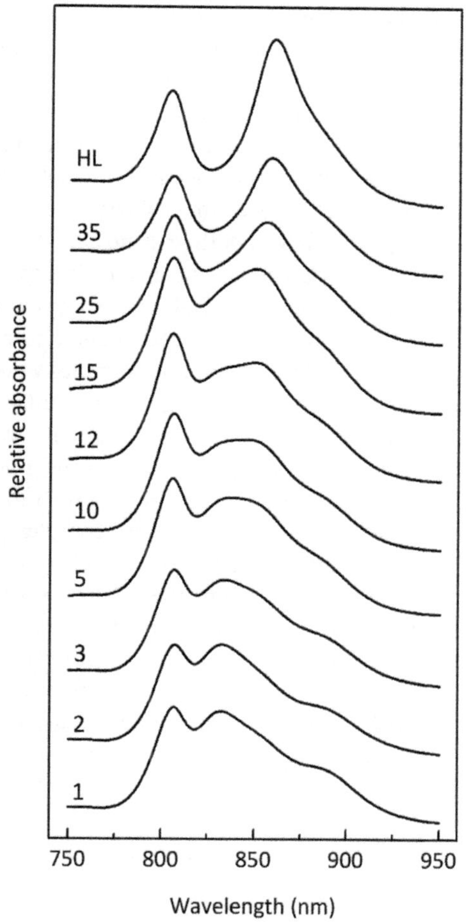

Fig. 3 Absorption spectra of membranes isolated from *Rps. acidophila* 7050 cells grown at progressively decreasing light intensities. The top, high-light spectrum corresponds to light conditions of approximately 150 μmol photons per s per m^2. As the light intensity decreases, the peak at ~850 nm progressively decreases whilst a new peak at ~820 nm becomes apparent.

bacteriophytochromes.[27] This is important as the purple bacterial light harvesting system is also subject to regulation by the wavelength of the incident radiation as well as its intensity.[28,29] The top absorption spectrum is from HL adapted cell membranes. It is dominated by peaks at 800 and ~850 nm that embody the typical HL form of LH2. There is a clear shoulder on the red side of the 850 nm band at about 880 nm that originates from the underlying LH1 complex (Fig. 1). As the incident light intensity is reduced the 850 nm absorption band gets smaller and appears to blue shift. A new peak at about 820 nm grows in as the light intensity decreases. Initially this is evident as the 'valley' between the 800 and 850 nm bands filling in but then a 820 nm band can be clearly seen at lower light intensities. The changes in the absorption spectrum can also be seen by comparing the relative heights of the 850 nm and 800 nm bands. As the light intensity is lowered the relative height of the 850 nm band strongly reduces. In all cases except at the highest light intensity the population of LH2 complexes present seems to be a mixture of different spectroscopic types. If the situation solely represents a progressive shift of the amount of B800–850 to B800–820 then the normalised difference spectra of the absorption spectra should reveal an isosbestic point somewhere between the 850 and 820 nm bands. A full set of difference spectra, where the absorption spectrum of each lower light intensity sample is subtracted from the HL spectrum, is shown in Fig. 4.

This set of difference spectra show that there is no clear isosbestic point (Fig. 4 inset), which suggests that the changes in the absorption spectra shown cannot be simply explained by a shift between two well-defined types of LH2. There must be a more complicated mixture of different spectroscopic forms of LH2 present.

To further investigate this finding, membrane samples were fractionated into LH2 and LH1 core complexes by sucrose gradient centrifugation following detergent solubilisation. Two such typical sucrose gradients are shown in Fig. 5A. The LH2 band is the major one about one third down the gradient and the LH1 core complex is the smaller band nearer the bottom of the tube. Two features are very clear. The ratio of the amount of LH2 to core complex present in the two examples shown is quite different. The LL tube has much more LH2 relative to the amount of core complex present. In addition, the colour of the LH2 band changes.

In the HL sample the LH2 band is orange/brown but in the LL tube the LH2 band is purple. This change has been documented before and is due to a change in the major type of carotenoid present. At HL, the major carotenoid in LH2 is rhodopin/rhodopin glucoside whereas at LL the major carotenoid is rhodopinal/rhodopinal glucoside.[24,30] Interestingly, rhodopinal is about 50% more efficient as

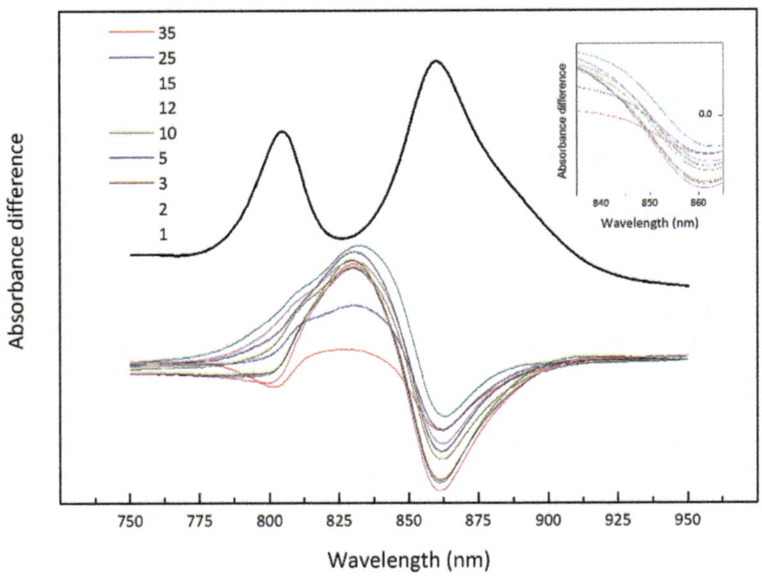

Fig. 4 Difference spectra of the NIR absorption from the *Rhodopseudomonas acidophila* 7050 membranes presented in Fig. 3. All absorption spectra were normalised at the Bchl a Q_x band, with the high-light membranes given in black. Each subsequent spectrum has been subtracted from this initial absorption spectrum to generate the set of difference spectra. The inset shows the possible isobestic region magnified for clarity.

 This journal is © The Royal Society of Chemistry 2018

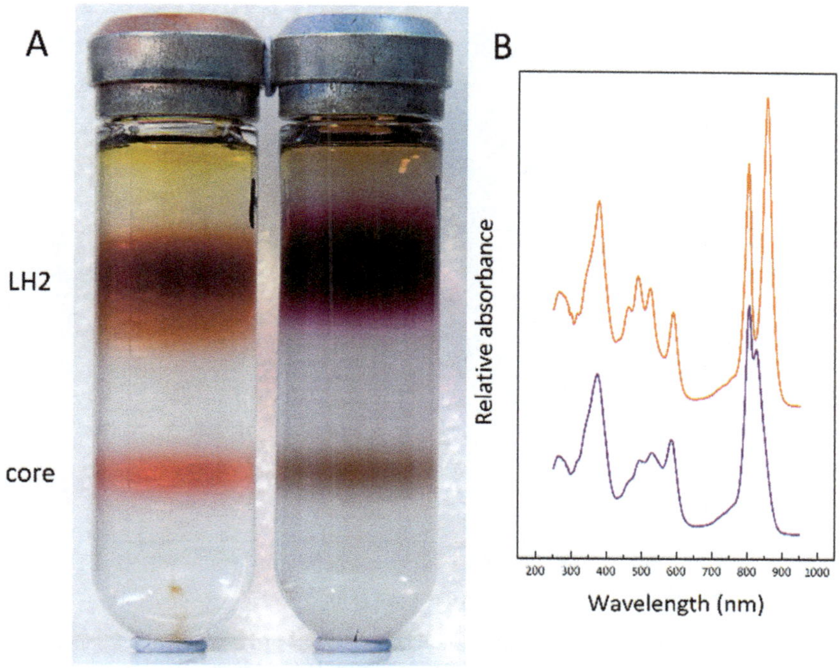

Fig. 5 A typical sucrose gradient centrifugation separation of solubilised membranes from cells grown at a high-light (A: left side) intensity and a low-light (A: right side) intensity. (B) Absorption spectra of the high-light (orange) and low-light (purple) purified LH2 complexes.

an accessory light harvesting pigment than rhodopin.[31] The colour change of these LH2 complexes is reflected in the change in the absorption spectra in the region between 450 and 550 nm where the carotenoids absorb. Other factors such as the carbon source on which the cells are grown also affect the response to growth at different light intensities.[30] *Rps. acidophila* 7050 cells were grown under identical conditions at the same LL intensity but in one case using succinate as the carbon source and the other in the presence of pyruvate. Growing with pyruvate rather than succinate shifts the light intensity response so that the cells have an absorption spectrum that is more like that of succinate cells grown at a higher light intensity.[30] Fig. 6 shows the resultant LH2 band spectra of the sucrose gradients of solubilised ICM isolated from succinate and pyruvate LL grown cells.

A solubilised ICM sample from succinate-grown LL cells was separated on a sucrose gradient and the isolated LH2 band further fractionated by ion exchange chromatography. Fig. 7 shows the absorption spectra of these different fractions that elute as the salt concentration increases. There appears to be a transition between several different forms of LH2 that cannot be simply explained by a mixture of the two previously described B800–850 (ref. 13) and B800–820 (ref. 16) complexes.

Considering the basic, repeating heterodimeric structure of a LH2 complex, when different pucBA genes contribute polypeptides to form the complete native structure, if these different polypeptides encode different site-energies for the

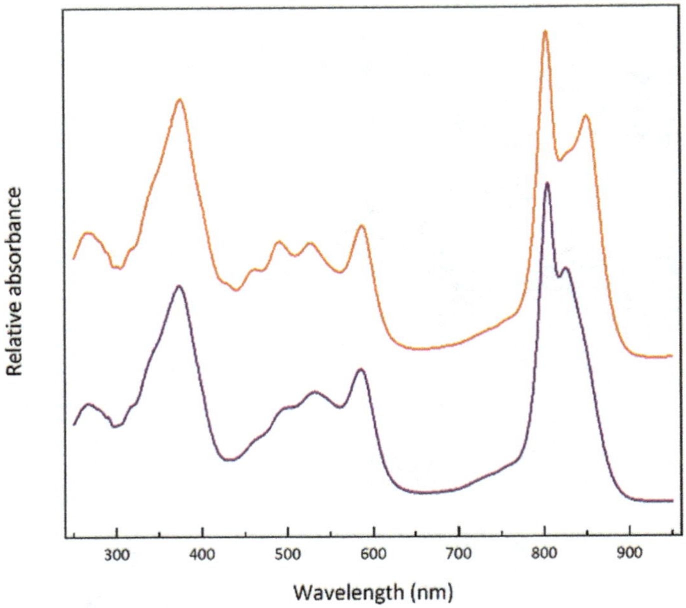

Fig. 6 Absorption spectra of the LH2 fraction taken from a sucrose gradient fractionation of membranes from *Rps. acidophila* 7050 cells grown at the same low light intensity (10 μmol photons per s per m^2), but with one where the cells were grown using succinate as the carbon source (purple) and with the other where the cells were grown using pyruvate as the carbon source (orange). The concentration of carbon in each medium was normalised to 20 mmol l^{-1}.

group of strongly coupled Bchl a molecules, then there are a large number of possible spectroscopic outcomes.[32,33]

Steady-state absorption of the membranes used in the transient absorption measurements

Steady-state absorption spectra of the HL and LL photosynthetic membranes of *Rps. acidophila* 7050 used in the transient absorption measurements, taken at room temperature and at 77 K, are shown in Fig. 8.

Low temperature clearly affects the spectra of each membrane sample. At 77 K the absorption spectra gain spectral resolution (the bands sharpen) and reveal additional details not seen in the room temperature spectra. For both HL and LL samples the absorption band at 903 nm becomes clearly recognizable. This peak is associated with the Q_y band of the LH1–RC core complex. In addition, for the LL membrane sample the broad absorption band between 820 and 880 nm, unresolved at room temperature, splits into two peaks with maxima at about 826 and 860 nm. The positions of those peaks suggest the presence of a mixture of at least two different spectroscopic forms of LH2. A clearer view of these differences between the two membranes is shown in Fig. 9 in which the 77 K absorption spectra are overlaid. This overlay shows an additional difference between the samples in the spectral shape of the Bchl a Q_x bands. The Q_x band is blue shifted and broader in the case of the LL membrane, which is consistent with the

 This journal is © The Royal Society of Chemistry 2018

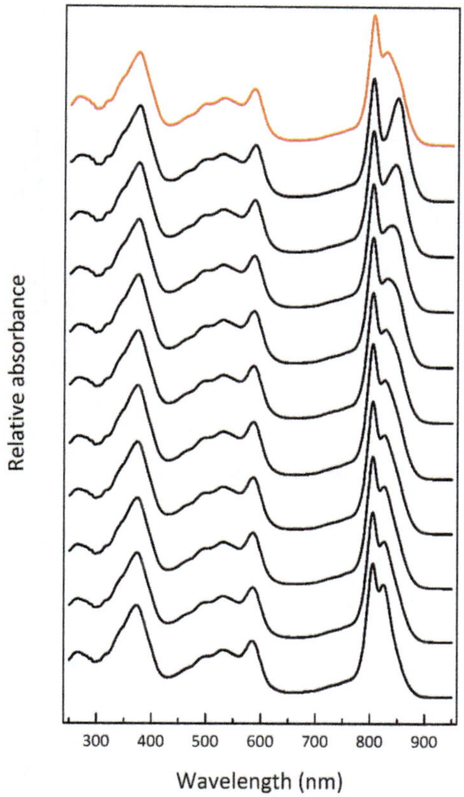

Fig. 7 A set of absorption spectra of the LH2 fractions eluted from an ion-exchange column loaded with the LH2 band (orange) from a sucrose gradient of solubilised membranes of cells grown at a light intensity of 15 μmol photons per s per m^2. The LH2 band was fractionated by ion exchange chromatography, as described in the Experimental section. It should be noted that this figure is representative only of the range of LH2 complexes that elute with increasing NaCl concentration and so comprise the LH2 band that was loaded onto the column. The fractions that elute at higher NaCl concentrations are the most abundant.

presence of a mixture of types of LH2 complexes, as seen in the fractionation studies described above.

Time-resolved absorption

Detailed information about the photo-physical processes that occur in the membrane after specific excitation (into the B800 band of the LH2 complexes) can be obtained from time-resolved absorption (TA) studies. Such a set of experiments are depicted in Fig. 10.

For easier interpretation of this data, the TA spectra are accompanied by a mirror image of the respective steady-state absorption (given as a dashed line). Just as seen in the case of the steady-state spectra, at 77 K the TA spectra gain a spectral resolution that cannot be seen at room temperature. The intermediate stages in time spectral evolution now also become visible. Since cryogenic data provide more insight into the photo-physical processes in the membranes, these

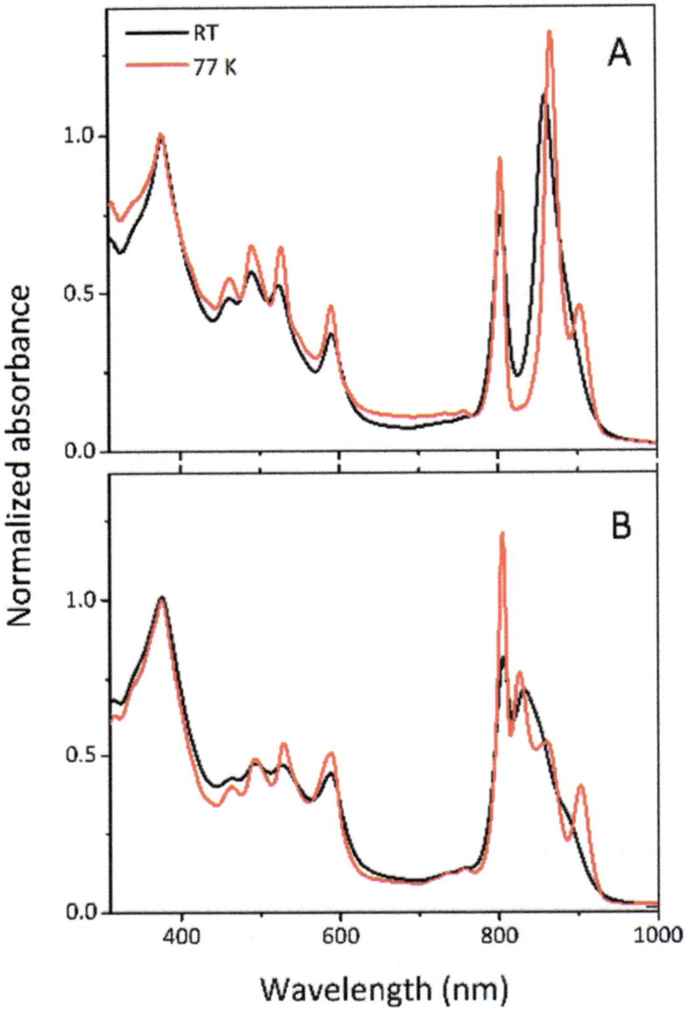

Fig. 8 Room temperature and 77 K steady-state absorption spectra of the photosynthetic membranes of *Rps. acidophila* 7050 cells grown under (A) high-light and (B) low-light intensity conditions. For better comparability, the spectra are normalized to unity at the maximum of the Soret band of Bchl a which should be least affected by low temperature.

will be described here but conclusions drawn from this can also be extended to the room temperature data. Excitation at 800 nm of the HL membranes leads to the instantaneous bleaching of two bands at 803 and 867 nm (black line). Those correspond to the bleaching of the Q_y bands of B800 and B850 Bchls in the well-characterised HL LH2 complex. The perfect match of the positions of the B850 band from the steady-state absorption spectrum and the observed bleaching band suggests that this excitonic level is not yet populated, otherwise it would be distorted by stimulated emission that would broaden and red-shift the observed "bleaching" band. Instead, excitation at 800 nm is also partially absorbed by the higher excitonic band of the "B850" exciton manifold that overlaps with the B800 band and which is responsible for efficient B800 → B850 energy transfer.[34,35]

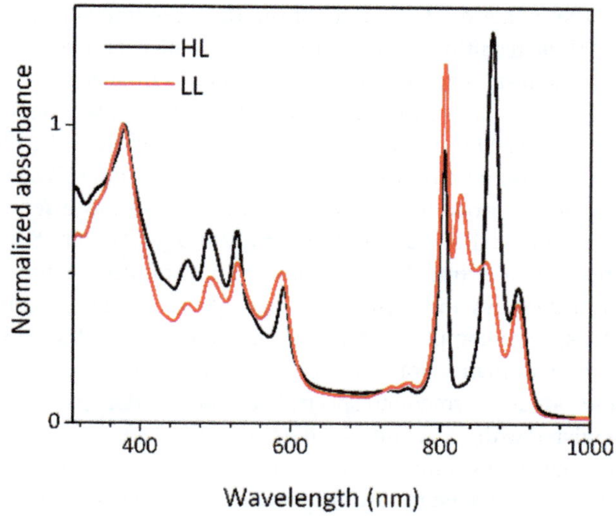

Fig. 9 Overlay of the 77 K absorption spectra of the photosynthetic membranes of *Rps. acidophila* 7050 cells grown under high-light (black) and low-light (red) intensity conditions.

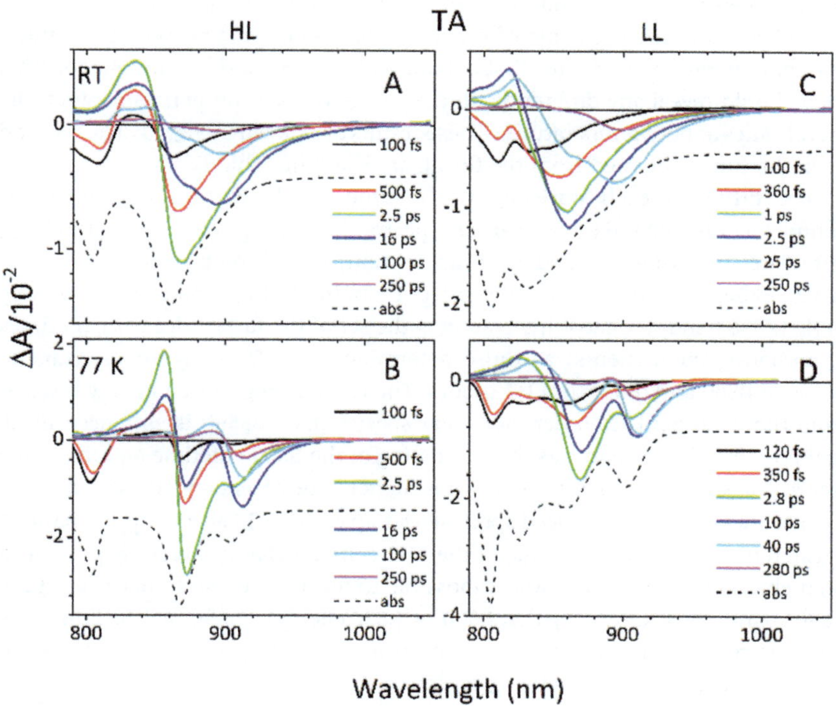

Fig. 10 RT and 77 K TA spectra of the (A and B) HL and (C and D) LL photosynthetic membranes of *Rps. acidophila* 7050 recorded at various times after excitation. The spectra were taken upon excitation at 800 nm corresponding to the Q_y band of B800 Bchl a. For comparison, a mirror image of the steady absorption (abs) of each membrane is overlaid with the TA spectra.

Population of the higher excitonic level of the B850 exciton will depopulate the ground state, thus resulting in the bleaching of all excitonic bands, which explains this initial fast bleaching in the B850 band. As time evolves bleaching of the B800 band decays and within ~1 ps all of the excitation is transferred to the B850 exciton (see the 500 fs and 2.5 ps TA spectra). At this stage the bleaching of the B850 band will also involve stimulated emission (SE) from this excitonic level (stimulation is by the probe light). This is clearly observed as a red-shift of the bleaching-SE band that shows a minimum at 872 nm (*vs.* 867 nm for the steady-state absorption). As the time delay increases to tens of picoseconds the excited B850 band of LH2 decays and excitation is passed to the LH1–RC core complex. Finally, the TA spectra completely recover back to the ground state in the sub-nanosecond time domain. For the LL membrane sample, one can clearly observe an additional intermediate spectral component that seems to be simultaneously populated with B800 and which decays within a couple of picoseconds. More details about the dynamics of those processes were obtained by fitting the TA data. First, single wavelength fitting was carried out on chosen kinetic traces corresponding to the changes seen at the maxima of B800, B850 (HL LH2) and B870 (LH1–RC) (HL membranes) and at 820 and 830 nm for the LL membranes (corresponding to the additional "B850-like" bands seen at LL). The fitting was carried out using a simple model based on a sum of exponentially decaying components convoluted by the instrument response function (IRF). In principle, the parameters obtained from the fitting show the effective decay lifetimes of the excited states of the pigments of interest (B800, B850, *etc.*) as well as the inter-pigment energy transfer rates that should be observed as rise components. This then should reveal any differences in the energy transfer properties between the membranes at both temperatures. These results are shown in Fig. 11. The rise and decay lifetimes obtained from the fitting are listed in Table 1.

The fitting at each wavelength gave multiple decay constants, which is completely understandable as various spectral features (ground state bleaching, stimulated emission, excited state absorption) of different molecular species (B800, B850, *etc.*) can coexist and overlap. From the fitting results of the recovery of the B800 band, the most interesting is the set of the fastest decaying lifetimes (dominating components) as those determine how efficiently energy transfer between B800 and 'B850' in LH2 occurs. The results suggest that energy transfer from the B800 Bchls to other molecular species (like 'B850') is enhanced in LL membranes. This is visible as the shortening of the B800* lifetime and this same trend is also noticeable at 77 K. This higher rate of energy transfer may be a consequence of the enhanced spectral overlap with the B830 absorption band of a different spectroscopic form(s) of the LH2 complex that is (are) only present in LL membranes. Note that it was impossible to resolve the rise component of the B830* band, thus suggesting that B800 → B830 energy transfer may be limited by the time resolution (~200 fs) of the spectrometer or may be due to a heterogeneous mixture of LH2 types being present.

The lifetimes associated with the rise components in the fitting of the signals associated with the bleaching of the B850 band correspond very well to the observed decay lifetimes of B800*, thus further confirming that efficient B800 → B850 energy transfer is indeed taking place. Subsequently, the majority of the B850* bleaching recovers with lifetimes ranging from 5.8 to 9.7 ps. Those lifetimes are typical for B850 → B870 (LH2 to LH1–RC) energy transfer.[34,36] Interestingly the

 This journal is © The Royal Society of Chemistry 2018

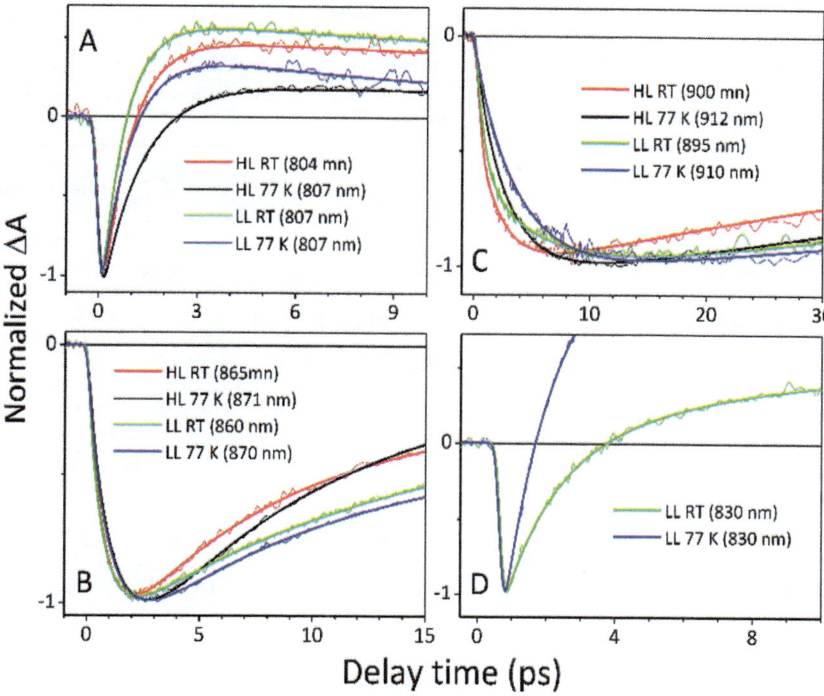

Fig. 11 Kinetic traces extracted from the maxima of the major spectroscopic features visible in the TA spectra of the studied membranes: (A) B800 band of LH2, (B) B850 band of LH2, (C) B870 band of the core and (D) "B830" band of an alternative LH2 at LL sample. The raw traces are accompanied by fits. All traces are normalized to unity at the signal minimum for better comparability.

rise components that are characteristic of the B800 → B850 energy transfer process also make a minor contribution to the population of signals that would be expected to be associated only with the bleaching of the B870 Bchls from LH1–RC. This is not an indication of any B800 → B870 energy transfer but rather reflects the fact that the transient signal associated with the B850* band also, however minimally, contributes to the absorption changes in this spectral range. Regardless of the fact that the steady-state absorption of the B850 band does not extend to such long wavelengths, the TA signal will also contain stimulated emission that overlaps, to some degree, with the B870 band. The TA spectra of the membranes, therefore, may always show some long-lived residual 'B850' bleaching associated with uncoupled LH2 complexes. A further complication arises from the possibility that some of the RCs may be closed during the measurements. This will unavoidably lead to substantial variation in the lifetime of B870*. This phenomena was observed in the present study as multiple kinetic components that are still associated with the same spectral species, B870*. The outcomes from the single wavelength fits were used to undertake some more advanced fitting, so-called global analysis that simultaneously fits the whole TA dataset. The results of such an analysis are given in Fig. 12.

Essentially, the effective lifetimes that are now associated with the EADS components are in a good agreement with those from the single wavelength fits.

Table 1 Rise and decay lifetimes obtained from single wavelength fitting of kinetic traces extracted at distinct spectral features (B800, B850, *etc.*) form TA data of HL and LL membranes at RT and at 77 K. The kinetic components were convoluted with IRF function mimicked by Gaussian with FWHM of 200 fs at room temperature and 250 fs at 77 K. Dominating components are marked by dashed line[a]

Sample	τ_R (ps)	τ_{R1} (ps)	τ_1 (ps)	τ_2 (ps)	τ_3 (ps)	Temperature
B800						
HL (807 nm)			0.84 ± 0.01	36.2 ± 1.5	751 ± 84	RT
HL (804 nm)			1.20 ± 0.01	35.7 ± 1.0		77 K
LL (807 nm)			0.71 ± 0.01	27.9 ± 1.2	177 ± 11	RT
LL (807 nm)			0.90 ± 0.01	8.3 ± 0.4	95 ± 8	77 K
"B830"						
LL (830 nm)			1.40 ± 0.04	6.2 ± 0.2	157 ± 3	RT
LL (830 nm)			1.57 ± 0.01	32.3 ± 0.7	166 ± 4	77 K
B850						
HL (865 nm)	0.76 ± 0.01		6.2 ± 0.1	34.3 ± 0.7	517 ± 31	RT
HL (871 nm)	1.10 ± 0.01		6.4 ± 0.1	23.0 ± 0.3	194 ± 17	77 K
LL (861 nm)	0.69 ± 0.01		9.7 ± 0.2	42.5 ± 0.5		RT
LL (871 nm)	0.97 ± 0.01		5.8 ± 0.1	29.1 ± 0.3	120 ± 3	77 K
B870						
HL (900 nm)	0.80 ± 0.01	3.5 ± 0.2	63 ± 1	353 ± 13		RT
HL (912 nm)	1.10 ± 0.01	6.4 ± 0.1	23.0 ± 0.3	194 ± 17		77 K
LL (895 nm)	0.80 ± 0.04	5.7 ± 0.2	87 ± 3	263 ± 7		RT
LL (910 nm)	0.73 ± 0.04	5.4 ± 0.1	207 ± 3			77 K

[a] $\tau_{R,R1}$ – lifetimes of the rise components, τ_{1-3} – lifetimes of the decay components, RT – room temperature, HL – high light membrane, LL – low light membrane. The rise and decay lifetimes that are coupled due to energy transfer between B800 → B850 and B850 → B870 are shaded.

The EADS provide more insight into the temporal evolution of the system as a whole, and allow the identification of the spectral change associated with each specific decay lifetime.

Back energy transfer

An interesting aspect of a whole membrane study is the probability of reversed energy transfer. This will depend on the extent of the spectral overlap of the

 This journal is © The Royal Society of Chemistry 2018

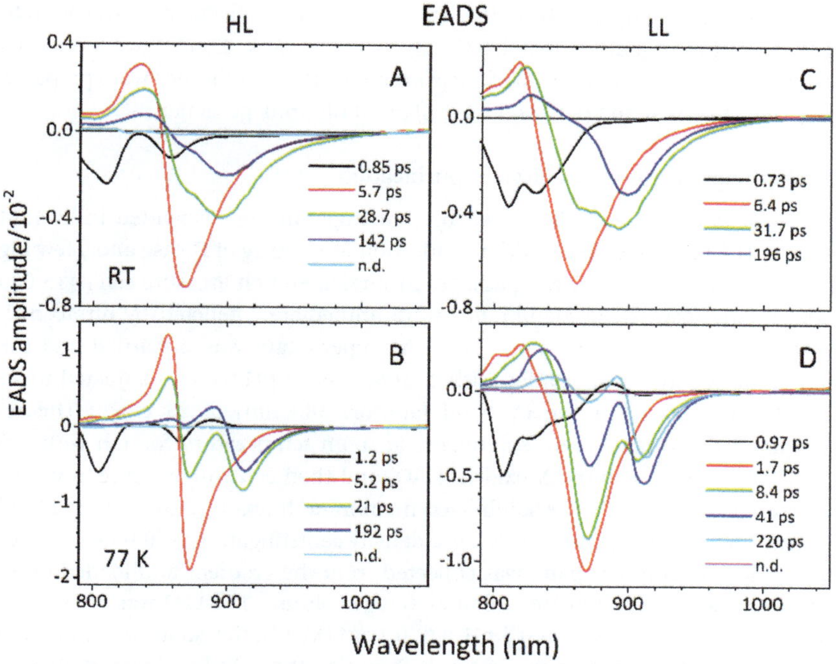

Fig. 12 Global analysis of the (A and B) HL and (C and D) LL membrane TA data with the application of a fitting model assuming the irreversible sequential decay of excitation from the shortest to longest decaying compartments.

emission spectrum of LH1 and the absorption spectrum of the longest wavelength band of the LH2 nearest to the excited LH1 complex. Unfortunately, the present experiments are unable to resolve any back transfers. One way to approach this problem would be to excite LH1 specifically and then to look for any transfer of excitation energy to the different 'B850' absorption bands in the membranes of cells grown at different light intensities. This type of approach has already been used.[35] However, what is really needed is an experimental protocol that is to be able to image the organisation of the LH complexes in individual membrane patches and which is then able to measure the energy transfer just from those small imaged regions. This approach would be able to avoid averaging over different arrangements of the LH complexes and thus would remove a lot of the current ambiguity. Unfortunately, at present there are no methods available to conduct these types of measurements.

Experimental

Cell growth

Cells of *Rps. acidophila* strain 7050 were grown anaerobically in C-succinate media in flat, glass bottles at 30 °C under anaerobic conditions at different light intensities as previously described.[18] HL cultures were placed between two incandescent bulbs at 100 μmol photons per s per m^2 and grown overnight. LL cultures were also grown under up to nine different conditions at 5, 10, 20, 25, 30,

40, 50, 75 and 100 cm away from the light source. These distances correspond to light intensities of (approximately) 35, 25, 15, 12, 10, 5, 3, 2 and 1 μmol photons per s per m^2. For the time-resolved measurements the membranes were prepared from cells grown at 100 (HL) and 15 (LL) μmol photons per s per m^2.

Membrane preparation and protein purification

Rps. acidophila cells were harvested by centrifugation, re-suspended in 20 mM Tris·Cl pH 8.0, homogenised with the addition of ~100 μg of DNase and a few mg of $MgCl_2$ and disrupted by two passages through a French Pressure cell (~15 000 psi). The photosynthetic membranes were immediately pelleted by ultracentrifugation (180 000 × *g*, 90 min, 4 °C). The supernatant was discarded and the membranes gently re-suspended with 20 mM Tris·Cl pH 8.0 and adjusted to an optical density (OD) at the Q_x band (approx. 580 nm) of 25 cm^{-1}. The re-suspended membranes were solubilised at room temperature for 1 h with 1% *N*,*N*-dimethyldodecylamine *N*-oxide (LDAO) and then centrifuged to remove any un-solubilised material. The solubilised membrane fraction (ODV = 15 loading) was fractionated using stepwise sucrose density centrifugation (150 000 × *g*, 4 °C, 16 h). The LH2 complex band was collected from the gradient and loaded on to a Q-Sepharose (GE Healthcare) anion exchange column. The LH2 was washed on the column with 20 mM Tris·Cl pH 8.0, 0.1% LDAO buffer solution and eluted with increasing concentrations of NaCl. Typically, the LH2 band would elute at around 100–150 mM NaCl.

Spectroscopic methods

The steady state absorption spectra shown in Fig. 1–7 were recorded using a Shimadzu UV-1700 spectrophotometer, with those shown in Fig. 8–9 recorded using a Shimadzu UV-1800 spectrophotometer. The spectroscopic measurements recorded at 77 K were carried out using a VNF-100 liquid nitrogen cryostat from Janis (Janis, USA) in 1 cm PMMA cuvettes. The samples were diluted in 60 : 40 v/v glycerol : buffer (20 mM Tris, 0.1% LDAO, pH 8.0), forming a fully transparent glass at cryogenic temperature. The absorbance of the samples in the Q_y region of Bchl a was set to ~0.5. The TA experiments were carried out using Helios, a femtosecond time-resolved and pump-probe absorption spectrometer (UltrafastSystems, LCC) coupled to a Spectra-Physics femtosecond laser system, as described in detail previously.[37] The samples were excited at 800 nm. The energy of the excitation beam was set to 50 nJ, corresponding to a photon flux of $\sim 3 \times 10^{13}$ photons per cm^2 per pulse (the beam was focused to a ~1 mm diameter spot on the sample and the laser repetition was set to 1 kHz), which was low enough to prevent a nonlinear response of the sample and excited singlet–singlet state annihilation.

Data analysis

Group velocity dispersion (temporal chirp) of the TA datasets was corrected using Surface Explorer software (UltrafastSystems LCC, Sarasota, FL, USA) by building a dispersion correction curve from a set of the initial times of transient signals obtained from single wavelength fits of the representative kinetics. Global analysis of the TA datasets was carried out by applying a sequential irreversible decay path model (A → B → C → D → …). In the sequential model, back-reactions are ignored on the assumption that the energy losses are large enough that the reverse reaction rates are negligible, however this may not be a true in a complex system such as a photosynthetic membrane. The spectral profiles obtained from this fitting of the TA datasets are termed Evolution Associated Difference Spectra (EADS). According to this model, a transient signal at any time delay and wavelength, $\Delta A(t,\lambda)$, can be decomposed to a superposition of the nth $C_i^{\text{S}}(t)$ $\text{EADS}_i(\lambda)$ products[38]

$$\Delta A(t,\lambda) = \sum_{i=1}^{n} C_i^{\text{S}}(t)\text{EADS}_i(\lambda) \tag{1}$$

where $C_i^{\text{S}}(t)$ are the time-dependent EADS concentrations defined by sequential decay. It should be noted that the complexity and structural inhomogeneity of the photosynthetic membrane precludes that decay of the excitation will occur in a simple sequential fashion. Therefore, the global fitting results should not be over-interpreted and assigned to spectroscopic signatures (bleaching of absorption, *etc.*) of specific pigment–proteins (LH2, LH1, *etc.*). Moreover, the EADS could also be mixed forms of a few independent spectra due to oversimplification of the fitting model. Global analysis was carried out using CarpetView software (Light Conversion Ltd., Vilnius, Lithuania). The concentration of the initial EADS was convoluted by the instrument response function (IRF) mimicked by a Gaussian with a full width at half maximum (FWHM) of ~200 fs (fixed fitting parameter).

Conclusions

Cells of *Rps. acidophila* strain 7050 produce more types of LH2 complexes than previously thought, and because these different spectroscopic forms are so similar it is not currently possible to individually isolate and purify them. Indeed with 12 members of the pucBA gene family present in the genome of *Rps. acidophila* 10050 (ref. 20), and with a similar number, possibly even more, present in the *Rps. acidophila* 7050 (ref. 19) genome, the actual apoprotein composition of these LH2 complexes could be even more complicated. Different apoprotein combinations might not even give different spectroscopic forms due to the heterogeneity of both of the different types of LH2 and their organisation in the photosynthetic membrane.[33] New experimental techniques are needed in order for definitive data to be obtained that can unequivocally explain the possible benefits that this species can obtain by regulating the type of LH2 that is made in response to changing environmental conditions.

Conflicts of interest

There are no conflicts to declare.

Acknowledgements

The authors gratefully acknowledge the support from the Photosynthetic Antenna Research Center, an Energy Frontier Research Center funded by the U.S. Department of Energy, Office of Science, Office of Basic Energy Sciences under the Award Number DE-SC 0001035.

References

1 R. A. Niederman and K. D. Gibson, in *The Photosynthetic Bacteria*, ed. R. K. Clayton and W. R. Sistrom, Plenum Press, New York and London, 1978, pp. 79–118.

2 J. Aargaard and W. R. Sistrom, *Photochem. Photobiol.*, 1972, **15**, 209–225.

3 S. Kaplan, in *The Photosynthetic Bacteria*, ed. R. K. Clayton and W. R. Sistrom, Plenum Press, New York and London, 1978, pp. 809–839.

4 B. Robert, R. J. Cogdell and R. van Grondelle, in *Light Harvesting Antennas in Photosynthesis*, ed, B. R. Green and W. W. Parson, Kluwer Academic Press, New York, 2003, vol. 5, pp. 169–194.

5 S. Scheuring and J. N. Sturgis, *Science*, 2005, **309**, 484–487.

6 J. N. Sturgis, J. D. Tucker, J. D. Olson, C. N. Hunter and R. A. Niederman, *Biochemistry*, 2009, **48**, 3679–3698.

7 C. S. Young and J. T. Beatty, in *Light Harvesting Antennas in Photosynthesis*, ed, B. R. Green and W. W. Parson, Kluwer Academic Press, New York, 2003, vol. 16, pp. 449–470.

8 R. J. Cogdell, T. D. Howard, N. W. Isaacs, K. McLuskey and A. T. Gardiner, *Photosynth. Res.*, 2002, **74**, 135–141.

9 B. Robert, in *The Purple Phototrophic Bacteria*, ed. C. N. Hunter, F. Daldal, M. C. Thurnauer and J. T. Beatty, Springer, Dordrecht, 2009, vol. 11, pp. 199–212.

10 R. J. Cogdell, A. Gall and J. Kohler, *Q. Rev. Biophys.*, 2006, **39**, 227–324.

11 S. L. Henry and R. J. Cogdell, in *Genome Evolution of Photosynthetic Bacteria*, ed. J. T. Beatty, Elsevier, Amsterdam, 2013, vol. 66, pp. 205–226.

12 T. H. P. Brotosudarmo, A. M. Collins, A. Gall, A. W. Roszak, A. T. Gardiner, R. E. Blankenship and R. J. Cogdell, *Biophys. J.*, 2011, **440**, 51–61.

13 G. McDermott, S. M. Prince, A. A. Freer, A. M. Hawthornthwaite-Lawless, M. Z. Papiz, R. J. Cogdell and N. W. Isaacs, Crystal structure of an integral membrane light-harvesting complex from photosynthetic bacteria, *Nature*, 1995, **374**, 517–521.

14 J. Koepke, X. C. Hu, C. Muenke, K. Schulten and H. Michel, *Structure*, 1996, **4**, 581–597.

15 A.-M. Carey, K. Hacking, N. Picken, S. Honkanen, S. Kelly, D. M. Niedzwiedzki, R. E. Blankenship, Y. Shimizu, Z. Y. Wang-Otomo and R. J. Cogdell, *Biochim. Biophys. Acta, Bioenerg.*, 2014, **1837**, 1849–1860.

16 K. McLuskey, S. M. Prince, R. J. Cogdell and N. W. Isaacs, *Biochemistry*, 2001, **40**, 8783–8789.

17 L. J. Cranston, A. W. Roszak and R. J. Cogdell, *Acta Crystallogr., Sect. F: Struct. Biol. Commun.*, 2014, **70**, 808–813.

18 Y. Z. Ma, R. J. Cogdell and T. Gillbro, *J. Phys. Chem. B*, 1997, **101**, 1087–1095.

19 Y. Z. Ma, R. J. Cogdell and T. Gillbro, *J. Phys. Chem. B*, 1998, **102**, 881–887.

20 D. Zigmantas, E. L. Read, T. Mančal, T. Brixner, A. T. Gardiner, R. J. Cogdell and G. R. Fleming, *Proc. Natl. Acad. Sci. U. S. A.*, 2006, **103**, 12672–12677.

21 G. Deinum, S. C. M. Otte, A. T. Gardiner, T. J. Aartsma, R. J. Cogdell and J. Amesz, *Biochim. Biophys. Acta*, 1991, **1060**, 125–131.

22 T. Kolubayev, N. E. Geacintov, G. Paillotin and J. Breton, *Biochim. Biophys. Acta*, 1985, **808**, 66–76.

 This journal is © The Royal Society of Chemistry 2018

23 R. J. Cogdell, I. Durant, J. Valentine, J. G. Lindsay and K. Schmidt, *Biochim. Biophys. Acta*, 1983, **722**, 427–435.

24 A. T. Gardiner, S. Takaichi and R. J. Cogdell, *Photosynth. Res.*, 1993, **38**, 159–167.

25 A. T. Gardiner, R. C. MacKenzie, S. J. Barrett, K. Kaiser and R. J. Cogdell, *Photosynth. Res.*, 1996, **49**, 223–235.

26 R. M. Mulvaney, PhD thesis, University of Glasgow, 2013.

27 E. Giraud, S. Zappa, L. Vuillet, J. M. Adriano, L. Hannibal, J. Fardoux, C. Berthomieu, P. Bouyer, D. Pignol and A. Vermegli, *J. Biol. Chem.*, 2005, **280**, 32389–32397.

28 M. Li, S. Noll and J. T. Beatty, *Curr. Microbiol.*, 2010, **61**, 429–434.

29 K. Evans, A. P. Fordham-Skelton, H. Mistry, C. D. Reynolds, A. M. Lawless and M. Z. Papiz, *Photosynth. Res.*, 2005, **85**, 169–180.

30 E.-A. Heinemeyer and K. Schmidt, *Arch. Microbiol.*, 1983, **134**, 217–221.

31 A. Angerhofer, F. Bornhauser, A. Gall and R. J. Cogdell, *Chem. Phys.*, 1995, **194**, 259–274.

32 T. H. P. Brotosudarmo, R. Kunz, P. Boehm, A. T. Gardiner, V. Moulisova, R. J. Cogdell and J. Koehler, *Biophys. J.*, 2009, **97**, 1491–1500.

33 M. Taniguchi, S. Henry, R. J. Cogdell and J. S. Lindsey, *Photosynth. Res.*, 2014, **121**, 49–60.

34 V. Sundstrom, T. Pullerits and R. van Grondelle, *J. Phys. Chem. B*, 1999, **103**, 2327–2346.

35 L. Luer, V. Moulisova, S. Henry, D. Polli, T. H. P. Brotosudarmo, S. Hoseinkhani, D. Brida, G. Lanzani, G. Cerullo and R. J. Cogdell, *Proc. Natl. Acad. Sci. U. S. A.*, 2012, **109**, 1473–1478.

36 S. Hess, M. Chachisvilis, K. Timpmann, M. R. Jones, G. J. S. Fowler, C. N. Hunter and V. Sundstrom, *Proc. Natl. Acad. Sci. U. S. A.*, 1995, **92**, 12333–12337.

37 D. M. Niedzwiedzki, T. Tronina, H. Liu, H. Staleva, J. Komenda, R. Sobotka, R. E. Blankenship and T. Polivka, *Biochim. Biophys. Acta, Bioenerg.*, 2016, **1857**, 1430–1439.

38 I. H. M. van Stokkum, D. S. Larsen and R. van Grondelle, Global and target analysis of time-resolved spectra, *Biochim. Biophys. Acta, Bioenerg.*, 2004, **1657**, 82–104.

Faraday Discussions

DISCUSSIONS

Bionanophotonics: general discussion

Amitava Chandra, Amitabha Chattopadhyay, Richard Cogdell, Ankona Datta, Arijit De, Shaina Dhamija, Murali Golla, Stefan Haacke, Mahesh Hariharan, John Helliwell, Roberto Improta, Ramapurath S. Jayasree, Mike Jones, Joshy Joseph, Tolga Karsili, Bern Kohler, Retheesh Krishnan, Imon Mandal, Dimitra Markovitsi, Himani Medhi, Padmaja P. Mishra, Priyadarshi Roy Chowdhury, Manas Sarangi, Ilme Schlichting, John Seddon, Amit Sharma, Atchimnaidu Siriki, Rajaram Swaminathan, Rienk van Grondelle, Reji Varghese, Ravindra Venkatramani and Anthony Watts

DOI: 10.1039/C8FD90007H

Richard Cogdell opened a general discussion of the paper by Amitabha Chattopadhyay: The analysis depends on the spatial resolution, so when you classify them as dimers or trimers they have to be within a certain volume, so can you say that they are really dimers or are they just very close to each other?

Amitabha Chattopadhyay responded: The analysis depends on spatial correlation of the photobleaching data. The size of the oligomers is deduced from fitting the model.

Amit Sharma asked: What is the size of these clusters when they are dimers or trimers? Would it not be possible to obtain the size through FRET studies?

Amitabha Chattopadhyay answered: The size of the oligomers will be in the nm range. FRET will provide proximity (distance) information, not size.

Rajaram Swaminathan said: I have two questions. First, what role does lipid composition play in the results shown? Will results be different with another lipid composition? Second, are there reports in the literature on GPCR oligomerization in artificial liposomes where composition can be controlled?

Amitabha Chattopadhyay responded: To answer the first question, results from our laboratory and a number of groups have shown that lipid composition could play a role in GPCR function and oligomerization. To the second question, there are not too many examples of GPCRs oligomers studied in reconstituted liposomes. This is due to a variety of reasons which include difficulty in purification, and more importantly, lack of other cellular ingredients (necessary for oligomerization) in liposomal systems.

Anthony Watts asked: Have you confirmed that EYGP does not influence the oligomerization state of the receptor? Have you any information on how oligomerization and cholesterol affects downstream signal processing, such as G protein coupling and bias coupling?

Amitabha Chattopadhyay replied: Yes – we have previously shown[1] that attachment of EYFP does not alter the affinity and G-protein coupling of the serotonin-1A receptor. Moreover, the type of GFP molecules used by us do not show oligomerization under these conditions. We have previously published how cholesterol affects downstream signaling.[2] We plan to monitor the effect of oligomerization on dowmstream signaling.

1 T. J. Pucadyil, S. Kalipatnapu, K. G. Harikumar, N. Rangaraj, S. S. Karnik and A. Chattopadhyay, *Biochemistry*, 2004, **43**, 15852–15862.
2 T. J. Pucadyil and A. Chattopadhyay, *Biochim. Biophys. Acta*, 2007, **1768**, 655–668.

John Helliwell communicated: Is there a GPCR serotonin crystal structure? If so is there a serotonin binding site *between* adjacent GPCRs, as well as to one GPCR? This would be an independent corroboration of the photobleaching image correlation spectroscopy method, which I assume is fairly new.

Amitabha Chattopadhyay communicated in response: The crystal structure of the serotonin-1A receptor is not available yet. However, crystal structures are available for a few other serotonin receptors. The crystal structure I showed in one of my slides where cholesterol molecules were shown to be bound between two monomers of a receptor was that of the β2-adrenergic receptor.

Priyadarshi Roy Chowdhury communicated: How do you identify the cholesterol binding sites in the serotonin1A receptor?

Amitabha Chattopadhyay communicated in reply: By examining the presence of CRAC motifs in the sequence of the receptor. Please note that these are putative cholesterol binding sites.

Imon Mandal opened discussion of the paper by Ramapurath S. Jayasree: Does the NIR absorption depend on the length of the nanorod? If it does, how will that affect the photothermal activity of the GNR and the overall drug release?

Ramapurath S. Jayasree answered: Yes, the NIR absorption depends on the aspect ratio of the nanorod. In this case, the absorption maximum was around 786 nm, which extends from 750 to 850 nm. A 786 nm laser would have been ideal. However, our laser has a wavelength of 808 nm and hence we used this for photothermal activity.

Bern Kohler asked: Can you comment more on the chemistry that you use to attach the drug to the gold nanorod? Could you also elaborate on the mechanism of drug release?

 This journal is © The Royal Society of Chemistry 2018

Ramapurath S. Jayasree replied: Thank you for the question. During the drug conjugation the formation of an ester bond is expected, which gets cleaved during drug release as the laser is applied. Drug release is hypothesized to occur by ester hydrolysis which is observed by the increase in fluorescence of the drug on laser irradiation.

Bern Kohler queried: The UV-visible absorption spectra in Fig. 1a in your paper appear to show a decrease in absorbance below 325 nm when the mitoxantrone drug is taken up by the gold nanorods, following treatment with folic acid. Both folic acid and mitoxantrone show strong absorption in the UV, so why doesn't the presence of both lead to the greatest absorption?

Ramapurath S. Jayasree answered: The reaction strategy adopted was such that conjugation of the drug was the final process in the system, which already has folic acid conjugated. The amount of drug used is half the concentration of folic acid so as to retain its identity. In the IR spectrum ester bond formation is visible in the final system; this ester bond may be formed between one of the carboxylic groups of folic acid and –OH of the mitoxantrone drug. This could be the reason for the decrease in absorption at 325 nm. From the peak at 275 nm, the presence of folic acid can be confirmed.

Anthony Watts asked: Are you sure that you have sufficient resolution to confirm co-localization of gold nanoparticles and nuclear marker, bearing in mind the Abbye diffraction limit problems when using co-localization of images?

Ramapurath S. Jayasree replied: Thank you for the question, we agree with your concern. The microscope had a resolution limit of 200 nm, which is nowhere near the size of the nanorod. Moreover, diffraction limit issues are prominent at red emission and hence need special attention in order to capture the image. Please note that we don't claim to visualize the co-localization of gold nanorods by performing the nuclear stains. We haven't visualized the rods as such inside the cell. But, we have considered the red fluorescence signal originating from the gold nanorods for ascertaining the co-localization of the material inside the cell. This is very clear from the images, which were acquired by optimizing the parameters and applying aberration corrections.

Atchimnaidu Siriki said: Your final target molecule which you used for photothermal treatment shows high fluorescence, how will that molecule be useful for photothermal treatment? Did you calculate the quantum yield for that final molecule? And what concentrations of final molecule were used for photothermal treatment?

Ramapurath S. Jayasree answered: The final material MTX@FPGNR is used for photothermal treatment in the bound stage which has comparatively low fluorescence (the result shown in the fluorescence spectrum is normalised). We didn't calculate the quantum yield. We are planning to do it soon, before doing any *in vivo* experiments. On laser irradiation, the drug is released from the gold nanorod and the inherent free electron transition of the rod gets exposed, facilitating its photothermal properties. We used a 1 mg ml^{-1} solution of the final product for photothermal therapy.

Reji Varghese commented: The efficacies of PTT and chemotherapy have to be separately investigated. What is the advantage of combined therapy *versus* individual therapy?

Ramapurath S. Jayasree answered: Thank you for the question. The efficacies of PTT and chemotherapy have been investigated separately in c6 glial cells and HeLa cells and results are given in the ESI† of our article. Fig. S6† gives an idea regarding the efficacy of chemotherapy alone while Fig. S7† gives an idea of PTT alone (GNR – represented by blue bars). It is clear from these results that the individual therapeutic approaches have less efficiency compared to the combined approach. The intention of having a dual modal therapeutic approach was to increase the efficiency, which we could achieve. I hope this response is clear, if not I will be happy to provide more explanation.

Joshy Joseph asked: Can you comment on the extent of anti-cancer drug loading onto these functionalized gold nanorods and their release with/without the laser irradiation?

Ramapurath S. Jayasree replied: Thank you for the question. The loading capacity of the system is approximately 675 μM. Without laser irradiation there is no drug release (see Fig. S2 in the ESI†). Drug release is qualitatively evaluated by measuring the increase in fluorescence of the drug (the cleavage of ester bonds formed during drug conjugation with GNR is predicted as we observe an increase in the fluorescence of the drug on laser irradiation, which initially got quenched on conjugation with the gold rods). Drug release with and without laser irradiation at two different pH values is given in Fig. S4.†

Himani Medhi communicated: This is a nice piece of work. This work will be helpful with further deep investigations if it can be applied directly to the target cancerous cells in a living system.

Ramapurath S. Jayasree communicated in reply: Thank you for the comment. We are working on it further to extend it to preclinical testing and if possible to clinical trials. Before which we have to collect the experimental data which are lacking, such as the quantum yield of the final system, stability, *in vitro* drug release profile, *etc.*

John Helliwell communicated: For the method described in Section 2.2.2 on page 4 of your article ("*in vitro* drug release study"), what percentage of the mitoxantrone is released from the gold nanorods on irradiation with the 808 nm laser for 1.3 s with 1 W power? This value would I presume be the best that can be achieved, and therefore it is important to know it, since *in vivo* I imagine the irradiation induced release of the drug will be less efficient as a process.

Ramapurath S. Jayasree communicated in response: In the case of *in vitro* drug release, the percentage drug release depends on the total amount of the final material used. When used in cells, more than 50% of (what we have used) drug release was observed. In the case of *in vivo* drug release, we agree with Prof. Helliwell that release will be less due to various reasons; but it can be controlled by suitably changing the total concentration of final material and the irradiation time. The concentration of material which would be used for *in vivo* studies will be optimized in future studies.

Joshy Joseph opened a general discussion of the paper by Ankona Datta: I have two questions. First, the authors have done theoretical calculations to find out the optimum distance (between amino acid residues) for placing the FRET pair for effective interactions. Do these calculations also account for the flexible spacer, which can change the effective distance between the FRET pair depending on different confirmations? Second, have the authors tried placing the fluorophore in the neighboring amino acids to check whether it is more/less effective than the computationally selected amino acid residues? This will probably ascertain the role of the flexible spacer.

Ankona Datta answered: To answer the first question: the donor tryptophan was a part of the peptide that we synthesized. The acceptor IAEDANS was conjugated to the peptide *via* a flexible linker. We have not included either the tryptophan or the flexible linker in our calculations which examine the original Gel 150–169 sequence. While the presence of the linker can change the effective distance between the FRET pair, we do find that the FRET efficiency increases as the peptide undergoes the coil–helix transition upon binding PI(4,5)P2 as predicted by the distance changes captured in the simulations. For the second question: this is a nice suggestion; we plan to do these experiments in the future. However we would also like to mention that the computations suggested multiple residue pairs like residues 11–18 and residues 10–18 which showed maximal inter-residue distance changes during peptide coil–helix transitions. In the present study, we selected residues 11–18 for our experimental investigations.

Padmaja P. Mishra asked: What is the smallest number of amino acids where you see this structural change?

Ankona Datta answered: The structural change is observed in the Gel 150–169 peptide sequence which is a 20 amino acid peptide. In the future, we could study shorter fragments of the peptide to check if the peptide can retain its helical form upon binding to lipids.

Arijit De commented: Very nice work! I guess the positions (11 and 18?) of the FRET pairs are on the same side of the helix – is that right? If so, during the coil to helix transition, when the helix starts to unfold, the angle between the transition dipoles of FRET pairs should also change which, apart from a change in the point to point distance, should change the FRET efficiency.

Ankona Datta answered: Thank you for appreciating our efforts. The residues 11 and 18 on the 20 amino acid Gel 150–169 correspond to Glutamine 160 and Glycine 167. These residues are on the same side of the helix. In order to evaluate the effect of the orientation of these residues on the FRET efficiency we also calculated the relative orientation of Glutamine 160 and Glycine 167 for 5000 conformations of the helical and coil peptide states obtained from the MD simulations. The relative orientation of residues was defined as the angle between the vectors normal to the plane containing the Cα and the backbone carbonyl group of the residues. The average orientation of Glutamine 160 and Glycine 167 peaked at around 30° for the helical conformers while the orientations for the coil conformers were random. This result has been summarized in our manuscript in Fig. 3 and indicates that the relative orientation of the two residues would favor increased FRET efficiency in the helical conformers. However, since we attached the acceptor *via* a flexible linker which is not present in the simulations, the effect of the favorable orientation cannot be directly assessed.

Anthony Watts asked: How sure are you that the peptide–membrane association is not a simple electrostatic association, and have you examined other anionic lipids? Your salt concentrations and ionic strengths are sufficiently high in the experiment that all charges are shielded.

Ankona Datta responded: We have examined the binding of the Gel 150–169 peptide with different anionic phospholipids in our previous work.[1] The binding affinity of the peptide is highest for PI(4,5)P2 (K_d 3 μM). Other anionic phospholipids like phosphatidylserine (K_d > 100 μM) and phosphatidylglycerol have much lower binding affinities to the peptide. We have performed all experiments at physiological pH 7.4 using 20 mM HEPES buffer and 100 mM NaCl. These results indicate that the peptide-membrane interaction is not a simple electrostatic association. We believe that the positively charged residues in the C-terminus of the peptide are sequentially arranged for optimal electrostatic interaction with PI(4,5)P2. However, the interactions of the hydrophobic residues near the N-terminus are also important for the selectivity. When we delete residues from the N-terminus we observe that the selectivity of the peptide toward PI(4,5)P2 is slightly lowered.[1] For example, the 13 amino acid peptide Gel 157–169 affords a K_d value of 1.4 μM with PI(4,5)P2 and 14 μM with PS. We have also performed molecular dynamics simulations on the helical conformer of the Gel 150–169 peptide in the presence of a lipid bilayer containing 0.74 mol% of PI(4,5)P2 on one leaflet using explicit solvent and 120 mM NaCl to mimic our experimental conditions. The simulation results indicated that the arginine residues at the peptide C-terminus formed stable salt bridge interactions with the PI(4,5)P2 headgroups throughout the entire timescale of the simulation at these ionic strength conditions, further corroborating our experimental observations.

1 S. Mondal, A. Rakshit, S. Pal and A. Datta, *ACS Chem. Biol.*, 2016, **11**, 1834–1843.

Anthony Watts asked: A unique PI(4,5)P2 binding site in your peptide would be a unique finding – is it found elsewhere?

Ankona Datta answered: The Gel 150–169 sequence is a PI(4,5)P2 binding site for the actin binding protein Gelsolin. This had been shown by Paul Janmey and his co-workers earlier.[1] Our studies on the Gel 150–169 peptide tagged to a polarity sensitive dye allowed the determination of K_d values for the interaction of the peptide with lipid vesicles containing different negatively charged phospholipids.[2] The results indicated highest affinity of the peptide toward PI(4,5)P2 followed by PI4P. Other negatively charged phospholipids like phosphatidylserine showed 30 times lower affinity to the peptide. Actin binding proteins like Villin and Profilin also have sequences that are similar to Gel 150–169 which are proposed to be phosphoinositide binding sites.[3] Whether these sites are selective toward PI(4,5)P2 can be investigated in the future.

1 W. Xian, R. Vegners, P. A. Janmey and W. H. Braunlin, *Biophys. J.*, 1995, **69**, 2695–2702.
2 S. Mondal, A. Rakshit, S. Pal and A. Datta, *ACS Chem. Biol.*, 2016, **11**, 1834–1843.
3 P. A. Janmey, W. Xian and L. A. Flanagan, *Chem. Phys. Lipids*, 1999, **101**, 93–107.

Amitabha Chattopadhyay commented: You could change the ionic strength. This would help in ruling out electrostatic factors.

Ankona Datta responded: The interaction of the Gel 150–169 peptide with phosphatidylinositol(4,5)bisphosphate (PI(4,5)P2) is primarily based on electrostatic interactions between lysine and arginine residues near the C-terminus of the peptide along with hydrophobic interactions with residues close to the N-terminus of the peptide.[1] In the present manuscript, we have also performed molecular dynamics simulations on the helical conformer of the Gel 150–169 peptide in the presence of a lipid bilayer containing 0.74 mol% of PI(4,5)P2 on one leaflet using explicit solvent and 120 mM NaCl to mimic the experimental conditions which utilize physiological ionic strengths. The simulation results indicated that the arginine residues formed stable salt bridge interactions with the PI(4,5)P2 headgroups throughout the entire timescale of the simulation. Previous work from our group[2] indicates that the binding affinity of the peptide is highest for PI(4,5)P2 (K_d 3 μM). Other anionic phospholipids like phosphatidylserine (K_d > 100 μM), phosphatidylglycerol, and phosphatidylethanolamine have much lower binding affinities to the peptide. We believe that the positively charged residues in the C-terminus of the peptide are sequentially arranged for optimal interaction with PI(4,5)P2. Further, the interactions of the hydrophobic residues near the N-terminus are also important for the selectivity. When we delete residues from the N-terminus we observe that the selectivity of the peptide toward PI(4,5)P2 is slightly lowered. For example the 13 amino acid peptide Gel 157–169 affords a K_d value of 1.4 μM with PI(4,5)P2 and 14 μM with PS.[2] As suggested by Prof. Chattopadhyay it would be interesting to perform experiments to study the effect of ionic strength on the salt-bridge interactions between the PI(4,5)P2 headgroup and the peptide.

1 I. Liepina, C. Czaplewski, P. Janmey and A. Liwo, *Biopolymers*, 2003, **71**, 49–70.
2 S. Mondal, A. Rakshit, S. Pal and A. Datta, *ACS Chem. Biol.*, 2016, **11**, 1834–1843.

Ilme Schlichting remarked: Your explanation of specificity relies on hydrophobic interactions of the N- of the peptide with the membrane and H-bonds/saltbridges of the C-terminus of the peptide with PI(4,5)P2. In holo-gelsolin these interactions are not possible since the peptide is part of the protein. Can you comment, please?

Ankona Datta responded: The interaction of the Gel 150–169 peptide with (PI(4,5)P2) is primarily based on electrostatic interactions between lysine and arginine residues near the C-terminus of the peptide along with interactions with hydrophobic residues close to the N-terminus of the peptide. The peptide used for all experiments and also for a sub-set of the molecular dynamics simulations had an amide moiety in the C-terminus. This would also be the situation when the peptide is embedded in the protein. Importantly, the stable salt-bridge interactions of the residues in the C-terminus with PI(4,5)P2 were with arginine side chains which would also be present in the protein. The C-terminus amide did not show any interactions with the PI(4,5)P2 headgroup. In the same vein, the hydrophobic residues toward the N-terminus of the peptide came closer to the membrane as a result of the binding of the C-terminus with the PI(4,5)P2 headgroup. Both the electrostatic and hydrophobic interactions were with the side chains of amino acid residues. These interactions could also occur in the holoprotein.

Tolga Karsili said: Constant temperature and pressure dynamics are known to under-sample low-energy conformations of proteins. Are you confident that the dominant conformations are sufficiently sampled in the propagation time window reported in your manuscript?

Ankona Datta answered: In our peptide simulations, we employ Langevin temperature control and Nosé Hoover Langevin piston pressure control which rigorously produce canonical sampling. Nevertheless, the conformation space for peptides is vast and rigorous sampling of all conformational states may require simulation trajectories spanning hundreds of microseconds.[1] However, the objectives of the peptide simulations in the present manuscript are far simpler: to sample inter-residue separations for two specific conformations (helix and coil forms) of the peptide. Circular dichroism experiments indicate that the Gel 150–169 peptide and similar peptides from phospholipid binding sites of actin binding proteins adopt a coil conformation in buffer in the absence of lipids.[2] In contrast, NMR experiments in organic solvents captured a helical conformation for Gel 150–169 which is the relevant lipid binding form for the peptide.[2] We, therefore directly sampled the helix form at 300 K and used high temperature (500 K) simulations to unfold the peptide and generate the extended coil form (Fig. 1 in the present manuscript). Subsequently we used 5000 conformations over 10 ns from both the coil and helix states to calculate and contrast average inter-residue distances (Fig. 2 in the present manuscript). The average distances obtained from simulations provided appropriate guidelines for the design of a peptide based fluorescence resonance energy transfer sensor. The experimental data from steady state and time-resolved fluorescence proved that the guidelines obtained from the computations were optimum. Hence, we can confirm that we have sufficiently sampled inter-residue distances in the two relevant conformations of the Gel 150–169 peptide in our simulations.

1 Y.-S. Lin and V. S. Pande, *Biophys. J.*, 2012, **102**, 315–324.
2 W. Xian, R. Vegners, P. A. Janmey and W. H. Braunlin, *Biophys. J.*, 1995, **69**, 2695–2702.

John Seddon opened a general discussion of the papers by Amitabha Chattopadhyay, Ramapurath S. Jayasree and Ankona Datta: Can you identify the common themes of the session?

Amitabha Chattopadhyay replied: Application of bionanophotonics to biomedical problems.

Ramapurath S. Jayasree added: Bionanophotonics is an emerging field in biomedical science.

Ankona Datta responded: Using biophysical and chemical tools to elucidate molecular mechanisms underlying biological processes and further using the insights obtained to develop tools for theranostic applications.

Amitava Chandra addressed Ramapurath S. Jayasree: You mentioned that the anticancer agents are incorporated into the cell by the folic acid pathway, could you talk more about this pathway? Is there any specific interaction that is related to the incorporation of the materials into the cell?

Ramapurath S. Jayasree replied: Thank you for the question. The concept of folate receptor mediated delivery of drugs and drug carrying systems is well established. There are several articles in this line.

Atchimnaidu Siriki said: Related to the drug release at pH 7 and pH 4 after treatment with an 808 nm laser, you have shown that release of the drug at pH 4 is very high, if you consider a cancer cell will it have that pH 4 in every organelle? And which part of the cancer cell is your final molecule entering?

Ramapurath S. Jayasree answered: It is well known that the extracellular region itself shows acidic pH in tumors.[1] When the cell organelles are considered, the lysosomal bodies of cells are more acidic and pH will be less than 4. In the cell uptake studies where we used the nuclear stains, we found that the drug enters more into the cytoplasm, though we expect entry into the nucleus as the drug used is expected to intercalate with DNA. A detailed time bound study in this direction is required to know more about this.

1 Y. Kato, S. Ozawa, C. Miyamoto, Y. Maehata, A. Suzuki, T. Maeda and Y. Baba, *Cancer Cell Int.*, 2013, **13**, 89.

Arijit De said: When you shine laser on the gold nanorod–drug (doxorubicin) construct, which is already inside the cancer cell, the cell can die either because of release of the drug or because of heating alone. My question is: can you pinpoint what is actually killing the cell, the free drug or the laser-induced heating? Now, as you said, there is a specific pH-dependent chemistry that leads to the cleavage of the ester bond releasing the drug; one can still wonder why that particular bond cleaves under irradiation, from the reaction dynamics point of view of unimolecular dissociation, but that is a fairly complicated process to have a simple answer.

Ramapurath S. Jayasree answered: Thank you for the good question. The whole purpose of using multimodal imaging and therapy is to increase the efficiency of both. In this study there are two simultaneous therapeutic approaches as well as two imaging approaches utilizing the same system. As you can see from Fig. S5 and S6 of the ESI† of our article the therapeutic effects of drug alone and PTT alone are much less than the combined effect, which fulfills the intended purpose of increasing the therapeutic effect. So, the cell death happens due to the effect of both drug release and laser induced heating.

Talking about the chemistry, the formation of the ester bond on drug conjugation is clear and is observed in the IR spectrum as well. The cleavage is predicted as we observe an increase in the fluorescence of the drug on laser irradiation, which initially got quenched on conjugation with the gold rods. Please note that MTX is the only molecule which has sufficient NIR fluorescence in the whole system. So we assume that the drug regains its identity on laser irradiation by ester hydrolysis, increasing the fluorescence. The acidic pH of the cells favors the reaction. Due to laser irradiation there may be dissociation of other bonds also. However, no major changes were observed and also a detailed study of this aspect is not considered in this study.

Mahesh Hariharan addressed Ankona Datta: I have a general question related to the FRET pair that is often used for probing biological environments. Do you also consider aggregation of peptides, where you can have energy transfer from one peptide to another peptide that might give wrong information?

Ankona Datta answered: It is true that aggregation of peptides can lead to intra-molecular FRET. In order to avoid aggregation affects we have performed all of our experiments at low μM concentrations of the peptide.

Mahesh Hariharan asked: How much influence will the acceptor/donor chromophore have in steady state emission when surrounded by the lipids?

Ankona Datta replied: In our studies we observe that the acceptor IAEDANS exhibits polarity sensitive emission. IAEDANS is placed close to the C-terminus which directly attaches to the PI(4,5)P2 headgroup through strong electrostatic interactions. Hence the steady state emission of IAEDANS is affected by the presence of lipids. However the donor tryptophan emission is not affected by the presence of lipids. The tryptophan residue is placed further away from the C-terminus towards the center of the peptide and both steady-state fluorescence and lifetime measurements indicate that the tryptophan emission is unaffected by the presence of lipids. Hence our conclusions on peptide conformational changes upon lipid binding have been drawn on the basis of the tryptophan emission data.

Rajaram Swaminathan said: I am interested in what is happening inside the living cell compared to *in vitro* measurements. Could you comment on the possibility of performing FRET measurements inside the cell?

Ankona Datta answered: In order to perform FRET experiments in living cells we would require FRET donor–acceptor pairs which absorb and emit in the visible wavelengths. We plan to explore such FRET pairs in the future.

Ravindra Venkatramani addressed Amitabha Chattopadhyay: Extrapolating the results in your paper, is it possible to systematically vary cholesterol concentration to selectively obtain monomer, dimer, trimer, and higher order oligomer populations for the GPCRs in membranes?

Amitabha Chattopadhyay responded: This is more involved and complicated. Different cell types have varying cholesterol content. Moreover, upon cholesterol replenishment, one is not sure whether cholesterol is put back from where it was depleted. These factors make the situation more complicated. A simple extrapolation is not possible.

Retheesh Krishnan addressed Ramapurath S. Jayasree: In your paper you have mentioned that the weak emission of PGNR and FPGNR are enhanced in the presence of PEG and folic acid. You have also mentioned that two strong emission peaks are present for the MTX@FPGNR. As the spectrum is normalized the strength of emission is not visible in it. You are using Raman to detect the delivery of the drug. Doesn't the strong emission of MTX@FPGNR affect your Raman measurements?

Ramapurath S. Jayasree answered: Thank you for the question. The data prior to normalization is shown here in Fig. 1. Please note that the fluorescence of the gold nanorods increases in PGNR and FPGNR. This increase is nominal with respect to fluorescence intensity. On addition of the drug mitoxantrone the fluorescence further increases. This could be due to energy transfer from gold to the drug (this we have not considered in detail and it is not explained in the paper). However, this increased fluorescence itself was not very strong as you see from the *y*-axis. Hence we could acquire Raman spectra without hindrance from

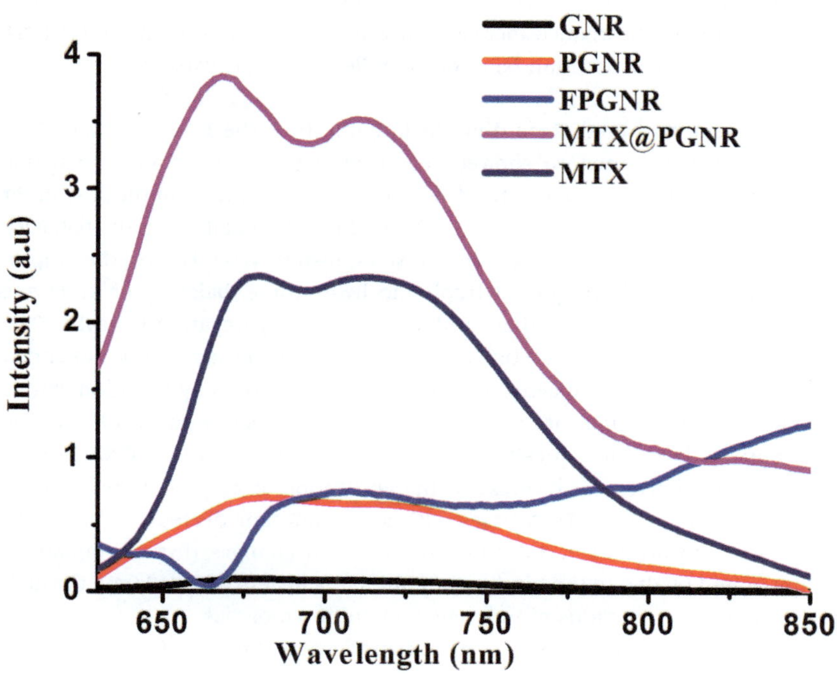

Fig. 1 Emission spectra of GNR, PGNR, FPGNR and MTX@FPGNR before normalization.

the fluorescence, initially. Moreover, the presence of GNR helps to enhance the Raman signal by many fold through the mechanism of SERS.

Imon Mandal opened a general discussion of the paper by Joshy Joseph: What is the molecular basis of the interaction (*e.g.* electrostatic, H-bonding *etc.*) between F-An and DNA molecules? How will the DNA absorption change if the pH of the F-An solution is altered?

Joshy Joseph replied: The functionalised fullerene (F-An) derivative containing aniline and fullerene subunits self-assembles in 10% DMSO–PBS mixture into 3–5 nm sized nanoclusters with the aniline groups projecting outside. Aniline remains neutral at physiological pH conditions and can bind to the grooves (preferably the minor groove) of DNA through H-bonding and van der Waals interactions. The initial interaction of F-An clusters with DNA is governed by the groove interactions of the pendent aniline moiety which might be sufficient enough to align the clusters along the grooves, facilitating further interaction of fullerenes through groove binding and π–π interactions.

We have carried out all the experiments reported in this paper in 10 mM phosphate buffer containing 2 mM NaCl, at pH 7.4. At higher pH and lower pH conditions the B-form DNA structure may get perturbed. At lower pH (pH < 5) the aniline moieties in F-An can get protonated and an additional, electrostatic mode of binding can be expected.

Bern Kohler said: I have two questions. First, how monodisperse are the fullerene clusters and what prevents their further aggregation? Second, you refer to the decrease in absorbance seen when the fullerene nanoclusters are disassembled as a hypochromic effect, but in DNA disassembly of stacked nucleobases leads to an absorbance increase. How does interaction with the CT-DNA modify the 336 nm absorption band of the fullerene nanoclusters?

Joshy Joseph answered: Regarding the first question, the TEM studies of F-An in 10% DMSO–PBS mixture showed the formation of almost mono-disperse nanoclusters with size ranging from 3–5 nm. Under similar conditions, other fullerene derivatives (F-Py and F-PTz) exhibited different self-assembly behavior (molecularly dissolved and larger nanoclusters, respectively). Hence, the amphiphilic structure and unique hydrophobic–hydrophilic balance of fullerene derivatives is responsible for their uniform clustered aggregation behavior. The clusters may be considered as a fullerene core on the inside and aniline appended triethyleneglycol group projecting outside. Therefore, further aggregation will be prevented due to the presence of the more hydrophilic periphery of the clusters.

To answer the second question, the sequential addition of CT-DNA to F-An solution showed a regular increase in the absorption at 260 nm corresponding to the absorption of the DNA bases with a concomitant decrease in the F-An nanoclusters absorption at 336 nm. The observed changes (hypochromism at 336 nm) could be due to the binding of F-An nanoclusters with CT-DNA through groove binding interactions of fullerene and aniline moieties.

The disassembly of CT-DNA/fullerene nanoclusters was not presented in this report. However, according to our previous report,[1] we observed that disassembly of F-An nanoclusters from DNA during DNA melting studies led to an increase in the absorption at 260 nm with concomitant enhancement in the tail-end absorption (400–600 nm).

1 S. K. Vittala, S. K. Saraswathi and J. Joseph, *Chem.–Eur. J.*, 2017, **23**, 15759–15765.

Reji Varghese asked: Is the space available in the groove (minor or major) sufficient enough for a cluster with a size of 3–4 nm to bind effectively?

Joshy Joseph answered: In this paper, we have used functionalised fullerene (F-An) containing aniline and fullerene sub-units which in the presence of 10% DMSO–PBS mixture preferably self-assemble into 3–5 nm sized nanoclusters with aniline groups projecting outside. As we mentioned in the paper, both aniline and fullerene are known to interact with DNA through groove interactions. In the case of fullerene clusters, based on the observed absorption changes and the induced circular dichroism (ICD) for CT-DNA/F-An complexes corresponding to the F-An absorption (336 nm), we proposed a possible groove binding of F-An clusters along the CT-DNA templates. The fullerene clusters (3–5 nm) as such may be slightly large to fit on the DNA grooves. However, the initial interaction of F-An clusters with DNA is governed by the groove interactions of the pendent aniline moiety which might be sufficient enough to align the clusters along the grooves facilitating further interaction of fullerenes through groove binding and π–π interactions.

Padmaja P. Mishra said: There are many experimental/computational efforts in this area where people have found unwinding or unzipping of DNA, while interacting with small charged molecules/particles. Do you find any such effect here? Does it bind to single stranded DNA?

Joshy Joseph responded: As reported previously, the unzipping or unwinding of DNA with fullerene can happen in the presence of external stimuli such as temperature or light. In the case of CT-DNA/F-An nanoclusters, morphological analysis revealed a perturbation in the highly networked structure of CT-DNA to form a nanofibrous structure. This is mainly due to the reduction in the DNA–DNA interactions subsequent to the groove binding of the F-An nanoclusters.

The UV-visible interaction studies of F-An nanoclusters with short single stranded DNA showed no change in the fullerene absorption indicating the absence of any kind of interactions with single strand DNA.

Murali Golla said: I have two questions. First, the CD signal of any DNA will depend on the conformation of the DNA. Generally CT-DNA show a bisignate CD signal, in your article after decoration of fullerene nanoclusters on CT-DNA, a positive CD signal was observed. Please could you explain what is the reason for there being only a positive CD signal? Second, after incubation of fullerene nanoclusters with CT-DNA, the entangled network was changed to a small network kind of morphology. Can you justify what is the reason for the change in morphology?

Joshy Joseph responded: In the present case upon increasing the concentration of F-An nanoclusters from 1 : 0 to 1 : 1, the CD spectra exhibited strong induced CD around 250–370 nm, which is the region corresponding to F-An nanocluster absorption. Since, F-An absorption is in line with that of DNA, the observed positive induced CD signal could be attributed to the chiral organization of the F-An nanoclusters on the CT-DNA template, facilitated by the groove binding interactions and possible π–π interactions between adjacently bound F-An nanoclusters.

To answer the second question, the CT-DNA/F-An complex exhibited a significantly different nanofibrous structure, where the highly networked structure of CT-DNA was broken down into smaller nanofibers with sizes of 50–100 nm. The change in morphology of CT-DNA upon F-An nanocluster binding could be attributed to the reduction in the DNA–DNA interactions subsequent to the groove binding of the F-An nanoclusters.

John Helliwell communicated: You state in your abstract that 10% DMSO was needed to dissolve the fullerene. What is the impact of 10% DMSO on the DNA (in the absence of fullerene)? Also, is "PBS" phosphate based saline?

Joshy Joseph communicated in reply: As stated in our manuscript, 10% DMSO–PBS solvent mixture was used in all experimental studies. Before carrying out DNA interaction studies, the stability of CT-DNA in the presence of 10% DMSO–PBS and PBS alone were compared using circular dichroism (CD) studies. The unchanged CD signal in the presence of 10% DMSO–PBS refers to the intact B-form of DNA and hence this solvent mixture was used for further DNA interaction studies with fullerene derivatives. "PBS" refers to phosphate buffered saline. In the present case we have used 10 mM phosphate buffer containing 2 mM NaCl at pH = 7.40.

Priyadarshi Roy Chowdhury communicated: What do you think about challenges in CT-DNA template analysis? What are the bottlenecks and how could these be overcome?

Joshy Joseph communicated in reply: The main challenge in the CT-DNA template analysis is the lack of information about nucleobase sequence structure and hence it is difficult to understand the specificity of binding. In addition, CT-DNA contains irregular lengths of nucleotide strands which cannot be used for the thermal denaturation studies. One can overcome these limitations by using specific oligonucleotide strands of known nucleotide sequences, in which case the exact nature of binding and accurate thermal denaturation can be studied.

Arijit De opened a general discussion of the paper by Richard Cogdell: Considering the molecular level understanding of energy transfer under low light (LL) conditions, a dark adapted strain shows shifting of energy levels in LH2 arresting the back-transfer of energy from LH1 (which is somewhat analogous to switching from carotenoid to chlorophyll energy transfer under LL conditions to chlorophyll to carotenoid energy transfer under high light (HL) conditions mediated *via* the xanthophyll cycle). On the other hand, the color change in LH2 is due to carotenoid incorporation. My question is: is there any specific role of these carotenoids or is the entire role still played by the chlorophylls alone, by tuning the coupling among themselves and with the protein, thereby shifting the absorption?

 This journal is © The Royal Society of Chemistry 2018

Apart from this "slow" change upon LL conditions, LH2 can also undergo "fast" conformation change (state transition) under strong pulse irradiation leading to dissipation of excess energy as heat, as revealed from the single complex studies by William Moerner and co-workers.[1] In this case, the LH2 complex remains the same (there is no carotenoid incorporation). So, conformation change may be a more desired pathway for quickly adapting to changing light intensities. Can you please comment on that?

1. G. S. Schlau-Cohen, Q. Wang, J. Southall, R. J. Cogdell, W. E. Moerner, *Proc. Nat. Acad. Sci. U. S. A.*, 2013, **110**, 10899–10903.

Richard Cogdell replied: The effect of the carotenoid on the absorption of the Bchl a molecules is rather small. But often with LH2 complexes they are not stably assembled in the absence of carotenoids. And with all proteins LH2 does show conformational flexibility and this can be best seen in single molecule experiments. There is no evidence in the purple photosynthetic bacteria that any such conformational changes are used to modulate the efficiency of light harvesting. I should remind you that these are anaerobic photosynthetic organisms and unlike plants do not have a problem with too much light as no oxygen is present.

Amitava Chandra said: I'm interested in knowing the molecular picture of LH1 and LH2 complexes. What molecules are they and what specific molecular changes are happening?

Richard Cogdell responded: There are X-ray crystal structures for these pigment protein complexes. References to these are provided in our paper.

Anthony Watts remarked: Why does the bacterium not simply change the density of LHC membrane to capture more or less light, rather than have such a complex adaptation in various components, their concentrations, and their composition resulting in different kinetics and efficiency of light capture?

Richard Cogdell responded: They do of course make more membrane and more photosynthetic units as the light intensity in which they are grown is reduced. However, the more subtle regulation where they change the spectroscopic form of LH2 that is made is because the changes in the spectroscopic type of LH2 allows the excited state to reside longer on LH1 at low light and it to escape and find another LH1 complex at high light. This provides more efficient use of the absorbed photons under both conditions.

Rajaram Swaminathan asked: I am curious about the shift in the absorption spectrum towards the red wavelengths. Can it be shifted to the blue side too?

Richard Cogdell answered: In the example shown in this paper the B850 band in the high light form of LH2 is actually blue shifted to 820 nm in the low light growth conditions.

Rienk van Grondelle added: Following on from Richard Cogdell's explanation for why there is a blue shift. There is also a degeneracy factor, so the possibility to be in the central state is already small which in fact rate limits the trapping process. Doubling the antenna size would reduce this even further and for that reason the added antenna is maybe blue shifted.

Shaina Dhamija asked: Is it possible to achieve pH-dependent spectral shift for LH2 (shifting the B850 band down to B830) instead of altering the light intensity?

Richard Cogdell answered: Most species of purple photosynthetic bacteria do not have LH2 complexes where the position of the B850 absorption band is pH sensitive in this way.

John Helliwell communicated: Can you please quantify the statement (on page 1 of your paper) that "the amount of ICM is a lot less"? On page 2, please can you quantify the statement "just a bit slower"?

Richard Cogdell communicated in reply: This is a very hard question to answer quantitatively. We are now doing some tomography to try to be more quantitative. The EM pictures only give a partial answer since the membranes appear not to be uniformly distributed throughout the cell. As to how much the LH2 to LH1 energy transfer step slows down on going from the HL to the LL LH2 complex, the rate slows from about 6 ps in the HL case to about 9 ps in the LL case. However since the Bchl a excited singlet state lifetime is about 1 ns these small changes have a very marginal effect on the overall energy transfer efficiency.

John Helliwell communicated: The method of varying the light intensity (described on page 4) is fine in the laboratory but in the natural environment absorbers of light such as water change the light spectrum as well as its intensity (a function of water depth and water type). Which environmental situation is the page 4 lab situation relevant to?

Richard Cogdell communicated in reply: The natural situation is so complicated because the light intensity varies, the spectral quality varies and of course there is the diurnal variation too. The lab situation is of course not the natural one, but it gives us the chance to dissect each effect individually. It is completely different to try to mimic the real natural system and far too complicated for me!

Priyadarshi Roy Chowdhury communicated: How do femtosecond dynamics affect the isolated LH2 complexes?

Richard Cogdell communicated in reply: The type of LH2 complexes we use are those that are the most stable. This means that they are not adversely affected by our femtosecond laser flash photolysis experiments.

Richard Cogdell addressed Joshy Joseph: One thing that is great about DNA is that you're not constrained by having single strand or double strand, you can produce origami structures. Have you thought about using DNA origami to produce ordered structures and place your fullerene derivatives on those? You could try to get well controlled patterns.

 This journal is © The Royal Society of Chemistry 2018

Joshy Joseph answered: Thanks for the excellent suggestion. We also have thought about the possibility of using fullerene nanoclusters for organizing higher order DNA structures. In future studies we will try to investigate these aspects.

Joshy Joseph asked Richard Cogdell: Do the light absorbing chromophores have any change in their chemical structure in the cells grown under low light and high light conditions? How about the sizes of the LH1 and LH2 complexes grown under LL and HL conditions?

Richard Cogdell replied: The bacteriochlorophyll molecules remain the same. Sometimes as I have shown here in this paper the type of carotenoid can change. I do not know of an example of where the ring sizes change dependant on the growth conditions.

Rienk van Grondelle addressed Joshy Joseph: I have a question about fullerene. Fullerene clusters in heterojunction solar cells are used to remove electrons and the size of the cluster is an important factor in the cell's efficiency. Is there a role for your fullerene clusters on DNA (*e.g.* electronic) in terms of electron transfer properties?

Joshy Joseph answered: Yes. Fullerene derivatives were extensively used in bulk heterojunction solar cells as acceptors. Also, it is reported that aggregation of fullerenes in the active layer of the device will deteriorate the device performance. In our previous report,[1] we observed that the nanowires obtained from the mutual assisted self-assembly of short DNA strands and fullerene clusters were semi-conducting in nature. Here, the conductivity is mainly arising due to the presence of fullerene clusters and hence confirms the positive role of fullerene clusters in the electron transfer properties in DNA.

1 S. K. Vittala, S. K. Saraswathi and J. Joseph, *Chem.–Eur. J.*, 2017, **23**, 15759–15765.

Mike Jones asked: You mentioned in your paper there are a variety of ways of condensing DNA. Are there advantages to using a fullerene as compared to other methods of packaging?

Joshy Joseph replied: Fullerene derivatives have been extensively studied for a variety of medical applications including their use as neuroprotective agents, HIV-1 protease inhibitors, bone-disorder therapy agents and X-ray contrast agents. The water soluble fullerenes find many advantages compared to other known DNA condensing agents: (i) fullerene is known to bind DNA through groove binding and the cationic fullerenes have an increased efficiency of DNA condensation due to the combined electrostatic, groove binding and π–π interactions; (ii) the high electron density of fullerene will be very helpful for the morphological studies of DNA/fullerene hybrids through HR-TEM analysis in the absence of staining agents; (iii) due to the semiconducting nature of fullerenes, the resultant nanostructures will have conductivity properties which might find applications in molecular electronics; and (iv) the high triplet quantum yield of fullerenes can be used for photo-induced DNA damage studies.

Amit Sharma asked Richard Cogdell: What applications can we learn from your work, *e.g.* in the field of bioenergy?

Richard Cogdell responded: Photosynthesis provides a blueprint of how to use solar energy to make fuels. When we understand the biological process at a deep mechanistic level the hope is we can use that information to make more efficient and robust systems for doing the same thing. I should say though that I do not believe that we should try to use the biological system directly as it is too fragile and has not evolved to efficiently make fuel for mankind.

Padmaja P. Mishra addressed Joshy Joseph: Fullerenes are known to produce singlet oxygen, which does damage in the DNA. Do you see any effects of singlet oxygen induced damage on your DNA during the interaction?

Joshy Joseph answered: Fullerene derivatives are known to produce singlet oxygen and hence are used as DNA photo-cleaving agents. In the current paper, we haven't done any photo-induced DNA damage studies.

Ilme Schlichting queried: Have you thought about using metal-doped fulllerenes?

Joshy Joseph replied: We haven't tried metal-doped fullerenes for the current studies.

Ilme Schlichting then asked Richard Cogdell: What is known about the light dependent transcription regulation?

Richard Cogdell answered: The regulation of all of these complexes is at the level of transcription. There is an all or nothing switch controlled by an oxygen sensor kinase and its regulator transcription factor. This is the RegA/RegB system. Then when the oxygen tension is low enough the extent of production of the LH complexes is controlled by both light intensity, by the AppA sensor kinase and its regulator transcription factor PpsR, and by light quality using a set of bacteriophytochromes. The AppA protein, which you determined an X-ray crystal structure for, is well characterised. Very little work has been done on the phytochrome regulation in this system.

Mike Jones said: In one case you see a carbon source dependent change in pigmentation. Do you know the biochemical basis for that?

Richard Cogdell responded: This effect was noted a long time ago by Karin Schmidt but it has never been really investigated. So the answer is we don't know.

Bern Kohler commented: In natural environments, low intensity light generally has a different spectral distribution than high intensity light. Have you looked into how spectral and intensity changes in illumination influence adaptation?

Richard Cogdell responded: In our experiments we make sure that the light intensity changes do not change the spectral composition of the incident light.

 This journal is © The Royal Society of Chemistry 2018

These bacteria do have bacteriophytochromes that can detect changes in the spectral composition but we have not studied that.

Mike Jones asked: You say there are 12 copies of the LH2 structural genes. How different are the protein sequences? Are all of them expressed? How much do we know about these multiple copies?

Richard Cogdell responded: Yes in *Rhodopseudomonas acidophila* there is this really large puc gene family. The protein sequences have subtle differences, but are all rather homologous. We have not been brave enough to get to grips with this level of complexity. Rather we have tackled the case of *Rhodopseudomonas palustris* where the puc gene family only has 5 members. In this case we have a full set of all the different puc gene deletions and will soon be able to answer your questions for that species.

Rienk van Grondelle remarked: I understand that these bacteria have amazing adaptability in terms of light harvesting capability. Why have plants lost this?

Richard Cogdell replied: You can never be sure that any answer to this type of question is actually correct. However I suspect that it is because plants typically are nearer the sun and too little light is often not a problem. For many plants, especially those here in India, quickly become light saturated soon after sunrise. Their problem is how to safely get rid of the excess light. They have developed multiple mechanisms to do this.

Amit Sharma said: Nanometals have been shown to improve photosynthetic efficiency in plants. Do you have any thoughts about the mechanism for this?

Richard Cogdell answered: I am sorry but I don't know what this refers to. Plasmonic effects can enhance light harvesting but I am not aware of any data to show this makes plants more efficient.

Ilme Schlichting opened a general discussion of the Concluding remarks lecture by Rienk van Grondelle: Just a comment. I fully agree with what you said. It would be very useful to calculate vibrational spectra using the proposed models and to compare whether they fit to the experimental spectra. However, this may not be so straight-forward in practice, as vibrational spectra code for a lot of details, such as for instance the chromophore–protein interactions. Accurate description of these interactions by computational models is still challenging, but the frequencies, as we know, are sensitive to these interactions. On the other hand, if one considers a (real) many-atom system there are many vibrational modes to consider, some of which have similar frequencies and some of which correspond to the chromophore and protein vibrations mixed. It is often unclear how to assign the computed frequencies to the experimental ones for such a complex system without additional experiments and computations, like for instance site-directed isotope labeling. Experimentally, site-specific isotope labeling is rather demanding...

Roberto Improta remarked: I strongly agree we should investigate the role of solvent molecules in excited dynamics with a lot of attention. We already have computational evidence of a strong coupling between the vibrational modes of the solute and those of the first solvation shell, which significantly modulate the spectral line shape. Vibronic coupling with solvent molecules rules a fundamental process as vibrational coupling and should modulate all the different dynamical features, including coherence, of excited state dynamics, especially for charge transfer states. The problems we face when investigating these kind of phenomena in systems as dinucleotides are connected to the existence of many different minima, and many key vibronic parameters exhibit a strong dependence on the geometry considered, making the treatment of the interplay between fast and slow degrees of freedom critical.

Bern Kohler addressed Rienk van Grondelle: In your remarks, you made a compelling case for the role that quantum coherence plays in promoting efficient charge separation in natural light harvesting complexes, but this has also been the subject of vigorous debates. Would you like to comment on the controversies for those outside the field?

Rienk van Grondelle replied: We/I have tried to quantitatively explain our observations on photosynthetic systems using correct physical models.[1] Such models typically include the coupling between electronic states giving rise to coherences. The environment was included as a classical bath. Since 2014, when we observed the long-lived coherences in photosynthetic reaction centers we have started to include specific vibrations into the quantum model for the electronic states. This implies building a set of quantized vibrational states on each of the electronic states. This leads to a reasonable but not perfect description of the observations. The people who criticize us typically do not do this. They have a very simple, according to me too simple, explanation of the observed spectroscopy, they never show a quantitative fit of their observations using a physical model of photosynthetic energy transfer and charge separation, which I find unsatisfactory.

1 V. I. Novoderezhkin, E. Romero and R. van Grondelle, How exciton-vibrational coherences control charge separation in the photosystem II reaction center, *Phys. Chem. Chem. Phys.*, 2015, **17**, 30828–30841.

Dimitra Markovitsi remarked: Fluorescence anisotropy, fluorescence decays and fluorescence spectra, recorded from the fs to the ns time-scales, for both genomic DNA and model oligonucleotides, have brought valuable information on: ultrafast energy transfer involving exciton states,[1–3] low energy (UVA) mixed $\pi\pi^*$/charge transfer states,[4–6] and High energy Emitting Long-lived Mixed (HELM) states.[7,8] It was shown that the collective behaviour of the bases in respect to these effects increases with the rigidity of the system which is affected by the duplex/G-quadruplex size and the ionic strength of the buffered solution. G-quadruplexes exhibit more pronounced collective effects compared to duplexes.[9]

1 D. Markovitsi, D. Onidas, T. Gustavsson, F. Talbot and E. Lazzarotto, *J. Am. Chem. Soc.*, 2005, **127**, 17130–17131.
2 D. Markovitsi, T. Gustavsson and I. Vayá, *J. Phys. Chem. Lett.*, 2010, **1**, 3271–3276.
3 I. Vayá, T. Gustavsson, T. Douki, Y. Berlin and D. Markovitsi, *J. Am. Chem. Soc.*, 2012, **134**, 11366–11368.
4 A. Banyasz, I. Vayá, P. Changenet-Barret, T. Gustavsson, T. Douki and D. Markovitsi, *J. Am. Chem. Soc.*, 2011, **133**, 5163–5165.

5 A. Banyasz, T. Douki, R. Improta, T. Gustavsson, D. Onidas, I. Vayá, M. Perron and D. Markovitsi, *J. Am. Chem. Soc.*, 2012, **134**, 14834–14845.
6 A. Banyasz, T. Gustavsson, D. Onidas, P. Changenet-Barret, D. Markovitsi and R. Improta, *Chem.–Eur. J.*, 2013, **19**, 3762–3774.
7 M. Huix-Rotllant, J. Brazard, R. Improta, I. Burghardt and D. Markovitsi, *J. Phys. Chem. Lett.*, 2015, **6**, 2247–2251.
8 I. Vayá, J. Brazard, M. Huix-Rotllant, A. Thazhathveetil, F. Lewis, T. Gustavsson, I. Burghardt, R. Improta and D. Markovitsi, *Chem.–Eur. J.*, 2016, **22**, 4904–4914.
9 P. Changenet-Barret, Y. Hua and D. Markovitsi, *Top. Curr. Chem.*, 2015, **356**, 183–202.

Stefan Haacke commented: I would be careful with the phrasing "quantum or optical coherence" because it leads to the misunderstanding that for vibronic coherences to occur one would need a coherent light source for their excitation. Vibronic coherence, *i.e.* vibrations with a defined phase relation, can exist in each single molecule that absorbs a photon. It is just that we use these coherent short pulse lasers, since we work on molecular ensembles and have to synchronise these to make the coherences among different molecules visible. I understand that you talk about superposition of quantum states, a coherent superposition, of the vibronic states in adjacent molecules in a light harvesting aggregate, and how this coherence survives energy relaxation, correct? Is the idea of "quantum coherence improves reaction efficiency" then simply due to the quantum entanglement of the vibronic wave functions of the initial and final states?

Rienk van Grondelle answered: I use "quantum coherence" on purpose because what we see is a quantum effect, this does not exist in classical physics. The major origin for speeding up the transition rate is 'delocalization' which is a true quantum effect. Because the frequency that couples the two (initial and final) states is very prominent in the spectral density the delocalization is also coupled to fast relaxation (in our model). We are working on a more detailed paper together with Pavel Maly, Doran Bennett and Alan Aspuru-Guzik.

Manas Sarangi asked: How does nature take care of the coherence and decoherence in photosynthetic systems?

Rienk van Grondelle responded: It is not a matter of 'taking care'. The coherences occur due to the strong coupling between the pigments, which with pigment–pigment distances of 1 nm are unavoidable and large. In photosynthetic antennae excitations are typically delocalized over several pigments and they do not decohere. In addition, different electronic states in the antenna or in the reaction center are resonantly coupled to some vibrations, typically 4 or 5, and as a consequence these electronic states are further delocalized. This delocalization will indeed decohere, although it is not precisely known how and how fast. This is a subject of intense investigation.

Arijit De said: I would like to add a comment following the last question about whether the noise, apart from (quantum) coherence, influences the energy transfer. Excitation delocalisation spanning over spatially separated chromophores leading to excitonic states is somewhat an obvious outcome of the close proximity of chromophores within an aggregate. To describe the robustness of such energy delocalisation, the "correlated fluctuation" model was proposed by Shaul Mukamel and co-workers which assumes the energy levels of two chromophores always remain close. So, we must always talk about exciton energies of dimers or oligomers and not site energies of monomers.

What is a more important problem to look at is how energy transfers between these excitonic states, how a coherent superposition between the excitonic states leads to efficient energy transfer and what is the role of environment (bath) induced fluctuations. In fact, there has been a great body of work, mostly from Martin Plenio's group, trying to address the second question. It has been argued, theoretically, that fluctuations are important to transport energy spatially. You may have a look at this fantastic paper[1] for more details.

1 F. Caruso, A. W. Chin, A. Datta, S. F. Huelga and M. B. Plenio, Highly efficient energy excitation transfer in light-harvesting complexes: The fundamental role of noise-assisted transport, *J. Chem. Phys.*, 2009, **131**, 105106.

Conflicts of interest

There are no conflicts to declare.

 This journal is © The Royal Society of Chemistry 2018

Faraday Discussions

PAPER

Photoinduced processes in nucleic acids and proteins: concluding remarks

Rienk van Grondelle

Received 26th February 2018, Accepted 27th February 2018
DOI: 10.1039/c8fd00049b

Light induced charge and energy transport in nucleic acids and proteins is the basis of fundamental biological processes such as photosynthesis, vision, DNA-photostability, DNA-photodamage and photosensing. This article summarises the concluding remarks given at the Faraday Discussions meeting on this topic. The specific themes arising from the meeting that are discussed herein include charge transfer in nucleic acids and proteins, excited state dynamics and topology, vibrations and dynamic disorder, proteins and charge transfer states, nanobiophotonics coupled to biomedical applications and photosynthesis.

1. Introduction

Light induced charge and energy transport in nucleic acids and proteins is the basis of fundamental biological processes such as photosynthesis, vision, DNA-photostability, DNA-photodamage and photosensing; for a review of this please see the introductory lecture given by Professor Ilme Schlichting (DOI: 10.1039/C8FD00058A). The most prominent chromophores in nucleic acids are nucleotides, retinal in rhodopsins, aromatic amino acids in proteins, and bacteriochlorophylls, carotenoids and bilins in photosynthesis. In addition there exists a variety of chromophores, often bound to specific proteins that play a role in sensing and in the regulation of DNA-expression (*e.g.* phytochromes). The relevant wavelengths of absorption for nucleotides are in the UV range (220–300 nm), while those for retinal are in the visible spectral range (300–700 nm), those for aromatic amino acids are in the UV-near UV region (220–320 nm), and those for the photosynthetic pigments chlorophyll *a*, bacteriochlorophyll *a*, bacteriochlorophyll *b* and bilins are in the ranges 300–700 nm, 300–900 nm, 300–1050 nm, 300–680 nm, respectively. Carotenoids (of which there are many) form a special class; they absorb light in the blue/green part of the spectrum (300/550 nm) to their S_2 state, while transitions to the lower energy S_1 state are forbidden or 'dark'. Excitation to the strongly allowed S_2 state determines the bright colours of carotenoids. Of course, light at wavelengths of below 300 nm is largely absorbed

Faculty of Science, Department of Physics and Astronomy, Vrije Universiteit Amsterdam, The Netherlands

by the ozone layer, and there is a lot of light at wavelengths of above 1000 nm that is not used by vision or photosynthesis.

The absorption of a solar photon by any of these chromophores causes an electronic transition. The strongly allowed transitions (*i.e.* those with a large transition dipole moment) are typically π–π^* transitions, while the weakly allowed transitions are typically n–π^* transitions or transitions to charge transfer (CT) states that have borrowed oscillator strength from the strongly allowed transitions. This Faraday Discussions meeting for a large part dealt with the fate of such an electronic excitation, *i.e.* how it is converted into the successful isomerization of retinal in vision or in bacteriorhodopsin, how it is converted into useful charge separation in photosynthesis, how the formation of damage in nucleic acids is avoided, and how proteins with a high density of aromatic or charged amino acids deal with UV excitations.

After considering the papers that were presented during the meeting, I concluded that the meeting concentrated on the following themes, which I will discuss sequentially followed by a few general remarks:

- Charge transfer in nucleic acids and proteins;
- Excited state dynamics and topology;
- Vibrations and dynamic disorder;
- Proteins and charge transfer states;
- Nanobiophotonics coupled to biomedical applications;
- Photosynthesis.

In this discussion I will follow the roadmap outlined by Professor Ilme Schlichting in her opening lecture (DOI: 10.1039/C8FD00058A). There is a need to understand function in order to potentially manipulate it. For instance, an understanding of specificity, (fluorescence) quantum yield, side reactions and colour tuning would all help in this regard. There is also a need to characterize in detail the photophysics and photochemistry, the influence of the protein matrix, steric constraints along all steps of the reaction coordinate, and the timing/coupling dynamics of a photoreaction and protein motion, while it might also be useful to look at the generality of the mode coupling model.

It is not always easy to check the consistency of structures and spectroscopy and calculation results using other techniques. It is known that the conditions of excitation matter, *e.g.* ultrashort laser flash, nano–microsecond flash, continuous illumination, high *vs.* low intensity, temperature and excitation colour. The interplay between electronic and nuclear motions can result in complex vibronic levels, thus complicating interpretations. Calculations (QM/MM, MD) would benefit from benchmarking with ultrafast transient intermediate structures in order to find the most appropriate approach. Finally, ground state/excited state/side reactions, as well as spectroscopically/structurally silent states, need to be taken into account, while low occupancy complicates analysis (extrapolated structure factors require lots of caution and control.

The new methods that were looked at were ultrafast time-resolved X-ray spectroscopy using a free-electron laser, femtosecond two-dimensional electronic spectroscopy (2D-ES), ultrafast two-dimensional vibrational spectroscopy (2D-IR) and quantum dynamics.

 This journal is © The Royal Society of Chemistry 2018

2. Charge transfer

When trying to understand energy and electron transfer in nucleic acids we must realize that DNA is basically a multichromophore aggregate, for which many pathways for excited state decay and charge separation, electron/hole migration and charge recombination are potentially simultaneously possible,while all pathways are effectively in equilibrium, with the free energy differences between the various pathways being small relative to kT. The pathway that is chosen in the end will depend strongly on the environmental effects that are controlling the relative energetics of each of the possible pathways. The outcome will depend on the interplay between the fast and slow degrees of freedom (dynamic disorder) (DOI: 10.1039/C7FD00195A).[1]

One important aspect that was broadly accepted by the community of scientists present at this meeting was the fact that in densely packed structures of aromatic molecules, a DNA duplex with nucleotides at a distance of 0.3 nm or a chlorophyll-protein with a densely packed set of chlorophylls and carotenoids separated by a nm or less, charge transfer following excitation is unavoidable and ultrafast. This implies that the electronic states that are excited by light are quantum mechanically mixed with CT-states, typically leading to effective charge transfer within a few hundred femtoseconds. For photosynthesis this is precisely what you want; excitation of the photosynthetic reaction center, either directly or *via* its light-harvesting antenna, leads to a trans-membrane charge separation with a close to 100% quantum yield (QY).[2] In DNA, however, this is not what you want; you want to get rid of the UV excitation as fast as possible, but the probability of charge separation occurring is high and is ultimately most likely the cause of DNA damage (DOI: 10.1039/C7FD00205J, DOI: 10.1039/C7FD00179G, DOI: 10.1039/C7FD00195A and DOI: 10.1039/C7FD00186J).[3–8] In rhodopsins, excitation of the retinal chromophore leads to an 'intra-retinal' charge separation (DOI: 10.1039/C7FD00207F, DOI: 10.1039/C7FD00200A and DOI: 10.1039/C7FD00198C), most likely involving the surrounding amino acids and water molecules. Such a complex charge transfer state could be the precursor for isomerization.[9]

Stacked DNA, both in single strand and double strand form, shows efficient charge transfer upon UV excitation.[3–5] The charge separation occurs on a sub-picosecond timescale and subsequent recombination takes place over 2–3 ps. Studying a variety of stacked dinucleotides yielded a logarithmic correlation between the rate of charge separation and the thermodynamics, which is a characteristic of incoherent electron transfer according to the Marcus model.[10] Excitation of an unstacked DNA/polynucleotide resulted in the excitation of two neighbouring bases but did not result in charge separation. In AT alternating duplexes, UV (266 nm) excitation with a small observed yield (1.5×10^{-3}) gave rise to electron ejection, producing adenine radicals (protonated and deprotonated) and forming AT-dimers (DOI: 10.1039/C7FD00179G).

As for single-stranded DNA, in DNA duplexes upon UV-excitation two π-stacked bases will create a pair of oppositely charged radical ions, which will trigger intrastrand proton transfer (Proton Coupled Electron Transfer (PCET)), whenever the driving force allows.[11,12] As a result, UV excitation will generate a distonic ion pair in which the spin and charge are separated on the two complementary strands (DOI: 10.1039/C7FD00195A).

An unresolved question in nucleic acid spectroscopy regards the presence of excitonic states, *i.e.* excited states that are delocalized over several neighbouring chromophores/bases. Based on the dipole strengths, distances and orientations, one can easily guess the strength of the dipole–dipole interaction between neighbouring bases to be around 100 cm^{-1}. Such an interaction strength would make the electronic states excitonic and delocalized over a number of bases (4–5). Such excitonic states mixed with charge transfer states would promote ultrafast charge separation.

What could the origin and nature of UV-induced DNA damage be (DOI: 10.1039/C7FD00179G, DOI: 10.1039/C7FD00193B, DOI: 10.1039/C7FD00201G, DOI: 10.1039/C7FD00188F and DOI: 10.1039/C7FD00202E)?[13,14] In the audience there was no doubt that charge transfer between neighbouring bases possibly coupled to proton transfer is the fundamental reason for oxidative damage done to DNA by UV light. Therefore, the reactivity of DNA excited states is determined by the probability of electrons and holes formed from the CT states to escape recombination. Nucleic acids must have been designed to minimize this probability, meaning that the electrons and holes must have remained localized on neighbouring bases after charge separation. The precise reason for this localization (*e.g.* topology, coupling to selected vibrations, polaron formation) is not known. 2D-ES experiments on selected nucleic acids (DOI: 10.1039/C7FD00201G) would be highly informative for learning more about the coupling between DNA-bases in a single or double strand, coupling to the environment and the possible pathways of charge transfer. During the meeting a variety of possible DNA products were discussed; adenine radicals, UV-induced base dimerization, ionization of G-quadruplexes *etc.* The latter serves as a trap for oxidative damage in the genome and may inhibit the activity of telomerase in cancer cells after the oxidation of its guanines by UV-light.

In photosynthesis, a large light-harvesting antenna collects solar excitations (DOI: 10.1039/C7FD00190H and DOI: 10.1039/C7FD00191F.[15] The antenna consists of membrane-bound and/or membrane-associated light-harvesting pigment proteins that contain bacteriochlorophyll and carotenoids, which are often densely packed with inter-chromophore distances of less than 1 nm. In these photosynthetic antennae, carotenoid S_2 transfers energy to chlorophyll on a femtosecond timescale, but also relaxes on the same timescale to a mixed S_1 state that is still capable of energy transfer to bacteriochlorophyll on a ps timescale.[16] Keto-carotenoids are special, as in these carotenoids, following excitation, the S_2 state relaxes to a mixed S_1-ICT state, which again is active in energy transfer on a ps timescale.[17]

The resultant bacteriochlorophyll excitations are transported on a very fast timescale to the photosynthetic reaction center, also a membrane pigment–protein, that upon excitation drives a transmembrane charge separation.[18] In the photosynthetic system the collective electronic states or Frenkel excitons[19] are coupled to charge transfer states which can easily be experimentally visualized using Stark spectroscopy. The charge separation occurs within a picosecond and is strongly coupled to certain vibrations that are quasi-resonant with the relevant energy gaps. This is discussed further towards the end of this review.

 This journal is © The Royal Society of Chemistry 2018

3. Excited state dynamics and topology

During the meeting a variety of contributions (DOI: 10.1039/C8FD00058A, DOI: 10.1039/C7FD00207F, DOI: 10.1039/C7FD00200A and DOI: 10.1039/C7FD00198C) discussed the process of photoisomerization in rhodopsins (rhodopsin, bacteriorhodopsin, proteorhodopsin *etc.*) that occurs after photoexcitation of the retinal chromophore. The retinal is bound to the protein *via* a Schiff-base that involves a strongly bound proton between the retinal and a nearby lysine residue. The general idea is that photoexcitation of the retinal induces a structural change in the chromophore that gives rise to a rearrangement of the C–C and C=C bonds. As a consequence of this rearrangement, a typical isomerization takes place in rhodopsin around the C14–C15 bond, the proton of the Schiff base escapes and initiates the formation of the signalling state (in rhodopsin). In bacteriorhodopsin the retinal isomerizes around the C13–C14 bond and the release of the Schiff-base proton starts the proton pumping process. In both rhodopsin and bacteriorhodopsin it is believed that the so-called Hydrogen-Out-Of-Plane (HOOP) wagging modes of the retinal are excited, which drives the retinal on its excited state surface to a conical intersection (CI) where isomerization may or may not take place.[9]

A remarkable result was presented during the meeting on mutants of the *Anabaena* sensory rhodopsin (DOI: 10.1039/C7FD00200A). The native retinal isomerizes over about 700 fs, but blue-shifted mutants could be produced, some of which isomerize over about 200 fs. It is not clear to me what the mutation does to the isomerization process. I find it difficult to imagine that the HOOP dynamics were affected at all by these mutations, so it must be something else. Therefore, I challenge the rhodopsin community to figure this out. In another presentation (DOI: 10.1039/C7FD00207F) the multi-functional/state role of water molecules in the rhodopsin structure was demonstrated. Protein bound waters exhibit a dynamic structural role; they play an important role in establishing the energy landscape that the excited retinal faces before it starts to isomerize. A water molecule close to the retinal and Schiff base was shown to become disordered on an ultrafast timescale (probably a few hundred fs), and this water molecule, or better still its charges and polarizability, may play a crucial role in the ultrafast isomerization process. It would be of great interest to demonstrate explicitly what the role of water in these ultrafast events could be, *i.e.* charge transfer, proton transfer or vibronic coupling?

In a very recent paper, Haacke and coworkers[20] showed that rhodopsin isomerizes according to a unique, ultrafast mechanism that preserves mode-specific vibrational coherence all the way from the reactant excited state to the primary photoproduct. Using quantum chemical simulations, they were able to show why the observed coherent nuclear motion depends critically on minor topological/structural variations capable of inducing strong electronic effects.

In conclusion, it is not totally clear to me what actually drives the isomerization of retinal in the rhodopsins. The HOOP modes are of a much higher frequency (10–20 fs) than the timescale of isomerization (200 fs). To me it seems that there could be a lower frequency vibrational mode that resonantly couples to the isomerization, and it would be very useful to study the process of isomerization using a variety of coherent spectroscopic techniques (2D-ES, 2D-IR, 2D-

Raman) in combination with ultrafast X-ray spectroscopy and quantum dynamics calculations.

4. Vibrations and dynamic disorder

Biological matter is by definition energetically disordered. This originates from the fast local motions (fs–ps) of atomic/molecular groups, slower (ns–μs) collective motions of amino acids/nucleotides/chromophores and slow (μs–s) motions of large parts of the biomolecule, for instance the re-orientation of an α-helix in a protein. This disorder appears as a broadening of the electronic transitions (due to the ultrafast dynamics), as multi-exponential kinetics (ns) or as 'static' disorder that can be probed using techniques such as spectral hole-burning, fluorescence line-narrowing and single molecule spectroscopy.[21] All of these dynamics are due to 'unavoidable' vibrations. Molecular bonds vibrate, collective atomic groups vibrate, α-helices vibrate and DNA vibrates. Consequently, the resulting energy landscape for a single biomolecule is complex and unique. When modelling dynamic processes in biomolecules, this complexity must be accounted for.

Disorder and vibrational dynamics in relation to fundamental biological processes were barely discussed during this meeting. This is remarkable since the potential energy surfaces of DNA excited states, DNA charge transfer, retinal isomerization and photosynthetic charge separation are typically presented with large displacements relative to the charge transfer or isomerization coordinate.[9,22,23] For photosynthetic reaction centers we have shown, using 2D-ES, that the dynamics of excitation energy transfer and charge separation are closely connected to a relatively small number of vibrations, whose major role is to delocalize the electronic states and thereby speed up the energy transfer and charge separation process by one order of magnitude.[2,24,25] It would be fantastic if similar results could be obtained for DNA charge transfer and retinal isomerization. The beauty of the 2D-ES experiment is that it shows precisely the mode of coupling between the initially excited state and the product state, including possible intermediates. Coupling this kind of information with 2D-IR, 2D-Raman and fs-X-ray spectroscopy would give a big boost towards our fundamental understanding of the elementary events in photoinduced processes in nucleic acids and proteins (DOI: 10.1039/C8FD00058A).

5. Spectroscopy of proteins

During the meeting there were several presentations of new spectroscopic features observed in dense solutions of proteins (DOI: 10.1039/C7FD00183E, DOI: 10.1039/C7FD00194K and DOI: 10.1039/C7FD00203C). Essentially, it was concluded that at high concentrations of certain proteins (proteins with a large number of Lys and/or Glu residues), new absorption bands appear in the 400–700 nm spectral region, which are ascribed to CT-states, mainly between Lys and Glu and the polypeptide backbone. A computational examination of other amino acids with charged side chains (Arg, Asp and doubly protonated His) and post-translationally phosphorylated amino acids (Ser, Thr and Tyr) indicated that all of these charged amino acids could present a donor–bridge–acceptor structure. It was proposed that the appearance of these new bands reflects the secondary structure and protein modifications, and could act as another fingerprint of

 This journal is © The Royal Society of Chemistry 2018

protein–protein interactions. One question to be asked is what are the fluorescence properties of these 'new' bands? In case they originate from unknown CT states, Stark spectroscopy, both in absorption and emission, could be very useful for further identification.

6. Bionanophotonics

I was personally very impressed by the work discussed in the talks given by Professor Chattopadhyay (DOI: 10.1039/C7FD00192D), Dr Jayasree (DOI: 10.1039/C7FD00185A), Dr Datta (DOI: 10.1039/C7FD00197E) and Dr Joseph (DOI: 10.1039/C7FD00196G), as well as the posters that described the application of light to induce or control molecular processes at the cellular level. From what I can see, there is great future potential for the application of 'photoinduced' technologies in medicine. An excellent example of this was seen in the talk given by Dr Jayasree (DOI: 10.1039/C7FD00185A), who used gold nanorods linked to an anti-cancer drug that could be applied for both imaging and therapy.

7. Photosynthesis

The light reactions of photosynthesis include two ultrafast processes: (i) excitation energy transfer among the pigments of the light-harvesting antenna followed by (ii) trans-membrane charge separation in the reaction center. Both the light harvesting antenna and the reaction center are pigment–protein complexes associated with the photosynthetic membrane, which in plants is the thylakoid membrane. Photosynthetic bacteria make do with a single reaction center that upon excitation drives cyclic electron transfer and builds up a proton gradient, whereas in plants there are two reaction centers, Photosystem 1 (PS1) and Photosystem 2 (PS2), which operate in series. Photosystem 2, after excitation and charge separation, oxidizes water to produce molecular oxygen. Photosystem 1 reduces $NADP^+$ to NADPH following excitation. Electron transfer from PS2 to PS1 is coupled to the transport of H^+ across the membrane. The resulting H^+ gradient is used to produce ATP. ATP and NADPH are required to fix CO_2 *via* the Calvin–Benson cycle.

At this meeting there were two 'photosynthesis' contributions, both relating to bacterial photosynthesis. In the first of these talks (DOI: 10.1039/C7FD00190H) bacterial photosynthetic units were placed on electrodes, supplied with electron donors and acceptors, thus forming a bio-solar cell. These so-called RC-LH1 complexes, in which the RC is surrounded by a circular LH1 antenna, naturally contain a small protein called PufX which creates a hole in the LH1-ring. This presentation was largely dedicated to understanding the role of PufX in the performance of the bio-solar cell. It was concluded that in the absence of PufX larger currents could be generated and the bio-solar cells were more stable. It is not the idea to use such cells for current generation but instead to drive specific biochemical reactions. In the second of these talks (DOI: 10.1039/C7FD00191F) the adaptation of the light-harvesting antenna of purple bacteria to different light intensities was investigated. For the purple bacterium *Rhodopseudomonas Acidophila* strain 7050, it was found that under low light conditions the peripheral light-harvesting complex LH2 was exchanged for a more 'blue' absorbing species. The reason for this blue-shift under low-light conditions was discussed. One possible reason is for the optimization of the efficiency of light capture by shifting the absorption spectrum of the antenna more to the visible region, not only by blue-shifting the bacteriochlorophyll absorption but also by incorporating redder carotenoids.

Finally, it is appropriate to return to the quantum-vibronic mechanism of energy transfer and charge transfer in photosynthesis. The model and conclusions are largely based on our work on the PS2RC using 2D-ES.[2,24–27] In this work we have shown that exciton transfer and charge separation are strongly coupled to a few specific low frequency vibrations. In the PS2RC (at least) two pathways of charge separation exist, one involving an exciton state largely localized on the active branch accessory chlorophyll and pheophytin, the second is an exciton state that is more or less localized on the 'special pair' of chlorophylls in the heart of the PS2RC. Our work has clearly demonstrated that these two pathways are coherently coupled through a 120 cm^{-1} vibration which shows up as a strong oscillatory cross peak in the 2D-ES frequency maps. In addition, the 'special pair' pathway is coupled to a real charge separated state *via* a 340 cm^{-1} vibration, which manifests itself *via* its coherent formation. I wish to emphasize that these coherences have nothing to do with the fact that we use ultrashort laser pulses to detect them. They are an intrinsic property of these reaction centers created by the delocalization of the relevant electronic states due to their mixing with resonant vibrations.

8. Conclusions

In conclusion, during this Faraday Discussions meeting, the participants extensively discussed the basic photophysics, photochemistry and photobiology of a variety of biological/biology related systems: DNA, rhodopsins, proteins and photosynthesis. Key considerations were given to ultrafast laser spectroscopy, charge transfer, conical intersections, topology, vibrations and functionality. It was remarkable to observe the tight connection between experiments and quantum chemical theory (hybrid QM/MM, TDDFT *etc.*). In my personal view it was a successful meeting, not only due to the quality of the presented work but also due to the participation of many young Indian scientists, including many young women. Personally I was impressed by the bridging between this area of fundamental science and possible new applications in medicine and medical therapy.

Conflicts of interest

There are no conflicts to declare.

References

1 R. Improta and V. Barone, *Angew. Chem.*, 2011, **50**(50), 12016–12019.
2 E. Romero, V. I. Novoderezhkin and R. van Grondelle, *Nature*, 2017, **543**(7645), 355–365.
3 C. Crespo-Hernandez, B. Cohen and B. Kohler, *Nature*, 2005, **436**, 1141–1144.
4 F. Santoro, V. Barone and R. Improta, *Proc. Natl. Acad. Sci. U. S. A.*, 2007, **104**, 9931–9936.

 This journal is © The Royal Society of Chemistry 2018

5 T. Takaya, C. Su, K. de La Harpe, C. E. Crespo-Hernandez and B. Kohler, *Proc. Natl. Acad. Sci. U. S. A.*, 2008, **105**, 10285–10290.
6 F. D. Lewis, R. L. Letsinger and M. R. Wasielewski, *Acc. Chem. Res.*, 2001, **34**, 159–170.
7 D. Markovitsi, D. Onidas, T. Gustavsson, F. Talbot and E. ELazzarotto, *J. Am. Chem. Soc.*, 2005, **127**, 17130–17131.
8 D. Markovitsi, T. Gustavsson and I. Vayá, *J. Phys. Chem. Lett.*, 2010, **1**, 3271–3276.
9 S. Gozem, H. L. Luk, I. Schapiro and M. Olivucci, *Chem. Rev.*, 2017, **117**(22), 13502–13565.
10 R. Marcus, Electron Transfer Reactions in Chemistry: Theory and Experiment (Nobel Lecture), *Angew. Chem.*, 1993, **32**(8), 1111–11211.
11 A. Zhang, K. de La Harpe, A. A. Beckstead, R. Improta and B. Kohler, *J. Am. Chem. Soc.*, 2015, **137**, 7059–7062.
12 A. Zhang, K. de La Harpe, A. A. Beckstead, L. Martinez-Fernández, R. Improta and B. Kohler, *J. Phys. Chem. Lett.*, 2016, **7**, 950–954.
13 A. Banyasz, I. Vayá, P. Changenet-Barret, T. Gustavsson, T. Douki and D. Markovitsi, *J. Am. Chem. Soc.*, 2011, **133**, 5163–5165.
14 A. Banyasz, T. Douki, R. Improta, T. Gustavsson, D. Onidas, I. Vayá, M. Perron and D. Markovitsi, *J. Am. Chem. Soc.*, 2012, **134**, 14834–14845.
15 G. D. Scholes, G. R. Fleming, A. Olaya-Castro and R. van Grondelle, *Nat. Chem.*, 2011, **3**(10), 763–774.
16 T. Polivka and V. Sundstrom, *Chem. Rev.*, 2004, **104**(4), 2021–2071.
17 K. Redeckas, V. Voiciuk, D. Zigmantas, R. G. Hiller and M. Vengris, *Biochim. Biophys. Acta, Bioenerg.*, 2017, **1858**, 297–307.
18 R. Van Grondelle and V. I. Novoderezhkin, *Phys. Chem. Chem. Phys.*, 2006, **8**(7), 793–807.
19 H. Van Amerongen, R. van Grondelle and L. Valkunas. *Photosynthetic Excitons*, World Scientific, 2000.
20 M. Gueye, M. Manathunga, D. Agathangelou, Y. Orozco, M. Paolino, S. Fusi, S. Haacke, M. Olivucci and J. Leonard, *Nat. Commun.*, 2018, **9**, 313.
21 T. Kondo, W. J. Chen and G. S. Schlau-Cohen, *Chem. Rev.*, 2017, **117**(2), 860–898.
22 C. T. Middleton, K. de La Harpe, C. Su, Y. K. Law, C. E. Crespo-Hernandez and B. Kohler, *Annu. Rev. Phys. Chem.*, 2009, **60**, 217–239.
23 I. Schapiro, M. N. Ryazantsev, L. M. Frutos, N. Ferre, R. Kind and M. Olivucci, *J. Am. Chem. Soc.*, 2011, **133**(10), 3354–3364.
24 E. Romero, R. Augulis, V. I. Novoderezkin, M. Ferretti, J. Thieme, D. Zigmantas and R. van Grondelle, *Nat. Phys.*, 2014, **10**(9), 677–683.
25 V. I. Novoderezhkin, E. Romero and R. van Grondelle, *Phys. Chem. Chem. Phys.*, 2015, **17**(46), 30828–30841.
26 V. I. Novoderezkin, E. Romero, J. Prior and R. van Grondelle, *Phys. Chem. Chem. Phys.*, 2017, **19**(7), 5195–5208.
27 E. Romero, J. Prior, A. W. Chin, S. E. Morgan, V. I. Novoderezhkin, M. B. Plenio and R. van Grondelle, *Sci. Rep.*, 2017, **7**, DOI: 10.1038/s41598-017-02906-7.

Poster titles

Facile synthesis of Fe/Ti layered double hydroxide for highly efficient photocatalysis of aqueous Rhodamine B and Congo Red, **P. Roy Chowdhury and K. G. Bhattacharyya**, *Gauhati University, India*

An Insight into the binding and photoinduced interactions of a psoralen derivative with G-Quadruplex DNA, **S. Paul and A. Samanta**, *University of Hyderabad, India*

Understanding mechanism of photoinduced conversion of monosulfiram to disulfiram, **A. Gupta**, *UDAI Pratap College Varanasi, India*

Exploring the ultrafast relaxation dynamics of the heme in the apoptosis process: Interaction dynamics of Cytochrome c with different liposomes, **C. Govind and V. Karunakaran**, *NIIST-CSIR, India*

Development of pump-probe Circular Dichroism at B23 beamline, Diamond Light Source, **A. Hughes, M. Panman, O. Berntsson, T. Javorfi and G. Siligardi**, *Diamond Light Source, United Kingdom*

Fluorescent naphthyl derivatives as preferential quadruplex DNA binders, **B. Kumari, S. Pany, P. I. Pradeepkumar and S. Kanvah**, *IIT Gandhinagar, India*

Kinetics and structural features of IMPDH: Possible target for treating *H. pylori* infection, **G. Purushothaman, K. Juvale, S. Ravi, V. Thiruvenkatam and S. Kirubakaran**, *Indian Institute of Technology Gandhinagar, India*

Columnar/lamellar packing in cocrystals of arylbipyridines with diiodoperfluorobenzene, **R. Ramakrishnan, A. R Mallia, N. M A., R. Sethy and M. Hariharan**, *IISER-Thiruvananthapuram, India*

Interplay between intersystem crossing and reverse intersystem crossing in bromoanthracene based biaryl crystals, **D. Sasikumar, K. Nagarajan, G. Gopan, R. Krishnan and M. Hariharan**, *IISER-Thiruvananthapuram, India*

Secondary electron acceptor to modulate the rate of charge recombination in DNA, **S. Deb, A. Mallia, R. Cheriya, V. Bhat, V. Vijayan and M. Hariharan**, *IISER-Thiruvananthapuram, India*

Threading intercalator for DNA condensation and efficient DNA delivery, **G. Thimmaiah, S. Pratihar and S. Y V.**, *Jawaharlal Nehru Centre for Advanced Scientific Research (JNCASR), India*

Crystalline triphenylamine substituted arenes: Solid state packing and luminescence properties, **N. M A., A. R. Mallia, R. Ramakrishnan and M. Hariharan**, *IISER-Thiruvananthapuram, India*

Engineered bacteriorhodopsin: a molecular scale condutance photoswitch, **J. Vinals and A. Watts**, *University of Oxford, United Kingdom*

Mutual self-assembly of fullerene nanoclusters and short DNA strands, **J. Joseph, S. K. V. and S. K. S.**, *CSIR-National Institute for Interdisciplinary Science and Technology, Thiruvananthapuram, India*

DNA photosensitization by benzophenone: A tale of two different solvent environments, **R. K. Venkatraman and A. J. Orr-Ewing**, *University of Bristol, United Kingdom*

Photon-induced protein stability and ligand binding using diamond B23 beamline for SRCD, **R. Hussain, T. Jávorfi, C. Hughes, A. Hughes and G. Siligardi**, *Diamond Light Source, United Kingdom*

Unveiling the potential of unfused bichromophoric naphthalimide to induce mitotic arrest by binding to tubulin, **S. K. Ghosh and R. Joshi**, *Visvesvaraya National Institute of Technology, India*

Fabrication of plasmonic chiral nanoassemblies on a helically twisted HPB-DNA ribbon, **M. Golla and R. Vargese**, *IISER-Thiruvananthapuram, India*

Self-Assembled Nucleobase-Arene Conjugate for Long-Lived Charge Separation, **A. M. Philip, F. Kuriakose and M. Hariharan**, *IISER-Thiruvananthapuram, India*

Aptamer DNA-BODIPY conjugates for selective cancer therapy, **A. N. Siriki and R. Vargese**, *IISER-Thiruvananthapuram, India*

Photoinduced phenomena in fluorescent proteins: Implications in photosynthesis, **S. Dhamija, B. Thakur, P. Guptasarma and A. K. De**, *IISER Mohali, India*

A new optical marker for proteins: Based on charge transfer transitions associated with charged amino acids, **I. Mandal, S. Paul and R. Venkatramani**, *Tata Institute of Fundamental Research, Mumbai, India*

Optical tracking of coil-helix transitions in the phospholipid binding domain of an actin severing protein, **A. Chandra, S. Mondal, R. Venkatramani and A. Datta**, *Tata Institute of Fundamental Research, India*

Light-dependent RNA interference with 2'-modified caged siRNA, **N. K. Sharma and A. Bollu**, *National Institute of Science Education and Research (NISER), India*

Metal-metal hybrid nano system: NIR imaging and therapy, **R. S. Jayasree, R. V. Nair and I. V. Nair**, *Sree Chitra Tirunal Institute for Medical Sciences and Technology, India*

Structural insights into light induced damage of human gamma-D crystalline, **B. Yorke and A. Pearson**, *University of Bath, United Kingdom*

The Faraday Discussions Poster Prize for the best poster was jointly awarded to Mr Vinayak Bhat of IISER-Thiruvananthapuram, India for his poster on "Secondary electron acceptor to modulate the rate of charge recombination in DNA", and Ms Beena Kumari of IIT Gandhinagar, India for her poster on "Fluorescent naphthyl derivatives as preferential quadruplex DNA binders".

 This journal is © The Royal Society of Chemistry 2018

List of participants

Mr Alfy Benny, *IISER Thiruvananthapuram, India*
Mr Vinayak Bhat, *IISER Thiruvananthapuram, India*
Mr Amarnath Bollu, *National Institute of Science Education and Research (NISER), India*
Mr Jack Busby, *Royal Society of Chemistry, United Kingdom*
Mr Amitava Chandra, *Tata Institute of Fundamental Research, India*
Professor Amitabha Chattopadhyay, *Centre for Cellular & Molecular Biology, India*
Professor Richard Cogdell, *University of Glasgow, United Kingdom*
Professor Carlos E. Crespo-Hernández, *Case Western Reserve University, USA*
Dr Ankona Datta, *Tata Institute of Fundamental Research, India*
Dr Arijit De, *IISER Mohali, India*
Miss Shaina Dhamija, *IISER Mohali, India*
Ms Anupriya E S, *IISER Thiruvananthapuram, India*
Mr Joseph Gibson, *Royal Society of Chemistry, United Kingdom*
Dr Angelo Giussani, *Università di Bologna, Italy*
Mr Murali Golla, *IISER Thiruvananthapuram, India*
Mr Chinju Govind, *NIIST-CSIR, India*
Dr Ashutosh Gupta, *Udai Pratap College Varanasi, India*
Miss Nikita Gupta, *Royal Society of Chemistry, India*
Professor Stefan Haacke, *University of Strasbourg - CNRS, France*
Dr Mahesh Hariharan, *IISER Thiruvananthapuram, India*
Mr Amal K Homer, *IISER Thiruvananthapuram, India*
Dr Ashley Hughes, *Diamond Light Source, United Kingdom*
Dr Roberto Improta, *Université Paris Saclay, France*
Mr Parin Jobanputra, *Royal Society of Chemistry, India*
Dr Mike Jones, *University of Bristol, United Kingdom*
Dr Joshy Joseph, *CSIR-NIIST, India*
Mr Ameen Jowhar E P, *IISER Thiruvananthapuram, India*
Mrs Sherin Joy, *Bangalore University, India*
Mr Sanath Raj K K, *IISER Thiruvananthapuram, India*
Ms Sajena K S, *CSIR-NIIST, India*
Mr Sandeepa K V, *CSIR-NIIST, India*
Dr Tolga Karsili, *Temple University, USA*
Professor Bern Kohler, *The Ohio State University, USA*
Dr Retheesh Krishnan, *Government College for Women, India*
Ms Beena Kumari, *IIT Gandhinagar, India*
Mr Pronab Kundu, *Jadvpur University, India*
Miss Anvy Kuriakose, *IISER Thiruvananthapuram, India*
Mr Mahil L, *IISER Thiruvananthapuram, India*
Professor Frederick Lewis, *Northwestern University, USA*
Miss Aiswarya M P, *IISER Thiruvananthapuram, India*
Miss Lijina M P, *IISER Thiruvananthapuram, India*
Mr Niyas M. A., *IISER Thiruvananthapuram, India*

Ms Imon Mandal, *Tata Institute of Fundamental Research, Mumbai, India*
Dr Dimitra Markovitsi, *Université Paris Saclay, France*
Professor P. Prasad Mishra, *Saha Institute of Nuclear Physics, India*
Ms Amalu Mohan M, *IISER Thiruvananthapuram, India*
Dr Michaela Muehlberg, *Royal Society of Chemistry, United Kingdom*
Miss Hridya P, *IISER Thiruvananthapuram, India*
Professor Pradeepkumar P.I., *Indian Institute of Technology Bombay, India*
Miss Sneha Paul, *University of Hyderabad, India*
Dr Ganga Periyasamy, *Bangalore University, India*
Mr Abbey Meprathu Philip, *IISER Thiruvananthapuram, India*
Mr Sumon Pratihar, *Jawaharlal Nehru Centre for Advanced Scientific Research (JNCASR), India*
Miss Gayathri Purushothaman, *Indian Institute of Technology Gandhinagar, India*
Miss Ambili R V, *IISER Thiruvananthapuram, India*
Ms Remya Ramakrishnan, *IISER Thiruvananthapuram, India*
Dr Priyadarshi Roy Chowdhury, *Gauhati University, India*
Ms Ananya S, *IISER Thiruvananthapuram, India*
Dr Ramapurath S Jayasree, *Sree Chitra Tirunal Institute for Medical Sciences and Technology, India*
Mr Sreehari S R, *IISER Thiruvananthapuram, India*
Dr Manas Sarangi, *IIT Patna, India*
Miss Devika Sasikumar, *IISER Thiruvananthapuram, India*
Dr Igor Schapiro, *The Hebrew University of Jerusalem, Israel*
Professor Gebhard F.X. Schertler, *Paul Scherrer Institut/ETH Zürich, Switzerland*
Professor Ilme Schlichting, *Max Planck Institute for Medical Research, Germany*
Mr Ebin Sebastian, *IISER Thiruvananthapuram, India*
Dr Javier Segarra Martí, *École normale supérieure de Lyon, France*
Dr Amit Sharma, *The Energy and Resources Institute, India*
Mr Mohammed Siddhique PK, *IISER Thiruvananthapuram, India*
Mr Atchimnaidu Siriki, *IISER Thiruvananthapuram, India*
Miss Kavya Smitha Pillai, *IISER Thiruvananthapuram, India*
Professor Rajaram Swaminathan, *Indian Institute of Technology Guwahati, India*
Miss Athira T John, *IISER Thiruvananthapuram, India*
Dr Arun Thomas, *IISER Thiruvananthapuram, India*
Mr Vishnu V, *IISER Thiruvananthapuram, India*
Miss Resmi V Nair, *Sree Chitra Tirunal Institute for Medical Sciences and Technology, India*
Professor Rienk van Grondelle, *Vrije Universiteit Amsterdam, Netherlands*
Dr Reji Varghese, *IISER Thiruvananthapuram, India*
Dr Ravi Kumar Venkatraman, *University of Bristol, United Kingdom*
Dr Ravindra Venkatramani, *Tata Institute of Fundamental Research, India*
Mr Javier Vinals, *University of Oxford, United Kingdom*
Professor Anthony Watts, *University of Oxford, United Kingdom*

This journal is © The Royal Society of Chemistry 2018